ORGANIC AND BIOLOGICAL CHEMISTRY

ORGANIC
AND BIOLOGICAL
CHEMISTRY

BRADFORD P. MUNDY
Colby College

MELVIN T. ARMOLD
Adams State College

JOHN R. AMEND
Montana State University

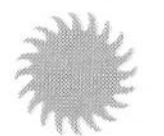

Saunders Golden Sunburst Series
SAUNDERS COLLEGE PUBLISHING
Harcourt Brace Jovanovich College Publishers

Fort Worth Philadelphia San Diego New York
Orlando Austin San Antonio Toronto
Montreal London Sydney Tokyo

Copyright © 1993 by Saunders College Publishing

All rights reserved. No part of this publication may be reproduced or transmitted in any form or by any means, electronic or mechanical, including photocopy, recording, or any information storage and retrieval system, without permission in writing from the publisher.

Requests for permission to make copies of any part of the work should be mailed to Permissions Department, Harcourt Brace Jovanovich, Publishers, 8th Floor, Orlando, Florida 32887.

Text Typeface: Times Roman
Compositor: Monotype Composition Co., Inc.
Acquisitions Editor: John J. Vondeling
Developmental Editor: Elizabeth C. Rosato
Managing Editor: Carol Field
Project Editor: Margaret Mary Anderson
Copy Editors: Linda Davoli and Andrew Potter
Manager of Art and Design: Carol Bleistine
Art Director: Anne Muldrow
Art Assistant: Caroline McGowan
Text Designer: Gene Harris
Cover Designer: Lawrence R. Didona
Text Artwork: J/B Woolsey and Associates
Layout Artists: Alan Wendt and Anne O'Donnell
Director of EDP: Tim Frelick
Production Manager: Charlene Squibb
Marketing Manager: Marjorie Waldron

Cover Credit: An Optical Photomicrograph of DNA Fibers by
© Phillip A. Harrington/Fran Heyl Associates

Printed in the United States of America

Organic and Biological Chemistry

0-03-029013-9

Library of Congress Catalog Card Number: 92-051014

3456 69 987654321

PREFACE

Audience

We wrote ORGANIC AND BIOLOGICAL CHEMISTRY to provide an introduction to chemistry for students pursuing allied health careers, including nursing and physical therapy. Students pursuing medical technology, agriculture, and home economics will also benefit from this broad introduction to chemistry. This text emphasizes basic chemistry and its applications. No background in chemistry is assumed and only introductory algebra is used in the development and application of chemical principles.

Objectives

We had three major objectives in mind when we wrote this text. First, we wanted a text that would progressively develop chemical principles and provide useful chemical concepts and information. Second, we wanted to provide the student with a broad range of examples that would increase their awareness of the involvement of chemistry in modern health care and in life. Third, we wanted a text that would contribute to a well-rounded liberal arts education by providing students with opportunities to enhance their problem-solving skills and comprehension of technical material.

Our text was examined carefully and repeatedly by ourselves and by several reviewers to check and remove any possible errors in content or calculations. We think we succeeded, but we encourage you to let us know of anything we might have missed.

Content and Scope

This text covers two major areas of chemistry, organic and biological chemistry. We included a brief chapter, Chapter R, that reviews and summarizes the more important aspects of inorganic chemistry.

Organic chemistry, Chapters 1 through 8, begins with an introduction to organic chemistry, which illustrates the unique features of carbon and organic chemistry. Chapters 2 through 8 develop the properties of important classes of molecules that are common in biochemistry.

Chapters 9 through 17, the biological chemistry portion of the text, describe structures and metabolism with considerable emphasis on human health care. Chapter 9, Introduction to Biochemistry, provides background information in cell biology and nutrition while introducing some general features of biomolecules. Chapter 17, Blood: The Constant Internal Environment, discusses blood chemistry and pulls together many topics developed earlier.

Features

Some students think that studying chemistry is a bit like hiking through a forest so thick that you cannot see your hand before you. Much of the time you press on, barely able to see a few feet ahead, hoping that the trail will lead somewhere worth visiting. We know how uncomfortable this can be for a student, and we have included many features to keep the trail visible and the destination in sight.

ORGANIC AND BIOLOGICAL CHEMISTRY contains the full complement of topics found in introductory texts. Many students in the introductory chemistry course, however, lack background in and appreciation of the physical sciences. We have added numerous pedagogical features to aid in the learning of chemical concepts and applications, as well as to stimulate interest and hold the student's attention. We know of no other text with this level of support integrated throughout each chapter. Because these features reinforce rather than build on each other, instructors and students can choose any or all of the features, as needed.

We intended to write a text that is clear, easily understood, well organized, and related to experiences that students have had or will have in the immediate future. Comments from reviewers indicate that we met these goals.

Our organization is traditional: Organic chemistry builds to biochemistry. We arranged the chapters and sections to aid in the selection or deletion of topics as needed to meet the needs of instructors and students.

Full color appears throughout this edition to organize material and to provide stronger pedagogical support. Our color assignments are often obvious and logical as well as consistently applied throughout the text. For example, water and aqueous solutions are blue, ATP in reactions is yellow, and tan emphasizes features introduced for the first time, particularly in the organic chemistry section. A complete color code appears at the back of the text.

In addition to the use of full color, we think our art program is exceptional among introductory chemistry texts. Most of the figures and many of the photographs for this edition are new. Chemical concepts and ideas are illustrated and reinforced with many superb figures as well as with numerous photographs. Perhaps our enthusiasm makes this sound a bit strong or presumptuous, but we think a glance through the book will convince you of the quality of our art program and of the level of depth and detail in our illustrations. See, for example, the following figures: 10.6, 13.9, and 13.18.

Worked examples that appear within each chapter illustrate the problem or concept, provide a method for solving the problem, and conclude with the correct answer.

New Terms are key words that are emphasized often for the student. They are clearly indicated throughout the text. We boldface each new key word or term the first time it is used in the text, and often repeat it as a margin note. Then, the key word or term is defined again in the New Terms list found within each chapter at the end of a section. New terms are listed in the Terms section at the end of each chapter, keyed to their corresponding sections. Finally, a list of all new terms appears in a glossary at the end of the text for easy reference.

At the end of nearly all sections are self-testing units called **Testing Yourself**. When students use these sets of questions and answers, they gain an immediate indication of how well they understand the content of the section just completed before they continue on in the chapter. By consistently using the Testing Yourself units, students gain a better grasp of concepts and are better prepared for the end-of-chapter exercises.

Answers to the odd-numbered exercises appear in the Appendix. (Answers to the even-numbered exercises appear in the Instructor's Manual.) Unique to our text is the keying of end-of-chapter exercises with the **Chapter Objectives** appearing at the beginning of each chapter.

We think applications of chemistry are a useful and necessary part of an introductory text. Examples of chemistry in health care and related disciplines are found in our **Chemistry Capsules**, which appear throughout the text and are usually accompanied by at least one illustration. You might find the following Capsules interesting: Correlation of Lipids and Health, Chapter 11; Enzymes and Medicine, Chapter 14; and Tay-Sachs Disease, Chapter 15.

Within the organic chemistry sections of our text, we created the **General Reaction Notecards**. They present the common, general reactions of organic compounds within the section where the reaction is presented.

Margin Notes appear throughout the text to reinforce learning and to stimulate interest. They also create a running outline of the chapter by highlighting important material.

Ancillaries

To supplement ORGANIC AND BIOLOGICAL CHEMISTRY, the following ancillaries are available:

- Experimental Chemistry: A Laboratory Manual to Accompany GENERAL, ORGANIC AND BIOLOGICAL CHEMISTRY, 2/e, by M. Armold, J. Amend, and B. Mundy. Thirty-one experiments in general, organic, and biological chemistry designed for students preparing for careers in nursing, health sciences, agriculture, and home economics.
- Instructor's Manual containing key objectives, multiple-choice questions, special project ideas, and the answers to the even-numbered, end-of-chapter exercises.
- Test Banks, both written and computerized in Macintosh and IBM formats.
- Student Companion with major objectives listed, additional Testing Yourself questions, multiple-choice questions, and a glossary of terms.
- 100 Overhead Transparencies in full color.
- Shakhashiri Videotapes by Bassam Shakhashiri (University of Wisconsin-Madison). Fifty 3- to 5-minute classroom experiments.
- Videodisc and Bar code Manual contains all the Shakhashiri demonstration and over 600 images drawn from various Saunders sources. Bar code manual allows easy access.

ACKNOWLEDGMENTS

Textbooks always represent team efforts. The staff at Saunders College Publishing provided very professional and enthusiastic support for the second edition. We wish to thank our Publisher, John Vondeling, our Developmental Editor, Elizabeth Rosato, our Project Editor, Margaret Mary Anderson, and our Art Director, Anne Muldrow. Special thanks to John Woolsey and Associates for taking our basic ideas and turning them into an excellent art program. We also want to thank Charles Winters for the many new photos in the second edition. Unless credited otherwise, all photos were taken by Charlie.

Our reviewers combed through our manuscript with a keen eye on content and provided us with many excellent suggestions for improving our text and removing errors. We thank the following reviewers of the second edition:

Grace Gagliardi, *Bucks County Community College*
Jane Toft, *Rochester Community College*
Miles D. Koppang, *University of South Dakota*
Mary Abraham, *University of South Dakota*
Shan Wong, *University of Lowell*
Robert Smith, *Skyline College*
Frank Milio, *Towson State University*
Robert Harris, *University of Nebraska-Lincoln*
Donald Williams, *Hope College*
Charles Bell, *Old Dominion University*
Rosemary Mollica, *The City University of New York, Brooklyn College*
Judith Kelley, *University of Lowell*
Josita Feighan, *Immaculata College*
Tom Grow, *Pensacola Junior College*
Richard Langley, *Stephen F. Austin University*
Hugh Akers, *Lamar University*
Sharmaine Cady, *East Stroudsburg State University*
Ronald Baumgarten, *University of Illinois at Chicago*
Suzanne Carpenter, *Armstrong State College*
Richard Pendarvis, *Central Florida Community College*
Ashis Basu, *University of Connecticut*
Robert Bohn, *University of Connecticut*
Chu-Ngi Ho, *East Tennessee State University*
Edith Rand, *East Carolina University*
Gretchen Webb-Kummer, *Modesto Junior College*
Edward Paul, *Brigham Young University*
Leland Harris, *University of Arizona*
Mary Campbell, *Mount Holyoke College*

Finally, we thank Joann, Margaret, Anita, and our children for their patience and understanding during the development and revision of the book.

Brad Mundy
Mel Armold
John Amend

Allied Health-Related Topics

The following topics, examples, and applications in this book will be of special interest to students in health-related career programs. (Note: Section numbers are given unless otherwise indicated.)

Medicine and Nursing

ABO blood group box in Section 17.5
Acidosis and alkalosis 17.3
Action of anesthetics 5.3
Anticoagulants 4.5
Artificial blood box in Section 2.7
Atherosclerosis 11.5
Beta keto acids and diabetes 7.5
Birth control compounds 3.5
Blood buffers 17.3
Blood clotting box in Section 14.5
Carcinogens 13.7
Diabetes test 6.5
Estrogen 4.5
Ether as anesthetic 5.3
Gas exchange 17.1
Gastric juices 15.2
Genetic diseases, origin 13.7
Genetic engineering 13.9
Heme 4.3
Hydrocarbons in medicine 2.7
Hyperglycemia and hypoglycemia 10.2
Hyperthermia and hypothermia box in Section 14.4
Immunity 17.5
Ketone bodies and ketosis 16.4 and 17.3
Kidney dialysis box in Section 17.2
Menstrual cycle Figure 11.8
Mouthwash 8.2
Nerve gases and organophosphates 14.4
Nitroglycerine for angina 7.3
Organohalogens as medicines Table 2.5
Peptide hormones 12.3
Phenylketonuria box in Section 16.5
Prostaglandins and leucotrienes 11.5
Pyrogenic 9.2
Regulation of blood nutrients 17.4
Sickle cell anemia box in Section 12.4
Steroid hormones 11.5
Tay-Sachs disease box in Section 15.6
Treatment of acne 3.7
Treatment of heavy metal poisoning 5.4

Nutrition and Food Science

Alcoholic beverages box in Section 5.1
Amines in decaying meat 8.1
Amino acids in nutrition box in Section 12.2
Antioxidants 6.6
Aromatic flavors 4.2
Aromatic hormones and vitamins 4.5
Artificial sweeteners box in Section 10.2
Carbohydrates and nutrition 10.5
Carbonyl compounds as odors and flavors 6.2
Carboxylic acids as food constituents 7.1
Cholesterol 11.5, 16.4, and box in Section 15.2
Cocaine 8.4
Digestion 15.2
Energy content of nutrients 11.3
Ergotism box in Section 8.4
Essential amino acids 16.5 and box in Section 12.2
Essential fatty acids 11.3 and 16.4
Esters as flavors 7.3
Fats and oils in North American diets 11.3
Flavorings and odors 5.1
General principles of nutrition 9.5
Lactose intolerance box in Section 15.2
Lactose synthesis box in Section 16.3
Lipids and health box in Section 11.5
Minerals 9.5 and box in Section 14.2
Nitrites as food preservatives 8.2
Organic acids as preservatives 7.6
Partial hydrogenation of oils 11.2 and 11.3
Phenolic flavorings 5.2
Preservatives (BHA and BHT) 11.3
Protein–calorie malnutrition box in Section 12.5
Starch and cellulose digestion 10.5
Sugars, dietary 10.4
Sulfur compounds as odors and flavors 5.4
Terpenes as flavors 3.7
Vitamin A and vision 3.7
Vitamin D synthesis box in Section 16.2
Vitamin K as antioxidant 6.6
Vitamins boxes in Sections 11.6 and 14.2

Health Risks

Addiction to terpenes box in Section 3.7
Alkylating agents and cancer 8.2
Aromatic compounds as carcinogens 4.4
Atmospheric oxidation of rubber 3.4
Benzpyrene from smoke 5.3
Carbon monoxide poisoning 2.6

CFCs and ozone layer depletion box in Section 2.6
Crack box in Section 8.4
Dioxin box in Section 5.3
Epoxides as carcinogens 5.3
Formaldehyde box in Section 6.2
Greenhouse effect box in Section 2.6
Harmful halogenated compounds 2.6
LSD and PCP 8.4
Morphine and heroin 8.4
Nicotine 8.4
Nitrosamines as carcinogens 8.2

Medical Laboratory and Radiology Technology

Electrophoresis of serum proteins Figure 12.15
Lucas test for alcohols 5.1
Plasma lipoproteins 15.2
Testing for aldehydes 6.5
Testing for glucose 10.3
Testing for intoxication 6.5
Testing for unsaturation 3.4

Drugs

Alkaloids as medicines 8.4
Ammonium salts as antiseptics 8.2
Anabolic steroids box in Section 11.5
Antibiotics 9.2
Anticancer drugs 14.4 and 16.6
Aspirin 3.7, 7.1, and 11.5
Barbiturates box in Section 8.4
Chloral hydrate 6.3
Compounds related to adrenalin 8.4
Contraceptives 11.5
Enzyme inhibitors as drugs 14.4
Enzymes as heart attack drugs box in Section 14.5
Heparin 10.5
Hydrocortisone 11.5
Laetrile 6.3
Liposomes for drug delivery 11.6
Phenols as antiseptics and medicines 5.2
Sulfa drugs 14.4
Synthetic heterocyclic compounds 8.4

BRIEF CONTENTS

CHAPTER R Electrons, Bonding, and Molecular Shapes R-1

CHAPTER 1 An Introduction to Organic Chemistry 1

CHAPTER 2 Alkanes and Cycloalkanes: Single-Bonded Hydrocarbons 17

CHAPTER 3 Unsaturated Hydrocarbons 62

CHAPTER 4 Aromatic Hydrocarbons 98

CHAPTER 5 Alcohols, Phenols, Ethers, and Thiols 122

CHAPTER 6 Carbonyl Group and Its Compounds: Aldehydes and Ketones 156

CHAPTER 7 Carboxylic Acids and Their Derivatives 187

CHAPTER 8 Amines and Amides 218

CHAPTER 9 Introduction to Biochemistry 246

CHAPTER 10 Carbohydrates 265

CHAPTER 11 Lipids 292

CHAPTER 12 Amino Acids, Peptides, and Proteins 322

CHAPTER 13 Molecular Basis of Heredity 352

CHAPTER 14 Enzymes 390

CHAPTER 15 Bioenergetics and Catabolism 416

CHAPTER 16 Anabolism 453

CHAPTER 17 Blood: The Constant Internal Environment 478

CONTENTS

CHAPTER R Electrons, Bonding, and Molecular Shapes R-1

R.1 An Introduction to the Chemistry of Carbon R-2
R.2 The Behavior of Electrons in Atoms R-3
R.3 Chemical Bonding R-10
R.4 Shapes of Covalent Molecules R-18
R.5 Molecular Dipoles: A Consequence of Molecular Shape and Electronegativity R-24
R.6 Behavior of Polar Molecules R-27
R.7 Hydrogen Bonding R-32

CHAPTER 1 An Introduction to Organic Chemistry 1

1.1 Introduction 2
1.2 Carbon's Electronegativity Makes It Unique 2
1.3 Introduction to Functional Groups 6

CHAPTER 2 Alkanes and Cycloalkanes: Single-Bonded Hydrocarbons 17

2.1 Structure and Physical Properties 18
2.2 Alkanes and Their Nomenclature 20
2.3 Alkyl Groups and Nomenclature 26
2.4 Cycloalkanes 35
2.5 Conformations of Alkanes and Cycloalkanes 40
2.6 Chemical Reactivity of Alkanes and Cycloalkanes 44
2.7 Health-Related Products Based on Hydrocarbon Structures 53

CHAPTER 3 Unsaturated Hydrocarbons 62

3.1 Introduction to the Unsaturated Hydrocarbons 63
3.2 Structure and Physical Properties 64
3.3 Nomenclature 68
3.4 Chemical Reactivity of Alkenes 72
3.5 Chemical Reactivity of Alkynes 84
3.6 Polyunsaturated Alkenes 86
3.7 Interesting Unsaturated Compounds 87

CHAPTER 4 Aromatic Hydrocarbons 98

4.1 Structure: Resonance and Electron Delocalization 99
4.2 Nomenclature 102
4.3 Aromatic Reactions: A Consequence of Overlapping Pi Bonds 107
4.4 Fused-Ring Aromatic Systems 113
4.5 Important Aromatic Hydrocarbons 113

CHAPTER 5 Alcohols, Phenols, Ethers, and Thiols 122

5.1 Alcohols 123
5.2 Phenols 135
5.3 Ethers 140
5.4 Thiols: Sulfur Equivalents of Alcohols 144

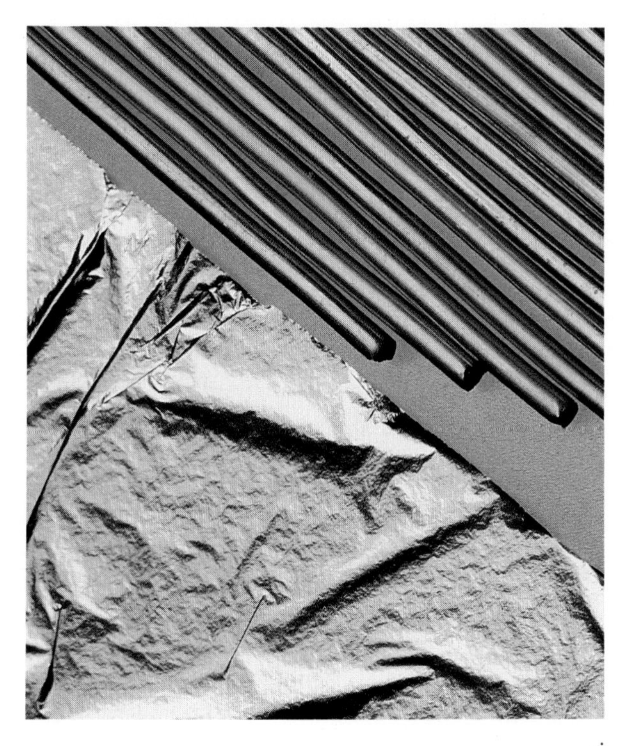

CHAPTER 6 Carbonyl Group and Its Compounds: Aldehydes and Ketones 156

6.1 Structure and Properties of the Carbonyl Group 157
6.2 Nomenclature of Carbonyl Compounds 159
6.3 Reactions of Carbonyl Compounds 164
6.4 Natural Examples of Acetals 170
6.5 Preparation Reactions of Carbonyl Compounds 170
6.6 Oxidation of Hydroquinones to Quinones 175

CHAPTER 7 Carboxylic Acids and Their Derivatives 187

7.1 Carboxylic Acids: Structure, Properties, and Nomenclature 188
7.2 Reactions and Preparations of Carboxylic Acids 193
7.3 Esters 200
7.4 Acid Anhydrides 209
7.5 Important Organic Acids and Acid Derivatives 211

CHAPTER 8 Amines and Amides 218

8.1 Amines: Structure, Properties, and Nomenclature 219
8.2 Reactions of Amines 222
8.3 Amides 229
8.4 Special Amines 232

CHAPTER 9 Introduction to Biochemistry 246

9.1 Introduction to Cells 247
9.2 Procaryotic Cells 249
9.3 Eucaryotic Cells 251
9.4 Biomolecules 253
9.5 General Principles of Nutrition 260

CHAPTER 10 Carbohydrates 265

10.1 Classification of Carbohydrates 266
10.2 Monosaccharides 269
10.3 Properties and Reactions of Sugars 277
10.4 Oligosaccharides 281
10.5 Polysaccharides 282

CHAPTER 11 Lipids 292

11.1 Classification of Lipids 293
11.2 Fatty Acids 294
11.3 Triacylglycerols 299
11.4 Saponifiable Lipids of Membranes 303
11.5 Nonsaponifiable Lipids 306
11.6 Liposomes and Membranes 313

CHAPTER 12 Amino Acids, Peptides, and Proteins 322

12.1 Protein Function 323
12.2 Alpha Amino Acids 323
12.3 Peptide Bonds and Peptides 331
12.4 Protein Structure 335
12.5 Properties and Classification of Proteins 345

CHAPTER 13 Molecular Basis of Heredity 352

13.1 Search for the Molecular Basis of Heredity 353
13.2 Nucleotides 355
13.3 Structure and Replication of DNA 359
13.4 RNA 368
13.5 Transcription 369
13.6 Translation 372
13.7 Mutagenesis 376
13.8 Regulation of Gene Expression 381
13.9 Genetic Engineering 385

CHAPTER 14 Enzymes 390

14.1 The Roles of Enzymes in Metabolism 391
14.2 Enzymes 391
14.3 Enzyme Specificity and Activity 396
14.4 Rates of Enzyme-Catalyzed Reactions 401
14.5 Regulation of Enzyme Activity 408

CHAPTER 15 Bioenergetics and Catabolism 416

15.1 Bioenergetics 417
15.2 Digestion, Absorption, and Transport 421
15.3 Carbohydrate Catabolism 429
15.4 Tricarboxylic Acid Cycle 434
15.5 Electron Transport and Oxidative Phosphorylation 437
15.6 Lipid Catabolism 441
15.7 Amino Acid Catabolism 446

CHAPTER 16 Anabolism 453

16.1 Introduction to Anabolism 454
16.2 Photosynthesis 455
16.3 Biosynthesis of Carbohydrates 460
16.4 Biosynthesis of Lipids 465
16.5 Biosynthesis of Amino Acids 472
16.6 Biosynthesis of Nucleotides 474

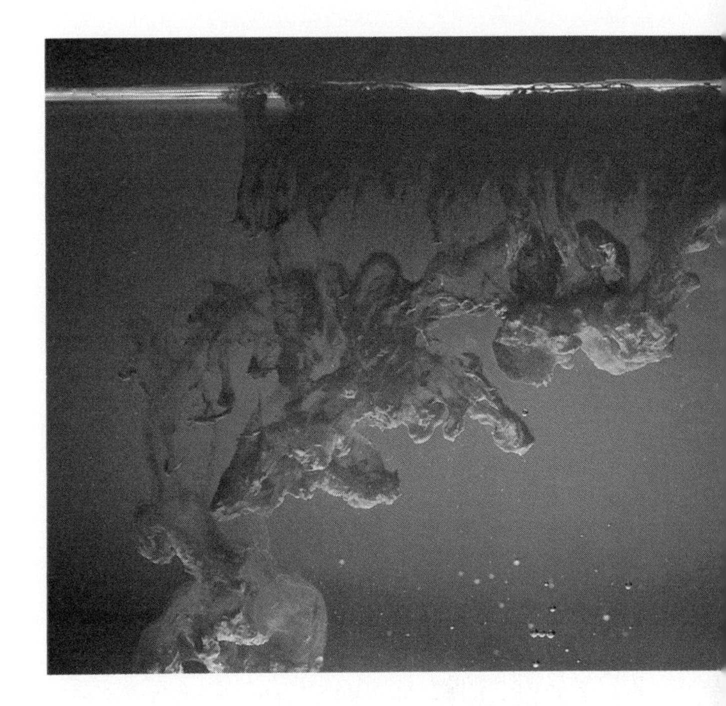

CHAPTER 17 Blood: The Constant Internal Environment 478

17.1 Gas Exchange 479
17.2 Removal of Wastes from the Blood 483
17.3 Regulation of pH 484
17.4 Circulation of Hormones and Their Role in Nutrient Maintenance 487
17.5 Blood and Immunity 492

Solutions to Odd-Numbered Exercises S.1

Glossary G.1

Photo Credits PC.1

Index I.1

ELECTRONS, BONDING, AND MOLECULAR SHAPES

This photo of a woman climbing a rock wall illustrates a stark contrast between two areas of chemistry: living and nonliving; organic and inorganic. The chemical structure of the sandstone is based on a framework of silicon atoms; it is inanimate and without life. In contrast, the chemical structure of the woman is based principally on molecules with a framework of carbon atoms. Carbon chemistry, the topic of this book, is the chemistry of life.

OUTLINE

R.1 An Introduction to the Chemistry of Carbon
R.2 The Behavior of Electrons in Atoms
R.3 Chemical Bonding
R.4 Shapes of Covalent Molecules
R.5 Molecular Dipoles
R.6 Behavior of Polar Molecules
R.7 Hydrogen Bonding

OBJECTIVES

After completing this chapter, you should be able to

1. Sketch shapes for *s* and *p* orbitals.
2. Use the electron filling series to predict the electron arrangement for neutral atoms up to atomic number 20.
3. State Hund's rule and its significance for predicting the arrangement of electrons in an atom's orbitals.
4. Sketch electron-dot models for neutral atoms.
5. Define *ionic* and *covalent bonding*.
6. Predict ionic charges of stable ions of elements up to atomic number 20.
7. Predict bond type by reference to the electronegativities of the elements involved.
8. Define the terms *dipole* and *molecular polarity*.
9. Predict the shape and molecular polarity of simple covalent compounds by use of electron-pair repulsion theory.

10. Explain how ionic compounds are dissolved in polar solvents.

11. Define the term *hydrogen bonding* and explain how the strength of the hydrogen bond is related to the strength of the molecular dipole.

R.1 An Introduction to the Chemistry of Carbon

This book is about the chemistry of living things. This chemistry involves only a very few elements, the same ones for plants and for people—mostly carbon, oxygen, hydrogen, and nitrogen, although many elements such as iron show up in trace amounts. Carbon, however, plays the central role.

The chemistry of carbon compounds is divided into two categories—organic chemistry and biochemistry. The carbon-based compounds studied in organic chemistry are derived from natural gas, petroleum, and coal—remnants of ancient vegetation and animals. Because the carbon-based molecules of prehistoric plants and animals have been stored for millennia, they have decomposed into simpler molecules. The molecules of organic chemistry are actually subsets of the original carbon-based compounds that formed the living organisms.

Organic chemistry affects each of us in countless ways. For the clothes we wear, the medicines we take, the preservatives we use, the herbicides and pesticides that protect our food supplies and the plastics that are used for containers and buildings, we are indebted to organic chemistry. We often find organic chemists making novel compounds that do "better" than nature. Wonder fibers such as those found in Goretex and Hollofil provide winter clothing that protects us against cold and water better than goose down. Goretex is also used to make NASA spacesuits. (Figure R.1)

Synthetic drugs target specific diseases and physiological conditions. Chemists are developing ways to make hydrocarbon fuels in the laboratory to reduce reliance on ever-scarcer foreign sources for petroleum. The use of chemicals from the reproductive system to control fertility has made a dramatic influence on the lives of many people. (Figure R.2)

(a)

(b)

FIGURE R.1
Goretex is used (a) in all-weather gear and (b) in high-tech NASA space suits.

FIGURE R.2
Birth-control methods have brought enormous changes to our society, allowing couples to plan when and if to have children.

Biochemistry, on the other hand, is the chemistry of living systems. Organisms and their chemical basis are fascinating topics, but they do not differ from the nonliving part of the world in any basic or fundamental way. All of the physical and chemical rules you have learned so far apply to biochemistry. Atomic structure and bonding are not different in biomolecules. The basic carbon structures and functional groups of organic chemistry are present in the molecules of living systems. Energy is vital to biochemical systems, and the rules that explain energy relationships in physics and chemistry are the same rules that govern energy relationships in biochemistry. Biochemistry differs from other areas of chemistry primarily in the complexity of its molecules.

To study and understand the molecules involved in organic and biochemistry requires that one recall some basic principles of atomic structure, electron distribution, and bonding. The behavior of electrons is really central to the chemical behavior of matter, because the outer part of all atoms is comprised of electrons, and it is by transferring and sharing these electrons that bonds are formed and molecules are built from individual atoms. (Figure R.3)

The first part of this chapter provides a brief review of some of the important ideas that we use to predict the shapes of the regions around the nucleus in which electrons are found. The second part of the chapter shows how the distribution of pairs of electrons around an atom determines the shape and solubility characteristics of many molecules.

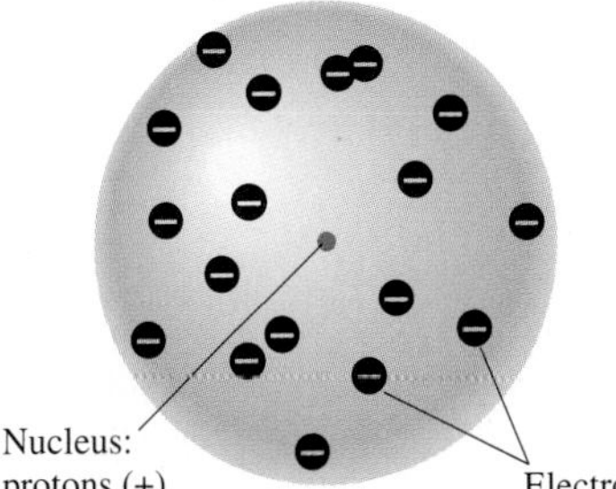

FIGURE R.3
Atoms are comprised of a tiny nucleus containing protons and neutrons, surrounded by a cloud of electrons. Electrons appear to be found at certain distances from the nucleus.

R.2 The Behavior of Electrons in Atoms

An atom-sized observer might tell us that, from his or her point of view, atoms appear to be mostly empty space and electrons, with a tiny nucleus containing protons and neutrons in the center. Suppose that a carbon atom could be enlarged to be a sphere as large as an ordinary gymnasium. The carbon nucleus with its six protons and six neutrons would be a tiny kernel in the center, about the size of a marble. The remaining space—mostly empty—would be used by six gnat-sized electrons as they buzz rapidly around.

There are two ways the gnat-sized electrons could use this space. They might fly randomly through it. Or they might be more territorial, each gnat or group of gnats staying pretty much in its own part of the gym. Our atom-sized observer reports that the latter is the case—the gnat-sized electrons are usually most stable when in groups of two, and there are a number of well-defined territories in which each pair of gnats spend most of their time. Gnats are seldom found in the no-man's-land between the territories.

Orbitals

♦ The space occupied by a pair of electrons is called an *orbital*.

Such is the case with electrons. Each pair of electrons is most likely to be found in its own region in space around the nucleus, and this territory is called an **orbital**. Orbitals never hold more than two electrons, and electrons are seldom found far outside their orbital. Orbitals may be classified in terms of energy—those in which an electron has the lowest energy are of the simplest shape, are closest to the nucleus, and are filled first as electrons are added to an atom. Some of these clouds are spherical, others are dumbbell-shaped, and still others are shaped like a four-leaf clover.

♦ The first major group of electrons has only one orbital, a *s* shape.

Electrons appear to form groups around the nucleus. (These groups are sometimes called *shells*.) The first group, closest to the nucleus, contains only two electrons. For this group of electrons, only one type or shape of electron orbital is possible. This orbital is of the simplest shape: a spherical orbital designated *s* (Figure R.4).

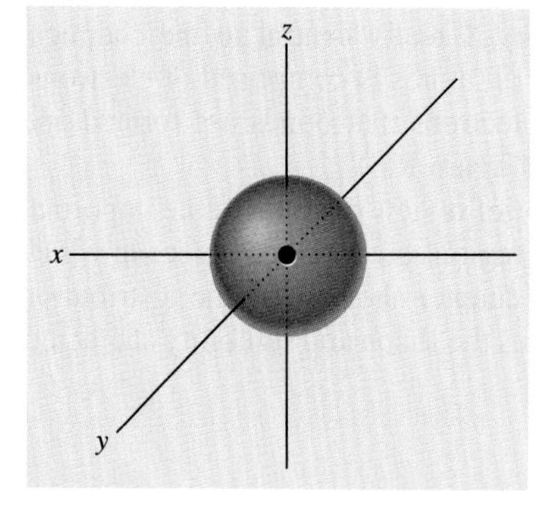

FIGURE R.4
For the first major electron group, only one **orbital** can exist. This orbital is spherically symmetrical. It is an *s* orbital.

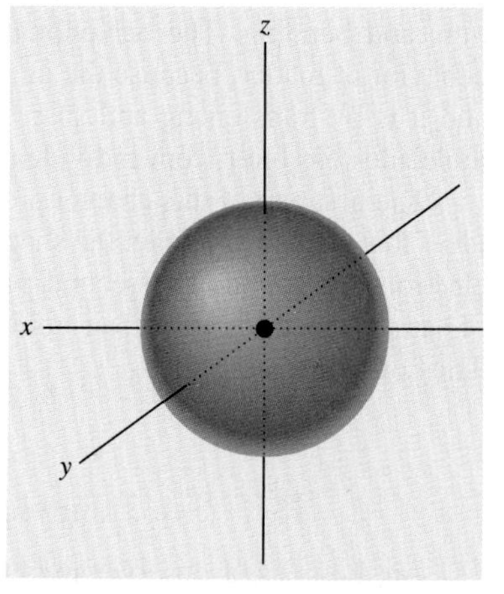

FIGURE R.5
The first of the orbitals in the second principal group of electrons is spherically symmetrical but larger than the *s* orbital of the first group.

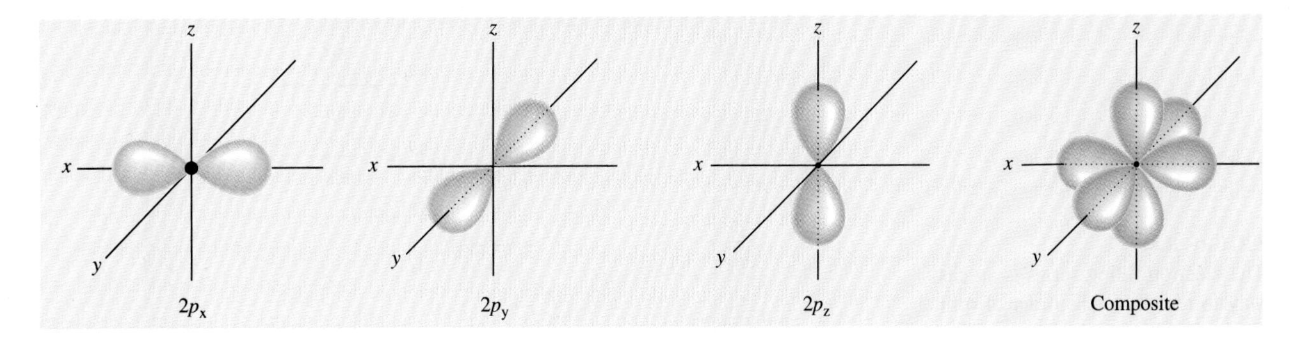

FIGURE R.6
Three *p* electron orbitals are possible. These orbitals are dumbbell-shaped and oriented at right angles to each other.

For the second principal group of electrons, two orbital shapes are possible. The first shape is spherically symmetrical. It is called a 2*s* orbital. (Figure R.5)

♦ The second major group of electrons has two orbital shapes possible, *s* and *p*.

The next set of orbitals in the second electron group have a *p* or dumbbell shape. Three *p* orbitals are possible. These orbitals are shown in Figure R.6. They are oriented on arbitrarily assigned *x*, *y*, and *z* axes. and are called p_x, p_y, and p_z orbitals. Note that they do not overlap. The second group of electrons will hold up to eight electrons.

♦ There are three *p* orbitals. These are oriented so that they do not overlap.

The third major group of electrons has three different types of orbitals. It has one *s* orbital, three *p* orbitals, and five *d* orbitals, four of which have shapes resembling a four-leaf clover. The *d* orbital electrons do not begin to fill until atomic number 21 in the periodic table.

♦ The third major group of electrons has *s*, *p*, and *d* orbitals.

As a shorthand, each orbital can be designated by a number that indicates its major electron group or distance from the nucleus, a letter describing its shape, and if necessary, a coordinate designation (*x*, *y*, or *z*, corresponding on a map to east-west, north-south, or up-down) describing its directional orientation. This coordinate designation is shown as a subscript to the shape indicator.

Thus an electron may be described as being in a 1*s* orbital (first major electron group, spherical orbital); a $2p_x$ orbital (second major electron group, dumbbell-shaped orbital on the *x* axis of the arbitrary coordinate system surrounding the nucleus); or a $4p_z$ orbital (fourth major electron group, dumbbell-shaped orbital, *z* axis). This notation is illustrated in Figure R.7.

♦ The common orbitals have spherical (*s*), dumbbell (*p*), or 4-leaf clover (*d*) shapes. Lowest electron energies are associated with the simplest shapes.

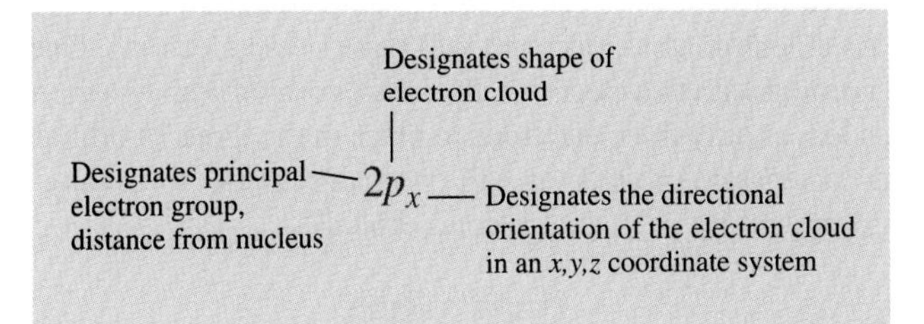

FIGURE R.7
Orbitals may be designated by a convenient notation that designates the distance of the major electron group from the nucleus, the shape of the orbital, and the directional orientation of this orbital with respect to a coordinate system whose origin is at the nucleus.

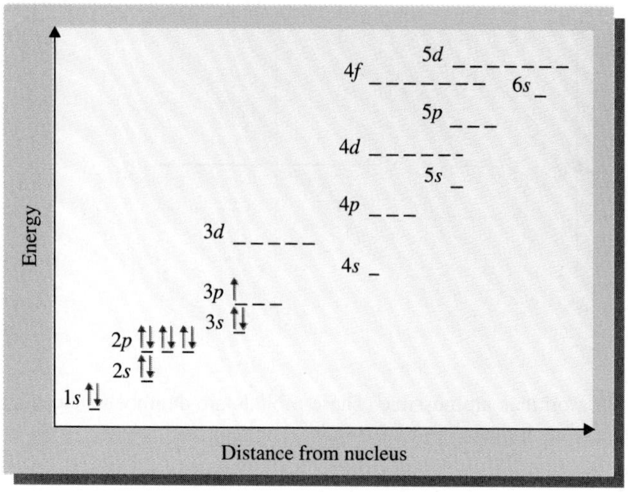

FIGURE R.8
The electron filling series is an easy way to remember the order of orbital filling. Orbitals each hold two electrons and fill from the lowest-energy orbitals first. The example shown is for aluminum, atomic number 13. Aluminum's 13 electrons are shown as arrows.

The Electron Filling Series

♦ *The electron filling series* is an easy way to predict the orbital filling sequence.

The electron filling series is an easy and useful way of remembering the sequence of orbital filling. This process of building up the electron structure of any atom assumes that electrons will first fill orbitals of lower energy. Each orbital will hold two electrons. These electrons differ slightly in energy, a difference caused by their direction of spin. Electrons always pair with another electron of opposite spin. The direction of electron spin is commonly indicated by a vertical arrow with the head up or down.

♦ The *atomic number* of an element tells us the number of protons in its nucleus and the number of electrons in a neutral atom.

A graphic way of presenting the electron filling series is shown in Figure R.8. The electron arrangement for aluminum, atomic number 13, is illustrated as an example. The nucleus of an aluminum atom has 13 protons, hence a neutral atom of aluminum has 13 electrons. Note the arrow symbolism for electrons presented in this figure.

♦ The *d* orbitals do not fill until the next higher group's *s* orbital is filled.

Note that the 4*s* orbital is actually lower in energy than the 3*d* orbitals. Because of this, the *s* orbital of the fourth electron group will fill before the *d* orbitals of the third group. A similar situation holds for the *d* orbitals in the fifth, sixth, and seventh major electron groups.

The electron arrangement for neutral atoms of hydrogen, helium, and lithium is illustrated in Figure R.9. The rules are simple: lowest-energy orbitals fill first, two electrons per orbital.

Figure R.10 presents an electron filling series diagram for nitrogen, atomic number 7. As a neutral atom, nitrogen will have seven electrons. One begins by filling the 1*s* orbital with two electrons, then the 2*s* orbital with two more electrons. This process leaves only three electrons to place in the three 2*p* orbitals. Do these electrons go in separate orbitals, or will two pair together to occupy one orbital and the third remain by itself in a second orbital?

Hund's Rule

This question is answered by **Hund's rule**, which states that electrons will not pair in orbitals of the same energy until each orbital is singly occupied. This

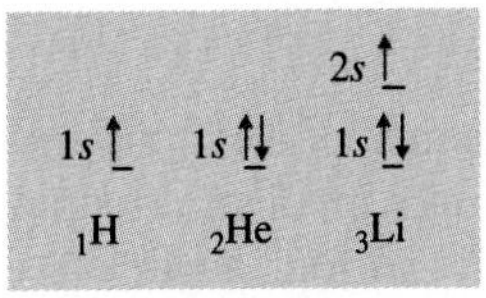

2s ↑

1s ↑ 1s ↑↓ 1s ↑↓

$_1H$ $_2He$ $_3Li$

FIGURE R.9
Electron energy diagrams for neutral atoms of hydrogen, helium, and lithium are illustrated in this figure. The number of electrons in a neutral atom is the same as the number of protons in the nucleus, as is indicated by the atomic number. It is normal practice to sketch only those orbitals actually occupied.

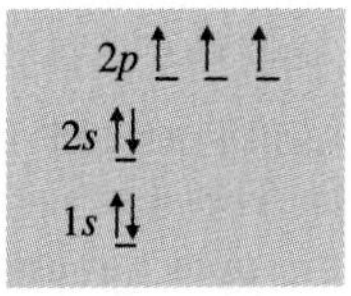

2p ↑ ↑ ↑

2s ↑↓

1s ↑↓

FIGURE R.10
An electron energy diagram for nitrogen. Note that the *p* orbital electrons are not paired.

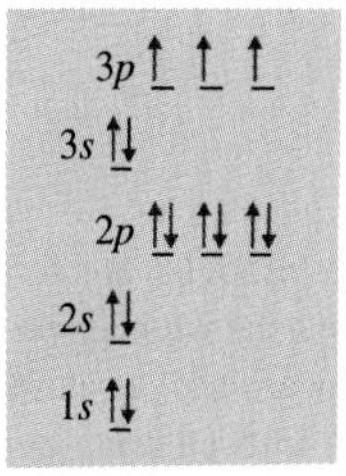

3p ↑ ↑ ↑

3s ↑↓

2p ↑↓ ↑↓ ↑↓

2s ↑↓

1s ↑↓

FIGURE R.11
An electron energy diagram for phosphorus. Note that the outer electron group arrangement for phosphorus is the same as that of nitrogen, without regard for the value of the major electron group: (s^2p^3).

rule is easily understood in terms of electron–electron repulsion. We recall that the three *p* orbitals differ in directional orientation, but not in energy. Since electrons carry a negative charge, it seems reasonable that a lower-energy situation would result if two electrons were in separate orbitals occupying different regions of space than if they occupied the same orbital. Thus the rule that electrons will not pair until each orbital of the same energy is occupied by one electron seems reasonable.

♦ Hund's rule states that electrons will not pair until each orbital of that energy is singly occupied.

Orbital Notation

In many cases it is convenient to express the electron configuration of an atom in a form that is more easily written than the electron filling series chart. In this case the orbital populations are simply listed, as in the case of nitrogen: $1s^22s^22p^3$. Phosphorus, atomic unumber 15, should have 15 electrons when in its neutral state. The electron filling series predicts that these electrons will be arranged as illustrated in Figure R.11. Note that the electron configuration of the outermost principal electron group (the third group for phosphorus—two *s* electrons and three *p* electrons) is the same as that for nitrogen! Note the relative locations of nitrogen and phosphorus in the periodic table. They are both in the same vertical column, or family of elements.

♦ Orbital notation is a convenient way to show the electron arrangement in an atom.

EXAMPLE R.1

Sketch an electron-energy diagram for a neutral atom of oxygen, $_8O$, and write its electron configuration in orbital notation.

Solution:

2p ↑↓ ↑ ↑ $_8O$

2s ↑↓

$1s^22s^22p^4$

1s ↑↓

Electron Dots

♦ Electron dot models are sometimes used to show the *outer-group* electron arrangement around an atom. They are quite useful in predicting chemical bonds.

Another useful method of indicating the arrangement of electrons around an atom is the **electron-dot** method. This approach simply involves arranging dots representing the outer-group electrons symmetrically in four quadrants around the symbol for the element. We must, however, sketch an electron-energy diagram to determine the number of outer-group electrons. The electron-dot method is quite useful in predicting chemical bonds. Since bonds form only with outer-group electrons of adjacent atoms, it has become common practice to indicate only the outer-group electrons in electron-dot diagrams. Note that electron-dot diagrams do not distinguish between *s* and *p* electrons.

EXAMPLE R.2

Construct an electron-dot model for a neutral atom of the element sulfur, $_{16}S$.

Solution: 3*p* ↑↓ ↑ ↓
3*s* ↑↓

2*p* ↑↓ ↑↓ ↑↓

2*s* ↑↓
1*s* ↑↓

Note: Electron-dot models only show outer group electrons.

$_{16}S$:Ṡ̤·

EXAMPLE R.3

Construct an electron-dot model for a neutral atom of the element argon, $_{18}Ar$.

Solution:

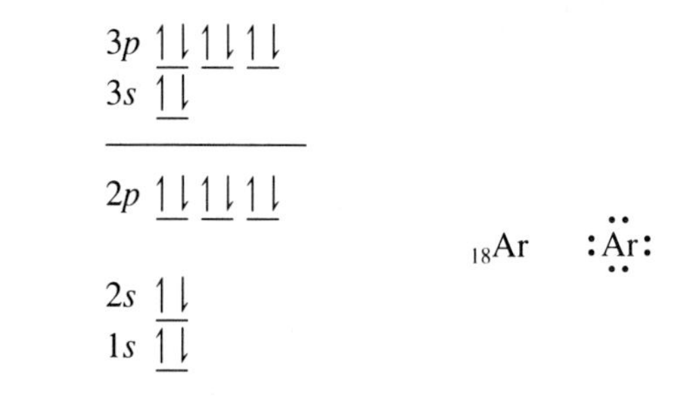

EXAMPLE R.4

Construct an electron-dot model for a neutral atom of the element neon, $_{10}Ne$.

Solution:

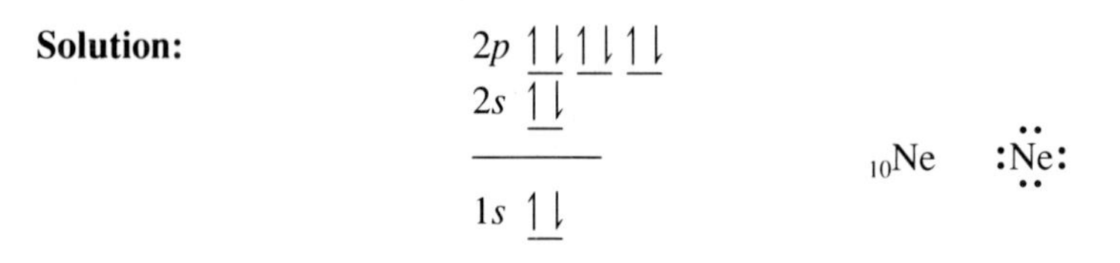

Note the similar configuration of outer-group electrons around the oxygen an sulfur atoms. Both oxygen and sulfur contain a filled *s* orbital, a filled *p* orbital, and two half-filled *p* orbitals. This similarity in electron arrangement results in similar chemical reactivity toward other atoms. From the point of view of an adjacent atom wishing to form a chemical bond, oxygen and sulfur look quite similar. Note also the similarity between argon and neon. These elements, members of the family called *noble gases*, are both chemically nonreactive.

♦ Elements in vertical columns of the periodic table have similar outer-group electron configurations. This is the reason for their similar chemical behavior.

Is it the number and arrangement of electrons in the outermost populated group of electrons that determines the chemical reactivity of the element? This is indeed the case, as we will see in the next section of this chapter.

NEW TERMS

orbital A region in space around the nucleus occupied by up to two electrons.

the electron filling series A system by which orbitals are filled with electrons, beginning with orbitals of lowest energy.

Hund's rule Electrons will not pair (occupy the same orbital) until all orbitals of that level have at least one electron. Electrons are negative, and will stay as far apart as possible as long as possible.

orbital notation A shorthand for indicating the orbital "addresses" of electrons. An atom with two 1*s* electrons, two 2*s* electrons, and one 2*p* electron could have its orbital population expressed as $1s^2 2s^2 2p^1$.

electron-dot model A model of the atom that indicates *only* the outer group electrons as dots surrounding the symbol for the atom. Useful principally for showing electron sharing and electron transfer, but does not specify the orbitals involved.

TESTING YOURSELF

Orbitals and the Electron Filling Series

1. A (*s*, *p*, *d*) orbital looks like a dumbbell.
2. According to the electron filling series and Hund's rule, how many *p* orbitals would be occupied in a neutral atom of oxygen (atomic number 8)? How many *p* orbitals would be filled (two electrons)?
3. The orbital notation for the outermost electron in a neutral atom of sodium is ______.
4. The electron configuration of a neutral atom of the element S is ______.
5. Sketch an electron-dot model of the molecule ammonia, NH_3.

Answers **1.** *p* **2.** 3, 1 **3.** 3*s* **4.** $1s^2 2s^2 2p^6 3s^2 3p^4$ **5.** H:N̈:H (with H below N; lone pair above N)

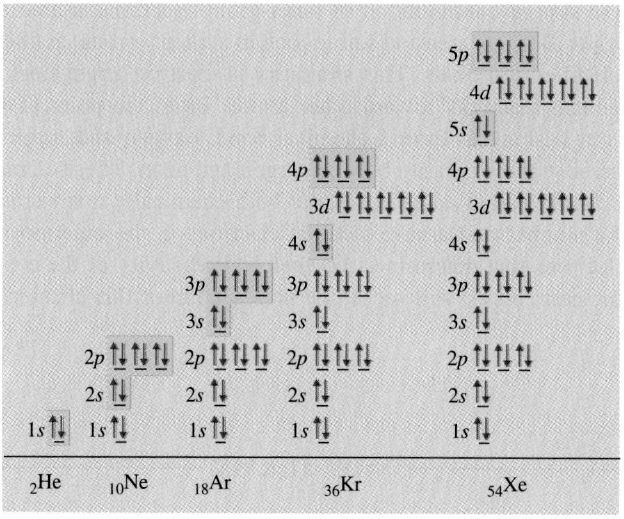

$_2$He	$_{10}$Ne	$_{18}$Ar	$_{36}$Kr	$_{54}$Xe
				5p ⇅⇅⇅
				4d ⇅⇅⇅⇅⇅
				5s ⇅
			4p ⇅⇅⇅	4p ⇅⇅⇅
			3d ⇅⇅⇅⇅⇅	3d ⇅⇅⇅⇅⇅
			4s ⇅	4s ⇅
		3p ⇅⇅⇅	3p ⇅⇅⇅	3p ⇅⇅⇅
		3s ⇅	3s ⇅	3s ⇅
	2p ⇅⇅⇅	2p ⇅⇅⇅	2p ⇅⇅⇅	2p ⇅⇅⇅
	2s ⇅	2s ⇅	2s ⇅	2s ⇅
1s ⇅	1s ⇅	1s ⇅	1s ⇅	1s ⇅

FIGURE R.12
Each of the noble gas elements has a complete complement of *s* and *p* electrons in its outer electron group.

R.3 Chemical Bonding

A Rule for Chemical Stability: Electron Structure of the Noble Gases

♦ The noble gas elements are chemically stable as individual atoms.

♦ The noble gas elements all have a full complement of outer-group *s* and *p* electrons.

All members of the noble gas family have one common characteristic: a complete complement of *s* and *p* electrons in their outermost group of electrons. (See Figure R.12.) Helium, an apparent exception to this rule, technically qualifies because a full complement of *p* electrons in group 1 is, of course, zero.

The fact that the chemically stable noble gas elements all have the same outer-group electron configuration suggests that this might also be a criterion that would provide for the chemical stability of other elements. How can an element such as sodium, with only one outer-group electron, or chlorine, with seven outer-group electrons, meet this criterion? Transfer of electrons from one atom to another is one way to meet this requirement.

Ionic Bonding: Electron Transfer Is One Route to Chemical Stability

♦ Atoms can transfer electrons from one to another to achieve a stable electron configuration.

Figure R.13 presents electron distribution diagrams for sodium and chlorine. According to our proposed criterion for chemical stability outlined earlier (a full complement of *s* and *p* electrons in the outermost electron group), neither atom should be chemically stable by itself: a fact supported by experimental evidence.

However, suppose that it were possible to remove one electron from the sodium atom. This would leave it with no electrons in its third electron group and would make its second electron group (consisting of two *s* and six *p* electrons) the outermost set of electrons. Sodium now qualifies for chemical stability, based on the rule proposed earlier. The electron configuration of the positively charged sodium ion then becomes the same as the noble gas neon.

But what about chlorine? Chlorine has two *s* electrons and five *p* electrons in its outermost group of electrons. Suppose that the single *s* electron removed from the sodium atom were given to the chlorine atom. This atom now would have two *s* electrons and six *p* electrons in its outer electron group: the full complement of *s* and *p* electrons required for chemical stability. The negatively charged chlorine ion now has the same electron configuration as the noble gas argon.

Electron transfer from sodium to chlorine is illustrated in the example below. Both atoms now have a full complement of *s* and *p* electrons in their outer group of electrons.

	Na		Cl
		3p	↑↓ ↑↓ ↑
3s	↑	3s	↑↓
2p	↑↓ ↑↓ ↑↓	2p	↑↓ ↑↓ ↑↓
2s	↑↓	2s	↑↓
1s	↑↓	1s	↑↓

FIGURE R.13
The electron configuration of sodium and chlorine. Neither element has the full complement of *s* and *p* electrons in its outer electron group required for chemical stability.

EXAMPLE R.5

Show the electron transfer of one electron from sodium (Na) to chlorine (Cl).

Solution:

	$_{11}Na^+$		$_{17}Cl^-$
		3p	↑↓ ↑↓ ↑↓
3s	__	3s	↑↓
2p	↑↓ ↑↓ ↑↓	2p	↑↓ ↑↓ ↑↓
2s	↑↓	2s	↑↓
1s	↑↓	1s	↑↓

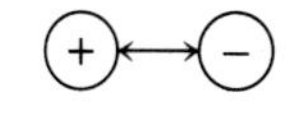

FIGURE R.14
Chemically stable charged atoms are called *ions*. Oppositely charged ions are attracted together by the difference in electrical charge. This is called an *ionic bond*.

What are the implications of such an electron transfer? Sodium, with 11 protons (11 + charges) in its nucleus, will now have only 10 electrons (10 − charges), for an overall charge of +1. And chlorine, with 17 protons in its nucleus, will now have 18 electrons, for an overall charge of −1. The transfer of electrons results in an unbalanced electrical charge at both atoms. Charged atoms are called **ions**. A positively charged ion is called a **cation**, a negatively charged ion an **anion**. The transfer of an electron from sodium to chlorine results in a full complement of outer group *s* and *p* electrons for each atom, but a +1 electrical charge will exist on the stable sodium ion, and a −1 electrical charge on the stable chlorine ion. These two ions will be held together by the electrical attraction between their unlike electrical charges. (Figure R.14.)

♦ An ion is a charged atom. Ions result when an atom has more or less electrons than the number of protons in its nucleus.

This type of chemical bond, an attractive force between ions of unlike electrical charge, is called an **ionic bond** or **electron transfer bond**.

♦ Ions of unlike electrical charge are attracted to each other. This is called an *ionic bond.*

Note that both the stable sodium ion and the stable chlorine ion have eight electrons in their outer group—an "octet." Before the discovery of orbitals, it was observed that a set of eight outer electrons appeared to confer chemical stability, and the **octet rule** for chemical stability was developed. The rule that a full complement of outer group *s* and *p* electrons results in chemical stability is more general in its application, however.

♦ Most atoms are chemically stable if they have an "octet" of electrons in their outer group.

Covalent Bonding: Electron Sharing Is Another Route to Chemical Stability

◆ Two atoms must have a large difference in their electron-attracting ability for electron transfer to occur.

The basic premise of ionic bonding is that electrons may be transferred from one atom to another, producing a stable noble gas electron configuration for both the resulting positive ion and the negative ion. The resulting ionic bond is caused by electrical attraction between the two oppositely charged ions. Ionic bonding is favored by very large differences in the electron-attracting ability of the component atoms, so that electrons can be easily transferred from one atom to the other.

Consider the opposite situation: that in which there is **no** difference in electron-attracting ability between the two atoms. An example of this is hydrogen gas, H_2. Hydrogen gas is composed of diatomic molecules, each containing two atoms of the simplest element known. The hydrogen atom does not have a stable electron configuration by itself, since it is short one electron of having a filled first electron group (Figure R.15). A hydrogen atom needs one more electron to be chemically stable.

An electron-transfer bonding model is not satisfactory for hydrogen gas, H_2. If one hydrogen atom were to remove an electron from another hydrogen atom, positive and negative hydrogen ions should result. Although this model predicts the correct formula for hydrogen gas, it provides a stable electron complement for only one of the two ions. Also, since there is no difference in the electron-attracting ability of the two hydrogen atoms, it is not reasonable to expect that one hydrogen should be able to remove an electron from the other.

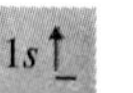

FIGURE R.15
The electron configuration of a hydrogen atom. It only has one electron.

However, if the two hydrogen atoms could *share* each other's electrons, this sharing would result in a full complement of electrons for each atom and, ac-

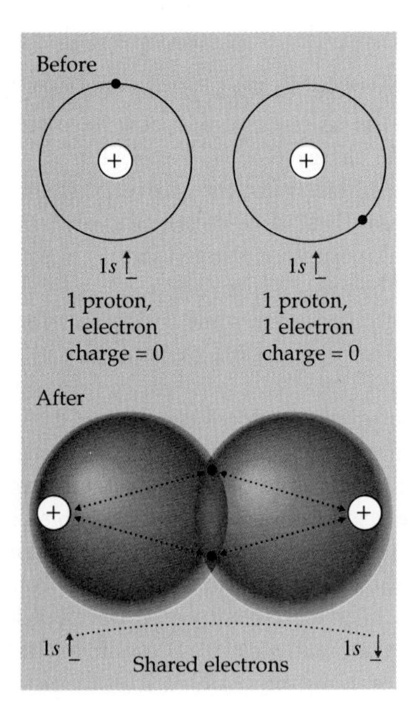

FIGURE R.16
An electron-sharing bond for hydrogen. By sharing electrons, both hydrogens achieve a stable electron configuration.

cording to our rule for chemical stability, a chemically stable situation for both atoms. Figure R.16 shows the overlap of the two 1*s* orbitals and a region of high electron density between the two hydrogen nuclei. The two electrons are shown as in a "flash" picture. *The common attraction of both positive nuclei for the shared negative electrons creates a* ***covalent*** *bond between the two atoms.*

♦ Atoms with small differences in electron-attracting ability share electrons to achieve a full set of *s* and *p* outer-group electrons.

Electron-sharing bonds of this type, where the region of maximum electron density is located along the axis between the two atoms, are called *sigma* bonds and are designated by the Greek letter σ. Since only two electrons may occupy a given orbital or region in space, any additional pairs of electrons shared between the two atoms must occupy space above and below or beside the sigma bond. These bonds are called *pi* (π) bonds. Sigma and pi bonding will be discussed further in the next chapter of this text.

♦ Covalent bonding is caused by a common attraction of two positive nuclei for one or more shared pairs of electrons.

♦ A sigma bond involves two electrons shared between adjacent nuclei. It is a linear bond.

Electronegativity and Bond Type

A convenient way to predict the ionic or covalent character of a chemical bond was suggested in the 1930s by Linus Pauling. Pauling assigned to each element an electron-attracting value he called **electronegativity**. These values ranged from a low of 0.7 for cesium to a high of 4.0 for fluorine. [The electronegativities of the elements are shown in Figure R.17.] Note that the electronegativity

FIGURE R.17
Electronegativities of the elements. Note that the electronegativity increases smoothly as one moves up and to the right in the periodic table.

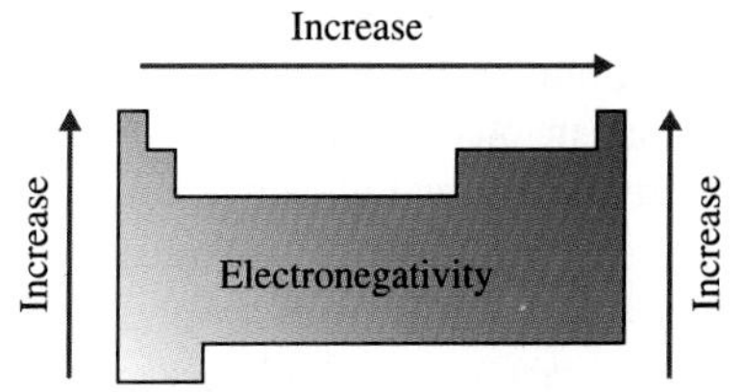

1	2	3	4	5	6	7	8	9	10	11	12	13	14	15	16	17	18
1 H 2.1																	2 He
3 Li 1.0	4 Be 1.5											5 B 2.0	6 C 2.5	7 N 3.0	8 O 3.5	9 F 4.0	10 Ne
11 Na 1.0	12 Mg 1.2											13 Al 1.5	14 Si 1.8	15 P 2.1	16 S 2.5	17 Cl 3.0	18 Ar
19 K 0.9	20 Ca 1.0	21 Sc 1.3	22 Ti 1.4	23 V 1.5	24 Cr 1.6	25 Mn 1.6	26 Fe 1.7	27 Co 1.7	28 Ni 1.8	29 Cu 1.8	30 Zn 1.6	31 Ga 1.7	32 Ge 1.9	33 As 2.1	34 Se 2.4	35 Br 2.8	36 Kr
37 Rb 0.9	38 Sr 1.0	39 Y 1.2	40 Zr 1.3	41 Nb 1.5	42 Mo 1.6	43 Tc 1.7	44 Ru 1.8	45 Rh 1.8	46 Pd 1.8	47 Ag 1.6	48 Cd 1.6	49 In 1.6	50 Sn 1.8	51 Sb 1.9	52 Te 2.1	53 I 2.5	54 Xe
55 Cs 0.8	56 Ba 1.0	57 *La 1.1	72 Hf 1.3	73 Ta 1.4	74 W 1.5	75 Re 1.7	76 Os 1.9	77 Ir 1.9	78 Pt 1.8	79 Au 1.9	80 Hg 1.7	81 Tl 1.6	82 Pb 1.7	83 Bi 1.8	84 Po 1.9	85 At 2.1	86 Rn
87 Fr 0.8	88 Ra 1.0	89 **Ac 1.1															

* Lanthanide series

58 Ce 1.1	59 Pr 1.1	60 Nd 1.1	61 Pm 1.1	62 Sm 1.1	63 Eu 1.1	64 Gd 1.1	65 Tb 1.1	66 Dy 1.1	67 Ho 1.1	68 Er 1.1	69 Tm 1.1	70 Yb 1.0	71 Lu 1.2

**Actinide series

90 Th 1.2	91 Pa 1.3	92 U 1.5	93 Np 1.3	94 Pu 1.3	95 Am 1.3	96 Cm 1.3	97 Bk 1.3	98 Cf 1.3	99 Es 1.3	100 Fm 1.3	101 Md 1.3	102 No 1.3	103 Lr 1.5

TABLE R.1

Percentage of Ionic Character in Relationship to Difference in Electronegativity

Difference in Electronegativity	Ionic Character (%)	Difference in Electronegativity	Ionic Character (%)	Difference in Electronegativity	Ionic Character (%)
0.1	0.5	1.2	30	2.3	74
0.2	1	1.3	34	2.4	76
0.3	2	1.4	39	2.5	79
0.4	4	1.5	43	2.6	82
0.5	6	1.6	47	2.7	84
0.6	9	1.7	51	2.8	86
0.7	12	1.8	55	2.9	88
0.8	15	1.9	59	3.0	89
0.9	19	2.0	63	3.1	91
1.0	22	2.1	67	3.2	92
1.1	26	2.2	70		

values are highest in the upper right of the periodic table and become less as one moves down and to the left in the table.

♦ Bond type may be predicted by computing the difference in electronegativity across the bond.

The ionic character of a chemical bond is predicted by calculating the **difference** in electronegativity of the two atoms involved. Large differences in electronegativity indicate a bond of high electron-transfer character, small differences indicate electron-sharing bonding. Table R.1 presents information that correlates difference in electronegativity with the percentage ionic or electron-transfer character of a single chemical bond.

♦ Polar bonds are covalent bonds in which the electrons are unequally shared.

Chemical bonds can be classified in three general categories, depending on their degree of ionic character. In general, bonds with low electronegativity differences (up to 0.5) are classified as **covalent**. Bonds with large electronegativity differences (1.7 or more) are arbitrarily classified as **ionic** and those of intermediate electronegativity difference as **polar covalent**. The dividing lines

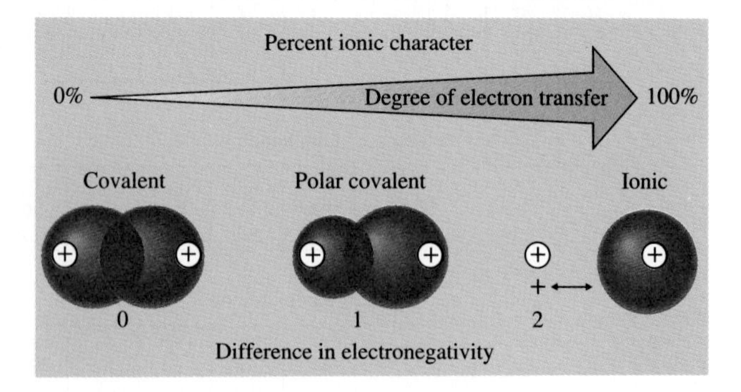

FIGURE R.18
The relationship between electronegativity difference and bond type is a gradually changing scale, with no clear-cut divisions. Bonds with small differences in electronegativity are classified as *electron-sharing* or *covalent*, bonds with large differences in electronegativity are classified as *electron-transfer* or *ionic*. Those bonds of intermediate electronegativity difference are classified as *polar covalent*.

between the three bond types are cloudy and subjective, however. The real situation, as illustrated in Figure R.18, is one of gradually increasing ionic character of the bond as the electronegativity difference between the two elements becomes greater. The only pure covalent bonds are those with zero difference in electronegativity across the bond—those bonds between atoms of the same element.

EXAMPLE R.6

Determine the type of bond in rubidium fluoride, RbF.

Solution:

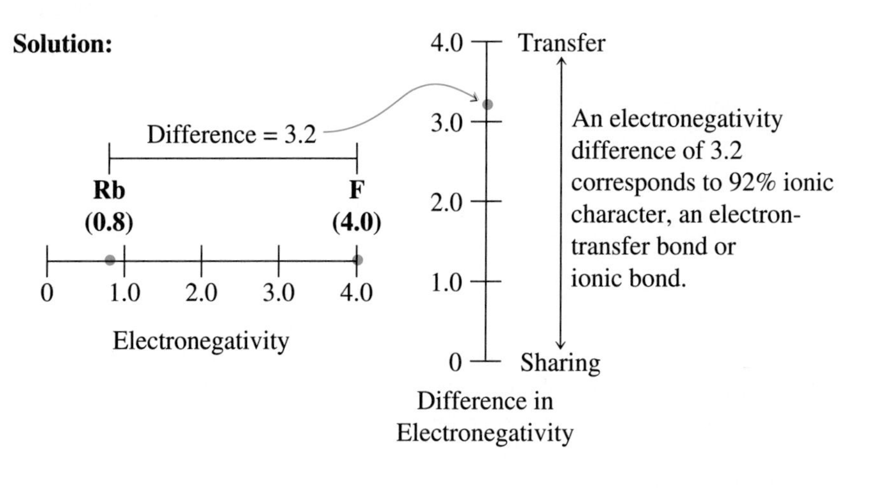

EXAMPLE R.7

Determine the type of bond in methane, CH_4.

Solution:

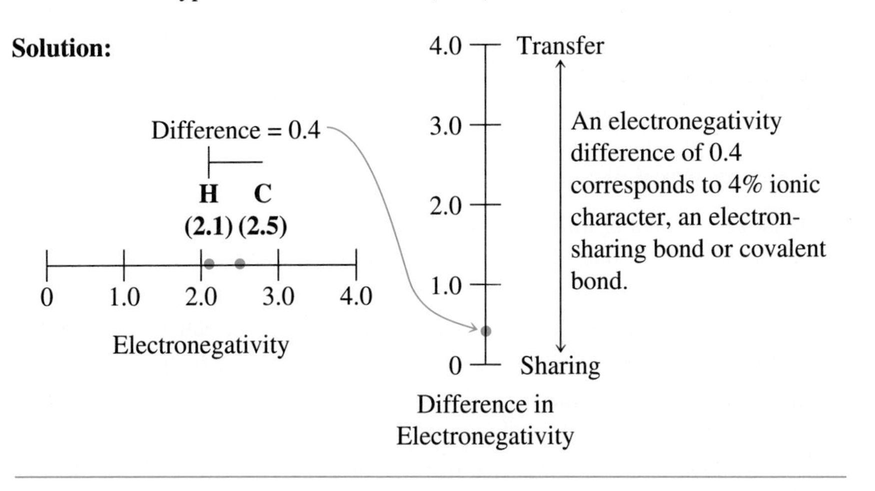

Some compounds, such as water, involve bonds that are intermediate between the electron-sharing or electron-transfer extremes of Pauling's scale.

EXAMPLE R.8

Determine the type of bond in water, H_2O.

Solution:

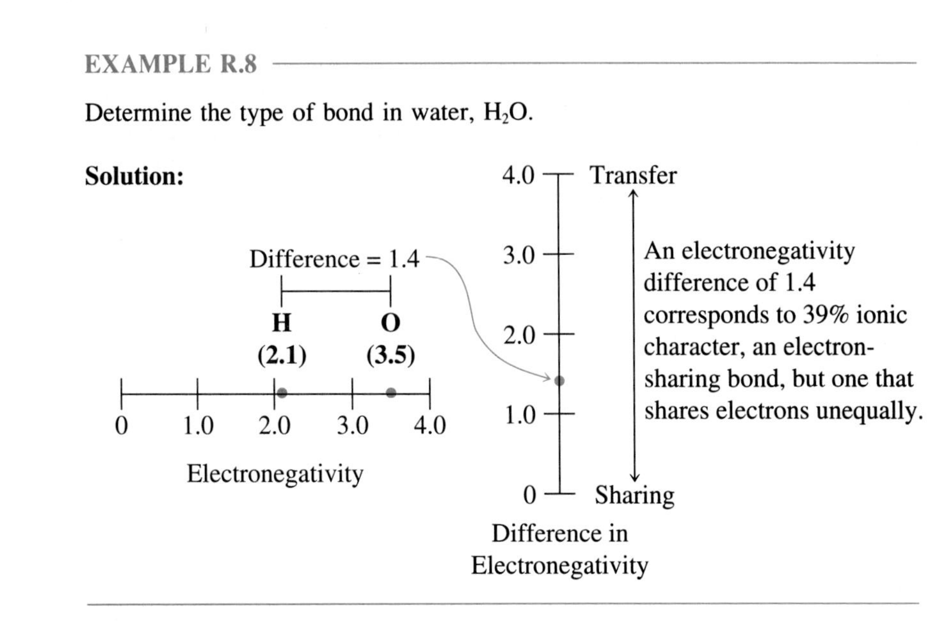

Such bonds are covalent, but with the electrons "shared" unequally. In the case of the hydrogen–oxygen bond found in water, the shared pair of electrons spend more time closer to the oxygen than to the hydrogen. This type of bond is called a **polar covalent bond** and results in the formation of an electrical **dipole** or region of unbalanced electrical charge. An electrical dipole is responsible for the unusual behavior of water. Dipoles will be discussed further in this chapter.

Although chemical bonds span the entire spectrum from complete electron transfer (ionic bonding) through intermediate polar covalent bonds to complete electron sharing (covalent bonding), *there are no clear-cut dividing lines between the bond types.* A similar situation exists in the growth of a person. A person can be classified as child, adolescent, and adult during his life, but it is difficult to draw clear-cut lines dividing these stages by age.

NEW TERMS

noble gases Atoms of this family are stable as individual atoms. They all have a full complement of outer-group *s* and *p* electrons. They are sometimes called the *inert gases.*

stable electron configuration A complete complement of outer-group *s* and *p* electrons. Often referred to as a "noble gas" electron complement.

electron transfer Transfer of an electron from one atom to another, generally to achieve a stable electron configuration.

ion A charged atom: an atom that has lost or gained electrons and no longer has the same number of electrons as it does protons in its nucleus. Positive ions are called **cations**, negative ions are called **anions**. Groups of atoms sometimes carry a charge; these are called *polyatomic ions.*

chemical bond An electrical force that holds atoms together.

ionic bonding A chemical bond formed by transfer of one or more electrons from one atom to another, and the resulting attraction between ions of unlike charge.

covalent bonding A chemical bond formed as two atoms share one or more pairs of electrons. The bond is due to the common attraction of each nuclei to the same pair(s) of electrons.

sigma bond (σ) Two electrons shared on the axis *between* the bonded atoms.

pi bond (π) Two electrons shared above and below or beside the axis between the bonded atoms.

electronegativity A measure of an atom's ability to attract electrons. One may predict the type of chemical bond that will form between two atoms by computing the difference in their electronegativities.

polar covalent bond A covalent bond in which electrons are shared unevenly. Except for bonds between atoms of the same element, with no difference in electronegativity, all covalent bonds have some polar character. For purposes of definition, bonds are considered to have polar properties when the electronegativity difference is 0.5 or greater.

dipole An electrical dipole is a region of unbalanced electrical charge. A dipole is a characteristic of all polar bonds. Dipoles also affect the solubility characteristics of a molecule, as we will see in the next chapter.

TESTING YOURSELF

Predicting Bond Type and Formulas

1. A stable ion's outermost group of electrons has a ______________.
2. By application of the electron filling series and your knowledge of the outer electron configuration of stable ions and elements, predict the charges of stable ions of the following elements:
   $_{3}Li$
   $_{8}O$
   $_{9}F$
   $_{19}K$
   $_{12}Mg$
   $_{15}P$
   $_{17}Cl$
   $_{19}K$
   $_{20}Ca$
3. The attractive force in a covalent bond is between the nuclei of the two atoms and the ______________.
4. What is the principal factor that determines whether a chemical bond will be ionic or covalent?
5. Use your knowledge of the electronegativity concept to classify the following compounds as ionic, polar covalent, or covalent:

Methane, CH_4 (natural gas)
Methyl alcohol

```
      H
      |
  H—C—O—H
      |
      H
```

Sodium iodide, NaI (iodized salt)
Stannous fluoride, SnF_2 (toothpaste)

6. Predict bond type for compounds of the following elements.
   Sodium and oxygen
   Hydrogen and sulfur
   Hydrogen and chlorine
   Carbon and fluorine

Answers **1.** A full complement of *s* and *p* electrons. **2.** Li = +1; O = −2; F = −1; K = +1; Mg = +2; P = −3; Cl = −1; K = +1; Ca = +2 **3.** Electrons shared between the two nuclei **4.** Difference in electronegativity between the two atoms in the bond **5.** Methane is covalent; methyl alcohol is covalent; sodium iodide is ionic; stannous fluoride is ionic **6.** Sodium oxide is ionic; hydrogen sulfide is covalent; hydrogen chloride is polar covalent; carbon tetrafluoride is polar covalent.

FIGURE R.19
The oil and vinegar to make this salad dressing do not mix well because of the different shapes of the oil and water molecules.

Nature Favors the Covalent Bond

Ionic compounds often come apart to their component ions when dissolved in water (this property is called *ionization*), but this behavior is not observed with most carbon-based molecules. On a planet in which water is one of the most abundant molecules, there is a definite advantage in having some molecules that do not come apart when they get wet.

The rocks and soil that make up the earth can afford to be slightly soluble in water. Living things, on the other hand, cannot afford to have their outer molecules come apart when they get wet. Because carbon is almost eactly in the center of the electronegativity scale, it forms covalent bonds with practically all other elements. For this reason, and because the four-bond character of carbon gives great flexibility in molecular design, all living things have covalently bonded molecules built on a carbon skeleton for their outer skin.

As we see next in this chapter, covalent bonding does not by itself keep a molecule from dissolving in water, but a covalent bond coupled with a symmetrical distribution of electrons around the atom will do the trick.

R.4 Shapes of Covalent Molecules

Figure R.19 shows something most of us have seen—a bottle containing salad dressing made from vinegar and oil. This salad dressing is made by mixing about two parts vegetable oil with one part vinegar (which is 95 percent water and 5 percent acetic acid) and a little salt added for flavor. As you can see in the figure, oil and vinegar do not mix. You have to shake the salad dressing vigorously before you pour it in order to get both oil and vinegar on your salad.

Figure R.20 shows another item common in the kitchen—mint extract, which contains oil of spearmint, oil of peppermint, and 90 percent ethyl alcohol. According to the label, it is 180 proof. (The proof of an alcohol solution is twice its percentage of alcohol.) Orange extract is only slightly less potent—160 proof. It contains 80 percent ethyl alcohol. The alcohol boils away during cooking, so it cannot be included as an intoxicant. It is included in such large proportions because the molecules responsible for flavor do not dissolve in water.

Figure R.21 shows another common occurrence in a kitchen, bacon grease being removed from a frying pan with soap and water. The soap is needed because the grease does not dissolve in water.

These three examples have something in common. What property of molecules permits one compound (ethyl alcohol) to dissolve with oil of spearmint, whereas other compounds (like vinegar and oil) do not mix? Grease does not dissolve in water unless soap is used. Why? The ability of one compound to mix with or dissolve into another is determined to a large degree by a property known as *molecular polarity,* which is related to the shape of each molecule and the symmetry with which electrons are distributed around it. The water, oil, and alcohol molecules behave differently because of their different shapes and symmetry of charge distribution.

In this section we will consider how to predict the shape of simple molecules and the symmetry of electrical charge distribution around these molecules. This symmetry will then be used to classify molecules into two groups—molecules with charge symmetry (nonpolar molecules) and molecules with nonsymmetric charge distribution (polar molecules). Polar molecules are attracted to each other, while nonpolar molecules move rather independently of each other. Some molecules have both polar and nonpolar parts. These molecules (as the alcohol and soap in the preceding examples) are able to mix freely with both polar and nonpolar molecules.

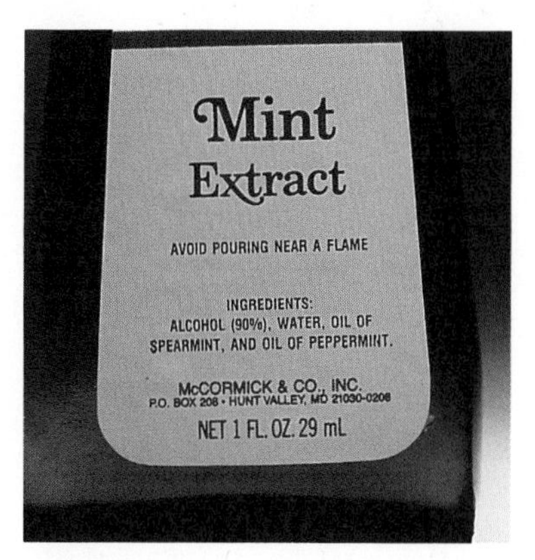

FIGURE R.20
Artificial flavor contains a large amount of alcohol. Note that, unlike the salad dressing, this material appears to be a homogeneous solution. The flavor molecules dissolve in alcohol.

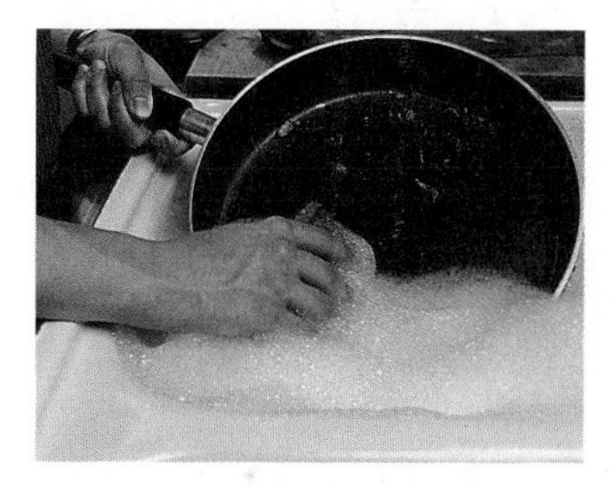

FIGURE R.21
Soap is an essential ingredient in dishwashing. It helps the grease dissolve in water.

Determining Molecular Shapes from Electron-Dot Models

Some quite reasonable predictions of molecular shape may be made by using two simple principles governing the behavior of electrical charges: unlike charges (+ and −) attract each other, and like charges (+, + or −, −) repel each other (Figure R.22). Electrons, being of like charges, repel each other. On the other hand, the unlike charges of nucleus (+) and electron (−) attract each other. The result is that whereas the positive charge of the nucleus of an atom holds the atom's electrons close by, the negative charges of the electrons cause them to stay as far apart from each other as possible. The result for stable molecules is that *pairs of electrons* around the bonded atoms place themselves as far apart as possible.

The name for this phenomenon is **valence-shell electron-pair repulsion (VSEPR)**. The term *valence-shell* refers to the outer-group electrons, and *electron-pair repulsion* describes the repulsion between these pairs of electrons.

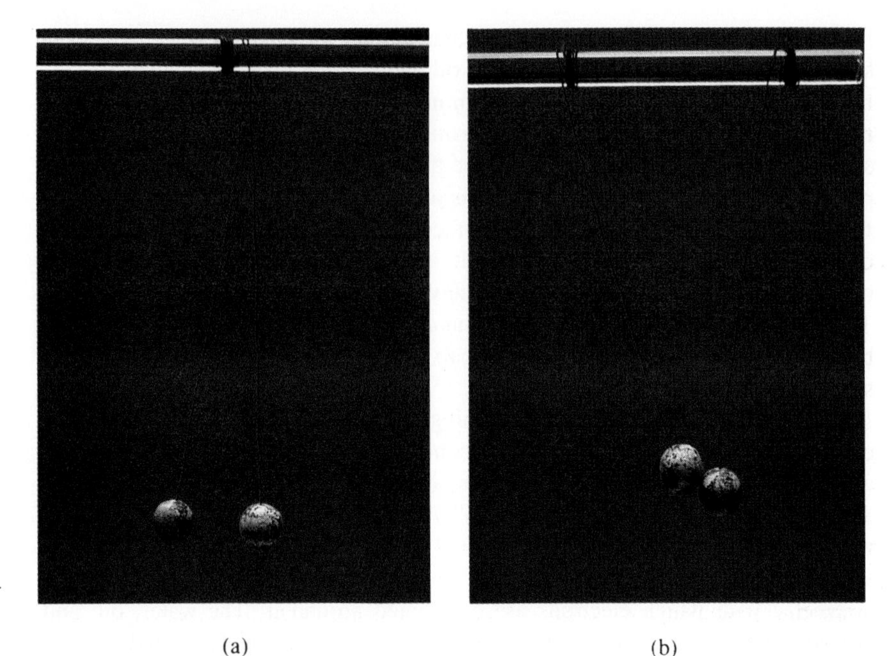

FIGURE R.22
While like electric charges repel each other (a), unlike electric charges attract each other (b). This principle causes pairs of electrons around an atom to place themselves in a symmetrical pattern as far apart from each other as possible.

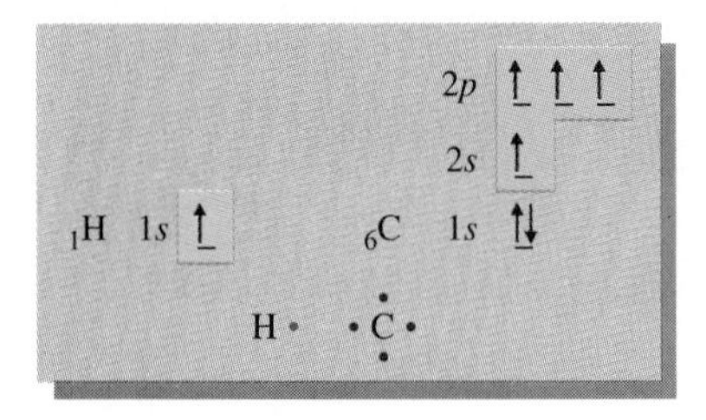

FIGURE R.23
Electron-dot models for neutral atoms of hydrogen and carbon reflect the electron complement of the atom's outer electron group.

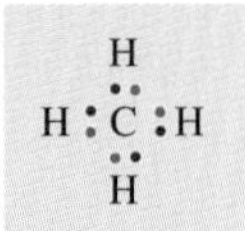

FIGURE R.24
An electron-dot model for methane. The carbon atom is surrounded by four pairs of electrons.

Consider the simple molecule methane (CH_4), commonly known as natural gas. Methane molecules each consist of one carbon atom and four hydrogen atoms. The electronegativity of carbon is 2.5; that of hydrogen is 2.1. The electronegativity difference is 0.4, resulting in a covalent bond. Figure R.23 shows electron filling series and electron-dot models for these two atoms. Note that, in the special case of carbon, the four outer electrons spread evenly across the four available orbitals.

Figure R.24 shows an electron-dot model of methane. Each carbon atom forms four single bonds with adjacent hydrogen atoms.

Consider now two types of forces acting on these pairs of electrons.

1. *Attractive forces:* Each pair of electrons is attracted both by the central +6 carbon nucleus, and its adjacent +1 hydrogen nucleus. This attraction holds the electron pair between the two nuclei. The common attraction of the two nuclei for the shared pair of electrons holds the two nuclei together, thus forming a chemical bond.
2. *Repulsive forces:* The four pairs of negative electrons, being of like electric charge, have strong repulsive forces between them. They move to get as far apart as possible.

Balloons can form useful models of molecules. Figure R.25 shows four balloons tied together in the center. The four balloons push against each other and move to get as far apart from each other as possible. The resulting geometric shape is called a **tetrahedron**.

The four electron pairs of the methane molecule likewise move to use the space around the carbon atom most effectively, each pair carrying its bonded hydrogen atom along. The result is a tetrahedral molecule with a carbon atom in the center (Figure R.26). Note that the electrical charge in this molecule is *symmetrically distributed.*

◆ The geometric configuration that places four pairs of electrons as far apart as possible is a tetrahedron.

The methane molecule has a mass of 16 amu. Methane boils at $-161°C$. Later in the chapter we will compare this value with that for some similar compounds.

Ammonia (NH_3) is a common molecule, similar in its structure to methane. The electronegativity of nitrogen is 3.0; that of hydrogen is 2.1. The bond between the nitrogen and carbon atoms is a polar covalent bond, with the bonding electrons shared a little closer to the nitrogen than to the hydrogen. Electron-dot models for nitrogen and hydrogen and an electron-dot model for the combination of these elements are shown in Figure R.27.

Note that, as with methane, the nitrogen atom in the ammonia molecule is surrounded by four pairs of electrons. As expected, these electrons arrange themselves as far apart as possible around the central nitrogen atom to form a tetrahedron (Figure R.28). There is a major difference between the methane and ammonia molecules, however. The distribution of electrical charge is symmetrical around the central carbon atom in the methane molecule, but this symmetry is not apparent in the ammonia molecule. As is shown in Figure R.28, an observer looking at the ammonia molecule from above first sees the negative charge of an unshared pair of electrons. Viewing the molecule from below, the observer sees three positive hydrogen nuclei. The ammonia molecule is far from symmetrical. Although the atomic mass of ammonia (17) is only 1 amu larger than that of methane, its boiling point is $-33°C$, about 130°C higher than methane. Why the difference in boiling point?

FIGURE R.25
A tetrahedron is a three-dimensional shape in which the four corners are as far apart as possible.

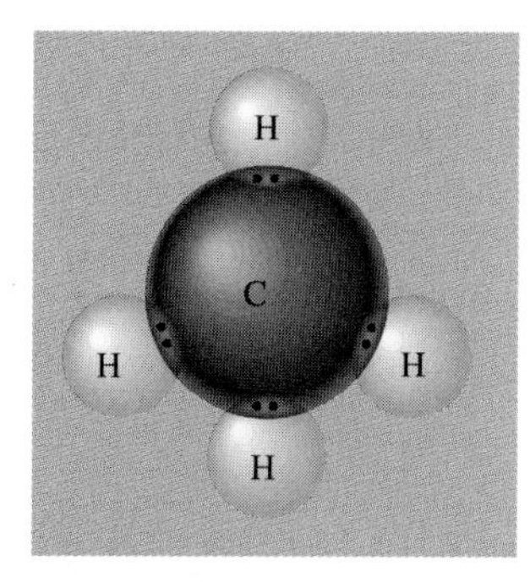

FIGURE R.26
Methane, CH_4, is a tetrahedral molecule. Note that the molecule is symmetrical.

$_1H$ 1s ↑
$_7N$ 1s ↑↓ 2s ↑↓ 2p ↑ ↑ ↑
H· ·N̈·
H:N̈:H
H

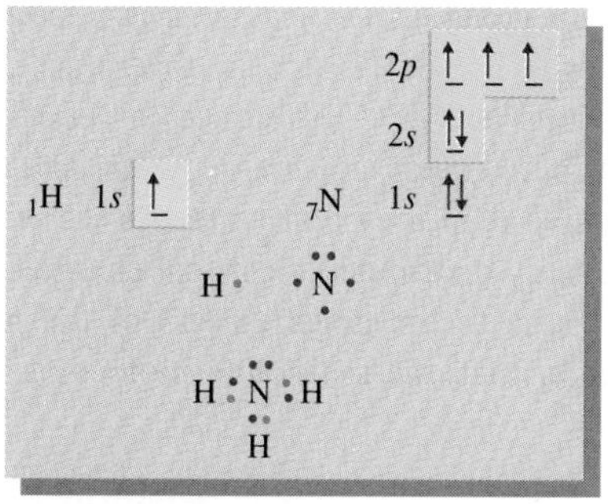

FIGURE R.27
An electron-dot model for the ammonia molecule. Note that the nitrogen atom has one electron pair that is not shared. This unshared pair occupies space just as do the bonding electron pairs.

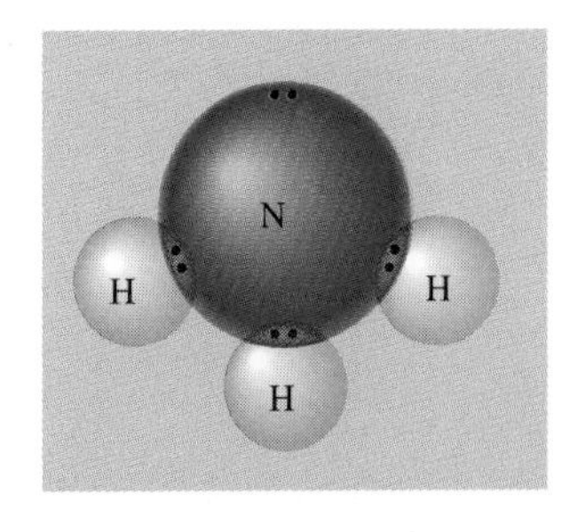

FIGURE R.28
The ammonia molecule is tetrahedral, with three of the four corners occupied by hydrogen atoms and the fourth corner occupied by an unshared pair of electrons.

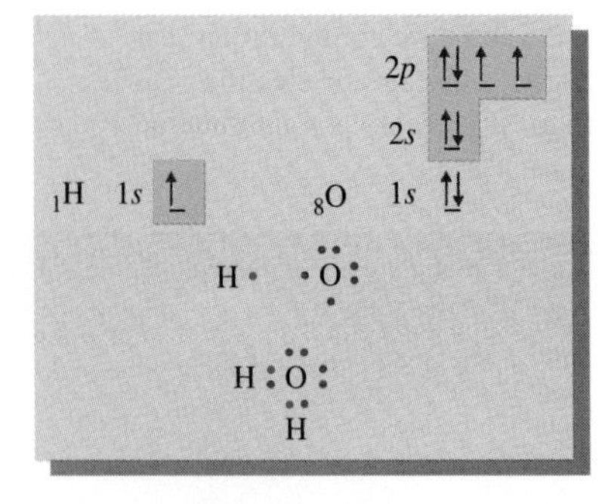

FIGURE R.29
An electron-dot model for a water molecule. Note that the oxygen atom has two unshared pairs of electrons.

TABLE R.2
The Boiling Point of Water Is Much Higher than One Would Predict from Its Molecular Mass

Compound	**Molecular Mass (amu)**	**Boiling Point (°C)**
Methane	16	−161
Ammonia	17	−33
Water	18	100

Figure R.29 shows an electron-dot model for a water molecule. The electronegativity of oxygen is 3.5; that of hydrogen is 2.1. The difference is 1.4 electronegativity units, resulting in a polar covalent bond. Again, the central atom is surrounded by four pairs of electrons. These electron pairs move as far apart from each other as possible, again forming a tetrahedron (Figure R.30). The water molecule is not at all symmetric. When viewing from the left in Figure R.30, an observer sees a negative charge from the unshared pairs of electrons. When viewed from the right, a positive charge from the exposed hydrogen nuclei is apparent.

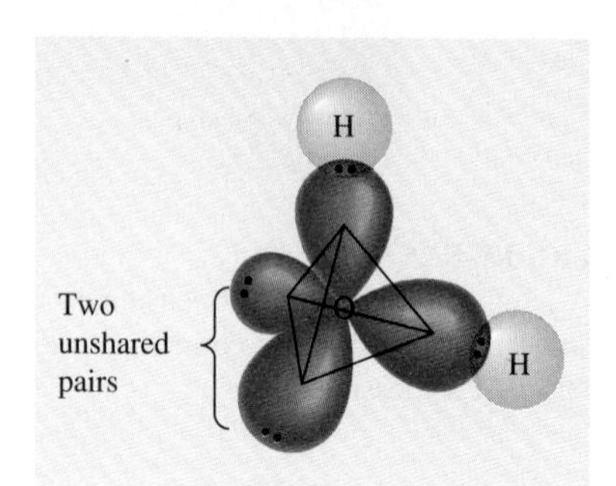

FIGURE R.30
The two hydrogen atoms and the two unshared electron pairs of a water molecule form a tetrahedron around the central oxygen atom.

Water, with a molecular mass of 18 amu, has a boiling point of 100°C. Although the methane, ammonia, and water molecules have very similar molecular masses, their boiling points are very different. In fact, there appears to be a direct correlation between the lack of symmetry of the molecule's electrical charge distribution and its boiling point. For molecules of similar molecular mass, the more symmetrical the distribution of electrical charge, the lower the boiling point. (Table R.2)

In general, the heavier a molecule, the more energy must be expended to vaporize it. The 11 percent difference in mass between methane and water hardly explains a 261°C difference in their boiling point. Because of their unevenly distributed electrical charge, water and ammonia behave as if they are much more massive than they really are. As we will see in the next section, molecules with unevenly distributed electrical charge can stick together—the positive end of the molecule to the negative end of the next—thus forming chains that are much heavier than the individual molecules.

Orbital Hydridization

♦ Symmetrical atomic orbitals are distorted as they combine during bonding. The process is called *hybridization.*

When a covalent bond forms between two atoms, the symmetrical orbital shapes of each unbonded atom's atomic orbitals are distorted, as the shared electrons are now held between the two bonded atoms, and the shape and orientation of the bonds shift to move the electron pairs as far apart as possible. This change in electron distribution that occurs during bonding is called **hybridization.**

Hybridized orbitals are named according to the set of orbitals from which they are formed. In the cases of methane, ammonia, and water, the set of four hybridized orbitals are formed from one atomic *s* orbital and three atomic *p* orbitals. They are thus called sp^3 orbitals. A central atom with sp^3 hybridization has tetrahedral geometry. Recall that the first bond between two atoms is called a *sigma* (σ) *bond.*

♦ An sp^3 hybridization results when an atom has a total of four electron pairs involved in single bonding and/or as nonbonded pairs.

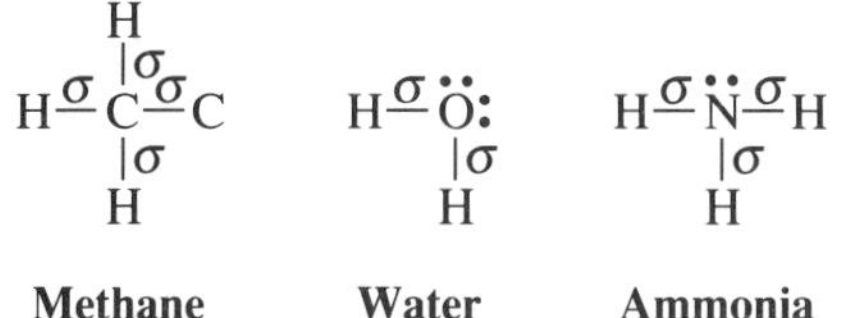

This is a linear bond, with the electron pair held between the two atoms. However, in some cases, two adjacent atoms must share more than one pair of electrons to reach their full noble gas complement of eight outer-group electrons.

Figure R.31 is an electron-dot model of the compound ethene (C_2H_4; sometimes called *ethylene*). This compound is a gas used in industry as the basis for many products (such as polyethylene plastic). It is also used to ripen bananas that were picked green. Note that, in the electron-dot model for this compound, each carbon atom shares two pairs of electrons with its neighboring carbon atom.

Each carbon atom is surrounded by three electron pairs that participate in sigma bonds, and one unhybridized *p* electron that is shared with a similar *p* electron in the adjacent carbon. Because the space between the atoms is occupied by the sigma-bonding electron pair, the next bond must use space above and below the sigma bond between the two carbon atoms. This overlap of unhybridized *p* orbitals is called a *pi* (π) *bond.*

The geometry of the system is determined by the number of sigma-bonding electron pairs (and, if present, by the number of unshared electron pairs). In this case, three sigma-bonding electron pairs determine the geometry (Figure R.32). When three atoms or unshared electron pairs surround a central atom, the resulting shape is a triangle. Because one *s* and two *p* orbitals have hybridized, the central carbon atoms are said to have sp^2 *hybridization.*

Some molecules, such as acetylene (C_2H_2), share three pairs of electrons. Figure R.33 shows electron-dot and hybrid orbital models of an acetylene molecule. In this case, each carbon atom forms a sigma bond to its adjacent hydrogen atom, and a sigma bond to the other carbon atom. Two electron pairs are involved in sigma bonding; the remaining electrons occupy overlapping *p* orbitals, forming

H H
:C :: C:
H H

FIGURE R.31
Ethane has a **double bond** between its two carbon atoms.

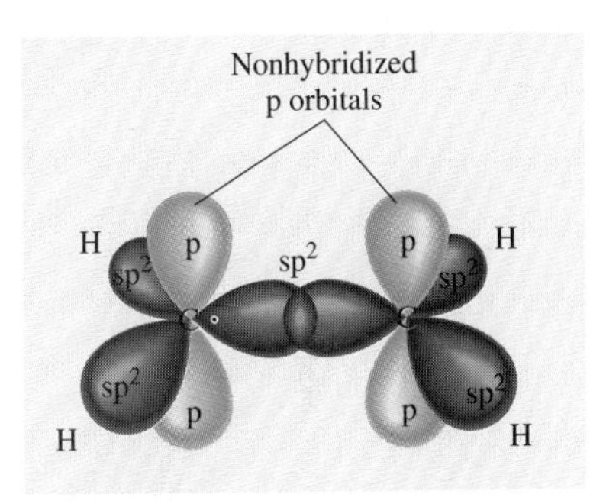

FIGURE R.32
When a double bond forms between two atoms, the first shared electron pair forms a sigma bond between the two atoms as two hybridized orbitals overlap. The second shared electron pair occupies overlapping, unhybridized *p* orbitals.

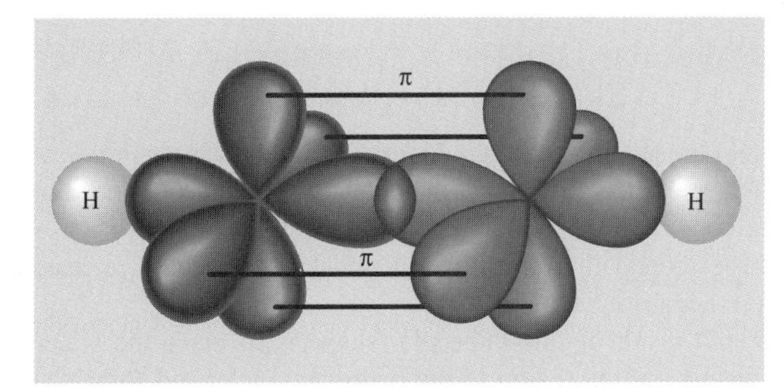

H••C⋮⋮C••H

FIGURE R.33
Acetylene is a linear molecule involving *sp* hybridization.

one pi bond above and below the sigma bond between the carbon atoms and one pi bond on either side of the carbon–carbon sigma bond. The hybridized orbitals are formed from a single *s* and *p* orbital; the hybrid is called an *sp* hybrid; such systems are linear.

NEW TERMS

valence-shell electron-pair repulsion (VSEPR) theory The idea that pairs of outer group electrons move as far apart as possible. It is used to predict the shape of the bonds around an atom.

tetrahedron A symmetrical geometric shape with four vertices and equal bond angles.

hybridization The rearrangement of bonding and nonbonding electrons around an atom that occurs during bonding.

TESTING YOURSELF

Covalent Molecules

Consider the compound carbon tetrafluoride, CF_4.

1. How many *bonding* pairs of electrons are found in the outer group of carbon's electrons?
2. How many *nonbonding* pairs of electrons are found in the outer electron group of carbon?
3. What is the hybridization of the carbon atom?
4. How many *bonding* pairs of electrons are found in the outer group of each fluorine?
5. How many *nonbonding* pairs of electrons are found in the outer electron group of each fluorine?
6. What is the hybridization of each fluorine atom?
7. What is the shape of the molecule?

Answers **1.** Four **2.** None **3.** sp^3 **4.** One **5.** Three **6.** sp^3 **7.** Tetrahedral

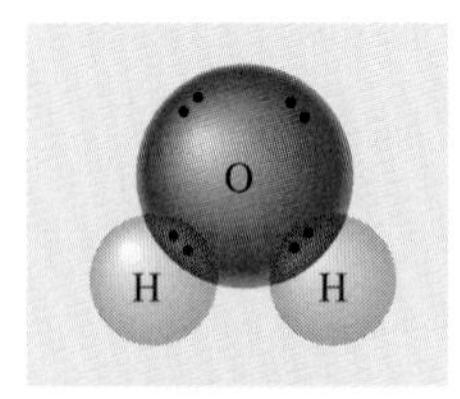

FIGURE R.34
The charge distribution in the water molecule is not symmetrical. The unshared electron pairs make the molecule negative on one side, whereas the two hydrogen atoms show a positive charge on the opposite side.

R.5 Molecular Dipoles: A Consequence of Molecular Shape and Electronegativity

The three-dimensional sketch in Figure R.34 illustrates an important property of water. The presence of hydrogen nuclei on one side of the molecule and unshared pairs of electrons on the other side results in an uneven distribution of electric charge around the molecule, an imbalance made more uneven by the greater electronegativity of oxygen.

The overall result of this molecular shape is to make the water molecule somewhat positive in the region of the hydrogen nuclei and somewhat negative in the region of the unshared electron pairs on the other side of the oxygen atom.

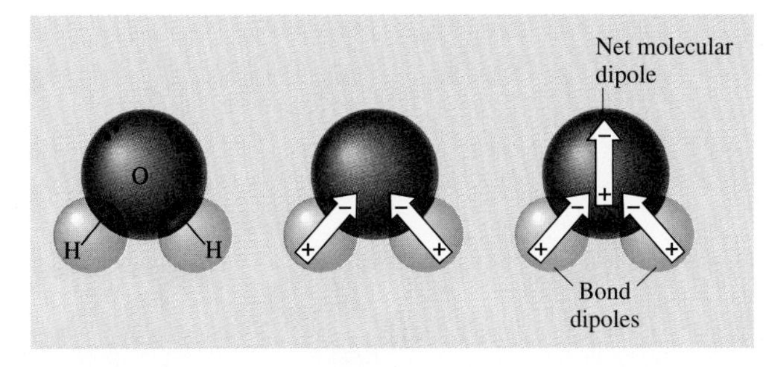

FIGURE R.35
The net molecular dipole for the water molecule is the sum of the individual H–O bond dipoles.

This unbalanced distribution of charge is called an *electric dipole* (Figure R.35); it is represented by an arrow pointing across the molecule from the region of extra positive charge and toward the region of extra negative charge. This unbalanced charge is said to form a **molecular dipole**, indicated by an arrow that points toward the region of extra negative charge. Since the water molecule has a strong molecular dipole, it is considered to be a **polar molecule**.

♦ If positive and negative charge is not arranged symmetrically in a molecule, a *molecular dipole* results.

The VSEPR and hybridization concepts can be used to predict the polar properties of any covalently bonded molecule. For example, ammonia (NH_3) is composed of nitrogen and hydrogen (Figure R.36).

Electron-pair repulsion theory predicts that ammonia's four pairs of electrons will form a tetrahedron as water does. The ammonia molecule's unshared pair of electrons give it a nonsymmetric charge distribution because there is no proton (hydrogen nucleus) to balance the negative charge of the unshared electron pair. The slightly polar N–H bond also contributes to this imbalance of electric charge. The dipole in the ammonia molecule, which is the sum of the individual bond dipoles, points toward the pair of unshared electrons.

♦ Electron-pair repulsion theory can be used to predict molecular polarity.

Methane is composed of carbon and hydrogen. Methane's four pairs of electrons form a tetrahedral structure, with a pair of electrons and a hydrogen nucleus at each corner (Figure R.37). Although very small dipoles exist across

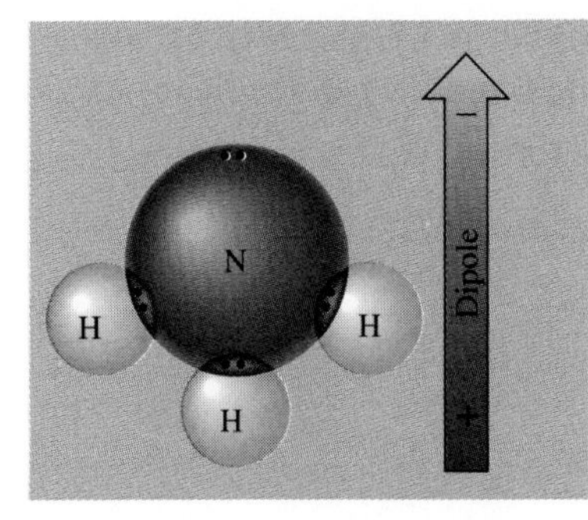

FIGURE R.36
Ammonia's unshared pair of electrons and its slightly polar N–H bond give it a nonsymmetrical charge distribution and make it a polar molecule. The net molecular dipole is the sum of the individual bond dipoles.

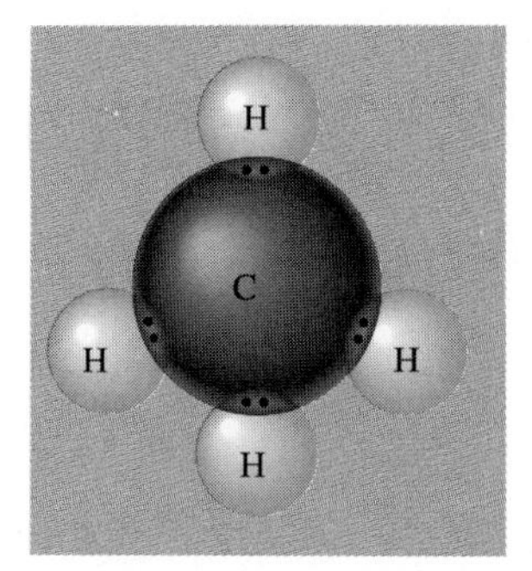

FIGURE R.37
The symmetrical distribution of electrical charge in methane makes it a nonpolar molecule. The individual bond dipoles are in opposite directions and cancel each others' effect.

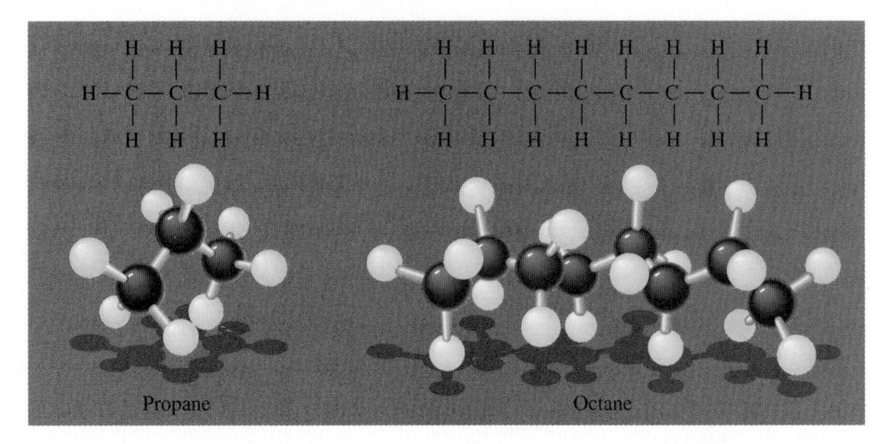

FIGURE R.38
As these examples of propane and octane show, all hydrocarbons have covalent bonds and symmetrical electron distribution and are nonpolar.

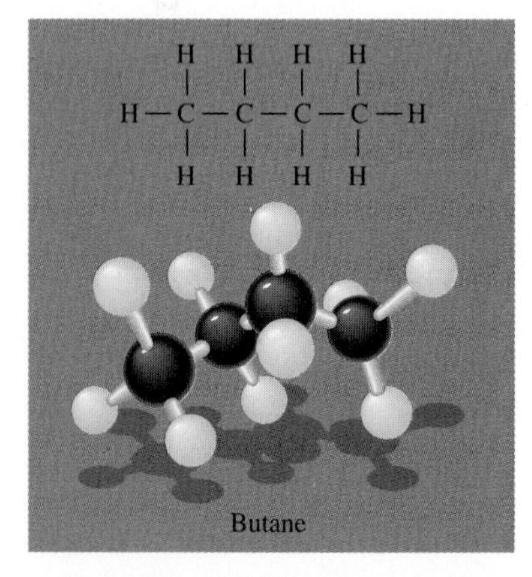

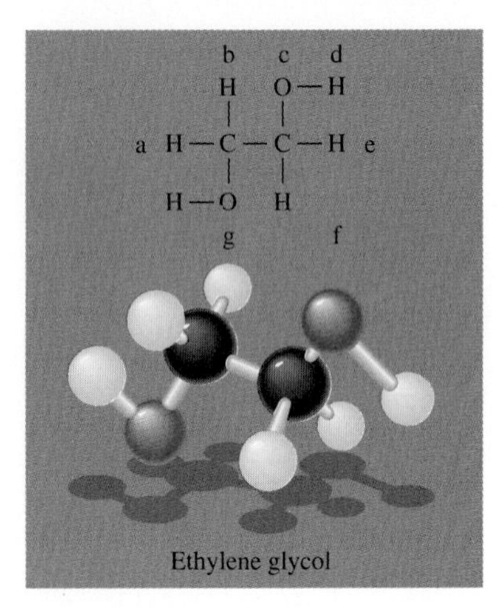

each of the H–C bonds, the symmetry of the structure causes their contributions to cancel. The symmetry of the electron distribution results in a balanced electrical charge. Thus, methane is considered to be a **nonpolar molecule**.

The C–H structure illustrated by methane can be extended by adding carbons and hydrogens to enlarge the molecule. Figure R.38 shows structural diagrams for propane (C_3H_8), a three-carbon hydrocarbon, and octane (C_8H_{18}), an eight-carbon hydrocarbon. Propane is a common fuel; octane is better known as a component of gasoline. Since the bonds are practically nonpolar and the electrical charge is symmetrically distributed, these molecules are considered to be nonpolar.

NEW TERMS

molecular dipole An unbalanced distribution of electrical charge across a molecule, giving one side a more negative charge and one side a more positive charge.

polar molecule A molecule with a molecular dipole.

nonpolar molecule A symmetrical molecule with no unbalanced electric charge.

TESTING YOURSELF

Predicting Molecular Polarity

1. Butane (C_4H_{10}) is a common fuel. Structural and ball-and-stick models of this molecule are shown here. Is this molecule polar or nonpolar?
2. Ethylene glycol ($C_2H_6O_2$), the major component in antifreeze, has the molecular structure shown in the models on the left. The dipoles in ethylene glycol point toward which of the lowercase letters shown there?

3. Carbon skeleton compounds sometimes bend completely around to form closed loops. The drawing on the right shows a compound of carbon, hydrogen, and oxygen that is important to the pine bark beetle. This molecule is a sex attractant used by the female pine bark beetle to lure the male beetle into her wings. It is used by forestry biologists to lure the male pine bark beetle into a trap, thus saving the pine trees from beetle damage. A model of this molecule are shown below the drawing. Does this molecule have any polar properties? If so, in which direction are they oriented? (Notice that since a dipole is caused by the presence of a pair of electrons not balanced by a bonding nucleus, we can expect that any atom with one or more unshared pairs of electrons contributes toward a molecular dipole. Notice that the presence of an oxygen atom is almost always a dead giveaway that a molecular dipole is present.)

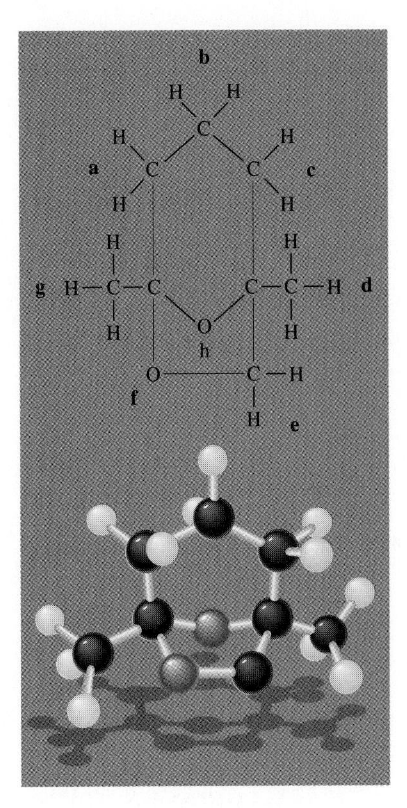

Answers **1.** Nonpolar **2.** c and g **3.** f and h

R.6 Behavior of Polar Molecules

The symmetry of the electron distribution around a molecule and the resulting molecular dipole (or lack thereof) strongly influence the physical and chemical behavior of the molecule. Consider again the water molecule (Figure R.39). The dipole in water is due to two factors: the presence of two pairs of unshared electrons and the highly polar O–H bond whose electrons reside considerably closer to the more electronegative oxygen than to the hydrogen.

As you might expect, because of the unlike charges at opposite ends, polar molecules are strongly attracted to one another. Figure R.40 illustrates the sort

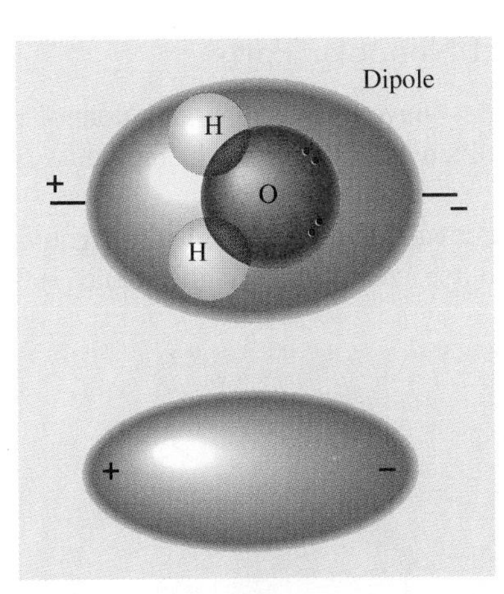

FIGURE R.39
A model of the polar water molecule. Polar molecules can be shown in a simplified manner as an oval or egg shape with positive and negative ends indicated by plus (+) and minus (−) charges.

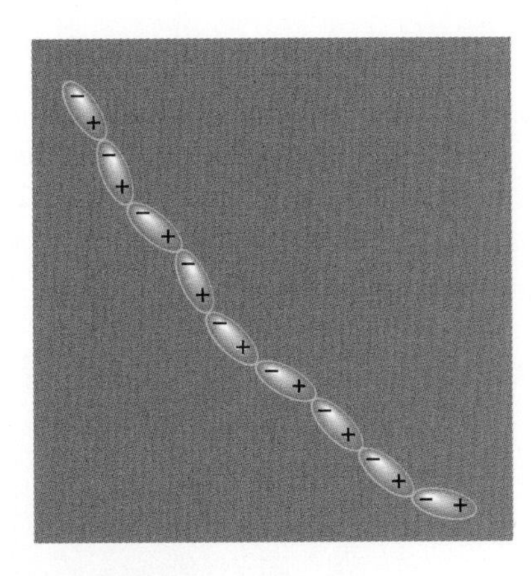

FIGURE R.40
Polar molecules are attracted to each other and often form loosely linked chains several molecules long.

♦ Polar molecules tend to form chains because of weak ionic forces between unlike ends of adjacent molecular dipoles.

of attraction that might exist between polar molecules. If the dipole is strong enough, the molecules actually form chains five to ten or more molecules in length, giving rise to significant changes in physical behavior, a phenomenon called *dipole–dipole interaction.*

Solubility and Molecular Polarity

The ability of one type of molecule to dissolve in another is related to the polarity of the molecules. The chain illustrated in Figure R.40 is not selective as to its membership. Formation of this chain requires only that each molecule have a partially negative end and a partially positive end, that is, they must be polar. For example, ammonia and water are both very polar molecules. Ammonia mixes very well with water—89.9 g of ammonia dissolves in 100 mL of water. We say it is very *soluble* in water.

♦ Molecules mix best with other molecules of like polarity.

Methane, on the other hand, is a nonpolar molecule. Since methane has no dipole, it cannot find a place in the chain. It is almost as if the polar molecules have formed an exclusive club. To join the club, a molecule must have an electric dipole. Methane, lacking this dipole, is squeezed out and excluded. Consequently, methane is very sparingly soluble in water—only 0.0064 g per 100 mL of water (Figure R.41).

Nonpolar molecules mix very well with other nonpolar molecules, however. Since no molecular dipoles are present, there is very little attraction or repulsion between their molecules. They move among each other attracted only by weak forces caused by induced charges between the molecules. These forces are called *van der Waals* or *London forces.*

Molecules with Polar and Nonpolar Parts

Some molecules—alcohols, for example—have both polar and nonpolar sections. A common compound used as a solvent and antiseptic in medicine is ethyl alcohol, C_2H_5OH (Figure R.42).

Ethyl alcohol has both covalent and polar covalent bonds. The electronegativity difference across a C–C bond is, of course, zero, making this bond 100

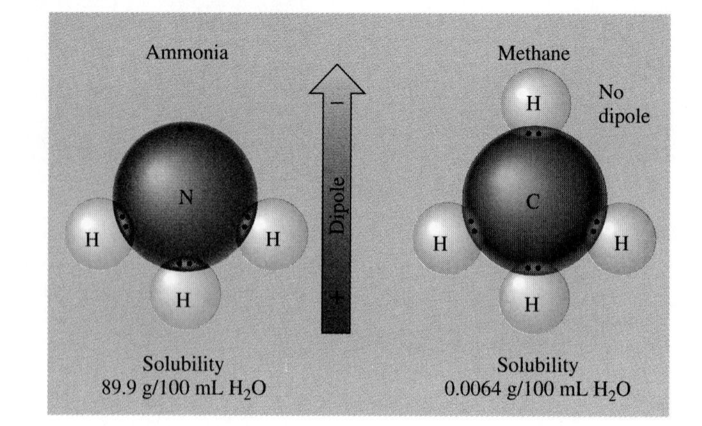

FIGURE R.41
Polar molecules, such as ammonia, are soluble in water, which is also a polar materials. Nonpolar compounds, such as methane, are not very soluble in water. The solubility of a compound is determined by the similarity of the symmetry of its electrical charge distribution to that of its solvent.

percent covalent. Carbon–hydrogen bonds have only 4 percent ionic character and therefore are essentially nonpolar; the C–O bond has only 22 percent ionic character and is polar; and the O–H bond, the most polar bond in the molecule, has only 39 percent ionic character. A model of this molecule is shown in Figure R.43. The polar O–H bond, and in particular the two unshared pairs of electrons, make the molecule very polar at the O–H end. However, from the C–O bond on down the carbon–hydrogen chain, the molecule is very symmetrical and nonpolar. This molecule has both polar and nonpolar properties and therefore is very soluble in both polar and nonpolar substances. (One end is soluble in polar materials, the other end in nonpolar materials.) For this reason, ethyl alcohol is a good solvent—a compound that can dissolve two mutually immiscible substances.

♦ Some molecules have both polar and nonpolar parts. These molecules will dissolve in both polar and nonpolar materials. They are good solvents.

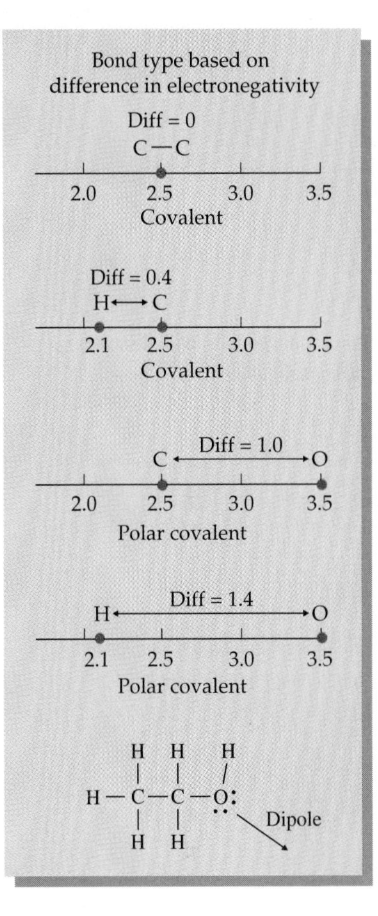

FIGURE R.42
Bond type and structure of ethyl alcohol. Note the direction of the dipole.

This property explains why ethyl alcohol serves as a solvent in many flavor extracts. Flavors, such as oil of spearmint and oil of orange, have large nonpolar molecules. By using the nonpolar end of a relatively large number of alcohol molecules (80 to 90 percent by volume), the nonpolar flavor molecules can be dissolved. The polar ends of the alcohol molecules then dissolve in water and thus mix the flavor into whatever is being cooked.

Soaps are another example of compounds that have both polar and nonpolar properties. Soaps and detergents consist of long nonpolar carbon–hydrogen chains with very polar groups attached to one end (Figure R.44). The nonpolar group (called *lipophilic*) mixes easily with nonpolar grease, and the polar end (called *hydrophilic*) remains dissolved in water. The effect is to tie the grease molecules to the water molecules, thus enabling us to rinse the grease away in the water. Without soap there would be no interaction between water molecules and grease molecules, and no cleaning action would be possible.

Dissolving Ionic Compounds

The process of dissolving ionic compounds in water is also related to the principle of molecular polarity. It is remarkable that a compound such as sodium chloride (NaCl)—in which the ionic bonds are so strong that a temperature of 801°C is required to melt the solid crystal—may be easily dissolved in water at room temperature. Sodium chloride, however, does not dissolve in nonpolar substances such as carbon tetrachloride.

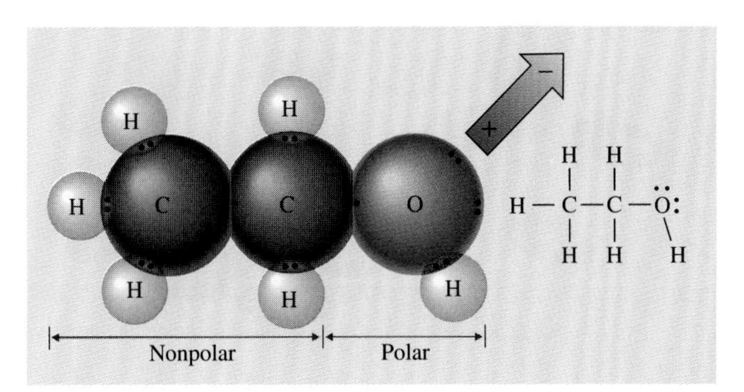

FIGURE R.43
The ethyl alcohol molecule has both polar and nonpolar parts. Alcohols all have the general structure R—OH, where R is a nonpolar hydrocarbon group.

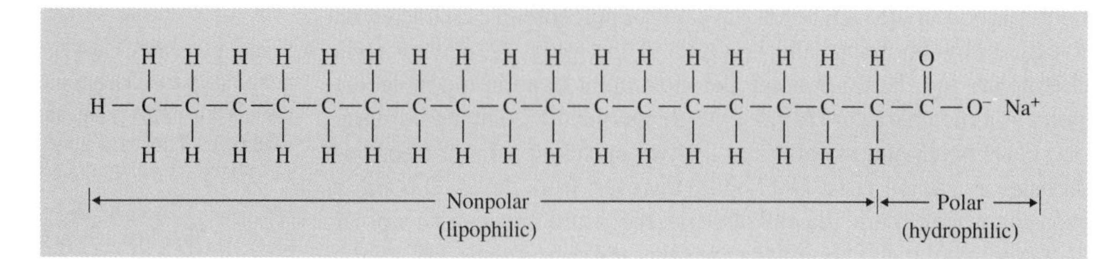

FIGURE R.44
The chemical structure of sodium stearate, a typical soap. This molecule has both polar (ionic, in this case) and nonpolar characteristics, and thus it dissolves in both grease and water.

♦ Ionic compounds will dissolve in polar solvents.

A sodium chloride crystal is held together by electrical attraction between the positive sodium ions and the negative chlorine ions. As is indicated by the high melting point, this attractive force is quite strong. If a crystal of sodium chloride is placed in water, however, the polar water molecules will surround the positive and negative ions, thus reducing the ionic attractive forces and permitting the ions to "float away" (Figure R.45). An ion thus surrounded by water molecules is called a **hydrated ion**. The resulting mobility of hydrated ions is demonstrated by the excellent electrical conductivity of water solutions of ionic compounds. Solutions that conduct electricity are called **electrolytes**.

The effect of molecular polarity on the solubility of ionic compounds is illustrated by the data in Figure R.46. The degree of solubility of an ionic compound determines the number of mobile ions, and thus the electrical conductivity of the solution. This experiment involves measurement of the electrical conductivity of saturated solutions of sodium chloride in each of four solvents—

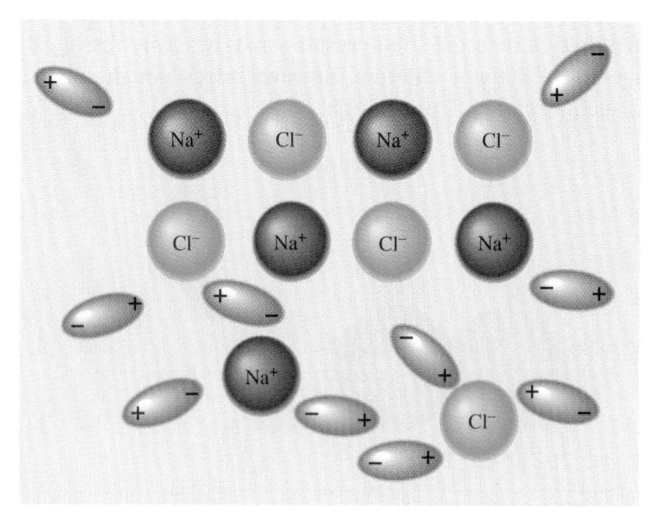

FIGURE R.45
Ionic compounds usually dissolve in polar solvents, such as water. Polar molecules surround the charged ions and cause them to float away.

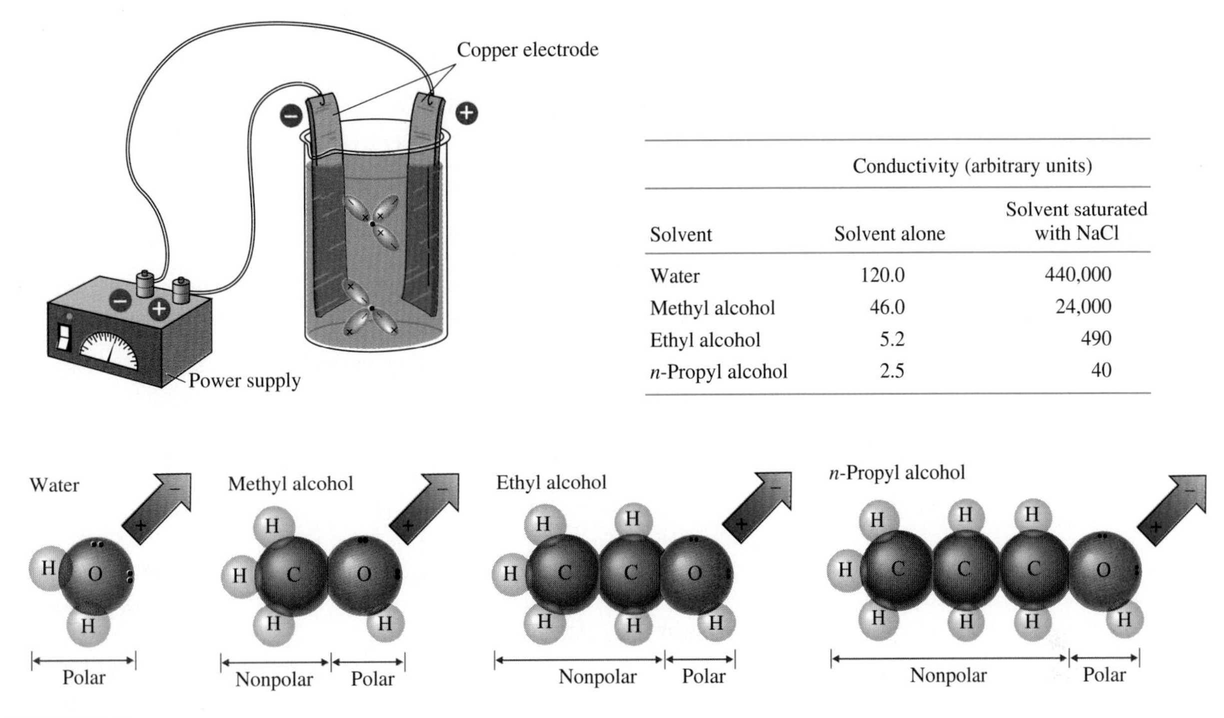

Solvent	Conductivity (arbitrary units) Solvent alone	Solvent saturated with NaCl
Water	120.0	440,000
Methyl alcohol	46.0	24,000
Ethyl alcohol	5.2	490
n-Propyl alcohol	2.5	40

FIGURE R.46
The solubility of an ionic compound is determined by the polarity of the solvent molecule. Highly polar solvents do the best job of dissolving ionic compounds. The apparatus illustrated in the upper left corner of this figure uses a battery to force electrons onto one electrode (−) and to draw them in from the other electrode (+). If ions are present in the solution, negative ions will give up electrons and positive electrons will accept electrons, and electron flow will be noted in the external circuit. The conductivity of the solution is a measure of its ionic content.

water, methyl alcohol, ethyl alcohol, and normal-propyl alcohol. These molecules are similar in that they each contain a polar O–H group. However, the degree of molecular polarity is decreased by the addition of progressively larger nonpolar C–H units in place of the second hydrogen found in water. This set of molecules provides a sequence of gradually decreasing polar character. As you might predict, the electrical conductivity due to dissolved ions decreases rapidly as the solvent molecule becomes less polar and less able to neutralize the attractive forces between charged ions.

NEW TERMS

hydrated ion An ion surrounded by water molecules.

electrolyte A solution that contains ions and conducts electricity.

TESTING YOURSELF

Molecular Polarity and Solubility

1. Of the three molecules shown below, which exhibits the most polar character?

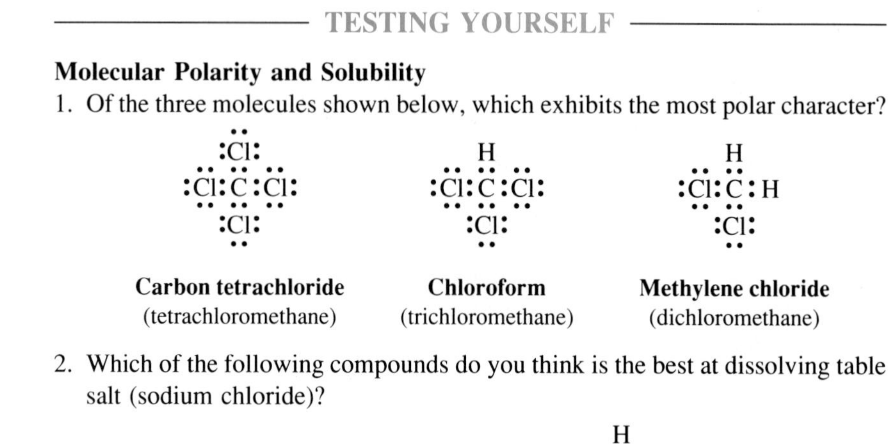

2. Which of the following compounds do you think is the best at dissolving table salt (sodium chloride)?

Water **Ammonia**

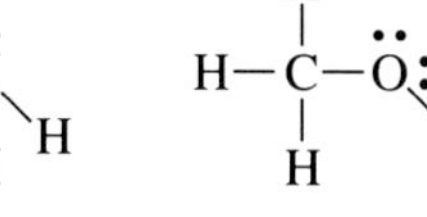

Methanol

Answers **1.** methylene chloride **2.** H_2O

R.7 Hydrogen Bonding

In covalent bonds there is an attraction from the positive nuclei of the bonding atoms toward one or more pairs of shared electrons. Another kind of bond exists that is considerably weaker and less substantial in that it does not involve shared electrons. It applies to only a few kinds of atoms, but where it operates it is an essential feature of life on our planet.

◆ Hydrogen bonding is a weak ionic bond between a hydrogen atom belonging to a polar molecule and the negative region of an adjacent polar molecule.

This bond is the **hydrogen bond**. Hydrogen bonding is a weak ionic bond that forms between oppositely charged parts of adjacent polar molecules. Hydrogen bonds are about 1/10 to 1/100 as strong as a normal covalent bond. It is only noticeable if the dipoles of the participating molecules are strong. Hydrogen bonding could be more accurately described as a permanent dipole–dipole attraction. The hydrogen bond is best exemplified by water, in which two pairs of unshared electrons cause a significant negative area on the side of the oxygen. On the other side, electrons shared between the hydrogens and the oxygen lie much closer to the oxygen because of the difference in electronegativity (H = 2.1, O = 3.5; difference = 1.4, for a 39 percent ionic character). As a result, the hydrogens appear as positive "bumps" on the side of the oxygen opposite its unshared electron pairs. The tendency of water's hydrogen atoms to be attracted to unshared electron pairs on adjacent water molecules is illustrated in Figure R.47.

Hydrogen bonding has several important effects on the behavior of water. First, it significantly increases water's boiling point because of the attraction of the water molecules to one another. A nonpolar molecule of methane (molecular weight 16) boils at about −160°C. Polar water (molecular weight 18), however, boils at 100°C, a 260°C difference.

Second, hydrogen bonding affects the freezing point of water. As water cools down, molecular motion becomes slower, and finally the water molecules align themselves with positive and negative ends together in much the same manner as you might expect for an ionic crystal. This regular arrangement is apparent if you take a close look at frost patterns (Figure R.48). However, a regular arrangement of water molecules in ice takes more space than the random arrangement in the liquid state, and ice is a little less dense than water. For this reason, unlike almost all other substances, the solid form of water is less dense than its liquid form. Ice floats on water, whereas most other liquids sink as they freeze.

Because of its strong electric dipole, water is unmatched in its ability to dissolve ionic compounds. Ionic compounds necessary to support human and animal life are transported dissolved in water in our blood. Sodium and potassium ions are necessary for nerve conduction. Sodium bicarbonate and carbonic acid regulate the acidity of blood. Many other ionic compounds necessary for life are dissolved in the blood.

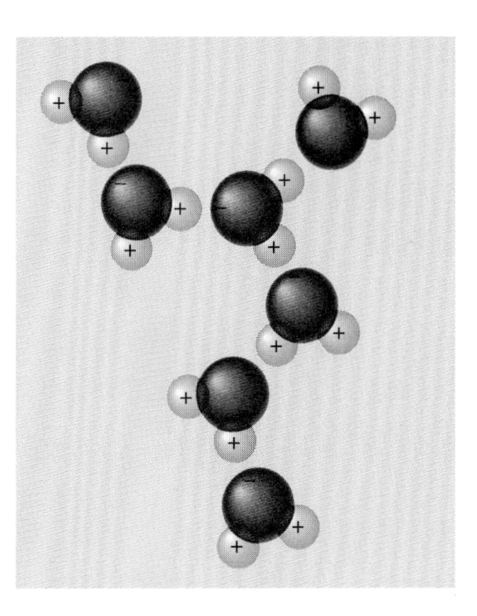

FIGURE R.47
Hydrogen bonding causes water molecules to form clusters of 9 to 12 molecules. The resulting higher effective molecular weight causes a large increase in boiling point compared with a nonpolar molecule of similar molecular weight.

NEW TERM

hydrogen bond A weak ionic bond between the negative region of a polar molecule and a positive hydrogen on an adjacent molecule.

Summary

This chapter had three purposes: To review the basic principles of atomic and electron structure that make it possible to predict the electron configuration of an atom; to review the principles governing ionic and covalent bonding; and to introduce methods of predicting shapes and polarities of covalent molecules by use of electron-pair repulsion theory.

Much of the material in this chapter can be summed up in six major concepts:

1. The electron filling series permits one to predict the electron configuration of any element, given its atomic number.
2. Atoms will seek to achieve chemical stability with a noble gas electron

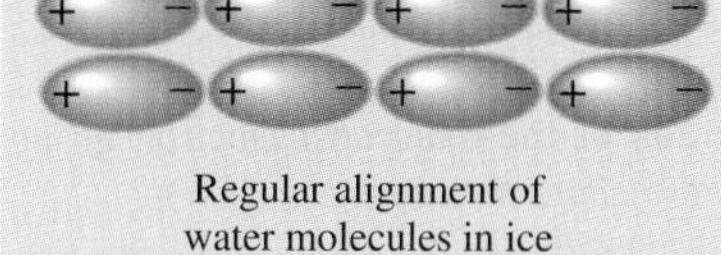

FIGURE R.48
The crystalline form of ice is due to the alignment of water dipoles. This is responsible for the strength of ice and the shape of frost patterns. It also makes ice float.

configuration (a full complement of *s* and *p* electrons in the outermost electron group) by either transferring or sharing electrons.

3. The type of chemical bond that will exist between two atoms can be predicted by computing the difference in electronegativity of the two atoms.
4. In covalently bonded molecules, electron pairs arrange themselves to be as far apart as possible. This is called *electron-pair repulsion theory*. When atomic orbitals overlap during covalent bonding, the resulting orbitals are said to be *hybridized*.
5. If the orbital hybridization results in an overall nonsymmetrical distribution of electrical charge across the molecule, an *electric dipole* results and the molecule is *polar*.
6. Interaction between molecules is determined by their polar properties. Polar molecules mix freely with other polar molecules. Nonpolar molecules mix freely with other nonpolar molecules. However, polar and nonpolar molecules do not dissolve in each other unless a third type of molecule having both polar and nonpolar parts is present to bring the two together.

Molecular polarity determines much of the physical and chemical behavior of a molecule. We will find it of considerable importance as we proceed with our study of organic and biological chemistry.

Terms

orbital (R.2)
the electron filling series (R.2)
Hund's rule (R.2)
orbital notation (R.2)
electron-dot model (R.2)
noble gases (R.3)
stable electron configuration (R.3)
electron transfer (R.3)
ion (R.3)
chemical bond (R.3)
ionic bonding (R.3)
covalent bonding (R.3)
sigma bond (R.3)
pi bond (R.3)
electronegativity (R.3)
polar covalent bond (R.3)
dipole (R.3)
valence-shell electron-pair repulsion theory (R.4)
tetrahedron (R.4)
hybridization (R.4)
molecular dipole (R.5)
polar molecule (R.5)
nonpolar molecule (R.5)
hydrated ion (R.5)
electrolyte (R.6)
hydrogen bond (R.7)

Exercises

The Electron Filling Series (Objectives 1 and 2)

1. What orbital will the outermost electron of a neutral atom of sodium occupy?
2. The electron distribution of which neutral atom is represented by the orbital notation shown below?

$$1s^2 2s^2 2p^6 3s^2 3p^4$$

3. The electron distribution of the outermost electron group of a neutral atom of the element chlorine is ($3s^2 3p^5$).
   a. What is the orbital notation for the electron distribution of a neutral atom of fluorine?
   b. What is the orbital notation for the electron distribution of a neutral atom of bromine?
4. Use the electron filling series to predict the electron distribu-

tion of the outermost electron group of neutral atoms of (a) oxygen, (b) sulfur, and (c) selenium.
What do these elements have in common?

5. Use the electron filling series to predict the electron distribution of the outermost electron group of neutral atoms of (a) magnesium, (b) calcium, and (c) strontium.
What do these elements have in common?
6. Potassium chloride is often used as a salt substitute for persons who should decrease their intake of sodium chloride. Use the electron filling series to predict the electron distribution of the outermost electron group of neutral atoms of sodium and potassium. What do these two elements have in common?

Prediction of Ionic Charge and Compound Formulas (Objectives 2 and 6)

7. Using the electron filling series and your rule for chemical stability, predict the charge of stable ions of

Li	B	O	F
Na	Al	S	Cl
K	Ga	Se	Br

Write the symbols for these elements as a table on your worksheet just as shown above. Show the predicted ionic charge in parentheses with the element—i.e., Li(+ 1).

8. The elements listed in Problem 7 are arranged in vertical columns as they are found in the periodic table. What generalization can be stated relating position in the periodic table to the charge of the stable ion of the element?
9. Predict the formula of a compound composed of stable ions of lithium and chlorine.
10. Predict the formula of a compound composed of stable ions of sodium and oxygen.
11. Predict the formula of a compound composed of stable ions of aluminum and fluorine.
12. Predict the formula of a compound composed of stable ions of gallium and bromine.

Electron Sharing and Covalent Bonding (Objectives 3–5)

13. When elements do not differ enough in electron-attracting ability for one atom to remove an electron from the other, the atoms will share electrons to achieve a stable electron configuration. Common attraction of both nuclei for the shared electrons holds the atoms together in a covalent bond.

Using your knowledge of the factors that determine an atom's electron-attracting ability, from the group of atoms listed below, decide

K N O

a. Which two elements are closest together in electron-attracting ability, and therefore which two will form a bond of the greatest covalent character.
b. Which two elements would form a bond of greatest ionic or electron transfer character.

14. Methyl alcohol is a covalently bonded compound with the structural formula

```
    H
    |
H—C—O—H
    |
    H
```

Sketch an electron-dot model for this molecule.

15. Diethyl ether, commonly used as tractor starting fluid, and once used as an anesthetic, has the structural formula

```
    H  H     H  H
    |  |     |  |
H—C—C—O—C—C—H
    |  |     |  |
    H  H     H  H
```

Sketch an electron-dot model of this molecule.

Electronegativity and Bond Type Prediction (Objective 7)

16. Use your knowledge of the electron filling series and electronegativity to predict the formula and bond type for compounds of the following elements.

	Bond Type	Formula
Aluminum and fluorine		
Sodium and oxygen		
Calcium and oxygen		
Magnesium and chlorine		
Strontium and oxygen		
Lithium and bromine		
Sodium and iodine		
Aluminum and oxygen		
Nitrogen and oxygen		
Nitrogen and hydrogen		

17. Using your knowledge of electronegativity, predict the type of bond that forms between carbon and each of the elements listed below: (See Figure R.17 on page R-13.)

Carbon and	Bond Type
Hydrogen	
Nitrogen	
Oxygen	
Fluorine	
Sulfur	
Chlorine	
Bromine	

18. Carbon has a unique position in the electronegativity table. Find the element that is farthest above carbon in electronegativity. What type of bond does carbon make with this element?
19. Find the element that is farthest below carbon in electronegativity. What type of bond does carbon form with this element?
20. Can carbon be expected to participate in ionic bonding?

Shapes of Covalent Molecules (Objectives 8 and 9)

21. Consider the compound ammonia, NH_3.
    a. What type(s) of bonds exist in this molecule?
    b. Draw an electron-dot model of this molecule.
    c. How many pairs of *bonding* electrons are there surrounding the nitrogen?
    d. How many pairs of *nonbonding* electrons are there surrounding the nitrogen?
    e. What is the shape of the nitrogen atom?
    f. Does the molecule have a dipole?
22. Consider the compound glycerol, commonly known as glycerine. Its molecular structure is sketched below:

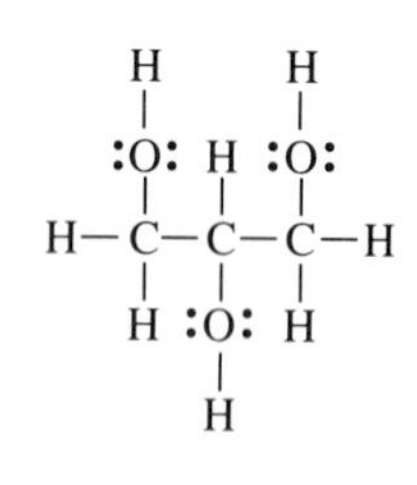

Glycerol

    a. What type(s) of bond are found in this molecule?
    b. How many pairs of electrons surround each carbon atom?
    c. What is the shape of each carbon atom?
    d. How many pairs of electrons surround each oxygen atom?
    e. What is the shape of each oxygen atom?
    f. Does the molecule have one or more polar regions? If so, where are they located?
23. Refer to the electron-dot model you drew for ammonia in Exercise 21. What is the hybridization of the outer electron group in the nitrogen atom in this compound?
    a. *s* b. *sp* c. sp^2 d. sp^3
24. What is the hybridization of the outer electron group of a carbon atom in glycerine (Exercise 22)?

Molecular Dipoles and Solubility (Objectives 9 and 10)

25. Electron-dot models of several molecules are sketched below. Classify these molecules as polar or nonpolar. If they are polar, show the location of their dipole.

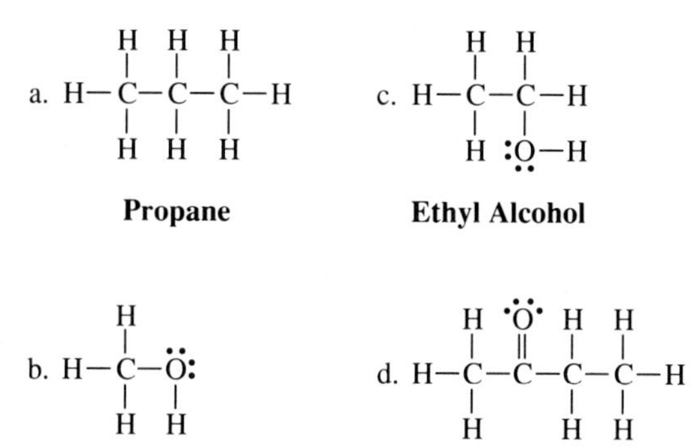

Propane

Ethyl Alcohol

Methanol

Methylethyl Ketone
(Fingernail-polish remover)

26. Which of the above compounds (a)–(d) is least soluble in water?
27. Which of the above compounds is most effective in dissolving salt?
28. Which of the above compounds have both polar and nonpolar parts, and therefore are good solvents to dissolve nonpolar materials in water?

Hydrogen Bonding (Objective 11)

29. What effect does hydrogen bonding have on the effective molecular mass of a molecule (increase, decrease, or no change)?

30. In general, what is the relationship between the molecular mass of a molecule and the energy required to make it boil?
31. Molecules that hydrogen-bond show (an increase, a decrease, no change) in boiling point compared with molecules of similar molecular mass that do not hydrogen-bond.
32. Compute the electronegativity difference across each bond in the following molecules, and decide which have the greatest polarity and therefore the greatest tendency for their molecules to hydrogen-bond.

H—Cl: H—C—O—H (with three H on C) H—N—H (with H on N)

Hydrochloric acid **Methanol** **Ammonia**

Unclassified Exercises

33. Use your knowledge of the electron filling series and chemical periodicity to predict the outer electron configuration of the element selenium, atomic number 34.
34. The ionic charge of a stable ion of gallium, atomic number 31 should be ________.
35. A compound comprised of sulfur and oxygen would probably have (a) ionic (b) covalent bonding.
36. A compound comprised of potassium and fluorine would probably have (a) ionic (b) covalent bonding.
37. Consider the compound ethylene glycol, commonly known as antifreeze. Its molecular structure is sketched below:

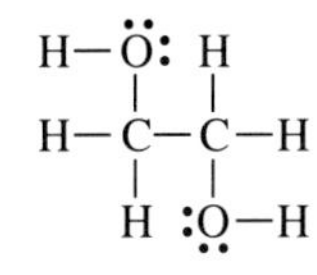

Ethylene Glycol

a. What types of bonds exist in this molecule?
b. Which atoms have *unshared* pairs of electrons?
c. What is the *hybridization* of each carbon atom: *s*, *sp*, sp^2, or sp^3?
d. How many dipoles are there in this molecule?
   i. none ii. one iii. more than one
e. Draw (an) arrow(s) across the molecule to show the position of the dipole(s).
f. Is this compound water-soluble?

38. Models of several molecules are sketched below. Classify these molecules as polar or nonpolar. If they are polar, show the location of their dipole.

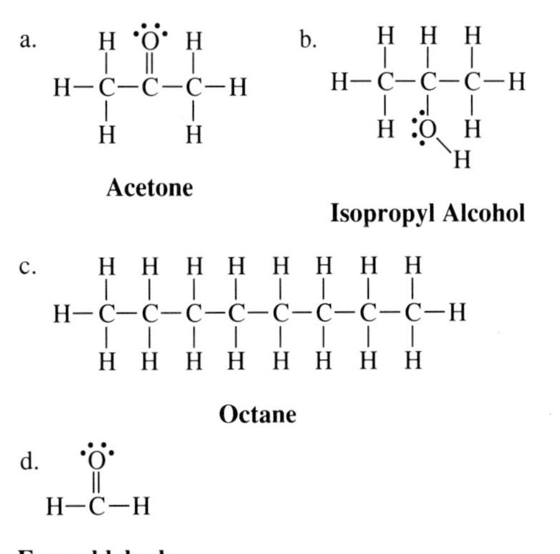

a. **Acetone**

b. **Isopropyl Alcohol**

c. **Octane**

d. H—C(=O)—H

Formaldehyde

39. Classify the following molecules as polar or nonpolar. Some molecules have both polar and nonpolar properties. In these cases, indicate the polar and nonpolar regions of each molecule.

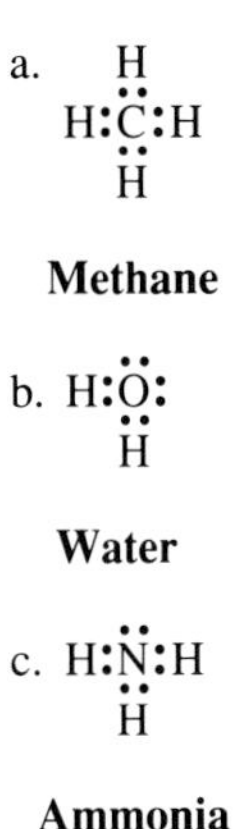

a. **Methane**

b. **Water**

c. **Ammonia**

40. Why are ionic compounds often soluble in polar solvents like water but not in nonpolar materials like gasoline?
41. Which of the molecules listed in Exercise 39 is the poorest solvent for table salt, sodium chloride?
42. Water molecules tend to stick together in clumps or chains because of weak bonds between the molecules. These bonds

are shown as dotted lines in the sketch below. What are these bonds called?

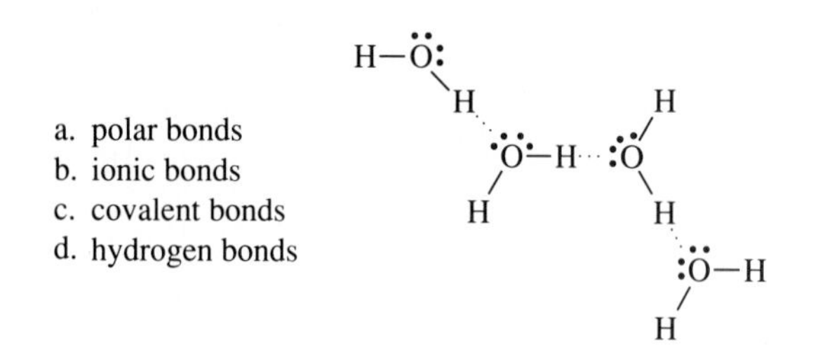

a. polar bonds
b. ionic bonds
c. covalent bonds
d. hydrogen bonds

43. What is the usual effect of hydrogen bonding on the boiling point of a compound?
  a. raises it
  b. lowers it
  c. no change

44. Which of the following molecules probably has the greatest tendency to participate in hydrogen bonding? Recall that the electronegativity across a bond is a good measure of the polar character of the bond.

  a. H—Ö: (with H bonded below O)

  Water

  b. H—S̈: (with H bonded below S)

  Hydrogen sulfide

  c. H—C̈l:

  Hydrogen chloride

45. Most solids are denser than their liquid counterparts, but water expands as it freezes. Why? What might be some environmental consequences if this did not occur and ice were denser than water?

AN INTRODUCTION TO ORGANIC CHEMISTRY

OUTLINE

1.1 Introduction

1.2 Carbon's Electronegativity Makes It Unique

1.3 Introduction to Functional Groups

OBJECTIVES

After completing this chapter, you should be able to

1. Identify common elements that form covalent bonds with carbon.
2. Define the term *alkyl group*.
3. Define the term *functional group*.
4. Classify organic compounds into families by identification of their functional groups—the hydrocarbons, alcohols, ethers, ketones, aldehydes, carboxylic acids, organohalogens, and amines.
5. Discuss the important relationship of petroleum to our everyday life.

Hydrocarbons protect fruit and vegetables. The waxy coating on the leaves of cabbage contains a hydrocarbon having 29 carbon atoms. Most plants have natural coatings rich in hydrocarbon materials.

1.1 Introduction

Carbon Compounds Are All Around Us

Compounds of carbon are all around us. The food we eat and the clothes we wear are two simple examples of our reliance on compounds of carbon. Although we may think of foods and natural fibers, such as wool and cotton, as being composed of compounds of carbon provided by nature, we must not forget about plastics and "artificial" fibers, such as rayon and nylon. These are also based on carbon; however, their origins were in petroleum. We describe petroleum as one of nature's forms of carbon compounds.

In the chapters that follow we will be examining the diversity of structural types associated with these compounds of carbon. It is exciting to marvel at the unique arrangements of atoms found in these compounds; it is also exciting to realize that chemists can duplicate almost all of the complex atomic arrangements that nature provides. The next few chapters will describe some of the important atomic arrangements common to even the most complex molecules; indeed, it will become evident that there are a few special arrangements of atoms (we will call these *functional groups*).

1.2 Carbon's Electronegativity Makes It Unique

◆ Carbon is essential to life. It is the major structural atom of the compounds found in living systems.

All living things have one basic characteristic in common. Whether from plant or animal, most molecules of living organisms are based on carbon. Consider the compounds shown in Figure 1.1. Glucose, a carbohydrate, is used by plants and animals as a source of quick energy. Acetic acid is an intermediate compound used in human metabolism and is the "sour" ingredient in vinegar. The amino

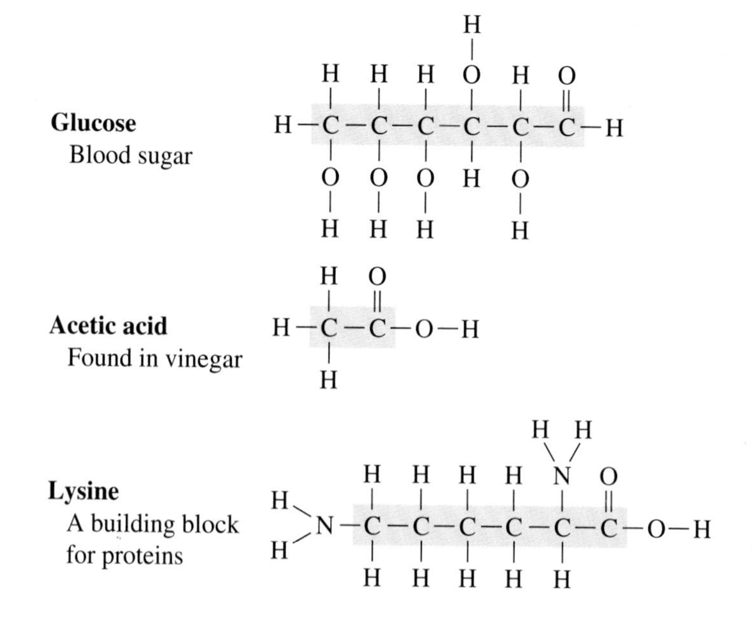

FIGURE 1.1
Three common compounds found in living things, all having something in common: they are built on a carbon skeleton.

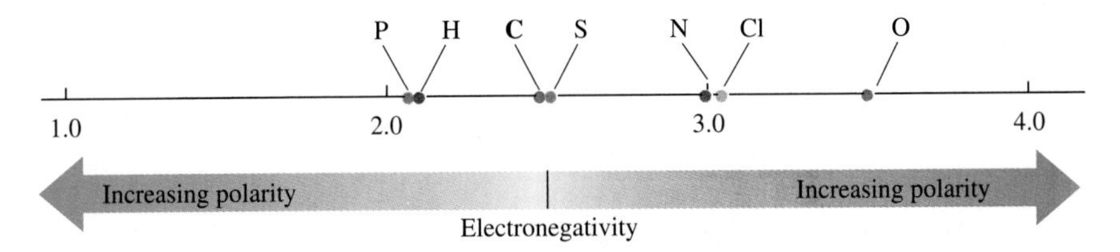

FIGURE 1.2
Because of its central location in the electronegativity scale, carbon forms covalent bonds with hydrogen, sulfur, nitrogen, chlorine, and oxygen, all of which are elements common to living things and abundant on the earth. The difference in electronegativity between two atoms determines the degree of polarity of the bond and, to a large extent, the physical and chemical behavior of the bond.

acid lysine is synthesized by plants and is used by plants and animals to make protein molecules. It is an "essential" amino acid for humans; we cannot synthesize it but we require it for protein synthesis. Notice that all three of these compounds are based on carbon.

Although carbon makes up less than 0.03 percent of the earth's crust, its importance to living systems makes it a very important element. The central position of carbon on the electronegativity scale (Figure 1.2) permits it to form covalent bonds with practically all elements. Because carbon normally has four bonds, there is an unlimited number of three-dimensional, covalently bonded molecular skeletons based on this atom. The number of known molecules based on carbon exceeds all other known molecules of all other elements. Living organisms are composed principally of molecules made of carbon, hydrogen, nitrogen, phosphorus, oxygen, sulfur, and (in the case of marine organisms) chlorine.

In the following sections we will preview the special arrangements of atoms that are found in organic chemistry. These special arrangements, which occur over and over again, are called **functional groups**. A functional group is a group of one or more atoms that may substitute for an alkyl group's missing hydrogen. Because of the electron distribution and electronegativity of its component atoms, it often dramatically changes both the physical properties and chemical reactivity of the parent hydrocarbon. Much of the chemical behavior of organic compounds can be directly related to functional groups.

Lewis Acid and Base Character from Electron-Dot Structures

There are several models for explaining the behavior of acids and bases. The Brønsted–Lowry model defines acids as molecules or ions that give up a hydrogen ion to an acceptor molecule, called a base. A more general acid–base model that is extremely useful in organic and biochemistry is the **Lewis theory**.

Any atom that has lone pairs of electrons associated with it can function as a **Lewis base**, an electron donor. This atom can be an individual atom, anion, or an atom in a molecule. Molecules containing double or triple bonds also function as Lewis bases because the electrons associated with multiple bonds are not

CHEMISTRY CAPSULE

Petroleum Is the Source of Our Chemical Industry

Where do we get the starting materials that allow us to create new fibers, materials, and medicines in the laboratory? To a large extent, these are nature's gift to us. Fossil fuels are what is left from the living chemistry of prehistoric times. Whether as peat, coal, oil, or natural gas, these materials are the building blocks upon which our modern world runs. It is somewhat ironic that the products of life from millions of years ago fuel our modern life (being burned to heat homes, provide electricity, or to fuel our engines). Because these fuels are the product of what was alive millions of years ago, we can appreciate why there is a limited, and nonrenewable supply of these precious materials.

FIGURE 1.A
Oil, coal, and peat have their origins in plant and animal life that lived long ago.

The most easily used of the fossil fuels is oil. Once the crude oil has been located, it must be extracted from the bowels of the earth. The oil is pumped from the deposits and carried to large industrial plants for purification. The petroleum can be separated into a number of different products through distillation, or different boiling point. In the next few chapters we shall see how boiling point and structure of these molecules are related.

TABLE 1.A

Hydrocarbon Fractions from Petroleum

Fraction	Number of Carbons	Boiling Range (°C)	Uses
Gas	1–5	<30	Fuel
Petroleum ether	5–7	30–90	Solvent
Gasoline	5–12	40–200	Motor fuel
Kerosene	12–16	175–275	Fuel
Fuel oil	15–18	250–400	Furnace fuel, diesel fuel, petrochemicals
Lubrications	>16	>350	Lubricants
Paraffins	>20	Melts	Candles, waxes
Pitch, tar		Residue	Asphalt

FIGURE 1.B
A petroleum refinery tower.

tightly attached to particular atoms and are available for reaction. In a similar definition, atoms (either alone, as cations, or in molecules) that accept electrons are classified as *Lewis acids*. By this definition we can include the Brønsted–Lowry acid, H^+, as a Lewis acid since it seeks electrons; the Brønsted–Lowry

CHEMISTRY CAPSULE (continued)

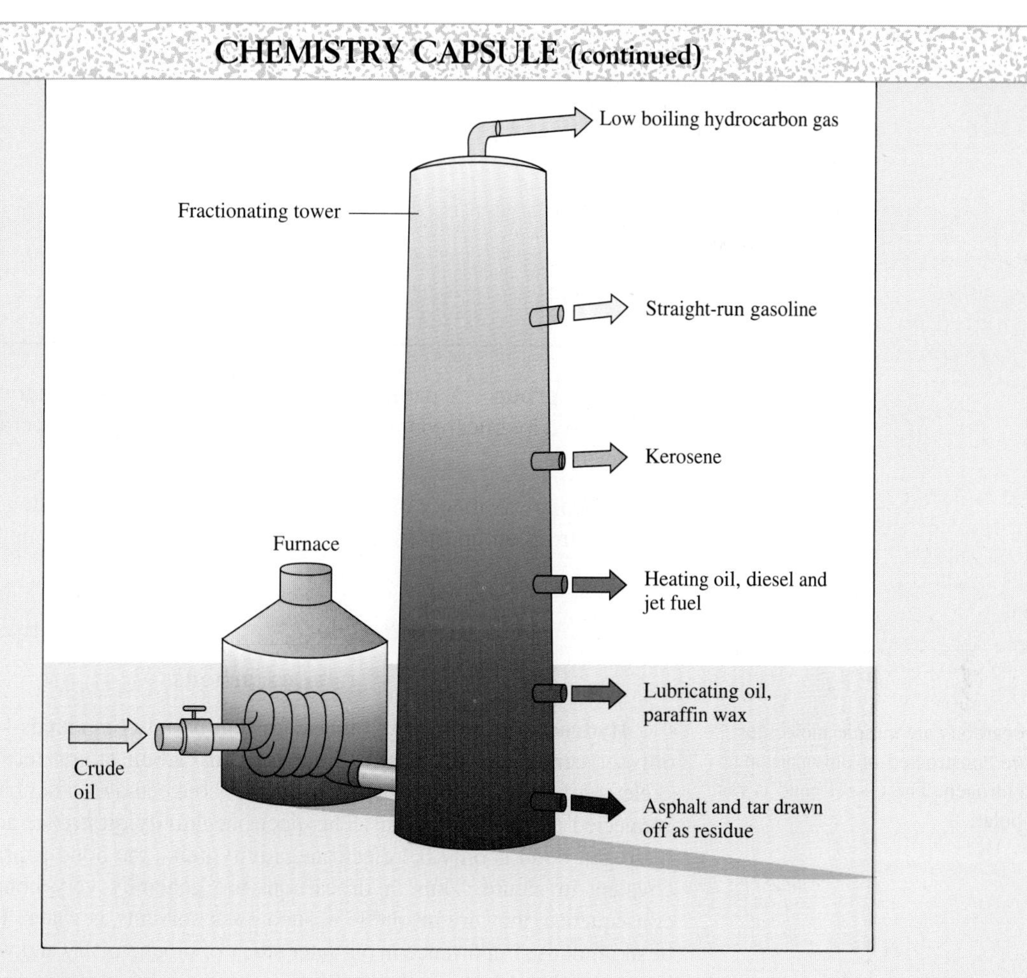

FIGURE 1.C
Distillation of crude oil separates the hydrocarbon material into classes based on rather broad ranges of boiling points. Because of the correlation of the boiling points with number of carbon atoms, we also notice that the fractions are roughly divided by carbon number.

base, OH^-, is an electron donor (Lewis base). We can also include anions such as Cl^- and O^{2-} as bases and cations such as Ag^+ and K^+ as acids. By this definition, water and ammonia function as Lewis bases because they have lone pairs of electrons. The Lewis theory of acids and bases will be seen in the chapters to follow as fundamental to much of the chemical behavior of organic and biological molecules. Examples of Lewis acid–base chemistry are seen in the following:

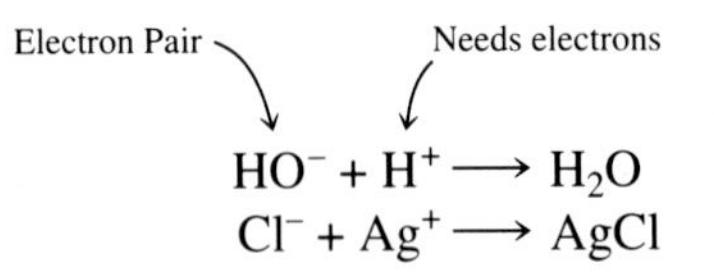

NEW TERMS

functional group A particular combination and arrangement of atoms; when attached to a hydrocarbon, unique physical and chemical properties are given to the molecule.

Lewis theory A theory of electron arrangement in a molecule to accommodate eight electrons around most atoms (hydrogen has two electrons).

1.3 Introduction to Functional Groups

Hydrocarbons

♦ Hydrocarbons are simple molecules that are constructed of only carbon and hydrogen. The C—H bond is not very polar.

Hydrocarbons are the simplest carbon-based compounds—they are made only of carbon and hydrogen. Four-bonded carbon atoms form the structural skeleton of the molecule, and all bonds not connected to other carbon atoms are connected to hydrogen atoms. The electronegativity difference across a carbon–hydrogen bond is only 0.4 electronegativity units; the bond is practically purely covalent in nature. Thus, hydrocarbons are generally very nonpolar, and as a consequence, they are insoluble in such polar solvents as water. This feature will be of immense importance in our later study of biochemistry and will be discussed in greater detail in later chapters.

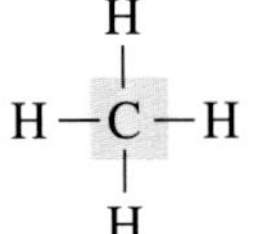

FIGURE 1.3
Methane (CH_4), commonly known as natural gas, is the simplest hydrocarbon.

Hydrocarbons take many forms, from the simplest, methane (Figure 1.3), to the very large, complex molecules found in crude oil and tars. In methane, a single carbon atom is bonded to four hydrogen atoms.

By adding other carbon atoms to the skeleton, larger hydrocarbon molecules can be created. Figure 1.4 shows several simple hydrocarbons. Note that all of the hydrocarbons discussed so far have single bonds between the carbon atoms, and their names end with the suffix *-ane*. The prefixes of each name reflect the

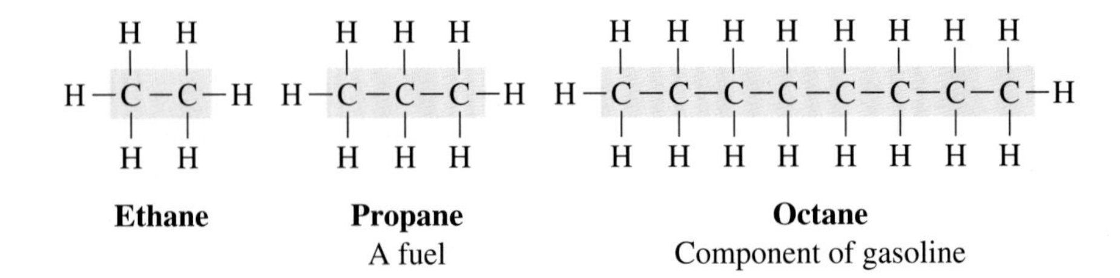

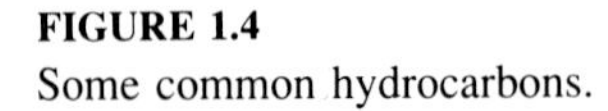
FIGURE 1.4
Some common hydrocarbons.

number of carbons in the chain. The family of single-bonded hydrocarbons is called the **alkane** family. They are also known as **saturated hydrocarbons** because the carbon atoms are unable to accept additional hydrogens.

Some hydrocarbons have double or triple bonds between the structural carbon's atoms. These are called **unsaturated hydrocarbons** because they are able to accept additional hydrogen or other atoms. Remember this fact by noting that saturated carbon systems have "full" bonds—that is, each of the four carbon bonds is attached to a different atom. Hydrocarbons with a double bond between adjacent carbons are called **alkenes**; those with a triple bond are called **alkynes** (Figure 1.5).

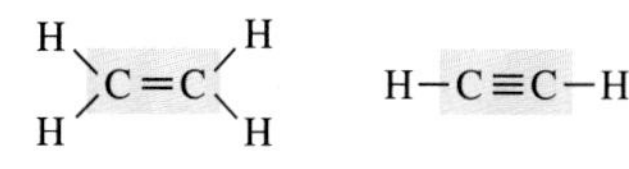

Ethene A simple alkene

Ethyne (acetylene) A simple alkyne

FIGURE 1.5
Hydrocarbons with double and triple bonds between structural carbons are called alkenes and alkynes, respectively. They are more reactive than their single-bonded counterparts.

Functional Groups Having Other Atoms

Hydrocarbons provide the basic structural unit for all molecules found in organic chemistry and biochemistry. The chemical and physical behavior of hydrocarbon molecules can be changed significantly by replacing one or more hydrogen atoms with functional groups made of other abundant elements—oxygen, hydrogen, chlorine, and nitrogen.

When a hydrogen atom is removed from a hydrocarbon, the resulting fragment is called an **alkyl group** (Figure 1.6). Alkyl groups are represented by the letter R in chemical structures. Alkyl groups have no stable existence, but they are a useful way of keeping some consistency in a system of organizing and naming compounds that could otherwise become quite complex.

Earlier in this chapter we identified several elements found in living organisms. These elements are listed in Table 1.1. They are all reasonably abundant on the earth, and all have electronegativity values close enough to carbon that they easily form covalent bonds with carbon.

From the set of elements in Table 1.1, the functional groups can be constructed that substitute easily for one of the hydrogen atoms in a hydrocarbon molecule,

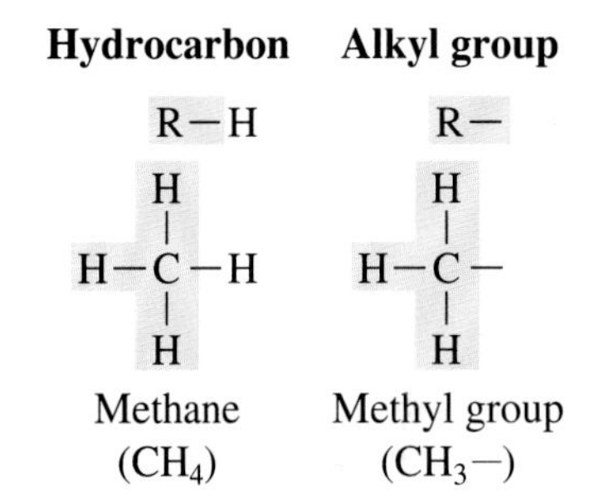

FIGURE 1.6
An alkyl group is a hydrocarbon with one hydrogen removed. The "open bond" is often filled with an atom or group other than hydrogen. Alkyl groups are often referred to by the symbol R—, where R represents any hydrocarbon group and the dash represents the open bond that awaits a new partner. Any hydrocarbon can thus be represented by the general form R—H.

TABLE 1.1
Common Elements Found in Living Organisms

Element	Electronegativity *Value*	Electronegativity *Difference from Carbon*	Usual Number of Bonds
Oxygen	3.5	1.0	2
Hydrogen	2.1	0.4	1
Carbon	2.5	0.0	4
Sulfur	2.5	0.0	2
Chlorine	3.0	0.5	1
Nitrogen	3.0	0.5	3
Phosphorus	2.1	0.0	3, 5

TABLE 1.2

Some Common Functional Groups and the Families of Compounds They Create

Functional Group −H (Replacement of −H)	General Formula (R−H)		HYDROCARBON Name
$-\ddot{\underset{..}{O}}-H$	$R-\ddot{\underset{..}{O}}-H$	R−OH	Alcohol
$-\ddot{\underset{..}{S}}-H$	$R-\ddot{\underset{..}{S}}-H$	R−SH	Thioalcohol (thiol)
$-\ddot{\underset{..}{O}}-R'$	$R-\ddot{\underset{..}{O}}-R'$	R−O−R′	Ether
$-\overset{:O:}{\overset{\Vert}{C}}-R'$	$R-\overset{:O:}{\overset{\Vert}{C}}-R'$	$R-\overset{O}{\overset{\Vert}{C}}-R'$	Ketone
$-\overset{:O:}{\overset{\Vert}{C}}-H$	$R-\overset{:O:}{\overset{\Vert}{C}}-H$	$R-\overset{O}{\overset{\Vert}{C}}-H$ R−CHO	Aldehyde
$-\overset{:O:}{\overset{\Vert}{C}}-\ddot{\underset{..}{O}}-H$	$R-\overset{:O:}{\overset{\Vert}{C}}-\ddot{\underset{..}{O}}-H$	$R-\overset{O}{\overset{\Vert}{C}}-OH$ $R-CO_2H$ R−COOH	Carboxylic acid
$-\overset{:O:}{\overset{\Vert}{C}}-\ddot{\underset{..}{O}}-R'$	$R-\overset{:O:}{\overset{\Vert}{C}}-\ddot{\underset{..}{O}}-R'$	$R-\overset{O}{\overset{\Vert}{C}}-OR'$ $R-CO_2R'$	Ester
$-\ddot{\underset{..}{X}}:$	$R-\ddot{\underset{..}{X}}:$	R−X	Organohalogen (X stands for F, Cl, Br, I)
$-\ddot{N}<$	$R-\ddot{N}<$	$R-NH_2$ $R-\underset{H}{\underset{\vert}{N}}-R'$ $R-N\genfrac{}{}{0pt}{}{R'}{R''}$	Amine
$-\overset{:O:}{\overset{\Vert}{C}}-\underset{..}{N}<$	$R-\overset{:O:}{\overset{\Vert}{C}}-\underset{..}{N}<$	$R-\overset{O}{\overset{\Vert}{C}}-N<$	Amide

* When more than one alkyl group is involved in a compound, the second and third groups are designated R′ and R″.

♦ The functional group concept allows for easy classification of compounds according to structure. This turns out to be also closely related to chemical behavior.

and in doing so the behavior of the molecule is significantly changed. This change in behavior is so pronounced that the newly formed compounds have been given new family names. Some common functional groups involving oxygen, chlorine, nitrogen, and sulfur are shown in Table 1.2. The remainder of the chapter surveys the common organic functional group families.

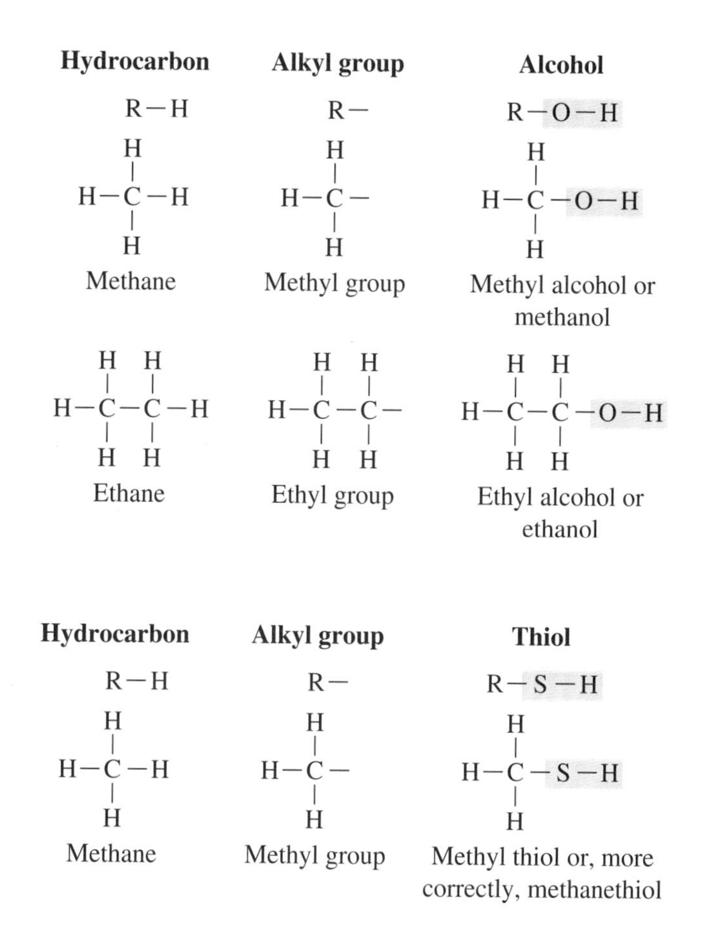

FIGURE 1.7
The formation of two simple alcohols by substitution of the —OH functional group. Notice how the alkyl group and the functional group combine to form the new class of molecules.

FIGURE 1.8
The creation of a simple thiol by the substitution of the —SH functional group.

Alcohols

Substitution of the —OH functional group for a hydrogen atom changes a hydrocarbon into an **alcohol**; thus, the generalized form for an alcohol is R—OH. Figure 1.7 shows the substitution of an alcohol group for hydrogen in two simple hydrocarbons; methane and propane.

♦ *Alcohols* are recognized by the general structure:

R—OH

Sulfur is in the same periodic family as oxygen, so it might reasonably be expected that a sulfur-containing compound similar to an alcohol could be found. As Figure 1.8 shows, this is indeed the case. The sulfur analogs of alcohols are called *thiols* and have the functional group —SH. The chemical and physical properties of alcohols and thiols will be discussed further in Chapter 5.

Ethers

An oxygen atom, which has two bonds, can form a bridge between two hydrocarbon groups. The resulting compound is called an **ether**. The ether functional group is shown in Figure 1.9. The ethers will be discussed further in Chapter 5.

♦ *Ethers* are recognized by the general structure:

R—O—R

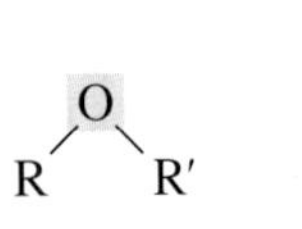

General formula of an ether

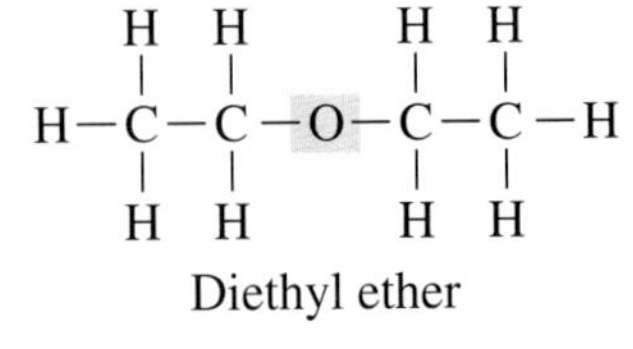

Diethyl ether

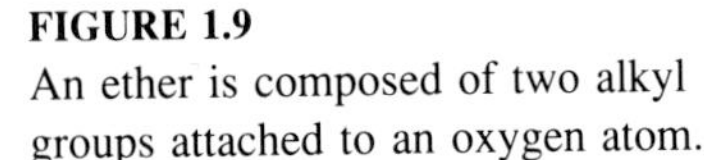

FIGURE 1.9
An ether is composed of two alkyl groups attached to an oxygen atom.

As you can see from the electron-dot structures in Table 1.2, the oxygen atom of ethers can be classified by the following generalized structure:

—O—

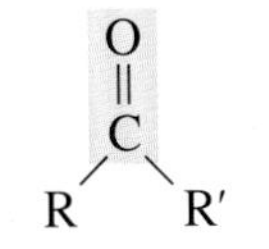

General formula of a ketone

Dimethyl ketone (acetone)

FIGURE 1.10
The ketone family is characterized by two alkyl groups connected to a carbonyl functional group.

Aldehydes and Ketones

Ketones are characterized by the **carbonyl** group, which is an oxygen atom double bonded to a carbon. The remaining two bonds of the carbon are available to connect to other fragments of the molecule. When each connects to an alkyl group, the resulting compound is called a **ketone** (Figure 1.10).

From application of the electron-dot structures in Table 1.2, we find that the carbonyl group can be generalized by the following structure:

Aldehydes are close relatives of the ketone family. Aldehydes, however, have an alkyl group and a hydrogen atom connected to the carbonyl group instead of two alkyl groups (Figure 1.11). The details of the chemistry of these functional groups will be discussed in Chapter 6.

Carboxylic Acid

Carboxylic acids are similar to the ketone–aldehyde families except that an —OH group is connected to one side of the carbonyl carbon atom (Figure 1.12). A **carboxyl group**, however, is *not* a carbonyl and a hydroxide; the combination

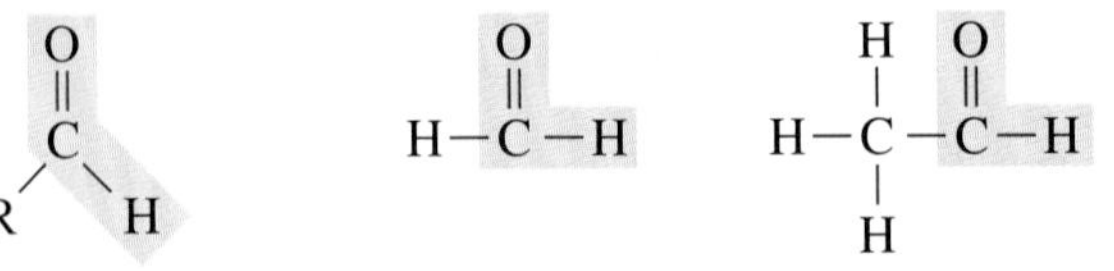

FIGURE 1.11
Aldehydes are characterized by an alkyl group and a hydrogen attached to a carbonyl functional group.

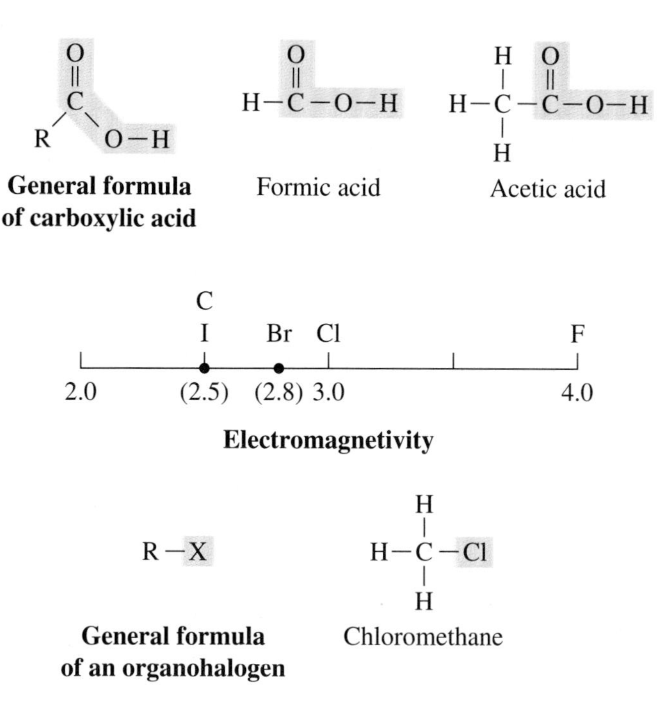

FIGURE 1.12
Carboxylic acids are characterized by the presence of the carboxyl group. An alkyl group or hydrogen atom is connected to the carboxyl group. The highlighted hydrogen atom is slightly ionizable and is responsible for the acidity of this class of compounds.

FIGURE 1.13
Chlorine is a member of the family of atoms (Group 7) known as the halogens. Although the atoms differ in electronegativity, they are all one electron short of a stable octet of electrons. The halogens can covalently bond to carbon. The organohalogens are compounds in which an alkyl group bonds to a halogen.

of one carbon atom, two oxygen atoms, and one hydrogen atom (COOH) forms a unique functional group having unique properties. The physical and chemical behavior of carboxylic acids will be discussed in Chapter 7.

Organohalogens

Twelfth in elemental abundance in the crust of the earth is chlorine, which forms one covalent bond with carbon. Chlorine is a member of the family of elements called *halogens*. Chlorine and the other members of the halogen family—fluorine, bromine, and iodine—can substitute for hydrogen atoms in a hydrocarbon to form a family of compounds known as **organohalogens** (Figure 1.13). The differences in electronegativity between carbon and the halogens are small enough that the bonds are still covalent. Note that more than one halogen can be connected to a carbon atom (Table 1.3). Thus, chlorinated hydrocarbons often contain more than one chlorine. If more than one carbon atom is present, these chlorines need not connect to the same carbon.

Amines

♦ *Amines* are recognized as the organic equivalents of ammonia and have one of these general structures:

```
R−N−H    R−N−R
  |        |
  H        H
     R−N−R
       |
       R
```

The *amino group* has nitrogen as its central atom. Nitrogen has a pair of unshared electrons that significantly affect the behavior of the group. The ammonia molecule is the parent compound of the amines. **Amines** are compounds formed by substitution of one or more alkyl groups for hydrogens in the ammonia (Table 1.4). An amine is the organic analog of ammonia, just as an alcohol can be considered as the organic analog of water.

TABLE 1.3
Chlorinated Hydrocarbons

Compound Name	Chemical Structure
Chloromethane	H | H—C—Cl | H
Dichloromethane	H | H—C—Cl | Cl
Trichloromethane (chloroform)	Cl | H—C—Cl | Cl
Tetrachloromethane (carbon tetrachloride)	Cl | Cl—C—Cl | Cl

TABLE 1.4
Amines

Compound Name	Chemical Structure
Ammonia	H—N—H | H
Methylamine	H H | | H—C—N—H | H
Dimethylamine	H H H | | | H—C—N—C—H | | H H
Trimethylamine	H | H—C—H H | H | | | H—C—N—C—H | | H H

The unshared electron pair on the nitrogen atom can serve as a hydrogen ion acceptor in the same manner that an unshared pair of electrons on the oxygen atom in water serves as a hydrogen ion acceptor to form hydronium ion. An amine is an "organic base" in the sense of a Lewis base—an amine is able to donate a pair of electrons.

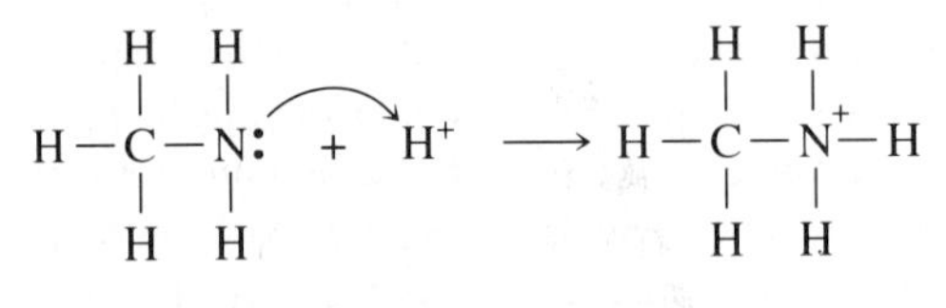

Amines will be discussed in further detail in Chapter 8.

Esters and Amides

The only two groups in Table 1.2 that we have not yet discussed are the esters and amides. Although very important to both organic chemistry and biochemistry, they are both simply modifications of the carboxylic acid functional group. Details of the chemistry of the esters and amides will be discussed under appropriate headings in Chapters 7 and 8, respectively. However, at this point it is appropriate to show the origins of these two functional groups:

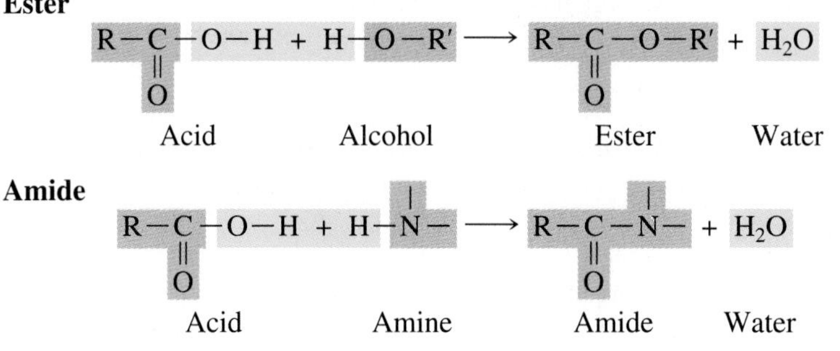

Examples of hydrocarbons. These materials are representative of the separation of crude hydrocarbon to get different ranges of carbon chain length.

NEW TERMS

hydrocarbon A compound consisting only of the elements hydrogen and carbon. Hydrocarbons are almost always nonpolar.

alkane A family of hydrocarbon compounds having only single bonds between carbon atoms in the molecular skeleton

saturated hydrocarbon A hydrocarbon containing only single carbon-to-carbon bonds

unsaturated hydrocarbon A hydrocarbon containing either double or triple carbon-to-carbon bonds

alkene A family of hydrocarbon compounds having at least one double bond between carbon atoms in the molecular skeleton

alkyne A family of hydrocarbon compounds having at least one triple bond between carbon atoms in the molecular skeleton

alkyl group A hydrocarbon with one hydrogen removed prior to bonding with another atom or group of atoms. Alkyl groups exist only on paper as a tool for naming and for explaining organic reactions. Alkyl groups are often represented by the symbol R.

alcohol An organic molecule of the form R—OH, where R is an alkyl group and the functional group is the —OH

♦ Alcohol
R—OH

ether A family of organic compounds formed when an oxygen atom serves as a bridge between two alkyl groups

♦ Ether
R—O—R′

carbonyl A family of organic compounds composed of a carbon double-bonded to an oxygen. The two remaining carbon bonds may be connected to other atoms or to alkyl groups.

♦ Carbonyl
$-\overset{\overset{\displaystyle O}{\|}}{C}-$

ketone A family of organic compounds formed when an alkyl group is connected to each of the two remaining carbon bonds of the carbonyl group

♦ Ketone
$R-\overset{\overset{\displaystyle O}{\|}}{C}-R'$

aldehyde A family of organic compounds formed when an alkyl group is placed on one of the carbon bonds of a carbonyl group and a hydrogen is placed on the other.

♦ Aldehyde
$R-\overset{\overset{\displaystyle O}{\|}}{C}-H$

♦ Carboxylic acid

$R{-}\overset{\overset{\displaystyle O}{\|}}{C}{-}O{-}H$

carboxylic acid An organic compound containing one or more carboxyl groups

♦ Carboxyl group

$-\overset{\overset{\displaystyle O}{\|}}{C}{-}OH$

carboxyl group A group of compounds composed of a carbonyl group with an —OH connected to one of the free carbon bonds. Carboxyls are the "trademark" of organic acids.

♦ Organohalogen

R—X

organohalogen An organic compound in which a halogen (fluorine, chlorine, bromine, or iodine) has replaced one or more hydrocarbon hydrogens

♦ Amine

$R{-}N{<}$

amine An organic compound in which a nitrogen is the central member of the functional group. Since nitrogen atoms have three bonds, only one of which connects to the parent hydrocarbon, it is possible to substitute additional alkyl groups on this nitrogen.

TESTING YOURSELF

Bonding and Functional Groups

1. Which of these elements makes the least polar bond with carbon: oxygen, hydrogen, nitrogen, or chlorine?
2. Which of the elements listed in question 1 make an ionic bond with carbon?
3. Structural diagrams for several compounds are shown below. Identify the family of organic compounds to which each belongs.

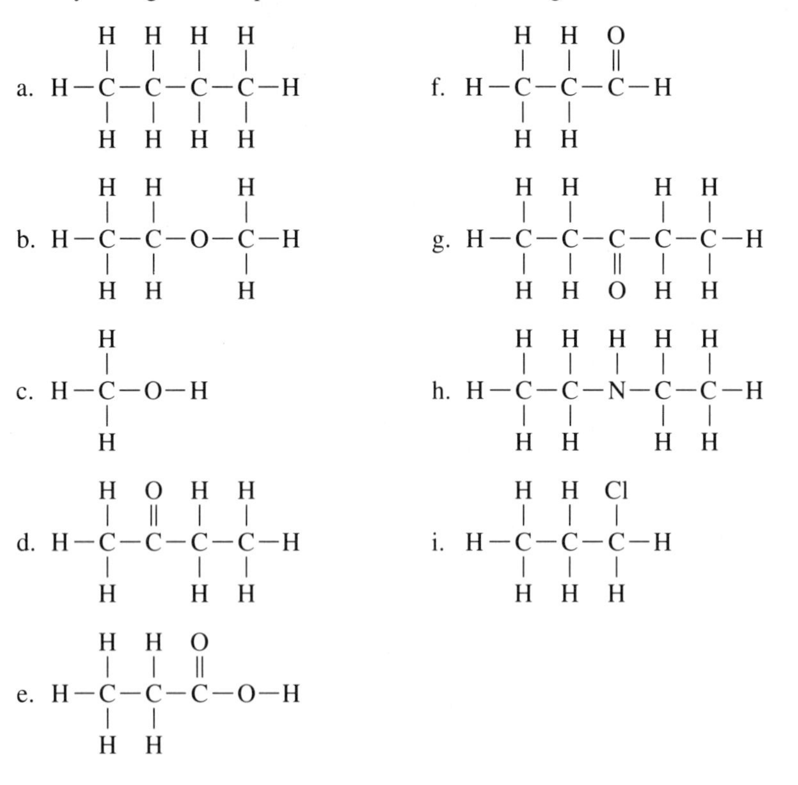

Answers **1.** hydrogen **2.** none **3. a.** hydrocarbon **b.** ether **c.** alcohol **d.** ketone **e.** carboxylic acid **f.** aldehyde **g.** ketone **h.** amine **i.** organohalogen

Summary

The chemical and physical behavior of a hydrocarbon molecule is drastically changed if one or more of its hydrogen atoms is replaced by special atoms or groups of atoms called *functional groups*. Carbon-based compounds can be classified into families of compounds of similar structure and reactivity based on the presence of these functional groups. The specific families of compounds surveyed in this chapter and to be discussed in detail in the chapters that follow include the hydrocarbons (alkanes, alkenes, alkynes, and aromatics), organohalogen compounds, alcohols, phenols, ethers, thiols, aldehydes, ketones, acids, esters, amines, and amides.

The shapes of molecules and functional groups can be predicted from the Lewis electron-dot structures. From the electron-dot structures, predictions about the Lewis acid–base character of the molecule can also be made. According to this theory, Lewis acids are electron acceptors, and Lewis bases are electron donors.

In the next chapter the simplest hydrocarbons, the alkanes, will be discussed. Highlights of that chapter will include the origins of nomenclature that will carry through the rest of the chapters, the concept of the alkyl group, and an introduction to the concept of molecular conformation—the preferred shapes of molecules.

Terms

functional group (1.2)
Lewis theory (1.2)
hydrocarbon (1.3)
alkane (1.3)
saturated hydrocarbon (1.3)
unsaturated hydrocarbon (1.3)
alkene (1.3)
alkyne (1.3)
alkyl group (1.3)
alcohol (1.3)
ether (1.3)
carbonyl (1.3)
ketone (1.3)
aldehyde (1.3)
carboxylic acid (1.3)
carboxyl group (1.3)
organohalogen (1.3)
amine (1.3)

Exercises

Bonding in Organic Compounds

1. If two atoms, A and B, are connected by a covalent bond, the electronegativity difference between the two atoms will influence the polarity of the bond. Thus, if B is more electronegative than A, the bond will be polarized:

$$\overset{+}{A} \longrightarrow \overset{-}{B}$$

Given the following bonds to carbon, which is most polarized in the direction C $\longrightarrow$ Z (where Z is another atom)?

a. i. C—Cl ii. C—N iii. C—S
b. i. C—C ii. C—O iii. C—N

Given the following bonds to carbon, which is more likely to appear as C ⟵ Z (where Z is another atom)?

c. i. C—Cl ii. C—P iii. C—N

d. Which of the following bonds is most polarized?

i. N—N ii. N—Cl iii. N—O

Functional Group Recognition (Objective 4)

2. Identify the functional groups in the following molecules:

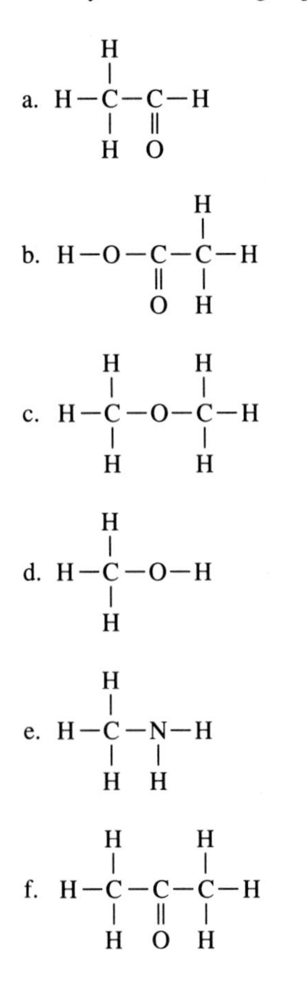

Unclassified Exercises

3. R—C(=Ö)—H (R—C—H with :O: double-bonded to C)

is a shorthand expression for a(n)

a. acid b. ketone c. aldehyde d. alcohol

4. An ether is best represented by which generalized structure?

a. R—O—R b. R—O—H

c. R—C(=O)—OH d. R—C(=O)—R

5. An amine is uniquely characterized by the presence of

a. hydrogen b. carbon c. oxygen d. nitrogen

6. An alkyne has what type of bond?

a. single b. double c. triple d. quadruple

7. An alcohol is represented by which generalized structure?

a. R—X b. R—OH c. R—O—R d. R—NH_2

8. A ketone is most similar to an

a. alcohol b. aldehyde c. acid d. amine

9. A ketone has how many R groups attached to the carbonyl group?

a. 0 b. 1 c. 2 d. 3

10. Cl—C(H)(H)—H

is classified as a(n)

a. organohalogen b. amine c. hydrocarbon d. alkene

ALKANES AND CYCLOALKANES: SINGLE-BONDED HYDROCARBONS

Combustion is a common reaction of hydrocarbons. Here we see the flame from methane combustion.

OUTLINE

2.1 Structure and Physical Properties
2.2 Alkanes and Their Nomenclature
2.3 Alkyl Groups and Nomenclature
2.4 Cycloalkanes
2.5 Conformations of Alkanes and Cycloalkanes
2.6 Chemical Reactivity of Alkanes and Cycloalkanes
2.7 Health-Related Products Based on Hydrocarbon Structures

OBJECTIVES

After completing this chapter, you should be able to

1. Describe structural features of the hydrocarbons.
2. Discuss how the structure and physical properties are related.
3. Identify simple alkane, cycloalkane, and organohalogen compounds.
4. Describe the orientations in space (conformations) of atoms in a molecule and how these orientations influence stability.
5. Name representative examples of each family.
6. Identify factors limiting reactivity of these compounds.
7. Predict products of some representative reactions of alkanes and cycloalkanes.

2.1 Structure and Physical Properties

Tetrahedral Shape

Based on the Lewis electron-dot structures, note that any atom having an arrangement of four atoms or electron pairs about it is called *sp^3 hybridized.* From this we know that the orientation of the bonds and electron pairs is tetrahedral, that is, they form a pyramid-like structure with four triangular faces. Thus, we find for the alkane hydrocarbons that the carbon atoms have tetrahedrally oriented bonds to hydrogen or to other carbon atoms.

As with atomic orbitals, the hybrid orbitals need two electrons per orbital. Since each of the hybrid orbitals has one electron, they share their electrons with four other atomic orbitals. In the case of methane, this means that one electron is accepted from each hydrogen atom.

◆ Three $2p$ orbitals of carbon mix with the $2s$ orbital to make four equivalent hybrid orbitals. These are said to be *sp^3 hybridized.*

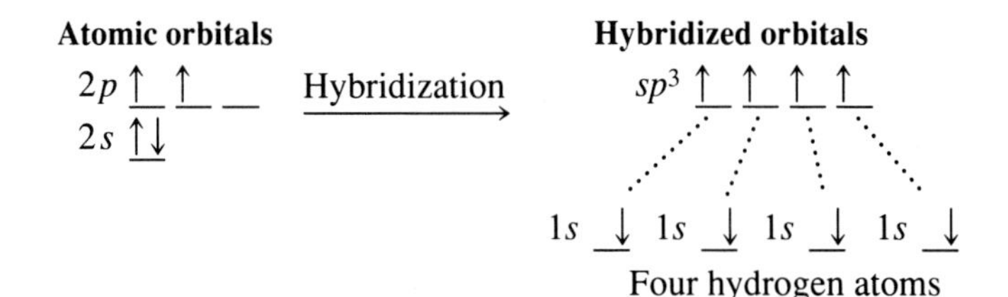

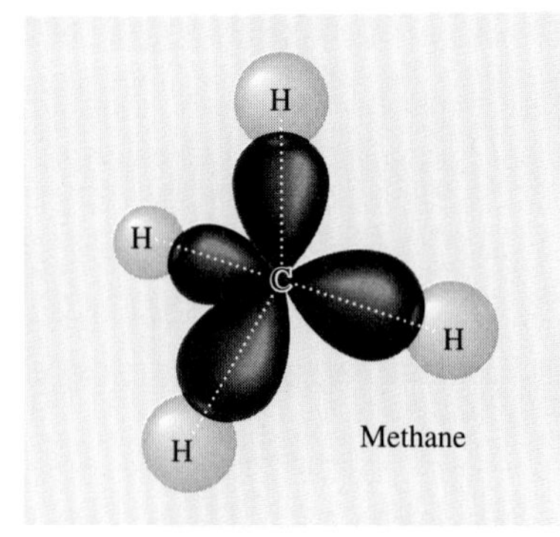

The Carbon–Hydrogen and Carbon–Carbon Bond

Carbon and hydrogen are very close in electronegativity. Bonds between these elements are therefore almost purely covalent. The hydrocarbon molecules are attracted to each other by *London forces*, attractive forces caused by *induced dipoles* (Figure 2.1). In these attractions, the positively charged atomic nucleus of one molecule attracts the electrons of another molecule. This induced dipole is only temporary, and the two molecules can then change roles. Since the London attractive forces are dependent on inducing a dipole in a molecule, the larger the molecule (the more electrons and protons present), the more important is the attractive force. Thus, the attraction between two molecules of pentane is greater than the attraction between two molecules of methane.

FIGURE 2.1
Schematic representation of a hydrocarbon. Because there is little difference in electronegativity between carbon and hydrogen, attractive forces between molecules containing only carbon and hydrogen have a unique origin. Induced dipoles attract these molecules to one another. The electron cloud associated with one molecule is attracted toward the nuclear charge of another. This "instantaneously" changes so that the two molecules change roles; there are no permanent dipoles. These attractive forces are called *London forces.*

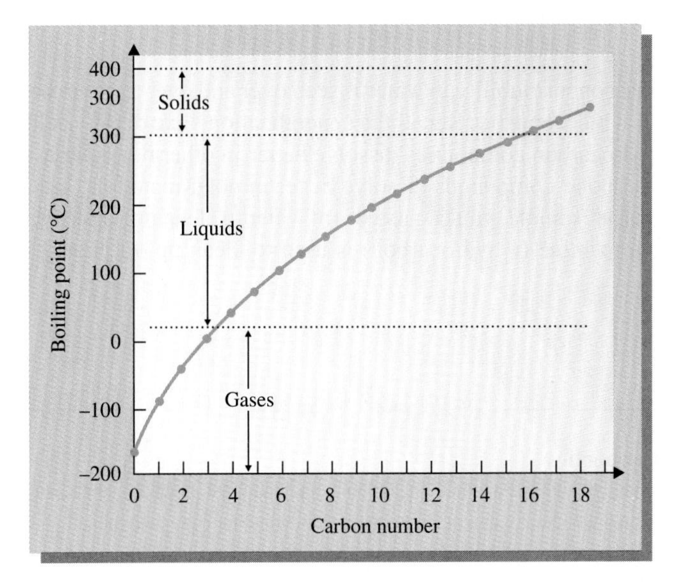

FIGURE 2.2
The boiling point of hydrocarbon compounds increases as the number of carbon atoms increases. This is because of the mutual attraction of the electrons of one molecule for the nuclei of the other molecule.

Boiling Point

A consequence of an increase in attractive force between the molecules is an increase in boiling point with increasing number of carbon atoms (Figure 2.2). Alkanes with fewer than five carbon atoms are gases at room temperature.

♦ In general, within a particular functional group, the boiling points of compounds increase with increasing carbon number.

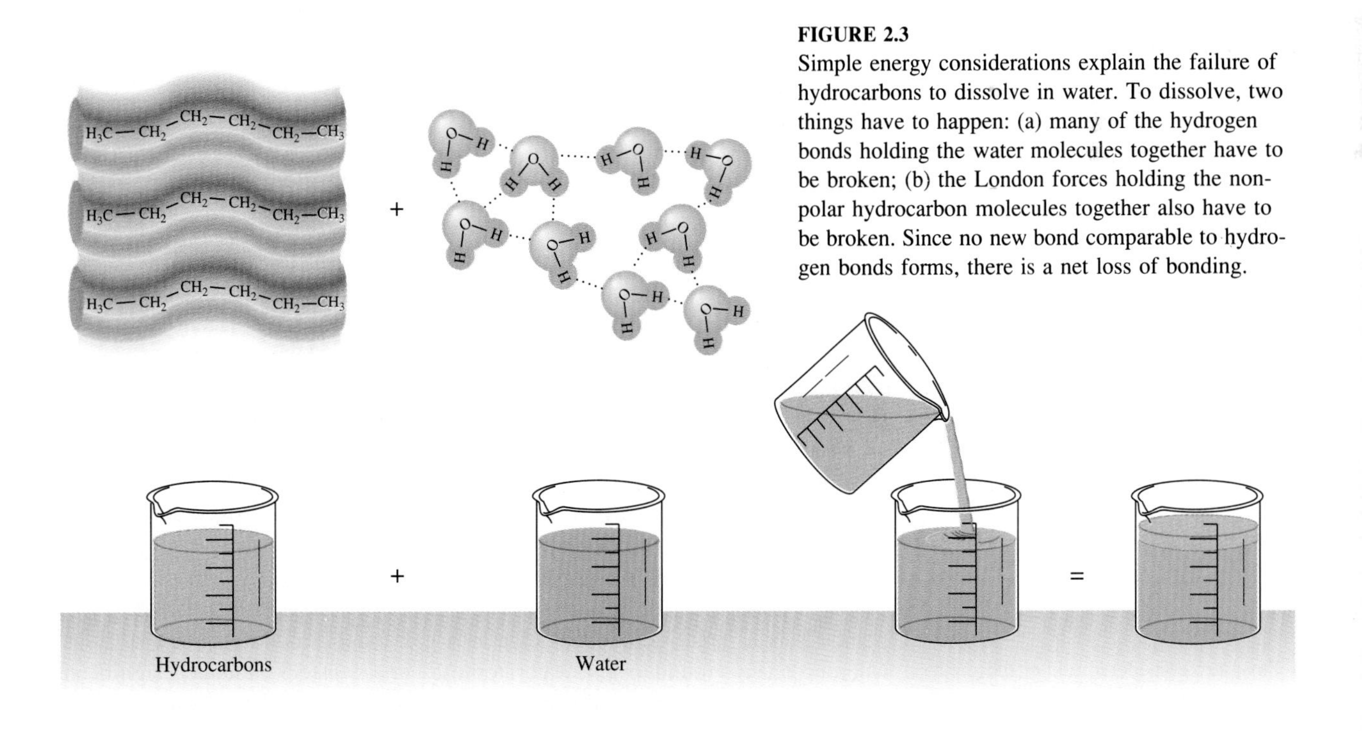

FIGURE 2.3
Simple energy considerations explain the failure of hydrocarbons to dissolve in water. To dissolve, two things have to happen: (a) many of the hydrogen bonds holding the water molecules together have to be broken; (b) the London forces holding the nonpolar hydrocarbon molecules together also have to be broken. Since no new bond comparable to hydrogen bonds forms, there is a net loss of bonding.

Solubility in Water

Hydrocarbon insolubility in water is seen over and over again in both organic chemistry and biochemistry. Recall the generalization "like dissolves like." Thus, nonpolar hydrocarbon compounds dissolve better in other nonpolar hydrocarbon materials and polar compounds dissolve better in polar materials, such as water. This is why, for example, you can use warm water to clean up water-based acrylic paints, but turpentine (a hydrocarbon) is used to clean up oil paints.

♦ Because of the nonpolar nature of the C—H bond, there is little reason to expect that hydrocarbons will be soluble in water.

Because oil floats on water, some mechanical control of oil spills is possible.

TESTING YOURSELF

Physical Properties

1. Discuss the physical states of Cl_2, Br_2, and I_2 (gas, liquid, and solid) in terms of the London forces attracting the molecules to one another.
2. Imagine that you are stacking logs. You have two kinds of logs to stack: straight logs (1 ft in diameter and 4 ft long) and T-shaped logs (1 ft in diameter, 3 ft long, with a 1-ft branch piece in the center). Which stack better: the piles of straight logs or the piles of T-shaped logs? Apply this same reasoning to molecules.

Answers **1.** The order of size, numbers of electrons and numbers of protons follow the trend: $I_2 > Br_2 > Cl_2$. Thus, the London attractive forces follow the same order. **2.** You could stack the straight logs better. There is more uniform surface contact between logs. The same holds for molecules; "odd shaped" molecules cannot maintain good contact with one another.

2.2 Alkanes and Their Nomenclature

The hydrocarbon family can be divided into four groups:

1. compounds that have single bonds between all carbons
2. compounds in which single-bonded carbon atoms form a ring
3. compounds that have one or more double bonds between adjacent carbons
4. compounds that have one or more triple bonds between adjacent carbons.

This chapter will consider hydrocarbons with single bonds between carbons. The next chapter will introduce carbon compounds with multiple bonds.

♦ Alkanes are hydrocarbons containing only single sigma bonds.

♦ Cows are considered one of the greatest sources of methane pollution.

The **alkane** family contains the simplest hydrocarbons. These are characterized by the formula C_nH_{2n+2}. All the carbon atoms have single bonds to other carbon atoms; all other carbon bonds are terminated with hydrogen atoms. Most of the common hydrocarbons have less than ten carbons. Methane and propane, two of the simplest hydrocarbons, are illustrated in Figure 2.4. Methane is the principal component of natural gas and occurs naturally in areas of decaying plants. It is a major component of marsh gas found around stagnant swamps. Microorganisms found in soil, water, termites, and livestock produce a significant amount of the methane found in our atmosphere. This could become a significant environmental problem. The only natural method to remove atmospheric methane is by *photooxidation*, a complex process of atmospheric oxidation using light energy.

The carbon atom in single-bonded alkane hydrocarbons always assumes a structure in which the carbon is at the center of a tetrahedron and the four pairs of shared electrons are at the vertices. The bond angles are about 109°. All alkanes have sp^3 hybridized carbon atoms. Look again at the structures of the methane and propane molecules in Figure 2.4. Notice that each carbon atom in the two molecules is tetrahedral in shape.

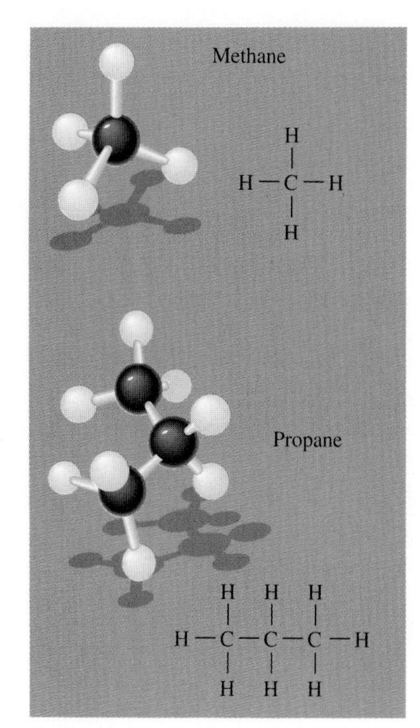

FIGURE 2.4
The alkanes are the simplest hydrocarbons. All carbons have single bonds to adjacent carbon or hydrogen atoms. Examples include methane and propane, both of which are used as heating fuels.

The Homologous Series

Hydrocarbons are usually named according to the number of carbon atoms in their longest continuous carbon chain. The number of carbons in the chain provides the stem name, and the suffix for all of the alkanes is always *-ane*. For example, see how the name of this five-carbon alkane relates to its structure:

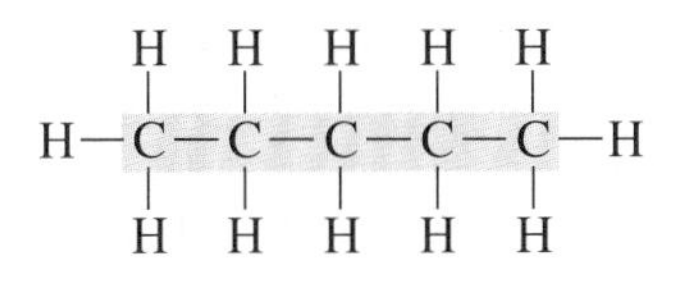

Stem name + *ane* = Pent*ane*

Names and formulas for the first ten alkanes are shown in Table 2.1. Most compounds we will study are derived from these ten basic hydrocarbons.

Figure 2.4 includes a "ball-and-stick" model of methane. This is a very useful representation of a chemical structure because it shows the shape of the molecule, whereas a **structural formula** shows only the connections of atoms. How do chemists, teachers, and students convey information about bonding in chemical formulas? A number of methods are available, depending on the information needed. Unfortunately, the ball-and-stick models, which are the most accurate, are also the least used because they require artistic drawings or photographs to do justice to the representation. What other ways can we show molecules?

Figure 2.5 shows several representations. A **perspective formula** is an attempt to couple the simplicity of a structural formula with the shape of a ball-and-stick model. A **condensed formula** is a way to simplify the structural formula. It is the easiest to write or type, but gives up the three-dimensional aspect. Finally, a **line formula** further simplifies the drawing of chemical structures, but it requires more "mental bookkeeping" because bonds to hydrogen are not shown. Each carbon has four bonds, and if the bond is to hydrogen, it is not shown. Line drawings are used extensively for ring compounds (discussed in a later section of this chapter).

Notice that the compounds shown in Table 2.1 differ only in the number of $—CH_2—$ groups inserted in the center of the carbon chain. A general formula for hydrocarbons with two or more carbon atoms can be written

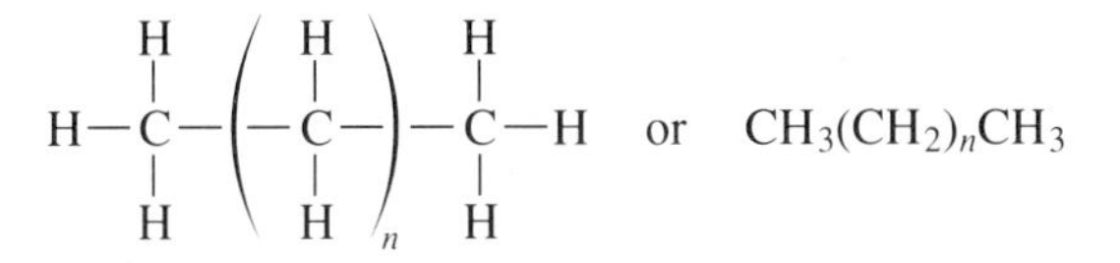

TABLE 2.1

Names and Structural Formulas for Simple Hydrocarbons Having up to Ten Carbons

Number of Carbons	Stem Name	Hydrocarbon Name	Structure
1	Meth	Methane	H | H—C—H | H
2	Eth	Ethane	H H | | H—C—C—H | | H H
3	Prop	Propane	H H H | | | H—C—C—C—H | | | H H H
4	But	Butane	H H H H | | | | H—C—C—C—C—H | | | | H H H H
5	Pent	Pentane	H H H H H | | | | | H—C—C—C—C—C—H | | | | | H H H H H
6	Hex	Hexane	H H H H H H | | | | | | H—C—C—C—C—C—C—H | | | | | | H H H H H H
7	Hept	Heptane	H H H H H H H | | | | | | | H—C—C—C—C—C—C—C—H | | | | | | | H H H H H H H
8	Oct	Octane	H H H H H H H H | | | | | | | | H—C—C—C—C—C—C—C—C—H | | | | | | | | H H H H H H H H
9	Non	Nonane	H H H H H H H H H | | | | | | | | | H—C—C—C—C—C—C—C—C—C—H | | | | | | | | | H H H H H H H H H
10	Dec	Decane	H H H H H H H H H H | | | | | | | | | | H—C—C—C—C—C—C—C—C—C—C—H | | | | | | | | | | H H H H H H H H H H

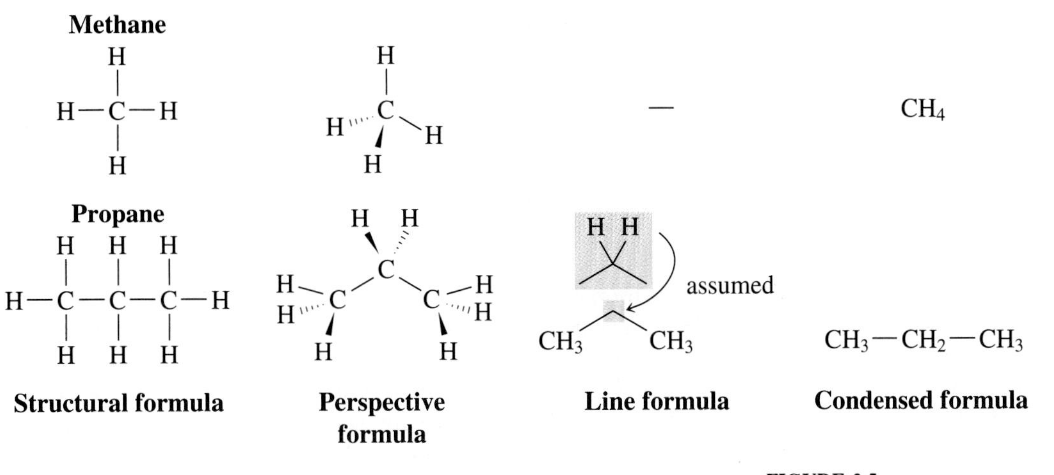

FIGURE 2.5
Several methods of representing chemical structures, using two alkanes as examples.

where $n = 1, 2, 3, \ldots$. A more general formula for any alkane is

$$C_nH_{2n+2}$$

where $n = 1, 2, 3, \ldots$.

Compounds that differ only in the number of $—CH_2—$ groups inserted in the carbon chain form a family group called a **homologous series**. Members of a homologous series are very similar in their chemical reactivity. But as more carbons are added and the size of the molecules increases, they exhibit gradually changing physical properties. We already saw (in Figure 2.2) that there is a regular relationship between the boiling point and the number of carbon atoms in the hydrocarbon chain.

♦ A series of compounds differing only in the insertion of a $—CH_2—$ group is called a *homologous series*.

Isomers

Until now we have assumed that all of a hydrocarbon's carbon atoms are connected in a simple unbranched chain. This need not always be the case. It is possible to assemble several different structures having different bonding patterns with the same set of atoms. Consider the case of the four-carbon hydrocarbon, butane, which can be represented three different ways:

```
  H  H  H  H
  |  |  |  |
H-C--C--C--C-H    CH3—CH2—CH2—CH3        /\_/CH3
  |  |  |  |                         CH3
  H  H  H  H
```

Structural **Condensed** **Line**

If butane contains four carbon atoms, then it must contain ten hydrogen atoms according to the formula

$$C_nH_{2n+2} \quad \text{where } n = 4$$

$$C_4H_{10}$$

Are there other arrangements of these atoms that can be made with this set of atoms? Consider the following four-carbon hydrocarbon:

♦ Isomers increase rapidly with carbon number

C_nH_{2n+2} n	# Isomers
4	2
6	5
8	18
10	75
20	366,310
40	62,481,801,147,341

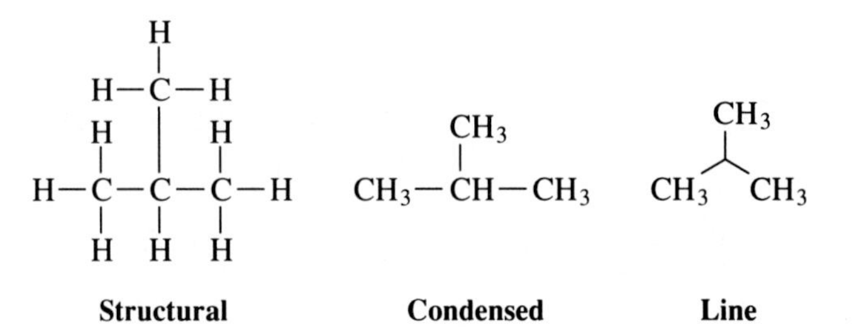

♦ *Isomers* are compounds having the same molecular formula but different atomic arrangements. These compounds differ in shape and properties, and are actually different compounds.

The formula of this branched compound is also C_4H_{10}. Molecules formed by different arrangements of the same atoms are called **isomers**. The linear and branched butane four-carbon compounds are both isomers of butane. The linear molecule is called *normal butane*. One accepted name for the second molecule is *isobutane*. The opportunities for isomerization are limited for butane due to its small number of carbon atoms. Simply having different shapes with the same bond connectivity does not change the molecule.

As the number of carbons in a hydrocarbon increases, the number of possible isomers increases very rapidly. In fact, isomerization is not only possible but common if the compound is larger than three carbons. For example, the five-carbon compound pentane has three isomers (a, b, and c).

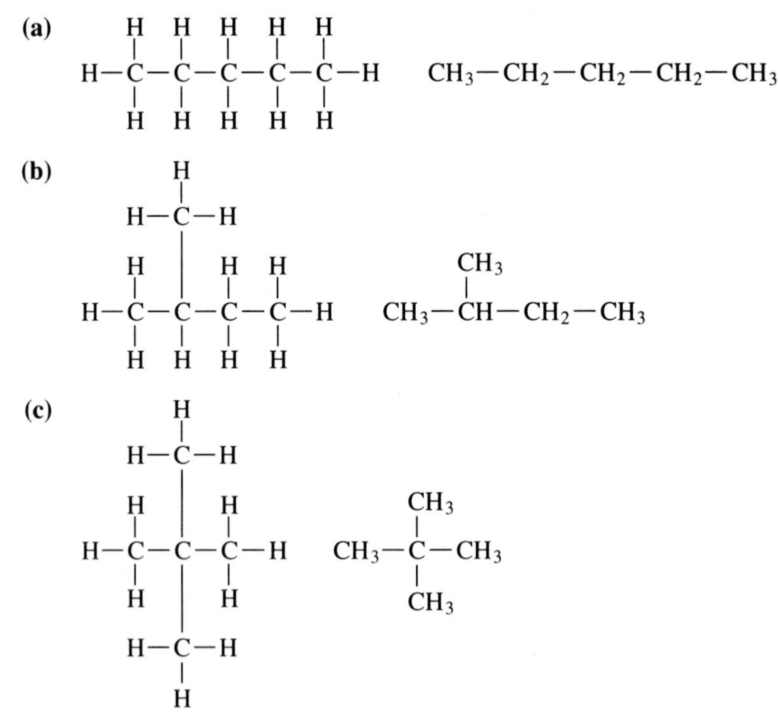

Three isomers of pentane

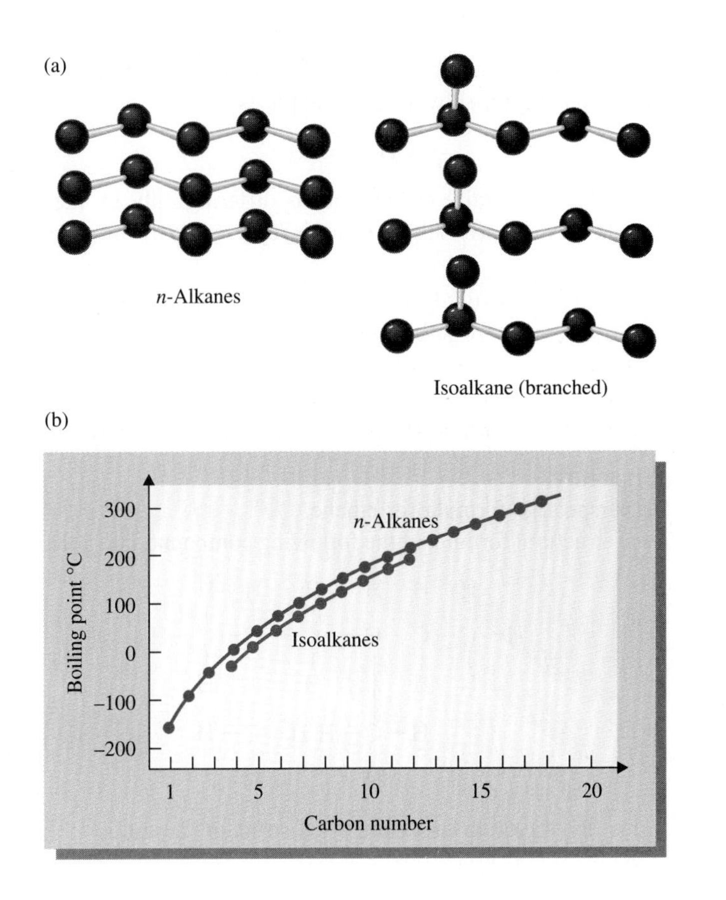

FIGURE 2.6
(a) Because of their stackable shapes, linear hydrocarbon molecules (such as *n*-alkane) interact with one another better than branched hydrocarbons (such as isoalkane). (b) The branched hydrocarbons thus have decreased London forces between molecules and resulting lower boiling points.

Because of the importance of London interactions to the boiling point of straight-chain hydrocarbons, they have higher boiling points than branched hydrocarbons (isomers). In the branched hydrocarbons there is less chance for continuous surface contact between molecules because the branched shapes are more "ball-like" and are less able to stack together. The effect of branching on boiling point can be seen in Figure 2.6.

♦ The more branched a molecule, the less it is able to interact with a neighboring molecule. This results in lower boiling points for liquids and lower melting points for solids.

NEW TERMS

alkane A family of hydrocarbons having only single carbon-to-carbon bonds characterized by the general formula C_nH_{2n+2}

structural formula A representation of a structure that emphasizes the bond connection between atoms

perspective formula A representation of a chemical structure that conveys the three dimensions of a ball-and-stick model and has some of the simplicity of the structural formula

condensed formula A condensed representation of a chemical structure that leaves out the vertical bonds and shows the whole structure set on one line, such as $CH_3CH_2CH_3$

line formula A simplified representation of a structural formula in which many of the C—H bonds are not shown. An example is

CH_3⁀CH_3

homologous series A family of compounds differing only by the number of CH_2 groups in the formula. The alkane series is represented by the general formula C_nH_{2n+2}

isomer Variations of a particular compound having the same molecular formula but different arrangements of atoms and bonds

TESTING YOURSELF

Chemical Structures

1. Draw the perspective formula for ethane.
2. Convert the structural formula of the following compound into a line structure.

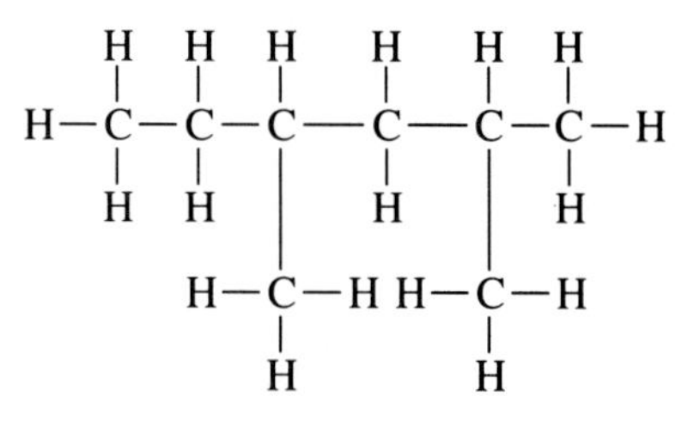

3. Convert this line structure to a condensed structural formula.

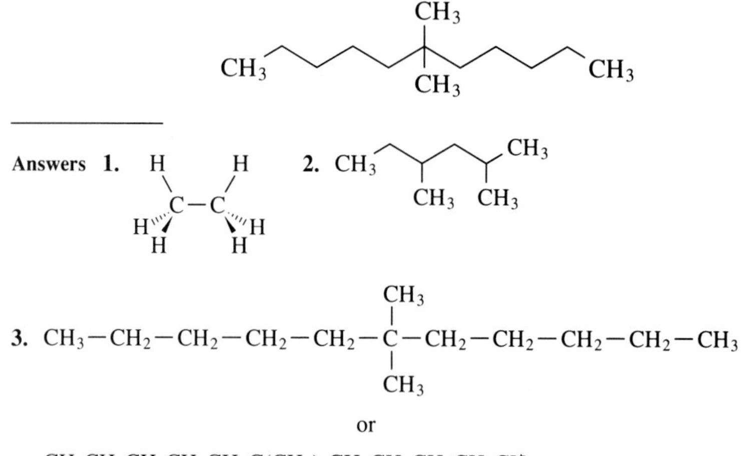

Answers 1. 2. 3.

$$CH_3-CH_2-CH_2-CH_2-CH_2-C(CH_3)_2-CH_2-CH_2-CH_2-CH_2-CH_3$$

or

$CH_3CH_2CH_2CH_2CH_2C(CH_3)_2CH_2CH_2CH_2CH_2CH_3$

2.3 Alkyl Groups and Nomenclature

Simple Alkyl Groups

Because so many potential isomers exist, there must be a logical system for naming them. The solution to the problem of naming isomers lies in the use of the alkyl group. As we showed in the previous chapter, an **alkyl group** is a hydrocarbon molecule minus a hydrogen. In place of the hydrogen is an "open

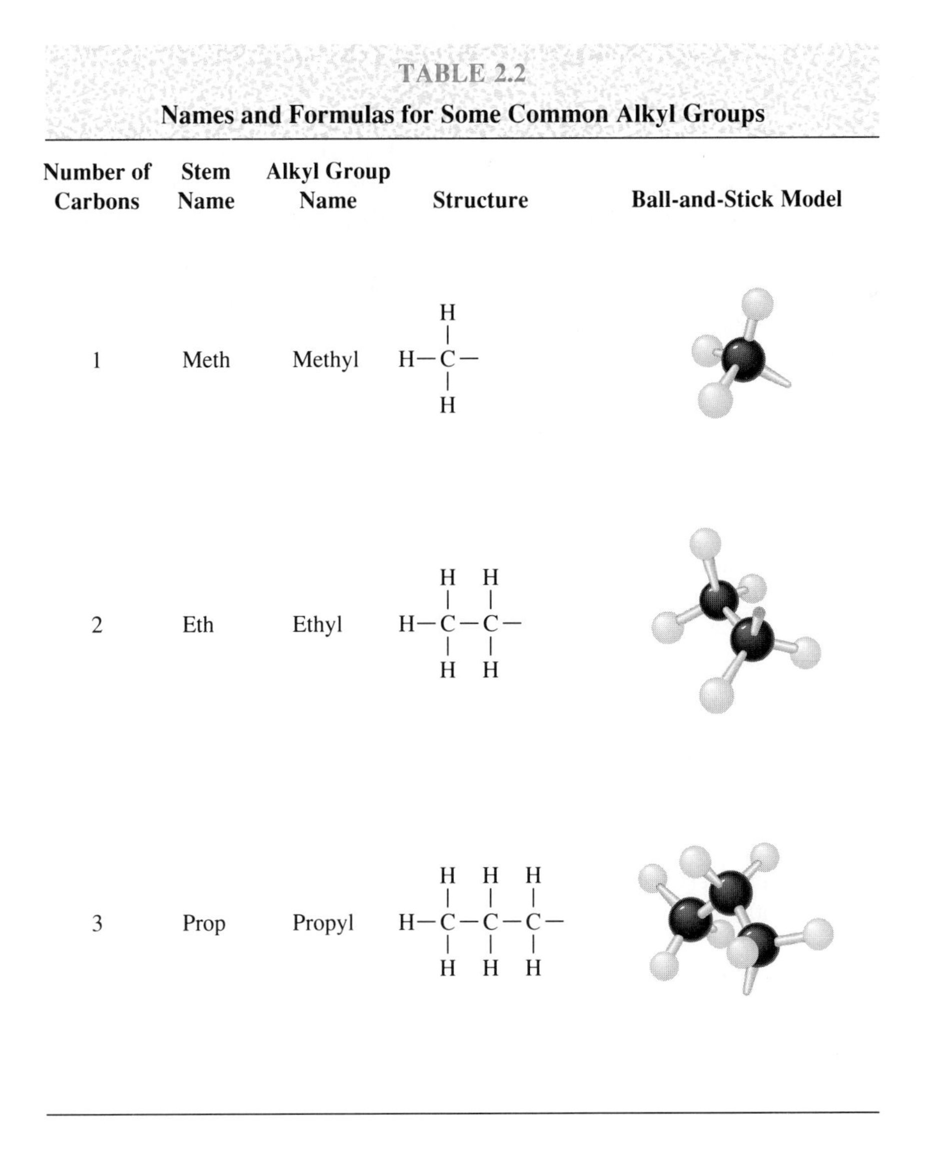

TABLE 2.2
Names and Formulas for Some Common Alkyl Groups

Number of Carbons	Stem Name	Alkyl Group Name	Structure	Ball-and-Stick Model
1	Meth	Methyl	H—C(H)(H)—	
2	Eth	Ethyl	H—C(H)(H)—C(H)(H)—	
3	Prop	Propyl	H—C(H)(H)—C(H)(H)—C(H)(H)—	

bond" that can be attached to other carbon chains. Alkyl groups carry the stem name of their parent alkane, but end with the suffix *-yl* instead of *-ane*. We can see in Table 2.2 that the alkyl group is directly related to the parent hydrocarbon alkane. From the names of the compounds in the homologous series, we know the names of the corresponding alkyl groups.

Starting with propane, we find it possible to make isomeric alkyl groups. Depending on the hydrogen removed, we can make different alkyl groups.

Primary, Secondary, and Tertiary Carbons

Carbons of an alkyl group are designated as **primary carbons** if they connect to only one additional carbon. Primary carbons are on the end of a chain. **Secondary carbons** bond to two adjacent carbons, and **tertiary carbons** bond to three adjacent carbons. **Quaternary carbons** are bonded to four other carbon atoms.

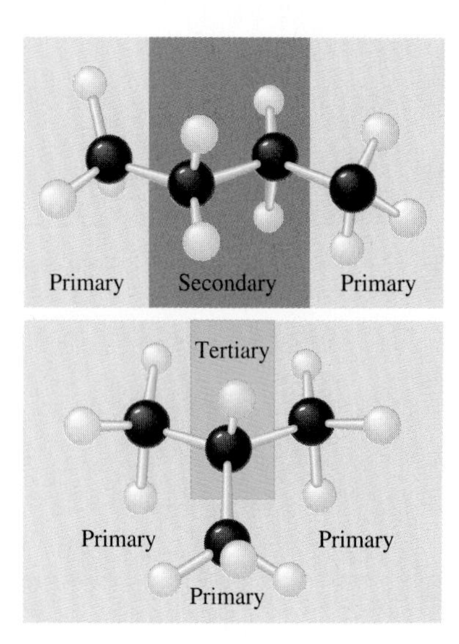

FIGURE 2.7
Carbon atoms are classified as primary, secondary, tertiary, or quarterrary, according to the number of other carbon atoms to which they are bonded, as shown in these two isomers of butane.

The carbons in the two isomers of butane shown in Figure 2.7 are indicated as primary, secondary, or tertiary. These are abbreviated *p*, *s*, and *t*, respectively.

Hydrogen atoms are classified according to the type of carbon to which they are attached and are given equivalent names. They are designated as *primary* if they connect to a primary carbon in any hydrocarbon chain, as *secondary* if they connect to a secondary carbon, or as *tertiary* if they connect to a tertiary carbon. In the following propane structure, the highlighted hydrogens are primary and all other hydrogens are secondary.

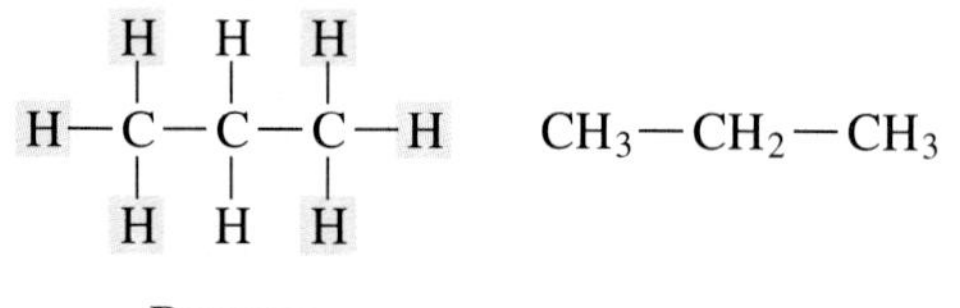

Propane

Suppose now that one of the primary hydrogens is removed to form the *normal propyl* group:

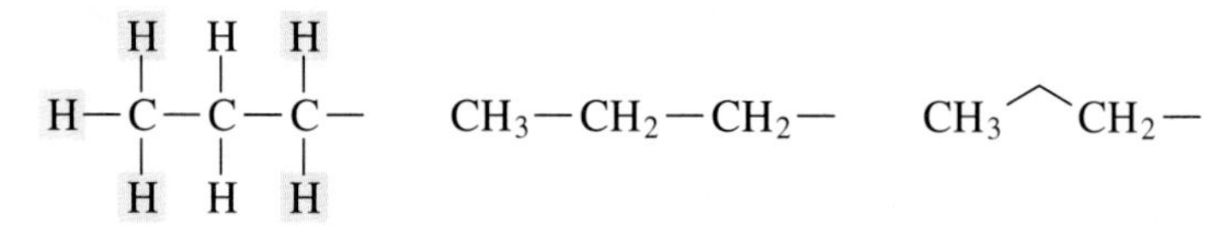

***n*-Propyl group**

The *n* in *n*-propyl stands for *normal*. If one of the secondary hydrogens is removed (the only other possibility for propane), the *isopropyl* group results.

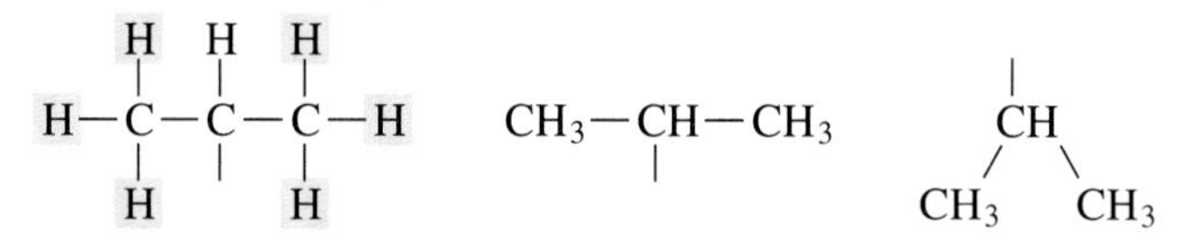

Isopropyl group

The situation is similar for butane. If one of the primary hydrogens is removed from butane, the *normal butyl* (or *n*-butyl) group results.

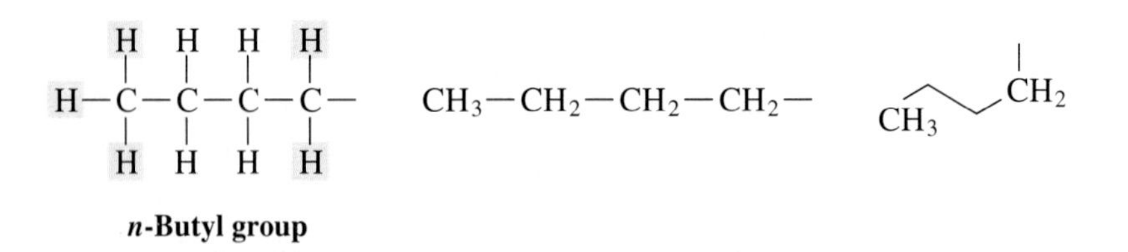

***n*-Butyl group**

Now recall that butane has two isomers, normal butane and isobutane. When one of the secondary hydrogens of *n*-butane is removed, the name isobutyl cannot be used because isobutane is a different compound. Instead, the resulting alkyl group is known as the *secondary butyl* (or *sec*-butyl) group.

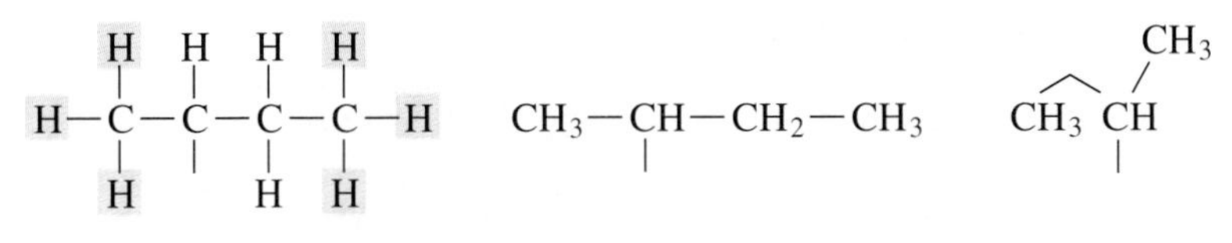

***sec*-Butyl group**

Consider now the case of isobutane:

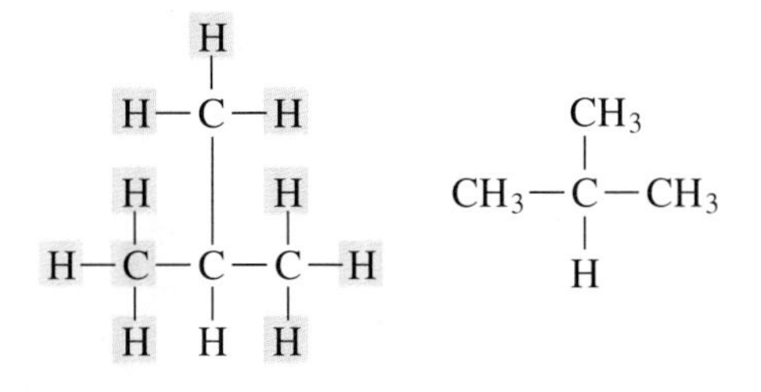

All of the hydrogen atoms in this compound are primary, except one, which is connected to a tertiary carbon. (It is the only hydrogen that is not highlighted.) If any primary hydrogen is removed, the *isobutyl* group results. This is the alkyl group (isobuty*l*) obtained form the alkane (isobut*ane*) by removal of a primary hydrogen.

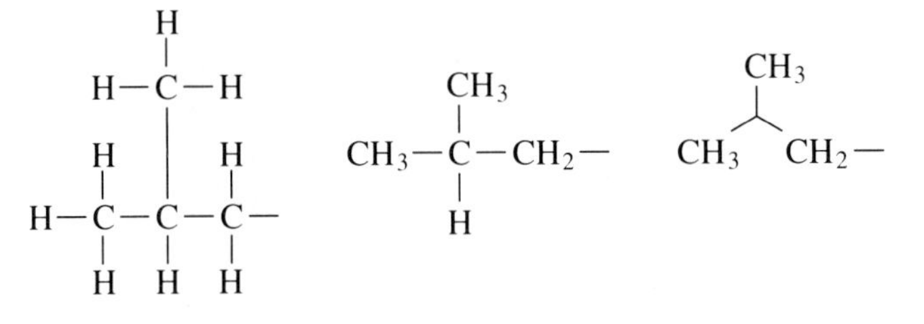

Isobutyl group

If, on the other hand, the sole tertiary hydrogen is removed, the *tertiary butyl* (or *tert*-butyl) group results.

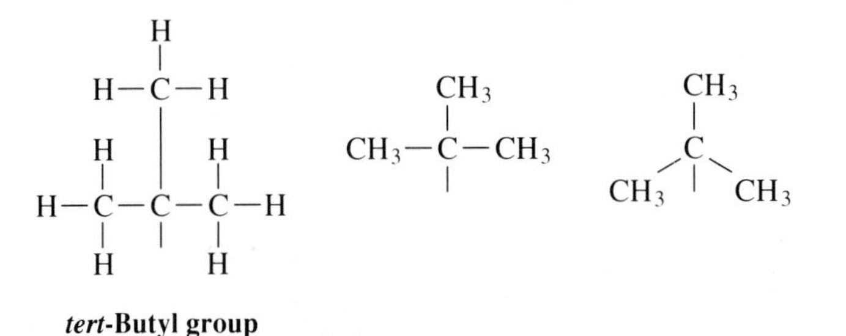

***tert*-Butyl group**

The alkyl groups derived from the isomers of propane and butane are summarized in Table 2.3.

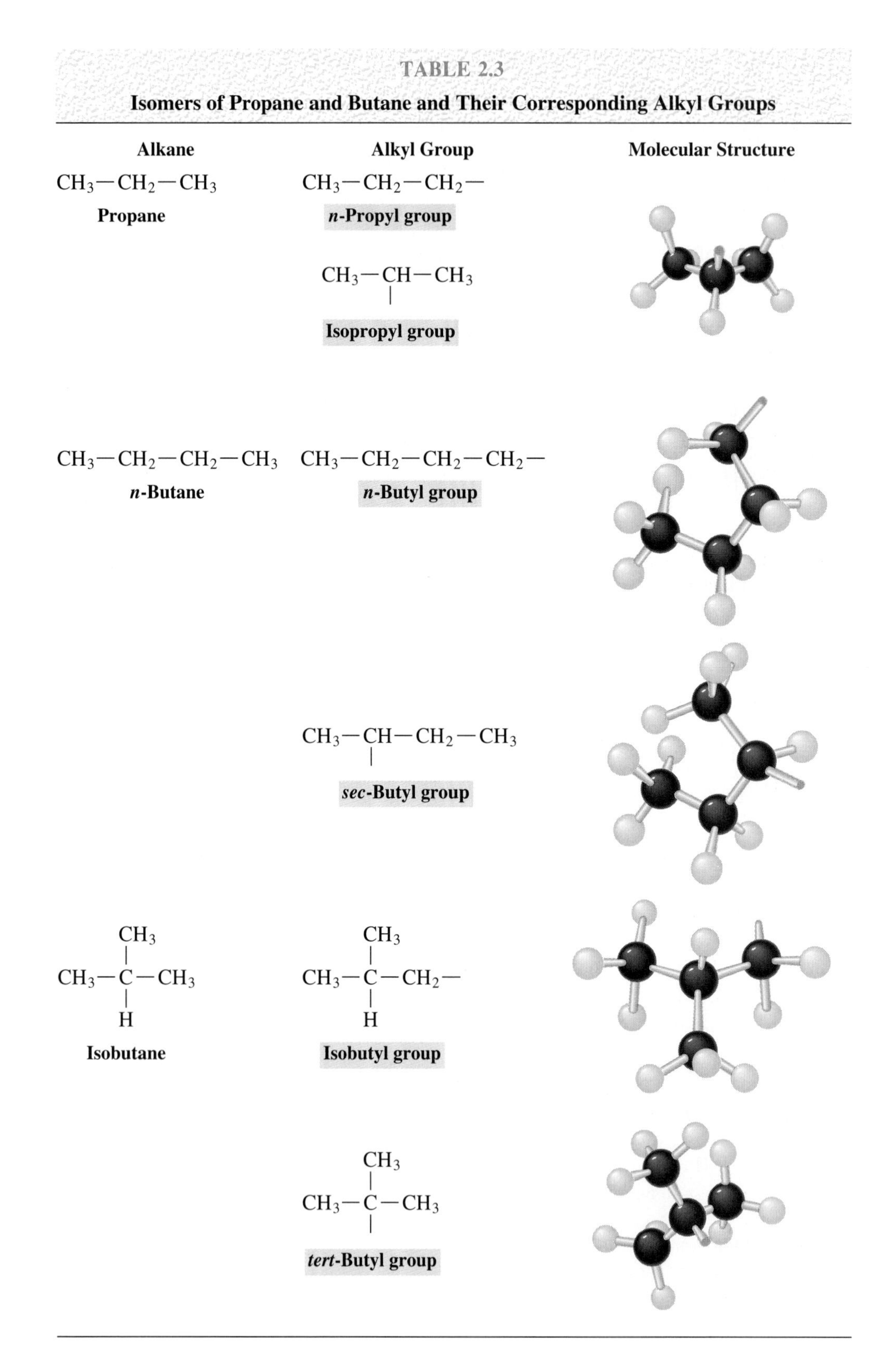

TABLE 2.3

Isomers of Propane and Butane and Their Corresponding Alkyl Groups

Alkane	Alkyl Group	Molecular Structure
$CH_3-CH_2-CH_3$ **Propane**	$CH_3-CH_2-CH_2-$ ***n*-Propyl group**	
	$CH_3-\underset{\mid}{CH}-CH_3$ **Isopropyl group**	
$CH_3-CH_2-CH_2-CH_3$ ***n*-Butane**	$CH_3-CH_2-CH_2-CH_2-$ ***n*-Butyl group**	
	$CH_3-\underset{\mid}{CH}-CH_2-CH_3$ ***sec*-Butyl group**	
$CH_3-\underset{H}{\overset{CH_3}{C}}-CH_3$ **Isobutane**	$CH_3-\underset{H}{\overset{CH_3}{C}}-CH_2-$ **Isobutyl group**	
	$CH_3-\underset{\mid}{\overset{CH_3}{C}}-CH_3$ ***tert*-Butyl group**	

General Rules of Nomenclature

Branched carbon skeleton compounds have alkyl groups substituted for hydrogens on one or more of the carbons of the parent straight-chain compound. The resulting compounds are named by a system developed by the International Union of Pure and Applied Chemistry (IUPAC). The IUPAC is a group through which chemists worldwide agree upon uniform ways of handling terms and communicating within their profession. The IUPAC system for naming hydrocarbons involves several simple steps. These steps will be illustrated by naming the two relatively simple carbon compounds shown below.

EXAMPLE 2.1

Name this hydrocarbon using the IUPAC system.

```
    H   H   H   H   H
    |   |   |   |   |
H — C — C — C — C — C — H      CH3—CH—CH2—CH2—CH3
    |   |   |   |   |                |
    H   |   H   H   H               CH3
        |
    H — C — H
        |
        H
```

(a) Find the longest *continuous* carbon chain. Number each carbon in the central chain, beginning at the end closest to the first substituent. The stem name is derived from the name of the parent alkane.

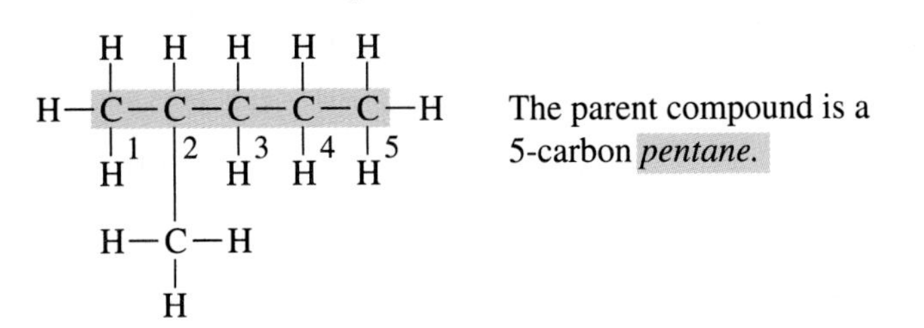

(b) Find each alkyl group side chain. Assign it a name and indicate to which carbon it is attached.

```
    H   H   H   H   H
    |   |   |   |   |
H — C — C — C — C — C — H
    |1  |2  |3  |4  |5
    H   |   H   H   H
        |
    H — C — H        Methyl group, attached to
        |            carbon number 2.
        H
```

(c) Number and name the substituents. If the same substituent occurs more than once, the number of each carbon of the parent alkane to which it is attached is given, and the number of substituent groups involved is indicated by a prefix such as *di-*, *tri-*, *tetra-*, and so on.

The compound has a methyl group attached to carbon number 2. The substituent called 2-methyl.

(d) Name the compound, beginning with the side chains in alphabetical order and ending with the name of the parent compound. Follow these rules:
  (i) *Always* use commas (,) between numbers.
  (ii) *Always* use hyphens (-) between numbers and words.
  (iii) Do not leave spaces in the name.
  (iv) When alphabetizing substituent groups, ignore the prefixes *di-*, *tri-*, and so on.
  (v) Prefixes such as *sec-*, *tert-* are not used in decisions of alphabetical order unless comparing the same group: *sec*-butyl > *tert* butyl.
  (vi) Use the lowest possible numbers.

Solution: The compound is thus named 2-methylpentane.

EXAMPLE 2.2

Name this branched hydrocarbon using the IUPAC system.

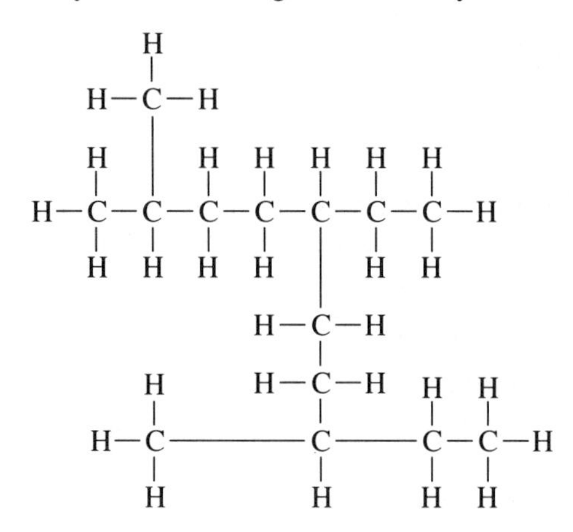

(a) Find the longest *continuous* chain. Number each carbon in the central chain, beginning at the end closest to the first substituent. Name the parent compound as a normal alkane. Note that the longest chain need not be straight on the drawing. The stem name of the compound is derived from the name of the parent alkane.

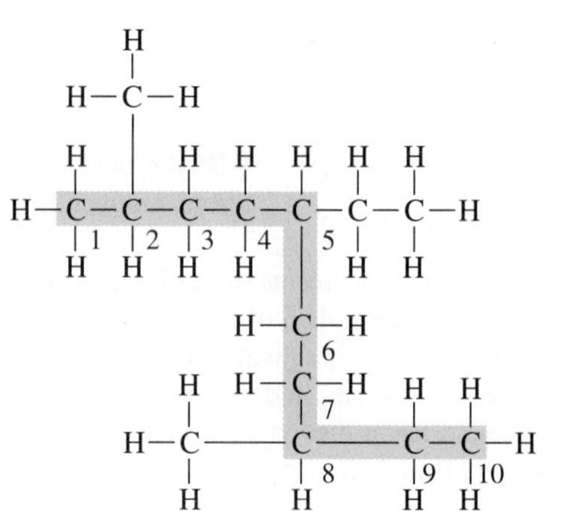

The longest continuous carbon chain in this compound is ten carbons in length; the parent compound is thus a *decane.* This chain is highlighted in the drawing. Identification of the longest chain requires care.

(b) Find each alkyl group side chain. Assign it a name and indicate to which carbon it is attached.

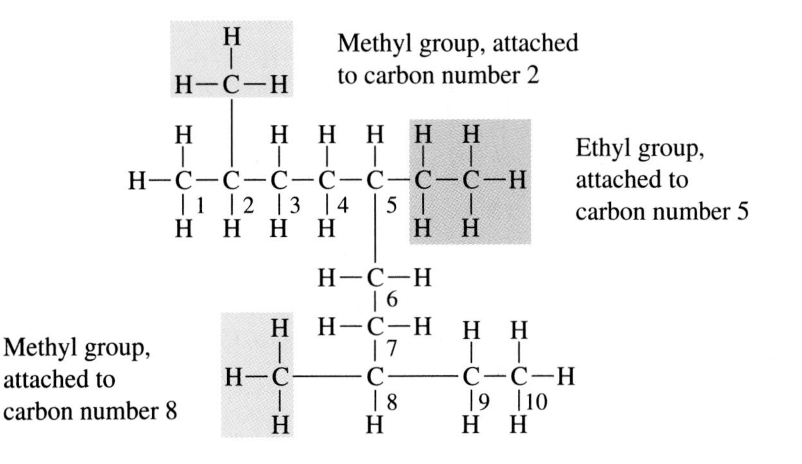

(c) Number and name the substituents.

This compound has methyl groups attached to carbon numbers 2 and 8, and an ethyl group attached to carbon number 5. The substituents are called

2,8-dimethyl and 5-ethyl

(d) Name the compound, beginning with the side chains in alphabetical order and ending with the name of the central chain.

Solution: Since *e* (ethyl) comes before *m* (methyl) in the alphabet, the compound is named 5-ethyl-2,8-dimethyldecane (Figure 2.8).

Recall that prefixes, such as *sec* and *tert*, are not used when alphabetizing substituent groups. The prefix *iso* is considered to be part of the alkyl group and is listed under *i* when alphabetizing.

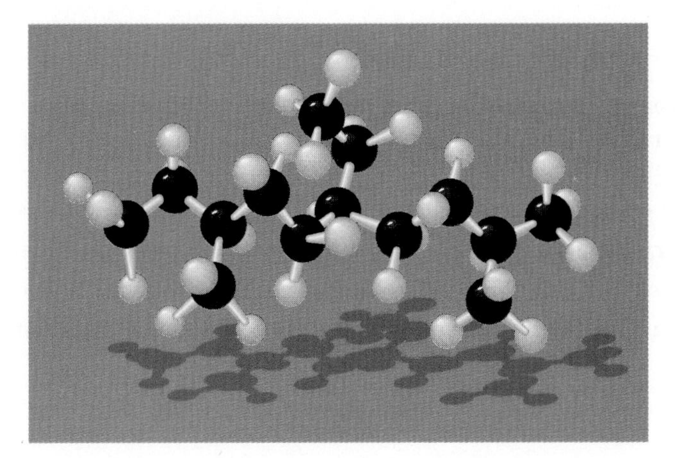

FIGURE 2.8
A ball-and-stick model of 5-ethyl-2,8-dimethyldecane.

The same process as shown in Examples 2.1 and 2.2 can be used to name the isomers of pentane. Note that each compound has the formula C_5H_{12}.

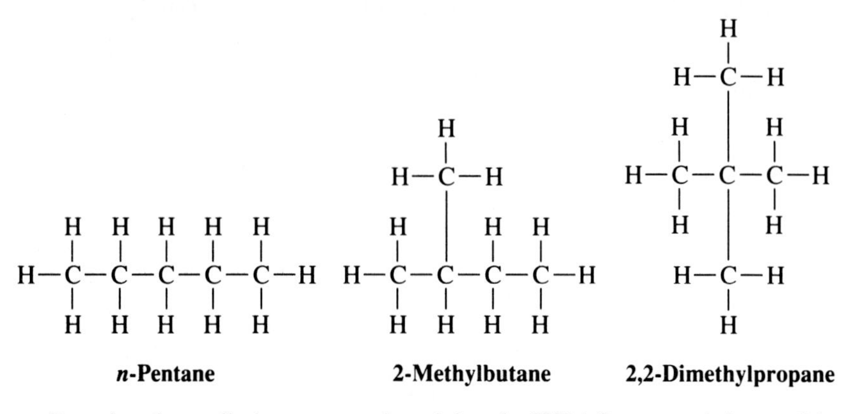

By using these alkyl groups, and applying the IUPAC system, it is possible to name a variety of more complex hydrocarbons. Consider the following examples (side chains are highlighted):

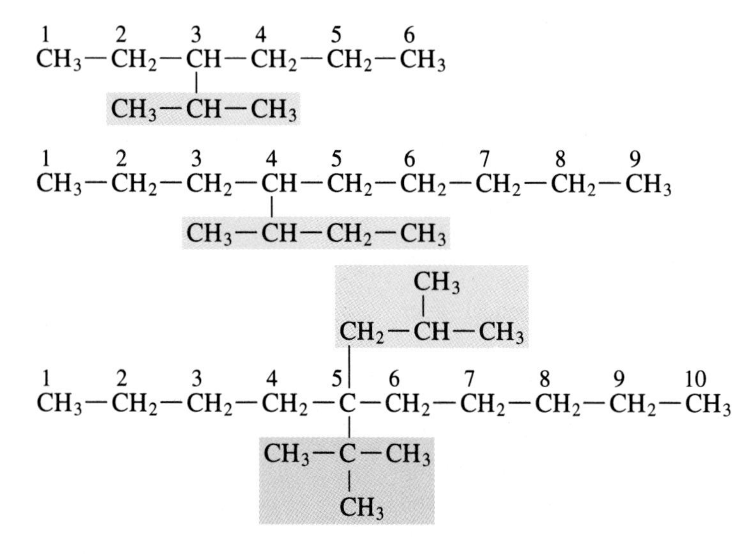

Long chain: hexane (six carbons)
Side chain: isopropyl group
Name: 3-isopropylhexane

Long chain: nonane (nine carbons)
Side chain: *sec*-butyl group
Name: 4-*sec*-butylnonane

Long chain: decane (ten carbons)
Side chains: isobutyl group and *tert*-butyl group
Name: 5-*tert*-butyl-5-isobutyldecane

NEW TERMS

alkyl group a hydrocarbon group made up of a hydrocarbon minus one of its hydrogen atoms. This group is named from the parent alkane by replacing the *-ane* ending with *-yl*.

primary carbon A carbon atom that is bonded to only one other carbon

secondary carbon A carbon atom that is bonded to two other carbons

tertiary carbon A carbon atom that is bonded to three other carbons

quaternary carbon A carbon atom that is bonded to four other carbons

TESTING YOURSELF

Naming Alkanes Using Alkyl Group Nomenclature

1. Give names for each of the following compounds.

a.
```
CH3—CH—CH2—CH3
     |
     CH3
```

b.
```
     CH3  CH3
     |    |
  H—C——C—CH2—CH3
     |    |
     CH3  CH3
```

c.
```
                   CH2—CH3
                   |
CH3—CH2—C——CH—CH2—CH2—CH3
                   |     |
                   CH3   |
                         |
               CH3—C—CH3
                         |
                         H
```

2. Draw structures for each of the following compounds.
   a. 4-ethyl-2-methylheptane
   b. 3,3,5-trimethyloctane
   c. 4-*tert*-butyl-4-isopropyloctane

Answers **1. a.** 2-methylbutane **b.** 2,3,3-trimethylpentane **c.** 3-ethyl-4-isopropyl-3-methylheptane

2. a.
```
CH3—CH—CH2—CH—CH2—CH2—CH3
     |          |
     CH3        CH2—CH3
```

b.
```
                  CH3
                  |
CH3—CH2—C——CH2——CH—CH2—CH2—CH3
                  |             |
                  CH3           CH3
```

c.
```
                       CH3
                       |
                CH3—C—CH3
                       |
CH3—CH2—CH2—C—CH2—CH2—CH2—CH3
                       |
                CH3—CH—CH3
```

2.4 Cycloalkanes

So far in this chapter we have considered straight-chain hydrocarbons as well as branched hydrocarbons having attached alkyl group side chains. Because of the tetrahedral shape of the single-bonded carbon atom, these are really "zigzag" in shape. A third possibility exists. Under proper conditions, alkane molecules

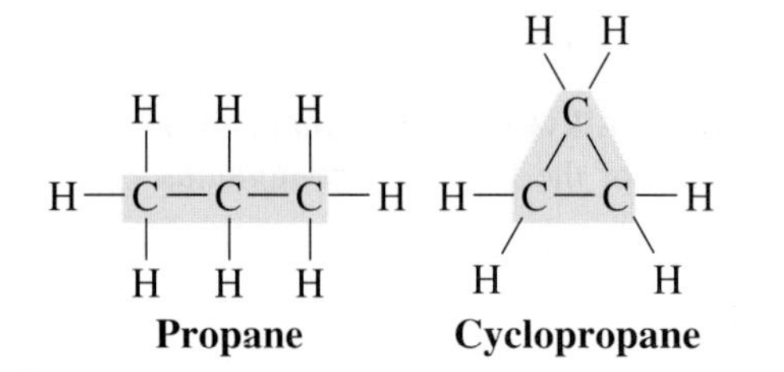

FIGURE 2.9
In the cycloalkane family, the skeletal carbon framework forms a ring.

can circle around to join head and tail, forming a ring. Compounds containing rings are quite common in nature.

Cycloalkanes are named according to the number of carbons in their ring in the same manner that normal alkanes are named according to the number of carbons in their longest continuous chain. Cyclic compounds have the prefix *cyclo-* preceding their name. A three-carbon cycloalkane, for example, is cyclopropane (Figure 2.9). A ten-carbon cycloalkane is cyclodecane.

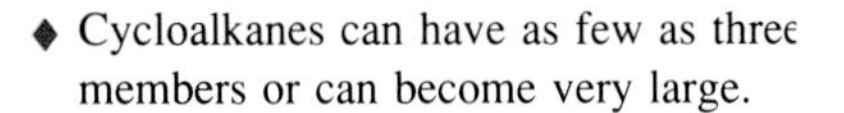
♦ Cycloalkanes can have as few as three members or can become very large.

Structure

Early thoughts about ring compounds were based on the assumption that they were planar compounds. From this assumption, only the amount of deviation from the stable 109° tetrahedral shape had to be considered to understand the strain built into a ring compound. Cycloalkanes have some special structural problems, since in the case of three- and four-membered rings, the 109° bond angle characteristic of the tetrahedral carbon cannot be achieved. In cyclopropane, for example, the three carbons form an equilateral triangle in which the bond angles are forced to be 60° (Table 2.4). As you might expect, the cyclopropane ring is highly strained. Larger cycloalkanes are less strained and are therefore more stable. Indeed, by applying this logic, you could predict that cyclopentane would be the most stable ring because its actual bond angle is very close to its required bond angle (Table 2.4). However, this is not observed.

♦ Except for cyclopropane, rings are *not* planar.

The reason that this simple theory does not work is because, aside from cyclopropane, none of the rings are planar. Table 2.4 shows the planar structures of several of the common rings, as well as more realistic ball-and-stick models.

The cycloalkanes lend themselves well to a shorthand method of representation based on combining the geometric shape of the ring with a line drawing. This **geometric structure** can be shown by an example.

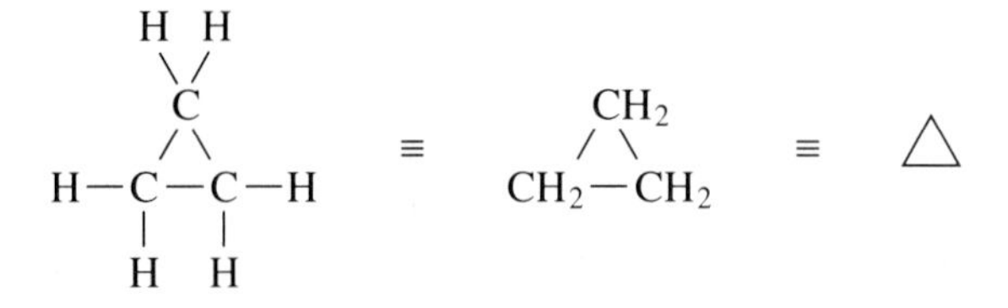

When this shorthand method is used, it is understood that there is a four-bonded carbon atom at each vertex of the figure and that each carbon has two hydrogens as well as two bonds to adjacent carbon atoms, for a total of four bonds.

Cycloalkanes can be represented by the general formula C_nH_{2n}, where n = 3, 4, 5, 6 . . . . Cyclopentane, for example, is C_5H_{10}.

TABLE 2.4
Some Common Cycloalkanes

Name	Required Bond Angle	Actual Bond Angle	Planar Structure	Ball-and-Stick Model
Cyclopropane	60°	60°		
Cyclobutane	90°	88°		
Cyclopentane	108°	105°		
Cyclohexane	120°	109°		

Nomenclature

When atoms other than hydrogen or alkyl groups are substituted for hydrogens on a cycloalkane, they have a well-defined position relative to other groups on the ring. If two groups are on the same side of a ring, they are said to be **cis** to each other. If they are on opposite sides of the ring, they are said to be **trans**. Substituted cycloalkanes can thus exist as cis or as trans isomers of the same molecular formula.

♦ Nomenclature of cycloalkanes includes cis and trans designations, where cis refers to groups on the same side and trans to groups on opposite sides.

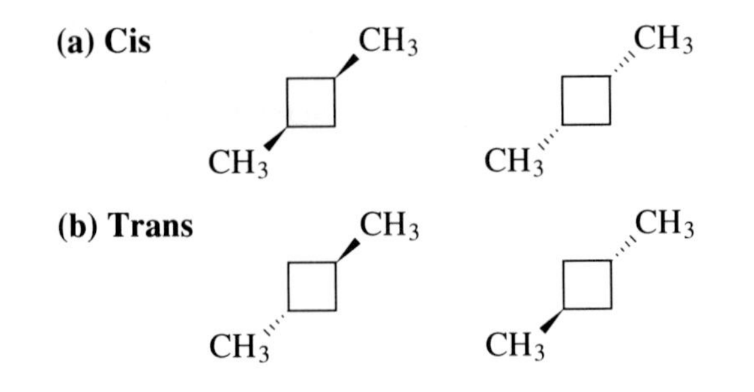

FIGURE 2.10
Two isomers of dimethylcyclobutane. (a) When groups are on the same side of the ring, the isomer is called a cis isomer. (b) When groups are on opposite sides of the ring, the isomer is called a trans isomer.

Figure 2.10 shows two isomers of 1,3-dimethylcyclobutane. In Figure 2.10(a), the two substituted methyl groups are on the same side of the molecule; this is the cis form of the molecule and is called *cis*-1,3-dimethylcyclobutane. In Figure 2.10(b) the methyl groups are on opposite sides of the molecule; this is the trans form and is called *trans*-1,3-dimethylcyclobutane.

A chemical shorthand is commonly used to indicate the position of substituents on a molecule. Bonds to groups "above" the ring are shown as solid wedges (▬), while bonds to groups "below" the ring are shown as dashed lines (····). This is shown in Figure 2.10.

The steps for naming substituted cycloalkanes are straightforward. When the cis and trans relationship of substituent groups is known, it is included in the name. Consider the following examples:

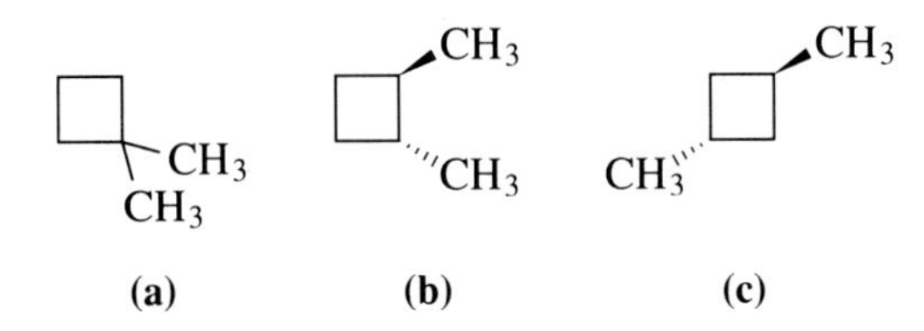

To name these cycloalkanes, first determine the parent hydrocarbon. Since all three of the compounds have four carbons in their ring, they are therefore cyclobutanes.

Next, you need to determine the alkyl groups and give each a name and number. Since there is no beginning or end of a ring, you start the numbering in such a way as to create the lowest possible numbers for the substituents. Compound (a) has two methyl groups on the first carbon. Thus, its name is 1,1-dimethylcyclobutane. Compound (b) has two methyl groups that are located on the first and second carbons. Note that they are trans to each other. This compound is thus called *trans*-1,2-dimethylcyclobutane. Compound (c) has two methyl groups that are located on the first and third carbons. They are also trans to each other. Its name is therefore *trans*-1,3-dimethylcyclobutane.

The nomenclature of cycloalkanes with different alkyl substitutions is shown in Figure 2.11. When compounds have more than one kind of alkyl group, these groups are listed in alphabetical order.

2-*sec*-butyl-5-*tert*-butyl-1-isopropyl-4-methylcycloheptane

If cis and trans information is not given in the example, it is not included in the name.

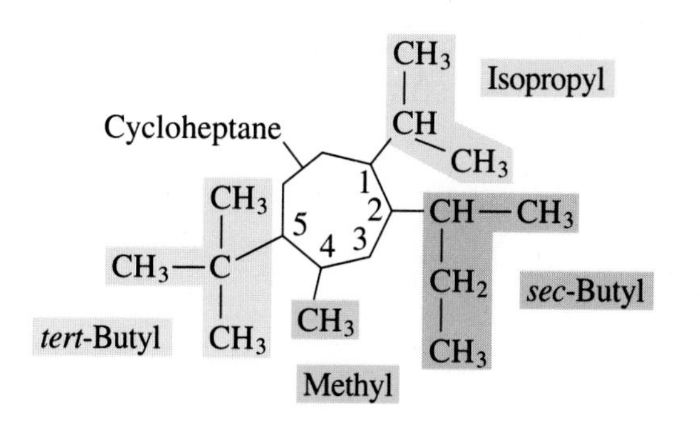

FIGURE 2.11
An example of nomenclature of a cycloalkane having a variety of alkyl substitutions.

NEW TERMS

cycloalkane A hydrocarbon compound with single carbon–carbon bonds, in which the skeletal carbons form a ring

geometric structure A geometric form representing a molecule; carbon atoms are assumed to be at each vertex and hydrogens are not shown

cis A prefix used to designate two similar groups on the same side of a molecule

trans A prefix used to designate two similar groups on opposite sides of a molecule

TESTING YOURSELF

Naming Cycloalkanes

1. Draw structures for each of the following molecules.
   a. 1-methyl-3-*n*-propylcyclohexane
   b. 1-isobutyl-1-isopropylcyclopentane
   c. 2-*tert*-butyl-1,5-diethylcyclooctane
2. Name the following compounds.

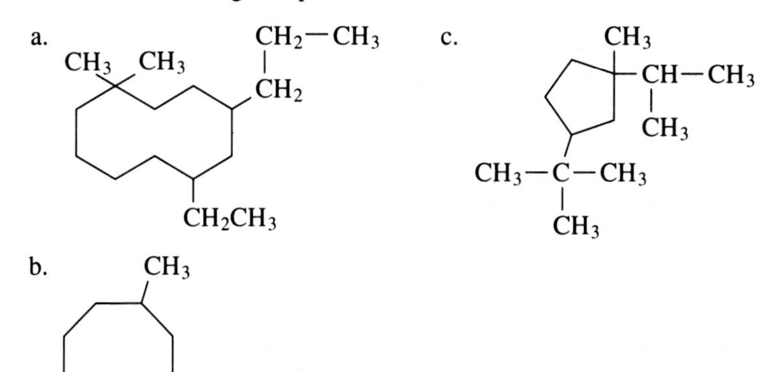

3. Identify the following substituents as cis or trans.

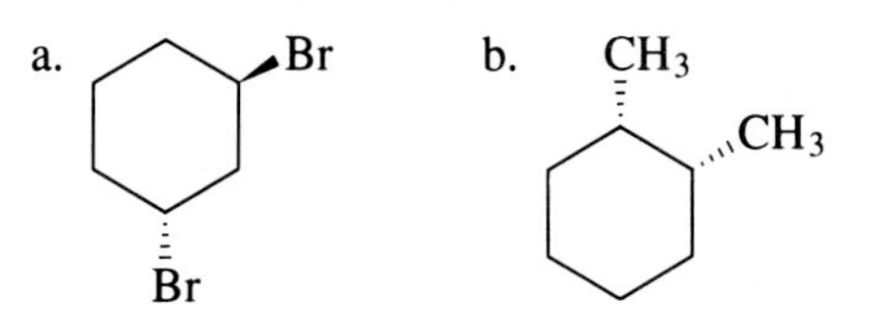

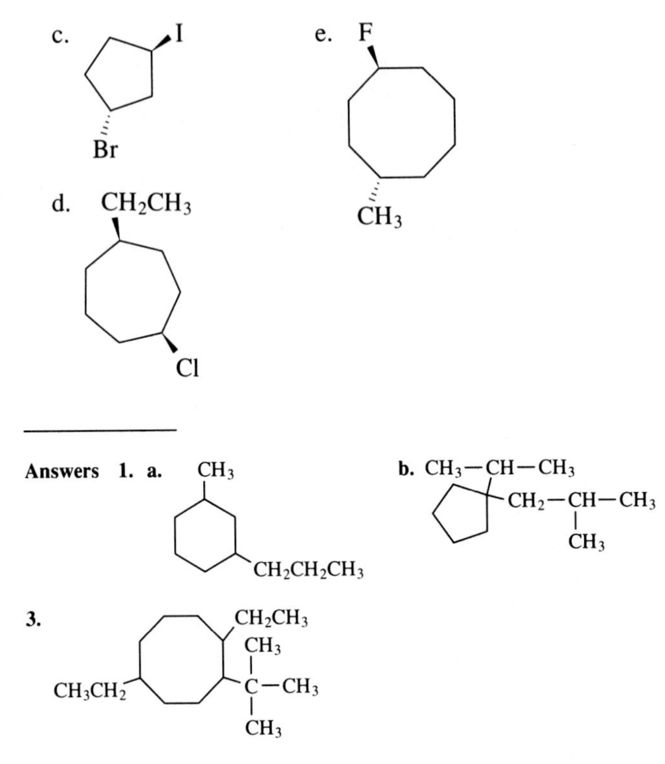

2. a. 6-ethyl-1,1-dimethyl-4-propylcyclodecane **b.** 1,3-dimethylcyclooctane **c.** 3-*tert*-butyl-1-isopropyl-1-methylcyclopentane
3. a. trans **b.** cis **c.** trans **d.** cis **e.** trans

2.5 Conformations of Alkanes and Cycloalkanes

Alkanes

♦ With free rotation about the C—C single bond, molecules can have their atoms in a number of different positions with respect to each other. These arrangements are called *conformations* and are *not* different compounds.

Because there is free rotation around a carbon–carbon single bond, atoms in a molecule can have different spatial orientations. These different orientations of the atoms are called **conformations**. Consider the molecule ethane. Several notably different conformations can be easily seen for ethane (Figure 2.12). When hydrogen atoms of ethane are directly behind one another, we call the conformation *eclipsed*. When the hydrogens are as far apart as possible, we call the conformation *staggered*.

You can see in Figure 2.13 that there are alternating staggered and eclipsed conformations every 60°. However, one hydrogen atom cannot be distinguished from another, so many conformations appear the same. This can be seen in an energy diagram, where the "front carbon" is allowed to spin and the interactions between hydrogen atoms are examined.

It should come as no surprise that some conformations of the atoms are better than others. Situations in which atoms or groups are close to one another are not as good as when the atoms or groups are farther away. If we consider a larger molecule, such as **butane**, we see that there are conformations in which hydrogens are close to hydrogens and other conformations in which methyl groups are close

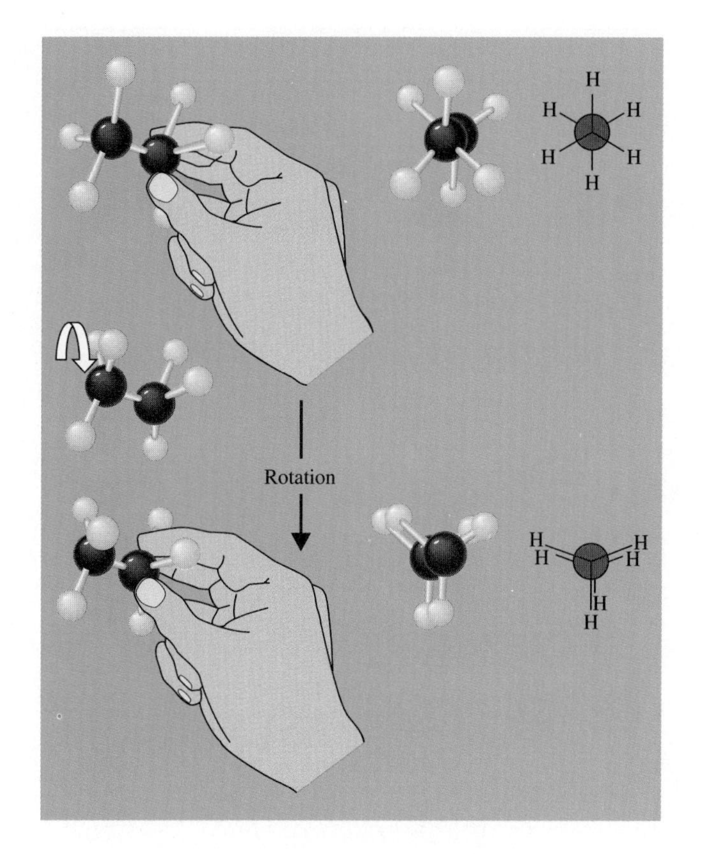

FIGURE 2.12
Conformations of ethane. Notice how the relative positions of the atoms change by simply rotating around the carbon–carbon bond. Since no bonds are broken, each rotation is simply a different orientation of the *same* molecule.

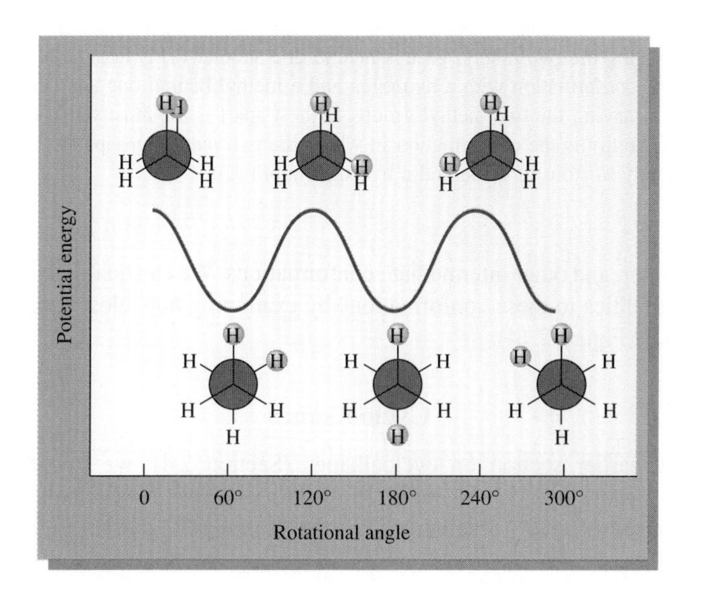

FIGURE 2.13
The relative energies of different conformations of ethane. When hydrogen atoms are directly behind one another, the conformation that has the highest energy is the least sable.

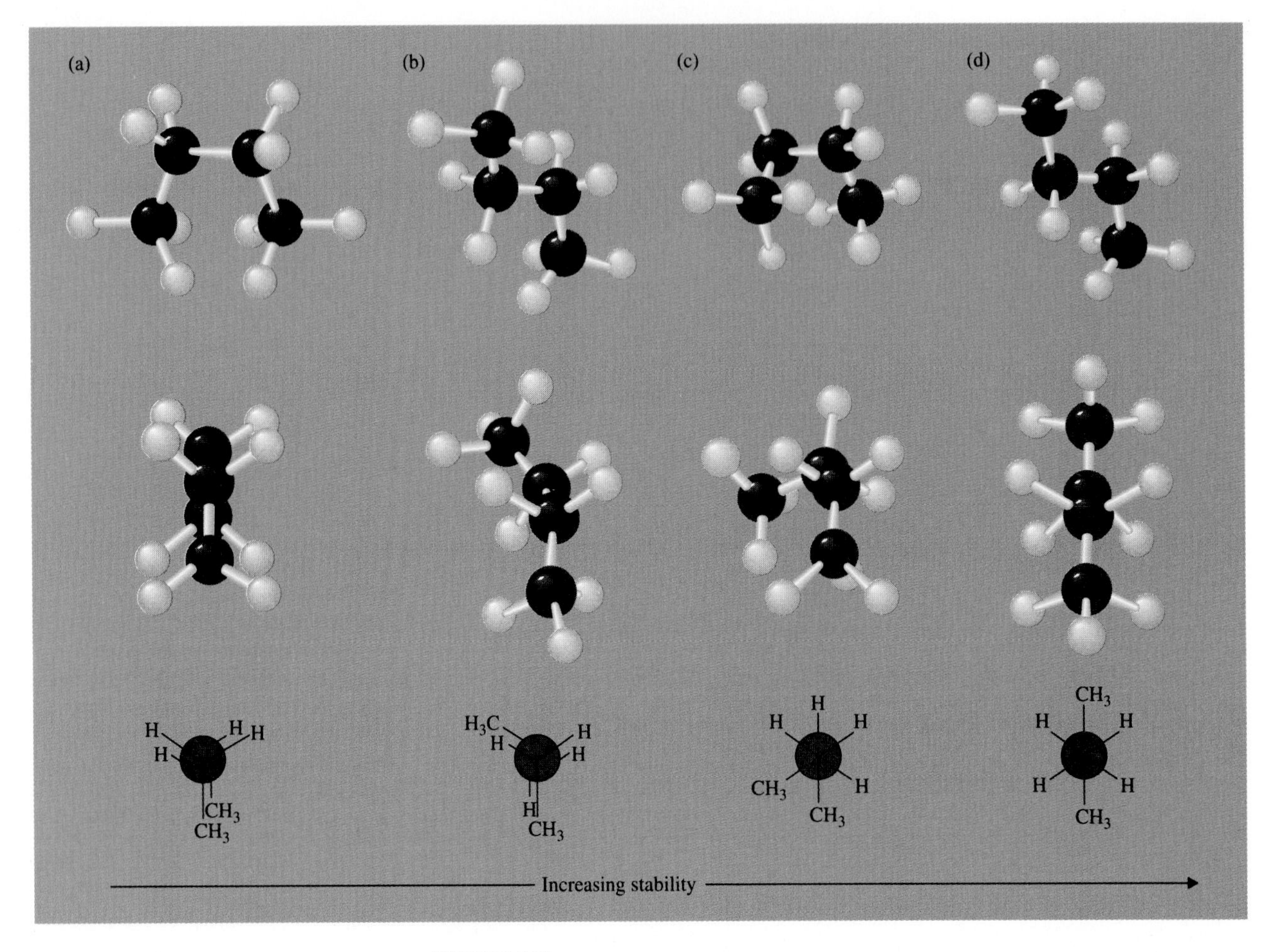

FIGURE 2.14
Some of the conformations of butane, showing a qualitative order of stability. The conformation having the two methyl groups behind one another (a) is least stable, followed by the conformation with a hydrogen and a methyl behind one another (b). The conformation having the two methyl groups farthest apart is the most stable (d). Of intermediate stability is the conformation in which the two methyl groups are close to but not directly in front of or behind a hydrogen atom (c).

to one another, and other intermediate conformations. We can qualitatively assign relative stabilities to these conformations by examining how close groups are to one another (Figure 2.14).

Cycloalkanes

In the earlier section on cycloalkanes (Section 2.4), we saw that rings were not planar. Just as for noncyclic hydrocarbons, the cycloalkanes also have preferred shapes—orientations of the atoms that give the most stable conformation.

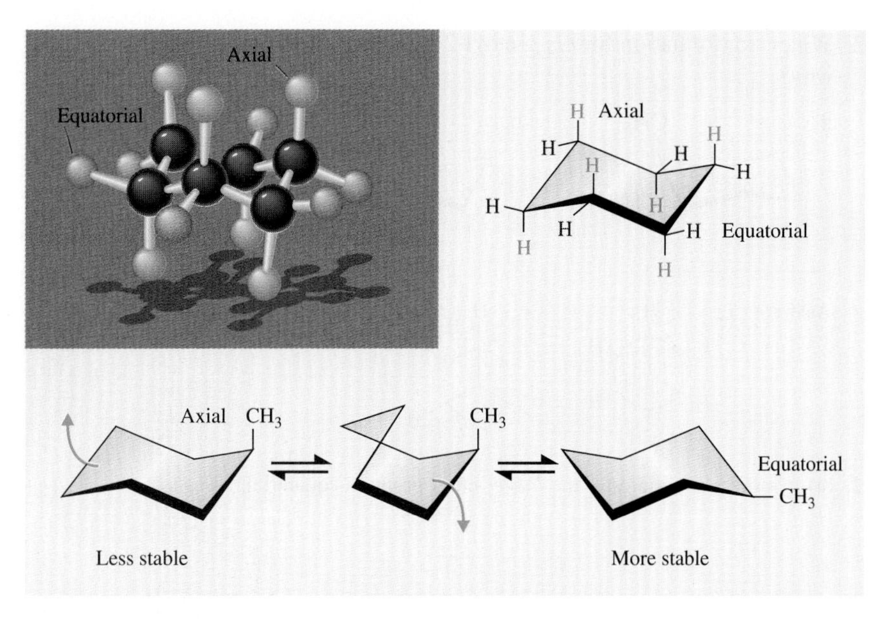

FIGURE 2.15
The chair conformation of cyclohexane. The chair form does not have any hydrogen atoms directly behind one another.

To better understand the effects of conformations on ring systems, let's look at the example of cyclohexane. For cyclohexane to maintain bond angles of 109°, it assumes what is called a *chair conformation* (Figure 2.15). A result of this shape is that all of the hydrogen atoms in cyclohexane assume a staggered arrangement. This results in two different orientations for the hydrogen atoms—*axial* and *equatorial*. These arrangements are shown in Figure 2.15. When only hydrogen atoms are involved on the carbons, there are six axial hydrogen orientations and six equatorial hydrogen orientations. If we replace a hydrogen with an alkyl group, this arrangement changes. In one conformation there is only one axial group, five axial hydrogens, and six equatorial hydrogens. In the other conformation, there are six axial hydrogens, five equatorial hydrogens, and one equatorial alkyl group. The conformation with equatorial alkyl groups is more stable.

♦ Drawings such as [Newman projection drawing] are called *Newman projections*.

♦ The cyclohexane molecule can undergo conformational changes such that groups attached to the ring have the more stable equatorial orientation.

NEW TERM

conformation The three-dimensional shape of a molecule emphasizing the relative orientation of atoms in space

TESTING YOURSELF

Conformations of Molecules

1. Looking at the C-1—C-2 bond of propane (right), draw its major conformations and show which is *least* stable.

2. Which conformation for *cis*-1-ethyl-2-methylcyclohexane is more stable, and why?

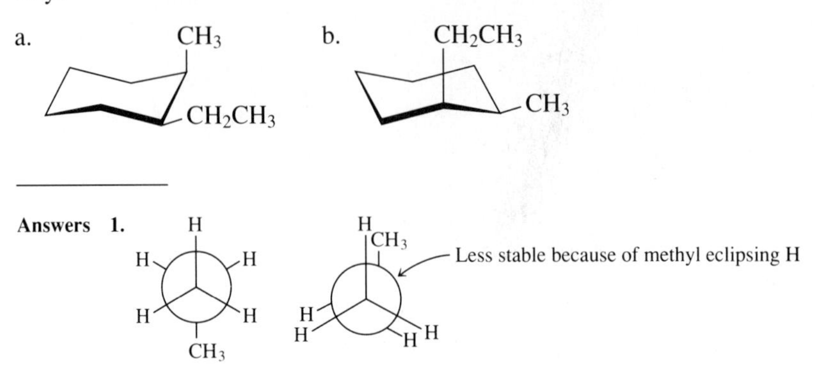

2. (a) is more stable because the *larger* ethyl group is equatorial.

2.6 Chemical Reactivity of Alkanes and Cycloalkanes

Effects of Sigma Bonding

◆ Alkanes and cycloalkanes have only sigma bonds and are thus relatively nonreactive.

The bonding of alkanes can be considered as very protected bonding. As we can see in Figure 2.16, the electrons involved in the bonding are shared directly between the atoms (recall that we call this a *sigma* bond). It is difficult for reagents to get to these electrons. There are also no "loose" electrons, such as nonbonding electrons, to participate in reactions. And without interaction with electrons, there is no reaction. Thus, we observe that alkanes are very resistant to normal chemical reactions.

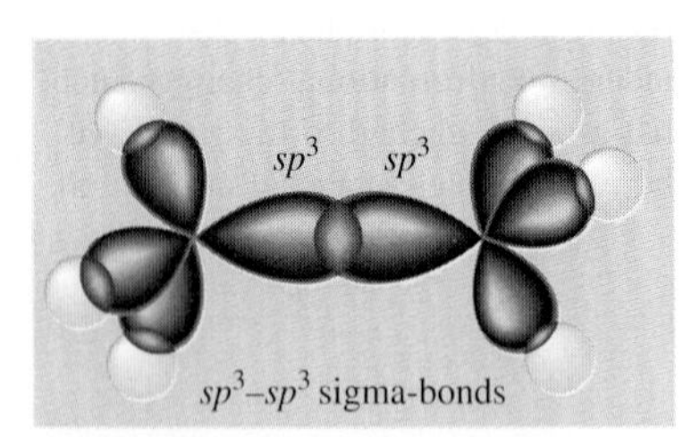

FIGURE 2.16
The chemical stability of the alkanes is due to the "protected" location of the shared bonding electrons between the bonded atoms. All electrons are involved in sigma bonding. There are no nonbonded or other available electrons to participate in reactions.

Combustion of Alkanes

About 90 percent of all petroleum products and natural gas is used for fuel—to heat our homes, to generate electricity, and to power our automobiles and aircraft. Propane, a common fuel for camp stoves and soldering torches, reacts with oxygen according to the following equation:

$$\mathrm{H{-}\underset{\displaystyle H}{\overset{\displaystyle H}{C}}{-}\underset{\displaystyle H}{\overset{\displaystyle H}{C}}{-}\underset{\displaystyle H}{\overset{\displaystyle H}{C}}{-}H} + 5\,O_2 \longrightarrow 3\,CO_2 + 4\,H_2O + \text{Heat }(\Delta)$$

where Δ indicates heat that is released from the reaction.

In a similar way, all alkanes undergo combustion, forming carbon dioxide (CO_2) and water as products. Since the combustion of every alkane follows the same general reaction, a generalized statement of the reaction that applies in all cases can be useful. To help you in your studies, simply remember the generalized reaction and apply it to the specific example in question rather than trying to remember a large number of specific examples. General reactions are presented in notecard format throughout the rest of the book.

GENERAL REACTION

Combustion of Alkanes

$$R{-}H + O_2 \longrightarrow CO_2 + H_2O + \text{Heat } (\Delta)$$

Examples

$$2\,H{-}\overset{\overset{\large H}{|}}{\underset{\underset{\large H}{|}}{C}}{-}\overset{\overset{\large H}{|}}{\underset{\underset{\large H}{|}}{C}}{-}\overset{\overset{\large H}{|}}{\underset{\underset{\large H}{|}}{C}}{-}\overset{\overset{\large H}{|}}{\underset{\underset{\large H}{|}}{C}}{-}H + 13\,O_2 \longrightarrow 8\,CO_2 + 10\,H_2O + \text{Heat } (\Delta)$$

$$CH_3{-}\underset{\underset{\underset{\underset{\large CH_3}{|}}{CH_2}}{|}}{CH}{-}CH_3 + 8\,O_2 \longrightarrow 5\,CO_2 + 6\,H_2O$$

If alkanes are burned with insufficient oxygen, as sometimes happens in automobile engines or heating systems that have insufficient draft, carbon monoxide (CO) is produced instead of carbon dioxide. In a balanced equation we can see that less oxygen is needed to form carbon monoxide than carbon dioxide (3.5 instead of 5 O_2).

$$2\,H{-}\overset{\overset{\large H}{|}}{\underset{\underset{\large H}{|}}{C}}{-}\overset{\overset{\large H}{|}}{\underset{\underset{\large H}{|}}{C}}{-}\overset{\overset{\large H}{|}}{\underset{\underset{\large H}{|}}{C}}{-}H + 7\,O_2 \longrightarrow 6\,CO + 8\,H_2O + \text{Heat } (\Delta)$$

It is very important that the production of carbon monoxide be avoided because of its harmful effects on humans.

Biological Combustion of Hydrocarbons

A large part of the food we eat contains fats and lipids (details of the structure and reactions of these molecules are discussed in later chapters). These types of molecules have a large hydrocarbon component as well as one or more functional groups as part of their structure. Since combustion of hydrocarbon fuels produces water as a by-product, it is not unreasonable to assume that the fuel we burn (or metabolize) in our body gives up a significant amount of water as a by-product of this metabolism. Many animals, including humans, dispose of excess water as perspiration and urine. However, many desert mammals use their metabolic water as a substitute for drinking water. The camel does not store water in its hump;

The Greenhouse Effect

Combustion of carbon fuel results in the formation of carbon dioxide. For every carbon atom burned, one molecule of carbon dioxide is produced. The natural abundance of carbon dioxide in our atmosphere has increased during this century as a result of the massive amounts of hydrocarbons burned as fuel for heating, power, and transportation. In the United States one third of all carbon dioxide produced is from automobile emissions. The increased levels of carbon dioxide contribute to the *greenhouse effect*.

This name aptly describes the results of the effect. As in a greenhouse, which captures and keeps heat within its glass roof and walls, the carbon dioxide (and other pollutant gases) in the atmosphere performs the same function. The short-wavelength solar energy from outer space is able to pass through the gas molecules in the atmosphere. But energy radiated from the earth back to outerspace is of longer wavelength. Thus, these wavelengths are absorbed by the polluting gas molecules and do not leave our atmosphere. This trapped energy remains as heat.

Until fairly recently we thought of carbon dioxide as the harmless by-product of combustion; we still rarely think of it as a deadly killer. Yet, on August 21, 1986, the West African nation of Cameroon felt the deadly effects of a deadly cloud of carbon dioxide belched up from the bottom of Lake Nios. Almost two thousand people and countless animals were suffocated as the heavier-than-air gas filled the town.

In our atmosphere carbon dioxide is the major contributor of the gases causing the greenhouse effect. The levels have been increasing rapidly since 1960. The concentration of methane, another greenhouse gas, has been steadily increasing, too. Indeed, during the past decade, atmospheric levels have been increasing by 1.5 percent per year. Fossil methane, created at the same time as petroleum products from ancient plants, is deposited in pockets of the earth's mantle. These gases escape during drilling and mining. By examining the ratios of carbon-14 to carbon-12 in the methane gases in our atmosphere, scientists believe about 15 to 20 percent comes from fossil sources. Modern methane comes from animal digestion processes, termites, and the vast wetlands and marshes of the world.

What are the long-term consequences of the greenhouse effect? Some scientists suggest the most dire of consequences: that our farmland will become deserts and the polar ice caps will melt, resulting in the flooding of our coastal cities. These researchers point to the measurable warming of our planet to support this view. In this view, the temperate regions of the earth would be most affected. Temperatures in some parts of the world may increase by as much as 5°C.

Others suggest that the earth's heating is localized and that there is in fact an overall cooling of the earth. Plants use a large amount of the excess carbon dioxide we are producing. However, there are estimates that the earth loses an acre of forest every second, primarily in tropical regions. It would seem prudent to recognize that these forests are important resources for many immediate needs as well as long-term storehouses for carbon. We need to ensure their survival. An interesting suggestion has been made that industrialized nations might consider renting rain forests from less developed countries!

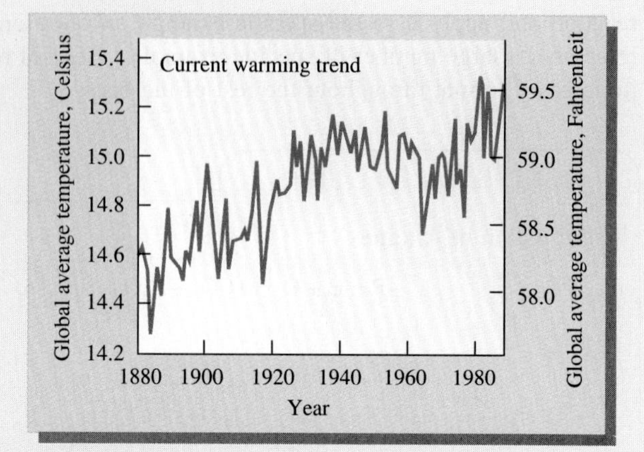

FIGURE 2.A
This graph shows the average global temperatures from 1880–1988. The average from 1950–1980 is used as a baseline. Notice the recent increase in temperature.

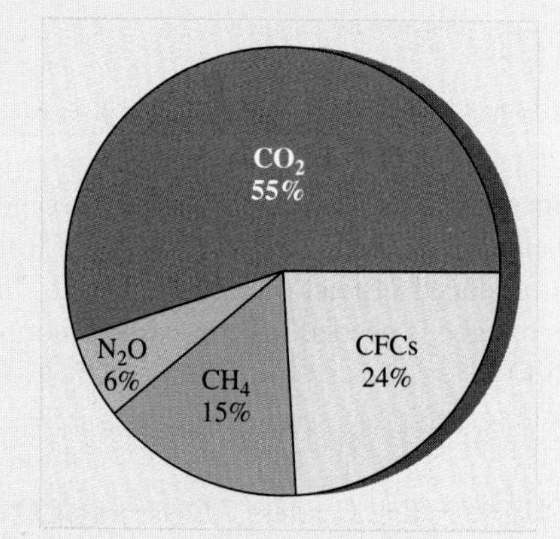

FIGURE 2.B
Of the major causes of the greenhouse effect, clearly carbon dioxide must be considered a major contributor.

Support for the notion that plants help reverse the heating trend is available. Palm Springs, California, is located in a desert area. As in many cities with large areas covered by concrete, the temperature of Palm Springs was regularly increasing. In the 1970s there was a trend toward construction of golf courses. Since that time, the temperature of that small oasis in the desert has dropped by 2 to 3°F.

No matter what the source or makeup of the gas, nor whether the problem is immediate or long range, there is

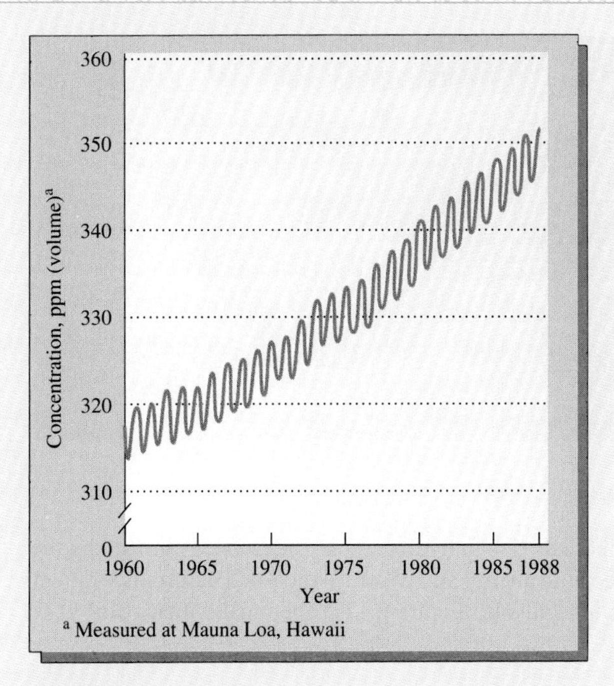

FIGURE 2.C
The levels of carbon dioxide in our atmosphere have been increasing rapidly over the past thirty years.

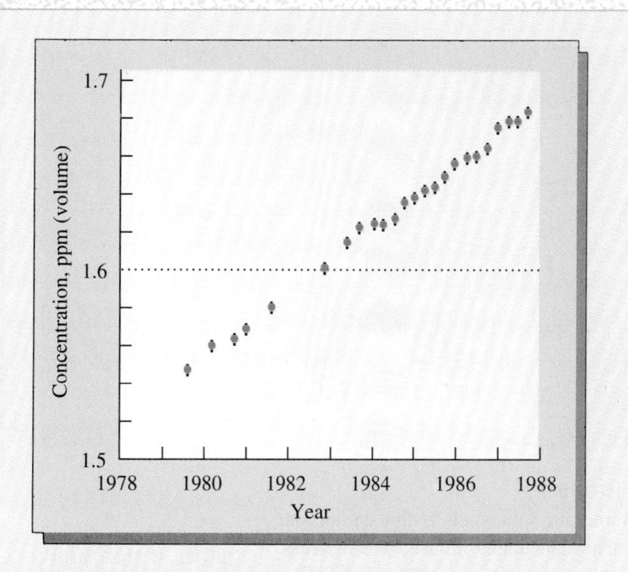

FIGURE 2.E
Methane concentrations have been steadily increasing.

no question that we need to study the greenhouse problem. We cannot go on forever polluting our air, land, and water without expecting a consequence for these actions. We cannot continue to burn fossil fuels and to destroy the tropical forests. We have an obligation to future generations to give them a future.

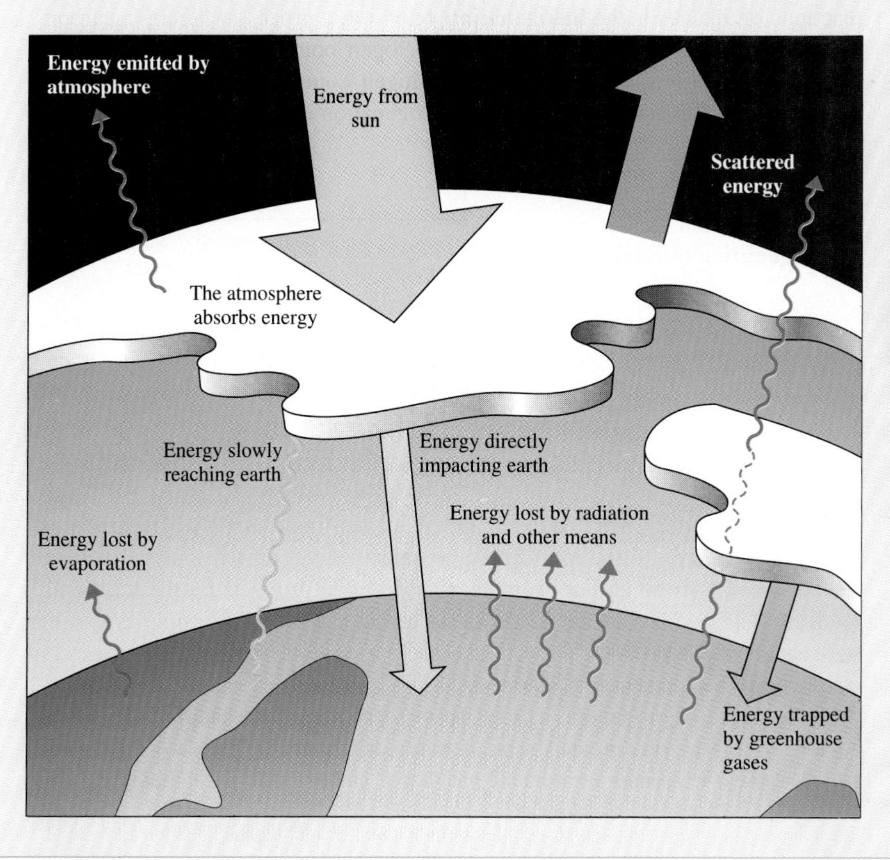

FIGURE 2.D
There are many factors controlling the heating and cooling of our planet.

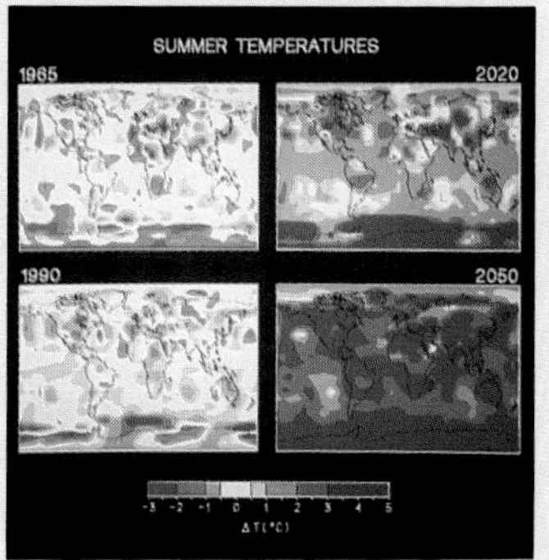

FIGURE 2.F
Analysis projection of change in summer temperature seems to support the notion of overall warming.

FIGURE 2.17
Some animals, such as this camel and Australian plains mouse, live in areas where water is scarce. These animals generate their own water as a by-product of metabolism.

rather, it stores fatty compounds rich in hydrocarbons that, when "burned" by the body metabolism, produce both energy and water. Camels and desert mice do not die easily when water is unavailable (Figure 2.17). They make their own!

Halogenation

In our discussion of the chemical reactivity of the alkanes (Section 2.6), we suggested that these hydrocarbons are considered chemically nonreactive. This is not completely true. Aside from combustion, one of the most important chemical reactions of alkanes is the **halogenation reaction**, a process whereby a carbon–hydrogen bond is replaced by a carbon–halogen bond. The products resulting from these reactions are called **organohalogen** compounds (sometimes called *alkylhalides*). The general reaction for halogenation can be stated as follows:

♦ Organohalogen compounds (R—X, where X represents a halogen atom such as F, Cl, Br, or I), are very important to the chemical industry.

GENERAL REACTION

Halogenation of Hydrocarbons

$$R{-}H + X_2 \xrightarrow[\text{or light}]{\text{Heat}} R{-}X + H{-}X$$

Example

$$CH_4 + Cl_2 \xrightarrow{\text{Heat}} CH_3Cl + HCl$$

The halogenation reaction has a limited but very important use. The limitation is primarily related to the indiscriminate nature of the process—a halogen can replace *any* C—H bond. For example, four different products for the reaction of methane (CH_4) with chlorine (Cl_2) can occur. Excess chlorine continues to react with each new by-product. The Cl_2 near each arrow indicates that the reaction is occurring in the presence of chlorine.

♦

$$CH_4 + Cl_2 \longrightarrow CH_3Cl + HCl$$
$$CH_3Cl + Cl_2 \longrightarrow CH_2Cl_2 + HCl$$
$$CH_2Cl_2 + Cl_2 \longrightarrow CHCl_3 + HCl$$
$$CHCl_3 + Cl_2 \longrightarrow CCl_4 + HCl$$

A mixture of products is always formed from the halogenation of alkanes. As long as this mixture is easily separated into its component parts (by distillation, for example), it creates no special problems, since most organohalogen compounds have some use. However, we can see the potential problems by considering the four products from the methane reaction. It takes little imagination to see the complexity that would result from the halogenation of a more complex hydrocarbon, such as pentane.

Nomenclature of Organohalogens

Organohalogen compounds have no special nomenclature. Indeed, the rules already established for hydrocarbons are used here as well. Since the four halogens are chlorine, bromine, iodine, and fluorine, the prefixes *chloro-, bromo-, iodo-*, and *fluoro-* are used if a halogen is part of a compound. They are given no special priority in the naming scheme. This is best shown by examples.

♦ The halogen imparts no special priority to the nomenclature process. A halogen is treated as an alkyl group.

Structure	Name
$CH_3-CH(Br)-CH_2-CH_2-CH_3$	2-*bromo*pentane (not 4-bromopentane)
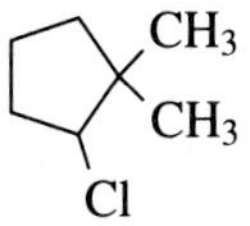 cyclopentane ring with two CH_3 groups on one carbon and Cl on the adjacent carbon	2-*chloro*-1,1-dimethylcyclopentane (*not* 1-chloro-2,2-dimethylcyclopentane)

As with all of the nomenclature developed to this point, we always strive for the lowest arrangement of numbers.

Artificial, or manufactured, organohalogen compounds have widespread uses in our everyday life. They are used widely as pesticides and herbicides, as solvents and refrigerants, as medicines and anesthetics, and as plastic and rubber. Examples of some of the common organohalogen compounds and their uses are listed in Table 2.5.

NEW TERMS

halogenation reaction A reaction of an alkane with a halogen that is catalyzed by heat or light, in which a C—H bond is replaced by a C—X bond

organohalogen A class of organic compounds characterized by carbon–halogen bonds.

TESTING YOURSELF

Chemical Reactivity and Organohalogens

1. Write a balanced equation for each of the following reactions:

a. $CH_3-CH_2-CH_3 + O_2 \longrightarrow$
Propane

b. $CH_3-C(CH_3)_2-CH_3 + O_2 \longrightarrow$
2,2-dimethylpropane, or commonly, neopentane

TABLE 2.5

Organohalogen Compounds

Pesticides and Herbicides

Name	Description	Structure
DDT	A powerful insecticide. Outlawed because of its toxic effects on humans and wildlife.	
Dieldrin	Insecticide for garden pests. It was once used to control DDT-resistant mosquitos. Manufacture and use has been discontinued.	
2,4-D	Ingredient in sprays to kill broadleaf plants such as dandelion. Makes the plant take up water too fast, causing a rapid growth of stem but no root.	OCH_2COOH

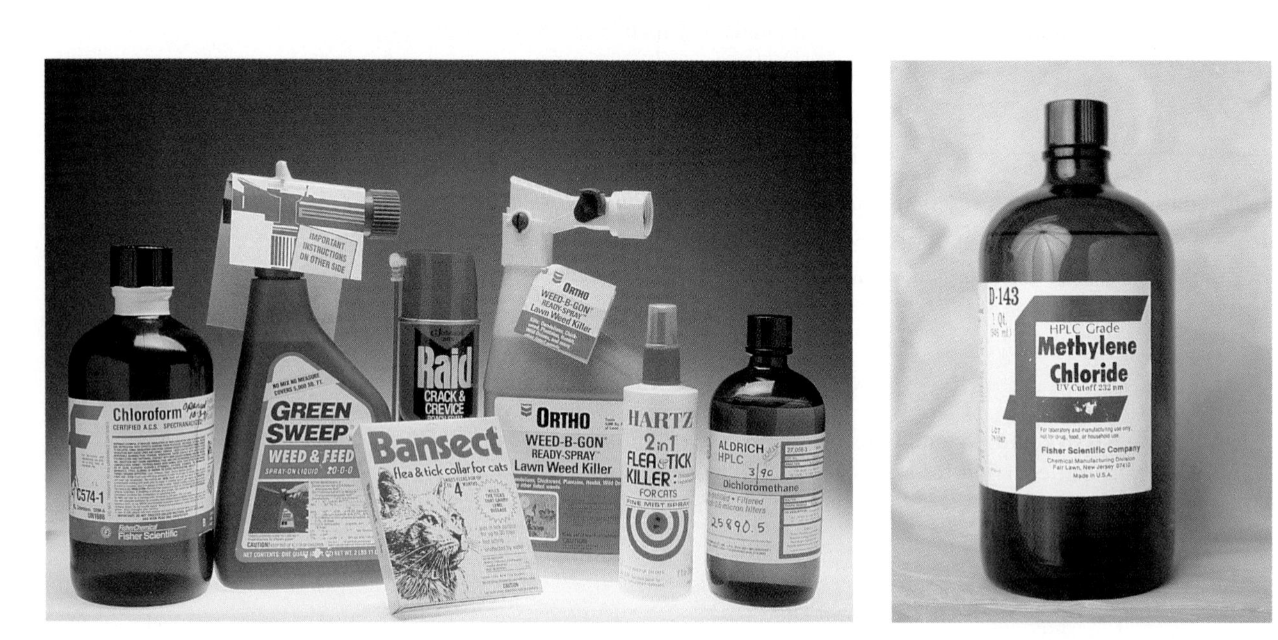

Organohalogen compounds are commonly found in garden supplies.

TABLE 2.5 (continued)

Organohalogen Solvents

Name	Description	Structure
Carbon tetrachloride	Once used as a cleaner, it is now known to be toxic and a potential cancer-causing agent (carcinogen).	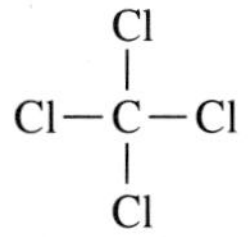
Chloroform	Once used as an inhalation anesthetic, chloroform (trichloromethane) is now known to be toxic when used this way and has been replaced by a variety of other compounds. It is now used mostly as a solvent.	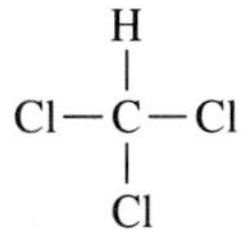
Freon-12 (dichlorodifluoromethane)	Refrigerant fluid and once used in propellant sprays. The latter use has been discontinued because of its potential harm to the atmospheric ozone layer. New refrigerant materials are being tested.	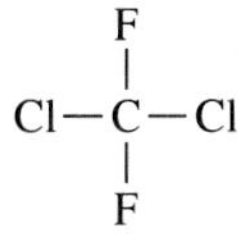
Dichloromethane	Commonly called methylene chloride. A very common solvent.	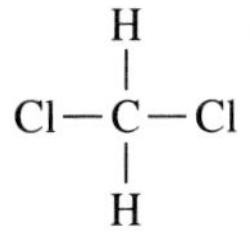
PCBs	The *p*oly*c*hlorinated *b*iphenyls (or PCBs) are used as insulating, nonflammable components in transformers and other electrical devices. They have proven to be carcinogenic.	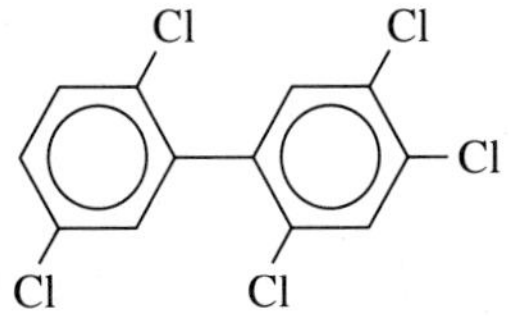

Medical Uses of Organohalogen Compounds

Name	Description	Structure
Halothane	A widely used inhalation anesthetic that has the added value of being nontoxic and nonflammable.	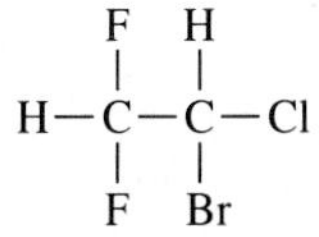
Tetrachloroethylene	An effective treatment for hookworm.	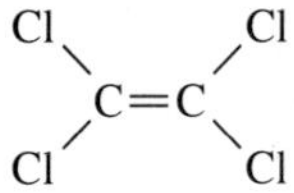
Mitotane	This compound is structurally similar to DDT and is used in the treatment of brain cancer.	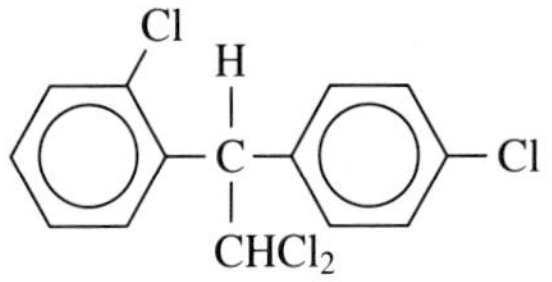

CHEMISTRY CAPSULE

Depletion of the Ozone Layer and CFCs

The stratosphere is one of the middle layers of our atmosphere and is found at distances of 15 to 50 km above the earth's surface. Within the stratosphere is a layer of ozone (O_3) that surrounds the earth. When we look into the sky, we imagine an unlimited expanse. Space goes forever. But the important space for our everyday life is much closer than we might imagine. The troposphere extends about 10 miles from the surface of the earth. The stratosphere extends an additional 20 miles or so. This is the region of space where the maximum ozone concentration is found. Ozone plays an important role in absorbing harmful ultraviolet (UV) radiation. If this protection decreased, we would be bombarded with sufficient additional UV radiation that the incidence of skin cancer would dramatically increase. This is the immediate important role of the ozone to us.

One of the leading culprits in the destruction of the ozone layer is the widely used class of compounds called *chlorofluorocarbons*, or *CFCs*. These are chlorinated and fluorinated hydrocarbons that are very useful in refrigerants, aerosols, cleaning agents, fire extinguishers, and insulation foams. On the surface of ice crystals in the clouds of the stratosphere, the CFCs tend to come apart to liberate chlorine atoms. As you recall from earlier discussions on bonding, the chlorine atom has seven electrons, and to fulfill the inert gas electronic structure, it reacts. In the atmosphere it reacts with ozone in the following way:

$$:\ddot{\underset{\cdot\cdot}{\mathrm{Cl}}}\cdot + O_3 \longrightarrow \mathrm{ClO}\cdot + O_2$$

$$\mathrm{ClO}\cdot + \mathrm{ClO}\cdot \longrightarrow Cl_2O_2$$

$$Cl_2O_2 + \text{Light energy} \longrightarrow ClO_2 + :\ddot{\underset{\cdot\cdot}{\mathrm{Cl}}}\cdot$$

$$:\ddot{\underset{\cdot\cdot}{\mathrm{Cl}}}\cdot \text{ repeats the process.}$$

This is called a *chain reaction* because a reactant is liberated each time, enabling the process to be repeated. You can see that because of the chain reaction property and the millions of tons of CFCs used each year, it will not take too long before this process has a measurable effect on our atmosphere.

There has been much interest in the popular press about the ozone "hole" over Antarctica. This observation is consistent with the theory that the reactions taking place require cold surfaces. Computer-enhanced satellite images of Antarctica show the depletion graphically. Much interest has been shown in the development of so-called safe CFCs; however, these cause the same effects, but at a rate of about one fifth of the presently used CFCs. The development of materials that can serve the same roles is a research priority in leading chemical companies.

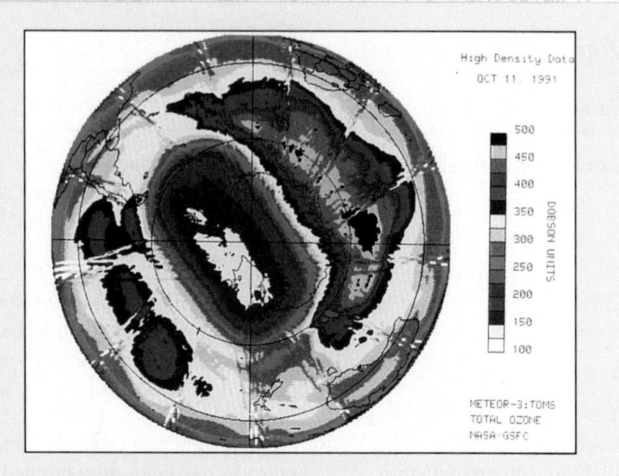

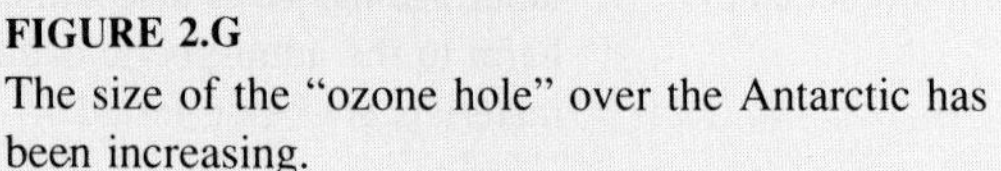

FIGURE 2.G
The size of the "ozone hole" over the Antarctic has been increasing.

The United States and 31 other nations have signed an agreement known as the Montreal Protocol, which mandates a freeze and then regular reduction in use of CFCs until they are eliminated. Twelve Western European countries that have been major producers of ozone-destroying chemicals have recently taken an aggressive attitude to help curb their production. They have completely banned both production and use over the next ten years. If the United States follows their lead, over two thirds of the total chemicals in this class will be removed by the year 2000. As optimistic as this appears, predictions are still not good for the atmosphere. It has been suggested that even if we stopped all use immediately, the ozone level would not reach 1985 levels until about 2050.

2. Draw and name all of the monobromination products from the bromination of pentane.
3. Draw and name all of the dibromination products from the bromination of pentane.

Answers **1. a.** $CH_3—CH_2—CH_3 + 5\ O_2 \longrightarrow 3\ CO_2 + 4\ H_2O$
b. $CH_3—C(CH_3)_2—CH_3 + 8\ O_2 \longrightarrow 5\ CO_2 + 6\ H_2O$

2.

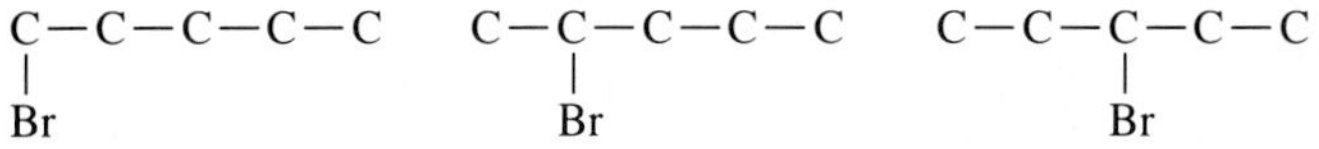

```
C—C—C—C—C        C—C—C—C—C        C—C—C—C—C
|                  |                  |
Br                 Br                 Br
```

1-Bromopentane **2-Bromopentane** **3-Bromopentane**

3.

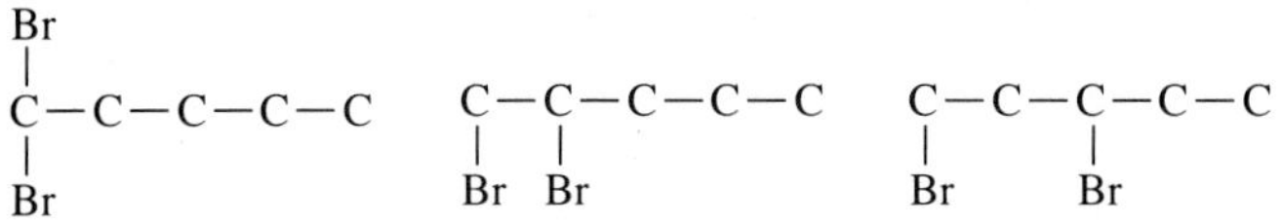

```
Br
|
C—C—C—C—C        C—C—C—C—C        C—C—C—C—C
|                |  |             |     |
Br               Br Br            Br    Br
```

1,1-Dibromopentane **1,2-Dibromopentane** **1,3-Dibromopentane**

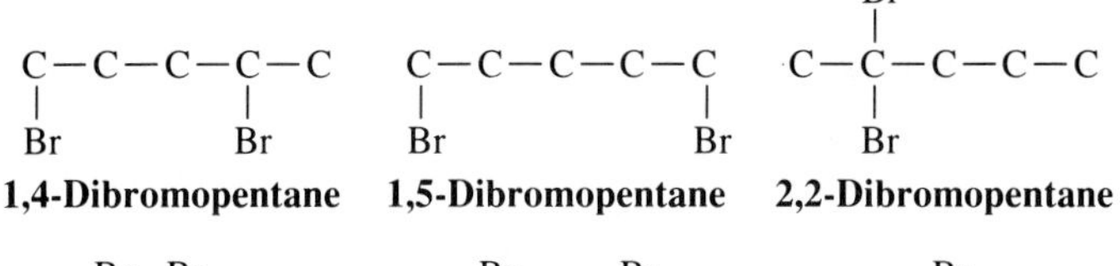

1,4-Dibromopentane **1,5-Dibromopentane** **2,2-Dibromopentane**

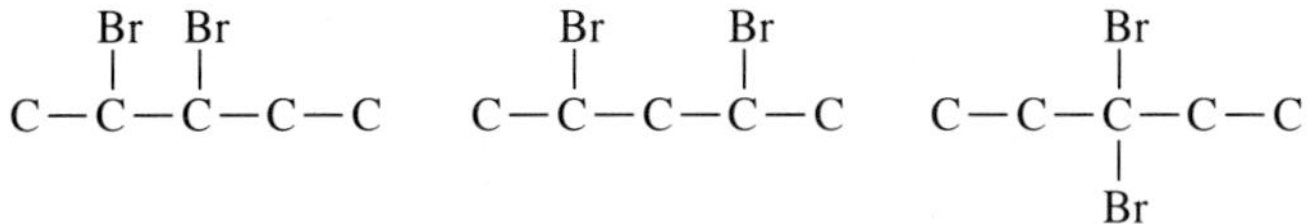

2,3-Dibromopentane **2,4-Dibromopentane** **3,3-Dibromopentane**

2.7 Health-Related Products Based on Hydrocarbon Structures

So far in this chapter we have surveyed the alkanes, cycloalkanes, and organohalogen compounds. All three of these groups are important in the medical and allied health fields. Thus, it is appropriate to conclude the chapter with a number of health-related applications.

Alkanes

The nonreactivity of hydrocarbons is evidenced by their scarcity in products for human use. However, it is precisely the nonreactivity and nonsolubility of high-boiling-point hydrocarbon fractions that make *petrolatum* (mineral oil) useful as a lubricant. Because it is an indigestible oil, there is only slight absorption from the intestinal tract. Yet the oil is able to soften stools in cases of minor constipation. Some fat-soluble vitamins dissolve in mineral oil; thus, prolonged use might deprive the body of this type of vitamin.

A major danger associated with ingesting liquid alkanes is that the hydrocarbon materials (lipids) of the cell membranes can be easily dissolved (like dissolves

CHEMISTRY CAPSULE

Artificial Blood

The great number of components in blood and its myriad functions make it highly improbable that a permanent substitute for blood could ever be found. However, we are often in need of short-term supplies of blood, and even with all of the care taken to maintain blood, there are many problems associated with blood transfusions. One problem is keeping a supply of fresh blood available for emergency use, since "old" blood is not as effective as an oxygen carrier. Although blood does many things other than transport oxygen, this is an essential function, and a synthetic blood substitute has been shown to be useful for short-term applications.

A salt solution of *perfluorodecalin* and *perfluorotripropylamine* has been marketed in Japan as Fluosol DA.

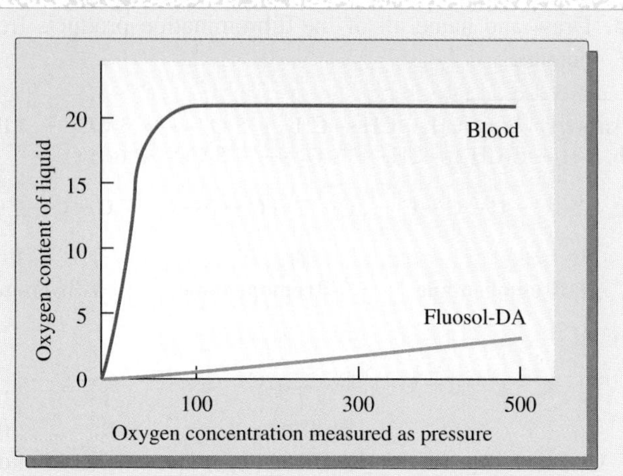

FIGURE 2.H
The fluorocarbons making up the artificial blood do not hold oxygen as efficiently as blood.

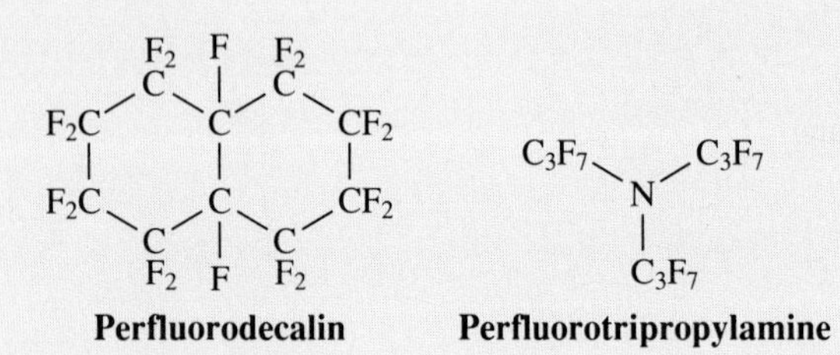

Although not as efficient as hemoglobin for transporting oxygen (see figure), the fluorocarbons are able to dissolve fairly substantial quantities of oxygen for transport.

Is this new product the answer to the problem of blood supply? It is a start. But problems remain and are now being studied. For example, what happens to the fluorocarbons in the body? Like Teflon, the molecules are supposed to be very resistant to chemical activity; however, chemical evidence shows that this is not true. Also, what effect do the small impurities in the synthetic material have on the body over the long term? Finally, is this *really* an effective substitute for blood? There are mixed clinical data.

One potentially useful outcome of the studies may be the application of these materials to cancer treatment. Cancer and tumor cells have low oxygen content, making it difficult to treat the cells with chemotherapy or radiation therapy. There are encouraging data to suggest that the artifical oxygen carriers (which are much smaller than hemoglobin) can be targeted to the cancer cells.

Whatever the outcome of these studies, it remains a positive sign of the ingenuity of science that we can even approach these questions. By trying to mimic the natural processes in the human body, we find out more about how it works. We can then try to find natural ways to assist body processes to fight disease and injury.

like). When this happens in the lung tissue, serious pneumonia-like health disorders often develop. This is the major reason that a person who has accidently swallowed hydrocarbon materials (for example paint thinner) should *not* be induced to vomit. It is bad enough in the stomach but worse if accidently inhaled.

Cycloalkanes

The most common health-related use of cycloalkanes can be found in the application of cyclopropane as an anesthetic. Cyclopropane is a fast-acting anesthetic having a wide margin of health safety associated with its use. Unfortunately, one danger is associated with its use—it is highly explosive. This has greatly restricted the use of cyclopropane as an anesthetic.

A cyclopropane derivative, *chrysanthemic acid*, is one of the basic structural units found in the *pyrethrins*. These are naturally occurring compounds found in the pyrethrum daisy and have very pronounced insecticidal activity. Indeed, most of the common insect sprays for household use have synthetic pyrethrins (or analogs) as the active ingredient.

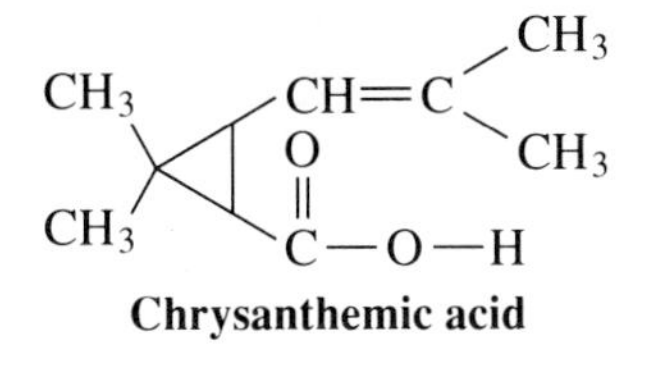

Chrysanthemic acid

Summary

Alkanes and *cycloalkanes* are the simplest of the organic hydrocarbons, having only C—C and C—H single bonds. These compounds have physical properties largely due to the London forces attracting molecules to one another. The electrons forming the bonds between the carbon atoms and between the carbon and hydrogen atoms are sufficiently tightly held that these molecules are not very reactive.

The regularly increasing series of alkanes is called a *homologous series*, and names for this series provide the basis for nomenclature of organic compounds. *Alkyl groups*, derived from the hydrocarbons, provide a unique method of naming complex organic structures. *Isomers* are molecules having the same formula (in the case of hydrocarbons, the same number of hydrogens and carbons), but different structures. They are conveniently named by combining the names of members of the homologous series and alkyl groups.

The shape of a molecule is reflected in the *conformation*—the direction in space for the bonds between C—C and C—H. Some conformations are of lower energy (and thus are more stable) than other conformations.

The simplest reaction of hydrocarbons is combustion (reaction with oxygen to form carbon dioxide and water) and *halogenation* (forming organohalogen compounds). The *organohalogen* compounds have extensive industrial use.

Reaction Summary

Combustion of Alkanes with Adequate Oxygen

$$R—H + O_2 \longrightarrow CO_2 + H_2O + \text{Heat } (\Delta)$$

Halogenation of Hydrocarbons

$$R—H + X_2 \xrightarrow[\text{or light}]{\text{Heat}} R—X + HX$$

Terms

alkane (2.2)
structural formula (2.2)
perspective formula (2.2)
condensed formula (2.2)
line formula (2.2)
homologous series (2.2)
isomer (2.2)
alkyl group (2.3)
primary carbon (2.3)
secondary carbon (2.3)
tertiary carbon (2.3)
quaternary carbon (2.3)
cycloalkane (2.4)
geometric structure (2.4)
cis (2.4)
trans (2.4)
conformation (2.5)
halogenation reaction (2.6)
organohalogen (2.6)

Exercises

Hydrocarbon Structure (Objective 1)

1. What is the hybridization of carbon atoms in the alkanes?
2. What is the orientation of atoms attached to carbon in the alkanes?
3. We often hear the term "straight-chain hydrocarbons." What is meant by this, and in what way is the phrase incorrect?

Physical Properties (Objective 2)

4. What is the force that attracts hydrocarbon molecules to one another?
5. Why is it difficult to dissolve hexane in water?
6. Hexane dissolves very readily in octane. Why?
7. 2,3-Dimethylbutane (six carbon atoms total) has a boiling point of 58°C, whereas hexane (six carbon atoms) has a boiling point of 69°C. Explain.

Isomers and Homologous Series (Objective 5)

8. What is a homologous series?
9. What is an isomer?
10. Draw and name all of the isomers of butane.
11. Draw and name all of the alkyl groups derived from the isomers of butane.
12. Assign the status (primary, secondary, tertiary, or quaternary) of each hydrogen atom on 2,2-dimethylbutane.
13. Assign the status (primary, secondary, tertiary, or quaternary) of each hydrogen atom on 2,3-dimethylbutane.
14. Draw all of the isomeric alkyl groups having the empirical formula, C_5H_{11}.
15. Draw and name all of the isomers of
    a. C_4H_{10}
    b. C_5H_{12}
    c. C_6H_{14}

Conformations of Alkanes (Objective 4)

16. What do we mean by the conformation of an alkane?
17. What is the preferred conformation of butane?
18. What is the preferred conformation of methylcyclohexane?
19. What is the orientation (axial or equatorial) of the methyl group in *cis*-1-methyl-2-isopropylcyclohexane?
20. Draw the two chair conformations of *tert*-butylcyclohexane. Which is more stable?

Alkane Nomenclature (Objective 5)

21. Give the IUPAC name for each of the following alkane hydrocarbons:

a.
$$\begin{array}{ccccccccc} & & & & CH_3 & & & & CH_2-CH_3 \\ & & & & | & & & & | \\ CH_3 & - & CH_2 & - & C & - & CH_2 & - & CH \\ & & & & | & & & & | \\ & & & & CH_3 & & & & CH_3 \end{array}$$

b.
$$\begin{array}{ccccccc} & & & & CH_3 & & \\ & & & & | & & \\ CH_3 & - & CH_2 & - & CH & - & CH_3 \end{array}$$

c.
$$\begin{array}{ccccccc} & & CH_3 & & & & CH_2-CH_3 \\ & & | & & & & | \\ CH_3 & - & C & - & CH_2 & - & CH \\ & & | & & & & | \\ & & CH_3 & & & & CH_2-CH_3 \end{array}$$

d.
$$\begin{array}{ccccccc} & & CH_3 & & & & \\ & & | & & & & \\ CH_3 & - & CH & - & CH_2 & & CH_3 \\ & & & & | & & | \\ & & & & CH_2 & - & CH_2 \end{array}$$

e.
```
      CH3
      |
CH3—C—CH2—CH3
      |
      CH3
```

f.
```
                 CH3
                 |
CH3—CH—CH2—C—CH3
      |          |
      CH3        CH3
```

g.
```
CH3—CH—CH2—CH3
      |
      CH2—CH3
```

h.
```
      CH3 CH3
      |   |
CH3—C———C—CH3
      |   |
      CH3 CH3
```

i.
```
CH3—CH2—CH—CH3
          |
          CH2—CH3
```

j.
```
      CH3          CH3
      |            |
CH3—CH   CH3—C—CH2—CH3
      |            |
      CH2—————————CH2
```

22. Give the IUPAC name for each of the following alkane hydrocarbons:

a.

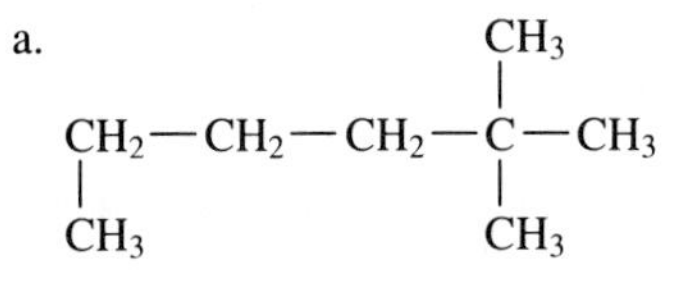

b.
```
CH3          CH3
|            |
CH—CH2—C—CH2—CH3
|            |
CH3          CH3
```

c.
```
CH3—CH2  CH2—CH2—CH3
      |   |
CH3—CH—CH—CH2—CH—CH2—CH3
                 |
                 CH3
```

d.
```
CH3—CH2          CH3
      |          |
CH3—CH—CH2—C—CH2—CH3
                 |
                 CH3
```

e.
```
CH3—CH2—CH2          CH3
          |          |
    CH3—C—CH2—C—CH3
          |          |
          CH3        CH3
```

f.
```
      CH3
      |
CH3—C——CH2   CH3 CH3
      |   |     |   |
      CH3 CH2—C——CH—CH3
                |
                CH3
```

g.
```
      CH2—CH3
      |
CH3—C—CH3
      |
      CH2—CH2—CH2—CH3
```

h.
```
CH3—CH2          CH3
      |          |
CH3—CH—CH2—C—CH3
                 |
                 CH3
```

i.
```
CH3—CH2—CH2—CH—CH3
               |
               CH2—CH2—CH3
```

j.
```
      CH3 CH3
      |   |
CH3—C———C—CH2—CH3
      |   |
      CH3 CH3
```

k.
```
              CH3
              |
CH3—CH2—C—CH2—CH3
              |
              CH3
```

23. Give the IUPAC name for each of the following cycloalkanes:

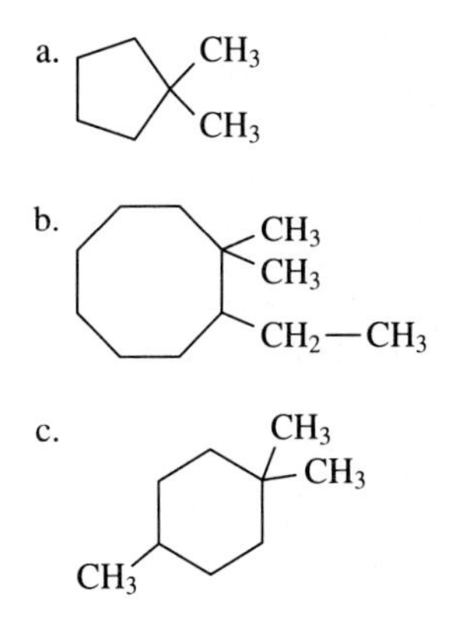

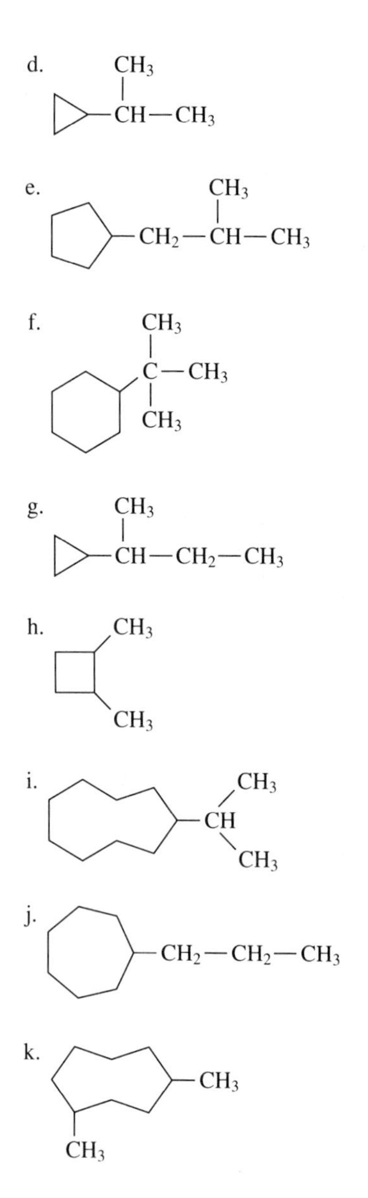

24. Give the IUPAC name for each of the following cycloalkanes. Include the cis and trans designations.

a.

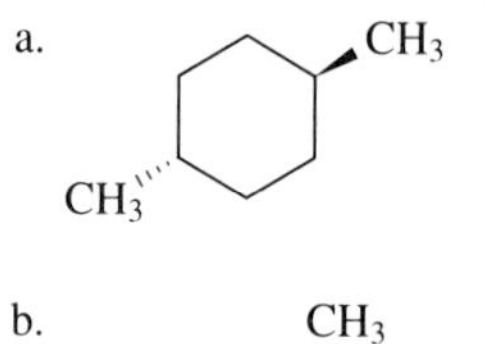

b.

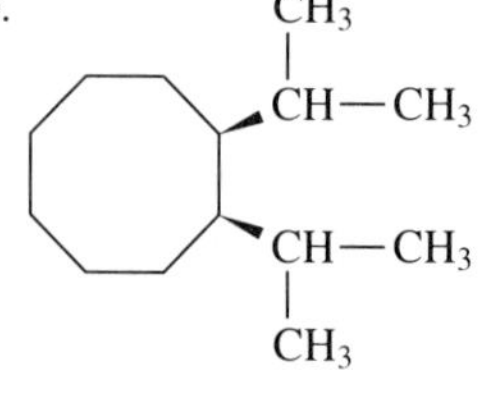

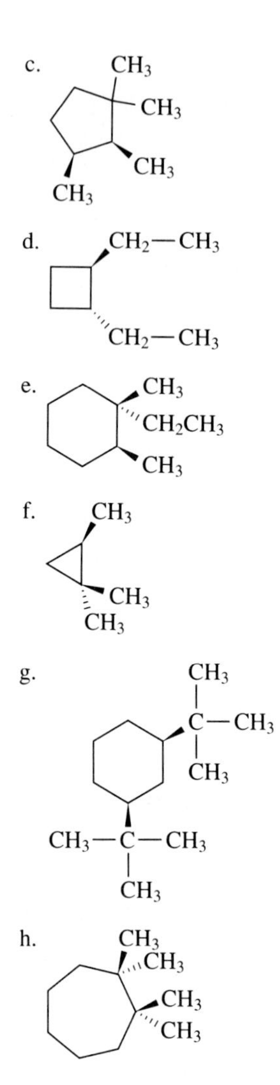

25. Give the IUPAC name for each of the following organohalogen compounds:

a. $CH_3-CH-CH_2-CH_3$
$\quad$ |
$\quad$ Br

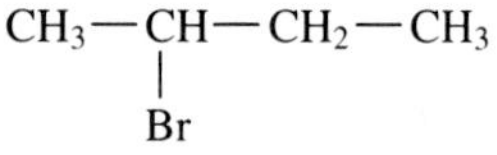

b.
$\quad$ Cl
$\quad$ |
$CH_3-CH_2-C-CH_3$
$\quad$ |
$\quad$ Cl

c.
$\quad$ Br $\quad$ Br
$\quad$ | $\quad$ |
$CH_3-CH-CH-CH_2-CH_3$

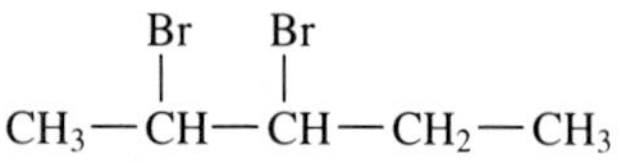

d.
$\quad$ CH_3 $\quad$ CH_3
$\quad$ | $\quad$ |
$CH_3-C-CH_2-C-CH_3$
$\quad$ | $\quad$ |
$\quad$ F $\quad$ CH_3

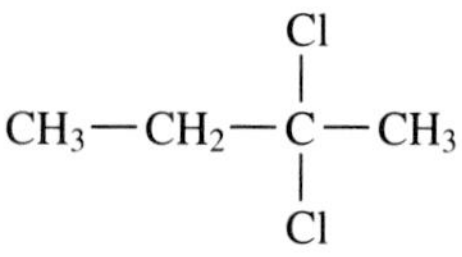

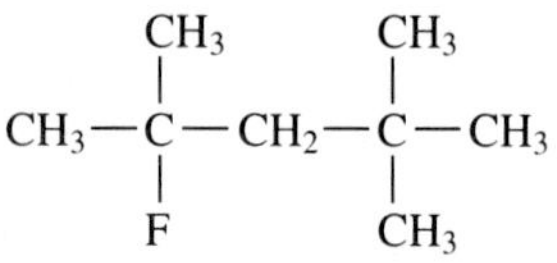

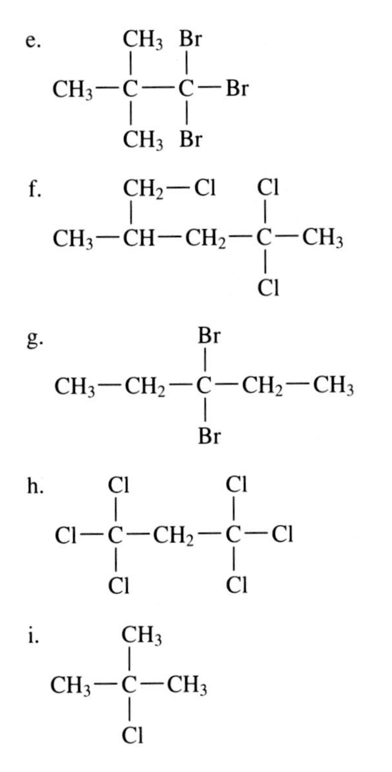

26. How many dibromobutane isomers can you identify? Draw each and provide a proper name.

Physical Properties of Organohalogens (Objective 2)

27. The boiling points of several 1-halogenbutanes are listed:

1-chlorobutane	78°C
1-bromobutane	101°C
1-iodobutane	130°C

Explain this ordering of boiling points.

Nomenclature (Objective 5)

28. Give the IUPAC name for each of the following halogenated cycloalkanes:

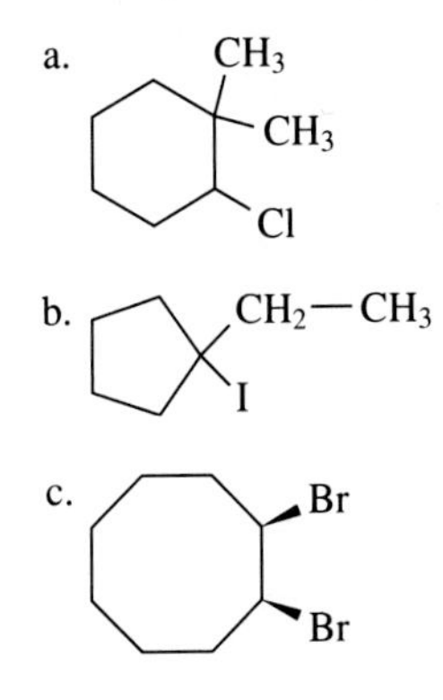

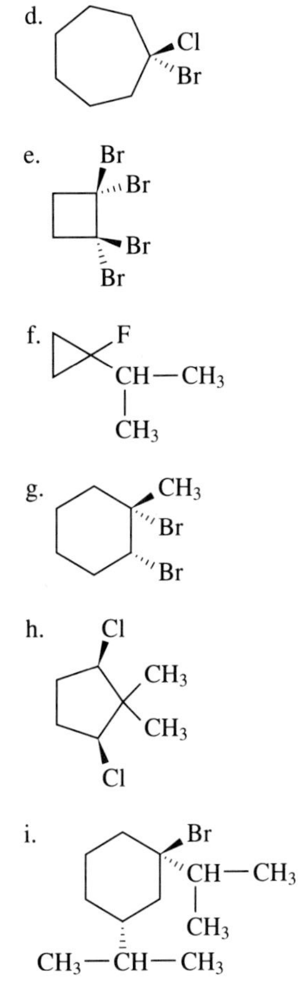

29. Give an IUPAC name for each of the following compounds.

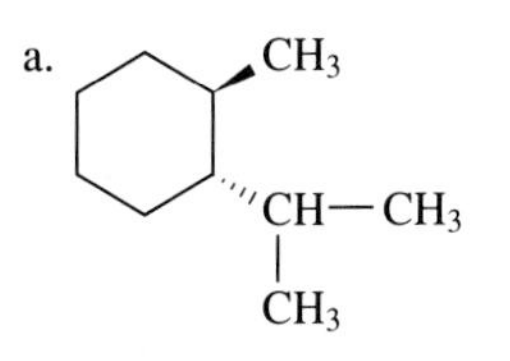

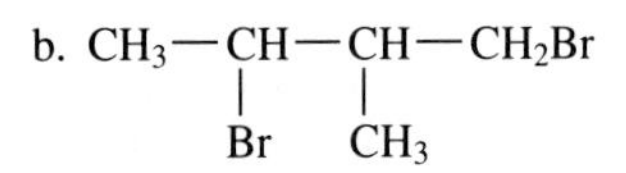

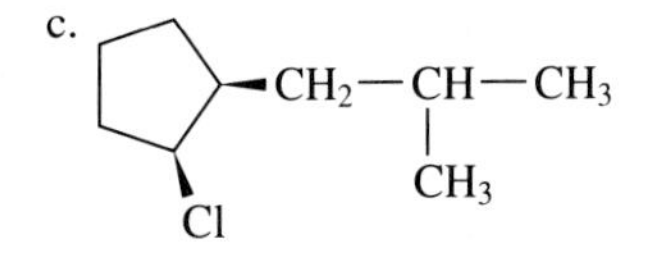

d. $CH_3-CH_2-CH(Br)-C(Br)(CH(CH_3)_2)-CH_2-CH_2-CH_3$

e.

f.

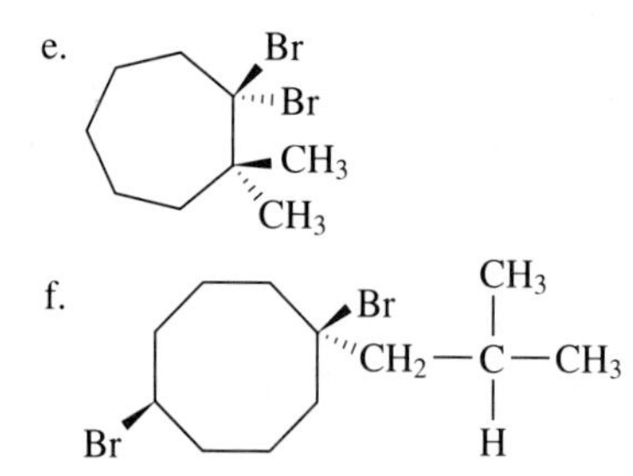

g.

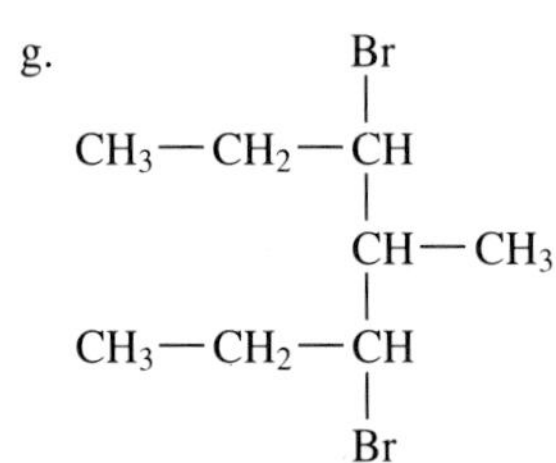

h.

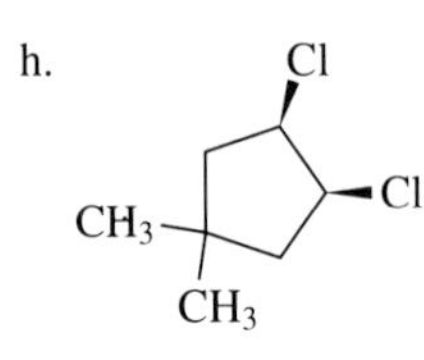

i.

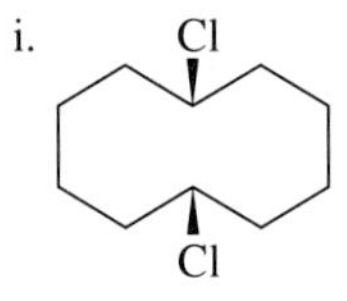

30. Draw structures corresponding to the following IUPAC names:
    a. 4-isopropyl-3,3-dimethyloctane
    b. *cis*-1,2-dimethylcyclohexane
    c. 2,2,3,3-tetramethylhexane
    d. 3-*tert*-butylheptane
    e. *trans*-2,3-diisopropyl-1,1-dimethylcycloheptane
    f. 1-*sec*-butyl-1-methylcyclohexane
    g. 1,1-diethylcyclopentane
    h. *cis*-1,3-diisopropylcyclohexane
    i. 4-*tert*-butyloctane
    j. 4-isopropyl-1,1,3,3-tetramethylcyclooctane
31. What is wrong with each of the following names?
    a. 2,4-diethylpentane
    b. *cis*-3,4-dimethylcyclohexane
    c. 2-isopropylbutane
    d. 1,5-diethylcyclopentane
    e. 2,2-dibromo-4,4,5,5-tetramethylhexane

Reactivity and Reactions (Objectives 6 and 7)

32. Why are alkanes and cycloalkanes generally unreactive?
33. What are the common products from combustion of any hydrocarbon?
34. Give products and balance the equations for the following reactions.
    a. Propane + $O_2 \longrightarrow$
    b. Isobutane + $O_2 \longrightarrow$
    c. Hexane + $O_2 \longrightarrow$
    d. Methylcyclohexane + $O_2 \longrightarrow$
    e. Cyclopentane + $Cl_2 \longrightarrow$
    f. Ethane + $Br_2 \longrightarrow$

Unclassified Exercises

35. Carbon atoms in alkanes and cycloalkanes (and in all carbon-containing compounds) always have how many bonds?
    a. 1. b. 2 c. 3 d. 4
36. The best name for

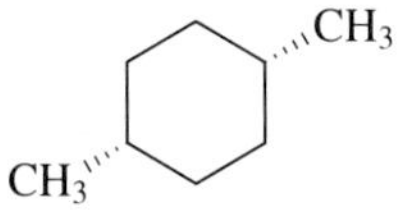

    CH_3 CH_3

    is
    a. *cis*-dimethylcyclohexane
    b. 1,4-dimethylcyclohexane
    c. *trans*-dimethylcyclohexane
    d. *cis*-1,4-dimethylcyclohexane
37. The best group name for

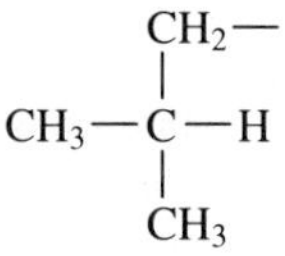

    CH_2-
    CH_3-C-H
    CH_3

    is
    a. isobutyl b. *tert*-butyl c. *sec*-butyl d. isopropyl
38. The best name for

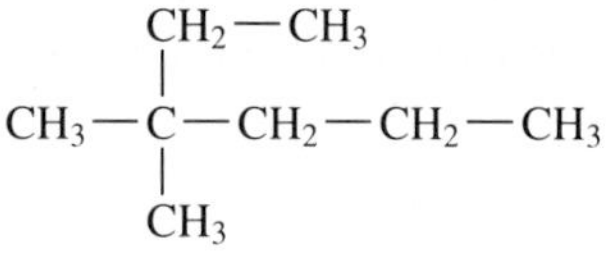

    CH_2-CH_3
    $CH_3-C-CH_2-CH_2-CH_3$
    CH_3

    is
    a. 4-ethyl-4-methylpentane
    b. 2-methyl-2-propylbutane
    c. 3,3-dimethylhexane
    d. 4,4-dimethylhexane
39. Hydrocarbons are
    a. insoluble in water
    b. composed of carbon and hydrogen
    c. both (a) and (b)
    d. none of these

40. The series of hydrocarbons of the general formula C_nH_{2n+2} is called
    a. hydrocarbon series
    b. increasing series
    c. homologous
    d. alkane series
41. Bonding in an alkane is
    a. ionic d. hydrogen bonding
    b. covalent e. both (b) and (c)
    c. sigma
42. The hydrocarbons are ______ in water.
    a. insoluble b. soluble c. reactive d. none of these
43. The major carbon compound formed from the combustion of a hydrocarbon in air is
    a. carbon dioxide b. carbon monoxide
    c. alkyl chains d. water
44. The general formula for a ring hydrocarbon is
    a. C_nH_n b. C_nH_{2n} c. C_nH_{2n+2} d. C_nH_{2n-2}

Synthetic polymers are formed from alkenes.

UNSATURATED HYDROCARBONS

OUTLINE

3.1 Introduction to the Unsaturated Hydrocarbons

3.2 Structure and Physical Properties

3.3 Nomenclature

3.4 Chemical Reactivity of Alkenes

3.5 Chemical Reactivity of Alkynes

3.6 Polyunsaturated Alkenes

3.7 Interesting Unsaturated Compounds

OBJECTIVES

After completing this chapter, you should be able to

1. Identify members of the alkene and alkyne families.
2. Discuss how the structure and properties of unsaturated hydrocarbons are related.
3. Explain how the double bond of an alkene results in specific stereochemical features.
4. Name representative members of the alkene and alkyne families.
5. Explain why unsaturated hydrocarbons are more reactive than the alkanes.
6. Predict products from common reactions of alkenes and alkynes.
7. Discuss features of synthetic and natural materials derived from alkenes and alkynes.

3.1 Introduction to the Unsaturated Hydrocarbons

The hydrocarbons introduced in the last chapter—the alkanes and cycloalkanes—are known as *saturated* hydrocarbons because all carbon–carbon bonds are single bonds and every carbon bond is connected to a different atom. When no more atoms can be placed on any carbon atom of the molecule, it is said to be ''saturated'' (Figure 3.1).

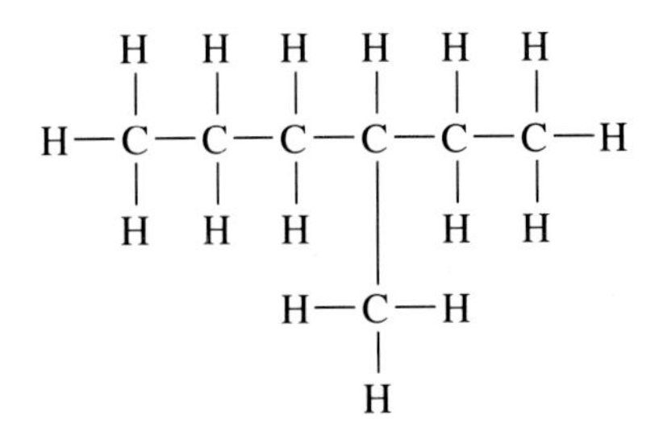

FIGURE 3.1
Saturated hydrocarbons are molecules that cannot accept any more atoms. Each carbon atom is bonded to four other atoms.

Another class of hydrocarbon molecules exists as well. These molecules do not have a different atom connected to every bond; they are known as **unsaturated** hydrocarbons. Figure 3.2(a) shows structural diagrams and ball-and-stick models of two hydrocarbons. One of these molecules is saturated; it is ethane. The second molecule is similar to ethane but has one important difference: it contains a double bond. It does not have enough hydrogen atoms to fill all four of the available carbon bonds. To achieve chemical stability, the unfilled orbitals share electrons with one another. This results in another bond between the carbon atoms. Organic compounds having a double bond between carbon atoms are called **alkenes**.

◆ *Alkenes* and *alkynes* are classified as *unsaturated hydrocarbons*. They have double and triple bonds, respectively.

Hydrocarbons are not limited to double bonds between carbons. If two less hydrogen atoms are present, a triple bond will result between the carbon atoms. Hydrocarbons with triple bonds are called **alkynes**. Examples of compounds containing triple bonds are shown in Figure 3.3.

Unsaturated hydrocarbons have characteristics that make them different from saturated hydrocarbons. The extra one or two bonds between carbon atoms hold its electrons quite loosely, and they are unprotected. These electrons, acting as Lewis bases, are available for reaction with Lewis acids. Molecules with double and triple bonds are likely to react at the site of their multiple bond. Thus, much of the chemical behavior of the alkenes and alkynes can be understood as examples of Lewis acid–base chemistry.

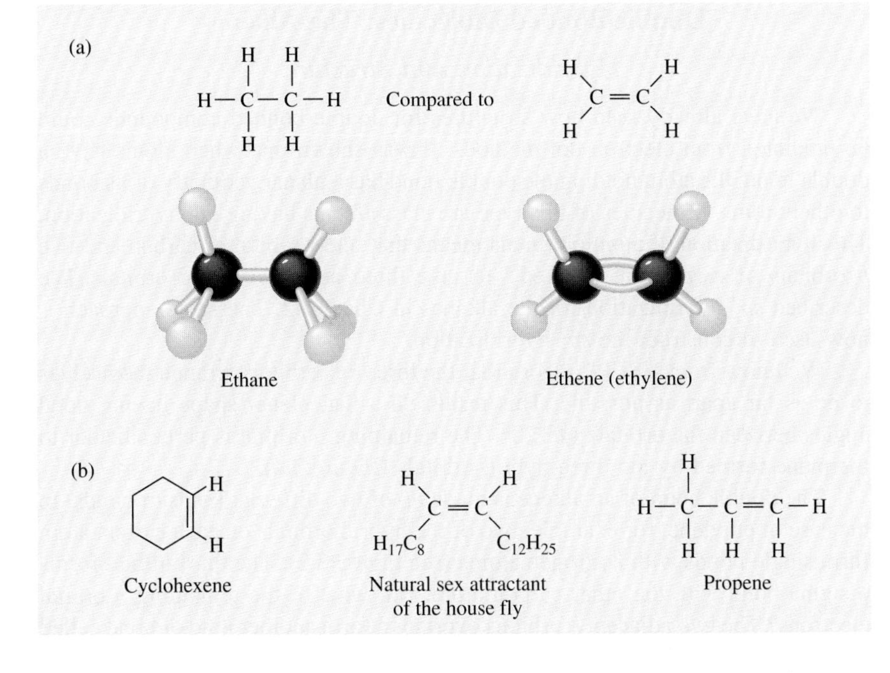

FIGURE 3.2
Alkenes (a) Ethane (on the left) is saturated (an alkane), whereas ethene (or ethylene) (on the right) is unsaturated. It is the simplest member of the alkene class of hydrocarbons. Ethylene is extensively used as an industrial intermediate and to help ripen fruit. (b) Other examples of artificial and natural alkenes.

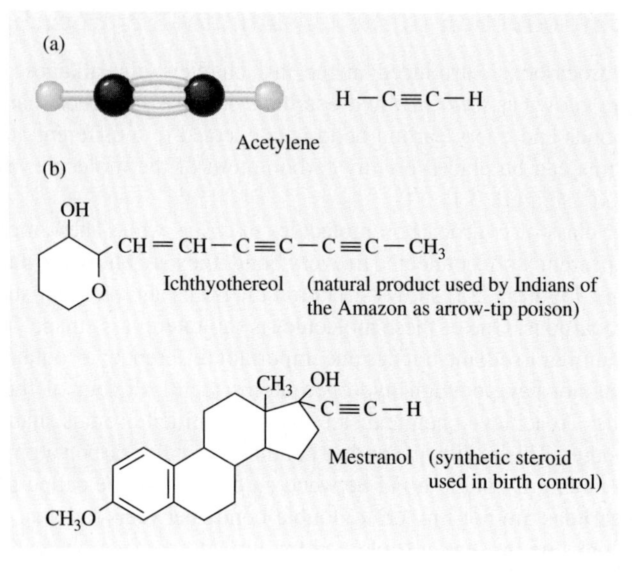

FIGURE 3.3
Alkynes (a) The simplest member of the alkyne series is acetylene. Note its triple bond. The alkyne functional group is not found to a large extent in naturally occurring molecules. (b) Examples of a natural and an artificial alkyne.

NEW TERMS

unsaturated One or more double or triple bonds exist between carbon atoms in the molecule

alkene A hydrocarbon having a double bond between carbon atoms

alkyne A hydrocarbon having a triple bond between carbon atoms

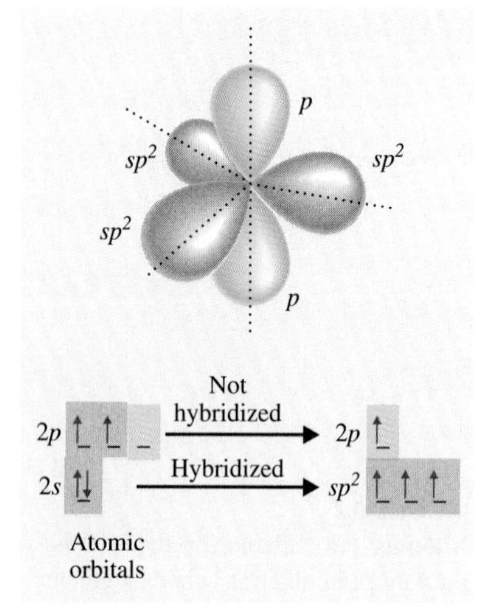

FIGURE 3.4
The sp^2 hybrid orbital. The three equivalent hybridized orbitals are in a plane with a bond angle of 120°. The p orbital is perpendicular to the plane formed by the hybridized orbitals.

3.2 Structure and Physical Properties

Double-Bonded Molecules: The Alkenes

Hybridization and Structure

We have already seen how structures for double-bonded compounds could be predicted from electron-dot methods. It was shown that when atoms have a double bond, the atom holding the double bond has a planar geometry of its bonds to other atoms. Molecules of this type are classsifed as having sp^2 hybridization. This hybridization of atomic orbitals means that one s orbital of carbon and two p orbitals of carbon are ''mixed'' to give three new hybridized orbitals. We described sp^3 hybridization for the alkanes in Chapter 2. The same approach is now used to construct the sp^2 hybridization.

We know from VSEPR theory that the three sp^2 orbitals arrange themselves to be as far apart as possible. This results in a flat, planar shape having bond angles between the orbitals of 120°. The remaining p orbital is perpendicular to the plane formed by the three hybrid orbitals (Figure 3.4).

The double bond of an alkene is made up of two different kinds of bonds. In the case of ethylene, for example, two of the hybrid orbitals on each carbon atom form single bonds with hydrogen atoms (see Figure 3.2). The third orbital forms a sigma bond with the other carbon atom. But what of the remaining p orbital electrons? The two adjacent p orbitals overlap to form another kind of bond called

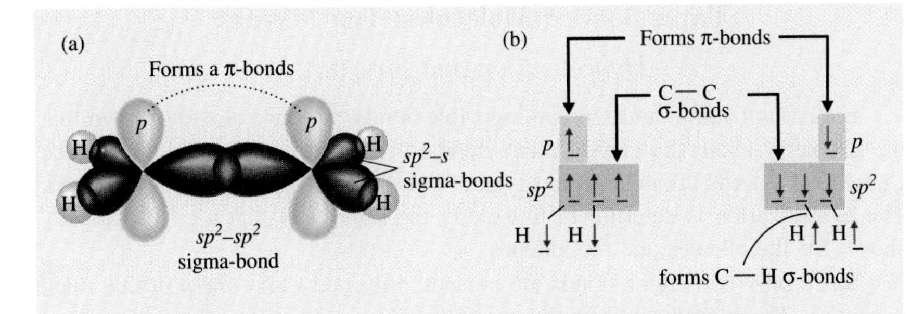

FIGURE 3.5
The two bonds of the alkene double bond are not the same. The double bond is made up of the sigma bond and one pi bond. The pi bond, formed by overlap of adjacent p orbitals, has its electrons more loosely held in space.

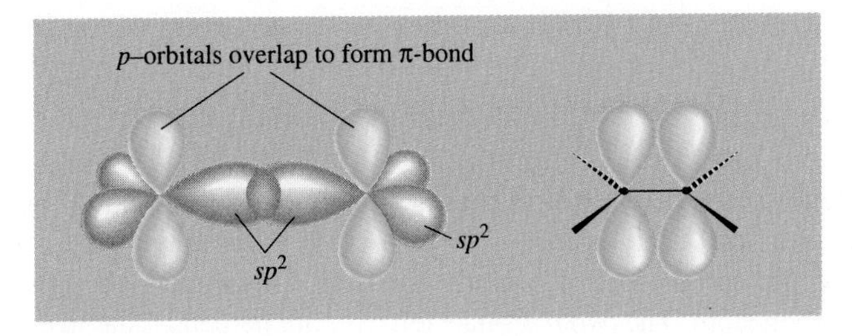

FIGURE 3.6
The C=C double bond gives a special rigidity to this portion of the alkene because of the overlap of the adjacent p orbitals to form the pi bond.

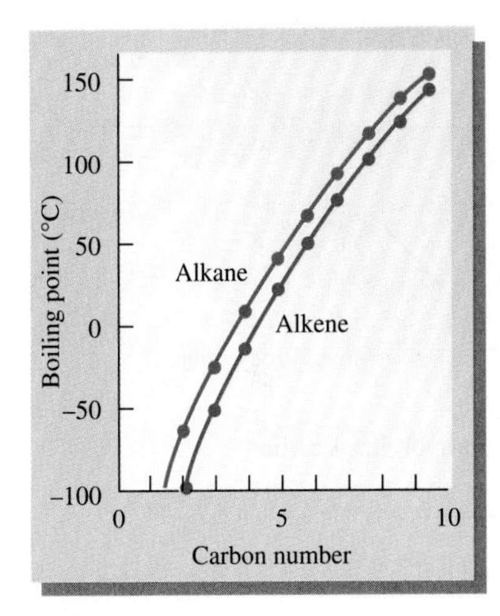

Boiling points of simple alkenes closely parallel those of the corresponding alkanes.

a **pi bond (π-bond)** (Figure 3.5). Since these electrons share a space between two atoms, they are attracted by both nuclei. This attractive force contributes to the increased strength of the double bond. However, since these electrons are not directly between the two nuclei, they do not contribute quite as much strength as do the sigma-bonded electrons. A double bond is stronger than a single bond, but not twice as strong.

Lack of Rotation About the Double Bond

Another characteristic of the double bond is implied by the ball-and-stick model of ethylene shown in Figure 3.2. Single-bonded molecules can rotate about the sigma bonds holding the atoms together, but this rotation cannot occur between double-bonded carbon atoms. (Figure 3.6). The additional pi bond firmly fixes the carbon atoms with respect to one another; there can be no free rotation about the carbon–carbon double bond.

♦ There is *no* rotation about a C=C bond.

Boiling Point and Carbon Number

The boiling points of a series of alkenes are quite similar to those for the corresponding alkanes. This is because the only attractive forces between the molecules are the London forces. Because we are comparing the same number of carbon atoms and only two fewer hydrogens, there is little difference between the alkanes and the corresponding alkenes.

Triple-Bonded Molecules: The Alkynes

Hybridization and Structure

♦ The carbon atoms involved in the pi bonds of the alkyne triple bond have *sp* hybridization.

For carbon compounds containing triple bonds, we have already shown that the geometry about the carbon atoms tends to be in a linear arrangement (see Chapter 1). Atoms involved in this type of bonding are said to be *sp*-hybridized. The hybridization is determined in exactly the same way that we have already shown for the alkanes and the alkenes.

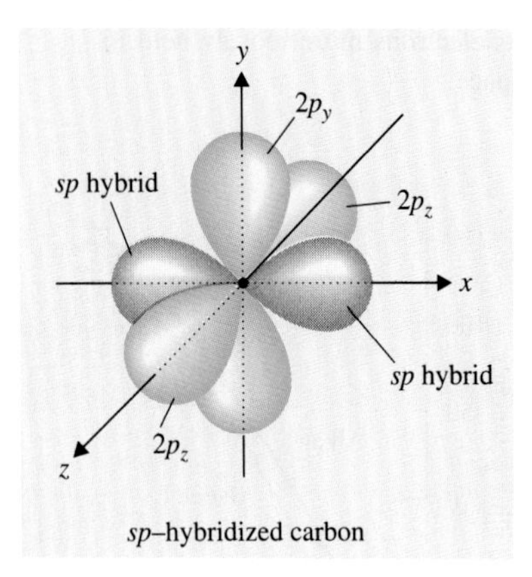

sp-hybridized carbon

Since only two sigma bonds are present, only one *s* and one *p* orbital must hybridize. The hybridization is thus *sp*,

$$2p\ \underline{\uparrow}\ \underline{\uparrow}\ \underline{\ \ }\ \xrightarrow{\text{Not hybridized}}\ 2p\ \underline{\uparrow}\ \underline{\uparrow}$$

$$2s\ \underline{\uparrow\downarrow}\ \xrightarrow{\text{Hybridized}}\ 2sp\ \underline{\uparrow}\ \underline{\uparrow}$$

Atomic orbitals

The geometric shape that places two pairs of electrons as far apart as possible is a line. For example, acetylene (H—C≡C—H) is a *linear* molecule.

One of the carbon hybrid orbitals bonds to a hydrogen, the other to the adjacent carbon atom. The two *p* orbitals left over are perpendicular to each other. These each contain an electron that can participate in pi bonding, as its orbital overlaps an adjacent *p* orbital. The triple bond comprises one sigma bond and two pi bonds, with the two pi bonds located at right angles to each other (Figure 3.7).

Triple-bonded molecules are much like double-bonded molecules, except that there are two pi bonds instead of one. The pi electrons are still on the "outside" of the molecule and are available for reaction. The pi bonds also restrict motion about the C—C sigma bond, although rotation of the entire linear molecule is still possible. Because of their linear geometry, triple-bonded molecules have no isomers. Only cyclic systems and alkenes have geometric isomers.

Boiling Points and Carbon Number

The linear alkynes have physical properties similar to the alkenes; however, they generally have slightly higher boiling points. This observation is nicely in

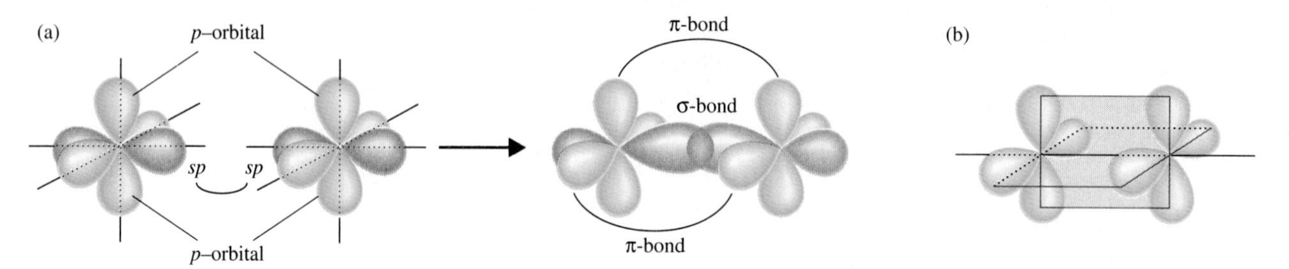

FIGURE 3.7
(a) Bonding in a triple-bonded molecule. The bonds consist of a single sigma bond and two pi bonds. (b) The lobes of the *p* orbitals overlap to form two pi bonds.

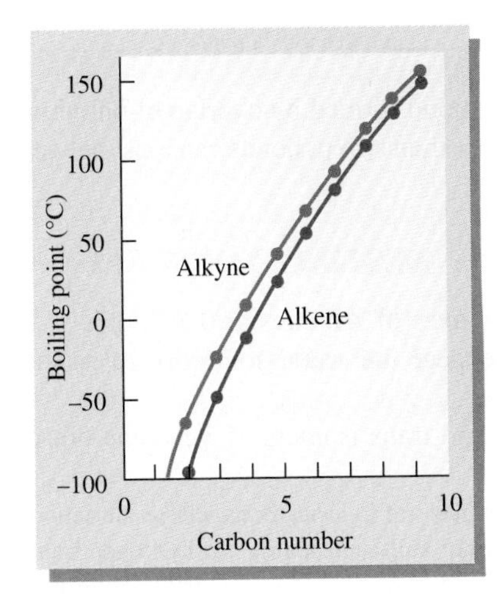

The boiling points of alkynes are generally a little higher than those of the alkenes.

accord with our discussions about London forces. Because the alkyne is linear and the alkene has a bent shape (120°), the alkynes are easier to stack next to one another. This phenomenon is similar to the lower boiling points of the branched hydrocarbons.

Isomers of Alkenes: Cis and Trans Relationships

One of the consequences of the geometry imposed by the bonding of alkenes is the existence of cis and trans isomers of alkenes. The same principles apply as for the cycloalkanes, for which totally free rotation around the C—C bond is also restricted. The terminology is the same, too. The cis isomer has similar groups on the *same side* of the pi double bond, the trans isomer has similar groups on *opposite sides* of the pi double bond (Figure 3.8). The cis and trans isomers are essentially different compounds.

♦ Cis alkenes have similar groups on the *same* side of the pi system:

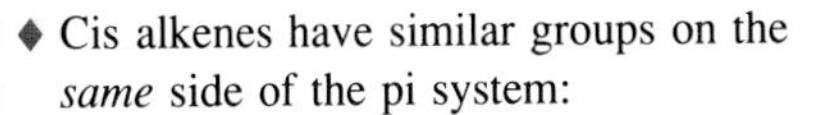

whereas trans isomers have similar groups on *opposite* sides:

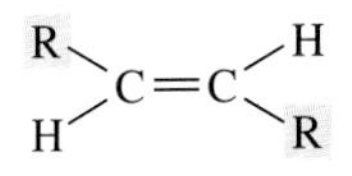

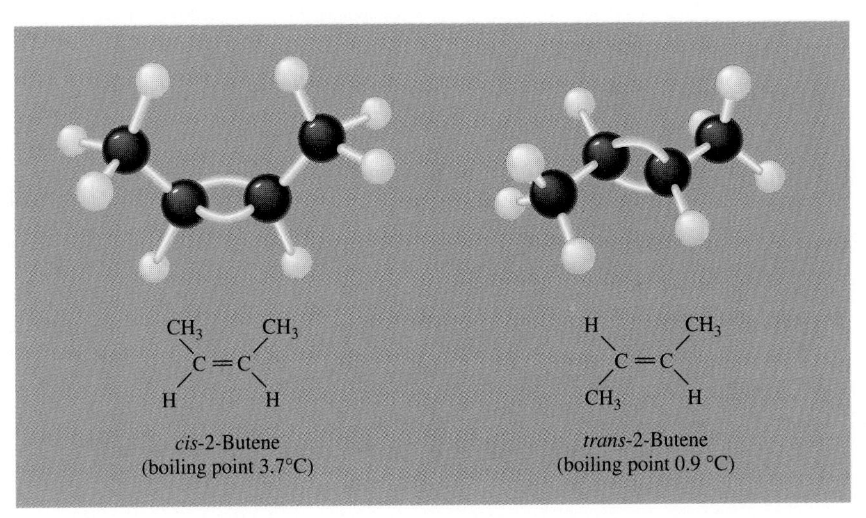

FIGURE 3.8
The geometric consequences of cis and trans isomers for alkenes. Since the pi bond prevents rotation about the C—C bond, the cis and trans isomers are truly different compounds and cannot be converted from one to the other by simple physical methods.

NEW TERMS

pi bond (π-bond) A bond formed by overlap of unhybridized *p* orbitals of two adjacent atoms. No more than two pi bonds can exist between two adjacent atoms.

TESTING YOURSELF

Properties and Structures of Alkenes and Alkynes

1. Why would you expect the water-solubility of octane and 1-octene to be similar?
2. Why are there cis and trans isomers of 3-hexene but not of 3-hexyne?

Answers **1.** Both molecules are hydrocarbons whose attractions are limited to London forces. For both there is not sufficient ability for hydrogen bonding to break up this arrangement of water. **2.** The alkyne C≡C triple bond is linear and thus cannot have isomers.

3.3 Nomenclature

Naming the Alkenes

Nomenclature for the alkenes is similar to that for the alkanes. Cis and trans isomers must be distinguished by including these terms in their names. As we have already seen for the cycloalkanes, the cis isomers have similar groups on the same side of the bond, but trans isomers have similar groups on opposite sides. In Figure 3.9 we show that in the nomenclature of alkenes, the cis or trans label is applied to the longest continuous carbon chain containing the double bond. Thus, if the longest continuous chain is on the same side of the pi bond, we call it a cis isomer; if it is on opposite sides, we call it a trans isomer.

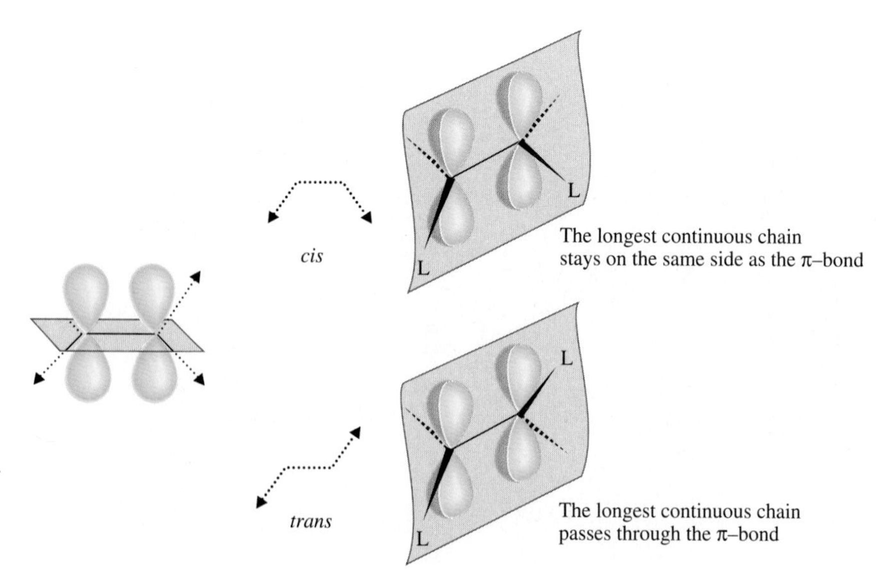

FIGURE 3.9
How alkene isomers are distinguished. The cis isomer has the longest continuous chain on the same side of the pi bond, but the trans isomer has this chain passing through the pi bond.

CHEMISTRY CAPSULE

Strange Hydrocarbons

It may be hard to believe, but organic chemists have a sense of humor too! As a result of interest in the preparation and chemical behavior of strained hydrocarbon systems, a number of unique compounds have been prepared. Because many of these have invoked some inner artistic image, they have received unusual names that express their strange-looking shapes. See if you can find the reason for the names in the few examples presented here.

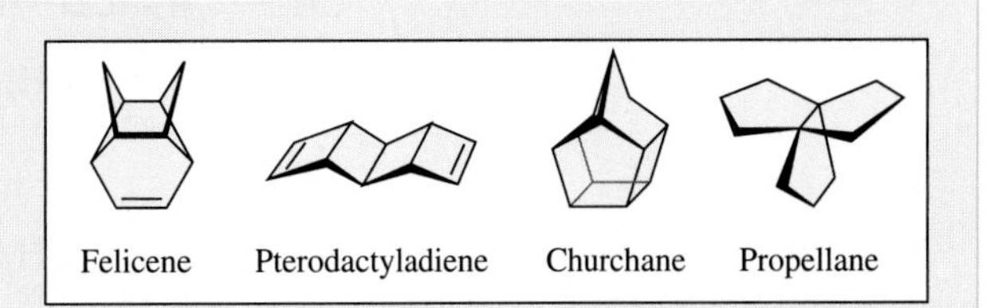

FIGURE 3.A
Chemists have a sense of humor. Common names are sometimes given to uniquely shaped molecules to express what the molecule looks like.

The rules for naming alkenes can be summarized as follows.

♦ Alkene nomenclature is similar to alkane nomenclature, except that there is an *-ene* ending, and the position of the double bond must be designated.

1. Find the longest carbon chain containing the double bond. This defines the parent compound.
2. Number the carbons from the end that places the double bond between the smallest possible numbers. Designate the location of the double bond with the lower of these two numbers.
3. If enough information is available about the molecule to decide if it is cis or trans, make this designation.
4. Name substituted alkyl groups, as was done for the alkanes.
5. Replace the *-ane* ending of the parent alkane with *-ene*.

EXAMPLE 3.1

Name this alkene.

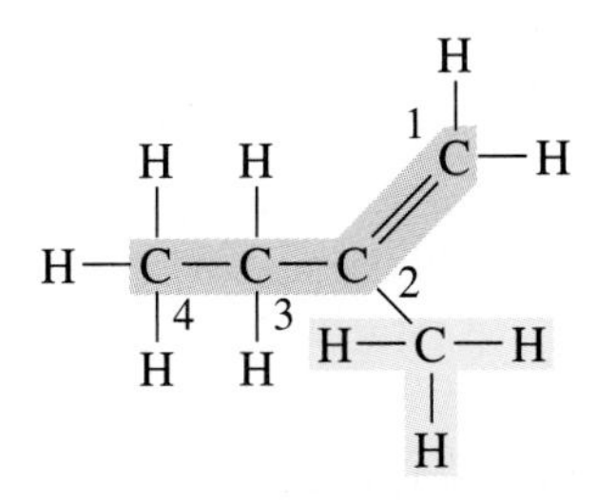

The longest chain that contains the double bond has four carbons. The parent compound is thus a butene. If we number from the left, the double bond is between carbons 3 and 4. If we number from the right, the double bond is between carbons 1 and 2. Numbering from the right gives the lowest number for the location of the double bond. The compound is a 1-butene. The methyl group is attached to carbon 2.

Solution: The name of this compound is thus 2-methyl-1-butene.

Listed here are some more alkene structures with their correct names.

$CH_3—CH=CH—CH_2—CH_3$ 2-Pentene

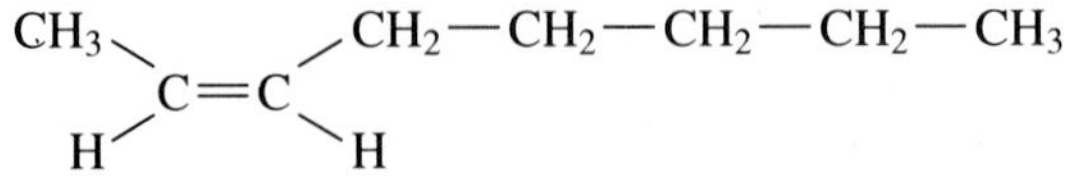

$$\begin{array}{c} CH_3 \quad\quad\quad CH_2—CH_2—CH_2—CH_2—CH_3 \\ C=C \\ H \quad\quad\quad H \end{array}$$ *cis*-2-Octene

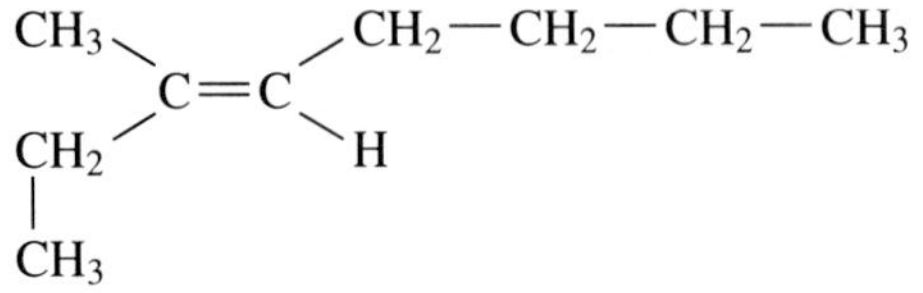

$$\begin{array}{c} CH_3 \quad\quad\quad CH_2—CH_2—CH_2—CH_3 \\ C=C \\ CH_2 \quad\quad\quad H \\ | \\ CH_3 \end{array}$$ *trans*-3-Methyl-3-octene

Polyunsaturated Alkenes

♦ Polyunsaturated molecules have more than one double or triple bond.

Often a carbon chain has more than one double bond. Multiple unsaturation is reflected in nomenclature as well as in unique chemical reactivity. A **polyunsaturated** compound has more than one double bond.

The number of double bonds can be described by the prefixes *di-*, *tri-*, *tetra-*, and so on. The name also has to tell *where* the double bonds are located. Nomenclature rules are very similar to those already defined for simple alkenes and are most readily demonstrated by an example.

EXAMPLE 3.2

Name this polyunsaturated alkene.

$$\begin{array}{ccccccccc} 1 & 2 & 3 & 4 & 5 & 6 & 7 & 8 & 9 \\ CH_3— & CH= & CH— & CH_2— & CH= & CH— & CH_2— & CH_2— & CH_3 \\ 9 & 8 & 7 & 6 & 5 & 4 & 3 & 2 & 1 \end{array}$$

The longest chain containing the double bonds has nine carbons, so the parent compound is nonene. Since there are two double bonds, we add *di-* to the name, to get nonadiene. The lowest numbers for the locations of the double bonds are 2 and 5.

Solution: Thus, the name for this compound is 2,5-nonadiene.

Listed here are some more polyunsaturated alkenes and their correct names.

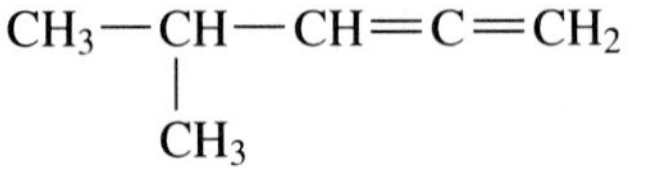

$$\begin{array}{l} CH_3—CH—CH=C=CH_2 \\ \quad\quad\; | \\ \quad\quad CH_3 \end{array}$$ 4-Methyl-1,2-pentadiene

$CH_3—CH=CH—CH=CH_2$ 1,3-Pentadiene

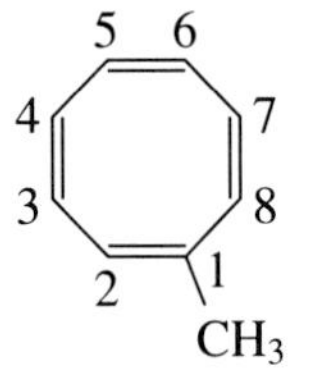

(cyclooctatetraene ring, positions 1–8, with CH_3 on C-1) 1-Methyl-1,3,5,7-cyclooctatetraene

Naming the Alkynes

The IUPAC rules for naming alkynes are similar to those for the alkanes and alkenes, except that the suffix *-yne* is used to indicate the presence of the triple bond. (See the rules listed for alkenes.) Briefly summarized, alkyne names are constructed this way:

♦ Alkyne nomenclature is similar to that for alkenes, except that there is an *-yne* ending.

Name = Substituents and locations + Location of triple bond

+ Parent compound + − *yne*

Listed here are some alkynes structure with their correct names.

H—C≡C—H	Ethyne (acetylene)
H—C≡C—CH_2—CH_2—CH_3	1-Pentyne
$\overset{1}{CH_3}$—$\overset{2}{CH}$(CH_3)—$\overset{3}{C}$≡$\overset{4}{C}$—$\overset{5}{CH_2}$—$\overset{6}{CH_2}$—$\overset{7}{CH_3}$	2-Methyl-3-heptyne

NEW TERM

polyunsaturated Molecules having more than one double or triple bond.

TESTING YOURSELF

Naming Alkenes and Alkynes

1. Name the following alkenes:

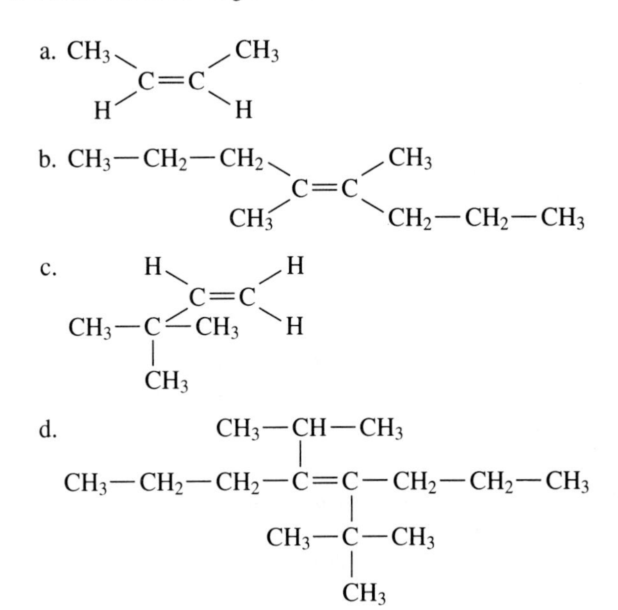

2. Name the following polyunsaturated alkenes:

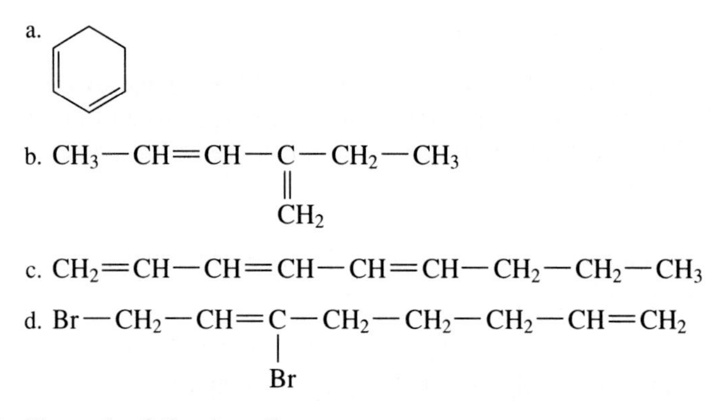

3. Name the following alkynes:

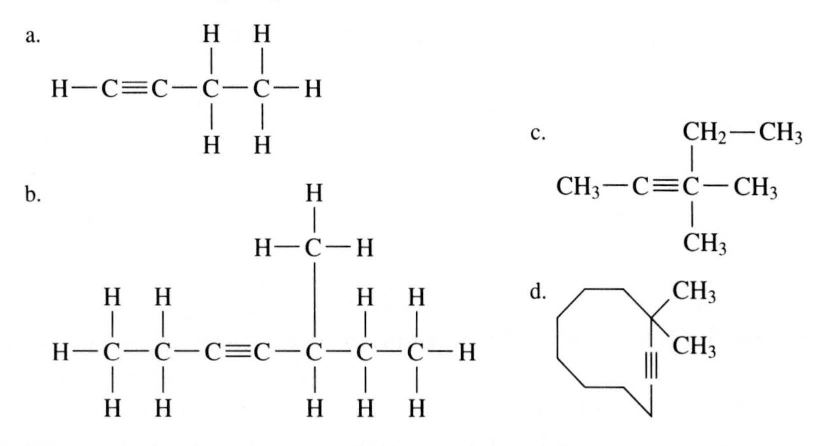

4. Why can't the cis and trans nomenclature be used for

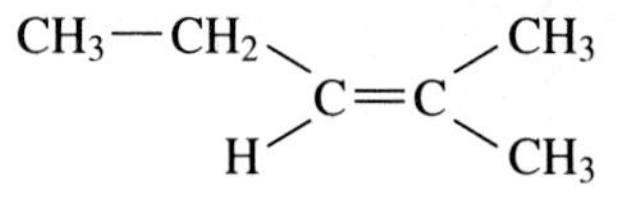

Answers **1. a.** *cis*-2-butene **b.** *trans*-4,5-dimethyl-4-octene **c.** 3,3-dimethyl-1-butene **d.** 4-*tert*-butyl-5-isopropyl-4-octene **2. a.** 1,3-cyclohexadiene **b.** 2-ethyl-1,3-pentadiene **c.** 1,3,5-nonatriene **d.** 6,8-dibromo-1,6-octadiene **3. a.** 1-butyne **b.** 5-methyl-3-heptyne **c.** 4,4-dimethyl-2-hexyne **d.** 3,3-dimethylcyclodecyne **4.** The longest continuous chain is the same.

3.4 Chemical Reactivity of Alkenes

The exposed pi-bonding electrons of the alkenes give this family of compounds enhanced chemical activity. We will examine four types of common reactions in this section.

◆ The pi electrons of the alkene and alkyne identify them as Lewis bases, and thus the unsaturated compounds are reactive to Lewis acids.

1. Combustion
2. Addition reactions, with symmetric or nonsymmetric reactants
3. Oxidation reactions, adding oxygen or hydroxide
4. Polymerization reactions, in which the molecules form extremely long and often cross-linked molecules

Combustion Reactions

We have already seen that one reaction characteristic of all hydrocarbons is combustion with oxygen. Since alkenes are hydrocarbons, it is no surprise to observe the same type of reaction. This reaction can be generalized as follows.

GENERAL REACTION

Combustion of an Alkene

$$C_nH_{2n} + 1.5_n\,O_2 \longrightarrow n\,CO_2 + n\,H_2O$$

Alkene Oxygen Carbon dioxide Water

Example

$$H_2C{=}CH_2 + 3\,O_2 \longrightarrow 2\,CO_2 + 2\,H_2O$$

Combustion produces carbon dioxide and water vapor. If insufficient oxygen is present, alkenes will also burn to produce poisonous carbon monoxide.

Addition Reactions

More important than combustion, however, is *addition* of other elements or groups to the double bond. These reactions divide into two classes: (1) reactions in which the same substituent is added to each carbon of the double bond, and (2) reactions in which different substituents are added to each carbon of the double bond. Reactions observed for the alkene family are also characteristic of the alkyne family of compounds. However, in the case of the alkynes, two pi bonds are available for reaction.

♦ In *addition* reactions, new atoms add to the double bond, converting it to a single bond.

Addition reactions are of the general form

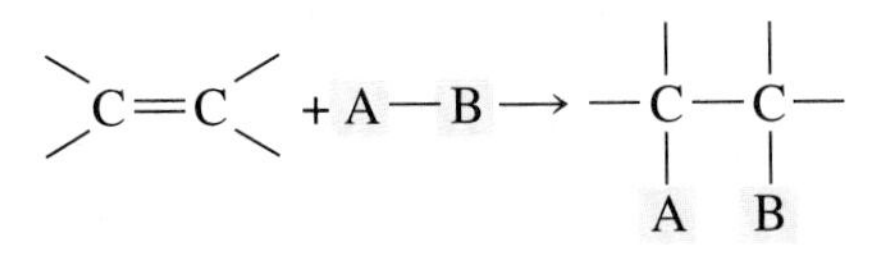

If the reactant A—B is symmetrical (for example, H—H or Br—Br), the addition to the double bond is symmetrical. For example, hydrogen gas reacts with an alkene (an unsaturated hydrocarbon) in the presence of a metal catalyst such as platinum (Pt) to produce a saturated hydrocarbon.

♦ When an alkene undergoes an addition reaction, the double bond is lost. Reactions take place at the double bond.

A Note About Writing Equations in Organic Chemistry

There are a number of ways to write reactions or equations in organic chemistry, but all should do the same thing: (1) show *what* reacted, (2) show special conditions, and (3) show products. Here are two general forms for the same reaction:

The hydrogenation of an oil to a fat is called *hardening*, because a liquid is converted to a solid.

$$A + B \xrightarrow{\text{Conditions}} C + D$$

$$A \xrightarrow[\text{Conditions}]{B} C + D$$

Note that the second reactant (B) can be shown either before the reaction arrow or above it. You will see it done both ways throughout this and other textbooks.

Addition of Hydrogen (Hydrogenation)

The reaction of ethylene (ethene) with hydrogen in the presence of a catalyst can be shown in two equivalent ways (according to the preceding discussion).

♦ Two ways of writing the same reaction.

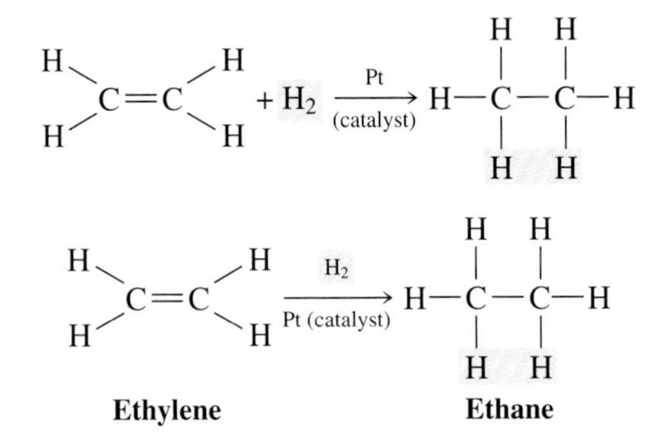

A reaction in which hydrogen is added to a double bond is called hydrogenation. **Hydrogenation** is used commercially to convert liquid oils to solid fats. As the number of double bonds is reduced, the melting point of the fat increases. Such a reaction, called *hardening*, is used to convert vegetable oils into margarine.

♦ The hydrogenation of an oil to a fat is called *hardening*, because a liquid is converted to a solid.

The hydrogenation reaction can be generalized as follows.

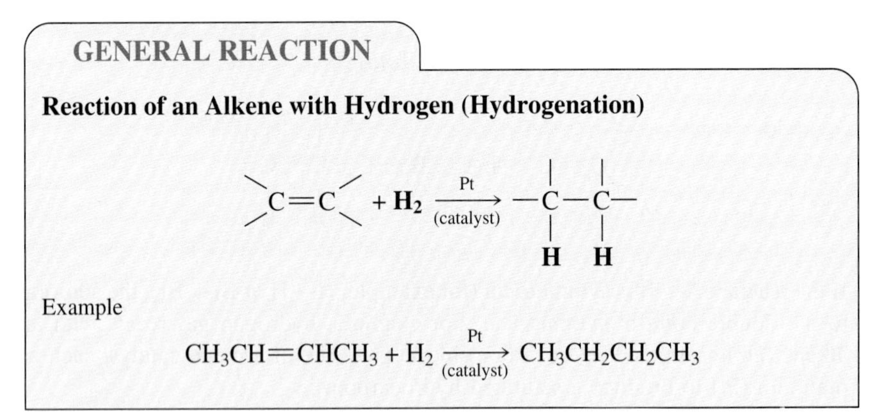

Addition of Halogens (Halogenation)

The double bond of an alkene reacts rapidly with chlorine or bromine at room temperature without a catalyst. Both chlorine and bromine are halogens, so this reaction is called a **halogenation** reaction. Thus, for example, we observe:

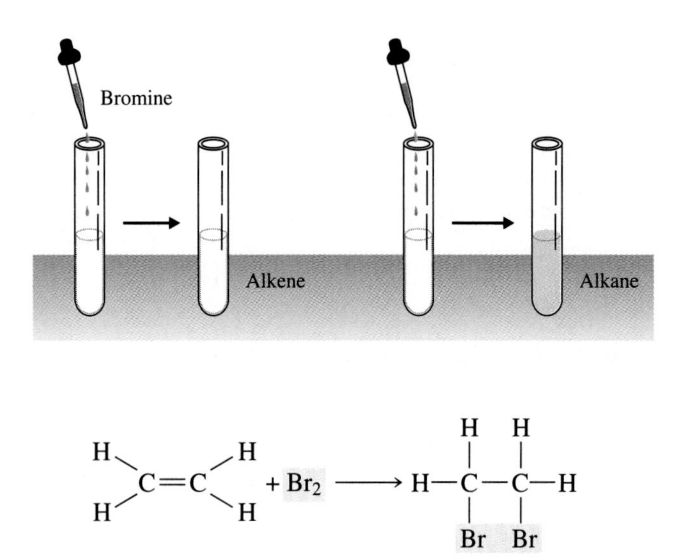

FIGURE 3.10
The addition of bromine to an alkene. The loss of the bromine color is an indication that the compound is an alkene; this observation is often used as a qualitative test for alkenes. If no double bond is present, such as in an alkane, the bromine color remains. The addition of bromine also gives a positive test for an alkyne.

$$H_2C{=}CH_2 + Br_2 \longrightarrow H{-}CH(Br){-}CH(Br){-}H$$

The relative reactivity of the halogens in this situation is

$$F > Cl > Br > I$$

Fluorine reacts explosively with alkenes, but iodine does not add to an alkene double bond under normal conditions. It makes sense, then, that chlorine (Cl_2) and bromine (Br_2) are the commonly used reactants.

Bromine, a red-brown liquid with dangerous fumes, is often used as a test reagent for identification of an alkene (Figure 3.10). It is usually prepared as a dilute solution in an inert solvent. A small sample of the suspected alkene is placed in a test tube, and a few drops of the deeply colored bromine solution are added. If a double bond is present, the bromine will add to this bond and the bromine color will disappear. If double bonds are not present, the bromine will not react and the solution will remain red-brown.

The halogenation reaction can be generalized as follows.

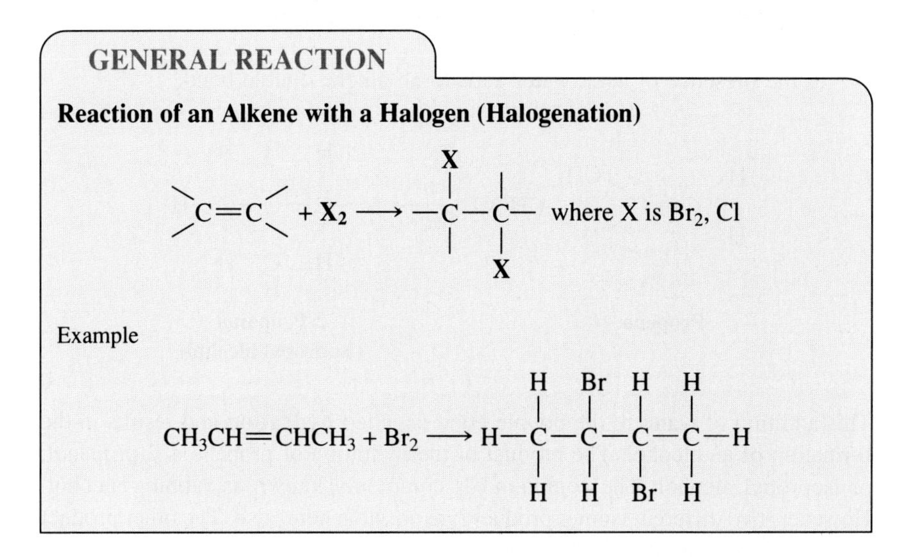

Addition of Hydrochloric Acid

Nonsymmetric reactants, such as hydrochloric acid (HCl), react with double bonds in the same manner as the symmetric additions of H_2 and Br_2 discussed earlier. However, this reaction has two possible isomer products. Consider the example of the reaction of HCl with propene.

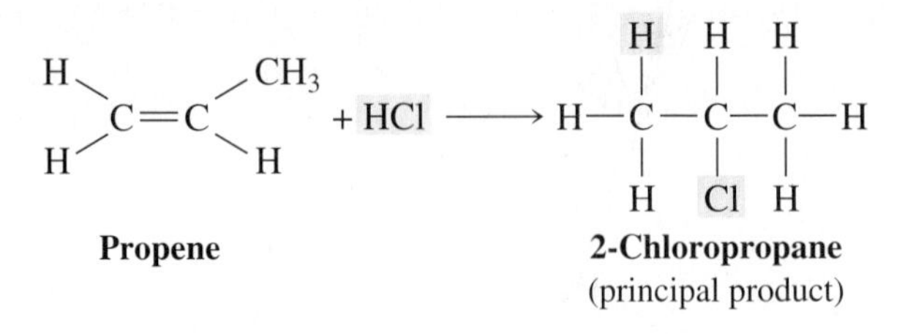

This reaction shows the chlorine connecting to the center carbon and the hydrogen from the HCl bonding to the outer carbon. The other possibility is for the chlorine to attach to the outer carbon and the hydrogen to the center carbon.

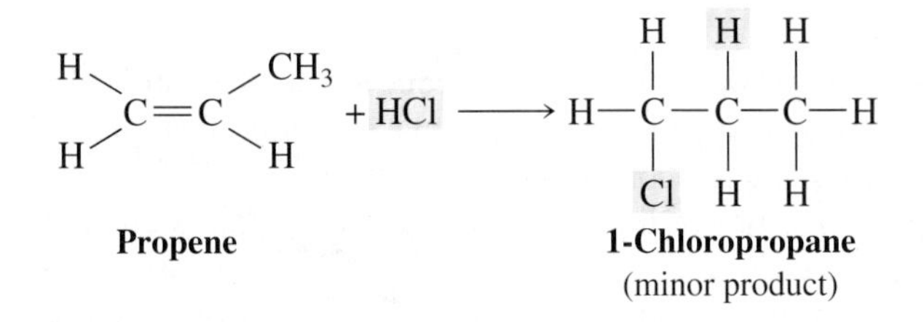

Both products shown above are possible, although the 2-chloropropane isomer predominates.

Addition of Water (Hydration)

In the presence of acid, water adds to an alkene double bond:

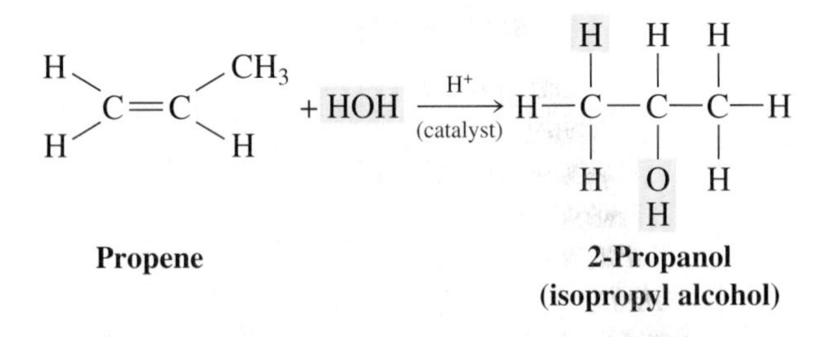

This addition of water to the double bond is called **hydration** and results in the formation of an alcohol. The product of the hydration of propene is 2-propanol, or isopropyl alcohol. The compound is commonly known as rubbing alcohol. However, two different isomer products are possible here, also. The other product is formed according to this reaction:

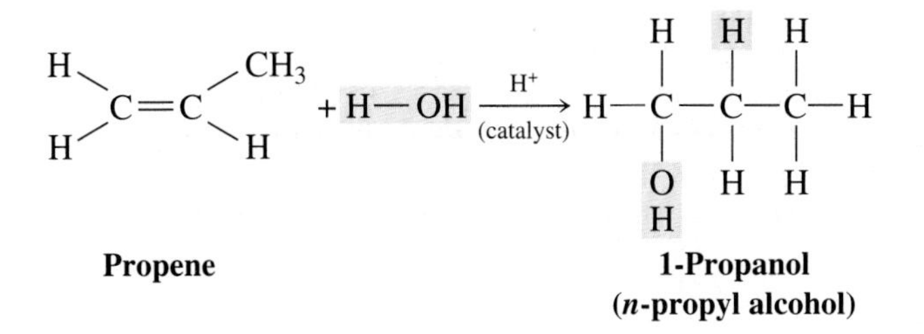

With hydration, the 2-propanol (isopropyl alcohol) predominates as the product. This type of reaction between ethylene and water is commercially used to prepare millions of gallons of ethanol (ethyl alcohol) each year.

Markovnikov's Addition Reactions

Organic chemists are often accused of being a little careless in the balancing of the chemical equations that describe organic reactions. There is, however, a reason for this. Inorganic chemists predict with reasonable certainty the products of their reactions. In organic chemistry, a number of reactions are almost always possible, and one can only indicate the relative probability that a certain product will be synthesized. The Russian chemist Vladimir Markovnikov (1838–1904) stated a general rule that helps us predict the structure of the principal product when a nonsymmetric molecule containing hydrogen is reacted with a double bond. Markovnikov observed that, when a nonsymmetric molecule such as HCl or H_2O adds to a double bond, the double-bonded carbon with the most hydrogens gets the additional hydrogen. **Markovnikov's rule** was a statement of fact based on observation. Markovnikov was unable to explain why the rule worked. However, he should not be judged too harshly. The electron was discovered only about 7 years before his death, and the ideas of orbitals and hybridization developed more than 20 years after after he had died. Markovnikov's rule is, however, still useful today.

◆ The H^+ in a reaction with a double bond adds to the carbon atom that has the most hydrogen atoms attached to it; this principle is called *Markovnikov's rule.*

GENERAL REACTION

Reaction of an Alkene with a Nonsymmetric Reactant (Markovnikov's Rule)

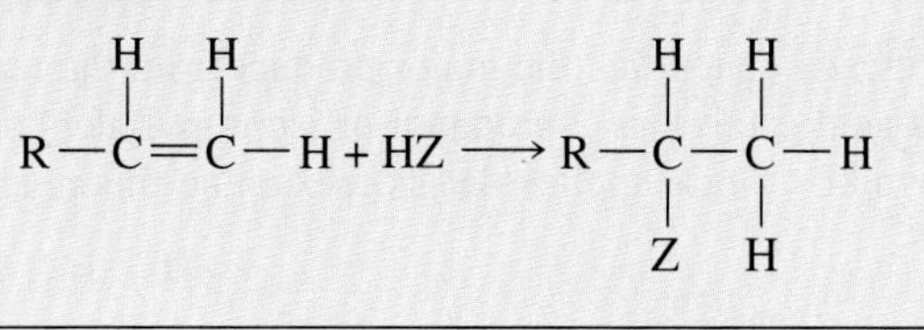

In this reaction, Z is Cl or OH (HZ is a "generic" reagent). Notice that the carbon that has the most hydrogens to begin with gets another.

We have discussed the notion that nonsymmetric reagents add to nonsymmetric alkenes in a particular way. We have given this concept the name *Markovnikov addition.* Let us now generalize these two reactions; the addition of HX (hydrohalogenation, or the addition of H^+ and X^-, where X = Cl, Br, I) and addition of water (hydration, or the addition of H^+ and OH^-).

GENERAL REACTION

Reaction of an Alkene with HX (Hydrohalogenation)

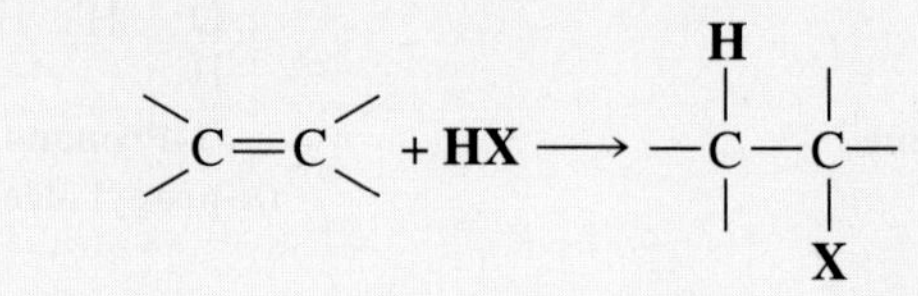

where X is Br, Cl. The H adds to the carbon bearing the most hydrogens.

Example

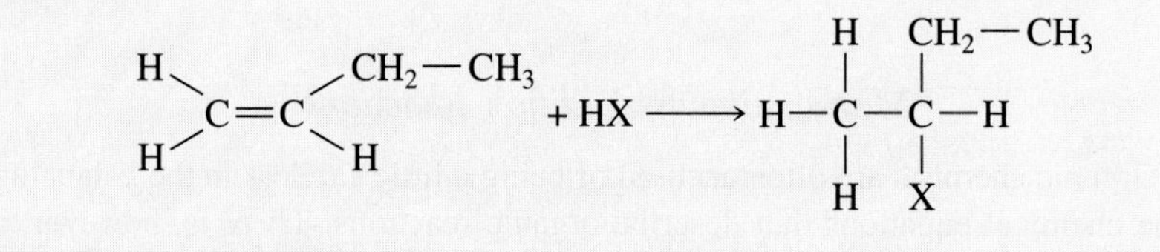

GENERAL REACTION

Reaction of an Alkene with Water (Hydration)

$$\text{>C=C<} + \mathbf{H{-}OH} \xrightarrow[\text{(catalyst)}]{H^+} -\overset{\mathbf{H}}{\overset{|}{C}}-\underset{\mathbf{OH}}{\underset{|}{C}}-$$

The H adds to the carbon bearing the most hydrogens.

Example

$$(H)_2C{=}C(H){-}CH_2{-}CH_2{-}CH_3 + H_2O \xrightarrow[\text{(catalyst)}]{H^+} H{-}\overset{H}{\underset{H}{C}}{-}\overset{CH_2-CH_2-CH_3}{\underset{OH}{C}}{-}H$$

♦ 2-propanol is the IUPAC name; but isopropyl alcohol is a very commonly used name.

We have already stated that the major product from the reaction of propene with water and acid catalyst is 2-propanol or isopropyl alcohol. As an example of how Markovnikov's rule applies, let us reexamine this reaction.

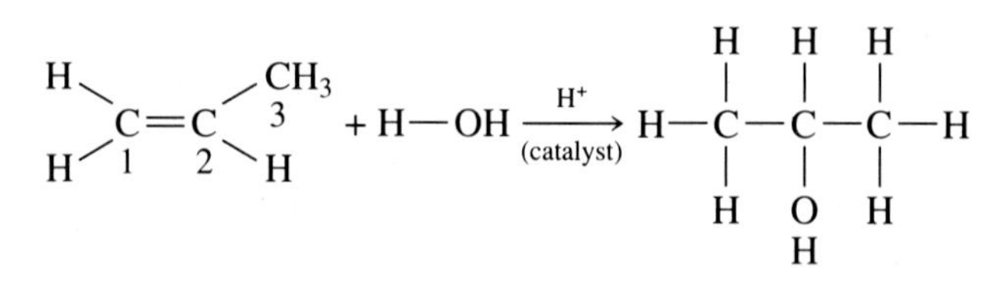

Propene **Isopropyl alcohol**

There are three carbons in the propene molecule; the double bond is between carbon 1 and carbon 2. Carbon 1 has two hydrogens attached to it; carbon 2 has one hydrogen and a methyl group attached to it. Markovnikov's rule predicts that the hydrogen from the reactant (H—OH) should go to the carbon that already has the most hydrogen (carbon 1). Carbon 2 must then get the —OH group.

We have seen that alkenes undergo addition reactions, both with symmetrical and unsymmetrical reagents. A few additional examples of the addition of unsymmetrical reagents is useful to illustrate the scope of these reactions. Remember that an unsymmetrical reagent can add to any alkene; however, if the alkene is unsymmetrical, we must apply Markovnikov's rule to ascertain the correct product.

EXAMPLE 3.3

Show the addition of water to fumaric acid to form malic acid

Solution:

$$\underset{\textbf{Fumaric acid}}{HOOC-CH=CH-COOH} + H-OH \xrightarrow{H^+} \underset{\textbf{Malic acid}}{HOOC-\overset{\overset{H}{|}}{C}H-\underset{\underset{OH}{|}}{C}H-COOH}$$

The addition of water to an alkene to form malic acid will be seen in more detail later in the biochemistry part of this book. This reaction is part of the *Krebs cycle* (or citric acid cycle, or tricarboxylic acid cycle). This is one of the important biochemical sequences in the human body that converts carbon compounds to energy through biological combustion, giving off carbon dioxide as a by-product.

EXAMPLE 3.4

Predict the product of the reaction of HBr with 2-methyl-1-pentene.

Solution:

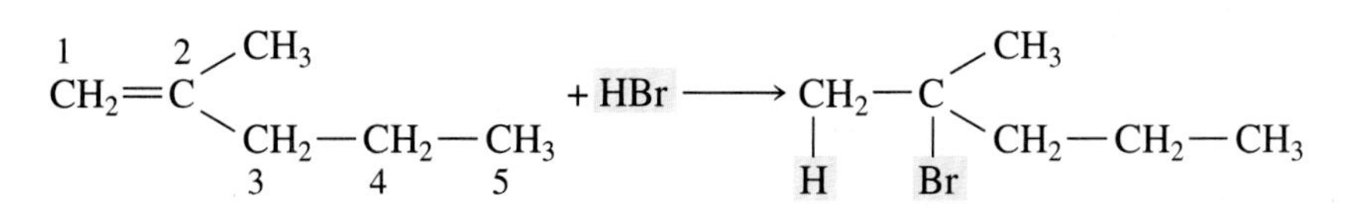

Notice that the alkene bond in Example 3.4 is between C-1 and C-2. The C-1 is directly attached to two hydrogen atoms, but C-2 is directly attached to two carbon atoms. Thus, C-2 is more substituted than C-1. From this we predict that the H^+ adds to C-1, whereas the Br^- attaches to C-2.

Oxidation Reactions of Alkenes

A third type of chemical reactivity in alkenes is oxidation. Double bonds are highly susceptible to attack by oxidizing agents, which include oxygen, ozone, potassium permanganate, and potassium dichromate. For example, rubber has a large number of carbon–carbon double bonds. The presence of sunlight acts as a catalyst to induce oxygen from the air to combine with these bonds.This process causes deterioration of rubber tires. Tires should not be exposed to direct sunlight during extended periods of storage because of this breakdown process (Figure 3.11).

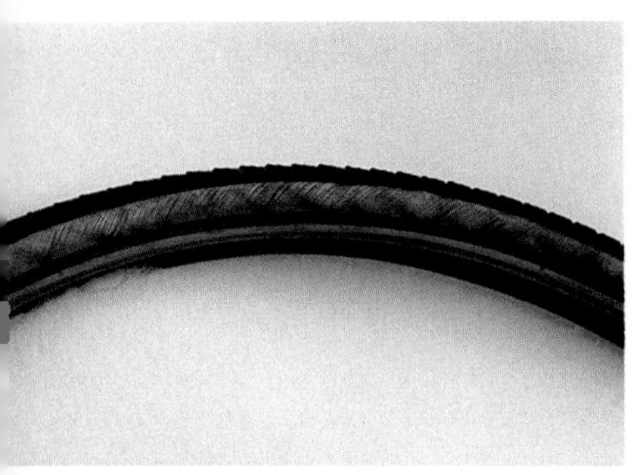

FIGURE 3.11
The "checking" or deterioration of tires is caused by oxidation of the double bonds in the rubber catalyzed by sunlight and atmospheric pollutants.

The type of oxidation that may occur depends on the strength of the oxidizing agent. In this important type of reaction type, it is known that, when a dilute water solution of potassium permanganate ($KMnO_4$) reacts with an alkene, —OH groups are added to the double bond.

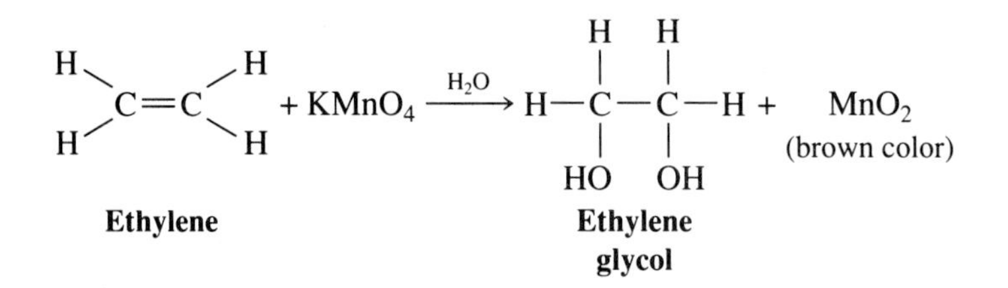

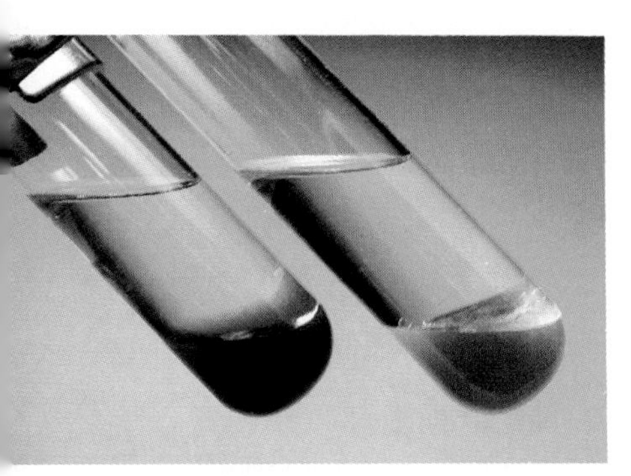

The purple of the permanganate solution is quickly discharged during the reaction with an alkene. The alkene is converted to a diol.

The result is a dihydroxy alcohol, commonly called *ethylene glycol*, a well-known antifreeze. The visual formation of the brown precipitate is sometimes used as a test for alkenes.

An oxidation reaction with an alkene can be generalized as follows.

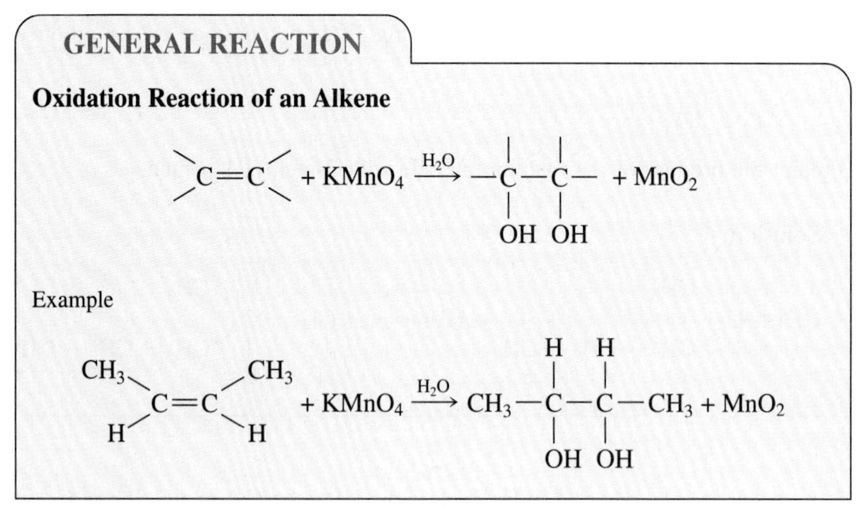

In Chapter 5 we will see another important oxidation reaction of alkenes, the epoxidation reaction.

TABLE 3.1
Some Common Polymers

Monomer	Polymer	Uses
Ethylene $CH_2{=}CH_2$	Polyethylene $-[CH_2-CH_2]_n-$	Plastic film, plastic pipe, plastic containers
Propylene $CH_3CH{=}CH_2$	Polypropylene $-[CH(CH_3)-CH_2]_n-$	Molded plastic objects, rope, synthetic fibers
Vinyl chloride $CH_2{=}CHCl$	Polyvinyl chloride (PVC) $-[CH_2-CHCl]_n-$	Plastic pipe, plastic containers, plastic rope
Dichloroethylene $CH_2{=}CCl_2$	Polydichloroethylene $-[CH_2-CCl_2]_n-$	Plastic food wrap, tubing
Acrylonitrile $CH_2{=}CHCN$	Polyacrylonitrile $-[CH_2-CH(CN)]_n-$	Orlon, fabrics
Tetrafluoroethylene $CF_2{=}CF_2$	Polytetrafluoroethylene $-[CF_2-CF_2]_n-$	Teflon

Polymerization of Alkenes

♦ Polymers are large, chained molecules that can be formed from alkenes.

Polymers are very large molecules made up of smaller and simpler units called **monomers**. Polymers are a classification (Table 3.1) that includes natural and synthetic molecules. The natural polymers, including starch, proteins, DNA and RNA, will be discussed in later chapters of this book. We now focus on the synthetic polymers, particularly those made from alkenes.

Many cooking utensils with "nonstick" surfaces are coated with a polymer such as Teflon.

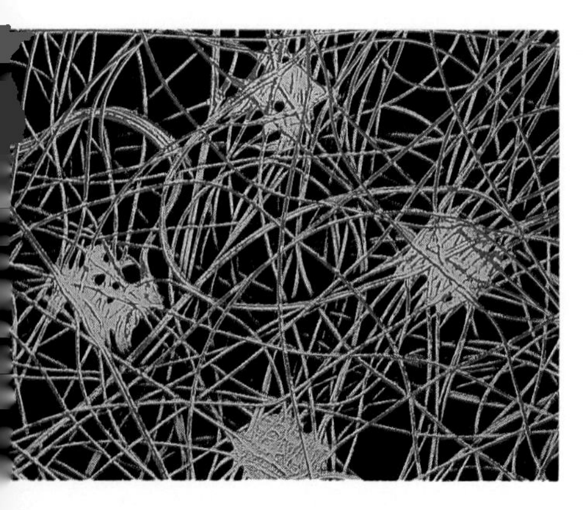

The rigidity of polypropylene can be seen in the complex molecular network.

The simplest polymer is polyethylene, which is made up of a large number of ethylene monomers. By convention we suggest that a large number (n) of ethylene molecules are connected to form a polymer.

$$CH_2{=}CH_2 \longrightarrow CH_3{-}CH_2{-}CH_2{-}CH_3 \longrightarrow CH_3{-}CH_2{-}(CH_2{-}CH_2)_n{-}CH_2CH_3$$

One 2-carbon unit | **Two 2-carbon units** | **(n + 2) 2-carbon units (polyethylene)**

Polytetrafluoroethylene is a remarkable material. Most of us know it as the coating on "nonstick" cooking utensils marketed as Teflon, among other names. Why is it so inert at high temperatures? In the previous chapter we saw that simple hydrocarbons are relatively nonreactive because all of their bonds are single bonds. There are no "loose" electrons. The C—C single bond is quite strong. Because of the high electronegativity of fluorine, the C—F bond is very strong. In many respects the polytetrafluoroethylene molecule looks like polyethylene, except that it has F atoms in place of H atoms. The F atoms form a protective cover over the internal backbone of carbon atoms. Thus, the normally strong C—C system is additionally protected.

Polymers are challenging nature in a number of ways. We have all noticed the way that plastic is replacing wood and metal in such things as furniture and construction materials. Down filling for jackets, once considered the ultimate in warmth and comfort, is rapidly being replaced with synthetic polymers. These materials can be processed to forms that provide the air-trapping spaces that provide the warmth of insulation. The new materials have the added benefits of not being allergenic. A scanning micrograph shows the structural similarities between natural and synthetic fibers.

Of current concern is how to dispose of plastics. Most plastics do not degrade naturally (biodegrade) in the environment. Many people believe that the best way to dispose of plastic is to burn it and recover the energy that was "sidetracked" when the petroleum was made into plastic, but obvious pollution problems are associated with this approach. A current area of interest is the preparation of new plastic materials that will biodegrade.

NEW TERMS

hydrogenation The addition of hydrogen to an alkene or alkyne. An H is added to both carbon atoms of the double or triple bond, and a pi bond is lost.

halogenation The addition of halogen to an alkene or alkyne. A halogen is added to both carbon atoms of the double or triple bond, and a pi bond is lost.

hydration The addition of water to an alkene or alkyne. An H is added to one of the carbon atoms of the double or triple bond, and an OH is added to the other carbon atom. A pi bond is lost.

Markovnikov's rule The generalization used to account for the way an unsymmetrical reagent adds to an unsymmetrical alkene. The positively charged reagent (often H^+) adds to the carbon atom directly attached to the greater number of hydrogens.

polymer A complex compound resulting from the end-to-end union of a large number of smaller units (monomers)

monomer The smallest repeating unit from which polymers are made.

TESTING YOURSELF

Reactivity of the Alkenes

1. Predict and name products for each of these hydrogenation reactions:

a. $$CH_3-\overset{H}{\overset{|}{C}}=\overset{H}{\overset{|}{C}}-CH_2-CH_3 + H_2 \xrightarrow[\text{(catalyst)}]{\text{Pt}}$$

b. $$(CH_3)_2C=C(CH_3)_2 + H_2 \xrightarrow[\text{(catalyst)}]{\text{Pt}}$$

c. $$CH_3-\underset{CH_3}{\overset{CH_3}{C}}-\overset{H}{\overset{|}{C}}=CH_2 + H_2 \xrightarrow[\text{(catalyst)}]{\text{Pt}}$$

2. Predict and name the products of the following reactions of alkenes with halogens:

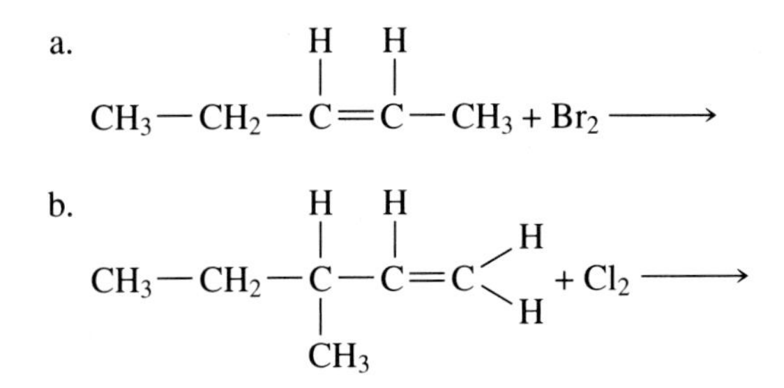

3. Predict products for the following reactions of alkenes with nonsymmetric reactants:

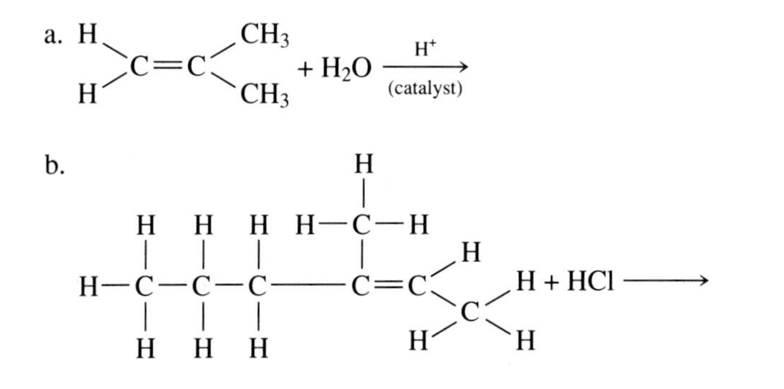

Answers **1. a.** pentane **b.** 2,3-dimethylbutane **c.** 2,2-dimethylbutane
2. a. 2,3-dibromopentane **b.** 1,2-dichloropentane

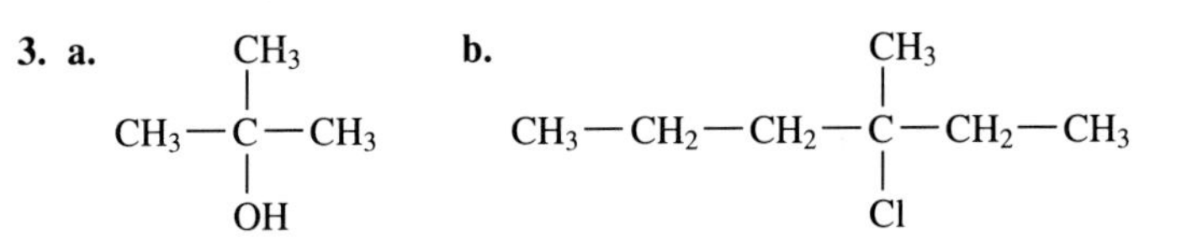

3.5 Chemical Reactivity of Alkynes

Alkynes undergo all of the reactions characteristic of pi-bonded hydrocarbons discussed in Section 3.4, except that there are twice as many pi bonds to react. (Recall that the triple bond has one sigma bond and *two* pi bonds.) Markovnikov's rule predicts the outcome of reactions with nonsymmetric alkyne reactants just as it does for the alkenes.

$$H-C\equiv C-CH_2-CH_2-CH_3 + 2\,HBr \longrightarrow CH_3-\underset{Br}{\overset{Br}{\underset{|}{\overset{|}{C}}}}-CH_2-CH_2-CH_3$$

1-Pentyne **2,2-Dibromopentane**

In the case of hydrogen addition in alkynes, it is possible, by use of selective catalysts, to stop an addition to a triple bond when only one of the pi bonds has reacted. This produces a double-bonded material with definite cis and trans properties. This procedure has value in artificial synthesis of biological compounds, such as insect sex attractants (pheromones).

EXAMPLE 3.5

Predict the product of the reaction of this alkyne in the presence of hydrogen.

Solution: $$C_5H_{11}-C\equiv C-CH_2-CH_2-\underset{\underset{H}{\|}}{C}-C_{10}H_{21} \xrightarrow[\text{(catalyst)}]{H_2}$$

$$\underset{C_5H_{11}}{\overset{H}{}}\!\!>C=C<\!\!\underset{CH_2-CH_2-\underset{\underset{H}{\|}}{C}-C_{10}H_{21}}{\overset{H}{}}$$

The product is the pheromone for the Douglas fir tussock moth. The pheromone is artificially produced and released into tree farms and woodlands to control the reproductiveness of this insect and thereby to limit the damage it causes.

The simplest alkyne, acetylene, is an extremely useful industrial chemical. Many other compounds are made using it as a starting material. Acetylene can be made by reacting carbon (coke) with lime (calcium oxide) in the presence of heat (Δ):

$$2\,C + CaO \xrightarrow[(2500°C)]{\Delta} CaC_2 + CO$$

Calcium carbide **Carbon monoxide**

When calcium carbide is placed in water, the following reaction takes place:

$$CaC_2 + 2\,H_2O \longrightarrow H{-}C{\equiv}C{-}H + Ca(OH)_2$$

The acetylene thus liberated can be burned or can be used as a raw material for synthesis of a more complicated compound.

$$2\,H{-}C{\equiv}C{-}H + 5\,O_2 \longrightarrow 4\,CO_2 + 2\,H_2O$$

The light of one kind of headlamp used by miners and cave explorers is given off by the combustion of acetylene.

Early automobile headlamps and miners' helmet lamps used this source of acetylene for their fuel. Miners had to give up these kinds of lanterns because they sometimes caused explosions in mines that were naturally filled with methane.

The alkynes are not as widely distributed in nature as are the alkenes. However, several synthetic materials having the alkyne functional group have had a major influence on society. Synthetic derivatives of the natural female sex hormones estrogen and progesterone are the major constituents of the widely used oral contraceptive known as the "pill." A common feature of these synthetic materials is the incorporation of an acetylene group.

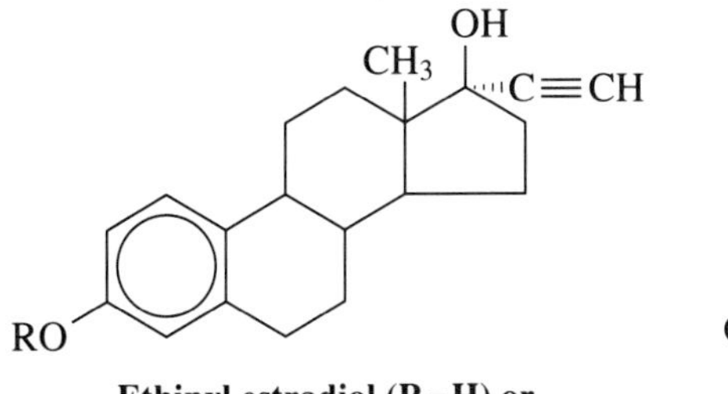

Ethinyl estradiol (R = H) or mestranol (R = CH_3)

OH, CH_3, C≡CH, O

Norethynodrel

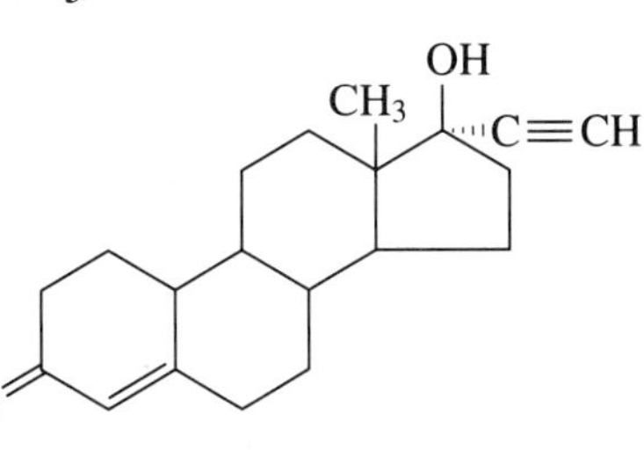

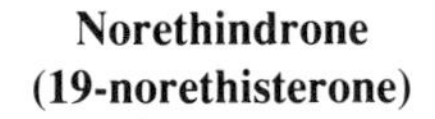

Norethindrone (19-norethisterone)

TESTING YOURSELF

Alkyne Reactions

1. Give products for the following reactions
   a. 2-hexyne + Br_2 (excess) $\longrightarrow$
   b. 1-octyne + H_2 $\xrightarrow{\text{(catalyst)}}$
   c. 1-octyne + HCl (excess) $\longrightarrow$

Answers 1. a. 2,2,3,3-tetrabromohexane **b.** 1-octene is the first product. Under controlled conditions, the reaction can stop here, but with excess hydrogen the alkene can be further reduced to octane. **c.** By Markovnikov's rule we would predict the H to add to C-1 and the Cl to C-2. The product from addition of two HCl molecules is 2,2-dichlorooctane.

3.6 Polyunsaturated Alkenes

A hydrocarbon containing more than one double bond is a *polyunsaturated alkene*. For most polyunsaturated alkenes, the chemical reactions already discussed can be easily applied. The major difference is that more reagent can be added (note the addition of *two* bromine atoms in the example below).

EXAMPLE 3.6

Show the products of these reactions with polyunsaturated alkenes.

Solution:

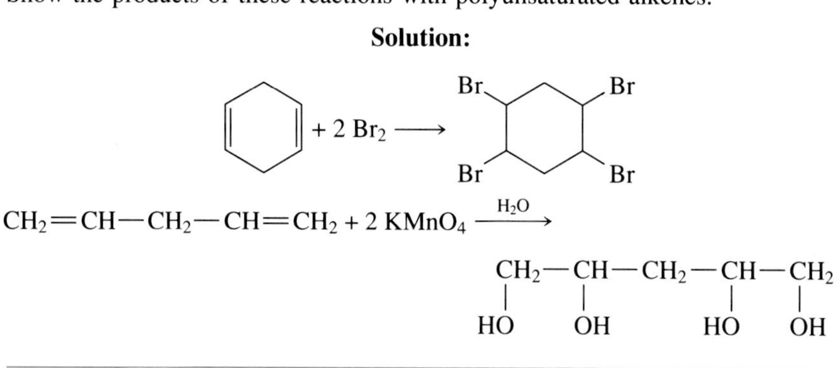

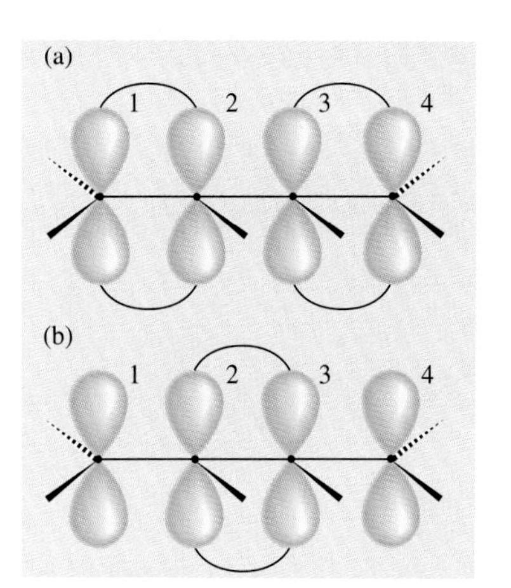

FIGURE 3.12
A conjugated polyene. (a) The orbitals of 1,3-butadiene. (b) A possibility of *p*-orbital overlap exists between the central two *p* orbitals.

♦ There is electron delocalization in polyunsaturated systems of alternating double and single bonds.

However, a special class of polyunsaturated compounds exists called the **conjugated polyenes**. *Conjugation* means that there are alternating double and single bonds. Why are these special? This is most readily seen by examining the simplest conjugated polyene, 1,3-butadiene. In Figure 3.12 we can see the orbitals of this molecule and gain insight into the special nature of a conjugated system.

It is easy to see the overlap of the *p* orbitals between C-1 and C-2 that give one of the double bonds and between C-3 and C-4 that give the second double bond. But why can't there also be overlap between C-2 and C-3? Actually, there is. The electrons in a conjugated system are *not* localized between specific atoms, but are rather spread out or *delocalized* over the entire pi framework. This delocalization of electrons leads to greater stability—a concept that will be seen again in the next chapter.

NEW TERM

conjugated polyene A polyene for which there are alternating double and single bonds

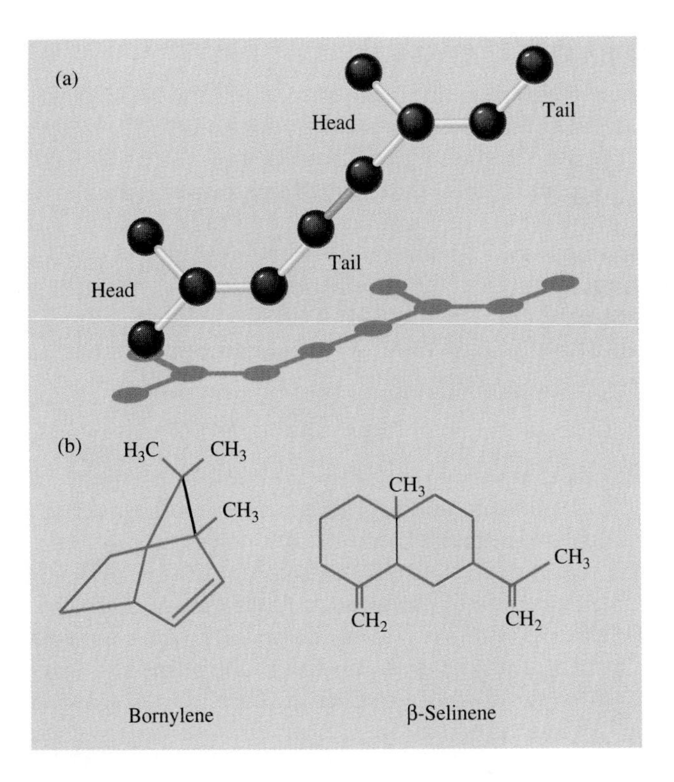

FIGURE 3.13
Structure of terpenes. (a) Isoprene units are often incorporated into terpenes in a head-to-tail fashion, as shown here. (b) Examples of two terpenes showing the head-to-tail arrangement of isoprene units.

3.7 Interesting Unsaturated Compounds

Many important natural products and biologically significant synthetic molecules have unsaturated compounds as part of their structure. A number of these molecules will be discussed in greater detail in other chapters. Presented here are a few molecules that represent some of the many unsaturated compounds found in daily life, including compounds that make up turpentine, vitamin A, fats and oils in our diet, and important chemical communicators called pheromones.

Terpenes

The large class of naturally occurring compounds known as **terpenes** are well represented as flavorings and fragrances. The terpenes are characterized by their carbon framework, which appears to be made up of repeating isoprene units. We should note here that it is *not* the positioning of double bonds that is important, but rather the general arrangement of carbon atoms. These are often found in a head-to-tail fashion (Figure 3.13).

A number of terpenes are common to our experience, including pine oil, lemon oil (Figure 3.14), and the oils in celery and ginger. A synthetic compound, isotretinoin, structurally related to vitamin A (shown below) is now marketed under the name Accutane. It is used for aggressive treatment of acne. Although

CHEMISTRY CAPSULE

Terpenes and van Gogh

The van Gogh legacy is the great art that he left behind. But we know that Vincent van Gogh was a troubled man who sometimes exhibited bizarre behavior. Recent medical sleuthing has come up with a "chemical" explanation for much of his behavior—he was addicted to terpenes. As a heavy drinker of absinthe flavored with wormwood, he ingested large amounts of the terpene *thujone*.

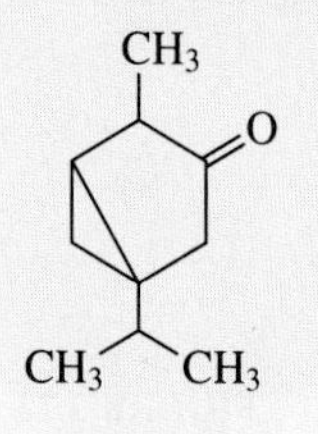

Thujone

This compound is now thought to have caused many of the disorders known to have troubled van Gogh. He apparently also became dependent on sniffing the odors of terpenes that he encountered every day in the turpentine so common to painters. This addiction finally reached the stage where he required camphor to be in his pillows and mattress in order to sleep. Camphor is also known to cause many of the same symptoms as thujone. As a result of his addiction to terpenes, he was even known to eat the paint and turpentine mixtures used in his art. Before he was finally committed to an asylum, it is reported that he had to be restrained from drinking a quart of turpentine.

FIGURE 3.B
Van Gogh, the artist.

the mechanism of action is not understood, the compound has been shown to be very effective in reducing skin oils responsible for acne. The product is a potent teratogen (causes abnormalities in the fetus if used during pregnancy). However, it is not mutagenic.

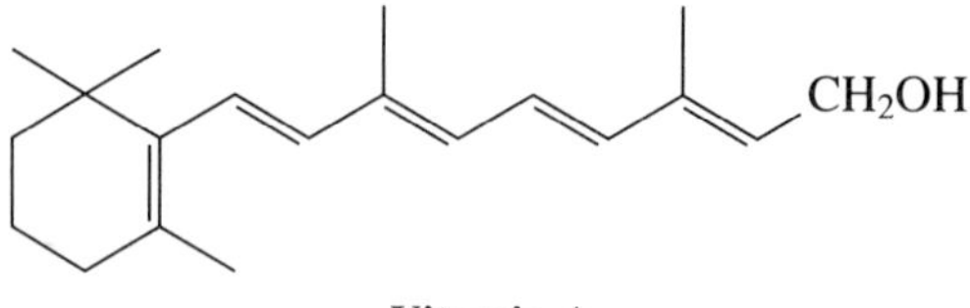

Vitamin A

Another analog, Retin-A, stimulates cell growth and is also used for treatment of acne. Some think it is also of cosmetic value in reducing wrinkles.

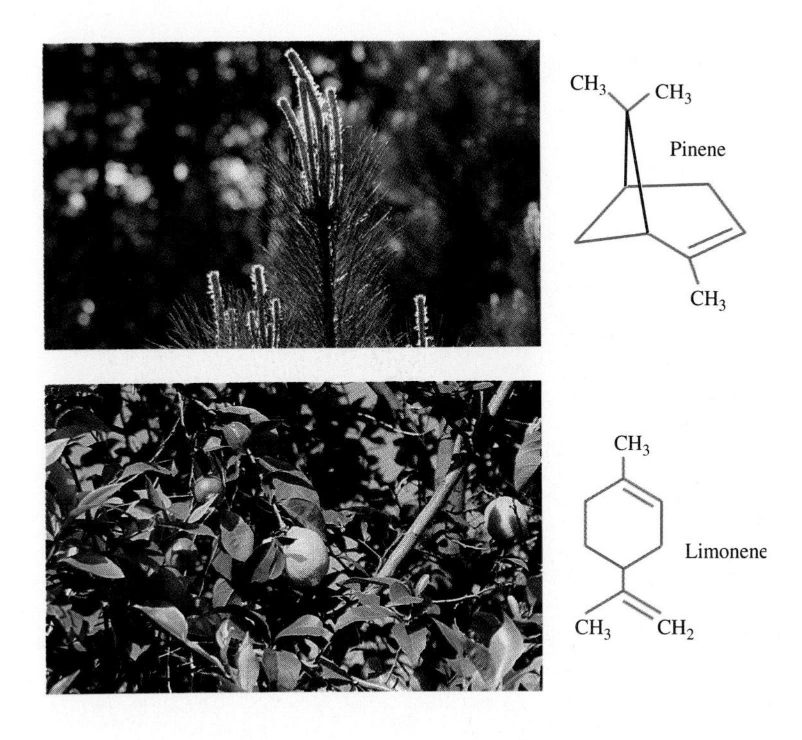

FIGURE 3.14
Two sources of common terpenes. (a) Pine oil (pinene). (b) Lemon oil (limonene).

Pheromones

Naturally occurring compounds used for communication by plants and animals are called **pheromones**. Some of these compounds signal for alarm, some for paths to food, and some for purposes of mating. Many of these pheromones are rich in unsaturation. An example of a pheromone is the sex attractant for the silkworm moth shown here.

$$CH_3{-}CH_2{-}CH{=}CH{-}CH_2{-}(CH_2)_8{-}CH_2OH$$

Note that this compound has only one double bond; however, even this solitary double bond makes all the difference between a compound that is biologicaly active and one that is not.

♦ The pheromone of the common house fly.

$$\begin{array}{c} CH_3{-}(CH_2)_6{-}CH_2 \quad\quad CH_2{-}(CH_2)_{11}{-}CH_3 \\ \diagdown \quad\quad \diagup \\ C{=}C \\ \diagup \quad\quad \diagdown \\ H \quad\quad\quad\quad H \end{array}$$

NEW TERMS

terpene A naturally occurring compound that contains isoprene units

pheromone A naturally occurring compound produced by an organism for the purpose of chemical communication

CHEMISTRY CAPSULE

Carrots Help You See

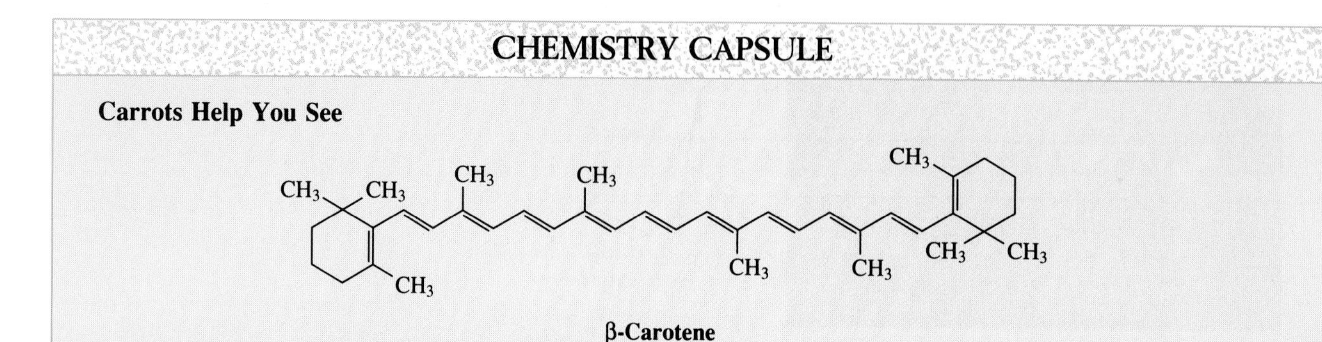

β-Carotene

A half cup of cooked carrots contains over 7000 IU (international units) of carotene, a compound that is very important to our well-being. Carotene is very efficient as an antioxidant; it even seems to decrease the incidence of lung cancer for smokers who take high doses of carotene.

In the liver, carotene is converted to vitamin A, which in turn, is converted to retinal, the major molecule that enables vision. Perhaps this relationship helps us to better appreciate the wisdom of the folk medicine concept of eating carrots to improve vision.

Light converts 11-*cis* retinal to all-*trans* retinal, and this conversion then enables the sensation of vision.

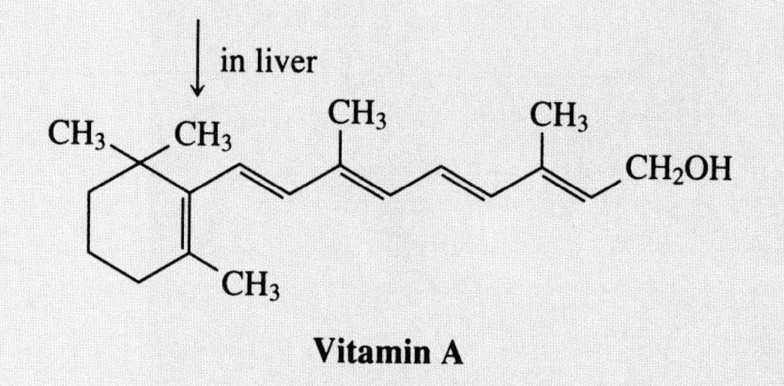

Vitamin A

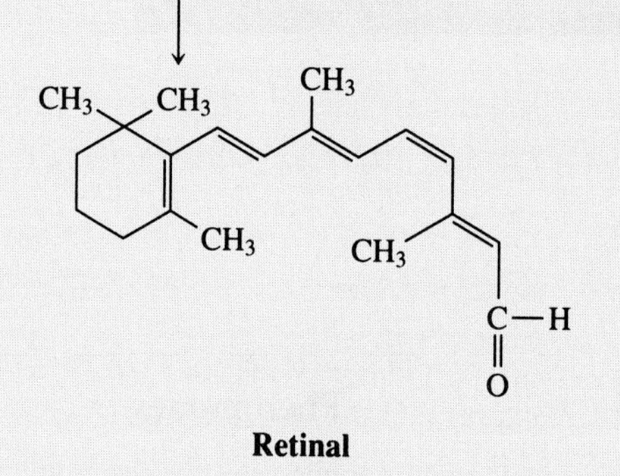

Retinal

FIGURE 3.C
Carrots and other common foods are rich sources of carotene. Carrots can really help you see!

Summary

An *unsaturated* molecule does not have enough atoms to provide four bonds for all of the carbon atoms. Carbon atoms in this situation form one or more multiple carbon–carbon bonds. There is only room enough between two carbon atoms for one sigma bond. When more than one bond forms between two adjacent atoms, the second and third bonds must use space above and below and on either side of the sigma bond. These bonds involve unhybridized *p* orbitals and are called *pi* (π) *bonds*. Since pi-bonding electrons are not shielded between the

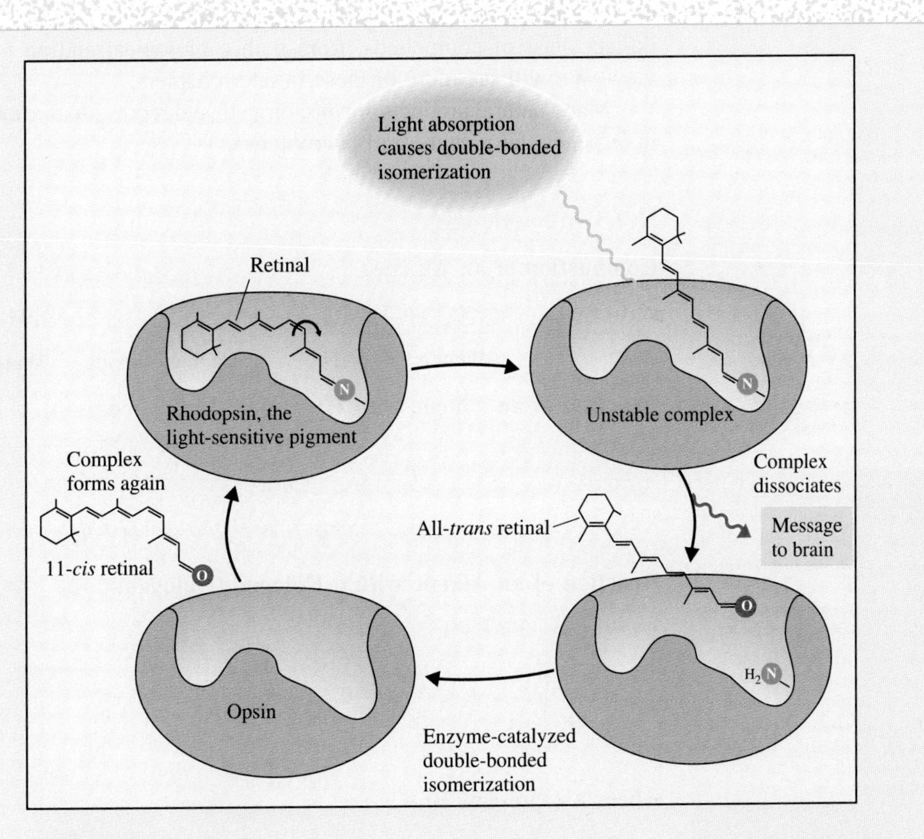

FIGURE 3.D
Light energy converts a cis double bond of retinal to a trans double bond as part of the process involved in vision.

bonding atoms but are highly visible from outside the molecule, they are readily susceptible to chemical attack by other molecules or ionic groups. Hydrocarbons with pi bonds are highly reactive compared with the sigma-bonded alkane family. Systems for naming double- and triple-bonded hydrocarbons (*alkenes* and *alkynes*, respectively) follow the same procedure as developed for the single-bonded alkane family, with the exception of procedures to designate and locate the multiple bond(s).

Polymerization is a characteristic of compounds that have one or more multiple bonds between adjacent carbon atoms. *Polymers* are an extremely important class of compounds, from both a biological and an industrial point of view. We will see more of these in later chapters.

Many natural products are unsaturated, and representative examples are found in terpenes, fatty acids, and pheromones.

Reaction Summary*

Combustion of an Alkene

$$C_nH_{2n} + 1.5n\ O_2 \longrightarrow n\ CO_2 + n\ H_2O$$

Alkene **Oxygen** **Carbon dioxide** **Water**

Reaction of an Alkene with Hydrogen (Hydrogenation)

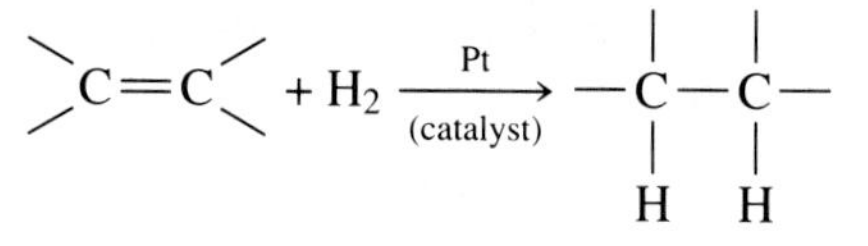

Reaction of an Alkene with a Halogen (Halogenation)

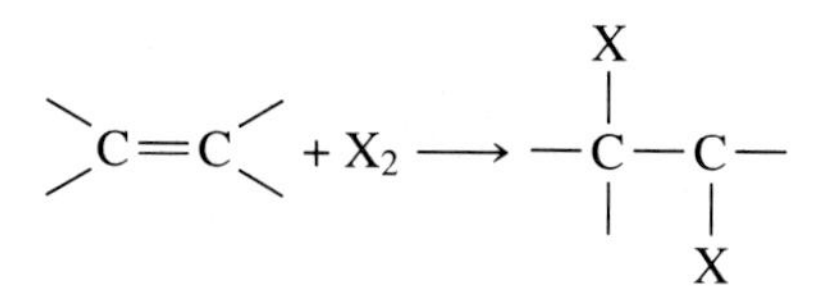

where X represents Br, Cl

Addition of HX to an Alkene

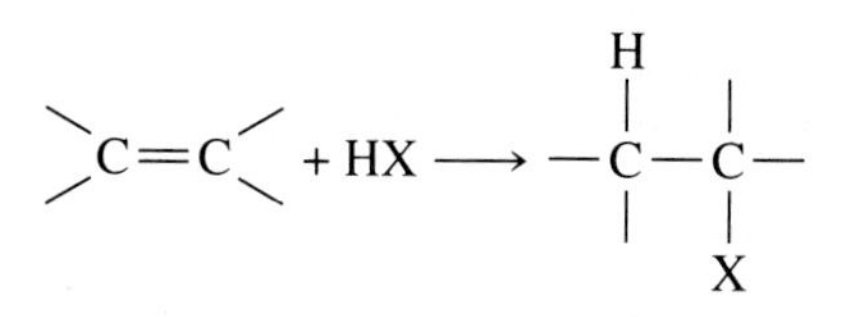

where X represents Br, Cl. The H adds to the carbon with the most hydrogens.

Addition of Water to an Alkene (Hydration)

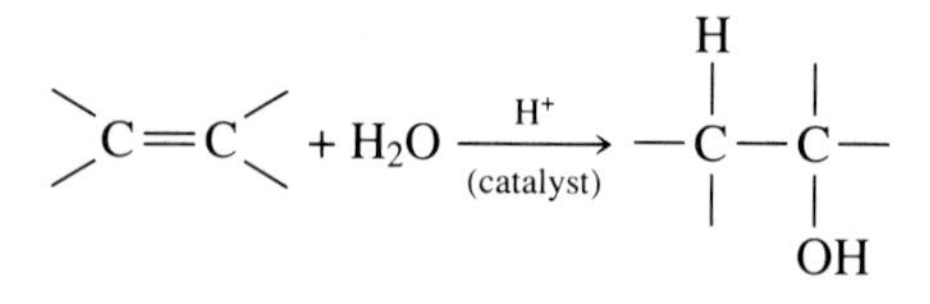

The H adds to the carbon with the most hydrogens.

---

* The reactions of the alkynes are the same as those of the alkenes, except that (1) 2 mol of reactant add to the triple bond and (2) the reaction of water does not apply to alkyne chemistry.

Terms

unsaturated (3.1)
alkene (3.1)
alkyne (3.1)
pi bond (π-bond) (3.2)
polyunsaturated (3.3)
hydrogenation (3.4)
halogenation (3.4)
hydration (3.4)
Markovnikov's rule (3.4)
polymer (3.4)
monomer (3.4)
conjugated polyene (3.6)
terpene (3.7)
pheromone (3.7)

Exercises

Functional Groups (Objective 1)

1. What is meant by an "unsaturated" hydrocarbon?
2. What are the differences between an alkene and an alkyne?

Structure of Unsaturated Hydrocarbons (Objective 2)

3. What is the hybridization of the carbon atoms in an alkene bond?
4. Describe the three bonds of the alkyne "triple bond."
5. What is the hybridization of the carbon atoms in the alkyne bond?
6. Describe the structural features of alkenes and alkynes.

Consequences of Unsaturation (Objective 3)

7. Explain why there is no rotation about a C=C double bond.
8. What do we mean by cis and trans isomers of alkenes?
9. Show how the cis and trans concept for alkenes and cycloalkanes is similar.
10. Why aren't there cis and trans isomers for alkynes?
11. Why can't cyclopentyne exist?

Nomenclature (Objective 4)

12. Give structures for the following compounds:
   a. cyclopentene
   b. *cis*-3-heptene
   c. 4-octene
   d. *trans*-2-pentene
   *e. *trans*-3,4-dimethylcyclohexene
13. Give an appropriate IUPAC name for each of the following compounds:

   a. 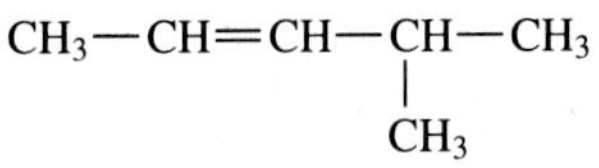 $CH_3-CH=CH-CH(CH_3)-CH_3$

   b. $CH_2=CH-CH_3$

   c. CH_3

   d.

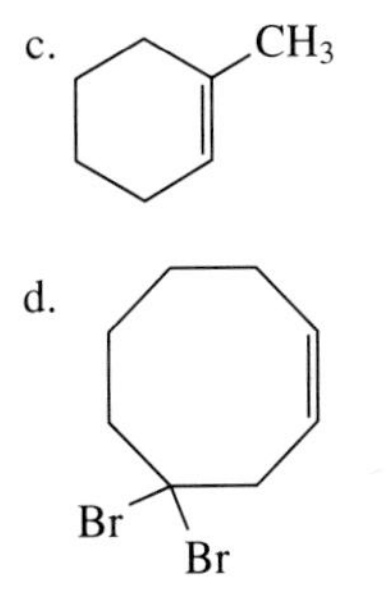

14. Give an appropriate IUPAC name for each of the following alkenes:

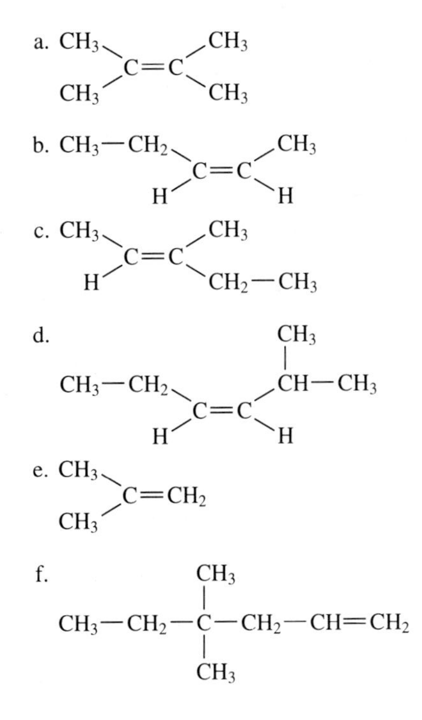

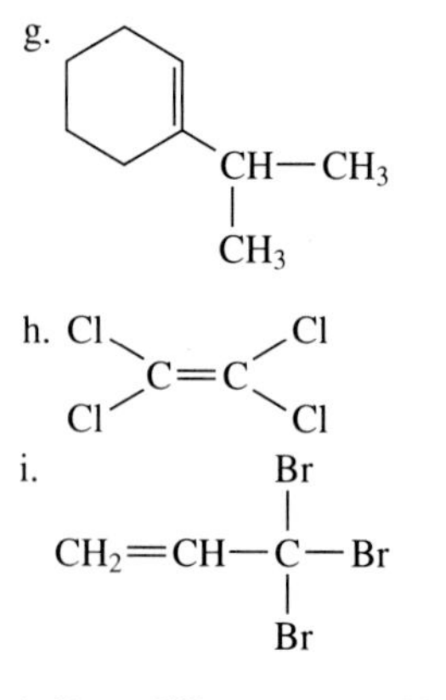

j. $Br-CH_2(H)C=C(H)CH_2-CH_2-CH_3$

15. Give a structure that corresponds to each of the following names:
    a. 5-methyl-*cis*-3-heptene
    b. 5,5-dimethyl-1-octyne
    c. 2-ethyl-1,3,3-trimethylcyclohexene
    d. 3,3-dimethyl-1-hexyne
    e. 5,5-dibromo-2-hexyne
    f. 1,1,1-trichloro-*cis*-3-octene
    g. *cis*-3,4-dimethylcyclohexene
    h. 3,4-dimethyl-*cis*-3-octene
    i. *trans*-2,2-dimethyl-3-nonene
    j. 2,2-dimethyl-2,2-4-decyne
    k. 3,3-dimethylcyclohexene
    l. 3,4-dimethyl-*trans*-3-hexene
    m. 1,3,4,4-tetramethylcycloheptene
16. Give an IUPAC name for each of the following compounds:

a. $CH_3-C\equiv C-C(CH_3)_2-CH_3$

b. $Cl-CH_2-C\equiv C-H$

c. CH_3-cyclopropenyl

d. $CH_3(Br)C=C(Br)CH_2-CH_2-CH_3$

e.

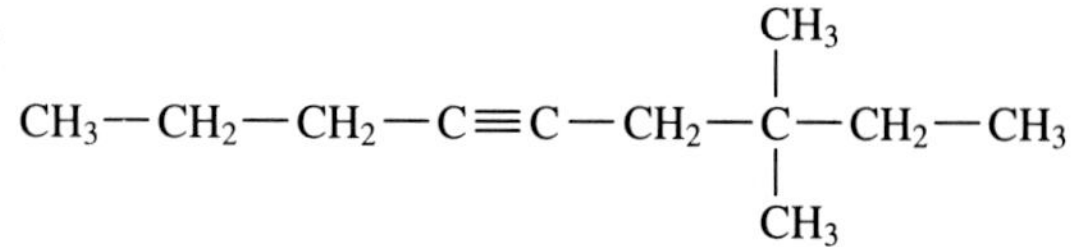

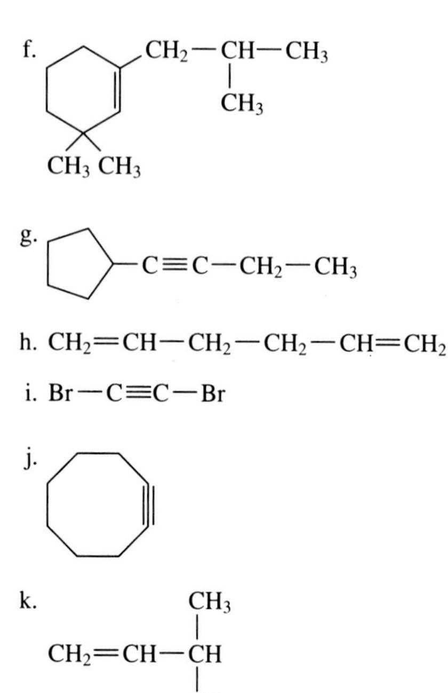

k. $CH_2=CH-CH(CH_3)-CH_3$

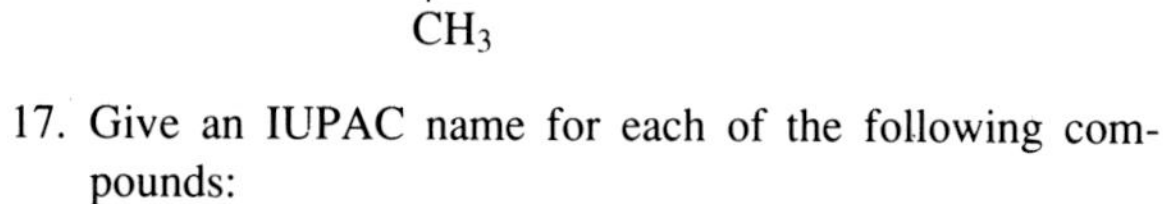

17. Give an IUPAC name for each of the following compounds:

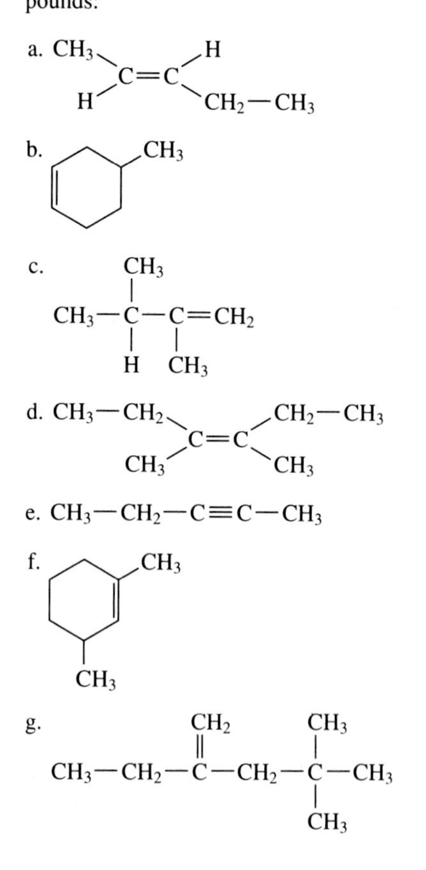

g. $CH_3-CH_2-C(=CH_2)-CH_2-C(CH_3)_2-CH_3$

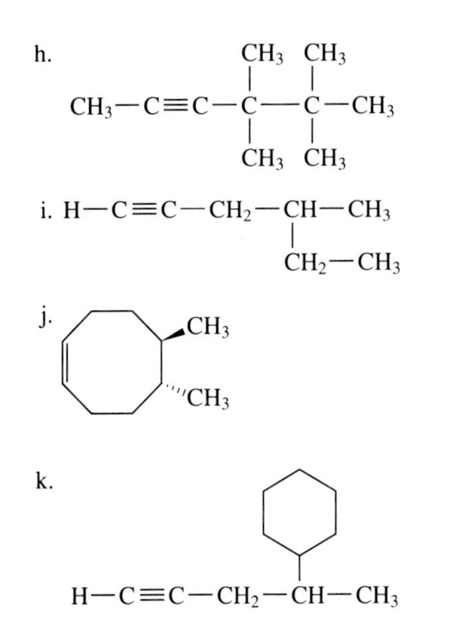

18. Give an IUPAC name for each of the following compounds:

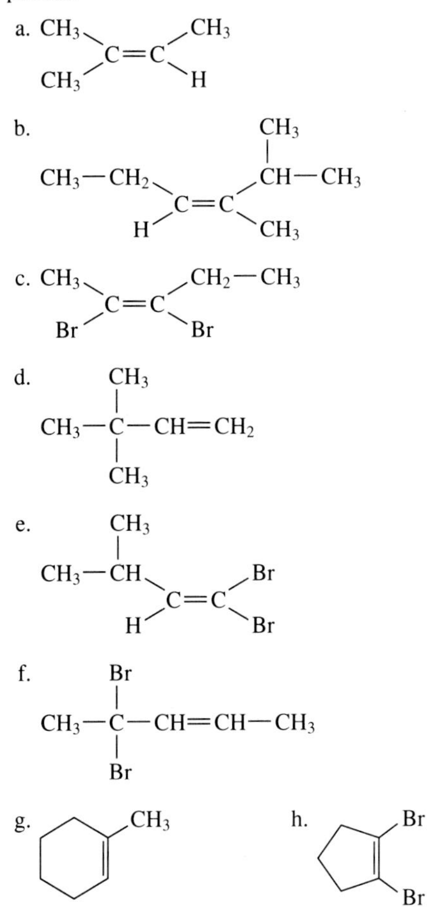

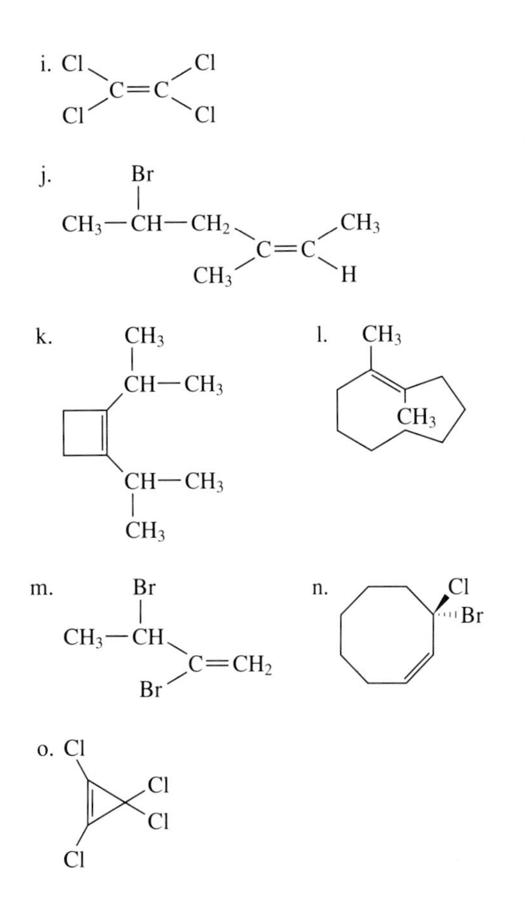

Reactivity and Reactions (Objective 5)

19. Why is ethylene more reactive to an acid than ethane?
20. Which type of bond, sigma or pi, is more reactive to an electrophile? Why?

Reactions (Objective 6)

21. What are the combustion products of an alkene?
22. What are the general products from reaction of any alkene with hydrogen and a catalyst?
23. With excess hydrogen and a catalyst, an alkyne is converted to what?
24. What is the general course of reaction of any alkene with bromine?
25. When hydrogen adds to an alkene, the carbon in the double bond changes hybridization from what to what?
26. What is Markovnikov's rule?
27. If H^+ reacts with 1-pentene, it becomes attached to carbon number what?
28. Any alkene reacts with dilute permanganate. What is the reaction at the site of the double bond?

29. Give the structures of products resulting from the following reactions:
    a. 4-ethyl-4-decene + HBr
    b. 1,2-dimethylcyclopentene + Cl_2
    c. 2-butyne + excess H_2 and catalyst
    d. cyclohexene + water and a trace of acid for catalyst
    e. cyclopentene + dilute permanganate
    f. 1,2-dimethylcyclohexene + hydrogen gas and catalyst

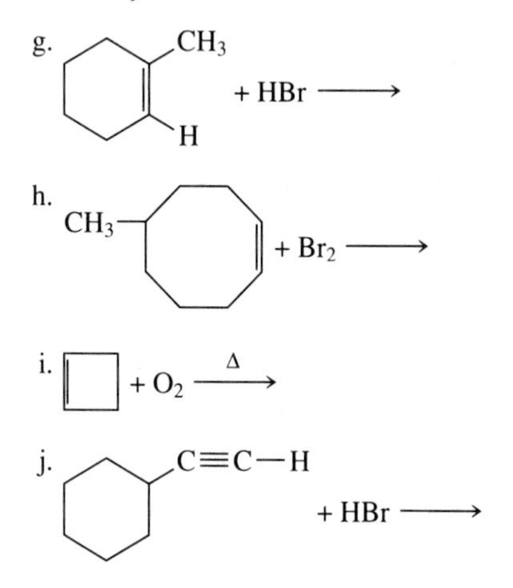

30. Give structures of products from the following alkene reactions:

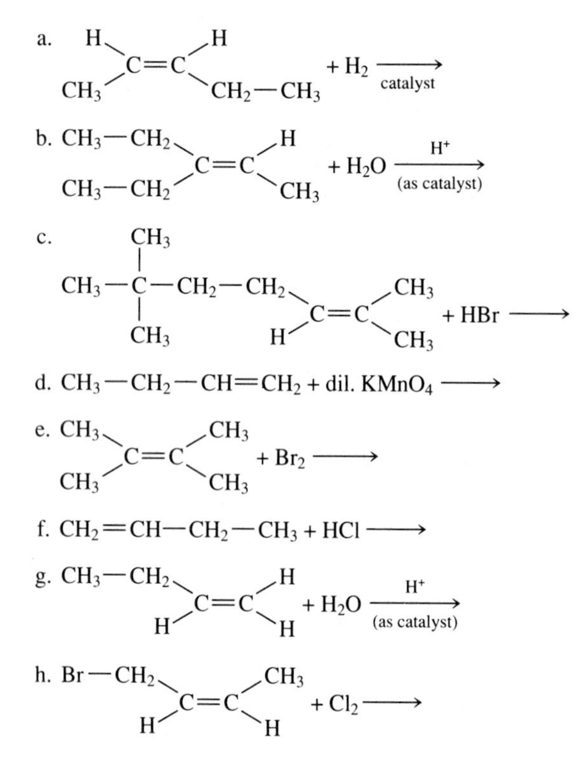

i. $CH_2{=}CH{-}CH_3 + H_2 \xrightarrow[\text{catalyst}]{}$

31. Give structures of products from the following cyloalkene reactions:

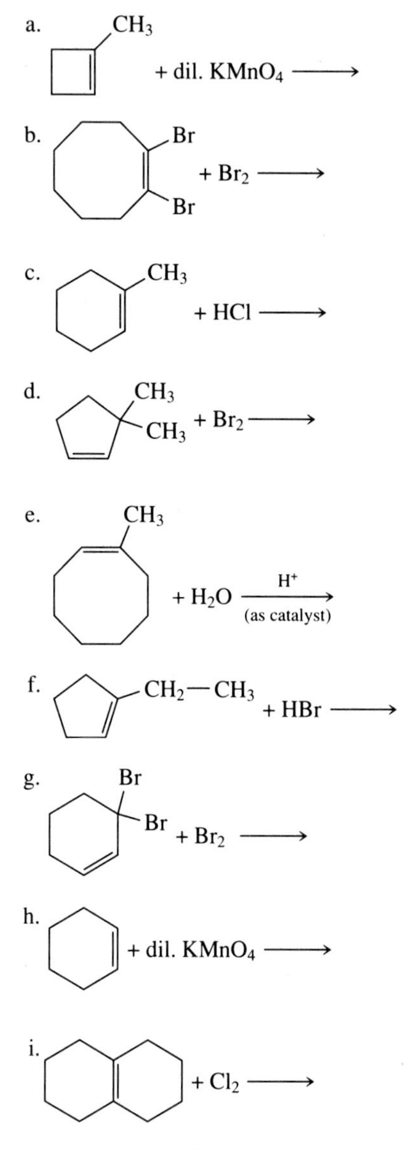

32. Give structures of products from the following alkyne reactions:
    a. $CH_3{-}C{\equiv}C{-}H$ + excess HBr $\longrightarrow$

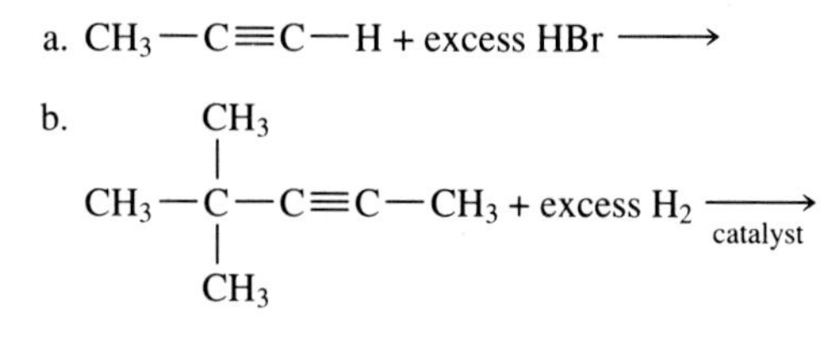

c. $CH_3-C\equiv C-CH_3$ + excess $Br_2 \longrightarrow$

d. $CH_3-CH_2-CH_2-C\equiv C-H$ + excess HCl $\longrightarrow$

e. $CH_3-C\equiv C-CH_3$ + excess $H_2 \xrightarrow[\text{catalyst}]{}$

f. $(CH_3)_2CH-C\equiv C-H$ + excess HBr $\longrightarrow$

Materials Based on Unsaturated Hydrocarbons (Objective 7)

33. What is a polymer?
34. If a polymer has the general structure

$$—[CCl_2—CCl_2]_n—$$

what is its basic building block?

35. Outline the isoprene units in the following terpenes:

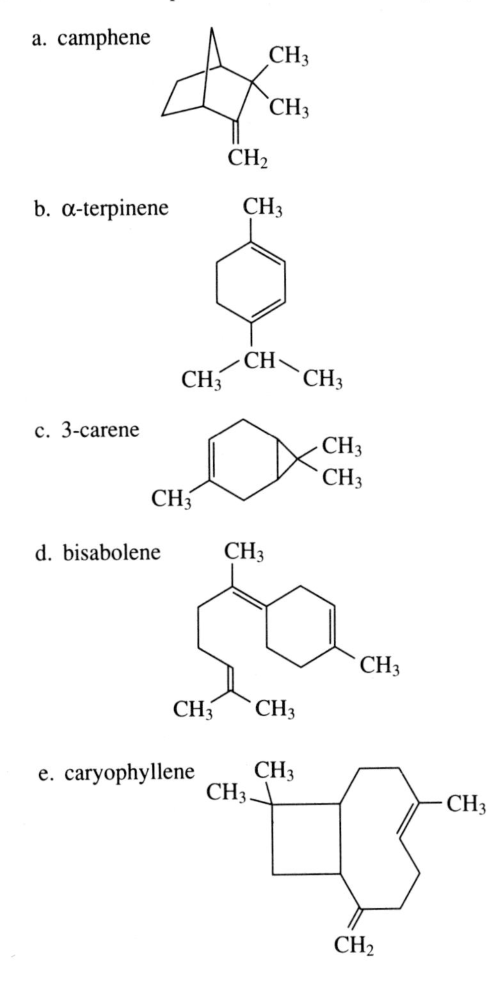

Unclassified Exercises

36. The carbon atoms involved in the double bond of an alkene have which hybridization?
   a. *sp* b. sp^2 c. sp^3
37. The most reasonable name for

$$(CH_3)(CH_3-CH_2)C=C(CH_2-CH_2-CH_3)(CH_3)$$

   is
   a. *cis*-2-ethyl-3-methyl-2-hexene
   b. *trans*-2-ethyl-3-propyl-2-butene
   c. *trans*-1,4-dimethyl-3-heptene
   d. *trans*-3,4-dimethyl-3-heptene
38. The higher reactivity of an alkene or alkyne, relative to an alkane, is due to
   a. sigma bonds b. pi bonds c. hydrogen bonds
   d. none of these
39. The bond angle associated with the hybrid orbitals of a carbon involved in a triple bond is
   a. 180° b. 120° c. 109° d. 45°
40. In the reaction of 1-butene with HCl, the H of the HCl becomes attached to carbon number
   a. 1 b. 2 c. 3 d. 4
41. If you have a polymer of the general structure

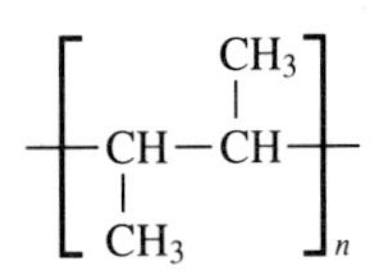

   the monomer is most likely to be:
   a. 1-propene b. 1-butene c. 2-butene d. 2-pentene
42. Combustion of an alkene with sufficient oxygen produces
   a. carbon dioxide and water
   b. carbon monoxide and water
   c. only carbon dioxide
   d. only carbon monoxide
43. Reaction of cyclohexene with bromine gives
   a. 1,2-dibromocyclohexene
   b. 1,1-dibromocyclohexane
   c. 1,2-dibromocyclohexane
   d. 1,1,2,2-tetrabromocyclohexane
44.

$$CH_3-C(CH_3)_2-C\equiv C-H$$

   is best named by IUPAC as
   a. isobutyl acetylene b. 2,2-dimethyl-3-butyne
   c. 3,3-dimethyl-1-butyne

AROMATIC HYDROCARBONS

Modern techniques allow us to look at benzene molecules.

OUTLINE

4.1 Structure: Resonance and Electron Delocalization

4.2 Nomenclature

4.3 Aromatic Reactions: A Consequence of Overlapping Pi Bonds

4.4 Fused-Ring Aromatic Systems

4.5 Important Aromatic Hydrocarbons

OBJECTIVES

After completing this chapter, you should be able to

1. Recognize conjugated double bonds, and explain why they differ in bond strength from normal double bonds.
2. Recognize resonance structures, and explain what is meant by *delocalization* of pi electrons.
3. Name some common aromatic compounds.
4. Predict products of representative aromatic reactions.

In the early years of investigating products from plants, fragrant materials such as clove, almond, and wintergreen were observed to produce interesting organic compounds. Because the structures of these distinctive-smelling compounds showed sufficient similarities, they were classified as **aromatic**. Later, however, many other compounds that were *not* fragrant were shown to have similar structural and chemical properties, and these were also classified as aromatic. Soon, the term had less to do with aroma than with classification by structural and chemical features; the terminology remains, however. In the following sections we shall present details to help us understand the reasons for placing the compounds in the same class and the chemical basis for the classification.

4.1 Structure: Resonance and Electron Delocalization

Benzene is the "parent" chemical compound of the aromatic family; it has the formula C_6H_6. A number of possible structures have been suggested for benzene, including the following:

◆ C_6H_6 is the formula for benzene.

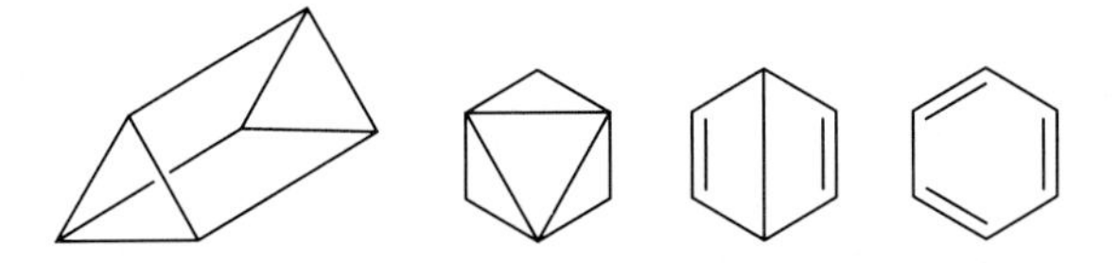

The structure on the far right, first proposed by Friedrich Kekule in 1865 after a dream, is closest to the structure as we now know it. Benzene represented in this way would really be 1,3,5-cyclohexatriene. This is nothing more than a cyclic triene, and we would expect it to behave as an alkene.

Between the time of Kekule and the present, there have been (and continue to be) many studies about benzene. As in many aspects of science, "right" or "wrong" answers are the products of many years of study and the work of many people. There was a time when people considered benzene as shown below:

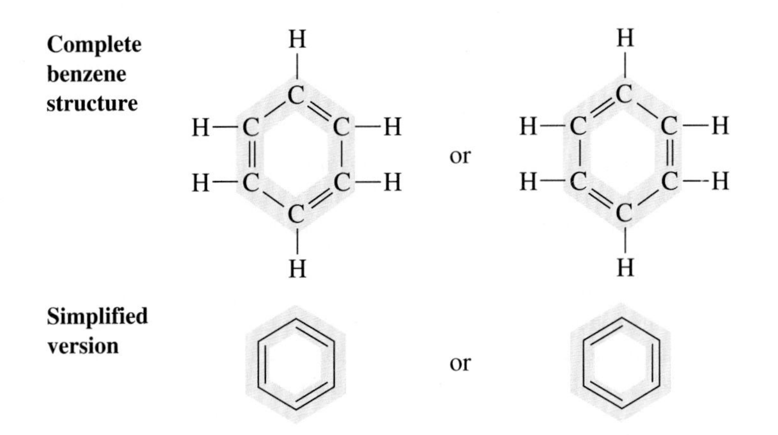

From the previous chapter, we might correctly assume benzene to have some additional stability because of the delocalization of electrons associated with the conjugation. However, benzene is *much* more stable than one might predict

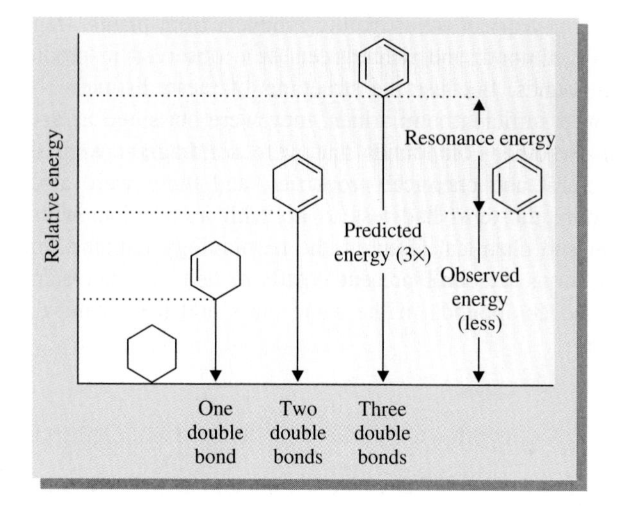

FIGURE 4.1
The resonance energy of benzene can be calculated from the predicted heat of hydrogenation. (See Section 4.1.)

from just considering it as a simple 1,3,5-cyclohexatriene. This extra stability is responsible for the lack of reactivity of benzene to reagents that normally react with alkenes. This extra stability is called **resonance energy**.

Figure 4.1 shows how the resonance energy is determined. There is a certain amount of heat liberated when cyclohexene is reduced to cyclohexane. You would expect three times as much heat to be liberated when 1,3,5-cyclohexatriene is reduced to cyclohexane (the predicted energy). But *less* than the predicted amount is actually observed. The difference between what is predicted and what is observed is the resonance energy. This energy is believed to be associated with the special stability achieved by the complete overlap of electrons in benzene's pi bond system.

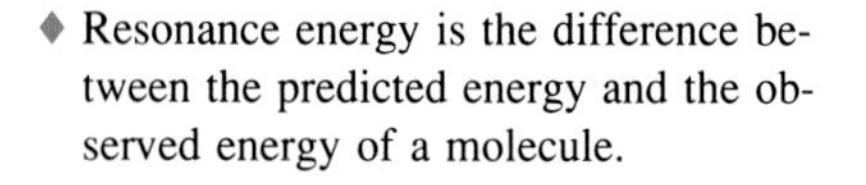

♦ Resonance energy is the difference between the predicted energy and the observed energy of a molecule.

When two different Lewis electron-dot structures can be drawn for a compound, the two structures are called contributing, or **resonance structures**, and it is understood *that the real structure is something between the two proposed structures*. Only compounds with alternating double bonds exhibit this property. When two or more resonance forms of a molecule can be drawn, the molecule is *always* more stable than would otherwise be predicted. Why is this so? We have shown that conjugated double bonds result from adjacent pi orbitals. The adjacent pi electron clouds form a *delocalized* orbital involving all of the pi-bonded carbon atoms. Such is the case with benzene. Benzene is much more resistant to reaction than might be expected because of the stability of this delocalized orbital (Figure 4.2). Thus, an aromatic compound must have a ring system with alternating double and single bonds. The total number of pi electrons must be a particular value to be aromatic. For simple aromatic compounds the total number of these electrons is 6, 10, 14, 18, . . . ($4n + 2$), where $n = 1, 2, 3, \ldots$. Considering benzene, we see that there are six pi electrons in the double bonds. Thus, the number of pi electrons, N is given by:

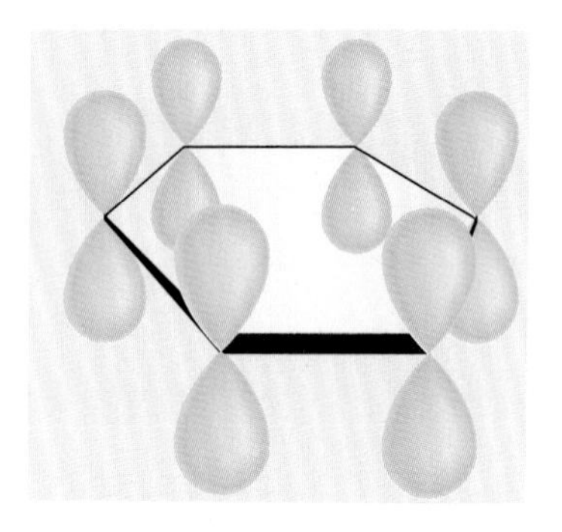

FIGURE 4.2
Alternating pi bonds in benzene produce a delocalized electron cloud of pi electrons.

$$N = 4n + 2$$
$$6 = 4n + 2$$

$$4 = 4n$$
$$1 = n$$

Since n here is a whole number, benzene fits the definition for an aromatic compound.

Benzene is best represented with the pi-bonded electrons shown as a ring.

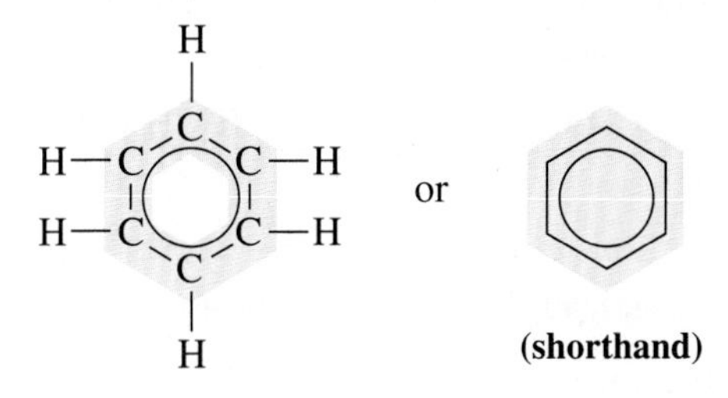

This representation is chemically more correct because the structure shows the electron delocalization, but many texts continue to use the traditional alternating double-bond resonance structure for convenience.

NEW TERMS

aromatic A class of ring compounds that have alternating double bonds and subsequent pi electron delocalization. An aromatic compound obeys the $4n + 2$ rule.

benzene The simplest aromatic hydrocarbon

resonance energy The energy due to delocalization of electrons

TESTING YOURSELF

Resonance and Aromatic Character

1. Draw the resonance structures for the bicarbonate ion

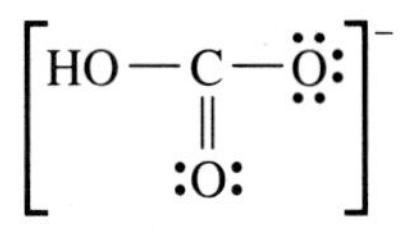

2. Show that cyclobutadiene

is not aromatic.

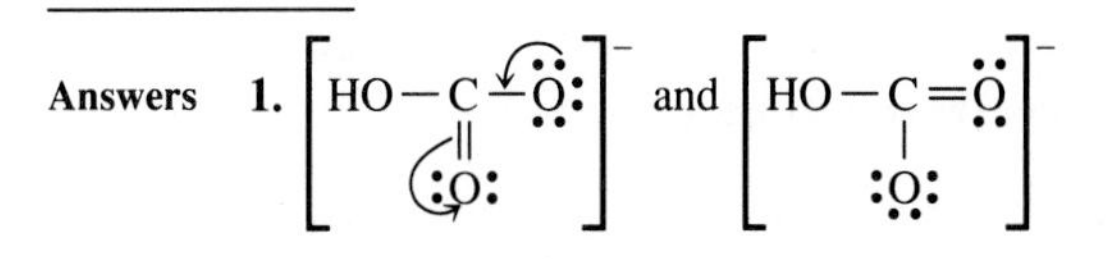

2. There are 4 π-electrons: $4 = 4n + 2$,
$4n = 2$, $n = \frac{1}{2}$, not a whole number; thus cyclobutadiene is *not* aromatic.

4.2 Nomenclature

Monosubstituted Benzenes

Since the "double bonds" of benzene are not located between specific carbon atoms, a different approach to nomenclature is used for aromatic compounds. This nomenclature often involves using the alkyl group name followed by the word benzene. For example,

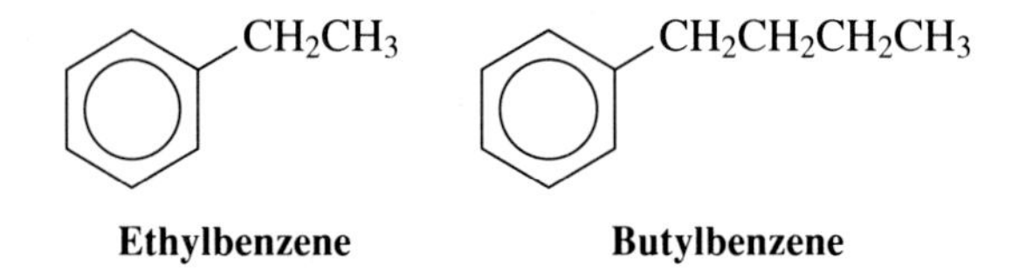

Ethylbenzene **Butylbenzene**

◆ Common names are important to the nomenclature of aromatic compounds.

A *monosubstituted* benzene is one in which *one* group is attached to the benzene ring. Common names take priority for some of these simple substituted benzenes. Two common monosubstituted compounds are *toluene* (methylbenzene) and *cumene* (isopropylbenzene).

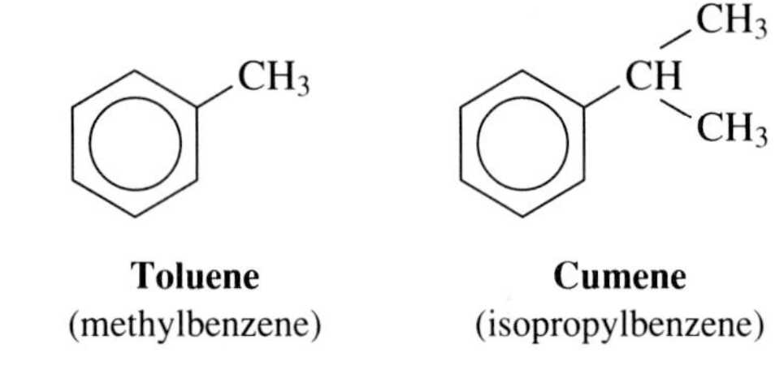

Toluene (methylbenzene) **Cumene** (isopropylbenzene)

The nomenclature of a number of other simple benzene derivatives is based on the name of the group attached to the benzene ring. The halogens, when attached to a benzene ring, are called *bromo* (Br), *chloro* (Cl), *iodo* (I), and *fluoro* (F). The NO_2 group is called *nitro*. When the SO_3 group is attached to a benzene ring, as in SO_3H, the resulting compound has the common name *benzenesulfonic acid*.

EXAMPLE 4.1

Name these monosubstituted benzenes.

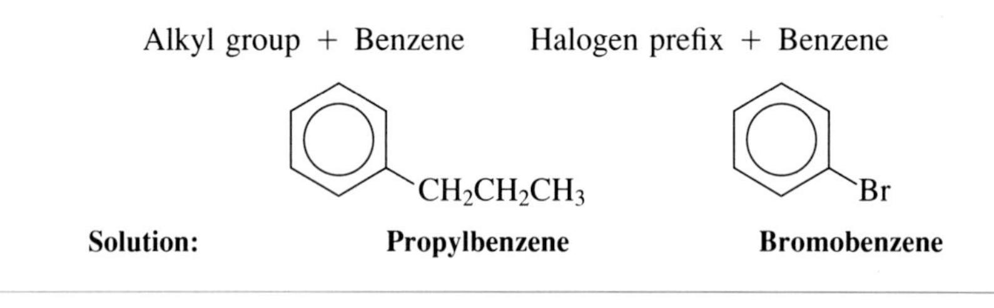

Solution: **Propylbenzene** **Bromobenzene**

Disubstituted Benzenes

In *disubstituted* benzenes two carbons have side groups attached. A unique system of nomenclature has evolved for these benzenes that requires one side group to be used to designate a *parent* aromatic compound. The second group is then assigned a position relative to the parent group. Two ways of assigning positions can be used.The first simply uses numbers, starting with the "parent" position as 1, and moving in the direction to give the second substituent the lowest possible number. For example,

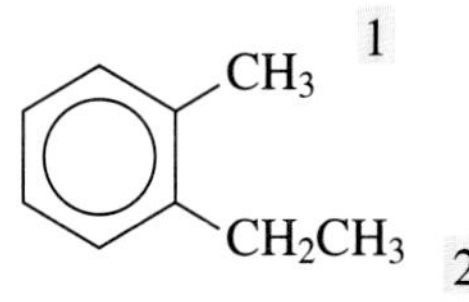

is known as 2-ethyltoluene (not 5-ethyltoluene or 2-methylethylbenzene). The methyl group (the parent compound is toluene) is designated as position 1.

The second method uses the prefixes **ortho- (*o*-), meta- (*m*-),** and **para- (*p*-)** to designate positions with respect to the parent group. These are shown in Figure 4.3. To use this nomenclature, the parent becomes the base name, and the second group is simply given a position and name using the prefix *o*-, *m*-, or *p*-. If one group is part of a *common* name, it becomes the parent.

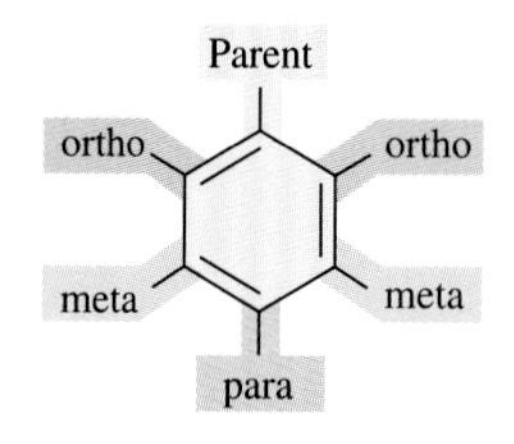

FIGURE 4.3
The prefixes ortho-, meta-, and para- can be used to designate position on a *benzene ring.*

◆ There is a move toward elimination of the *o*-, *p*-, *m*- method in favor of numbers. This makes computer searching easier. But, use of the *o*-, *p*-, *m*- nomenclature remains common.

EXAMPLE 4.2

Name these disubstituted benzenes:

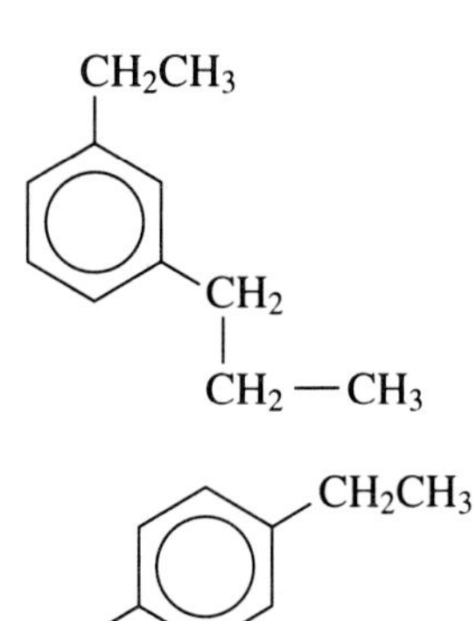

Solution:
The correct name is 3-ethylpropylbenzene, or *m*-ethylpropylbenzene. Note that we used the propylbenzene as the parent compound. We could have used the ethylbenzene, and then the name would be 3-propylethylbenzene, or *m*-propylethylbenzene.

4-Methylethylbenzene, or *p*-methylethylbenzene, is an incorrect name because "methylbenzene" has the common name *toluene*. Thus, the correct name is either 4-ethyltoluene, or *p*-ethyltoluene.

Polysubstituted Benzenes

When more than two side groups are attached to the benzene ring, as in *polysubstituted* benzenes, numbers are used to show the position relative to the parent compound. As in previously discussed rules of nomenclature, count around the ring in such a way as to get the lowest possible carbon numbers for the substituents.

EXAMPLE 4.3

Name these polysubstituted benzenes:

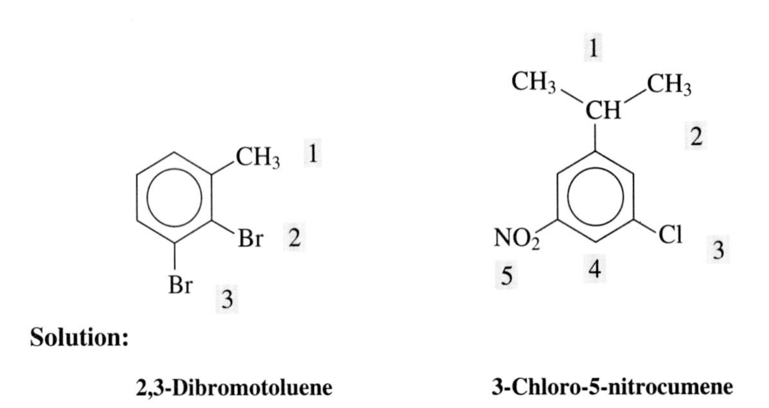

Solution:

2,3-Dibromotoluene **3-Chloro-5-nitrocumene**

Benzaldehyde is responsible for the strong odor of almond.

Additional Nomenclature

Sometimes the benzene ring appears as a substituent group on a larger compound. When the benzene ring is used this way, it is called a **phenyl** group. Another common benzene derivative is the **benzyl** group, which is a phenyl group with a CH_2 attached.

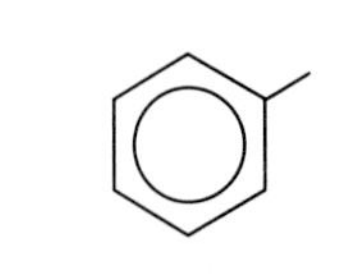

Phenyl-

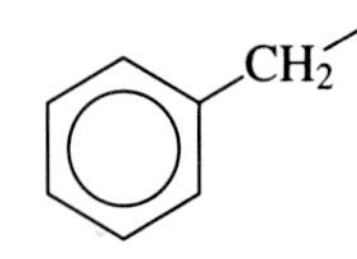

Benzyl-

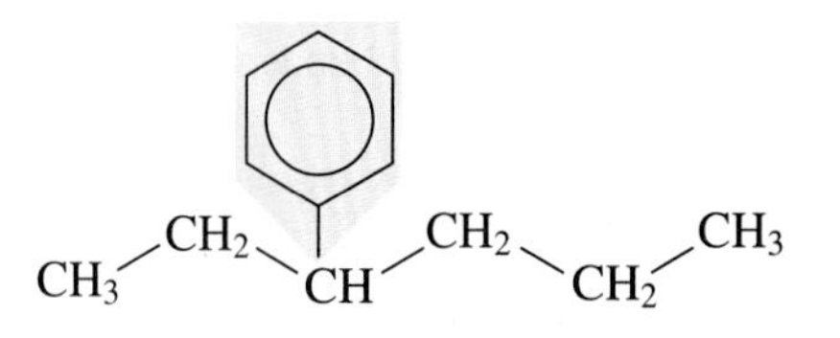

3- Phenylhexane

1,1-Dimethyl-2- benzylcyclohexane

A summary of aromatic nomenclature can be seen in the following examples.

***p*-Bromotoluene**

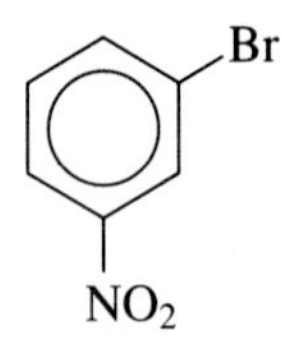

***m*-Nitrobromobenzene**
(or *m*-bromonitrobenzene)

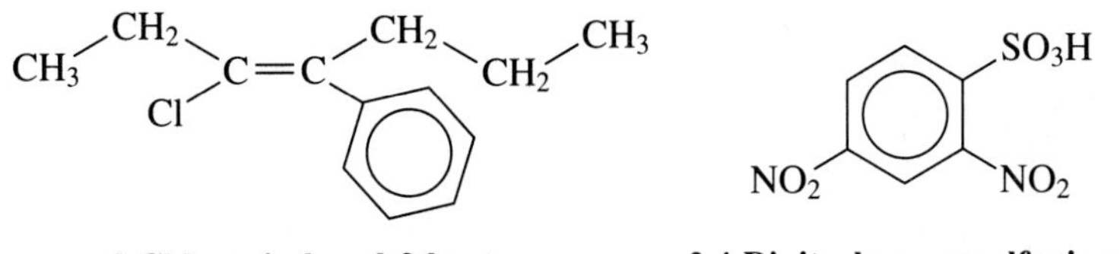

3-Chloro-4-phenyl-3-heptene **2,4-Dinitrobenzenesulfonic acid**

Now that we know the rules of nomenclature for the aromatics, we can give the names and structures of the major components of the oils of clove, almond, and wintergreen that were mentioned at the beginning of this chapter. (Note that eugenol is a common name.)

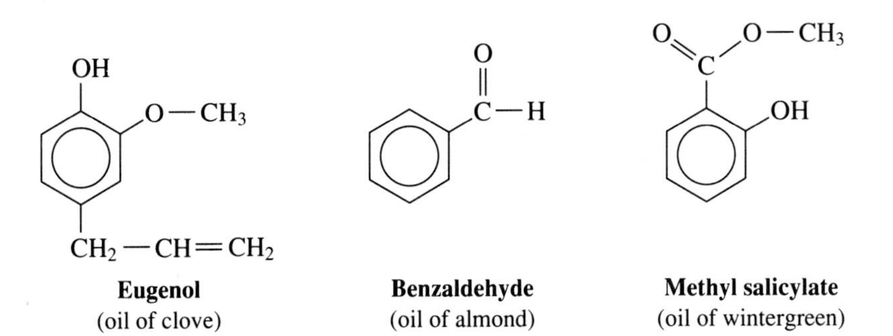

Eugenol (oil of clove) **Benzaldehyde** (oil of almond) **Methyl salicylate** (oil of wintergreen)

NEW TERMS

ortho- (*o*-) A prefix used to designate substituent position on a benzene ring. The ortho position is immediately adjacent to the parent substituent.

meta- (*m*-) A prefix used to designate substituent position on a benzene ring. The meta position is second from the parent substituent.

para- (*p*-) A prefix used to designate substituent position on a benzene ring. The para position is across from the parent substituent.

phenyl A benzene ring when used as an alkyl group attached to a larger molecule

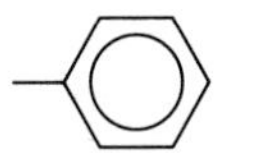

benzyl The alkyl group derived from toluene by loss of a hydrogen from the methyl group

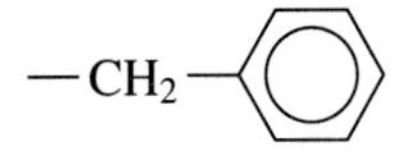

TESTING YOURSELF

Nomenclature

1. Name these substituted benzene compounds:

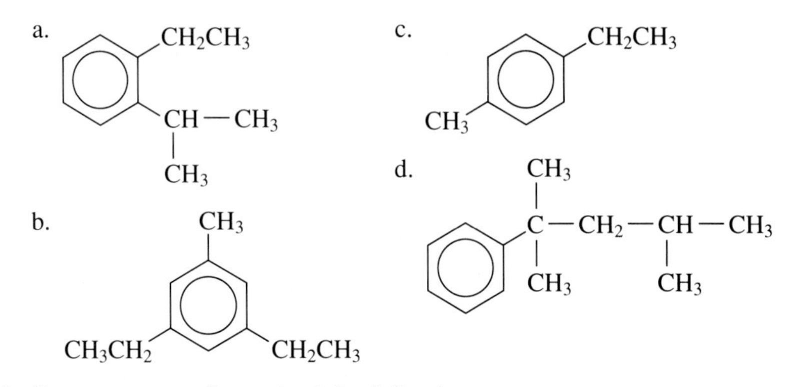

2. Draw structures for each of the following names:
   a. *o*-butylethylbenzene
   b. 2,3-dimethyltoluene
   c. 4,5-dibutyl-2-ethylcumene
3. Name these compounds, which have substitutions other than alkyl:

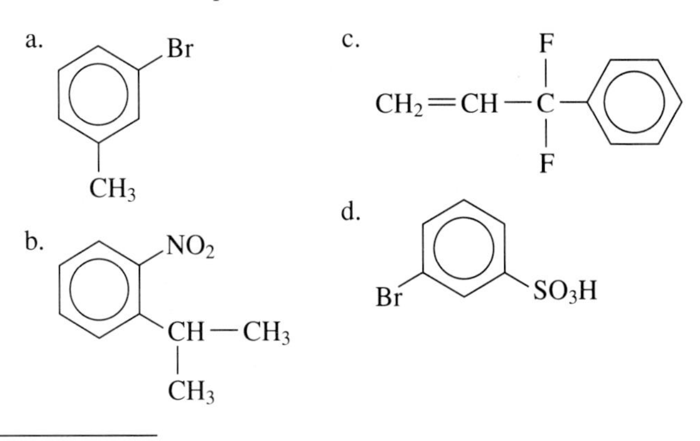

Answers **1. a.** *o*-ethylcumene, or 2-ethylcumene **b.** 3,5-diethyltoluene **c.** *p*-ethyltoluene, or 4-ethyltoluene **d.** 2-phenyl-2,4-dimethylpentane

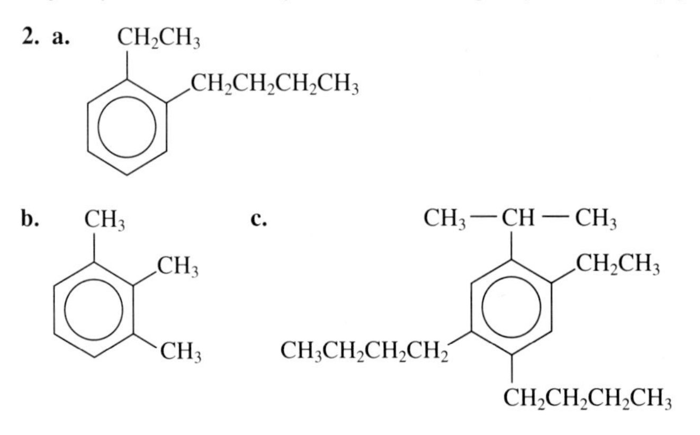

3. a. *m*-bromotoluene **b.** *o*-nitrocumene, or 2-nitrocumene **c.** 3,3-difluoro-3-phenylpropene **d.** *m*-bromobenzenesulfonic acid

4.3 Aromatic Reactions: A Consequence of Overlapping Pi Bonds

Substitution Reactions

The delocalized nature of pi-bonding electrons in benzene makes it unreactive to normal alkene reactions. Most of the alkene reactions discussed in the previous chapter are *addition* reactions. The alkene double bond is lost as a result of the reaction. If benzene were to participate in an addition reaction, its double bond(s) would be lost, and pi-electron delocalization would no longer be possible. The resulting compound would no longer be aromatic (Figure 4.4).

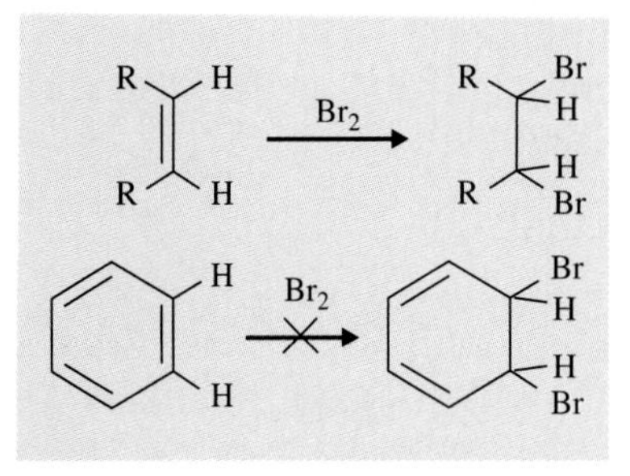

FIGURE 4.4
Alkenes generally react by addition reactions. The double bond of an aromatic compound does not do this because if there was an addition across one of the double bonds, the benzene ring would lose its resonance stability.

Lewis acids are *electrophiles* (abbreviated E^+), which means electron "lovers." They attack the pi electrons of the aromatic ring. However, rather than adding to the ring, the electrophile *substitutes* for a hydrogen atom. The replacement of H by E^+ is the reason that most aromatic reactions are called **electrophilic substitution reactions**.

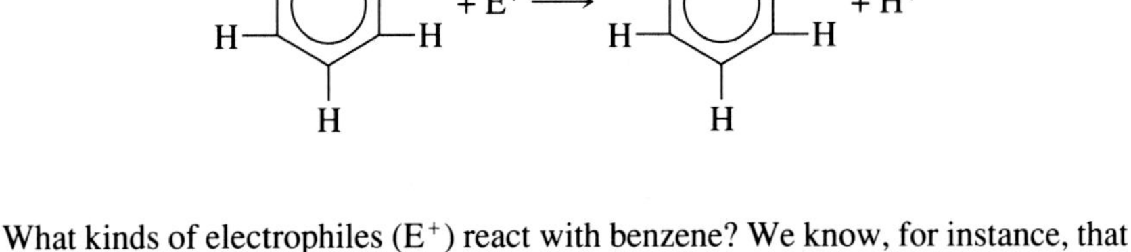

What kinds of electrophiles (E^+) react with benzene? We know, for instance, that Br_2 *does not* react with benzene. Yet, Br_2 and $FeBr_3$ are an effective source of Br^+. There are special reaction conditions required for aromatic reactions. Table 4.1 lists some of the special reaction conditions needed to generate the electrophiles.

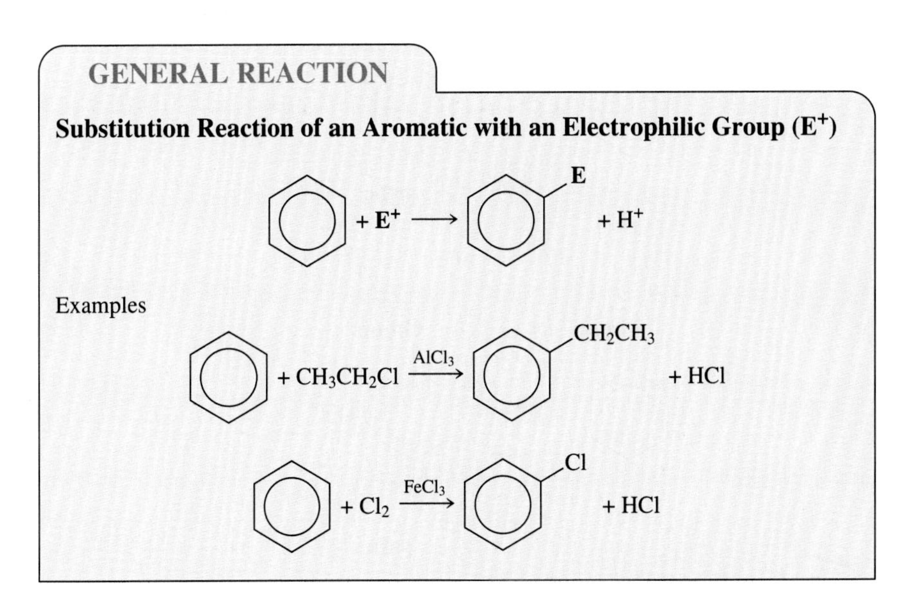

TABLE 4.1
Reaction Conditions and Electrophiles

Reaction Conditions	Identity of Electrophile
$X_2 \xrightarrow[\text{(catalyst)}]{FeX_3}$	^+X
$H_2SO_4 \xrightarrow{\Delta}$	$^+SO_3H$
$HNO_3 + H_2SO_4 \longrightarrow$	$^+NO_2$
$RX \xrightarrow[\text{(catalyst)}]{AlCl_3}$	^+R (only for methyl and ethyl)

Substitution of a Ring Already Having Substituents

Often a benzene ring already has one or more groups attached to it before it is reacted. For example, in a reaction of toluene with some E^+ reagent, where would the E^+ substitute?

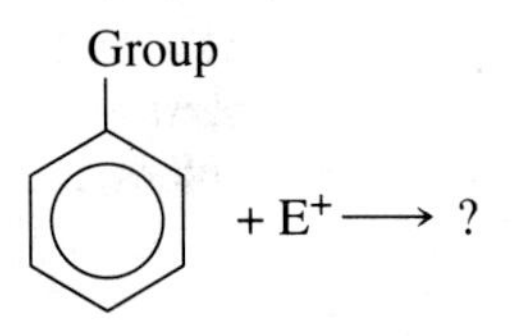

♦ The group already present on a substituted benzene ring directs the position of attack by a new reagent.

After many years of observation and testing, chemists have determined that the group or groups already on the ring directs where the E^+ substitutes and that these groups can be classified into two categories. One category of groups always directs the incoming E^+ reagent to an ortho and/or para position—these are called *ortho–para directors*.

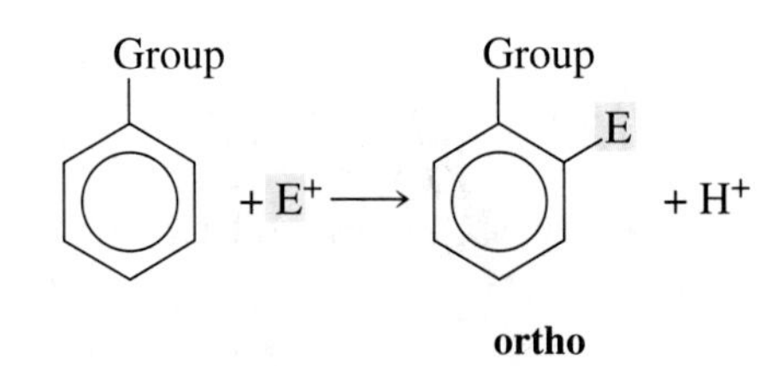

or

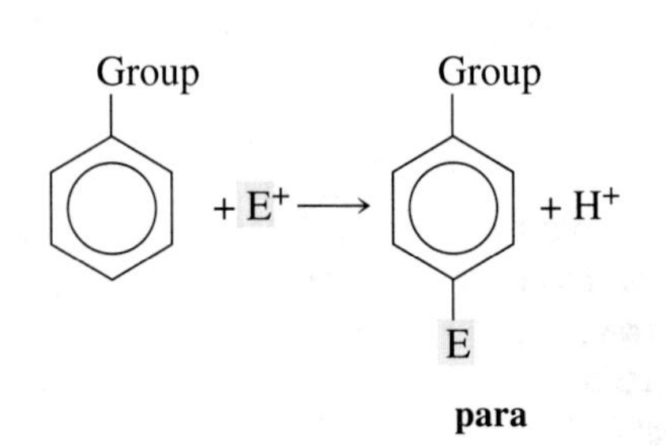

Except for alkyl groups, the ortho–para-directing groups have one or more pairs of unshared electrons associated with the atom directly attached to the aromatic ring.

The second type of group directs toward the meta position and is called *meta directors*.

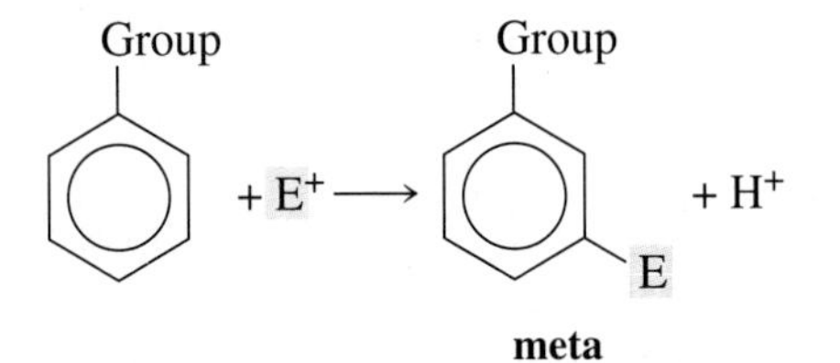

In the meta-directing groups, a common characteristic is that the atom directly attached to the benzene ring is multiply bonded to an electronegative atom. Table 4.2 lists a number of these common **directing groups**.

When reactions are carried out on benzene rings that are already substituted, there are two steps we must think about: (1) what is the E^+ and (2) where will it substitute? The first question is answered in exactly the same way that was done in the previous section—the reagents that are used determine the identity of E^+. Once the identity of E^+ is known, we have only to decide what kind of substituent there is on the ring. This tells where the E^+ will attach itself.

TABLE 4.2
Directing Groups on a Benzene Ring

Symbol	Name or Prefix
Ortho–para-directing groups	
—R	Alkyl
—OH	Hydroxy
—OCH_3	Methoxy
—X	Halogen
—NH_2	Amine
Meta-directing groups	
—NO_2	Nitro
—C(=O)—OH	Carboxyl
—C(=O)—R	Ketone
—C(=O)—H	Aldehyde
—SO_3H	Sulfonic acid
—CN	Nitrile

EXAMPLE 4.4

Show products for the following reactions:

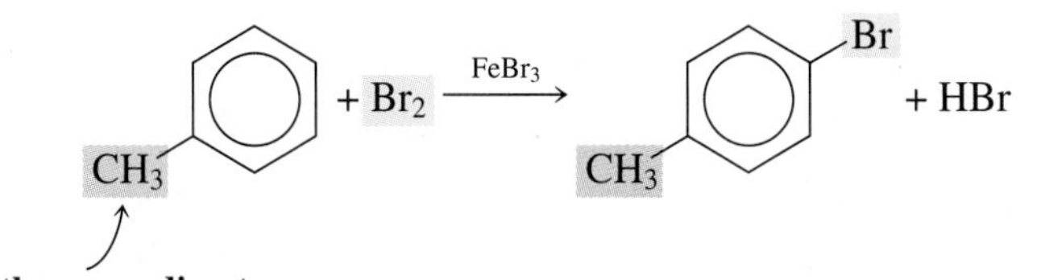

Ortho–para director

You would also be correct in giving the product as

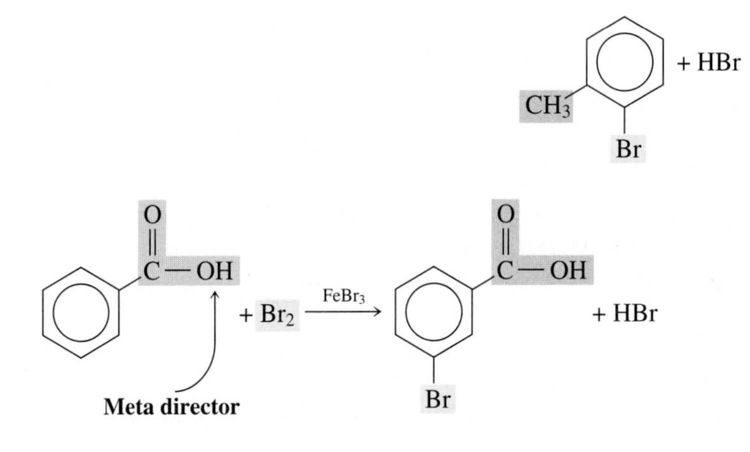

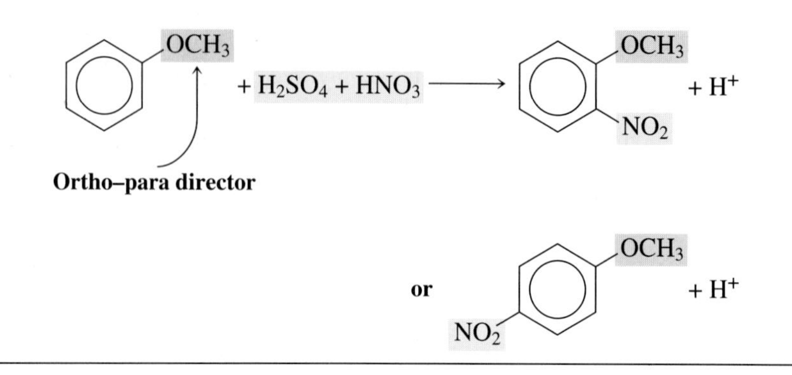

Phenol is a powerful disinfectant.

Oxidation Reactions

The aromatic ring does not readily undergo oxidation. When ring oxidation does occur, particularly in biological oxidation reactions, the products are often *phenols* (see Chapter 5 (Section 5.2) for details). Phenols are benzene rings with hydroxyls attached and many are quite toxic.

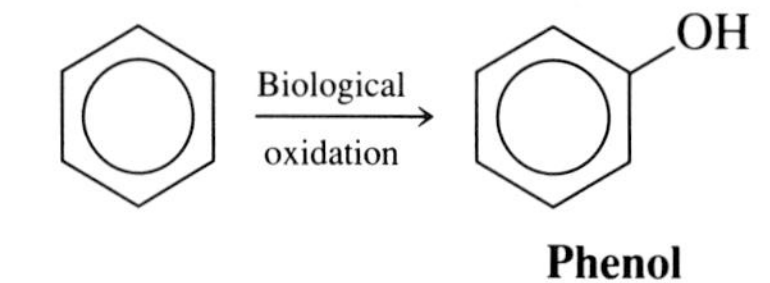

Alkyl-substituted benzene rings undergo side-chain oxidation to benzoic acid and its derivatives. Benzoic acid derivatives are less toxic than phenol derivatives. Thus, we find that toluene is less harmful to us than benzene because the metabolic oxidation product of toluene is benzoic acid.

♦ Carbon side chains on an aromatic ring are readily oxidized.

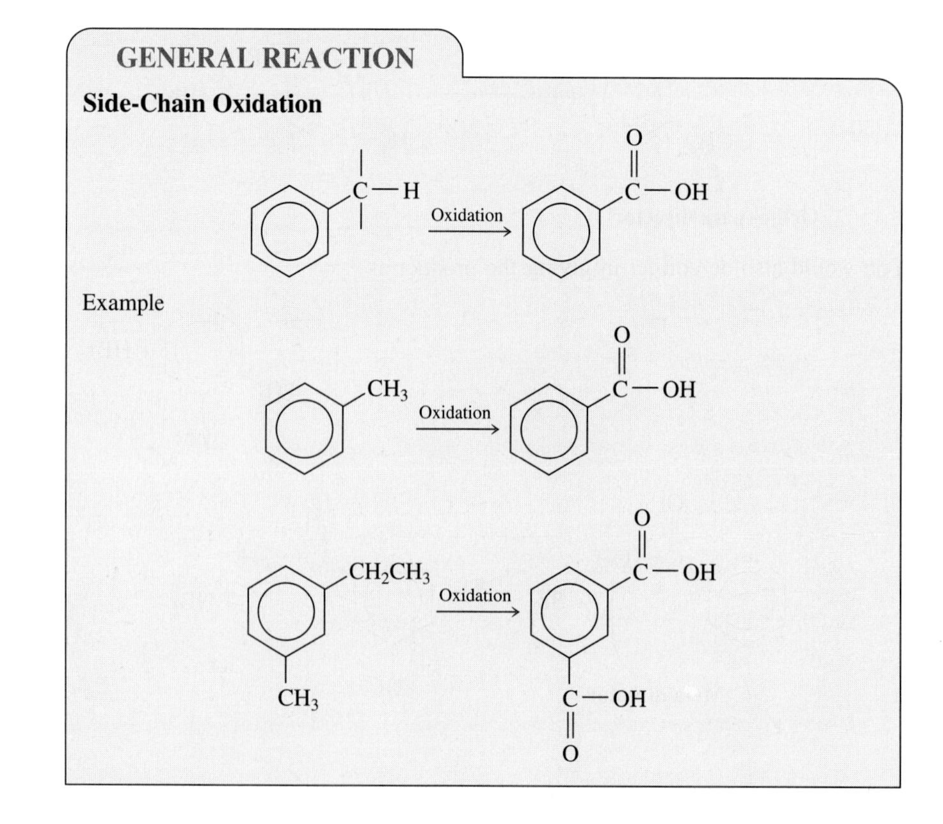

CHEMISTRY CAPSULE

Interesting Aromatic Compounds

A fundamental question for organic chemists has been the size of a ring that can maintain alternating double bonds and still provide special "aromatic" properties. Novel hydrocarbons have been constructed to test these questions, and in each case the application of the $4n + 2$ pi electron rule has been found to hold true.

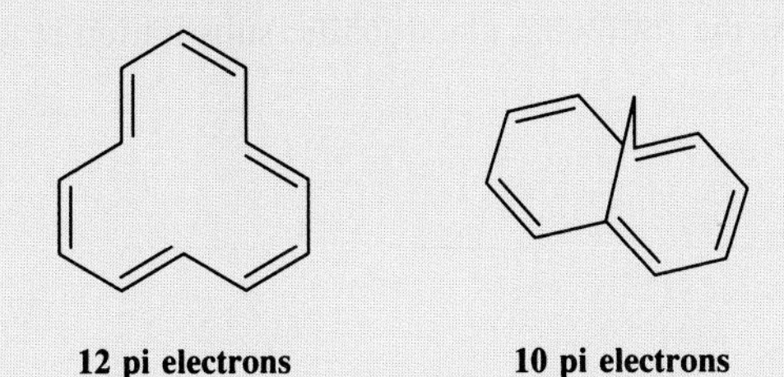

12 pi electrons

$4n + 2 = 12$

$4n = 10$

$n = 2.5$

(not a whole number)

10 pi electrons

$4n + 2 = 10$

$4n = 8$

$n = 2$

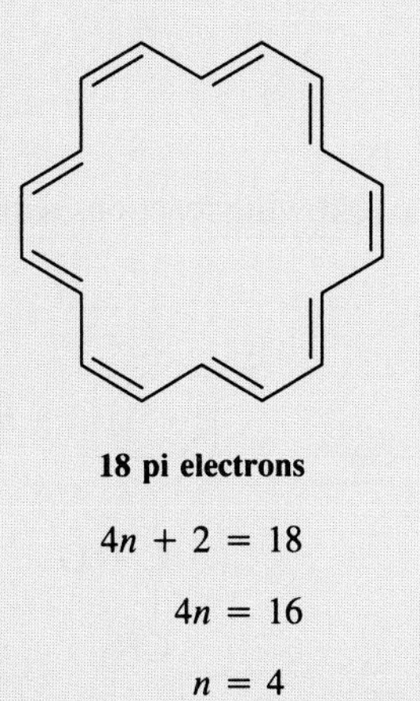

18 pi electrons

$4n + 2 = 18$

$4n = 16$

$n = 4$

It was thought that the rule would fail for molecules having $n > 5$. Recently, the rule was tested, and it is now known that it holds for larger values of n. The test molecules also happen to have some biological and potential medicinal significance.

Chlorophyll and heme are complex aromatic (porphyrin) systems that contain nitrogen and are able to attach to metals, such as magnesium and iron. There has been interest in putting other metals in the cavities of these porphyrins, but the available space is too small for many metals of interest.

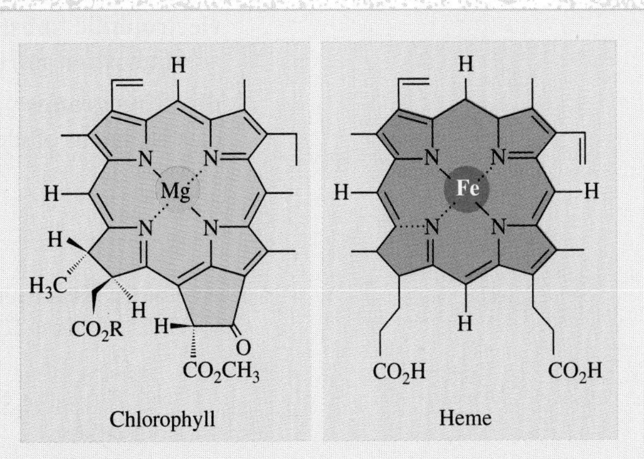

FIGURE 4.A
The basic structural units of heme and chlorophyll are similar. A highly structured aromatic system is found to be essential for both plants and animals.

Recently additional space in the center of the molecule was provided when chemists synthesized *texaphyrin*, a molecule that can incorporate cadmium. (The name comes from Texas, the "Lone-Star" State.)

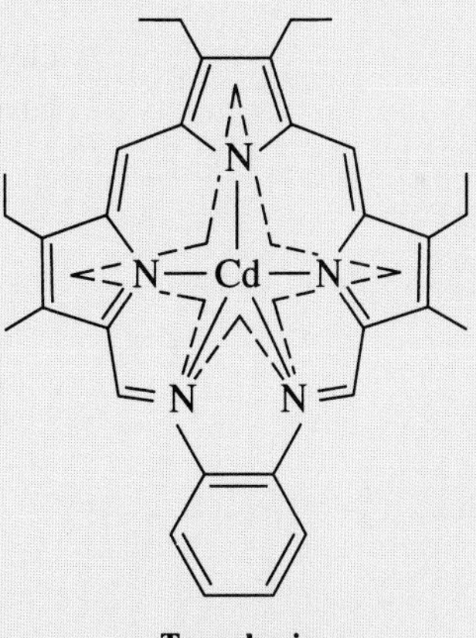

Texaphyrin

The cadmium system absorbs more light in the red region of the spectrum. This spectral region is very interesting because it can excite oxygen to a special, highly reactive state called *singlet oxygen*. Singlet oxygen has the ability to kill tumor cells, but it has been difficult to get this oxygen into the tumor cells. A possible future use of molecules such as texaphyrin might be to incorporate them into tumor cells, then excite them to help transfer energy to oxygen. This might provide a unique way to supply singlet oxygen to tumor cells.

Source: *New Scientist* (1989), Feb., p. 32.

NEW TERMS

electrophilic substitution reaction Reaction in which an electrophile (E^+) substitutes for an H on an aromatic ring

directing groups The groups already on a benzene ring that direct the position of attachment of electrophiles

TESTING YOURSELF

Aromatic Reactions

1. Give products for each of the following electrophilic substitution reactions:

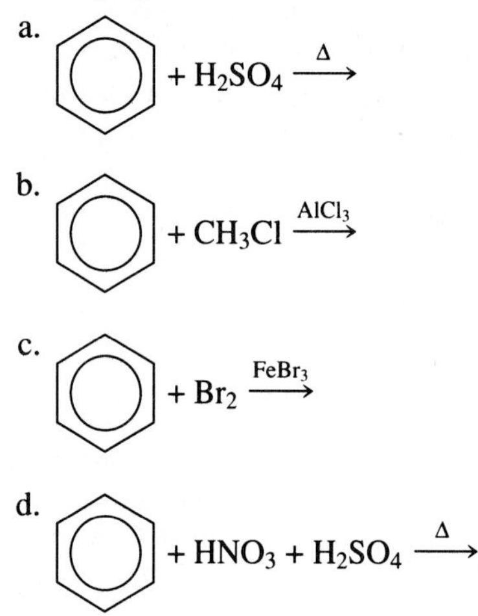

2. Give the products expected from the following reactions with rings that already have substituents:

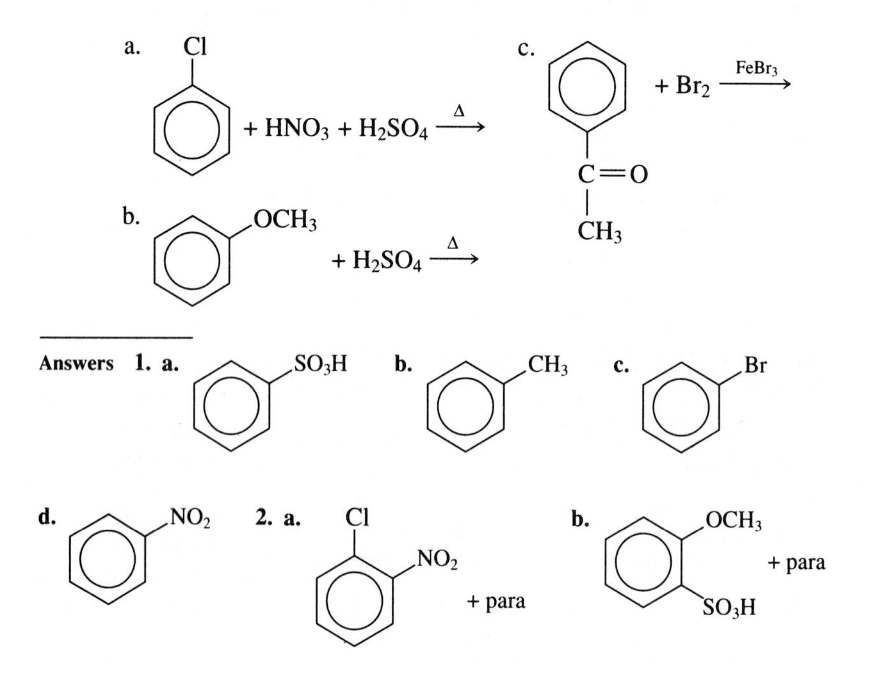

c.

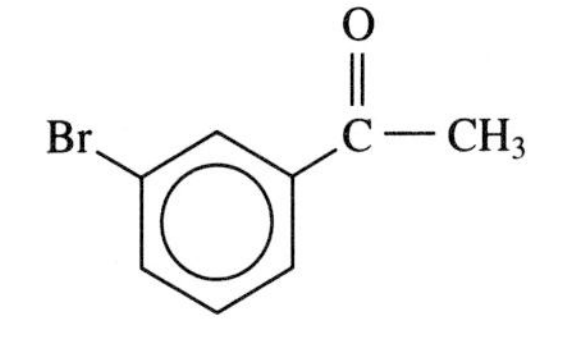

4.4 Fused-Ring Aromatic Systems

Aromatic character is due to the overlap of adjacent pi-bonding electrons in a ring compound. Although the six-carbon system is most common, linked rings and other ring-sized aromatics are also found in nature. As with mono rings, the major characteristic determining aromatic character (and thus substitution rather than addition reactivity) is the presence of $4n + 2$ pi electrons in the ring. For example, note the number of pi electrons in these fused-ring compounds:

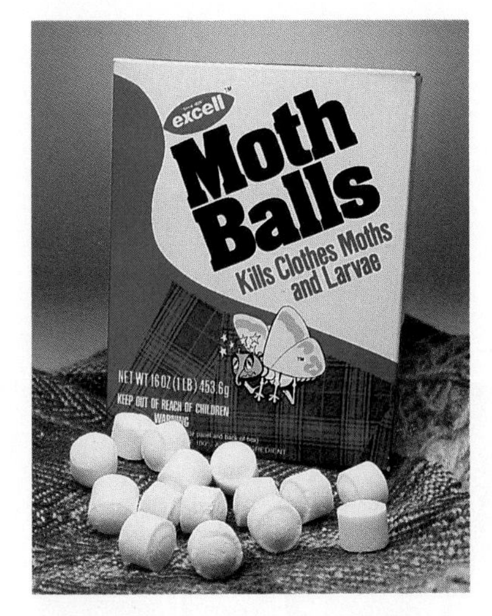

Naphthalene is the major ingredient of moth balls.

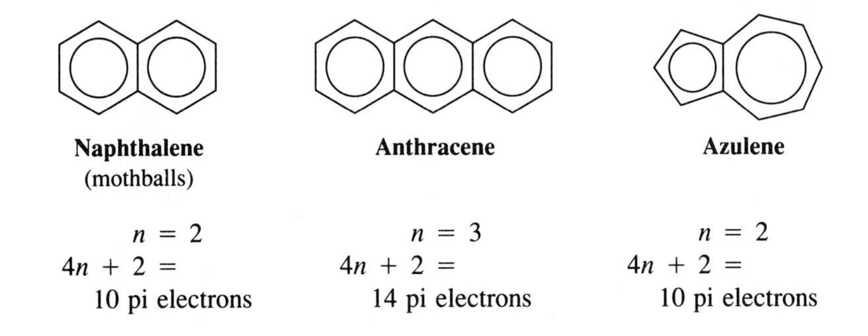

A number of fused-ring aromatic compounds are known to be toxic or carcinogenic (cancer causing). Often these compounds are "bent" aromatics, that is, they are multiple rings fused together in such a way as to not be linear. For example,

♦ Some fused-ring aromatic hydrocarbons have been found to cause cancer in humans

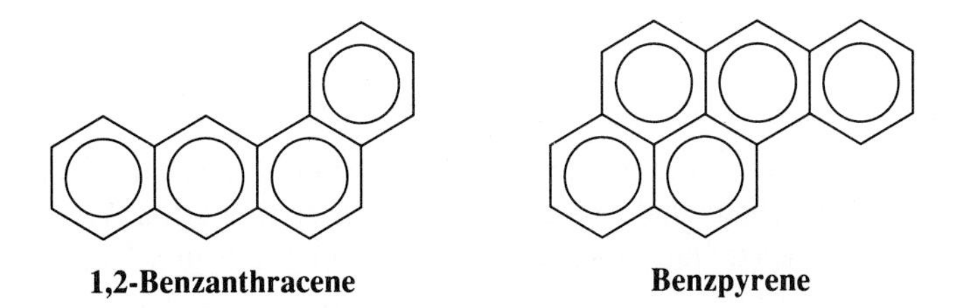

Benzypyrene is one of the major substances found in cigarette smoke and has been implicated as a potent carcinogen. It should be understood that benzpyrene itself is not the problem. However, it is easily converted by the body's oxidation processes to a compound that tightly binds to DNA and causes mutations.

4.5 Important Aromatic Hydrocarbons

One of the most notable examples of a biological aromatic compound is *estradiol*, a major female sex hormone.

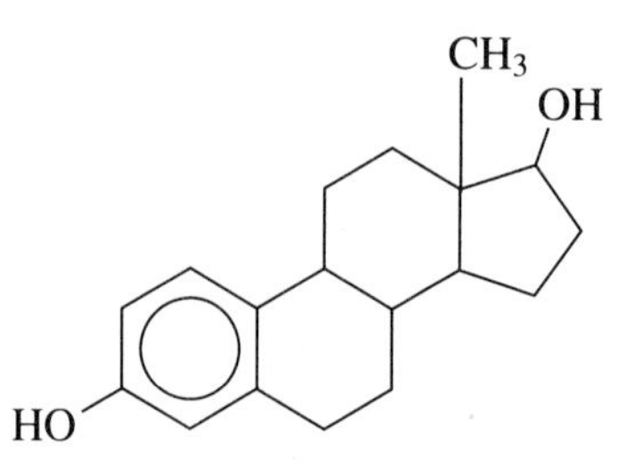

Estradiol

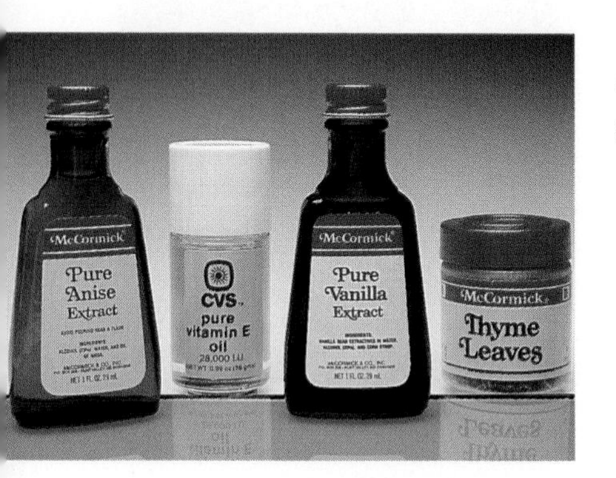

Many aromatic compounds were first recognized in spices and natural preservatives.

Flavorings are well represented by aromatic compounds. Three examples were presented at the introduction of this chapter (clove, almond, and wintergreen). Additional examples include the flavorings from anise, vanilla, and thyme.

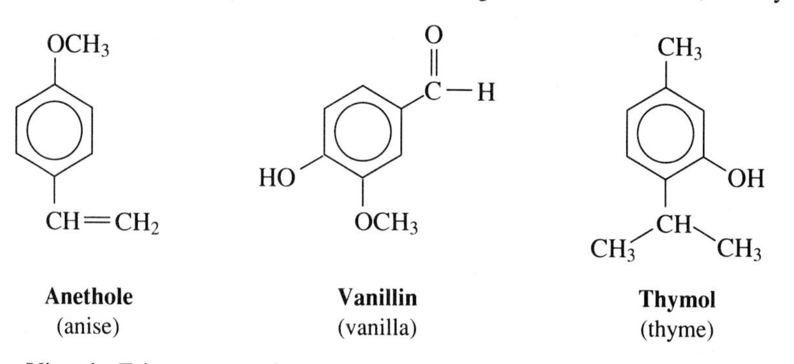

Anethole (anise) | **Vanillin** (vanilla) | **Thymol** (thyme)

Vitamin E is an aromatic compound with the chemical name α-tocopherol. It is an excellent antioxidant and has been shown to have antisterility properties in some animals.

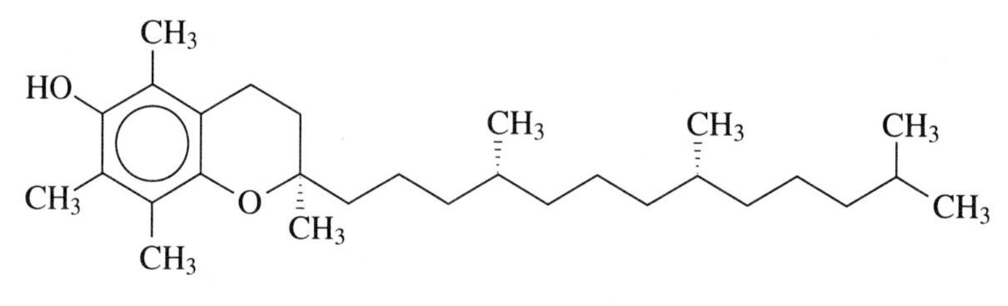

α-Tocopherol

Warfarin is an aromatic compound that is a potent anticoagulant. One of the major uses of this compound and its derivatives is as rat poison. When mice or rats ingest foods laced with this ingredient, they tend to hemorrhage and die.

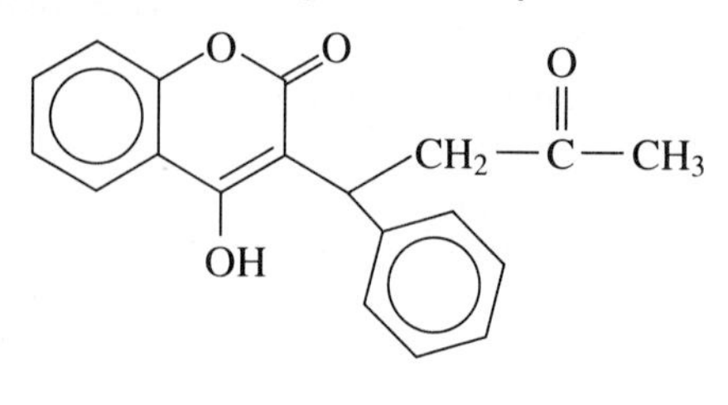

Warfarin

CHEMISTRY CAPSULE

Bucky Balls—the New Excitement

The laser-induced formation of "Bucky Balls" has created a new excitement in chemistry. The discovery of Buckminsterfullerene has ushered in a new form of carbon into the world of chemistry. Until the recent discovery of the uniquely shaped Bucky Ball, (named for the American architect, R. Buckminster Fuller, who designed the geodesic dome, which this carbon structure resembles), carbon was thought to exist in two elemental forms: diamond and graphite. Now a series of ball-shaped carbon compounds, of which the symmetrical C_{60} compound is an example, are being studied.

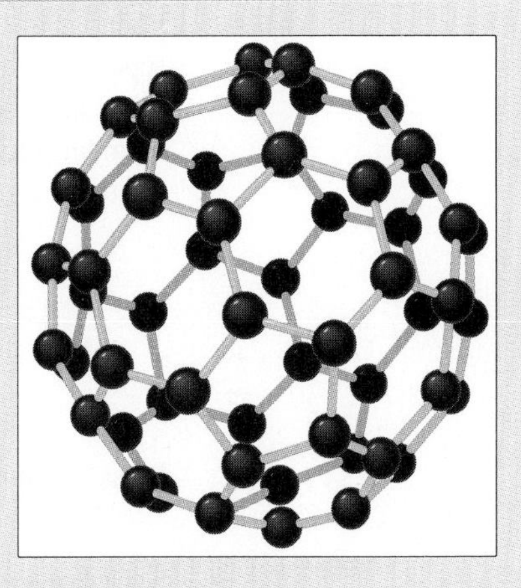

FIGURE 4.B
The "Bucky Ball," short for the soccer ball-shaped aromatic compound buckminsterfullerene, is a 60-carbon atom system. The benzene rings are connected by a pentagon-shaped skeleton.

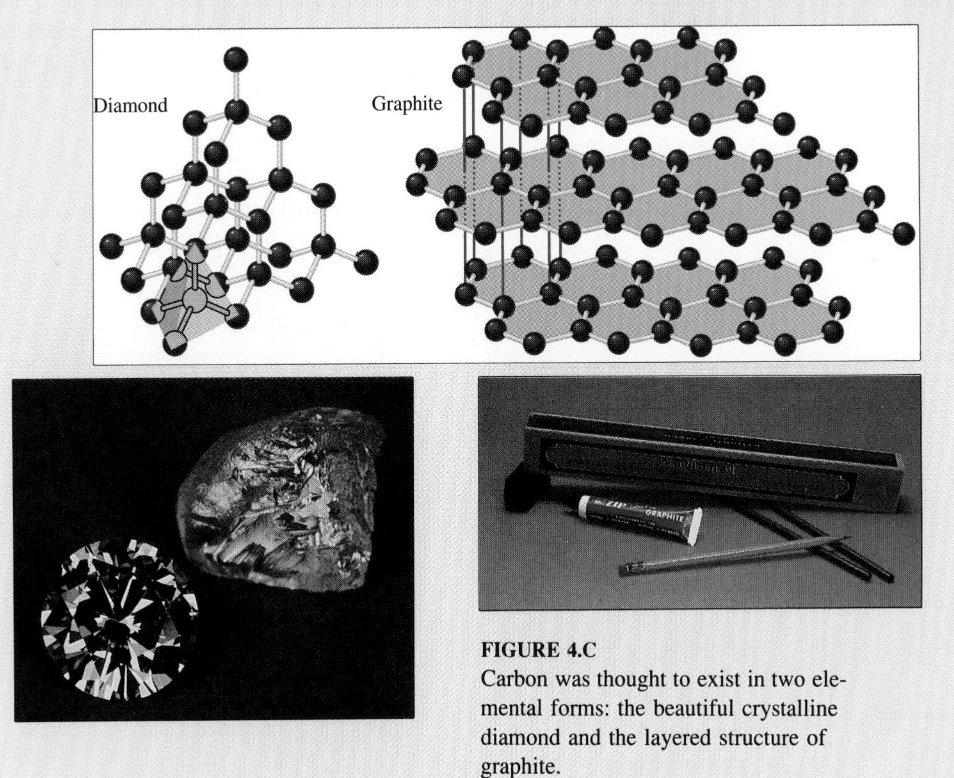

FIGURE 4.C
Carbon was thought to exist in two elemental forms: the beautiful crystalline diamond and the layered structure of graphite.

From these few examples it is readily apparent that —OH or —OR groups connected to an aromatic ring are quite common for biologically active compounds. In the next chapter we will be discussing in detail compounds of this type, and additional examples will be provided.

Later, we will encounter a number of aromatic ring compounds that have elements other than carbon making up the ring. For example, an analog of benzene containing nitrogen is called *pyridine*.

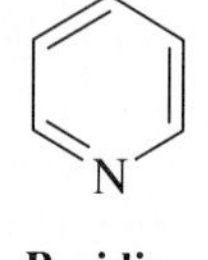

Pyridine

Ring compounds containing atoms other than carbon are called *heterocyclic*.

Summary

Conjugated double bonds give a special stability, called *resonance energy*, to molecules that contain them. This is particularly demonstrated in *aromatic* molecules. These are ring systems with alternating double and single bonds that obey the rule that there are $4n + 2$ pi electrons. Because of the special stability of these molecules, they do not undergo simple addition reactions to the double bonds. Instead, the molecules undergo aromatic *substitution reactions*, where an electrophile (E^+) substitutes for a hydrogen. Another unique reaction of an aromatic molecule is *side-chain oxidation*.

Reaction Summary

Substitution Reaction of an Aromatic with an Electrophilic Group (E^+)

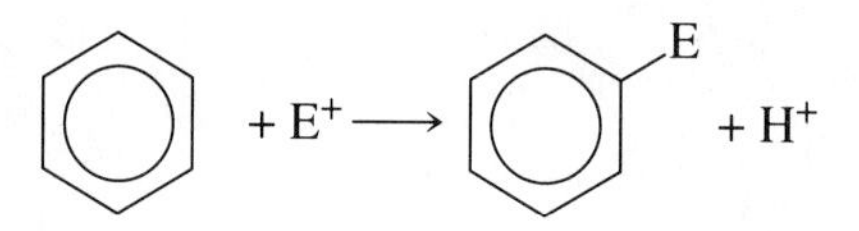

Side-Chain Oxidation

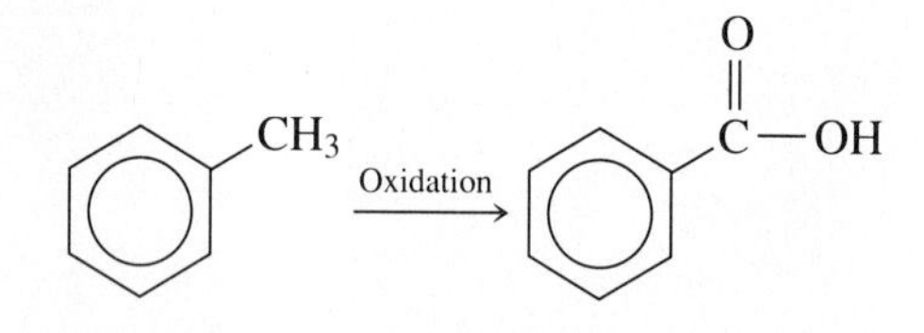

Terms

aromatic (4.1)
benzene (4.1)
resonance energy (4.1)
ortho- (4.2)
meta- (4.2)
para- (4.2)
phenyl (4.2)
benzyl (4.3)
electrophilic substitution reaction (4.3)
directing group (4.3)

Exercises

Resonance (Objective 2)

1. What do we mean by "delocalized" electrons?
2. What are resonance structures?
3. Why is part a below a better representation for benzene than b?

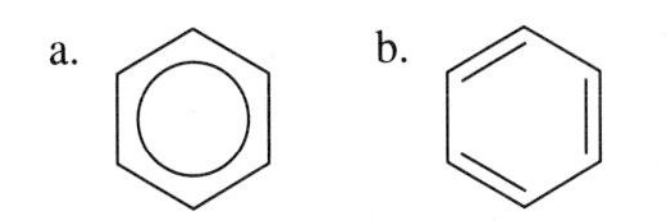

Nomenclature (Objective 3)

4. Draw structures corresponding to the following names:
   a. *o*-ethyltoluene
   b. *p*-ethyltoluene
   c. *m*-bromocumene
   d. 2,3-dichlorotoluene
   e. *m*-iodoethylbenzene
   f. *p*-bromochlorobenzene
   g. *m*-chlorotoluene
   h. 2,4,6-trinitrotoluene
5. Draw structures for the following common aromatic compounds:
   a. toluene
   b. cumene
   c. nitrobenzene
6. Name the following compounds

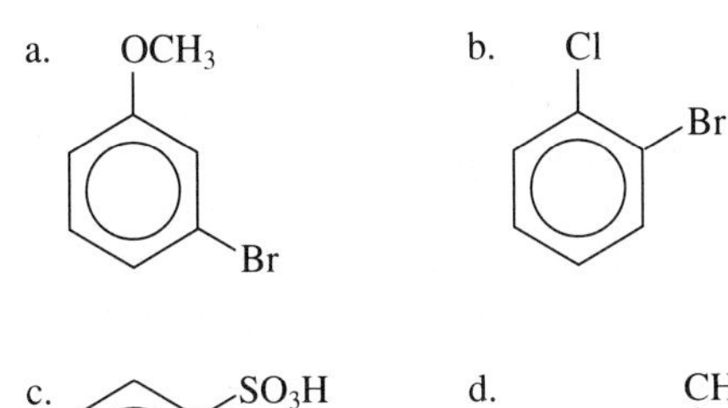

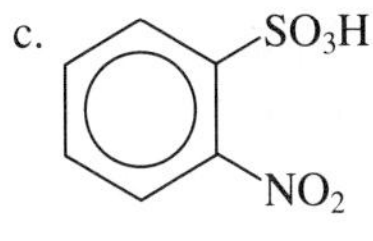

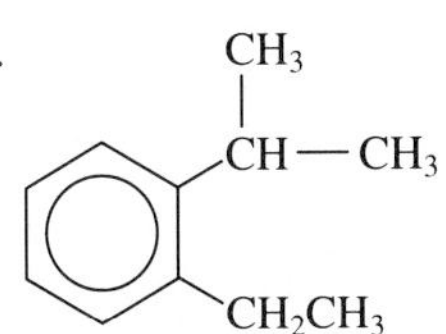

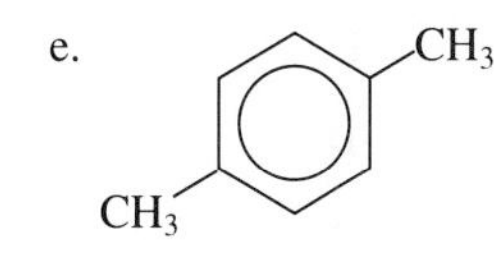

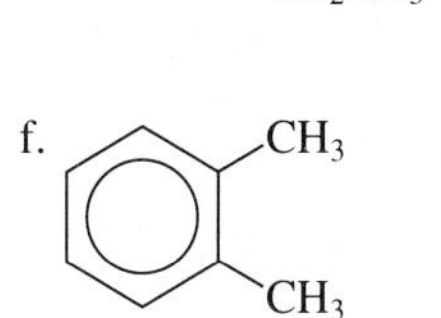

7. Draw structures corresponding to the following names:
   a. *o*-bromocumene
   b. 3,5-dinitrotoluene
   c. *p*-chloronitrobenzene
   d. *p*-ethylbenzenesulfonic acid
   e. *o*-propyltoluene
   f. 4-benzyl-1-octene
   g. 1,3-dibromobenzene
   h. *m*-cyanotoluene
   i. 2,3-dibromotoluene
   j. 3-phenylcyclohexene
8. Name the following compounds:

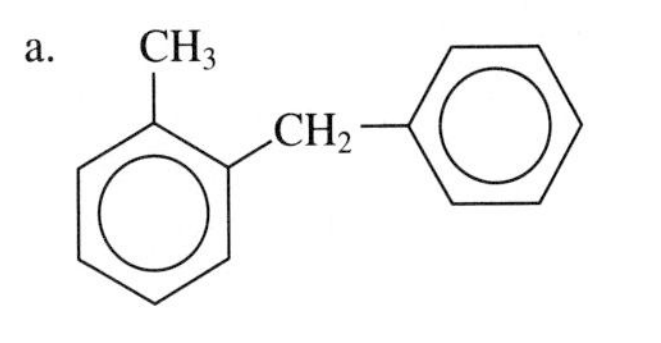

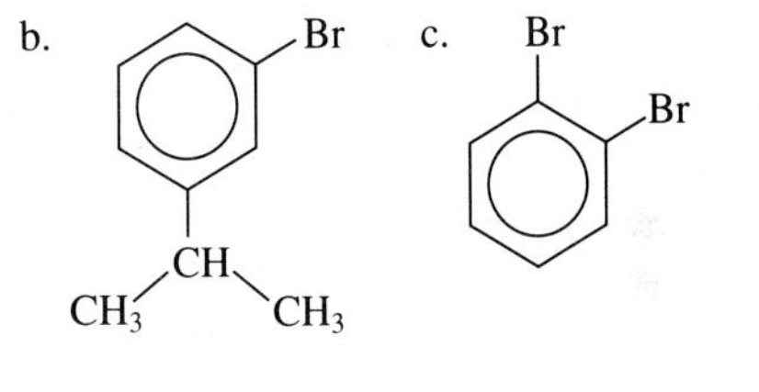

d. CH_2CH_3 ... I

e. CH_3 ... Cl

9. Give names for the following aromatic compounds:

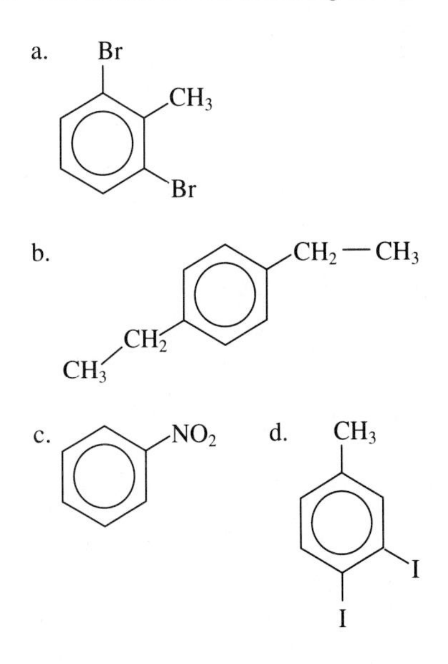

10. Give an appropriate name for each of the following compounds:

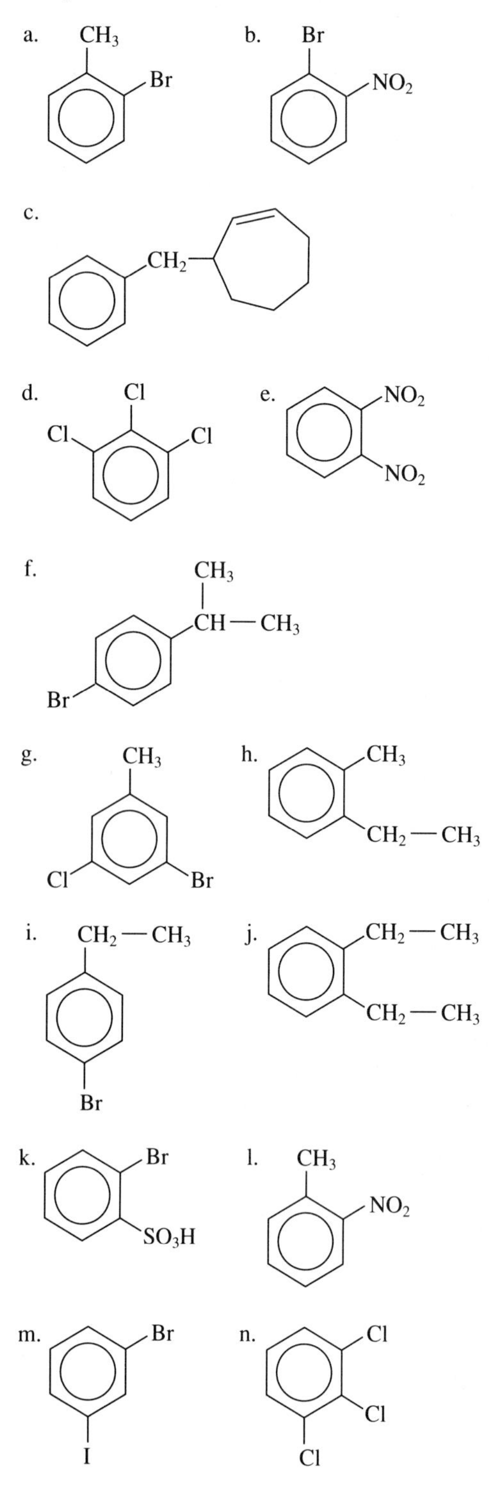

Resonance (Objective 2)

*11. Draw resonance structures for the carbonate ion

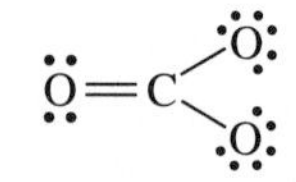

12. Draw resonance structures for naphthalene.

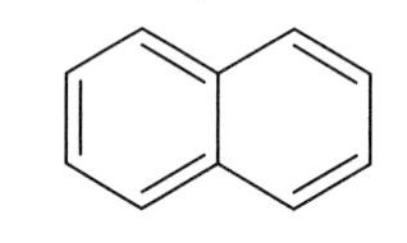

13. What is resonance energy?
14. Explain the difference between the symbolism $\longleftrightarrow$ and $\rightleftarrows$.

Reactions (Objective 4)

15. Benzene derivatives undergo substitution reactions rather than addition reactions, as we saw with alkenes in the previous chapter. Explain.
16. What do we mean by ortho–para directors?
17. Show on the generic-substituted benzene ring

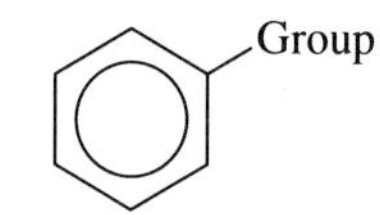

   a. What is the ortho position of toluene?
   b. What is the para position of nitrobenzene?
   c. What is the meta position of ethylbenzene?
18. List the following groups as ortho–para- or meta-directors:
   a. methyl b. nitro c. hydroxy
   d. bromo e. ethyl f. sulfonic acid
19. Label the kind of directing group each of the following belongs to:

   a. $CH_3—CH_2—$
   b. $CH_3—C(=O)—$
   c. $—C\equiv N$
   d. $Br—$
   e. $HO—$
   f. $—NO_2$
   g. $—O—CH_3$
   h. $—SO_3H$
   i. $—C(=O)—H$

20. Show the electrophilic agent (E^+) that is formed from each of the following reagents:
   a. $Br_2 + FeBr_3$
   b. H_2SO_4 and heat (Δ)
   c. $H_2SO_4 + HNO_3$
   d. $CH_3CH_2Cl + AlCl_3$

21. What is meant by side-chain oxidation?
22. Give the structure of the major product(s) from each of the following reactions:

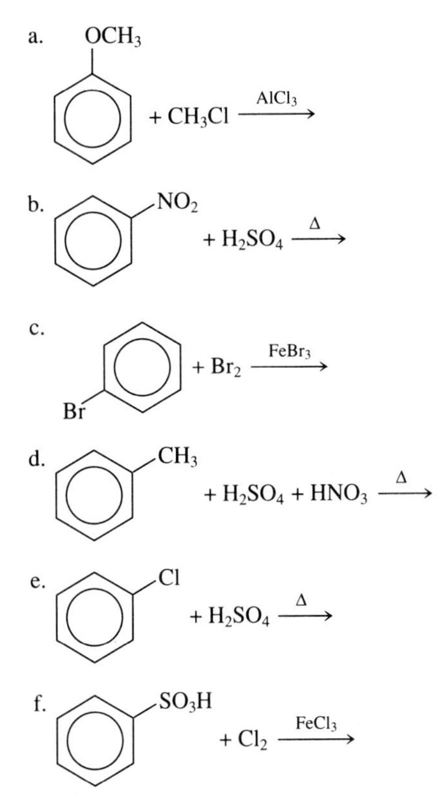

23. Give the structure of the major product from each of the following reactions:

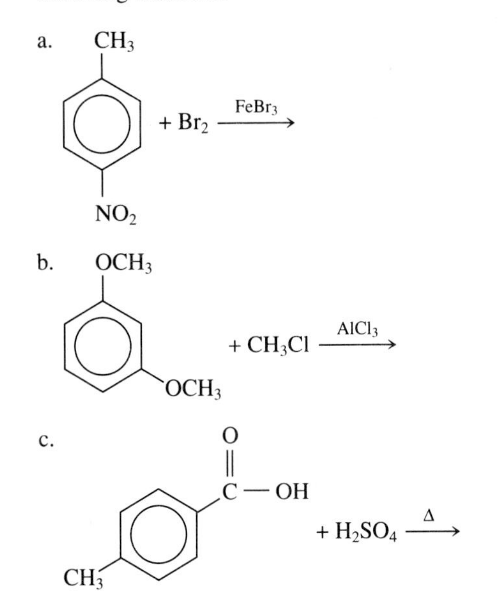

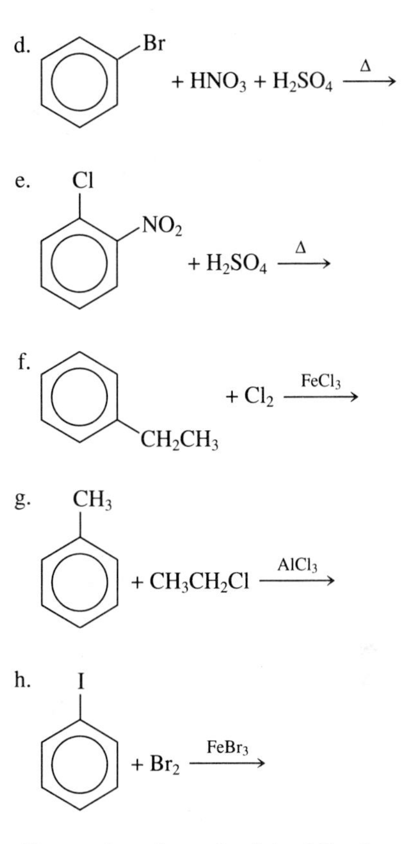

24. Give products for each of the following reactions:

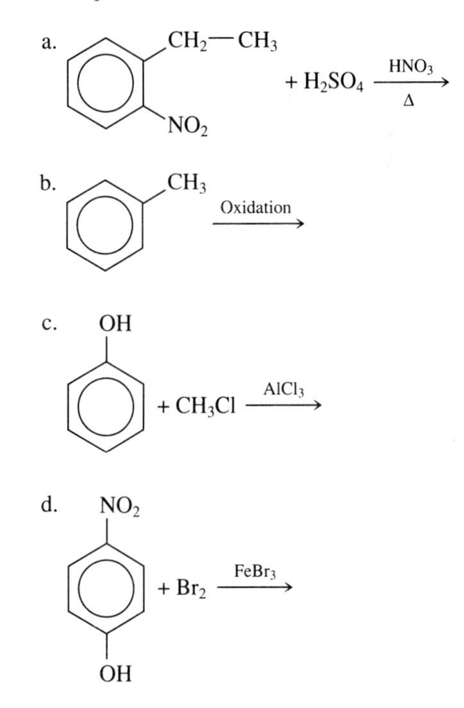

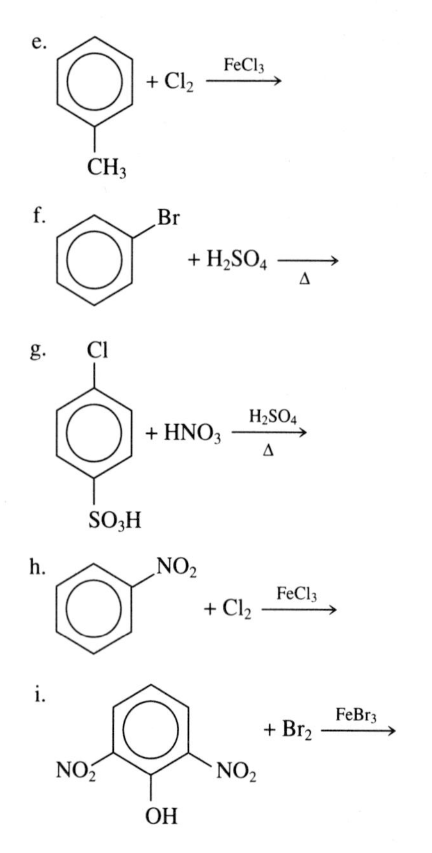

25. Give major products for each of the following reactions:

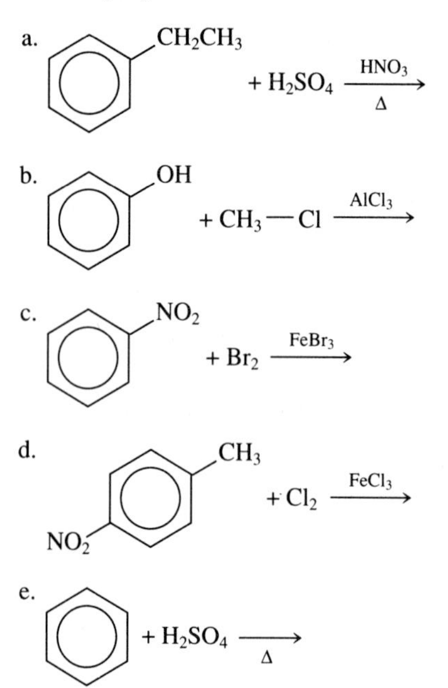

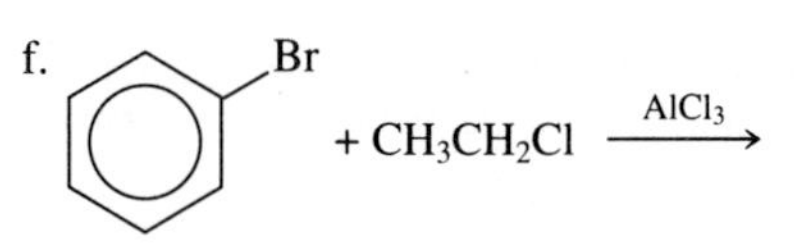

Unclassified Exercises

26. a. 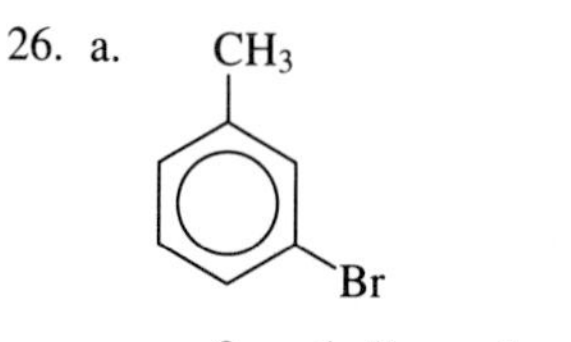

is best named

   a. 3-methylbromobenzene c. *p*-bromotoluene
   b. *o*-methyl-*p*-bromobenzene d. *m*-bromotoluene

b. Benzene derivatives undergo what type of reactions?
   a. aromatic electrophilic substitution
   b. aromatic nucleophilic
   c. electrophilic addition
   d. nucleophilic addition

c. When a benzene ring is used as an alkyl group, it is called
   a. benzyl
   b. aromaticyl
   c. phenyl
   d. none of these

d. When the reactants $HNO_3 + H_2SO_4$ and heat are used with benzene, the resulting product is
   a. nitrobenzene c. sulfobenzene
   b. benzenesulfonic acid d. no reaction

e. For a reaction of ethylbenzene, the ethyl is considered what type of directing group?
   a. ortho b. ortho–para c. meta d. ortho–meta

f. 2,3-Dichlorobromobenzene is

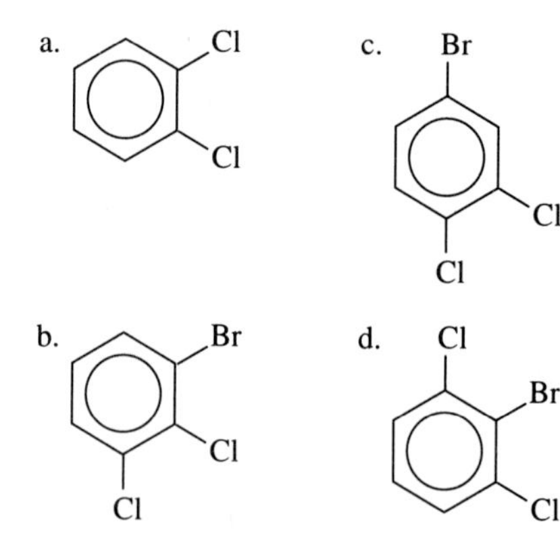

g. Which of the following is *not* aromatic?

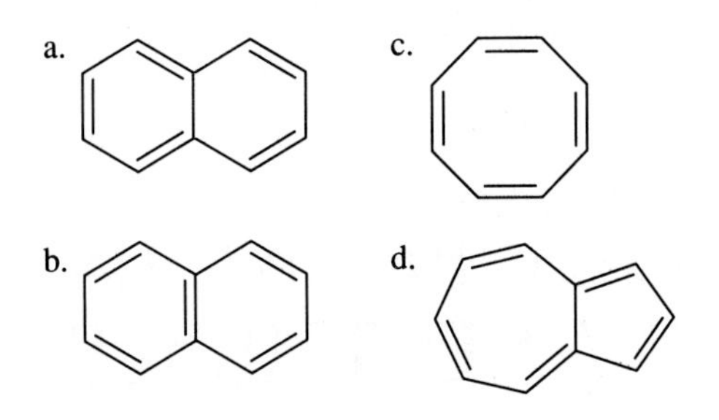

h. The carbon atoms in a benzene ring are ______ hybridized.
a. *sp* b. sp^2 c. sp^3

i. Which of the following resonance structures is most likely *not* to contribute to the stability of azulene?

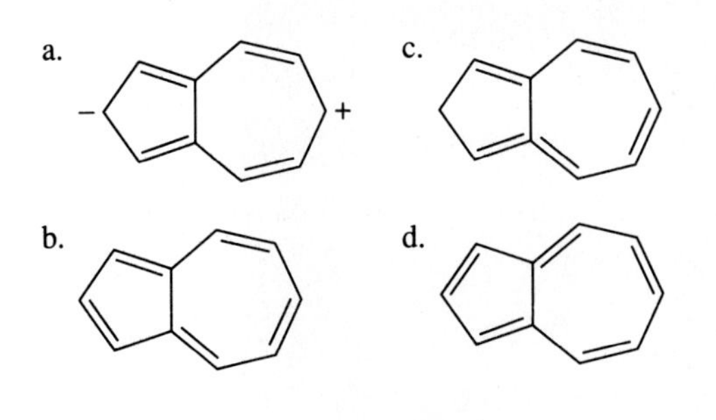

Sulfur compounds in marigolds are responsible for their unique odor.

ALCOHOLS, PHENOLS, ETHERS, AND THIOLS

OUTLINE

5.1 Alcohols
5.2 Phenols
5.3 Ethers
5.4 Thiols: Sulfur Equivalents of Alcohols

OBJECTIVES

After completing this chapter, you should be able to

1. Discuss how the structures of alcohols, thiols, and ethers are related to physical properties.
2. Identify and name representatives of the alcohol, phenol, thiol, and ether families.
3. Predict products from common reactions of these functional groups.

Oxygen is the most abundant element in the earth's crust, and water is one of the most abundant oxygen-containing molecules. In this chapter we introduce the oxygen-containing functional groups most related to water: those found in *alcohols, phenols*, and *ethers*. Alcohols and phenols are compounds in which one of the hydrogens of water is replaced with a carbon-containing group. Ethers are organic compounds that result from replacing both of the hydrogens with carbon-containing groups. Also included here are the *thiols*, which can be considered as the sulfur equivalent of alcohols; they are based on hydrogen sulfide (the sulfur equivalent of water). The relationship between water and these functional groups can be summarized as follows:

◆ There are structural similarities among water, alcohol, and ethers:

H—O—H Water

H—O—R Alcohol

R—O—R Ether

H—O—H $\xrightarrow{\text{Replace an H with R}}$ R—O—H $\xrightarrow{\text{Replace the second H}}$ R—O—R′

Water **Alcohol** **Ether**

(If R is aromatic,
this is a phenol.)
(If OH is replaced by
SH, this is a thiol.)

Reaction of sodium metal with ethanol gives sodium ethoxide and hydrogen. Ethanol is much less acidic than water, so its reaction with sodium is much less vigorous than the reaction of water with sodium.

5.1 Alcohols

Structure and Physical Properties

From application of the rules for Lewis electron-dot structures, an **alcohol** can be characterized by this general structure:

R—O—H

where R can be an alkyl or cycloalkyl (but not aromatic) group and —OH is the hydroxyl group. The oxygen atom of the alcohol has the same bonding characteristics that we have seen for water: sp^3 hybridization with two lone pairs of electrons.

Hydrogen Bonding and Boiling Point

For all of the functional groups discussed so far, a similar small increase of boiling point has occurred as the carbon number increases. We now begin to see a change in this relationship. The alcohol functional group (—OH) dramatically influences the physical properties of alcohols. First, notice the influence of the —OH group on boiling point (Figure 5.1). The curves for alcohol and alkane converge as carbon number increases. To explain this, let's reexamine an analogous molecule—water.

Although having a molecular weight of only 18, water is a liquid at room temperature and must be heated to 100°C at atmospheric pressure to be converted to a gas. In contrast, compare these properties with those of a similar molecular weight hydrocarbon. Methane, with a molecular weight of 16, is a gas at room temperature and doesn't liquify until it reaches the very low temperature of − 162°C. It is apparent that the —OH in the water plays a special role in its high boiling point. We would expect the similar structure of an alcohol to be reflected

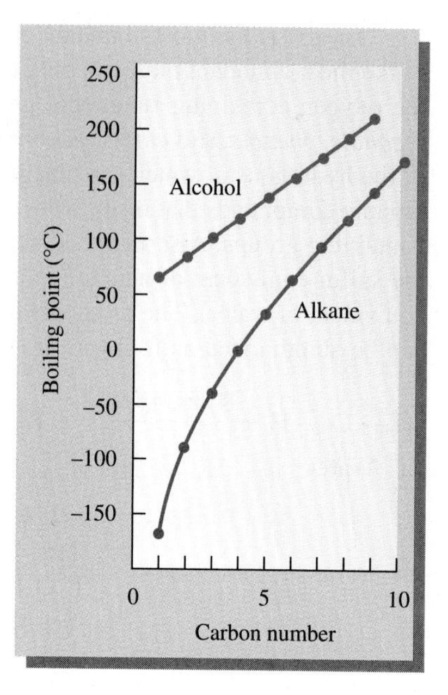

FIGURE 5.1
The relationship between carbon number and boiling point for the alcohols compared with the alkanes. For the lower carbon number alcohols the —OH group plays an important role through its hydrogen bonding and dominates the physical properties. As the carbon number increases, an alcohol more resembles an alkane and the influence of the —OH group decreases. The physical properties of long-carbon-chain alcohols more closely resemble the parent hydrocarbon.

in similar unique properties for alcohols. And this is in fact the case. The hydrogen-bonding capabilities of water and alcohols are responsible for their similarities in physical properties (Figure 5.2).

♦ Hydrogen bonding in alcohols influences the boiling points of the lower-carbon-number alcohols.

Energy is needed for the alcohol molecules to free themselves from the weak hydrogen-bonding forces that hold them together in the liquid state. Thus, alcohols have higher boiling points than we would expect based on the molecular weight.

It is apparent from Figure 5.1 that the hydroxyl group plays a more important role in determining the boiling points of the small-carbon-chain alcohols than for the longer chained alcohols. As the carbon chain becomes longer, the hydroxyl portion loses its *relative* importance and the molecule behaves more and more like a hydrocarbon. This ability of long hydrocarbon chains to mask functional group properties is repeated in a number of biologically important molecules.

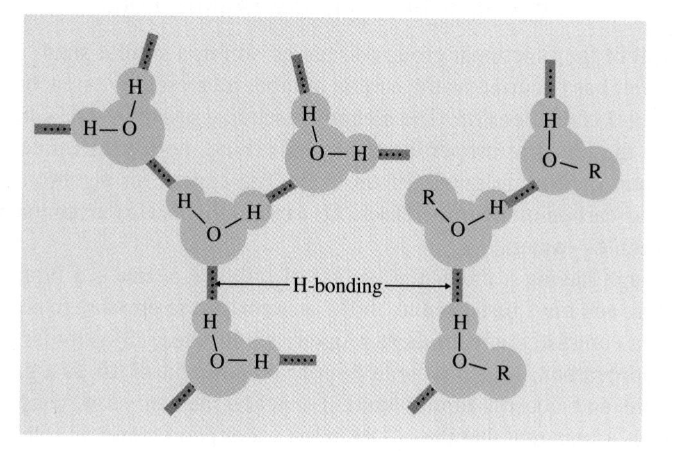

FIGURE 5.2
The structures of water and alcohols are very similar, as is their tendency to hydrogen-bond. Because there is only one hydrogen–oxygen bond per alcohol molecule, there is less hydrogen bonding.

This often allows a polar portion of a molecule to function in a hydrocarbon-like environment. We will see a number of examples of this interplay between hydrocarbon (nonpolar) character and polar functional groups throughout our study of biochemistry.

Solubility: Like Dissolves Like

Since alcohol so closely resembles water in its ability to participate in hydrogen bonding, it is no surprise that water and alcohol have similar solubility properties. Figure 5.3 shows the influence of hydrogen bonding on the ability of alcohols to dissolve in water. The compounds methanol (CH_3—OH) and ethanol (CH_3CH_2—OH) have infinite solubility in water. Just as the hydroxyl group has its main effect on the boiling points of the smaller alcohols, it also influences the

♦ Ethanol is the accepted IUPAC name. However, ethyl alcohol is the commonly used name.

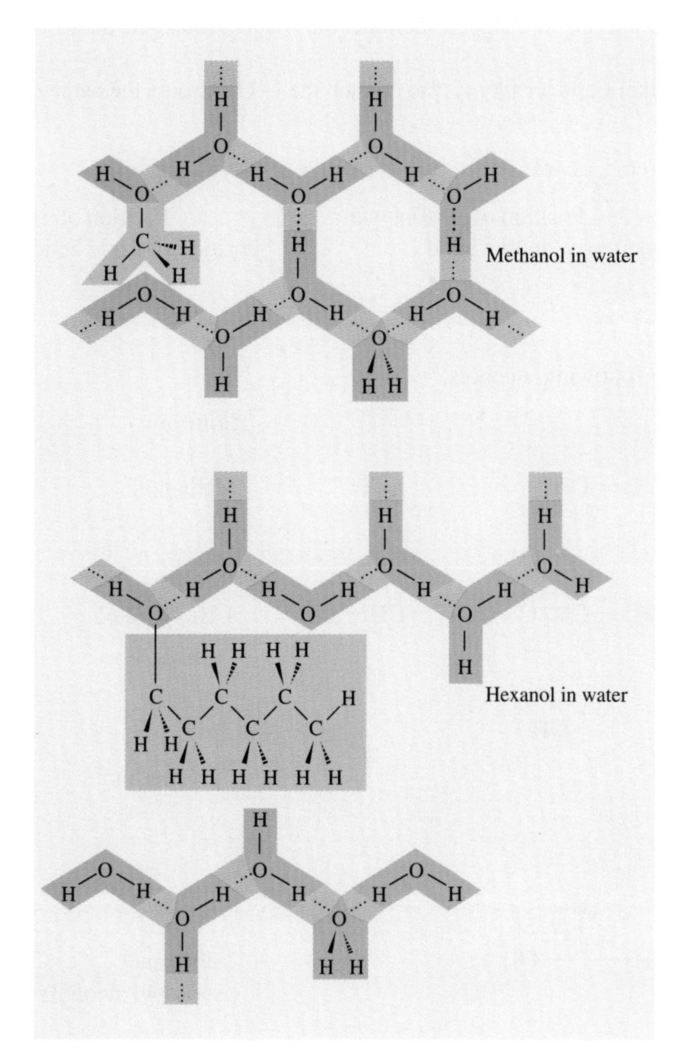

FIGURE 5.3
The polarity and the hydrogen bonding (dashed lines) of the —OH group helps an alcohol to dissolve in water. As the number of carbon atoms in the carbon chain increases, an alcohol starts to resemble a hydrocarbon and the solubility decreases until it becomes insoluble in water.

solubility of these lower carbon alcohols more. As the hydrocarbon portion of the alcohol increases in size, the alcohol starts to behave more like a hydrocarbon and thus becomes more insoluble in water (Figure 5.3).

Nomenclature of Alcohols: The Priority of the Hydroxyl Group

◆ In the nomenclature of alcohols, the *-ol* ending is used and a number is given for the position of the —OH group. This group takes on a numbering priority.

In previous chapters, we have developed organized ideas and rules for naming simple hydrocarbons and organohalogen compounds. With these rules, unambiguous names can be given to complex molecules. The same ideas can be applied to naming alcohols as well, with only one major change: the —OH group takes *priority* over alkyl groups or halogens in nomenclature. This simply means that a way must be found to give the position of the —OH group special significance. Here are the basic nomenclature rules for alcohols:

1. When the —OH group is present in a hydrocarbon, the *-e* ending of alkan*e*, alken*e*, and alkyn*e* is replaced by *-ol*.
2. The numbering is assigned such that the —OH group gets the lowest possible number.
3. If an alkene and an alkyne also contain the —OH group, the name is expressed as:

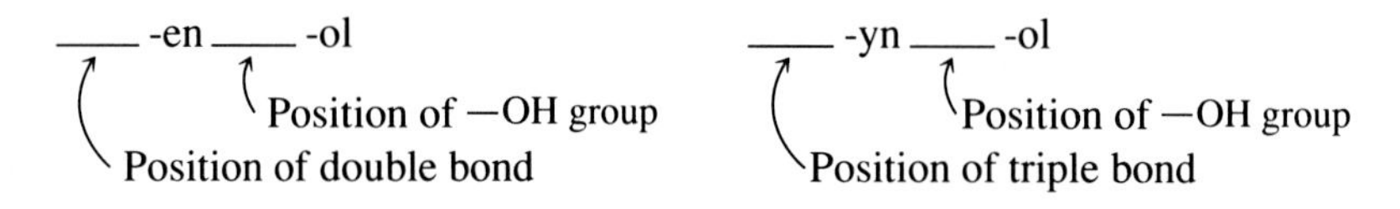

EXAMPLE 5.1

Name the following alcohols.

Solution:

$CH_3—OH$ — Methanol

$$\begin{array}{ccccccccccc} & & & & & & & & OH & & \\ & & & & & & & & | & & \\ CH_3 & — & C & \equiv & C & — & CH_2 & — & CH & — & CH_3 \\ 6 & & 5 & & 4 & & 3 & & 2 & & 1 \end{array}$$

4-Hexyn-2-ol

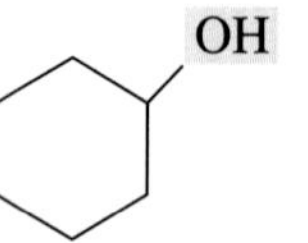

Cyclohexanol

$$\begin{array}{ccccc} & & H & & \\ & & | & & \\ CH_3 & — & C & — & OH \\ & & | & & \\ & & CH_3 & & \end{array}$$

2-Propanol
(isopropyl alcohol)

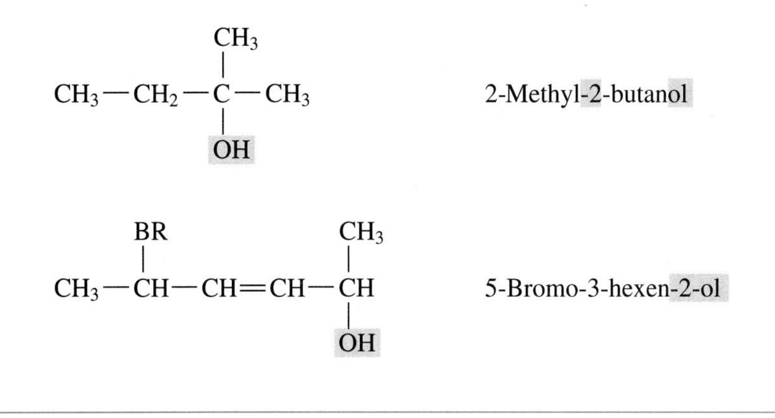

Some common artificial alcohols as well as a number of naturally occurring alcohols are shown in Table 5.1. Notice that some of them are known by a common name as well as an IUPAC name.

Reactions of Alcohols

As our study has progressed from the chemistry of alkanes to the alkenes, alkynes, and aromatics, the chemical reactivity was shown to be related to the availability of electrons, or the *Lewis basicity*. This is a direct reflection of the Lewis base character of many of the functional groups. We will see this property repeated over and over again in our study of organic chemistry and biochemistry.

In the reactions of the alkenes, alkynes, and aromatics, the electrophiles (represented by E^+ in Chapter 4) attack the electrons in the pi bonds. This behavior initiates most of the chemical reactivity of the unsaturated hydrocarbons. If an alcohol is considered to be a Lewis base as a result of the presence of lone pairs of electrons on the oxygen, it follows that alcohols will react with Lewis acids (electrophiles).

At this point it is appropriate to point out the ability of an alcohol to also function as a Brønsted–Lowry acid. By giving up the hydrogen attached to the —OH group (much like water can give up H^+), an alcohol can function as an acid.

$$R—O—H \rightleftarrows R—O^- + H^+$$

As with water, the equilibrium favors the undissociated alcohol.

Dehydration: An Elimination Reaction

♦ Alcohols can be prepared by the hydration of alkenes. The reverse reaction, loss of water, is called *dehydration.*

When an alcohol reacts with a strong acid (often H_2SO_4), **dehydration**, or the elimination of water, occurs and an alkene is formed.

TABLE 5.1

Examples of Common and Naturally Occurring Alcohols

Name	Description	Structure
Common Alcohols		
Methanol (wood alcohol)	A solvent and an intermediate for other compounds. Several billion pounds are converted to formaldehyde each year.	$CH_3—OH$
Ethyl alcohol (grain alcohol)	The fermentation alcohol (sometimes called *ethanol*) is used in alcoholic beverages and as a solvent.	$CH_3—CH_2—OH$
Butyl alcohol (isomer mixture)	A large quantity of butyl alcohol is made each year, and is used as a solvent and plastizer.	$C_4H_9—OH$
Natural Products		
Menthol	Found in peppermint oils. Used in throat sprays, cough crops, and inhalers.	CH_3; OH; CH; CH_3 CH_3
Cholesterol	Necessary for the preparation of steroids. Found in all body tissues, especially the brain. A main constituent of gallstones.	CH_3; CH_3 CH; CH_2; CH_3; CH_2 CH_2; CH; CH_3 CH_3; HO
Grandisol	Boll weevil pheromone	CH_3; $CH_2—CH_2OH$; CH_3; C; ‖; CH_2

EXAMPLE 5.2

Show the reaction product of these alcohols with H_2SO_4.

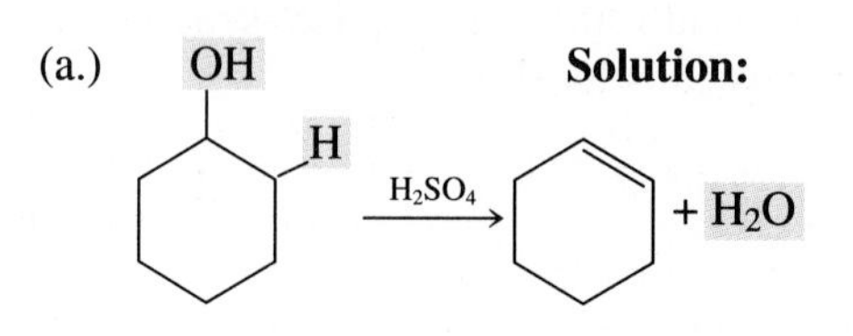

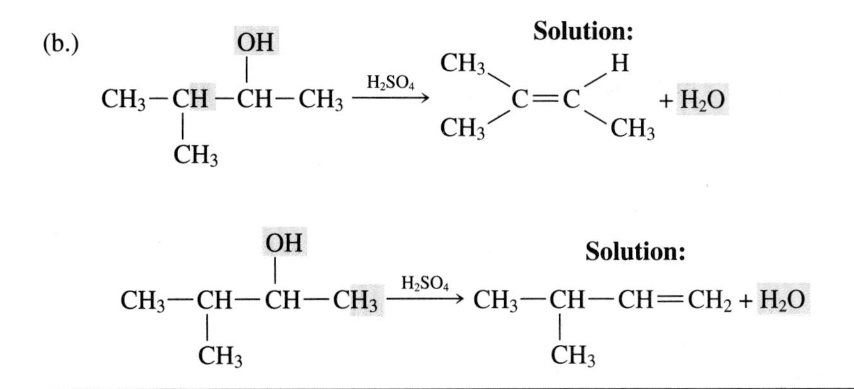

Note that the second product for Example 5.2b is *not* found in the actual reaction. In general, it is observed that the *more substituted* alkene is favored in dehydration reactions. What do we mean by more substituted? Examine the atoms directly attached to the double-bond carbon atoms—the fewer the hydrogens, the more substituted the alkene.

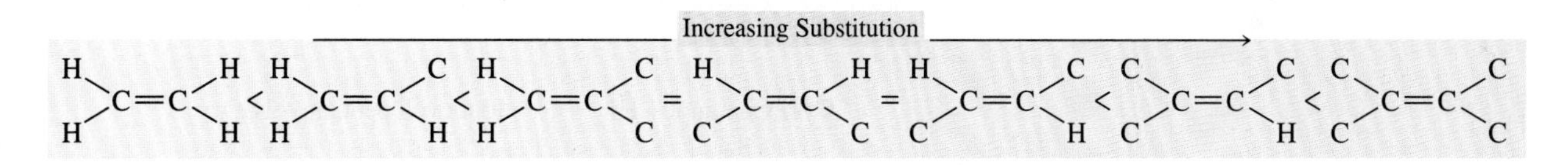

One of the important steps in the biological conversion of organic materials to carbon dioxide and water is the dehydration of citric acid. Essentially all organic compounds are converted to citric acid as part of human and animal metabolism.

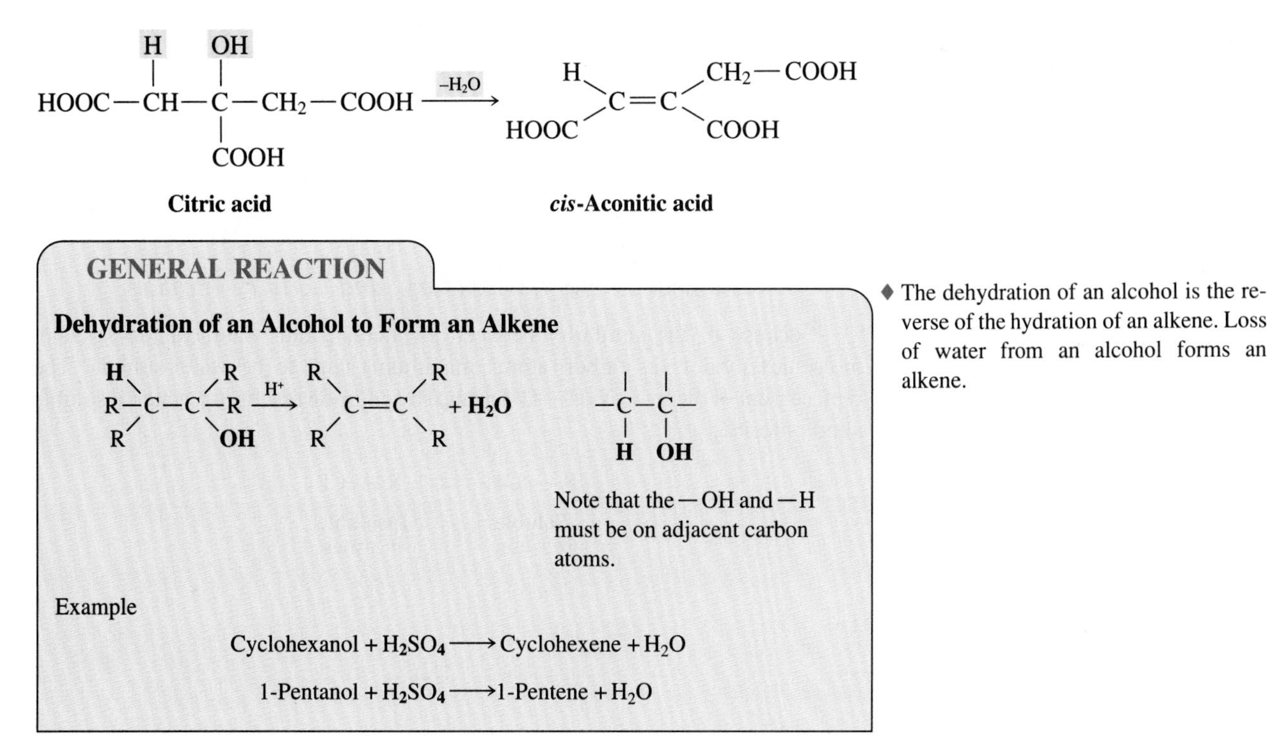

◆ The dehydration of an alcohol is the reverse of the hydration of an alkene. Loss of water from an alcohol forms an alkene.

Formation of Alkyl Halides

♦ An alcohol can be converted to an organohalogen.

Alcohols are often used as intermediates for preparing alkyl halides (organohalogen compounds):

$$R—OH \longrightarrow R—X$$

This reaction can be accomplished by reacting the alcohol with HX. Hydrogen iodide and hydrogen bromide have been found experimentally to be the best hydrogen halides for this reaction.

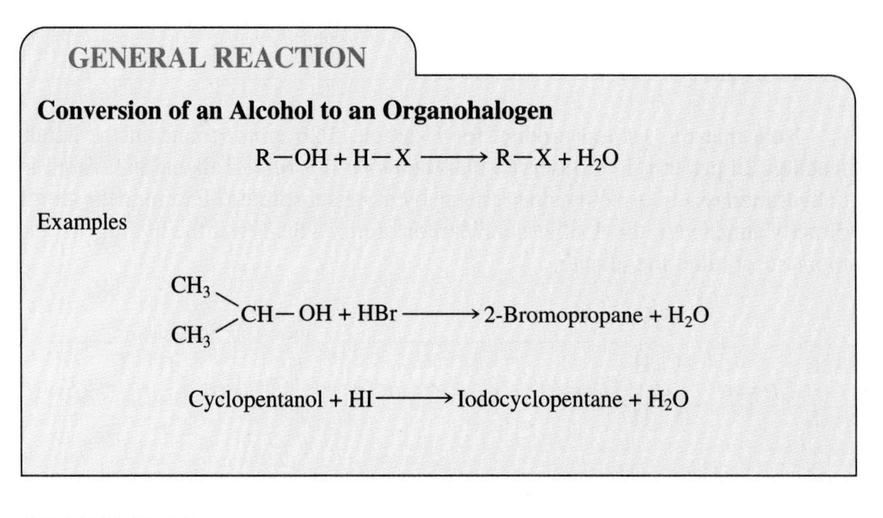

EXAMPLE 5.3

Show the alkyl halide products of these reactions:

Solutions:

$$CH_3—CH_2—OH \xrightarrow{HI} CH_3—CH_2—I + H_2O$$

$$CH_3—CH(CH_3)—OH \xrightarrow{HCl} CH_3—CH(CH_3)—Cl + H_2O$$

Testing for Primary, Secondary, and Tertiary Alcohols

Because of the similarity between an alcohol and water, smaller alcohol molecules (about six carbon atoms maximum) tend to be water-soluble. The replacement of the hydroxyl (—OH group) by chloride results in a water-insoluble alkyl chloride.

$$R—OH \longrightarrow R—Cl$$

Soluble in water	**Insoluble in water**

CHEMISTRY CAPSULE

Alcoholic Beverages

The fermentation of sugar generates ethyl alcohol, or ethanol. The first examples of alcoholic beverages were probably the result of natural yeasts acting on the sugar in honey to form mead. There is evidence that this drink was known as long as 10,000 years ago. It is reasonable to suspect that the early interest in alcohol had its origins in the observation that fermented foods kept better—they did not spoil. Even the ancients had problems with maintaining supplies of pure water, and fermented juices were found less likely to make a person sick. The ancient Egyptians wrote of the use of beer and wine as medicine. Later, as humans were able to control this process, specialized methods were developed to make drinks.

Natural fermentation generally does not produce ethyl alcohol contents greater than 15%, as this is about the limit of tolerance for the yeast to alcohol. To obtain greater alcohol content, the ethyl alcohol must be distilled from the solution. Since ethyl alcohol has a lower boiling point than water, this provides greatly purified ethyl alcohol.

Most alcoholic beverages have unique flavors and qualities. These flavors are generally due to substances other than ethyl alcohol, which in its pure form has no flavor. The unique tastes come from the materials fermented, the addition of flavorings, or the wooden containers that the beverage is aged in. Here are a few of the common groups of alcoholic beverages.

Table Wines These wines, sometimes called *still wines*, are widely produced throughout countries of the temperate zone, generally from various varieties of grapes. Many of these wines are named for the locations in which they were first made. Examples include Burgundy, Bordeaux, Claret, Tokay, Rhine, Zinfandel, Sauterne, Chablis, and Chianti. Wine can also be prepared from other fruits and berries. Most table wines have alcohol contents in the range of 10 to 14%.

Sparkling Wines Champagne (from the Champagne area of France) and other similar wines contain carbon dioxide.

Dessert Wines Sometimes classified as *fortified wines*, these wines have additional alcohol added to them, bringing the alcohol content to about 20%. Representatives of these wines include Sherry, Port, Muscatel, Madeira, and Marsala.

FIGURE 5.A
Many alcoholic beverages have their beginning in fruits and grains.

Aromatized Wines Addition of various herbs and spices gives wines such as Vermouth and Dubonnet their unique flavors and odors.

Beer Beer is formed by the fermentation of barley with the addition of hops for flavoring. Lager beer generally has an alcohol content of 3 to 6%, and stout has a content of 4 to 8%.

Liquors These can contain up to 50% alcohol (100 proof) as well as a variety of flavorings.

Brandy Distilled from wines, the generic name *brandy* includes cognac, absinthe, and Benedictine.

Whiskey These include a number of beverages prepared from fermented corn, rye, and barley. Bourbon is a corn-fermented product; Scotch is primarily fermented from barley.

Rum Prepared from fermented molasses and sugar cane

Gin Essentially an ethyl alcohol solution flavored with juniper

Vodka Prepared from fermented potatoes; essentially an unflavored solution of ethyl alcohol.

Sake The Japanese drink from fermented rice (also classified as a wine)

♦ The Lucas test differentiates among small-carbon-chain alcohols by differences in reactivity. The formation of two layers is the indication of a positive test.

The **Lucas test** uses the differences in solubility between small-molecular-weight alcohols and the corresponding organohalogen compunds (alkyl halides) to test whether an alcohol is primary, secondary, or tertiary. (Notice that the alkyl group of the alcohol is the alkyl group found in the organohalogen compound.) A *primary alcohol* is one in which the carbon bearing the —OH group is attached to only hydrogens or one alkyl group. A *secondary alcohol* has the same carbon atom attached to two other alkyl groups, and a *tertiary alcohol* has the same carbon atom attached to three other alkyl groups. A mixture of HCl and $ZnCl_2$ is called the *Lucas reagent* and reacts with an alcohol according to the following reaction:

$$R—OH \xrightarrow[\text{(Lucas reagent)}]{HCl + ZnCl_2} R—Cl$$

A tertiary alcohol reacts almost instantly. A secondary alcohol takes 5 to 10 min to react, and a primary alcohol does not react at all within 10 min.

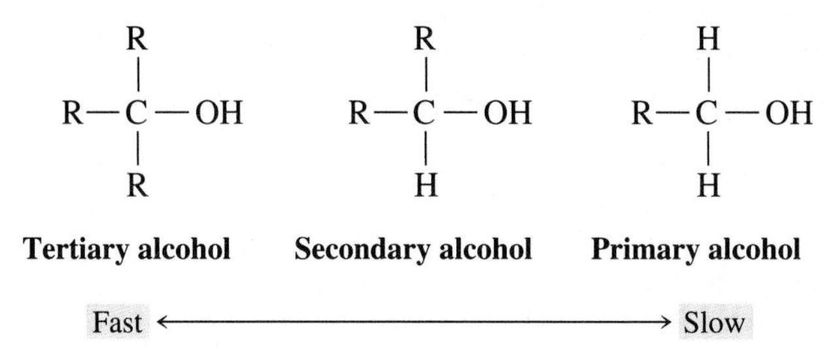

Special Alcohols: Polyhydroxylic Compounds

♦ Alcohols with more than one —OH group are called *polyhydroxylic*. Compounds with two —OH groups are often called *glycols*.

There are many compounds that contain more than one —OH group per molecule. These kinds of compounds are named in the same way as ordinary alcohols except that each —OH is given a number to show its position, and the prefixes *di-*, *tri-*, and so on are used to designate the number of —OH groups. Common examples include ethane-1,2-diol (commonly called ethylene glycol) and propan-1,2,3-triol (commonly known as glycerol or glycerine).

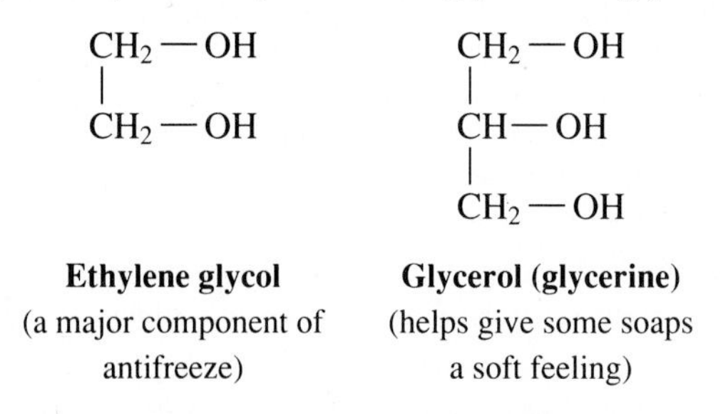

Ethylene glycol (a major component of antifreeze)

Glycerol (glycerine) (helps give some soaps a soft feeling)

An important characteristic of these polyhydroxylic compounds is their water-solubility. This property is expected since they closely resemble water, and the large ratio of —OH to hydrocarbon negates much of the nonpolar hydrocarbon character. The water-solubility of ethylene glycol makes it possible to add it to water, in any proportion, for antifreeze. Also, its high boiling point (187°C) prevents it from evaporating from the hot water in a car's radiator.

The common name **glycol** is used for alcohols having two —OH groups. The antifreeze properties of glycols are not unique to human use. Many butterflies produce a mixture of glycols, which circulate in their body fluids before cold weather arrives. This "butterfly antifreeze" allows them to survive the winters.

The process of fermenting sugars to produce ethyl alcohol is an important reaction of the carbohydrates. In a later chapter, the details of the *fermentation* process (a reaction carried out by yeast) will be discussed. Because of the potential use of ethyl alcohol as a fuel (such as in Gasohol, a mixture of one part ethyl alcohol and nine parts gasoline), as a disinfectant, and as a component in alcoholic beverages, it is important to recognize here that grain or ethyl alcohol is readily produced by fermentation of glucose and other carbohydrates.

$$\text{Glucose} \xrightarrow{\text{Fermentation}} \text{Ethyl alcohol}$$

NEW TERMS

alcohol A functional group characterized by the general formula R—O—H

dehydration The formation of an alkene from an alcohol as a result of the loss of water

Lucas test A test for identifying whether an alcohol is primary, secondary, or tertiary by its rate of conversion to a chloroalkane

glycol A dihydroxylic alcohol, that is, one with two —OH groups

TESTING YOURSELF

Alcohols

1. Name the following simple alcohols:

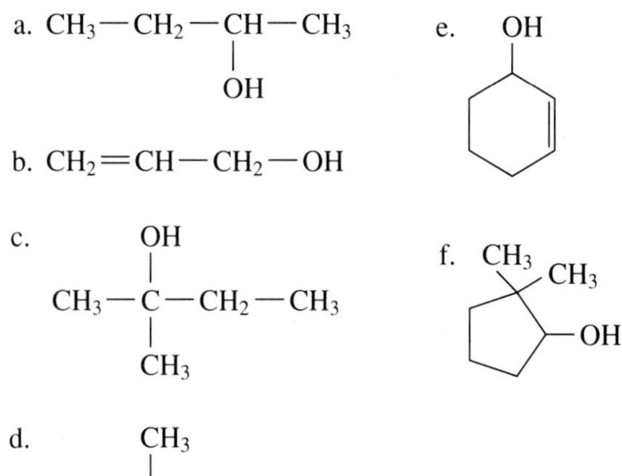

2. Draw structures for each of the following compounds:
   a. 5,5-dibromo-3-hepten-2-ol c. 4,4-dichlorocyclooctanol
   b. 3-hexyn-2-ol

3. Give the major product for each of the following dehydration reactions:

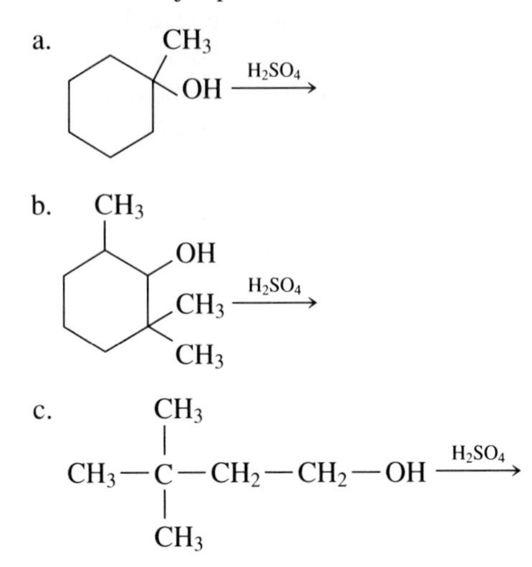

4. Give products for each of the following:

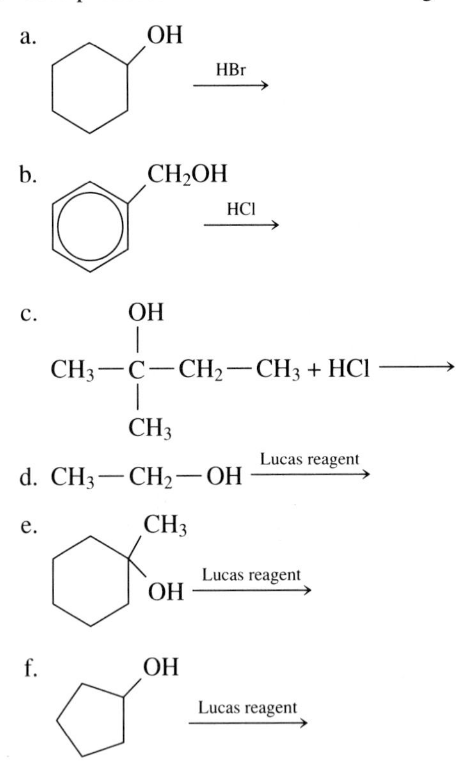

Answers **1. a.** *sec*-butyl alcohol or 2-butanol **b.** 2-propen-1-ol **c.** 2-methyl-2-butanol **d.** 2-methyl-3-pentyn-1-ol **e.** 2-cyclohexenol **f.** 2,2-dimethylcyclopentanol

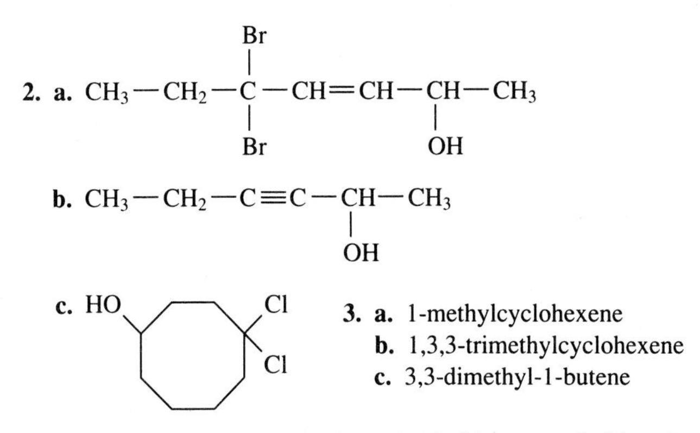

3. a. 1-methylcyclohexene
b. 1,3,3-trimethylcyclohexene
c. 3,3-dimethyl-1-butene

4. a. bromocyclohexane **b.** benzyl chloride **c.** 2-chloro-2-methylbutane **d.** No reaction within 10 min. **e.** Almost immediate conversion to the 1-chloro-1-methylcyclohexane and formation of two layers. **f.** The formation of two layers (the product, chlorocyclopentane, is insoluble in water) takes between 5 and 10 min.

5.2 Phenols

Structure

The **phenols** appear similar to alcohols but are identified by the R group being an aromatic ring. The most common molecule of the phenol class of compounds is phenol itself.

♦ *Phenols* are characterized as a benzene ring with an —OH group attached.

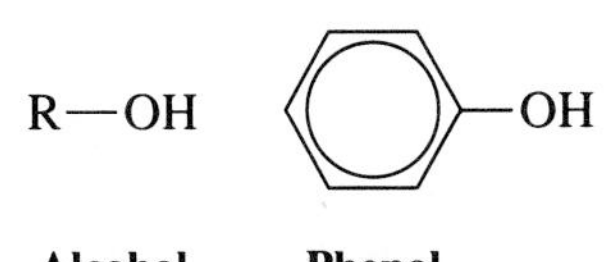

Carbolic acid is an old name for phenol, and although it does not receive much modern usage, the term is still seen on occasion. It might be of interest to note that the IUPAC has suggested that phenol be named *benzenol*; however, this has not met with much acceptance among practicing chemists.

Nomenclature: Use Aromatic Rules

We have already seen the extensive IUPAC nomenclature for aromatic compounds. The same ideas as applied to toluene (methylbenzene) are used for the nomenclature of phenols. Phenol is used as the parent compound, and other compounds are named as derivatives of phenol. Thus, either the ortho, para, and meta nomenclature can be used or the numbering methods. In all cases, the positions of new groups are relative to the position of the —OH group of phenol.

♦ Nomenclature of phenols follows the rules of aromatic chemistry.

TABLE 5.2
Comparison of Alcohol and Phenol Acidity

Alcohol Reactions	Phenol Reactions
$2\ R—OH + 2\ Na \longrightarrow 2\ R—O^-Na^+ + H_2$*	$C_6H_5—OH + Na \longrightarrow C_6H_5—O^-Na^+$
$R—OH + NaOH \longrightarrow$ No reaction	$C_6H_5—OH + NaOH \longrightarrow C_6H_5—O^-Na^+$
$R—OH + NaHCO_3 \longrightarrow$ No reaction	$C_6H_5—OH + NaHCO_3 \longrightarrow$ No reaction

* This reaction is more correctly an oxidation–reduction reaction.

EXAMPLE 5.4

Name the following phenols:

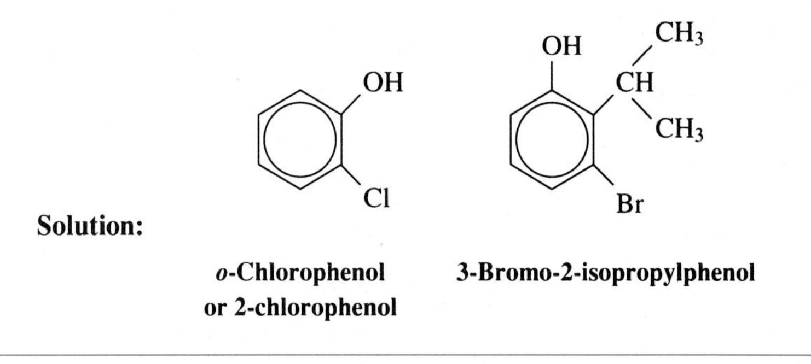

Solution:

***o*-Chlorophenol or 2-chlorophenol** **3-Bromo-2-isopropylphenol**

Reactions of Phenols

Acidity of the —OH Bond: A Qualitative Look

♦ Phenols are much more acidic than alcohols.

An important difference between alcohols and phenols can be found in the relative acidities of the —OH hydrogen. In fact, phenol was once called *carbolic* acid to reflect its substantial acidity. This is qualitatively demonstrated by the comparisons shown in Table 5.2. Alcohols and phenols both react with metallic sodium to produce anions and hydrogen gas. Sodium hydroxide (NaOH) is not a strong enough base to remove the hydrogen from an alcohol, but it is strong enough to do so from phenols. Sodium bicarbonate ($NaHCO_3$), a very weak base, does not react with either an alcohol or a phenol.

Because the —OH group in phenol is attached to a carbon that has sp^2 hybridization, phenols do *not* undergo dehydration or conversion to alkyl halides as observed for alcohols. Thus, at this point, it seems that only acidity distinguishes the —OH group of phenols from that of alcohols. However, as we will see in the following section, the presence of the aromatic ring on phenols allows some interesting aromatic substitution reactions of phenols.

Aromatic Substitution Reactions

Phenols react like other aromatic compounds except that they are strikingly more reactive. The phenolic —OH group is a powerful activator (ortho–para director) of the benzene ring for aromatic electrophilic substitution reactions, and aromatic substitution reactions of phenols tend to take place very easily. The general reaction for an aromatic substitution reaction of phenol is shown here, where E^+ can be any of the electrophiles discussed in the previous chapter.

♦ Phenols undergo aromatic substitution reactions, with the —OH group acting as a powerful ortho–para director.

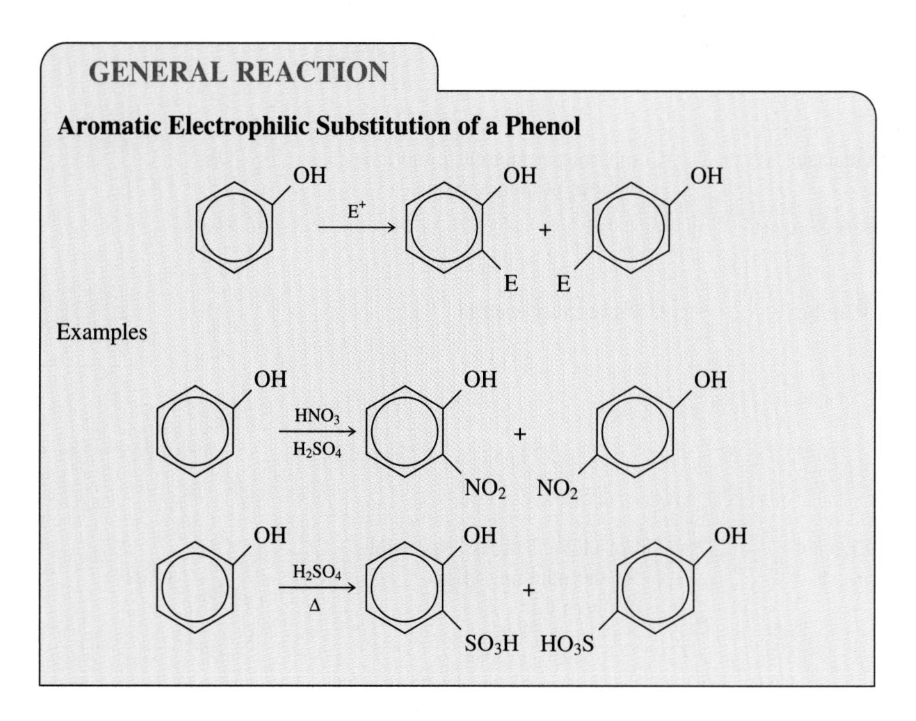

Special Phenols

Phenols are found in a variety of natural and synthetic products. Phenolic materials can be responsible for burning and blistering (as in poison ivy) and also can be used as highly effective antiseptics. One of the earliest uses of phenol as an antiseptic occurred in 1867 when Joseph Lister found that a solution of phenol killed bacteria. Examples of phenols are given in Table 5.3.

NEW TERM

phenol Any aromatic ring with an —OH group is classified as a phenol. Phenol is also the name for the simplest member of this class.

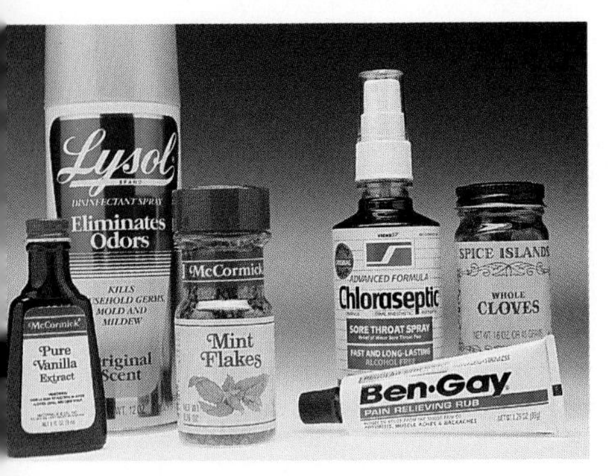

The —OH alcohol group is found in the menthol of mint, and the —OH phenol group is found in oil of clove. These groups are incorporated into cleaners, disinfectants, and analgesic agents. The —OH group is associated with many interesting compounds.

TABLE 5.3
Examples of Common Phenols

Name	Description	Structure
Methyl salicylate	Oil of wintergreen	OH; $C-OCH_3$; O
Urushiol	Constituent of poison ivy	OH; OH; $C_{15}H_{31}$
Guaiacol	Common expectorant found in a number of cough syrups.	OH; OCH_3
Vanillin	Component of vanilla	OH; OCH_3; C; O; H
Eugenol	Oil of clove. This is responsible for the numbing effect of clove oil.	OH; OCH_3; $CH_2-CH=CH_2$
Thymol	Constituent of mint	CH_3; CH; CH_3; CH_3; OH
2-Phenylphenol	Constituent of Lysol	OH

TESTING YOURSELF

Phenols

1. Name the following phenols:

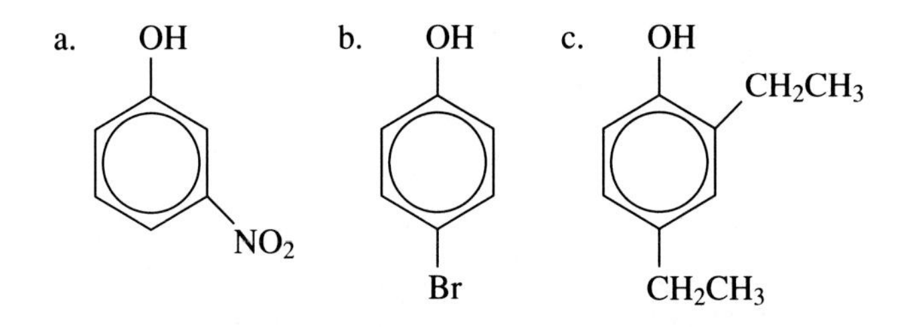

2. Which is more acidic, 2-propanol or *m*-nitrophenol? (Draw both.)
3. Give products for the following phenol reactions:

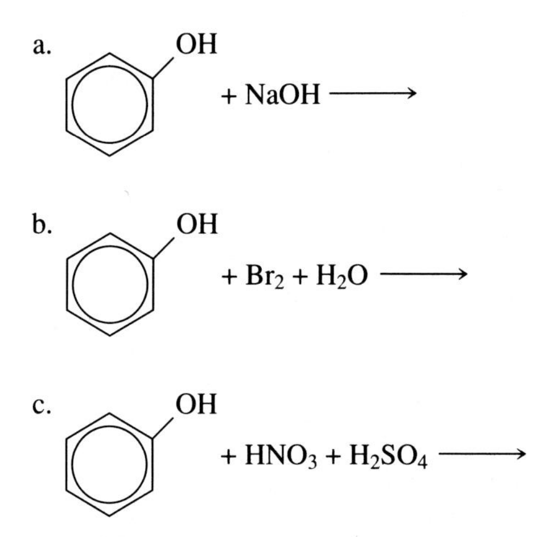

Answers **1. a.** *m*-nitrophenol (3-nitrophenol) **b.** *p*-bromophenol (4-bromophenol) **c.** 2,4-diethylphenol **2.** Phenols are always more acidic. The structures are

$$CH_3-\underset{\displaystyle OH}{\underset{|}{CH}}-CH_3$$

Isopropyl alcohol

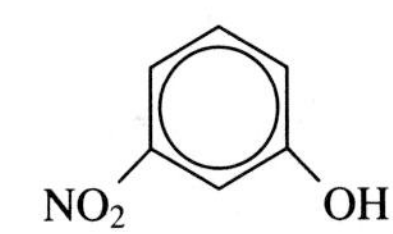

m-Nitrophenol

3. a.

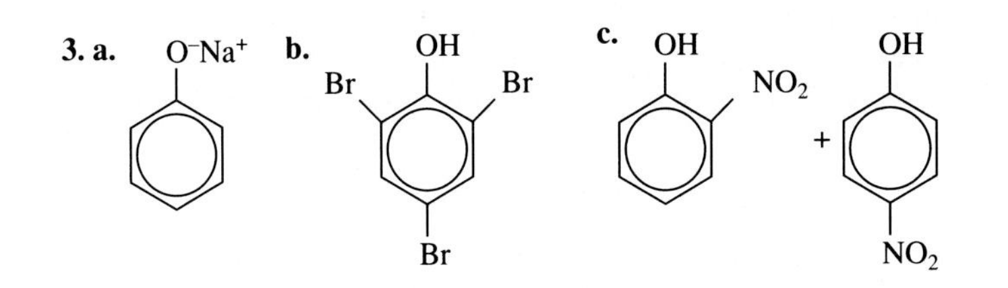

5.3 Ethers

Structure

◆ Ethers resemble polarized hydrocarbons.

R—O—R H—O—H

An **ether** can be considered to be an alkane in which an oxygen atom has been substituted for one of the $—CH_2—$ groups.

$$\underset{\textbf{Alkane}}{R—CH_2—R} \qquad \underset{\textbf{Ether}}{R—O—R}$$

They are perhaps more properly considered as analogs of water, in which both hydrogens have been replaced by R— groups. Thus, ethers have the unique ability to have some polarity associated with the oxygen parts of their structure while having solubility and reaction characteristics more closely resembling the hydrocarbons. This makes them very good solvents.

What is the result of the substitution pattern found in ether relative to water? First, since the oxygen is more electronegative than the carbon, there is a polarization of the C—O bond toward oxygen. This polarization leads to a slight overall polarity for the ether. Since the ethers are so similar to hydrocarbons and have no O—H bonds (and hence no hydrogen bonding between ether molecules), their boiling points are quite similar to those of hydrocarbons (Figure 5.4).

◆ The boiling points of ethers are similar to the corresponding hydrocarbons and lower than the alcohols of similar formula. This is due to the lack of hydrogen bonding between ether molecules.

Because ethers possess a great deal of hydrocarbon character, they are able to dissolve many hydrocarbons; at the same time, the slight polarity of the oxygen allows for weak hydrogen bonding with alcohols or water.

$$R_2O \cdots H—O—R \qquad R_2O \cdots H—O—H$$

Thus, an ether is usually more water-soluble than its corresponding hydrocarbon. A useful rule of thumb is that compounds having a ratio of four carbons per oxygen or less are water-soluble. Thus, $CH_3—O—CH_3$ is miscible with water in all proportions, whereas $CH_3—CH_2—O—CH_2—CH_3$ is much less soluble (10 g per 100 g of water).

Nomenclature

Compounds of the general structure

$$R—O—R \qquad Ar—O—R \qquad Ar—O—Ar$$

are classified as ethers, where Ar— is a symbol for a general aromatic group. Although there is a systematic IUPAC nomenclature, it is more common to name these compounds according to the alkyl (or aromatic) groups on each side of the oxygen. For example,

$$CH_3—O—CH_3$$

is called *dimethyl ether*. Unsymmetrical ethers are named in the same manner.

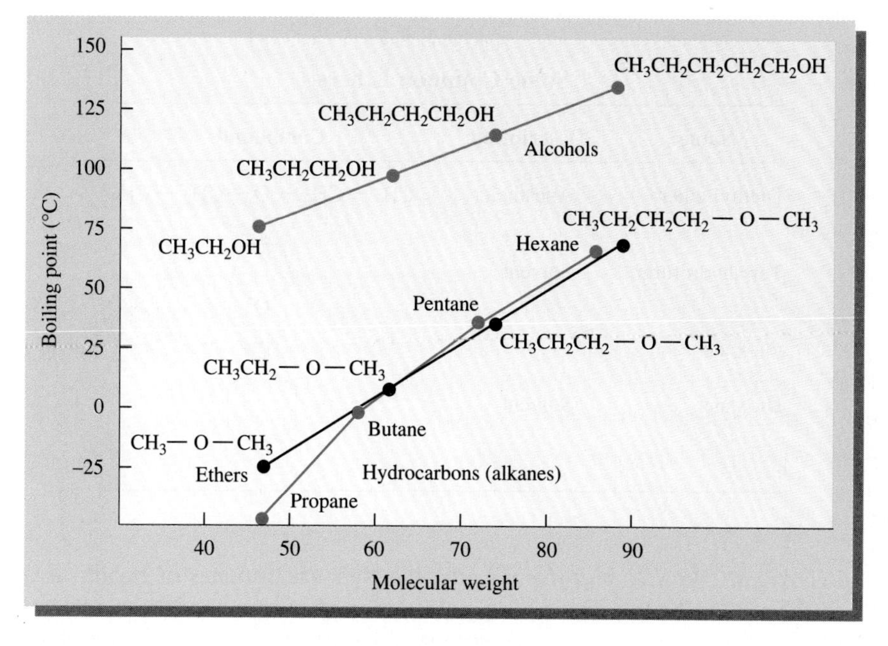

FIGURE 5.4
The relationship between molecular weight and boiling points of ethers, hydrocarbons, and alcohols of similar molecular weights. Note the higher boiling points of alcohols, caused by the hydrogen bonding. These data show that hydrogen bonding is not present in ethers and that their physical and chemical properties more closely resemble the alkanes.

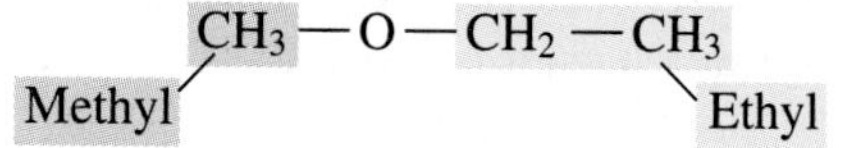

Thus, this compound is called ethyl methyl ether. Since a benzene ring is called *phenyl* when it is used as a substituent,

$$C_6H_5-O-CH_2-CH_2-CH_3$$

is called *phenyl propyl ether*. Three very common ethers and their names are given in Table 5.4.

Special Ethers: The Epoxides

An **epoxide** is the ether analog of a cyclopropane ring.

◆ An *epoxide* is a cyclic ether related to cyclopropane.

Cyclopropane **Epoxide of ethylene**

Epoxides are very important intermediates in many industrial and laboratory procedures. They also play a very important role in biological processes.

Epoxides and Carcinogenic Compounds

In the last chapter we introduced the idea that biological oxidation of aromatic compounds results in harmful products. There is a wealth of experimental evidence suggesting that biological oxidation of aromatic compounds occurs by way

TABLE 5.4
Some Common Ethers

Name	Description	Compound
Diethyl ether	Anestheic	$CH_3—CH_2—O—CH_2—CH_3$
Tetrahydrofuran	Solvent	
Dioxane	Solvent	

◆ Diethyl ether is commonly known as ethyl ether.

of epoxides. The *NIH shift* (named after the National Institutes of Health) is a name applied to the conversion of an aromatic compound to a phenolic compound through an epoxide intermediate. Once the epoxide is formed, the aromatic character is lost (recall the resonance stability). To regain the energy, the ring is rearomatized. In doing so, a phenol is formed.

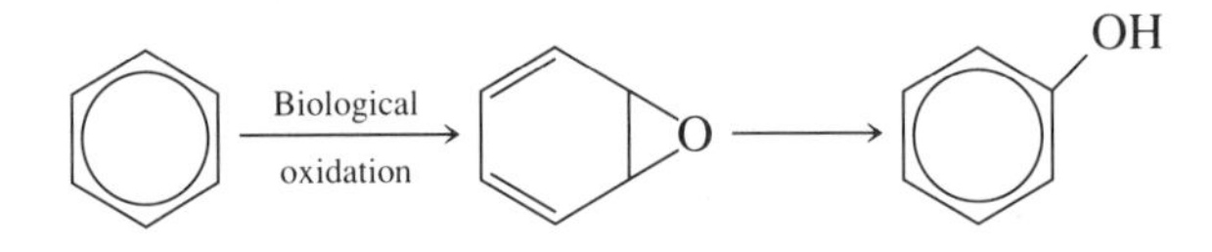

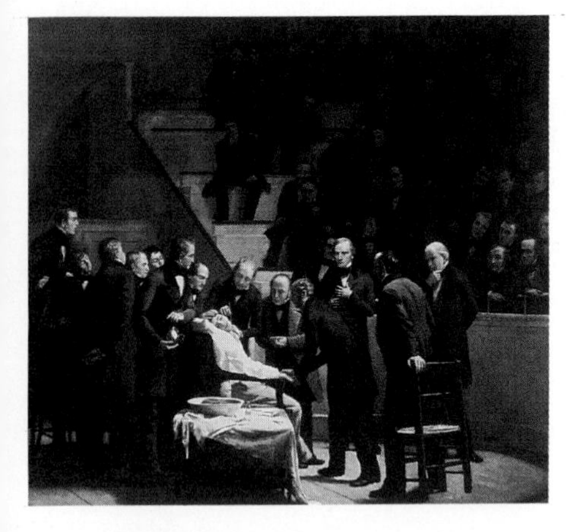

FIGURE 5.5
This painting by Robert Hinckley shows the first use of ether as an anesthetic in 1846. The patient in the picture, Gilbert Abbott, was having a tumor removed from his neck by Dr. John Collins. The ether is being administered by the dentist, W. T. G. Morton, who discovered the anesthetic properties of ether.

There is evidence that the cancer-causing, or **carcinogenic**, properties of aromatic compounds are directly linked to the epoxides of these compounds. For example, 1,2-benzpyrene is formed as a product of incomplete combustion of aromatic compounds (cigarette smoke, fat from cooking meat dropping onto hot coals, and so on) and is known to be a potent carcinogenic compound. In the body it becomes epoxidized and the *product* of this reaction causes the problems.

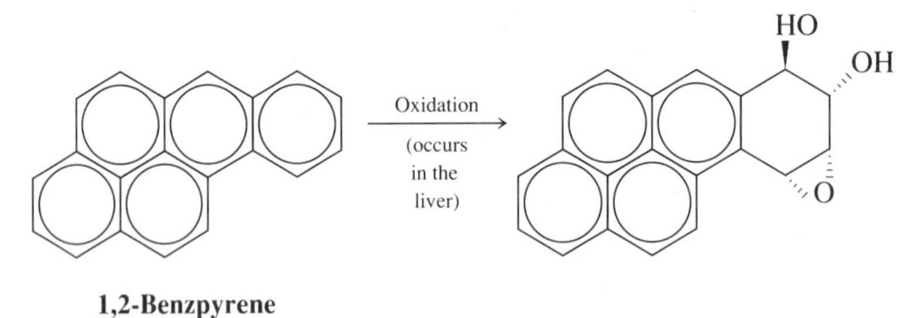

Ether as an Anesthetic

From the earliest use of diethyl ether as a general anesthetic in 1846 (Figure 5.5), there has been constant interest in developing new, safer anesthetics and in trying to understand how they work.

CHEMISTRY CAPSULE

Dioxin

Dioxin is an ether with the following structure:

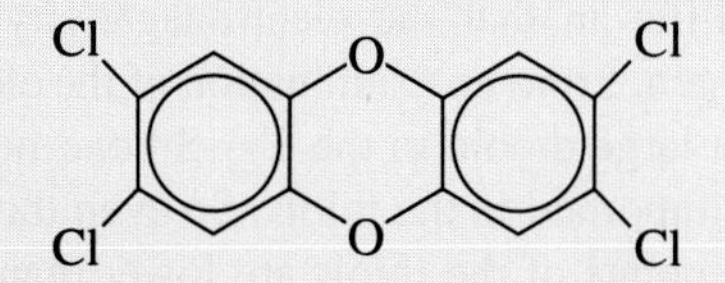

Although dioxin is widely thought of as a serious health and environmental problem, there are some differences of opinion. A by-product of the manufacturing process for making herbicides (including Agent Orange), dioxin has been around for about 40 years. The molecule is quite stable and is strongly held to soil. However, in direct sunlight it is more rapidly decomposed. The chemical details of the decomposition are not fully understood.

Dioxin is highly toxic to some experimental animals. Indeed, a lethal dose for a guinea pig is only about 0.6 μg (microgram) per kilogram of body weight. However, for a hamster it requires almost 2000 times more for the same level of toxicity. Reproductive problems have also been noted for experimental animals. Dioxin is also a proven carcinogen in rats.

Many symptoms of exposure to dioxin have been noted for humans as well, including dermatitis, aches, digestive disorders, effects on the nervous system, and some psychological effects. However, in a recent review of dioxin, the long-term human impact was examined and found to be nonexistent.* For example, careful monitoring of workers exposed to high levels in 1949 showed no long-term effects, including birth defects or chromosomal damage. Clearly more studies are needed.

What is the important lesson from this? Chemicals are always going to be with us. Some are harmful, some are not. Federal and local agencies are looking out for our welfare and have made the determination that chemicals harmful to animals are assumed to be harmful to humans. This may not always be the case; however, it is much better to err on the side of increased caution.

* From Fred H. Tschirley, *Scientific American* (1986), Vol. 254, No. 3, p. 29.

The electrical signals that are involved in transmission of nerve impulses require the presence of sodium ions (Na^+). These ions must pass through cell channels. When certain kinds of molecules (such as ether) are dissolved in the hydrocarbon layer (or fat layer) of the cell, the channel does not open on the arrival of an impulse and sodium ions are blocked. There is a direct relationship between a compound's solubility in hydrocarbon and its effectiveness as an anesthetic. Ether and chloroform are more effective as anesthetics than nitrous oxide or carbon tetrafluoride because they are more soluble in the fat layer of the cell.

♦ Anesthetic ability can be related to solubility in hydrocarbon.

NEW TERMS

ether An organic functional group characterized by the bonding

$$R—O—R$$

epoxide A three-membered, oxygen-containing ring

carcinogenic Capable of inducing the formation of cancer cells

5.4 Thiols: Sulfur Equivalents of Alcohols

Structure

♦ Thiols are the sulfur equivalent of alcohols;

R—S—H

♦ Sulfur is less able to hydrogen-bond than oxygen, so boiling points are lower.

Thiols are similar in structure to alcohols except that oxygen is replaced by sulfur. Figure 5.6 shows a comparison of the boiling points of the thiols with the alcohols. Their properties differ in that the electronegativity of sulfur is considerably less than that of oxygen. From an examination of the electronegativity table, you would not expect a large dipole in the C—S bond nor would you expect that hydrogen bonding is important in the thiols. So even though sulfur is heavier than oxygen, the boiling points of the thiols are lower than those of the alcohols.

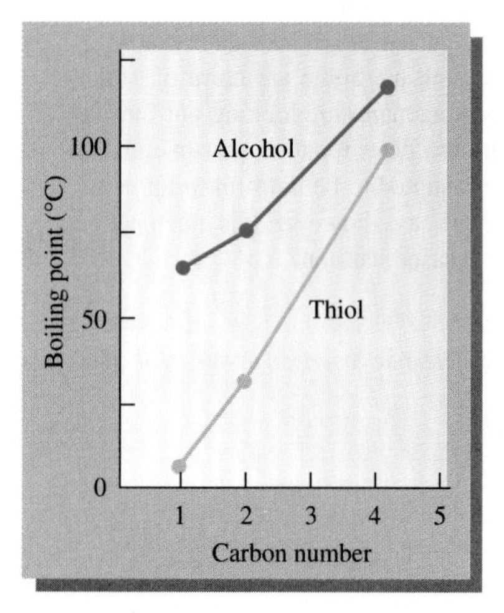

FIGURE 5.6
Comparison of boiling points of thiols and alcohols. Thiols of equivalent carbon number boil at lower temperature than the corresponding alcohols. There is less hydrogen bonding in thiols than in alcohols.

Nomenclature

The rules for naming the sulfur analogs of alcohols are exactly like those for alcohols, except that the suffix *-ol* is replaced by *-thiol*. Several examples demonstrate this nomenclature. The *e* on the alkan*e* is retained.

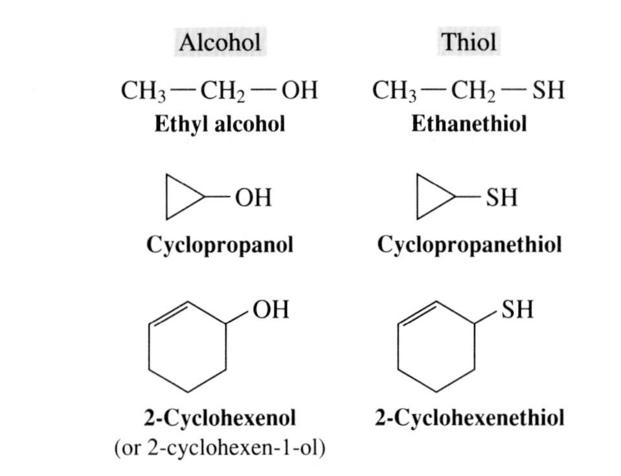

(Notice in 2-cyclohexenol that 1- can be "understood" to be part of the nomenclature if not listed.)

An older nomenclature system describes thiols as **mercaptans**. Using this system of nomenclature, the compounds shown above are named ethylmercaptan, cyclopropylmercaptan, and 2-cyclohexenylmercaptan, respectively. In biochemistry, particularly in the chemistry of amino acids, the —SH grouping found in some amino acids is referred to as the *sulfhydryl group*.

Reactions

Oxidation

♦ Oxidation of thiols produces disulfides.

Of particular interest and importance to our later understanding of biochemistry is the oxidation of a thiol to a **disulfide**, a compound with two sulfurs as its functional group.

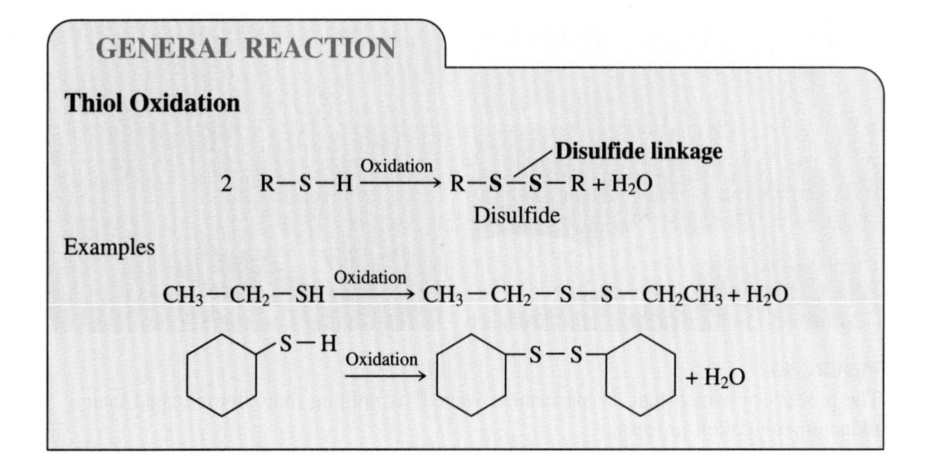

A thiol can be oxidized to a disulfide by any of a number of oxidizing agents, including oxygen. In a reverse reaction, it is also possible to reduce a disulfide to a thiol.

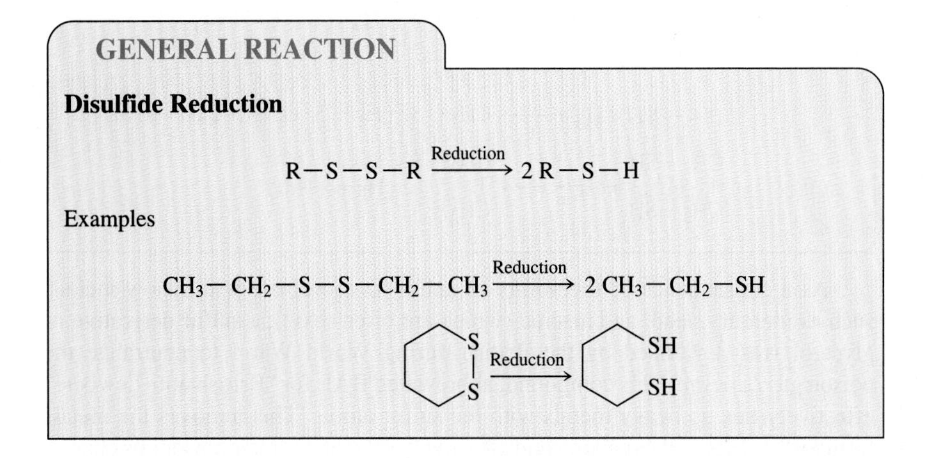

Disulfide linkages are very important in the human body for maintaining the correct structures of proteins and enzymes. A protein found in hair has a large number of disulfide linkages, and by reducing these linkages, free thiol functional groups can be created. If the orientation of these free thiol groups is changed, then new disulfide linkages are created when hair is reoxidized. These processes are responsible for hair taking on a new shape as is done in permanent waving (Figure 5.7).

Thiol Reactions with Mercury

One of the underlying reasons for the toxicity of mercury is its affinity for sulfur. Mercury salts react with thiols to form very stable compounds.

♦ The sulfur of thiols reacts with heavy metal ions such as Hg, As, and Pb.

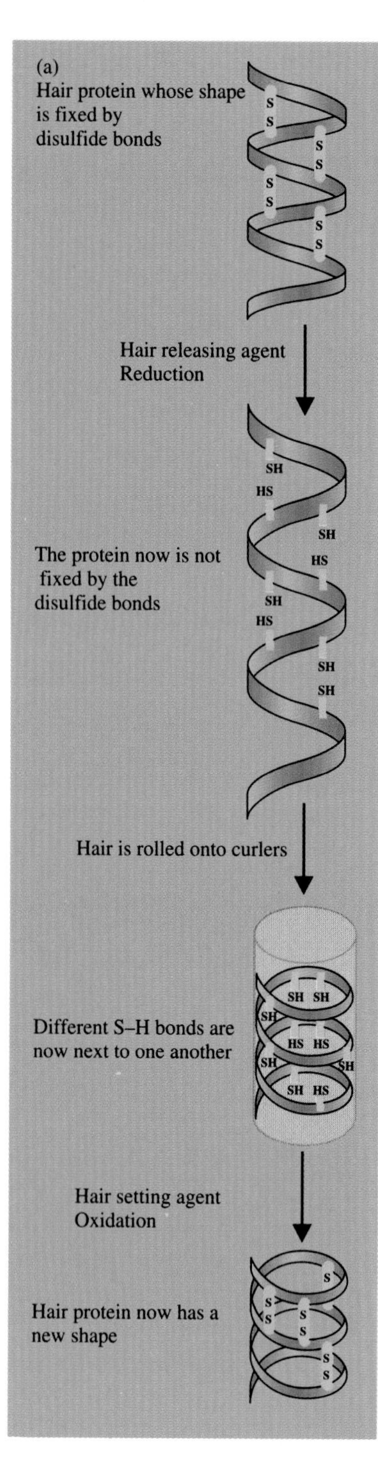

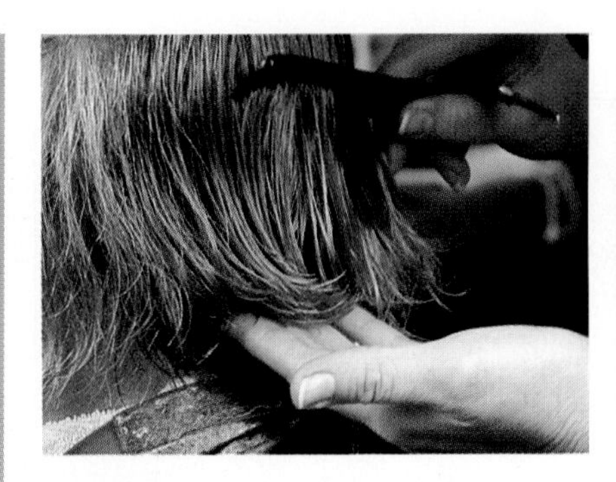

FIGURE 5.7
The process by which hair is "permanent waved" is nothing more complicated than a reduction–oxidation process.

GENERAL REACTION

Reaction of Mercury with a Thiol

$$2\,R{-}S{-}H + Hg^{2+} \longrightarrow R{-}S{-}Hg{-}S{-}R + 2\,H^{+}$$

Examples

$$2\,CH_3{-}SH + Hg^{2+} \longrightarrow CH_3{-}S{-}Hg{-}S{-}CH_3 + 2\,H^{+}$$

$$\begin{array}{l} CH_2{-}SH \\ | \\ CH_2{-}SH \end{array} + Hg^{2+} \longrightarrow \begin{array}{l} CH_2{-}S\diagdown \\ | \qquad\quad Hg \\ CH_2{-}S\diagup \end{array} + 2\,H^{+}$$

As a consequence of the ability of sulfur to coordinate with heavy metals, such as mercury, lead, and arsenic (which are often toxic), a sulfur derivative of glycerol was developed by the British during World War I to neutralize the poison gas *Lewisite*. This compound, abbreviated BAL for "British anti-Lewisite" effectively ties up heavy metals with the sulfur atoms. This removes the metals from further biological reaction and effectively neutralizes their poisonous character. BAL is still used for treating severe cases of arsenic and mercury poisoning and is currently marketed as *dimercaprol*.

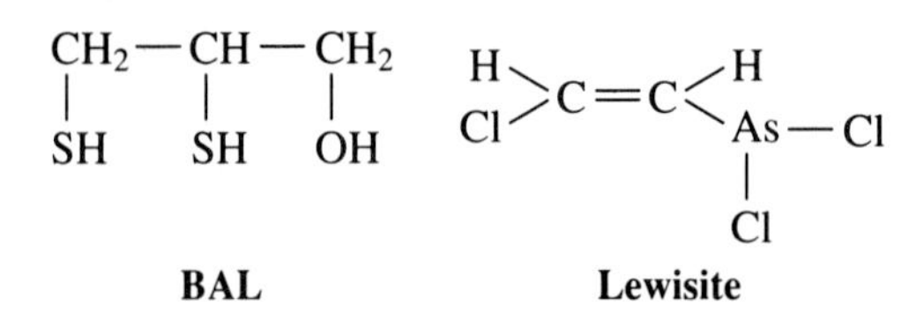

Special Thiols

Sulfur-containing compounds are abundant in nature. There is often an association of sulfur with some disagreeable odors: the smell of hydrogen sulfide (H_2S)

CHEMISTRY CAPSULE

Garlic and Onions

In 1609, it was written

> Garlic then have power to save
> from death
> Bear with it though it maketh
> unsavory breath,
> And scorn not garlic like some
> that think
> It only maketh men wink
> and drink and stink.*

Garlic and onion contain special thiols that have been suggested as having medicinal value. Many folk medicines have regularly employed these plants in their cures. There is a basis for this widespread use. Extracts of garlic and onions have been shown to have antifungal and antibacterial properties, as well as acting as an anticoagulant. In the writings of the ancients, garlic was prescribed as a therapeutic for ailments as diverse as headaches and tumors. No less an authority on antibiotics than Louis Pasteur studied these properties of garlic. Extracts of garlic juice can be diluted to less than one part in 100,000 and still actively inhibit growth of a number of bacteria.

* From Sir John Harrington, in the "Englishman's Doctor."

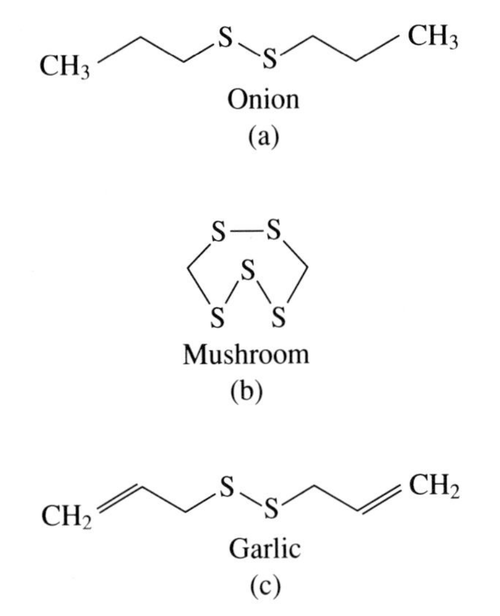

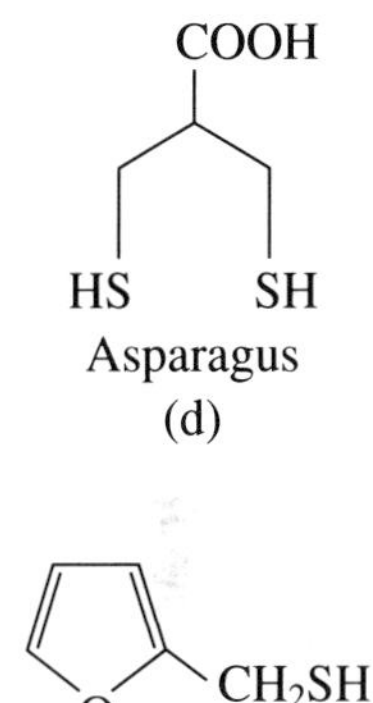

FIGURE 5.8
Some common natural substances that are rich in sulfur-containing compounds.

from rotten eggs, the essence of skunk, and the odor of garlic and onion. However, lest we always think of sulfur in a negative way, we should also be aware that the clean smell of the marigold and the aroma of coffee are both due to sulfur-containing compounds (Figure 5.8).

NEW TERMS

thiol The sulfur analog of an alcohol in which oxygen is replaced by sulfur

disulfide The product of thiol oxidation, having the structure R—S—S—R

TESTING YOURSELF

Thiols

1. Give an appropriate name for each of the following compounds:

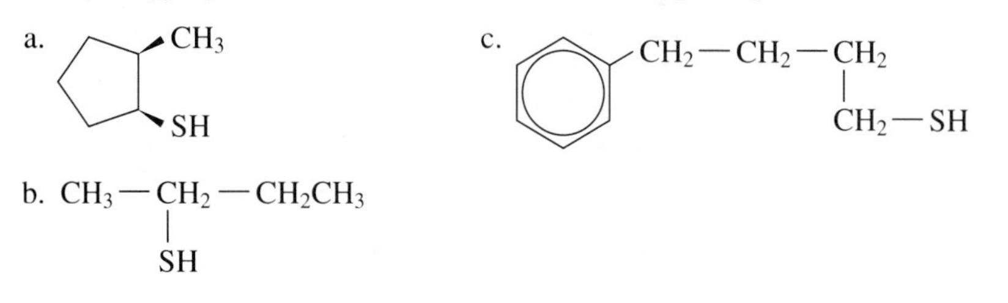

2. Suggest products for each of the following reactions:

a. $CH_3-CH_2-SH \xrightarrow{\text{Oxidation}}$

b. $\underset{\displaystyle SH}{CH_2}-\underset{\displaystyle SH}{CH}-CH_2OH + Hg^{2+} \longrightarrow$

Answers **1. a.** *cis*-2-methylcyclopentanethiol **b.** 2-butanethiol **c.** 4-phenylbutanethiol

2. a. $CH_3-CH_2-S-S-CH_2-CH_3$ **b.** $CH_2-CH-CH_2OH$, with each of the first two carbons bonded to S, and both S atoms bonded to Hg (ring: CH_2—S—Hg—S—CH)

Summary

Alcohols and *phenols* are characterized by the presence of the *hydroxyl group* as the important functional group. This group is responsible for the solubility of small- and medium-sized alcohols in water. Phenols are more acidic than alcohols. Alcohols can lose water through *dehydration* to yield alkenes and can be converted to organohalogen compounds. Phenols do not undergo either of these reactions. *Ethers* have the oxygen placed between two carbon atoms. The ether analog of cyclopropane (*epoxide*) is an important functional group in biological reactions. *Thiols* are the sulfur analogs of alcohols.

Reaction Summary

Dehydration of an Alcohol to Form an Alkene

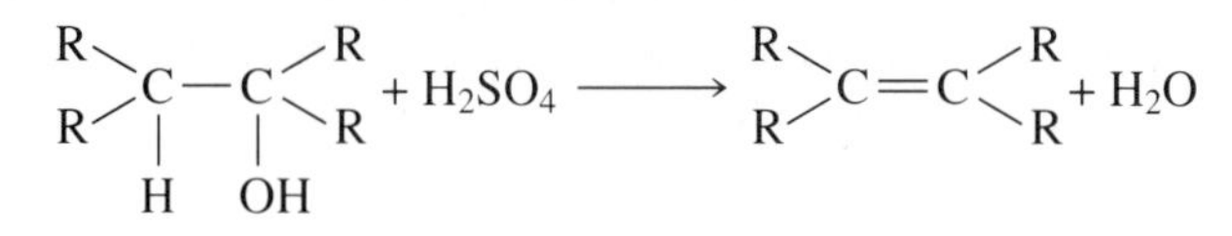

Conversion of an Alcohol to an Organohalogen

$$R—O—H + HX \longrightarrow RX + H_2O$$

Aromatic Electrophilic Substitution of a Phenol

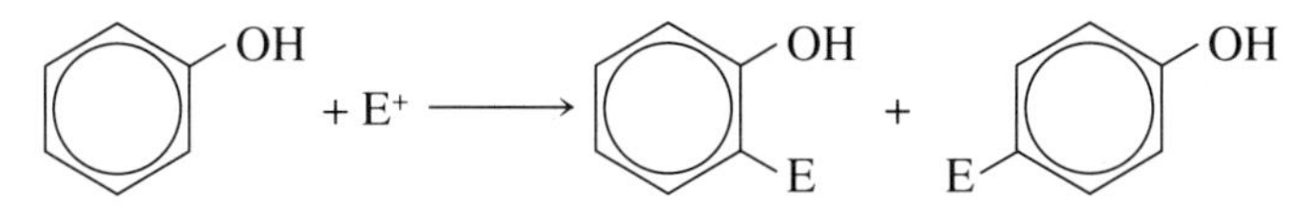

Thiol Oxidation

$$2\,R—S—H \xrightarrow{\text{Oxidation}} R—S—S—R + H_2O$$

Reaction of Mercury with a Thiol

$$2\,R—S—H + Hg^{2+} \longrightarrow R—S—Hg—S—R + 2\,H^+$$

Terms

alcohol (5.1)	ether (5.3)
dehydration (5.1)	epoxide (5.3)
Lucas test (5.1)	carcinogenic (5.3)
glycol (5.1)	thiol (5.4)
phenol (5.2)	disulfide (5.4)

Exercises

Alcohols

***Structure and Physical Properties* (Objective 1)**

1. Although of higher molecular weight, methanol has a lower boiling point than water. Explain.
2. Discuss why the solubility of alcohols decreases rather rapidly after pentanol.
3. Ethyl alcohol is totally soluble in water, but ethane is not. Why?
4. Even though a bromine is "heavier" than an OH group, methanol has a higher boiling point than bromomethane. Explain.

***Nomenclature* (Objective 2)**

5. Draw structures for the following alcohols:
   a. 2-methylcyclohexanol
   b. 2,3-dimethyl-2-octanol
   c. 4-isopropyl-3-decanol
   d. *cis*-1,2-dimethylcyclohexanol
   e. 3-hexen-2-ol
   f. 2-pentyn-1-ol
6. Give IUPAC names for the following simple alcohols:

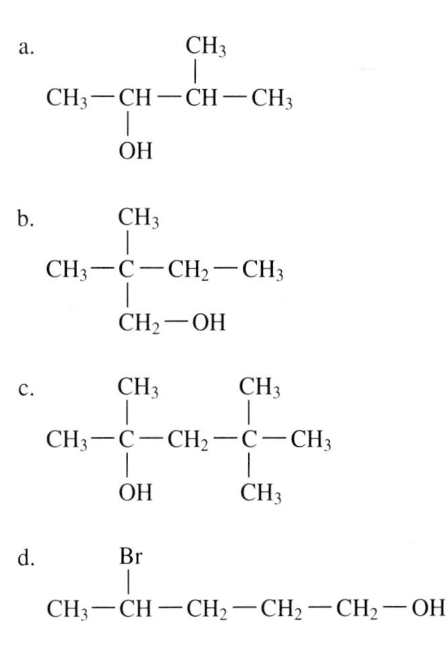

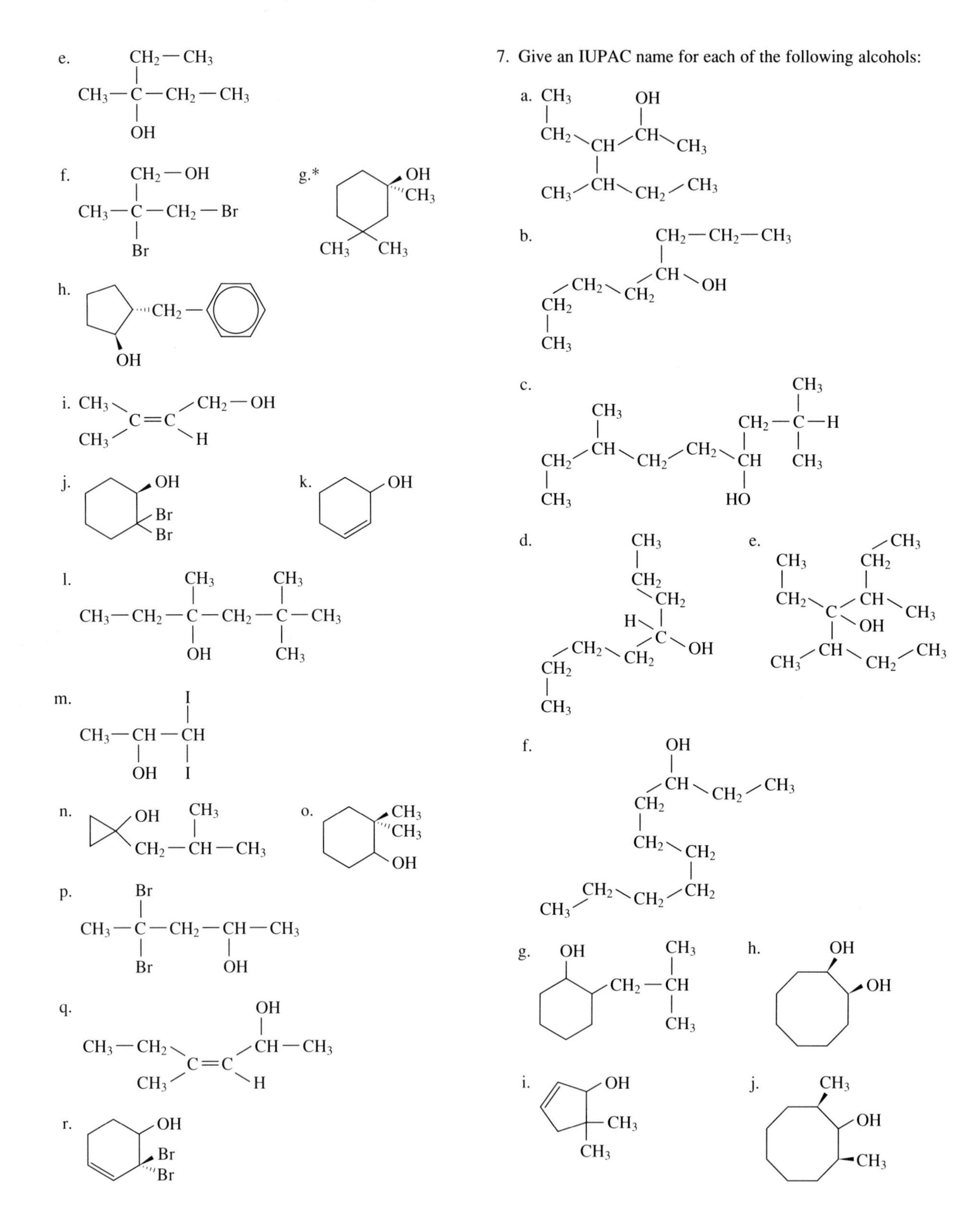
7. Give an IUPAC name for each of the following alcohols:

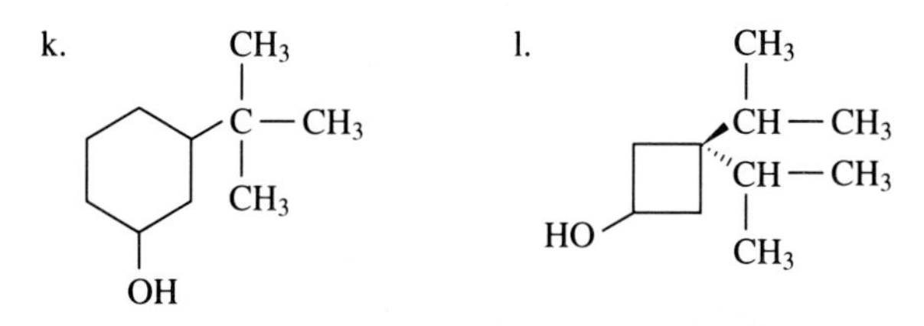

8. Identify each of the following as a primary, secondary or tertiary alcohol:

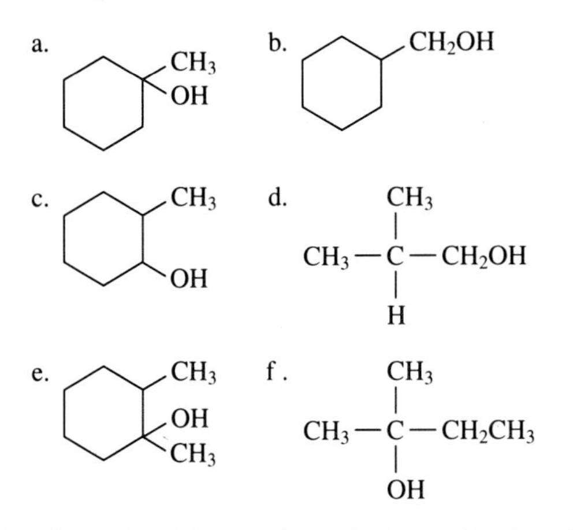

9. Give an IUPAC name for each of the following alcohols:

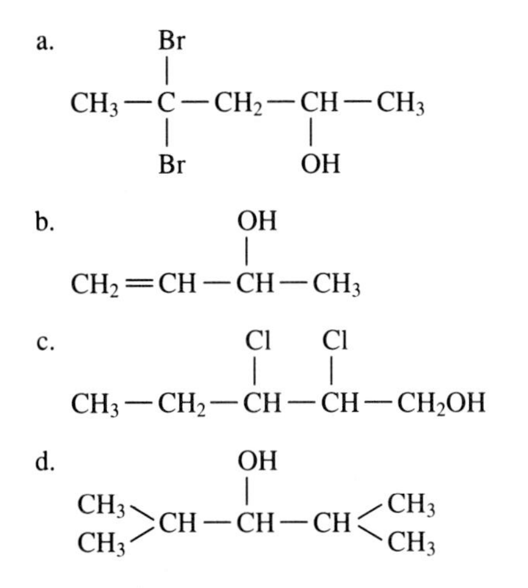

Reactions (Objective 3)

10. How can an alcohol be considered as both an acid and a base?
11. Give products for the following reactions of alcohols:

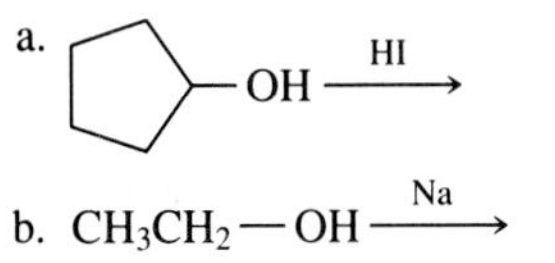

b. $CH_3CH_2—OH \xrightarrow{Na}$

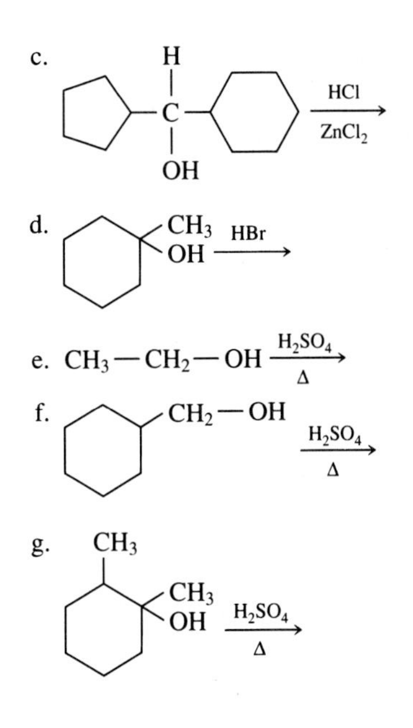

Phenols
Nomenclature (Objective 2)

12. Give a name to each of the following phenols:

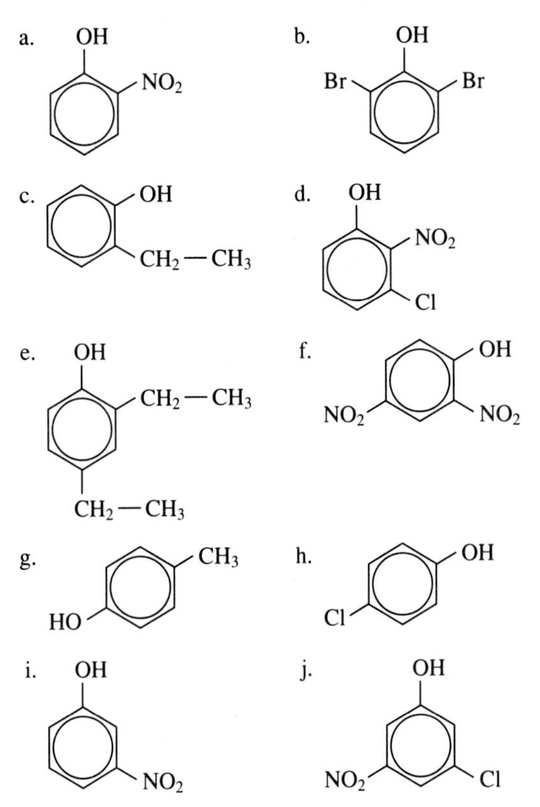

Reactions (Objective 3)

13. Why are phenols more acidic than alcohols?
14. Suggest major products for each of the following reactions:

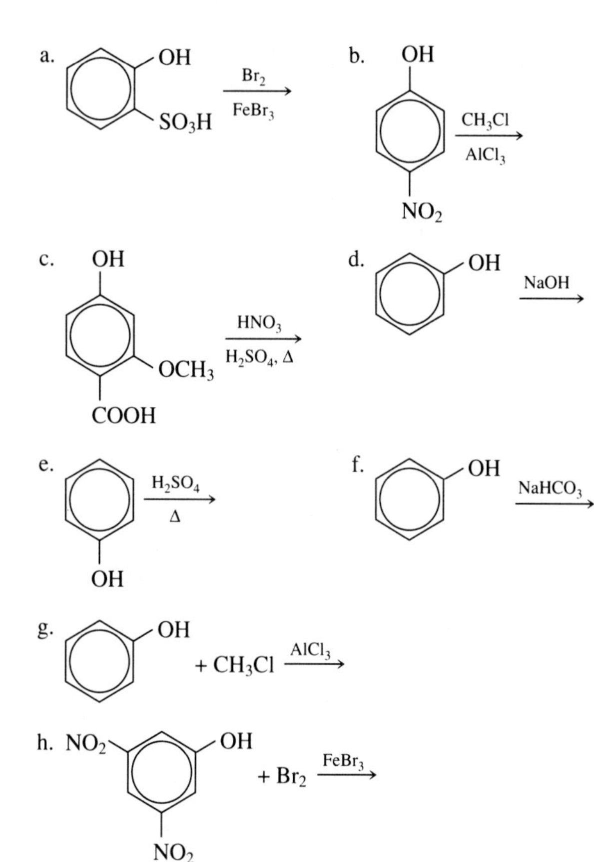

Ethers

Structure and Physical Properties (Objective 1)

15. Dimethyl ether has the same molecular formulas as ethyl alcohol but has a much lower boiling point. Explain.
16. Explain why alcohols are more water-soluble than ethers.
17. Given the formula C_3H_8O, we can draw both alcohols and ethers. Show all the isomeric structures.

Nomenclature (Objective 2)

18. Give structures for the following ethers:
    a. isopropyl methyl ether
    b. *sec*-butyl ethyl ether
    c. *tert*-butyl phenyl ether
    d. dipropyl ether

Thiols

Structure and Physical Properties (Objective 1)

19. Thiols have lower boiling points than alcohols of the same carbon number. Explain.

Nomenclature (Objective 2)

20. Give names to the following thiols.

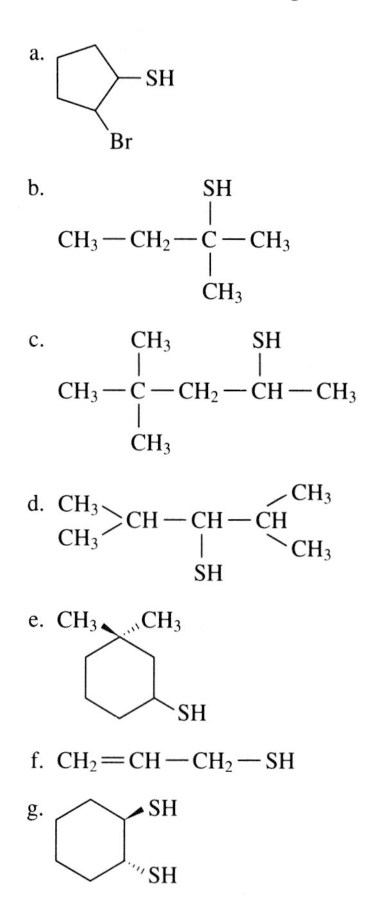

Reactions (Objective 3)

21. Suggest products for the following reactions:

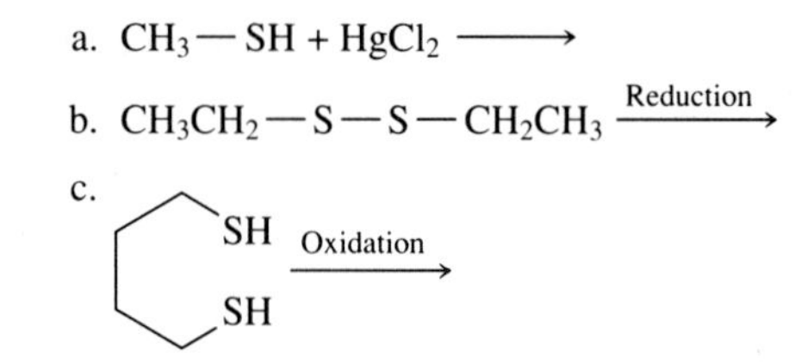

d.

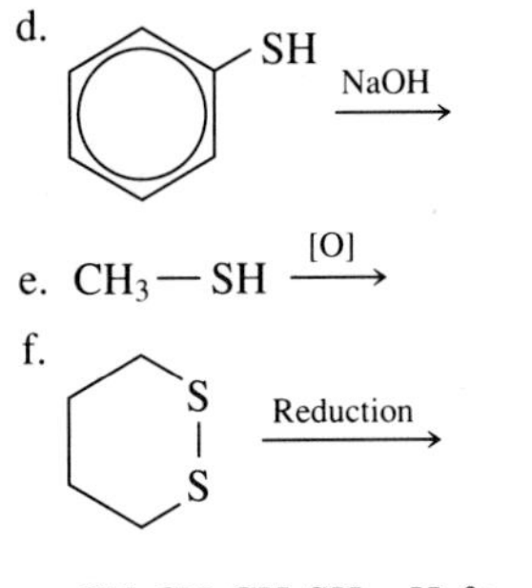

e. $CH_3-SH \xrightarrow{[O]}$

f.

g. $CH_3CH_2CH_2SH + Hg^{2+} \longrightarrow$

General Review Exercises

Structure and Physical Properties (Objective 1)

22. For each of the following compounds, identify the alcohol, phenol, thiol, or ether functional group:

a.

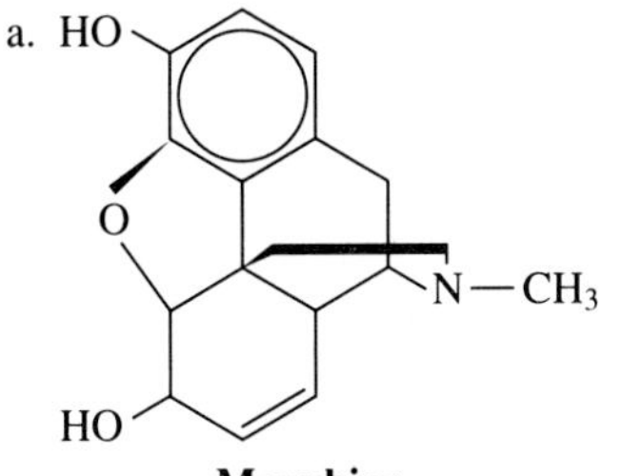

Morphine

b.

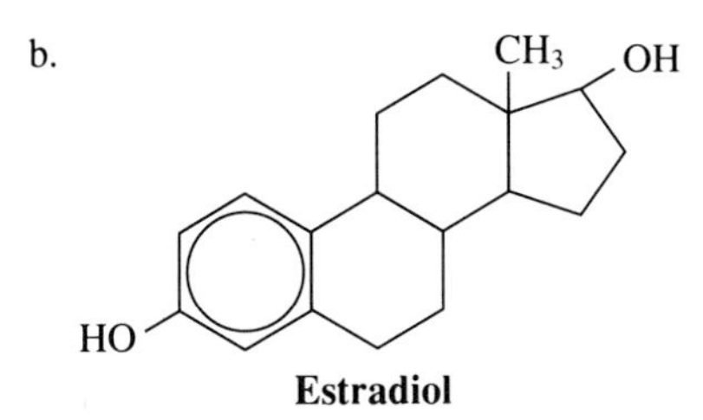

Estradiol

c.

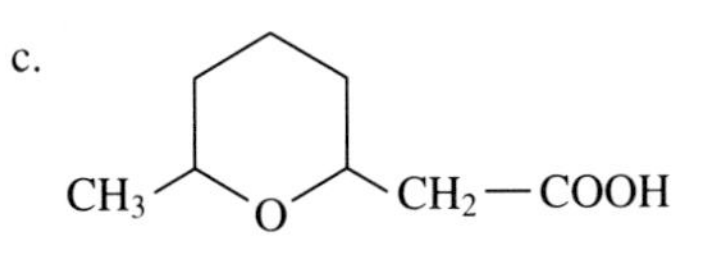

Civet constituent

d.

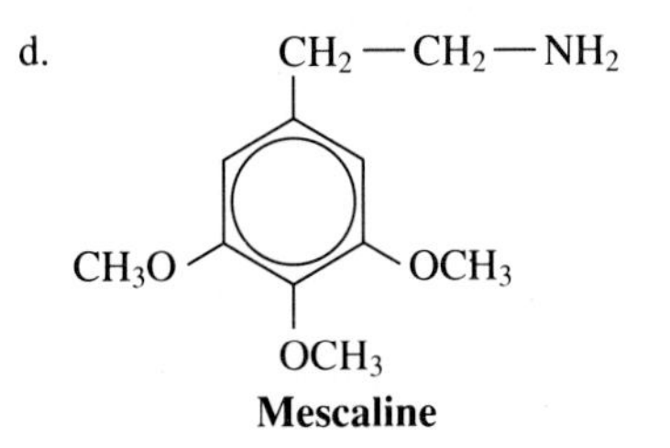

Mescaline

23. Draw a general structure for each of the following:
    a. phenol
    b. disulfide
    c. ether
    d. thiol
    e. epoxide
    f. tertiary alcohol
    g. primary alcohol
    h. secondary alcohol

Nomenclature (Objective 2)

24. Draw structures for each of the following compounds:
    a. 2-chloro-3-hexanol
    b. *cis*-2-methylcyclohexanol
    c. *trans*-1,3-cyclohexanediol
    d. *p*-bromophenol
    e. isobutyl methyl ether
    f. 3-cyclohexenethiol
    g. 2,3-dimethyl-2-pentanol
    h. benzyl isopropyl ether
    i. 3,4-dimethylphenol
    j. 2-bromo-2-methyl-3-hexanethiol
    k. 3-bromo-2-heptanethiol
    l. *sec*-butyl isobutyl ether
    m. 4-methylcyclooctanol
    n. 3,3-dibromopropan-1-ol
    o. 2,4,6-trinitrophenol
    p. 1-hexene-5-yne-3-ol

25. Give an IUPAC name to each of the following compounds:

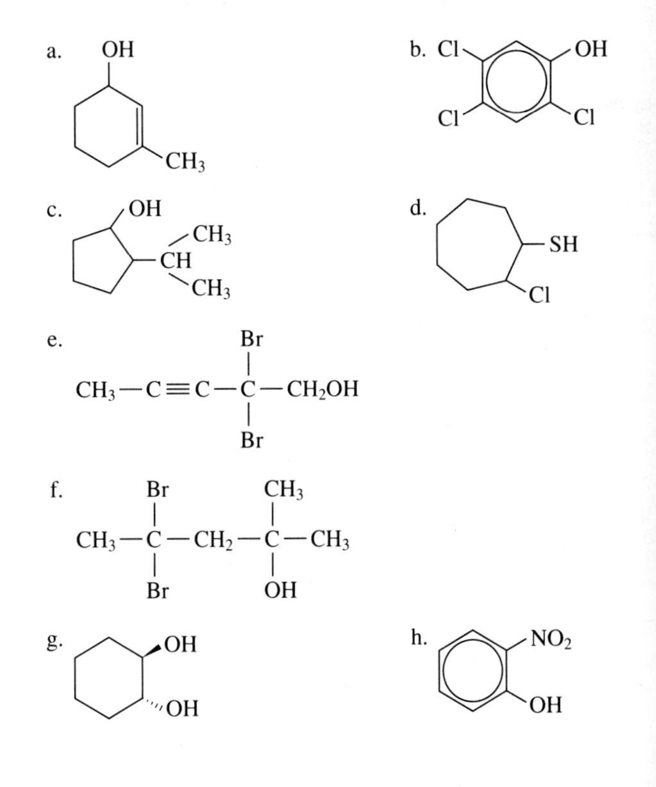

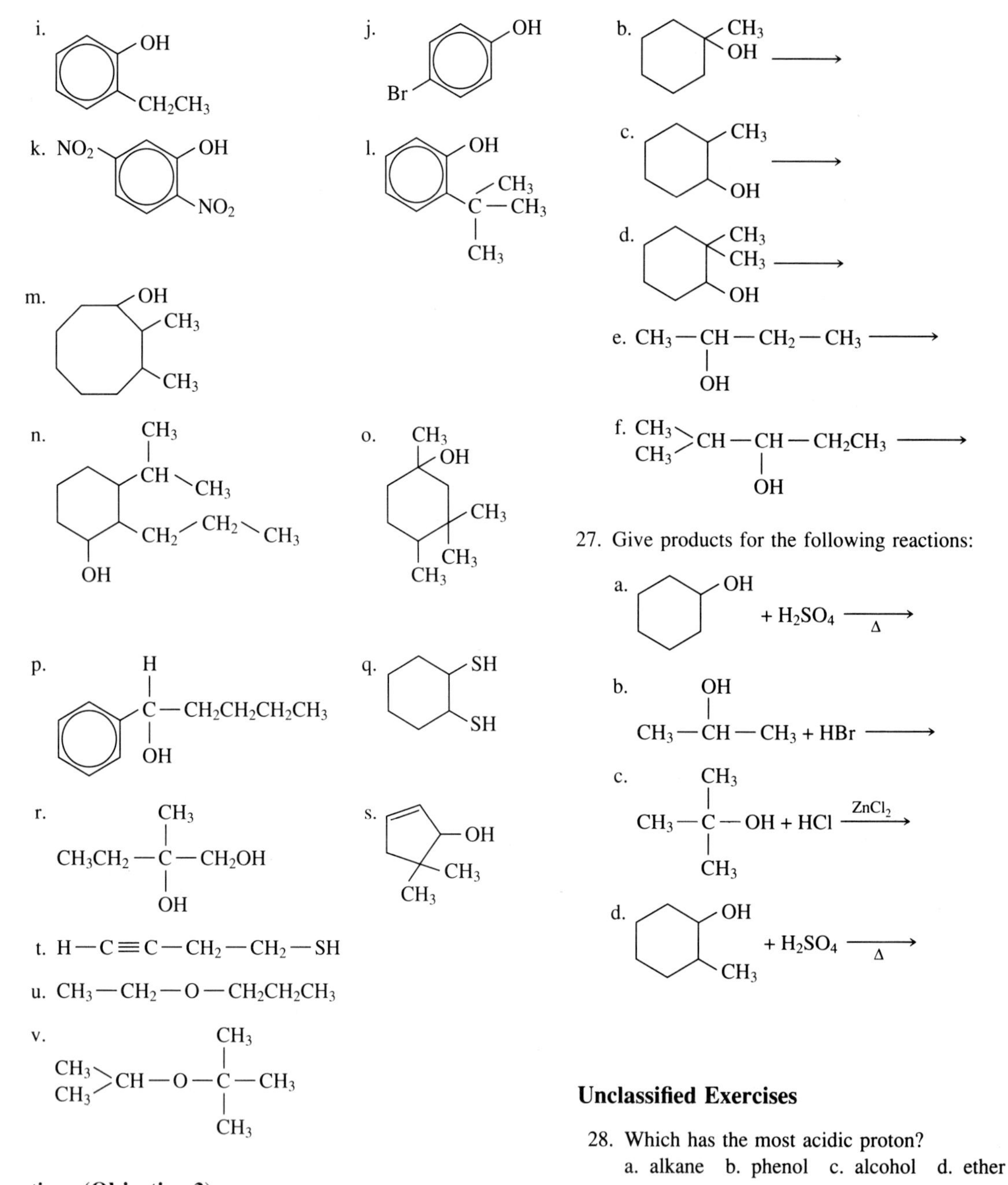

27. Give products for the following reactions:

a. OH + H_2SO_4 $\xrightarrow[\Delta]{}$

b. $CH_3—CH(OH)—CH_3 + HBr \longrightarrow$

c. $CH_3—C(CH_3)_2—OH + HCl \xrightarrow{ZnCl_2}$

d. OH, CH_3 + H_2SO_4 $\xrightarrow[\Delta]{}$

Reactions (Objective 3)

26. Give the major dehydration product from reaction of each of the following alcohols to sulfuric acid:

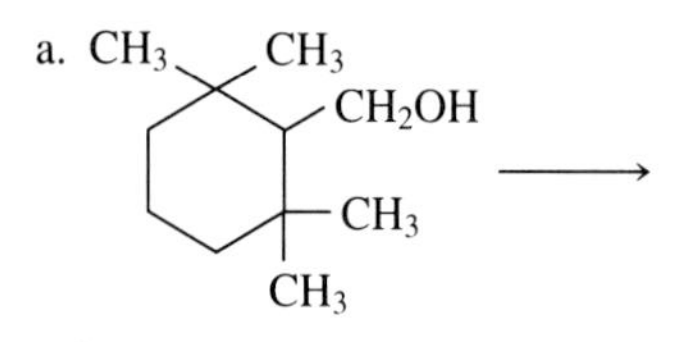

Unclassified Exercises

28. Which has the most acidic proton?
   a. alkane b. phenol c. alcohol d. ether

29. OH, CH_3

   is best named
   a. *cis*-4-hydroxy-3-methylcyclohexene
   b. *trans*-4-hydroxy-3-methylcyclohexene
   c. *cis*-2-methyl-3-cyclohexenol
   d. *trans*-2-methyl-3-cyclohexenol

30. In the following dehydration reaction, what is the major product?

31.

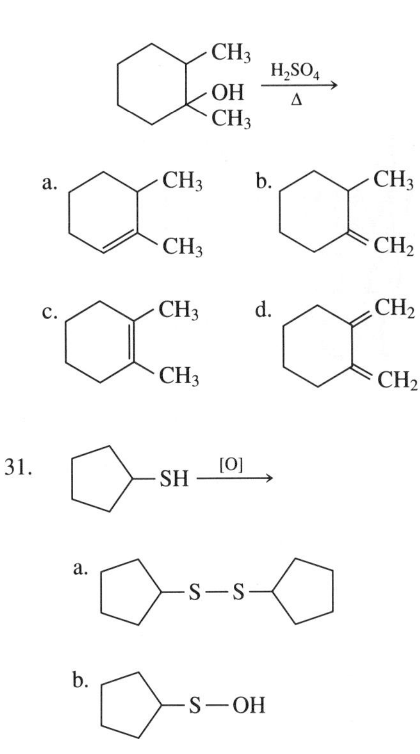

c. No reaction

32. The most water-soluble material is most likely

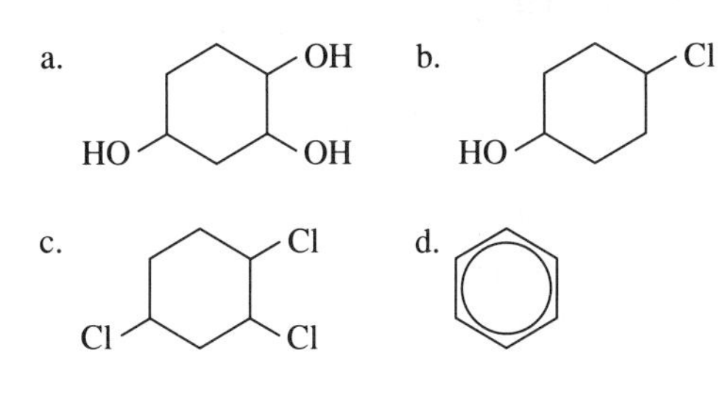

33. What is the most reasonable product from the following reaction of the phenol?

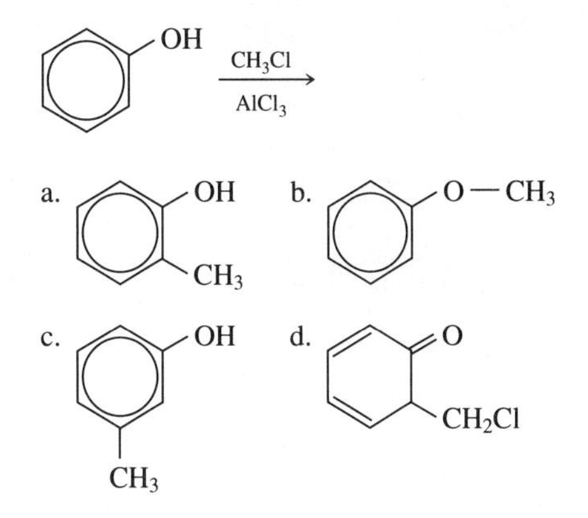

34. Assuming that each is water-soluble, which of the following alcohols reacts most quickly to the Lucas test?

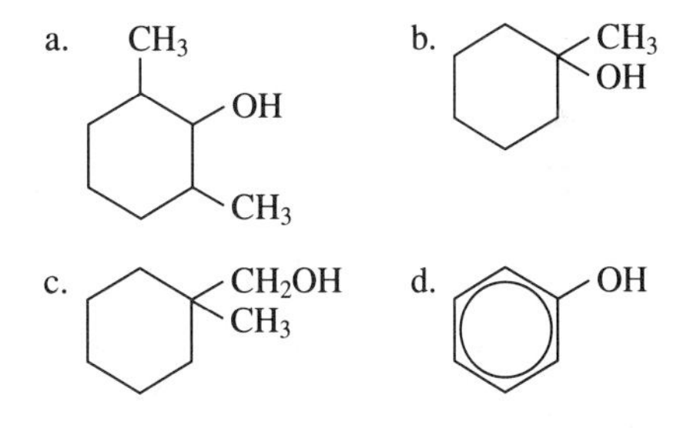

35. The high boiling points of alcohols compared with the corresponding alkane hydrocarbons are due to
   a. hydrogen bonding
   b. heavy oxygen atoms
   c. water-solubility
   d. none of these
36. Which of the following functional group series is ranked according to *increasing* boiling points?
   a. diethyl ether, ethane, ethyl alcohol, ethanethiol
   b. ethane, ethyl alcohol, diethyl ether, ethanethiol
   c. ethane, diethyl ether, ethanethiol, ethyl alcohol
   d. diethyl ether, ethane, ethanethiol, ethyl alcohol
37. The acidities of what functional group can be understood by resonance concepts?
   a. alcohols b. phenols c. thiols d. ethers

CARBONYL GROUP AND ITS COMPOUNDS: ALDEHYDES AND KETONES

The female bark beetle prepares chambers beneath the bark. During this process she gives off the mating pheromone.

OUTLINE

6.1 Structure and Properties of the Carbonyl Group
6.2 Nomenclature of Carbonyl Compounds
6.3 Reactions of Carbonyl Compounds
6.4 Natural Examples of Acetals
6.5 Preparation Reactions of Carbonyl Compounds
6.6 Oxidation of Hydroquinones to Quinones

OBJECTIVES

After completing this chapter, you should be able to

1. Explain how the reactivity of the carbonyl group is similar to the reactivity of an alkene and discuss how it is different.
2. Give names to representative members of the aldehyde and ketone families.
3. Predict products of common reactions of carbonyl compounds.
4. Show how carbonyl compounds can be prepared from alcohols.
5. Recognize the structures of acetals and show where they come from.

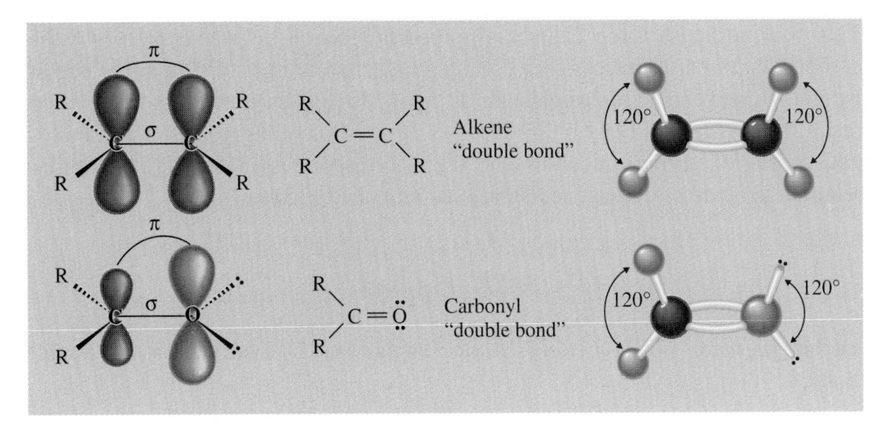

FIGURE 6.1
A comparison of an alkene and a carbonyl group. Although both functional groups have a double bond, the polarization of the carbonyl due to the higher electronegativity of oxygen results in a very different chemical behavior for the carbonyl group. The larger orbitals for the oxygen in the carbonyl group represent its greater electron density.

6.1 Structure and Properties of the Carbonyl Group

Structure

In many respects, the **carbonyl group** is similar to the alkene functional group. Whereas an alkene has two carbon atoms joined by a double bond, a carbonyl group has a carbon and an oxygen atom attached by a double bond (Figure 6.1). The difference between the chemical behavior of these two functional groups is due to the greater electronegativity of the oxygen atom. The double bond is polarized *toward* the oxygen atom; thus, the carbonyl group undergoes many reactions as a result of the carbon atom bearing a partial positive charge and the oxygen having a partial negative charge. This polarization can be represented by resonance structures (Figure 6.2).

◆ The *carbonyl group* is structurally similar to an alkene.

◆ The difference in electronegativity between an oxygen and a carbon gives the carbonyl a polarization.

The carbonyl group is common to both the **aldehydes** and **ketones**. The general structures of these compounds are given below, where R is used to designate an alkyl or cycloalkyl group. The symbol R′ simply suggests that the side groups may be (but are not necessarily) different. The shorthand notations Ar and Ar′ designate aromatic groups.

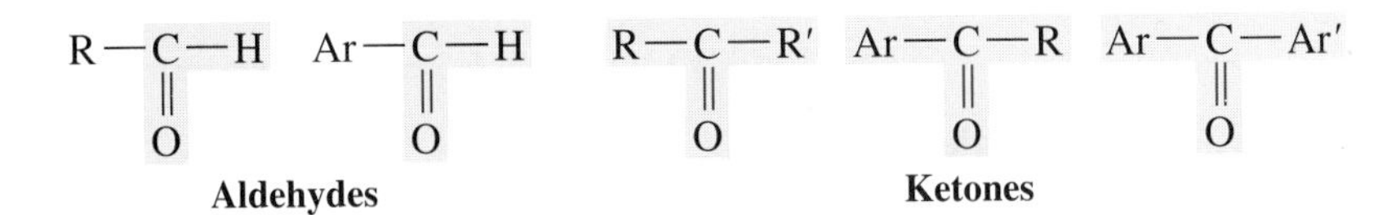

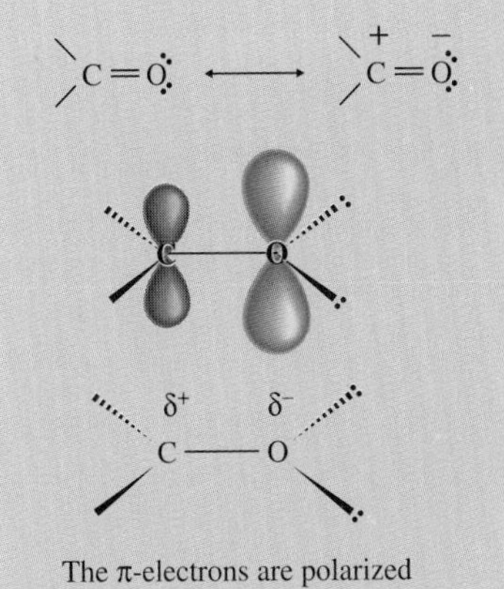

FIGURE 6.2
Resonance structures of the carbonyl group. The polarization of the carbonyl group can be represented by a resonance structure that has a positive charge on the carbon atom and a negative charge on the oxygen atom.

Physical Properties: Dipole–Dipole Interaction

In the previous chapter the dramatic influence of intermolecular hydrogen bonding of alcohols on their boiling points was observed. The carbonyl group, although polarized, cannot participate in the same kind of interaction because there are no O—H bonds. However, as a result of the polarity of the carbonyl group, a weak attraction occurs:

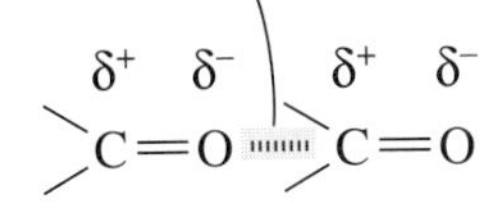

δ^+ and δ^- mean relative positive and negative charge

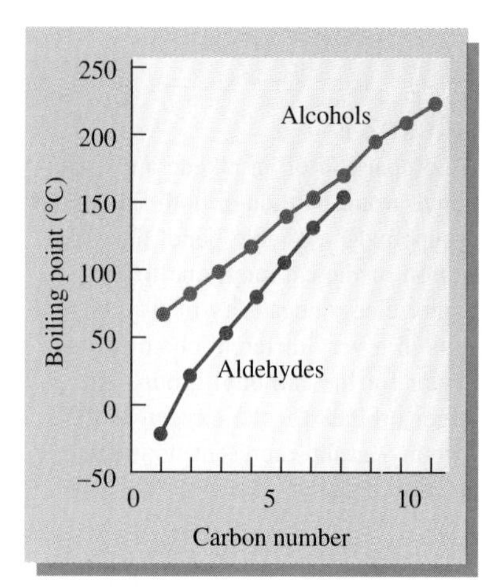

FIGURE 6.3
The boiling points of alcohols and aldehydes versus carbon number. The dipole–dipole interactions are not nearly as important as hydrogen bonding in determining boiling point. Again notice that as the carbon number increases, the importance of the functional group in dominating physical properties decreases.

This weak attractive force, called a **dipole–dipole force**, helps to explain why the carbonyl compounds have higher boiling points than alkanes of similar molecular weight. However, the intermolecular attractive forces are much less than would result from hydrogen bonding. Figure 6.3 shows the boiling points of a homologous series of primary alcohols and their corresponding aldehydes. A similar relationship exists between secondary alcohols and ketones.

NEW TERMS

carbonyl group The functional group characterized by a carbon–oxygen double bond:

$$>C=O$$

aldehyde A class of carbonyl-containing compounds of the general structure

$$Ar-\underset{\displaystyle \overset{\|}{O}}{C}-H \qquad R-\underset{\displaystyle \overset{\|}{O}}{C}-H$$

The carbonyl group is always on a terminal carbon atom.

ketone A class of carbonyl-containing compounds of the general structure

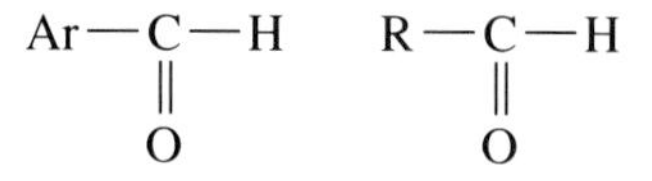

$$R-\underset{\displaystyle \overset{\|}{O}}{C}-R' \qquad Ar-\underset{\displaystyle \overset{\|}{O}}{C}-Ar' \qquad Ar-\underset{\displaystyle \overset{\|}{O}}{C}-R$$

The carbonyl group is not on a terminal carbon atom.

dipole–dipole force An attractive force between two molecules as a result of the polarity of molecules

TESTING YOURSELF

Structure and Properties

1. Show the Lewis electron-dot structure for formaldehyde,

$$\begin{matrix} H \\ & >C=O \\ H \end{matrix}$$

2. Which would have the stronger dipole–dipole interaction, H—Cl ········ H—Cl or H—I ········ H—I? Why?

3. Why does HI have a higher boiling point than HCl?

Answers **1.** $H_2C=\ddot{\underset{\cdot\cdot}{O}}:$ **2.** H—Cl, because Cl is more electronegative and the H—Cl bond is more polarized than the H—I bond. **3.** The molecular weight of HI is greater than that of HCl.

6.2 Nomenclature of Carbonyl Compounds

Aldehydes are named using the same general rules already described for alcohols, with the exception that the *-ol* ending of the alcohol is replaced by *-al*. Because of the structure of an aldehyde, the carbonyl group is always on a terminal carbon (carbon number 1), and all other substituents are numbered from this position.

◆ In nomenclature, aldehydes have an *-al* ending and ketones have an *-one* ending

EXAMPLE 6.1

Name this aldehyde:

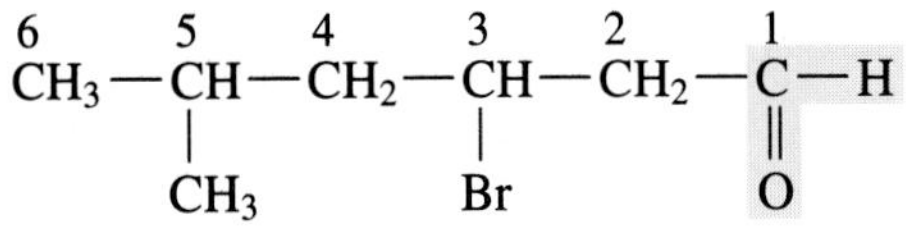

```
6     5     4     3     2     1
CH3—CH—CH2—CH—CH2—C—H
      |           |           ||
     CH3          Br          O
```

Solution:

3-Bromo-5-methylhexanal

Ketones also follow similar nomenclature; the *-ol* ending is replaced by *-one*. The carbonyl group takes priority in naming over all other functional groups studied so far. If there is also an alcohol group present as a substituent, the OH group is given as *-hydroxy*.

EXAMPLE 6.2

Name this ketone:

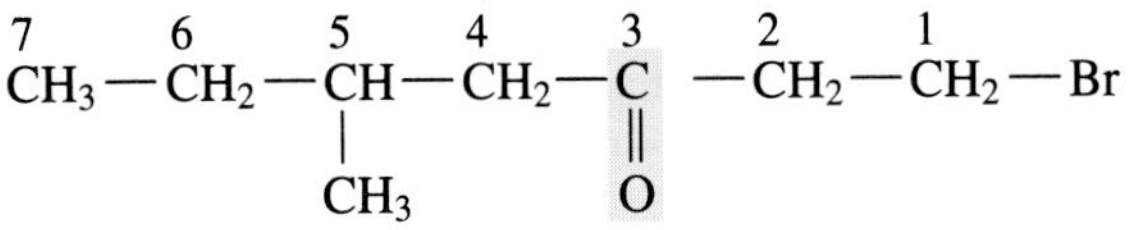

```
 7     6     5     4     3     2     1
CH3—CH2—CH—CH2—C —CH2—CH2—Br
             |           ||
            CH3          O
```

Solution:

1-Bromo-5-methyl-3-heptanone

Listed here are a variety of examples of aldehyde and ketone structures along with their correct names. Make sure you understand how each was named before continuing on.

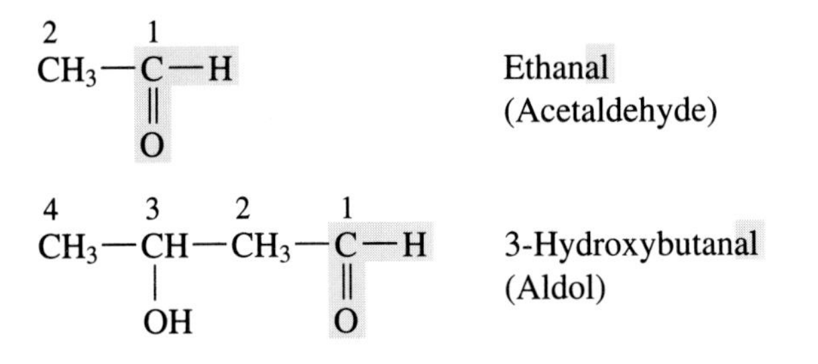

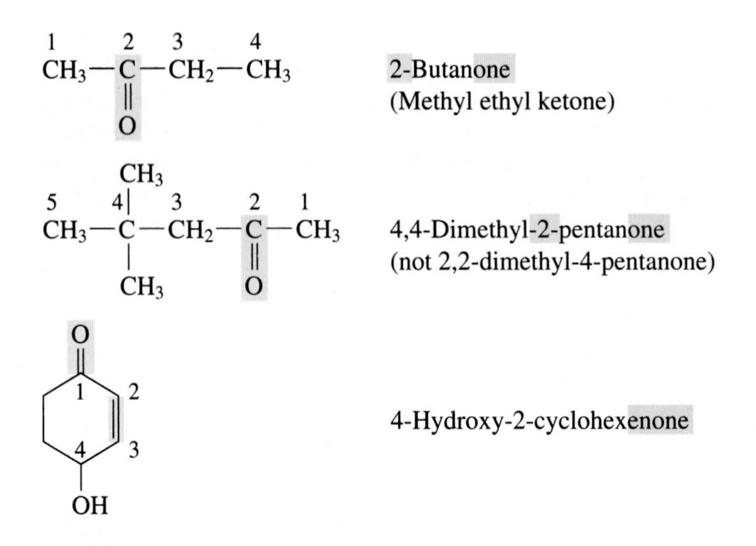

♦ In nomenclature the >C=O group has higher priority than the —OH group.

Notice that the carbonyl group has a higher priority than the alcohol. Always start numbering the carbons of a ring compound at the attachment point of the highest priority functional group.

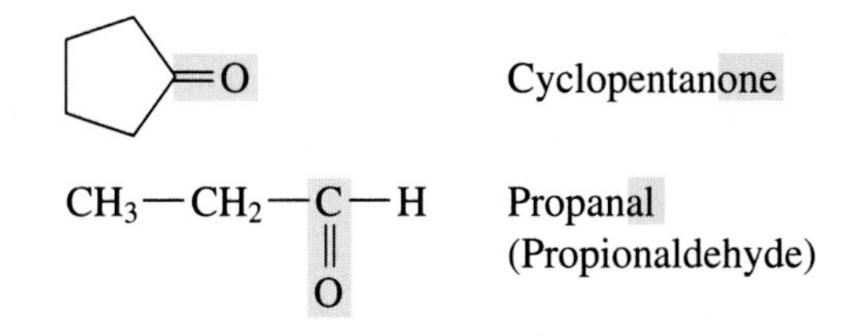

There are also several carbonyl compounds (including those in parentheses in the lists above) whose common names have attained sufficiently wide acceptance that they should be learned as the "official" names (Table 6.1).

TESTING YOURSELF

Nomenclature

1. Give the correct name of each of the following:

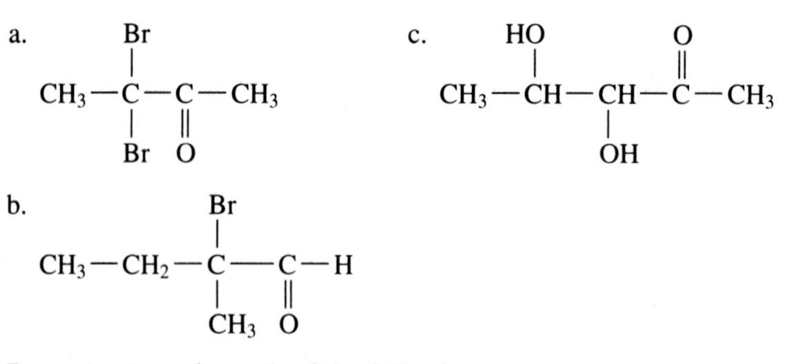

2. Draw structures for each of the following:
   a. *m*-nitrobenzaldehyde c. 2-cyclopentenone
   b. 2-bromocyclohexanone

CHEMISTRY CAPSULE

Formaldehyde

Although formaldehyde is actually a gas at room temperature, most people who have worked with formaldehyde have used it in a diluted, aqueous solution called *formalin*. In the past, formalin was commonly used in all biology labs as a preservation solution for biological specimens. Many foams, insulations, and bonding materials also contained formaldehyde. Much of its use has now declined with the recent finding that it is a suspected carcinogen.

A common industrial use of formaldehyde is found in the polymerization reaction with phenol to form *Bakelite*. Bakelite is a special polymer used for coatings and plastics.

FIGURE 6.A
Bakelite is an important polymer based on formaldehyde and phenol. It is found in a number of common products: the base of the radio tube (*left*); a rotor cap (*middle*); and a distributor cap (*right*).

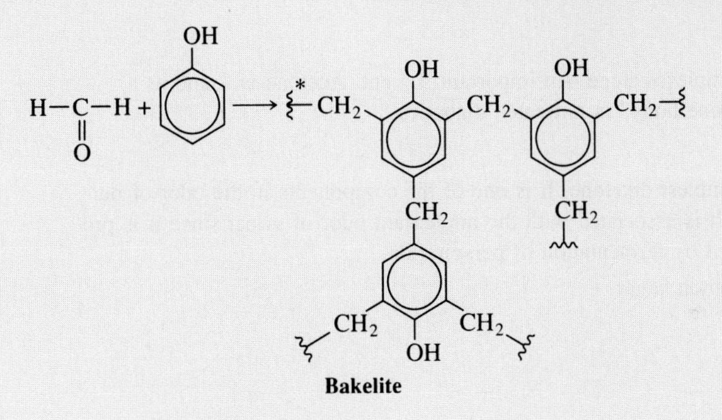

Bakelite

* Wavy line (⌇) indicates a structure of unknown or uncertain configuration.

Answers **1. a.** 3,3-dibromo-2-butanone **b.** 2-bromo-2-methylbutanal **c.** 3,4-dihydroxy-2-pentanone

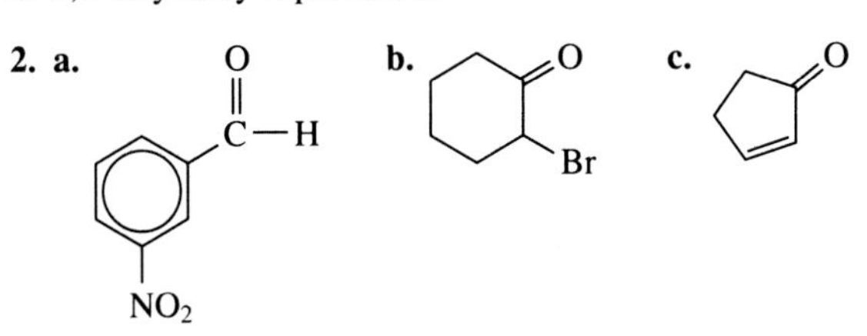

TABLE 6.1

Common Aldehydes and Ketones

Name	Description	Structure
Formaldehyde	The simplest aldehyde. A 40% water solution (formalin) is useful for preserving biological samples. It is a component of wood smoke, and the smoking of foods depends on formaldehyde to kill bacteria.	$H-\overset{\overset{O}{\|}}{C}-H$
Acetaldehyde	A simple aldehyde (sometimes called ethanal) formed from the oxidation of ethyl alcohol. It is responsible for the "hangover" from drinking. It is associated with the nutty flavor of sherry wine.	$CH_3-\overset{\overset{O}{\|}}{C}-H$
Benzaldehyde	The simplest aromatic aldehyde. It contributes to the aroma of almonds.	$C_6H_5-\overset{\overset{O}{\|}}{C}-H$
Cinnamaldehyde	A fragrant compound of cinnamon.	$C_6H_5-CH=CH-\overset{\overset{O}{\|}}{C}-H$
Acetone	The simplest ketone. An important solvent. Acetone is found as a "ketone body" in untreated diabetics.	$CH_3-\overset{\overset{O}{\|}}{C}-CH_3$
Butanedione	The simplest diketone. It is one of the components in the odor of butter. It is associated with the unpleasant odor of sweat since it is produced by fermentation of perspiration.	$CH_3-\overset{\overset{O}{\|}}{C}-\overset{\overset{O}{\|}}{C}-CH_3$
Glucose	A common sugar	$H-C=O$ $H-C-OH$ $HO-C-H$ $H-C-OH$ $H-C-OH$ CH_2OH
Glyceraldehyde	An intermediate in carbohydrate metabolism	$H-C=O$ $H-C-OH$ CH_2OH

TABLE 6.1 ***(continued)***

Name	Description	Structure
Dihydroxyacetone	An intermediate in carbohydrate metabolism.	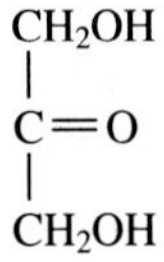 CH_2OH–C=O–CH_2OH
Camphor	From the camphor tree. Used in medicine, although most is used to make film and celluloid.	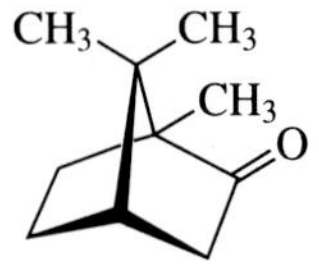 CH_3, CH_3, CH_3, O
Civetone	A natural scent from the civet cat used in perfumes.	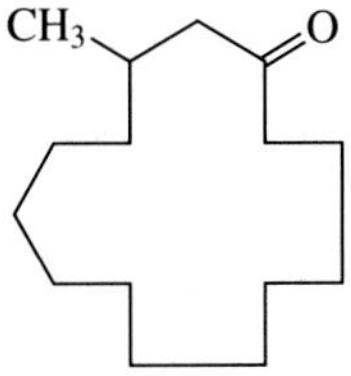 CH_3, O
Jasmone	A compound obtained from the jasmine flower. The fragrance is used for perfumes.	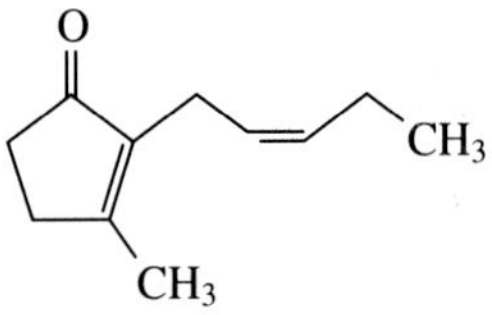 O, CH_3, CH_3
Vanillin	From the vanilla bean (both a phenol and an aldehyde)	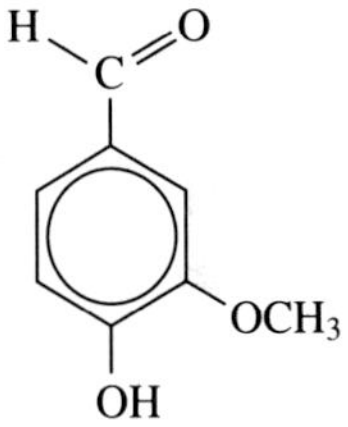 H, C, O, OCH_3, OH
Progesterone	A female sex hormone	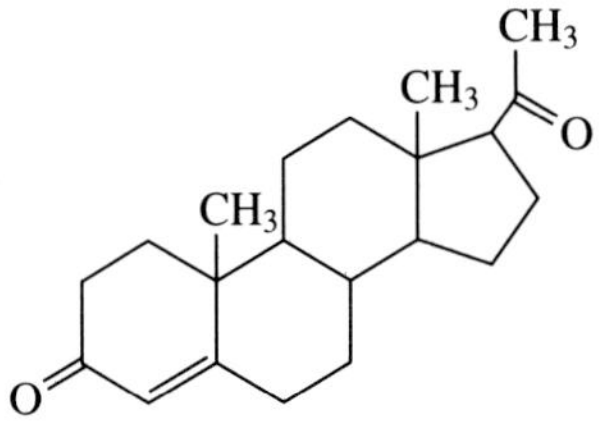 CH_3, CH_3, CH_3, O, O
Testosterone	A male sex hormone	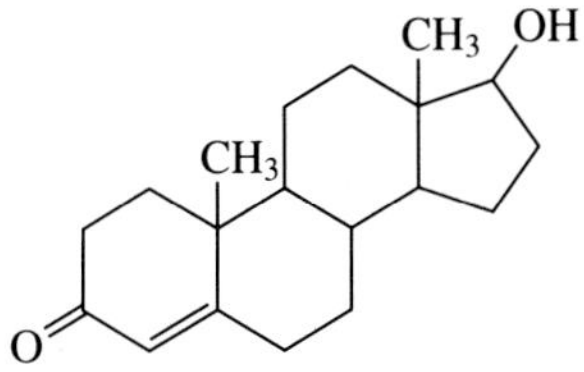 CH_3, OH, CH_3, O

6.3 Reactions of Carbonyl Compounds

Additions to the Carbonyl: Double-Bond Chemistry

♦ The addition reactions of a carbonyl can be related to the reactions of an alkene.

As for all of the functional groups studied so far, the carbonyl group exhibits unique and important chemical behavior. We will now examine some of the important reactions of the carbonyl groups, along with examples to show their importance. Recall the addition of E^+ and Nu^- (an electrophile and a nucleophile) to an alkene. In this same addition to a carbonyl group, it is easy to predict where the E^+ and Nu^- attach. Since the carbonyl group has a partial negative charge on the oxygen, the E^+ preferentially adds to the oxygen atom. In turn, the negatively charged nucleophile preferentially adds to the partially positively charged carbonyl carbon atom. This can be seen in the following general reaction.

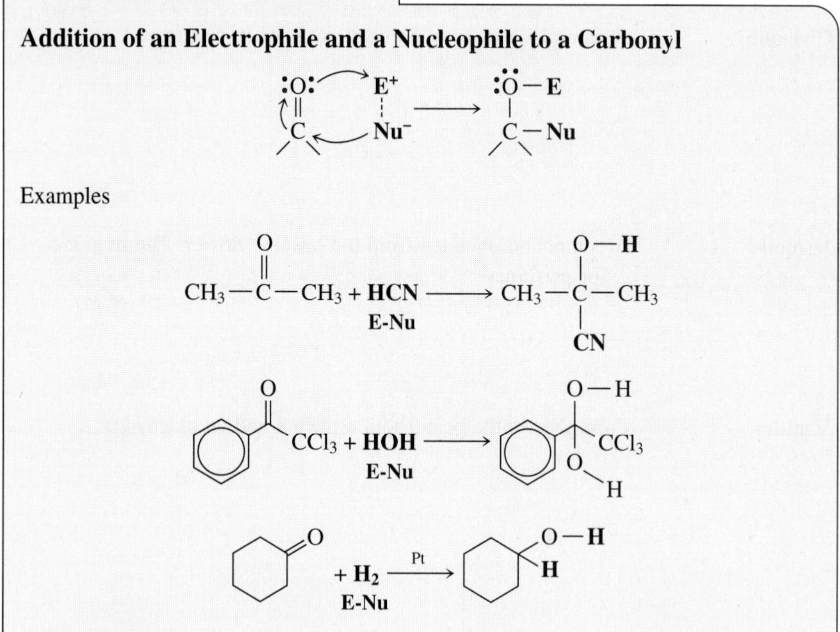

♦ Addition of HCN to a carbonyl gives a *cyanohydrin*.

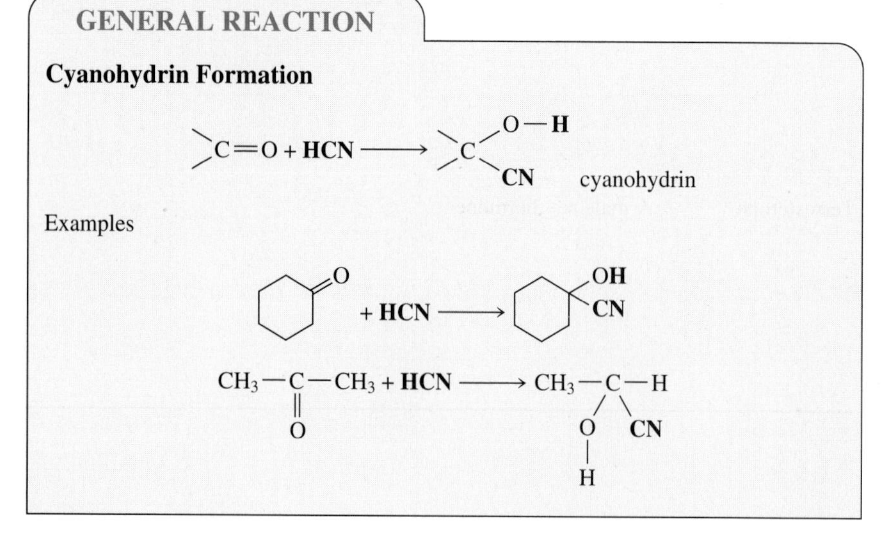

FIGURE 6.4
A millipede produces a compound that readily comes apart to form benzaldehyde and hydrogen cyanide (HCN). The hydrogen cyanide is a very toxic gas.

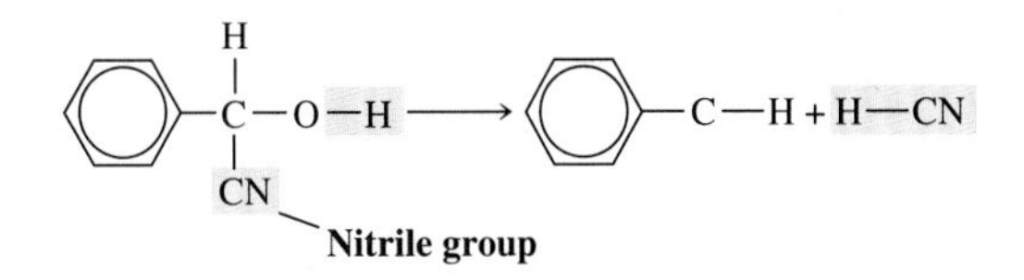

Cyanohydrin as an Insect Defense Mechanism

Several insects make use of the reversible nature of the addition of HCN to the carbonyl group in their defense methods. For example, the millipede, *Apheloria corrigata*, uses the hydrolysis of *mandelonitrile*, a **cyanohydrin**, to ward off attackers (Figure 6.4). The insect uses an enzyme to catalyze the reaction that produces hydrogen cyanide. This deadly gas inhibits the ability of any organism to use oxygen, resulting in death. There are a number of organic compounds containing the cyanide group, —C≡N. These are often classified as *nitriles*.

Cyanohydrin in Laetrile

The reputed anticancer drug, laetrile, contains a complex compound called *amygdalin*. This compound has been known for many years as a constituent of bitter almond. Hidden in this complex structure is the cyanohydrin of benzaldehyde. Laetrile has not been demonstrated to be effective as a cancer treatment, and its continued use can result in chronic poisoning by HCN.

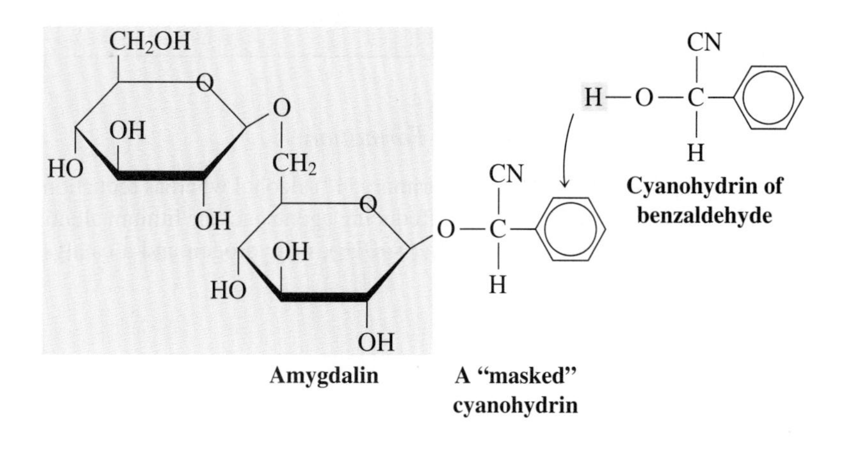

Hydrate Formation

♦ Addition of water to a carbonyl gives a *hydrate*.

A **hydrate** of a carbonyl group results from addition of water to the carbon–oxygen double bond, much as we saw in Chapter 3 for the addition of water to the carbon–carbon double bond. Because of the electronegativity difference between carbon and oxygen, the H adds to oxygen and the OH adds to carbon. It is generally difficult to obtain a hydrate, since the equilibrium favors the reverse reaction, but certain aldehydes give a stable product. The most well-known example is the reaction of *chloral* with water to form *chloral hydrate*.

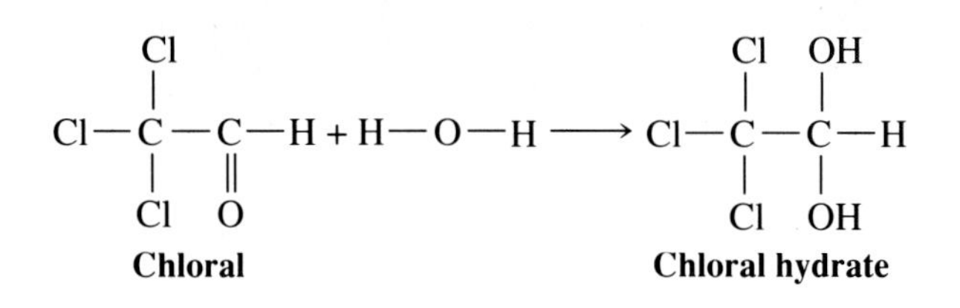

Chloral hydrate, sometimes known as "knockout drops," is used as a hypnotic and in drug-withdrawal treatments.

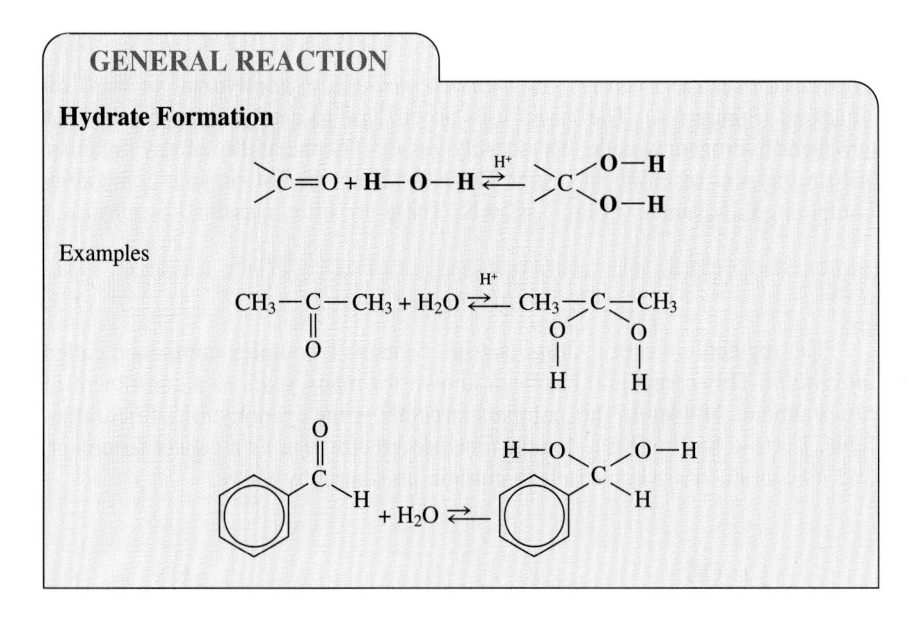

Alcohol Formation

♦ Reduction of a carbonyl gives an alcohol.

Reduction of an aldehyde gives a primary (1°) alcohol whereas reduction of a ketone gives a secondary (2°) alcohol. Reducing agents include lithium aluminum hydride ($LiAlH_4$), sodium borohydride ($NaBH_4$), or hydrogen and a catalyst.

GENERAL REACTION

Carbonyl Reduction

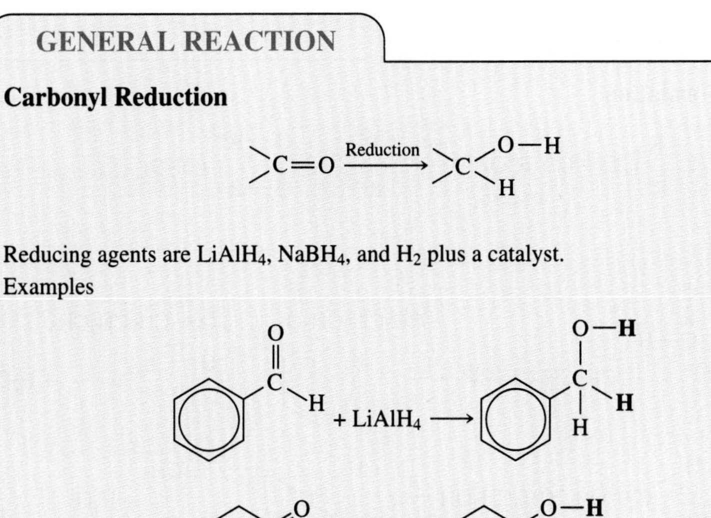

Reducing agents are $LiAlH_4$, $NaBH_4$, and H_2 plus a catalyst.

Examples

$$\text{cyclohexanone} + H_2 \xrightarrow[\text{(catalyst)}]{\text{Pt}} \text{cyclohexanol}$$

$$CH_3-\overset{\displaystyle O}{\overset{\|}{C}}-H + NaBH_4 \longrightarrow CH_3-CH_2-O-H$$

♦ Reduction of a carbonyl group is the result of adding two hydrogen atoms.

Formation of Acetals and Ketals

We have seen several general reaction types demonstrating that the carbonyl carbon is readily attacked by nucleophiles.

If a carbonyl compound is treated with excess alcohol in the presence of a trace of acid, aldehydes will give an **acetal** product, whereas ketones will give a **ketal** component.

♦ *Acetals* and *ketals* are formed from the reaction of a carbonyl compound with an alcohol. Most modern reports will no longer use the word "ketal."

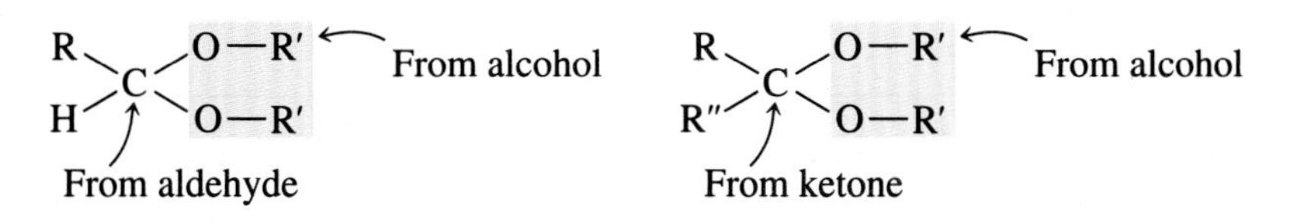

There has been a recent trend away from using the two different names, and correct nomenclature in the future will not include the name ketal. All molecules of this general type will be classified as acetals.

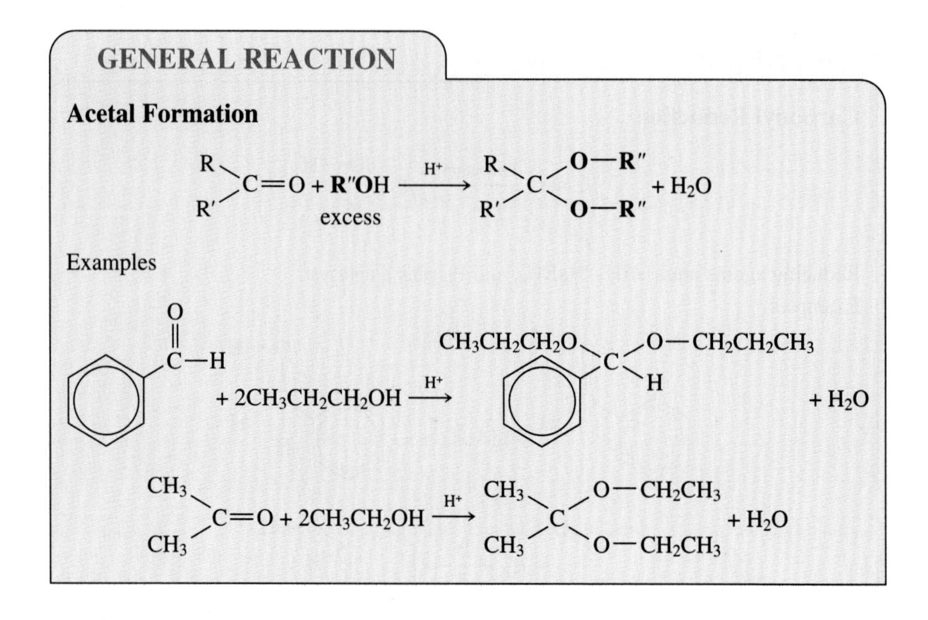

In later chapters we will notice that the reaction of a carbonyl compound actually takes place in two steps. If there is not an excess of alcohol, only one molecule will add. This gives a *hemiacetal*.

$$>C=O + R—O—H \xrightarrow{H^{\oplus}} >C<^{O—H}_{O—R}$$

Carbohydrates use an internal alcohol to form cyclic hemiacetals:

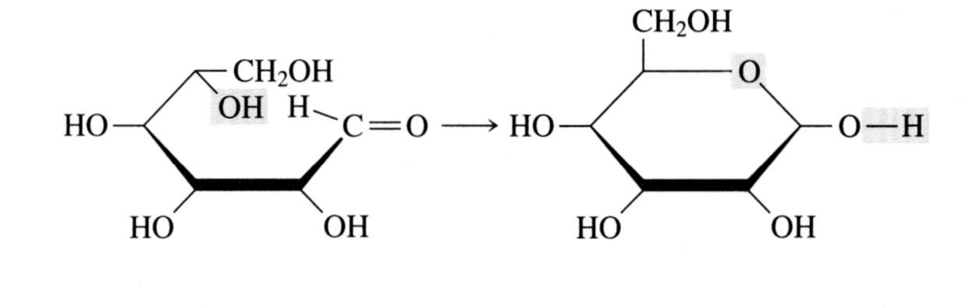

NEW TERMS

cyanohydrin A carbonyl derivative made by the addition of H—C≡N to C=O

hydrate A carbonyl derivative, usually unstable, made by the addition of water to a carbonyl

acetal The product formed from the acid-catalyzed reaction of an aldehyde with two equivalents of an alcohol

ketal The product formed from the acid-catalyzed reaction of a ketone with two equivalents of an alcohol. This is now considered under the generic term, acetal.

TESTING YOURSELF

Carbonyl Reactions

1. Give products for the following addition reactions:

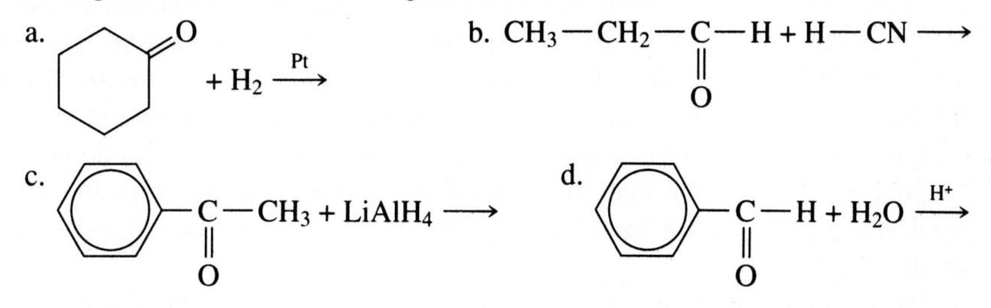

2. Draw the structure for the major product expected from each of the following reactions:

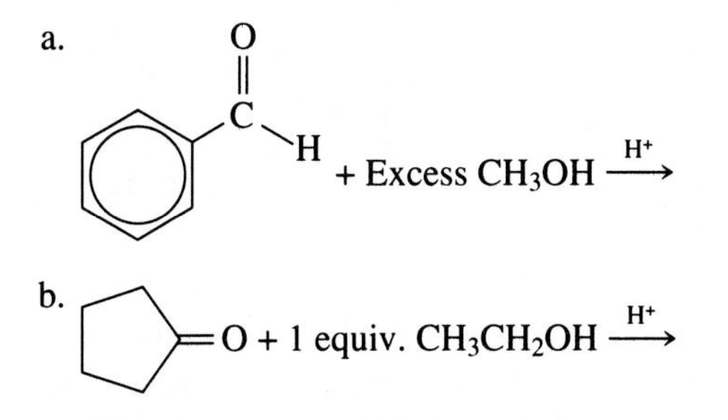

3. In the following general reaction what will happen to the equilibrium if a large excess of alcohol is used?

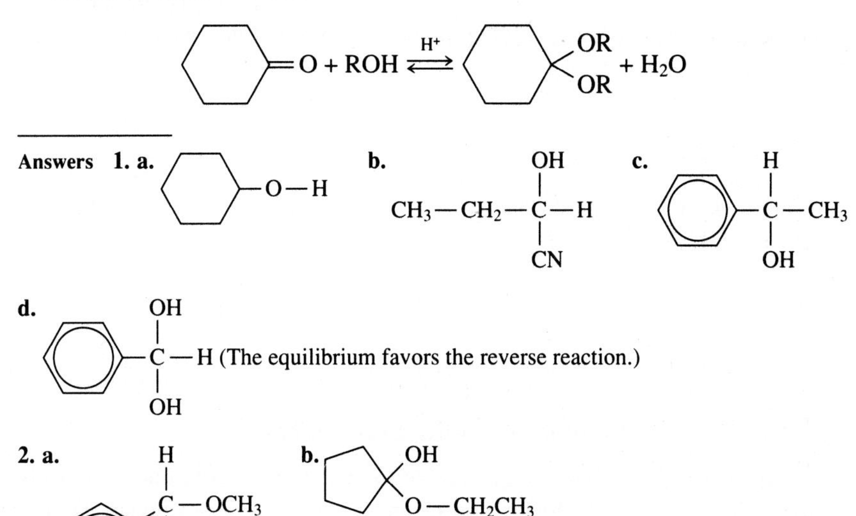

3. The equilibrium will be shifted towards products.

FIGURE 6.5
The beetle-damaged trees stand out in the western forests.

6.4 Natural Examples of Acetals

Pheromones: Examples of Natural Acetals

Pheromones are chemicals produced by an organism to communicate information to other members of the same species. Some species of bark beetles have acetals as their *sex pheromones*, chemicals that aggregate and excite insects for mating. The female pine bark beetle *Dendroctonus brevicomis* releases a compound, given the trivial name of brevicomin, to attract male beetles to her "nuptial chamber" in the tree. The tree rarely survives these massive attacks of beetles, and many of the forests of the northwest have been devastated by the beetles (Figure 6.5).

If the structure of brevicomin is dissected, the origins of the acetal can be seen. This approach has been taken as part of its successful laboratory synthesis.

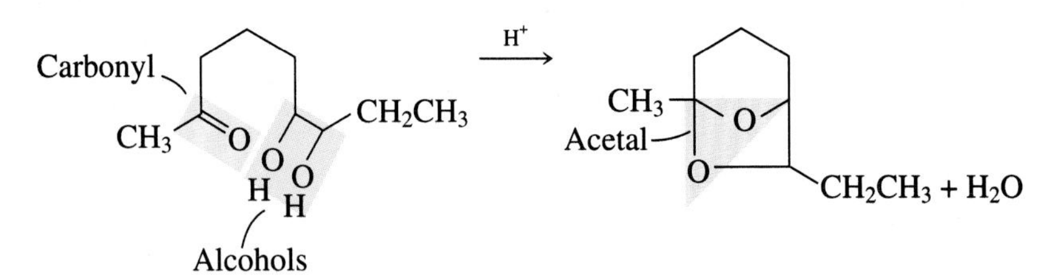

CH3 O O CH2CH3

An aggression pheromone of the house mouse

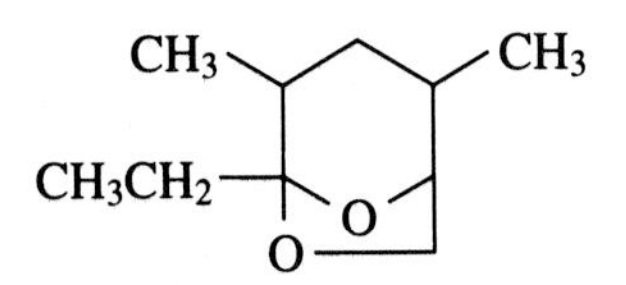

Pheromone of the Dutch elm beetle

FIGURE 6.6
The common house mouse and the beetle responsible for spreading Dutch elm disease both have pheromones based on an acetal structure.

A pheromone that induces aggressive behavior in the house mouse (*Mus musculus*) has been recently characterized and is surprisingly similar to brevicomin. A bicyclic acetal is also found as the sex pheromone of the Dutch elm beetle, *Scolytus multistriatus* (Figure 6.6). This beetle has been largely responsible for the widespread destruction of the stately elm trees of the eastern part of the United States. The beetles themselves do not kill the elm trees, but carry a fungus that does.

NEW TERMS

pheromones Chemicals produced by organisms to communicate with other members of the same species

6.5 Preparation Reactions of Carbonyl Compounds

Oxidation–Reduction Reactions

♦ Oxidation of primary alcohols gives aldehydes and of secondary alcohols gives ketones. Tertiary alcohols are resistant to oxidation.

Aldehydes and ketones are most often formed by the **oxidation** of alcohols. Primary alcohols form aldehydes, and secondary alcohols form ketones. Tertiary alcohols are resistant to oxidation. These general oxidation patterns are shown in Figure 6.7. Typical oxidizing agents include a host of chromium reagents, including CrO_3 and H_2CrO_4. Sometimes just the oxygen in air can carry out an oxidation. In the remainder of this chapter, a "generic" oxidizing agent in a reaction will be designated as [O] over the reaction arrow. The oxidation of ethyl

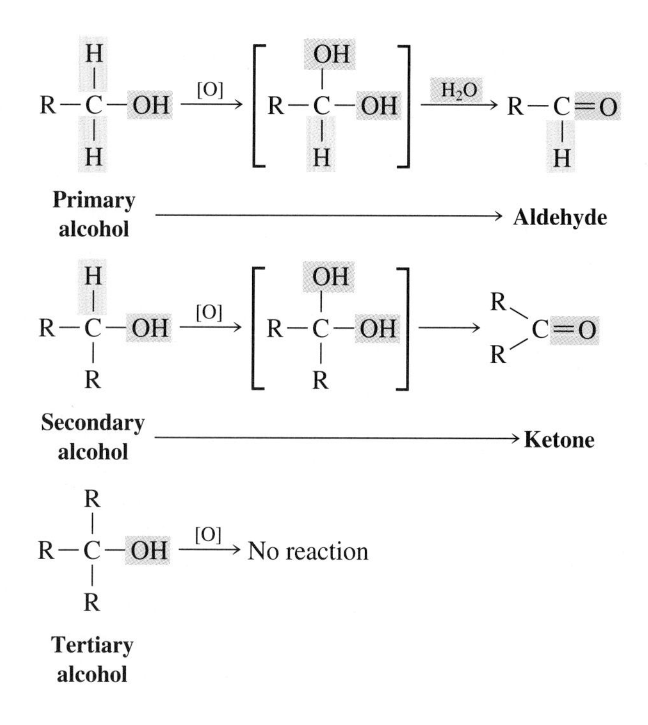

FIGURE 6.7
A generalized oxidation scheme. This scheme is not intended to show *how* an oxidation takes place. Rather, the purpose is to clearly show the need for hydrogens attached to the carbon if oxidation is to take place.

alcohol forms the basis of the breath tests given by police to drivers suspected to be driving while intoxicated. In the road test, a standard solution of orange-yellow dichromate ion ($Cr_2O_7^{2-}$) is the oxidizing agent and the alcohol is the reducing agent. The alcohol is oxidized while the chromium is reduced to a green-colored Cr^{3+} ion. The change in color is measured, and the level of alcohol in the driver is determined.

♦ The police analysis of alcohol content relies on the fact that the ethyl alcohol ingested becomes rapidly equilibrated with water in the body. Ethyl easily diffuses through cell membranes, and the partial pressure of ethyl alcohol in the breath is a very good measure of the amount in the blood.

The reverse reaction, the **reduction** of carbonyl compounds to alcohols, is the method used for many alcohol preparations. For these reactions, typical reducing agents include sodium borohydride ($NaBH_4$), lithium aluminum hydride ($LiAlH_4$), and H_2/Pt. Note that the catalytic reduction of a carbonyl group is similar to the catalytic reduction of an alkene.

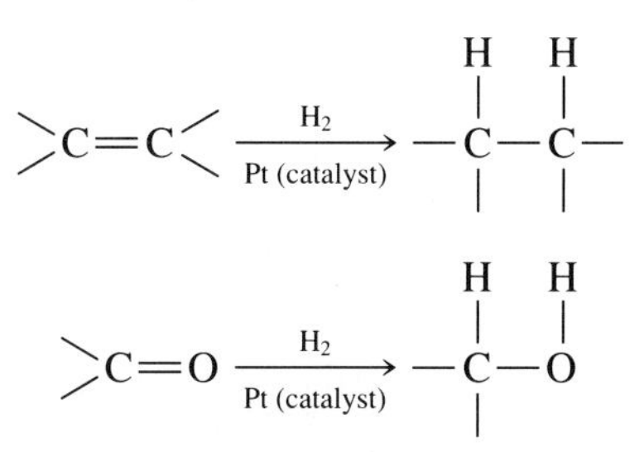

Just as the general symbol [O] is used for an oxidizing agent, the general symbol [H] is used for a reducing agent. Aldehydes are reduced to primary alcohols, and ketones are reduced to secondary alcohols. This is the reverse of

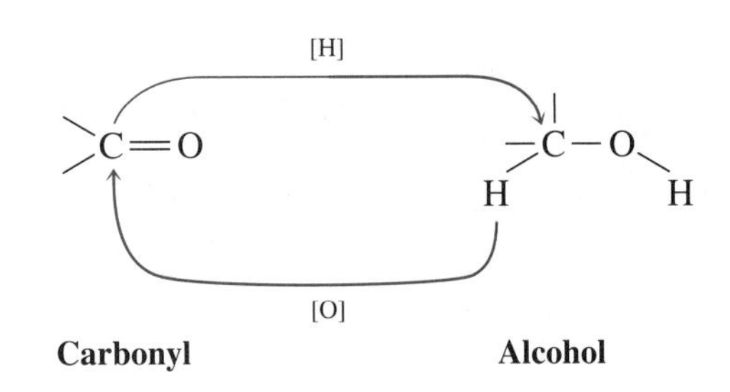

FIGURE 6.8
The reversible nature of the oxidation–reduction process. A carbonyl compound can be reduced to an alcohol and an alcohol can be oxidized to a carbonyl compound.

the oxidation process. Thus, the reversible nature of these oxidation–reduction reactions is apparent (Figure 6.8).

Two important points should be reviewed and reinforced at this time.

1. Every time an oxidation is carried out, there must also be a reduction.
2. In the complete sense, an oxidation is a loss of electrons and a reduction is a gain of electrons.

This will become more apparent in other specific oxidation reactions. However, the concept is very important for biological processes because electron movement in the living organism operates by the oxidation–reduction process. For example, in the important oxidation–reduction processes that keep the body operating, a carbonyl reduction takes place after strenuous exercise. When we become so winded from exertion that we feel we can't continue, we often feel muscle discomfort. This is due to the need for more oxygen to help in the conversion of NADH to NAD^+ (the acronym for the biological oxidizing agent of nicotinamide adenine dinucleotide). If more oxygen is not available, the body finds another way to make the conversion—it uses the NADH to reduce pyruvic acid:

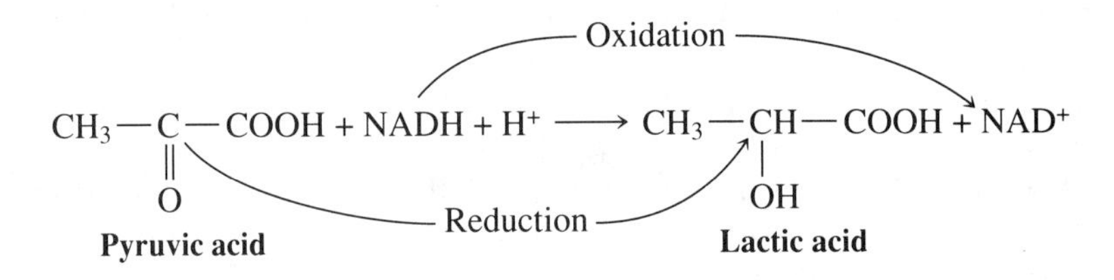

The buildup of this *lactic acid* in the muscle tissue is responsible for the muscle pain we experience when exercising heavily.

Tests for Aldehydes

♦ Oxidation reactions are easy tests for aldehydes, since aldehydes are easily oxidized to acids.

Aldehydes are reactive toward further oxidation, but ketones are resistant. This difference in reactivity towards oxidizing agents provides a test to determine whether a carbonyl compound is an aldehyde or a ketone. In general, the oxidation of an aldehyde can be shown as:

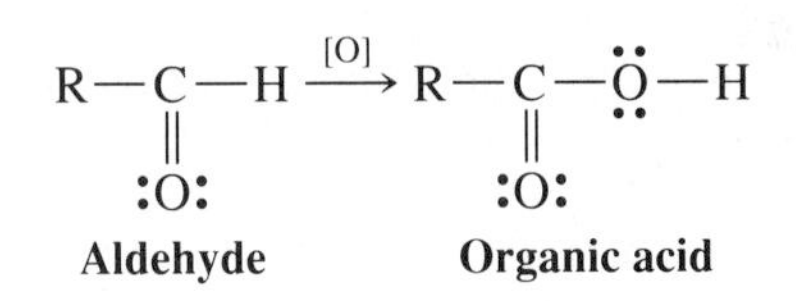

As already mentioned, for every oxidation that takes place, there must also be a reduction. This idea of *coupled reactions* is extremely important for biological processes.

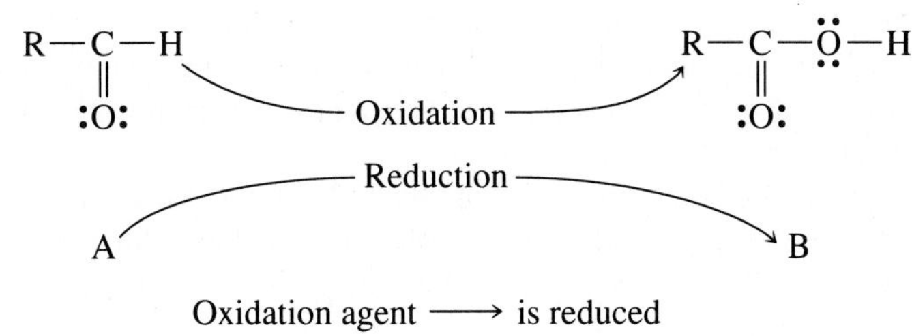

Oxidation agent ⟶ is reduced

Tollen's Test

The Tollens test is a simple test for aldehydes. In this procedure, the silver ion in $Ag(NH_3)_2^+$ is the oxidizing agent.

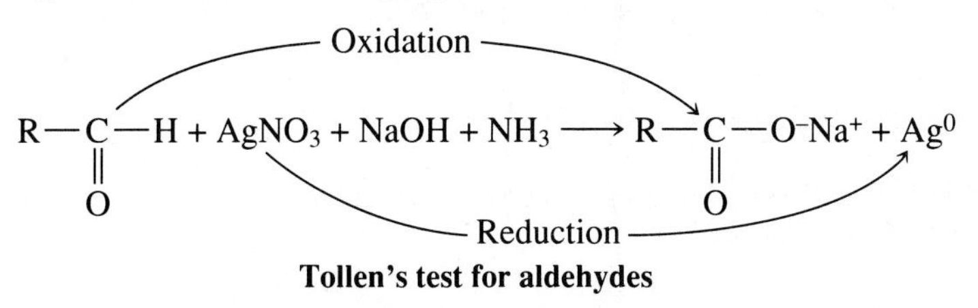

Tollen's test for aldehydes

A positive result is easily observed for this test because a perfect film of silver coats the inside of the test tube. In this example the Ag^+ has gained electrons (been reduced) to form metallic silver, and the aldehyde has been oxidized to an acid.

Benedict's or Fehling's Test for Aldehydes

A very similar idea of oxidation–reduction is applied in **Benedict's** or **Fehling's tests**. In these visual tests, the deep blue color of Cu^{2+} disappears as the aldehyde is oxidized to an acid and the copper is reduced to Cu^+. The copper(I) ion gives a very characteristic brick-red color as it forms an insoluble copper(I) oxide. The two tests differ only in the identity of the complex anion that is associated with the Cu^{2+}. Citrate is the anion of citric acid; tartrate is the anion of tartaric acid. Organic acids are discussed in the next chapter.

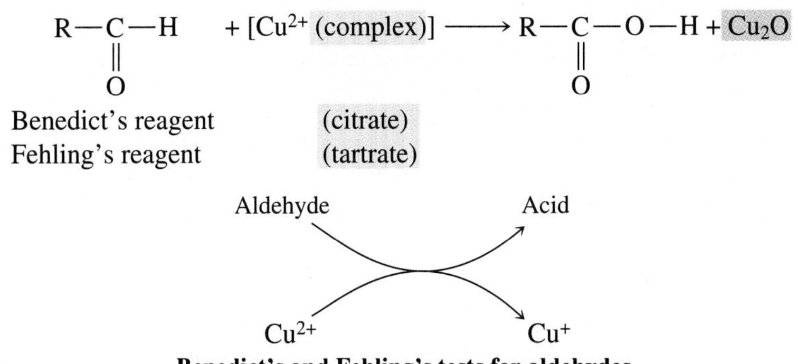

Benedict's and Fehling's tests for aldehydes

We saw from Table 6.1 that glucose (a common sugar) contains an aldehyde group. One of the simple tests for diabetes is an analysis for glucose in the urine. This is easily done with reagents that give certain color changes. As the concentration of copper (II) ions (blue color) decreases and the color of Cu(I) ion increases (red color), solutions pass through a color change continuum. This change can be used as an indication of the amount of sugar in urine (oxidation of glucose) and can help screen for diabetes.

Color

Percentage glucose 0.5% ⟷ 2% (or greater)

The preparation of wine vinegar is simply an example of the oxidation of an alcohol (ethyl alcohol) to an aldehyde (acetaldehyde) and then further to an acid (acetic acid, also called ethanoic acid). The same process occurs in our bodies after alcohol consumption.

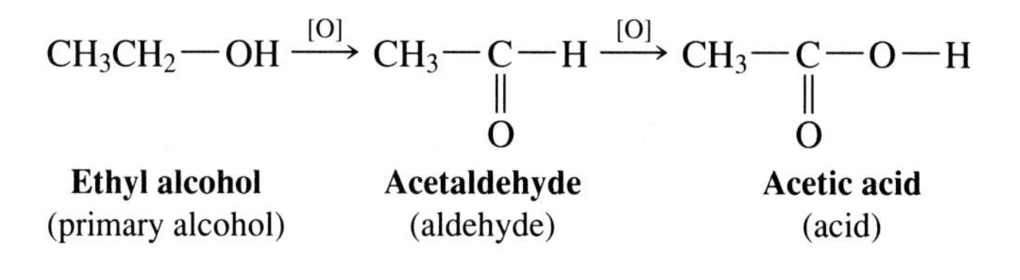

NEW TERMS

oxidation The loss of electrons; in carbonyl chemistry, it relates to the reaction:

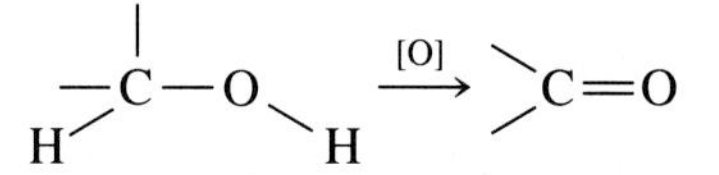

reduction The gaining of electrons; in carbonyl chemistry, it refers to the process:

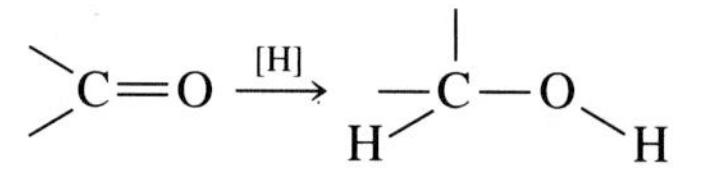

Tollen's test A test for aldehydes using a silver ion as the oxidizing agent. As the aldehyde is oxidized to an acid, the silver ion is reduced to free silver.

Benedict's or Fehling's test A test for aldehydes using copper ion as the oxidizing agent.

TESTING YOURSELF

Reactions of Carbonyl Compounds

1. Give products for each of the following oxidation reactions:

a.

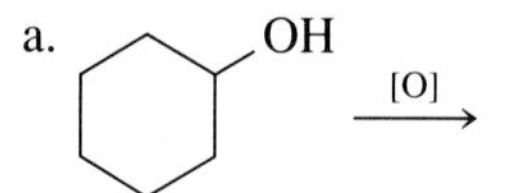

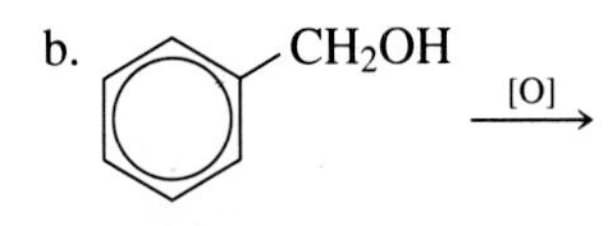

c.

$$CH_3-\underset{\underset{OH}{|}}{\overset{\overset{CH_3}{|}}{C}}-CH_3 \xrightarrow{[O]}$$

2. Give products for each of the following reduction reactions:

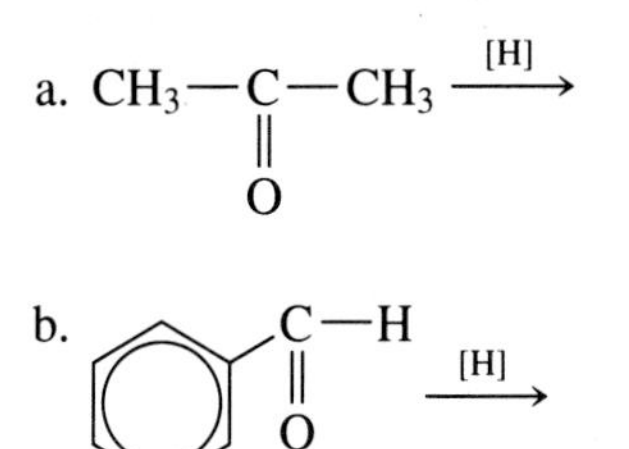

3. Show the acetal resulting from the following reactions:

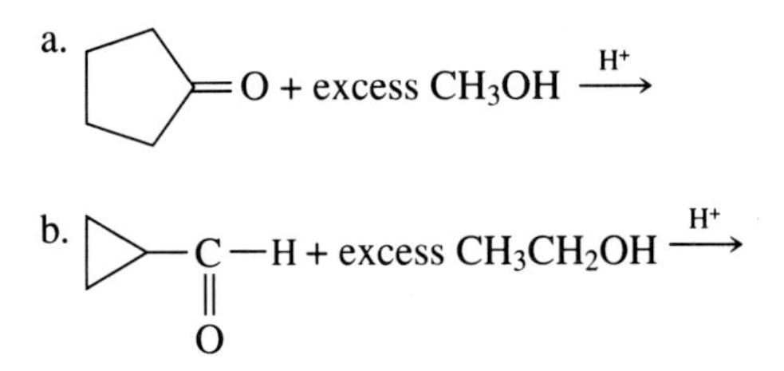

Answers **1. a.** cyclohexanone **b.** benzaldehyde **c.** no reaction
2. a. isopropyl alcohol **b.** benzyl alcohol

3. a. OCH_3, OCH_3 **b.** $-C-H$, CH_3CH_2O OCH_2CH_3

6.6 Oxidation of Hydroquinones to Quinones

Phenols do not readily oxidize; however, *p*-hydroxyphenol, commonly known as **hydroquinone**, can be easily oxidized to **quinone**. Oxidation (and its reverse, reduction) is common to biological systems. Hydroquinone is often added to commercial products to inhibit their oxidation, since it is more reactive to

♦ Quinones are formed from the oxidation of hydroquinones.

CHEMISTRY CAPSULE

Antioxidants

Many foods rich in fats and oils become spoiled easily because of oxidation (particularly unsaturated oils because the double bonds are very reactive). Although the reaction processes by which these reactions take place are quite complex, simply stated, atmospheric oxygen causes these reactions. To prevent the reactions from taking place, *antioxidants* are sometimes added to food as preservatives. Antioxidants are compounds that inhibit oxidation by molecular oxygen (autoxidation). Phenolic compounds and hydroquinone compounds are especially useful in this regard.

The use of antioxidants to preserve food is not new, only the understanding of their chemical action is. Since ancient times, spices and herbs have been used to help in the safe storage of food. It is interesting to note that many of these natural materials are pleasant smelling—indeed, the *aromatic* spices such as clove, oregano, thyme, rosemary, and sage contain phenolic compounds that were among the very early compounds associated with the aromatic class of hydrocarbons. Even the process of smoking meats provides antioxidants to the food because formaldehyde and phenolic compounds are found in wood smoke. Antioxidants are also present in animal tissue to prevent oxidation of the natural fats in the body. There is some thought that the "aging process" is accelerated by the oxidation of natural double bonds.

You need not become concerned that the "sea" of oxygen around us is destructive, because the complex reactions involved in autoxidation processes do not use oxygen as it naturally occurs in the atmosphere. However, the pollutants in the environment are capable of initiating some of these reactions, and this is one of the reasons why there is so much concern about atmospheric contamination.

In addition to food spoilage, there is another common autoxidation problem. Most of us drive a car and have probably heard of the fuel delivery system getting "gummed up." This is precisely what happens when the hydrocarbons in fuel undergo autoxidation. Under the high-temperature conditions in the combustion chamber, the lubricating oil undergoes this process, accelerating damage to the engine. Natural rubber, synthetic rubber, and most plastics are attacked by oxygen in much the same way. We find that when antioxidants are added to these materials, the structural damage resulting from oxidation is greatly reduced.

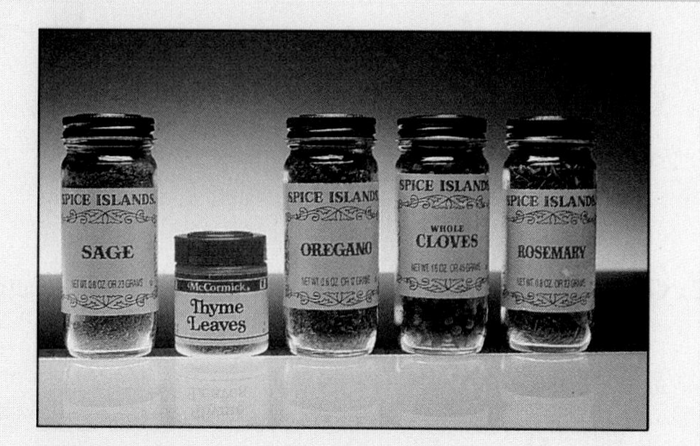

FIGURE 6.B
Many natural spices have important antioxidation properties, which have contributed to their use over the centuries. Not only do these materials improve taste, they also preserve foods.

oxidizing agents than most other functional groups. Compounds are added as antioxidants to many snack foods, fats, and oils. The common antioxidants include BHT (butylated hydroxy toluene) and BHA (butylated hydroxy anisole).

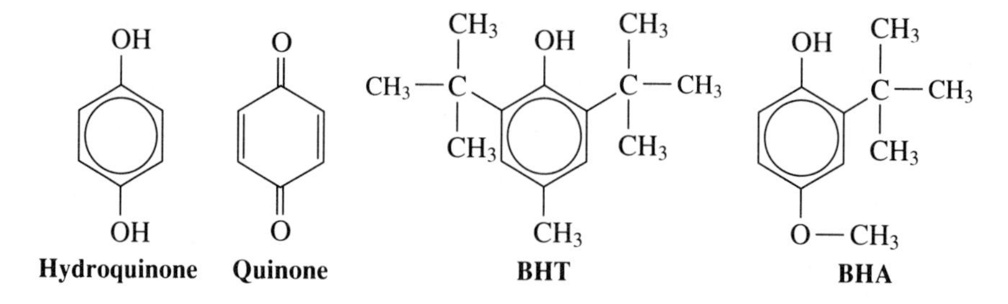

The reaction of an oxidation of hydroquinone and the reverse reduction of a quinone are shown below.

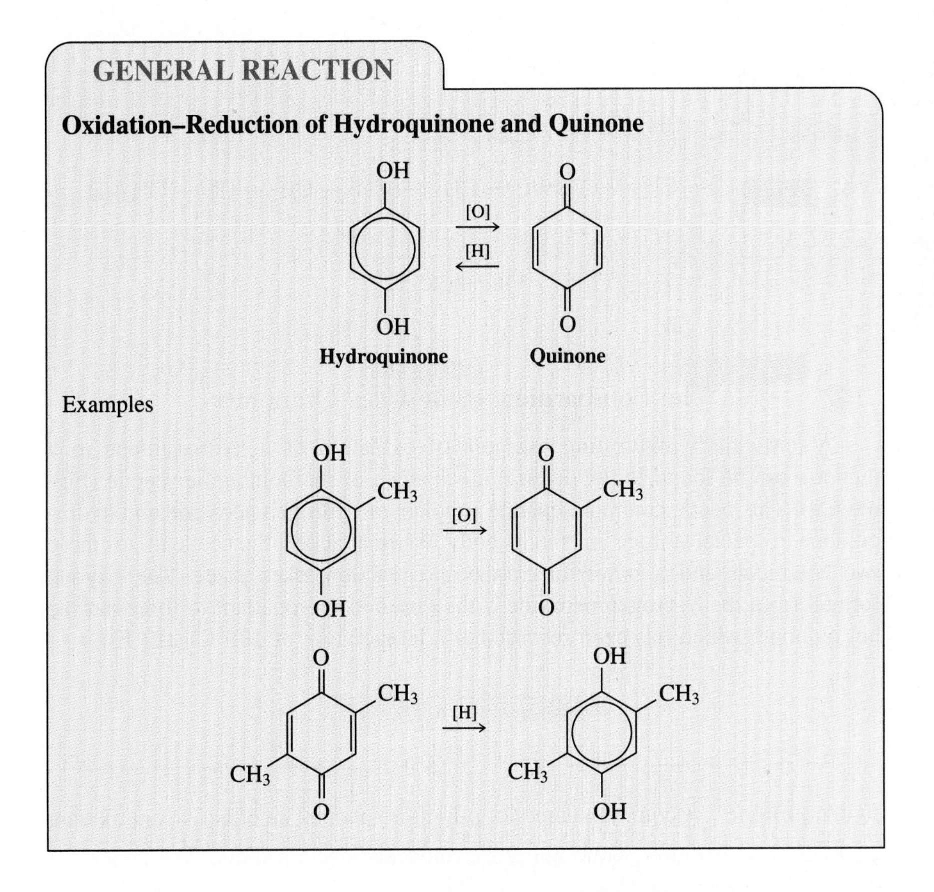

Quinones in Our Everyday Life

As part of the process of respiration (biological oxidation), oxygen must be converted to water. If we keep track of the electrons involved in this process, we can see that the oxygen has been reduced. This requires that something else has been oxidized. In the biochemistry chapters of this book, it will be noted that carbon compounds are oxidized to carbon dioxide.

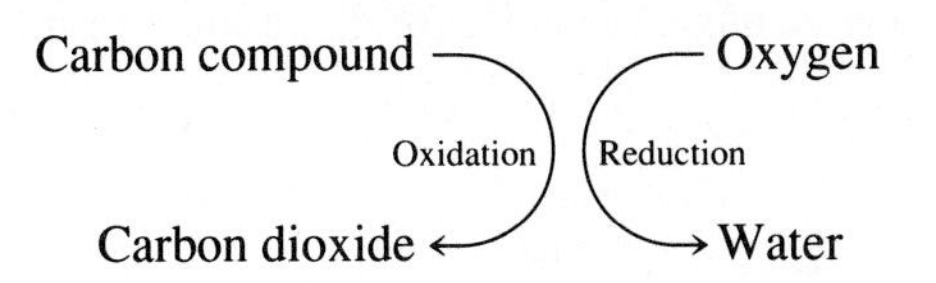

The electrons from the oxidation process get carried down a chain of secondary oxidations and reductions, including an $Fe^{2+} \rightleftharpoons Fe^{3+}$ equilibrium and a hydroquinone $\rightleftharpoons$ quinone equilibrium. In a section on photosynthesis in Chapter 16, the same conversion will be seen, but in the reverse sense. Water will be oxidized to oxygen, and the electrons will be used to reduce carbon dioxide into organic compounds.

FIGURE 6.9
The bombardier beetle has a unique defensive spray that has its origins in the hydrogen peroxide oxidation of hydroquinone to quinone. The exothermic nature of the reaction and the formation of an oxygen-based propellant system make this a very effective defense.

Vitamin K is important in the maintenance of our blood-clotting ability. Because of its obvious hydrocarbon nature, Vitamin K is a fat-soluble vitamin; it is also a quinone.

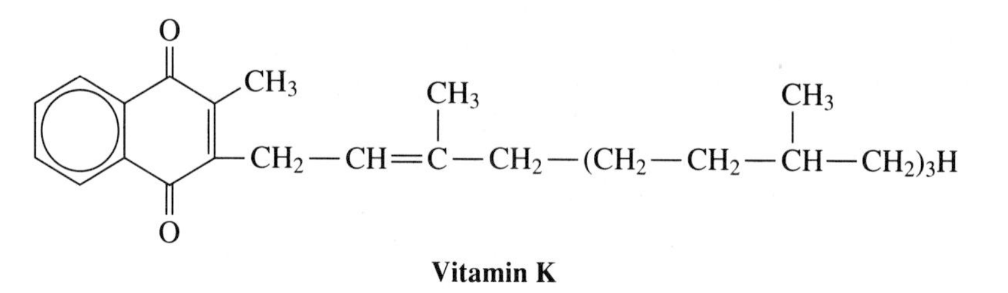

Vitamin K

The Bombardier Beetle Uses Quinones

A particularly interesting example of oxidation of a hydroquinone to a quinone can be found in the defense mechanism of the bombardier beetle (Figure 6.9). This beetle contains separate supplies of hydrogen peroxide and hydroquinone in special storage units in its body. When attacked, the beetle mixes these two chemicals and a powerful exothermic reaction takes place. The oxygen formed from the hydrogen peroxide is then used as a propellant to force out the hot quinone, which has been recorded at a temperature of 100°C (212°F)!

NEW TERMS

hydroquinone A common name for *p*-hydroxyphenol, an effective antioxidant

quinone A common name for 2,5-cyclohexadiene-1,4-dione, the oxidation product from hydroquinone.

TESTING YOURSELF

Hydroquinones and Quinones

1. Show what happens to hydroquinone when it is oxidized.
2. What might be a reasonable structure for the product resulting from reduction of vitamin K?

Answers **1.** **2.**

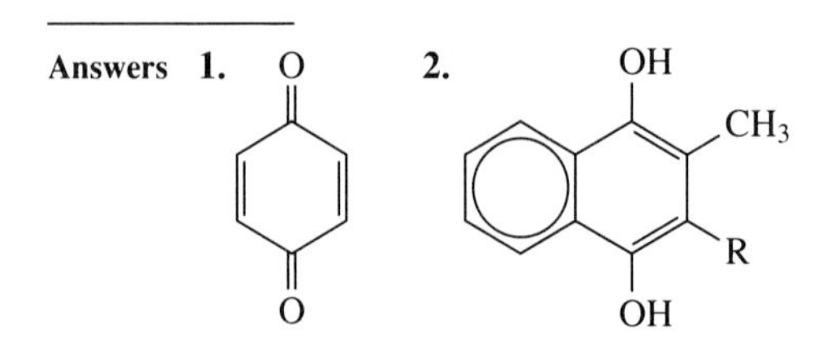

Summary

The greater electronegativity of oxygen dominates the chemical behavior of the carbon–oxygen double bonds of the *carbonyl group*. In reactions, the oxygen atom appears as though it has a negative charge and reacts with electrophiles. The carbon atom of the carbonyl group appears as though it has a positive charge and reacts with nucelophiles. Reactions of the carbonyl double bond include *reduction, hydration, cyanohydrin* formation, and acetal formation. In general, the addition reactions of the carbonyl group resemble those of alkenes, with an important difference due to the electronegativity differences between the oxygen and the carbon.

Acetals are important derivatives of the carbonyl group and are formed by the reaction of the carbonyl compound with alcohol in the presence of an acid catalyst.

Reaction Summary

Addition of an Electrophile and Nucleophile to a Carbonyl

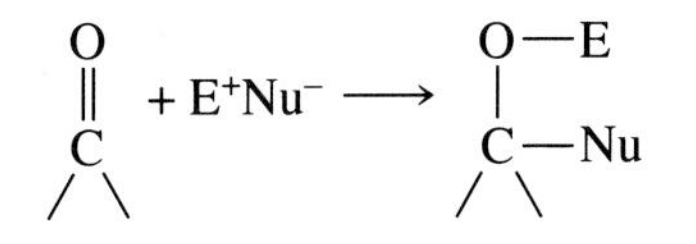

Cyanohydrin Formation

$$\rangle C{=}O + HCN \longrightarrow \rangle C(OH)(CN)$$

Hydrate Formation

$$\rangle C{=}O + HOH \underset{}{\overset{H^+}{\rightleftharpoons}} \rangle C(OH)(OH)$$

Carbonyl Reduction

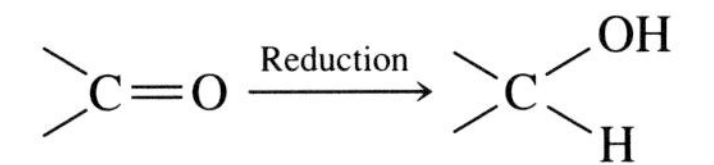

Acetal Formation

$$\rangle C{=}O + \underset{\text{excess}}{ROH} \xrightarrow{H^+} \rangle C(O{-}R)(O{-}R) + H_2O$$

Alcohol Oxidation
(1° and 2° alcohols)

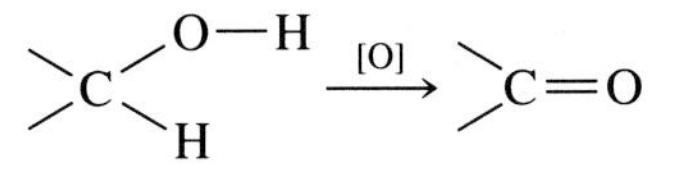

Aldehyde Oxidation

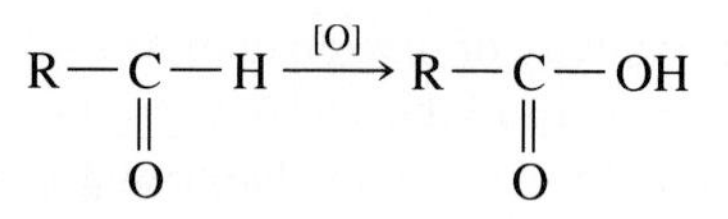

Oxidation of Hydroquinones

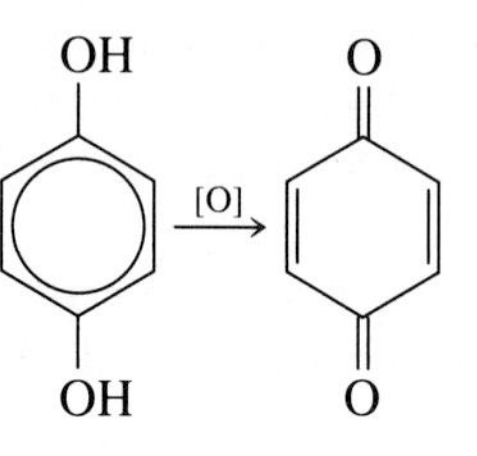

Reduction of Quinones

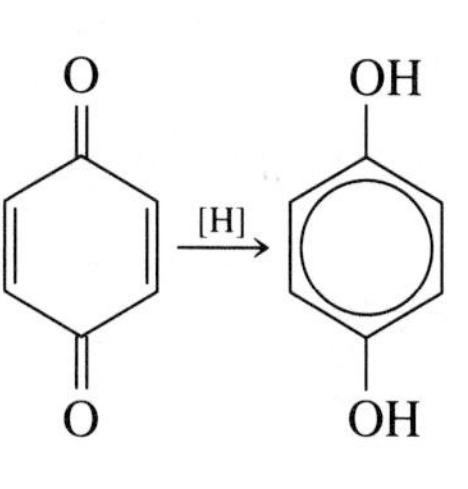

Terms

carbonyl group (6.1)
aldehyde (6.1)
ketone (6.1)
dipole–dipole force (6.1)
cyanohydrin (6.3)
hydrate (6.3)
acetal (6.3)
ketal (6.3)
pheromones (6.4)
oxidation (6.5)
reduction (6.5)
Tollen's test (6.5)
Benedict's or Fehling's test (6.5)
hydroquinone (6.6)
quinone (6.6)

Exercises

Structure and Physical Properties (Objective 1)

1. Compare the structures of acetone (2-propanone) and isobutylene (2-methylpropene). Which has the greater dipole? Why?
2. For each of the molecules in Exercise 1, show where an H^+ becomes preferentially attached.
3. Rank the following compounds according to boiling point (lowest first):
   a. pentanal
   b. 1-chloropentane
   c. 1-pentanol
   d. pentane
4. Why don't ethylene molecules interact with each other as well as molecules of formaldehyde?
5. Describe why aldehydes have lower boiling points than primary alcohols.
6. Why might you expect methyl ethyl ketone (2-butanone) to be less water-soluble than *sec*-butyl alcohol (2-butanol)?

Nomenclature (Objective 2)

7. Give names to each of the following aldehydes:

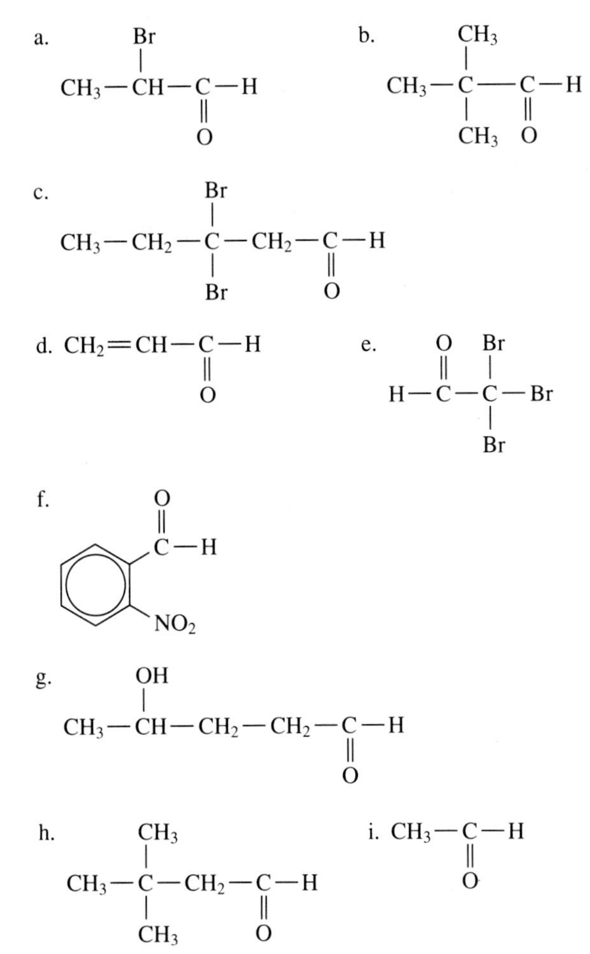

8. Give names for the following aldehydes:

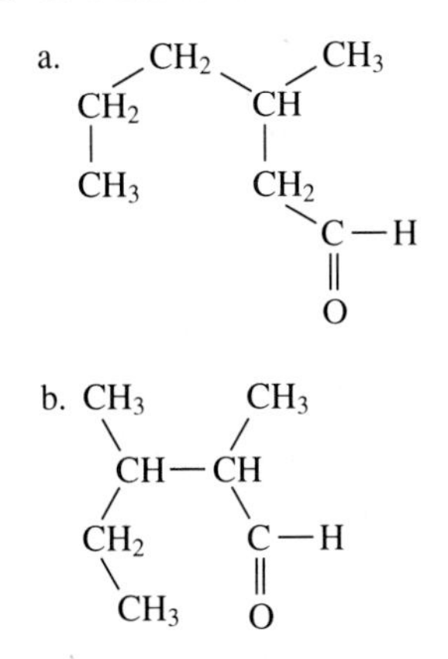

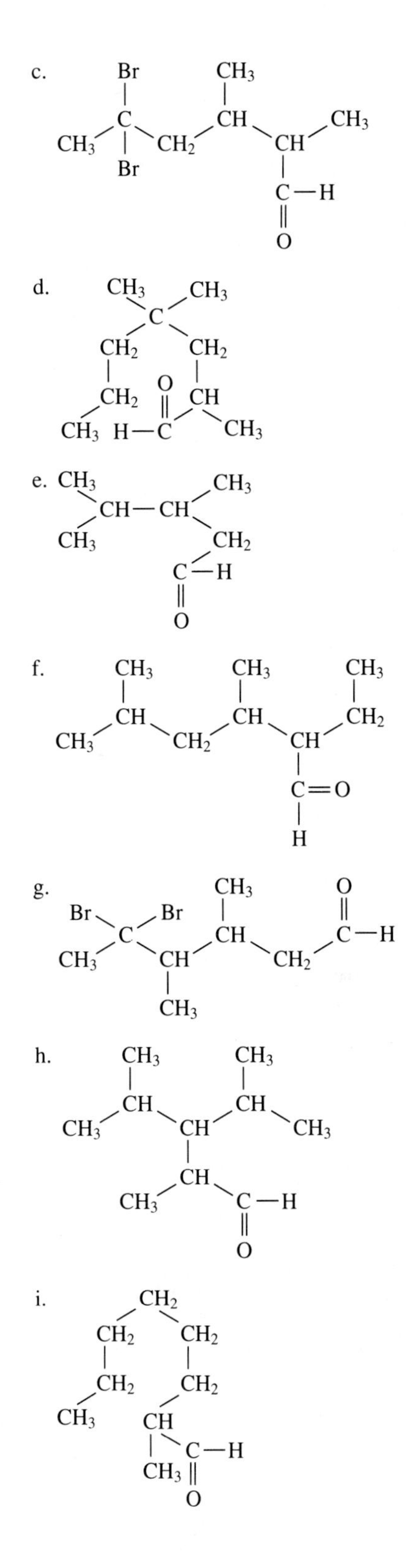

9. Give names for the following ketones:

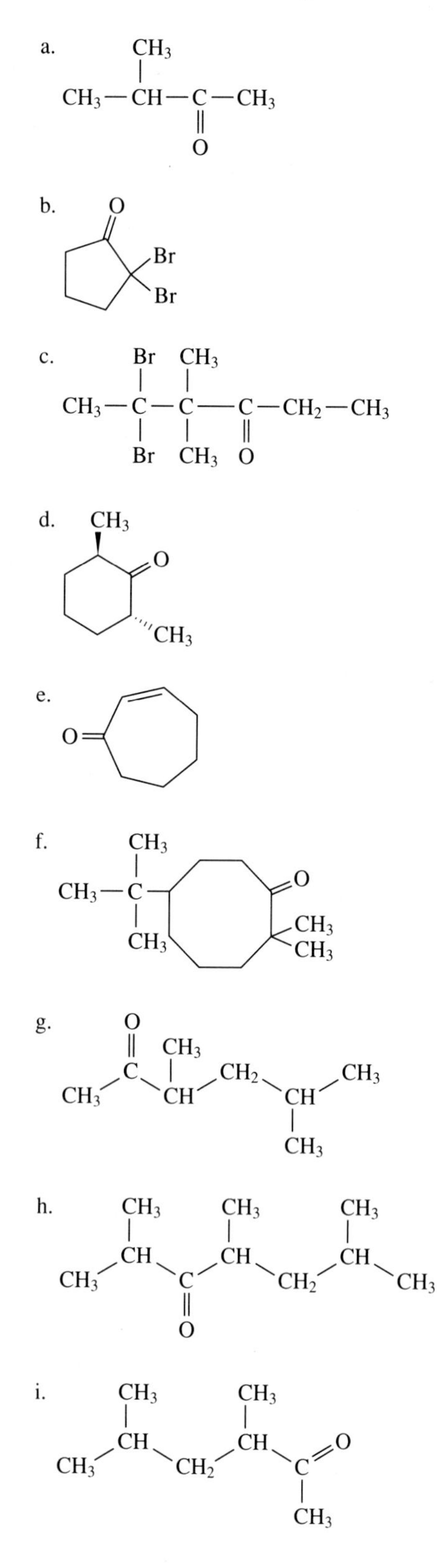

10. Give names for the following ketones:

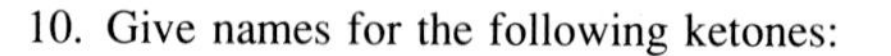

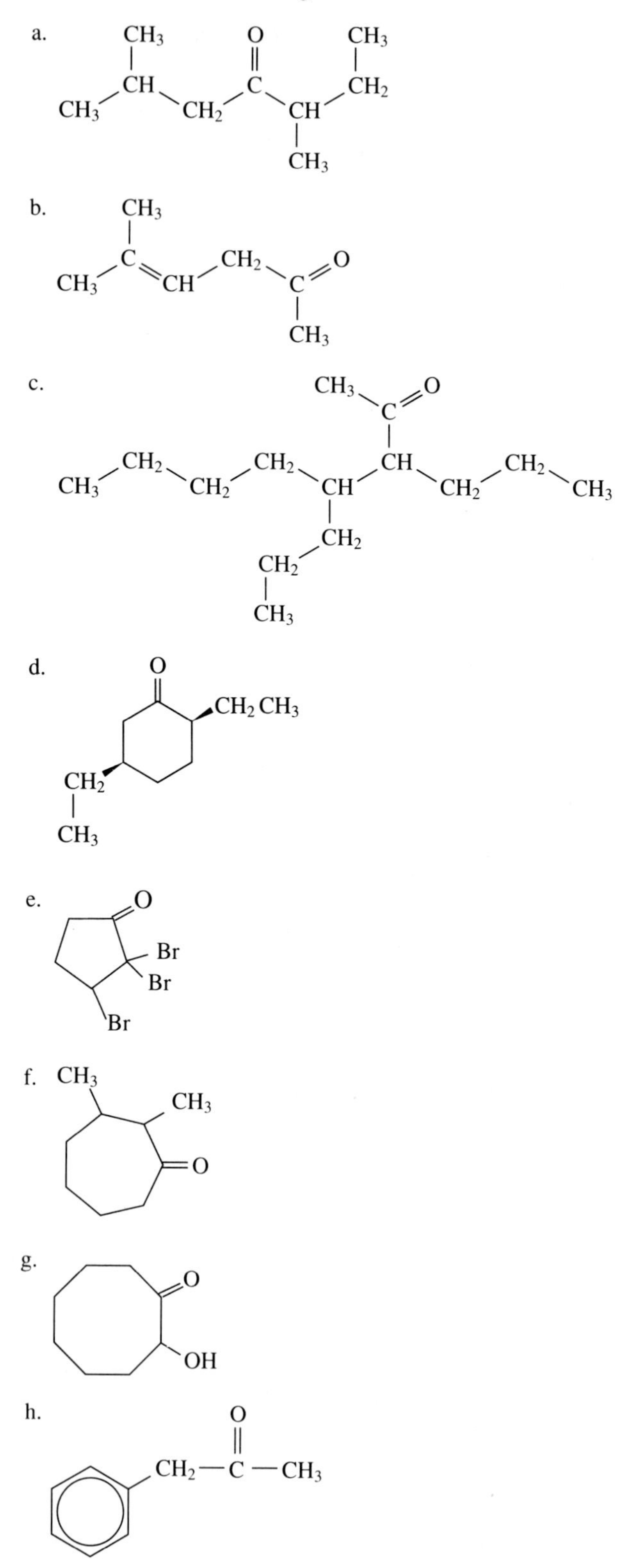

i.

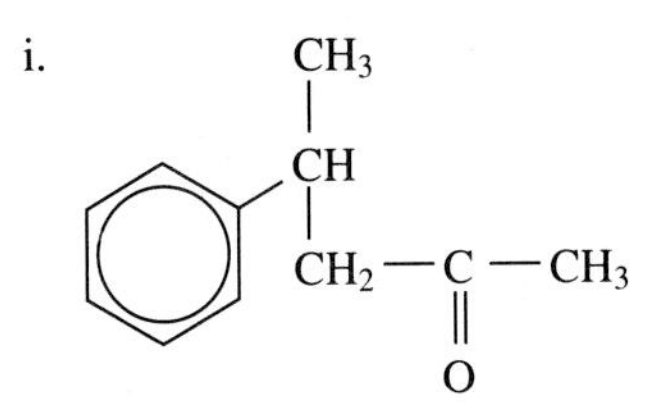

11. Give names for the following aldehydes and ketones:

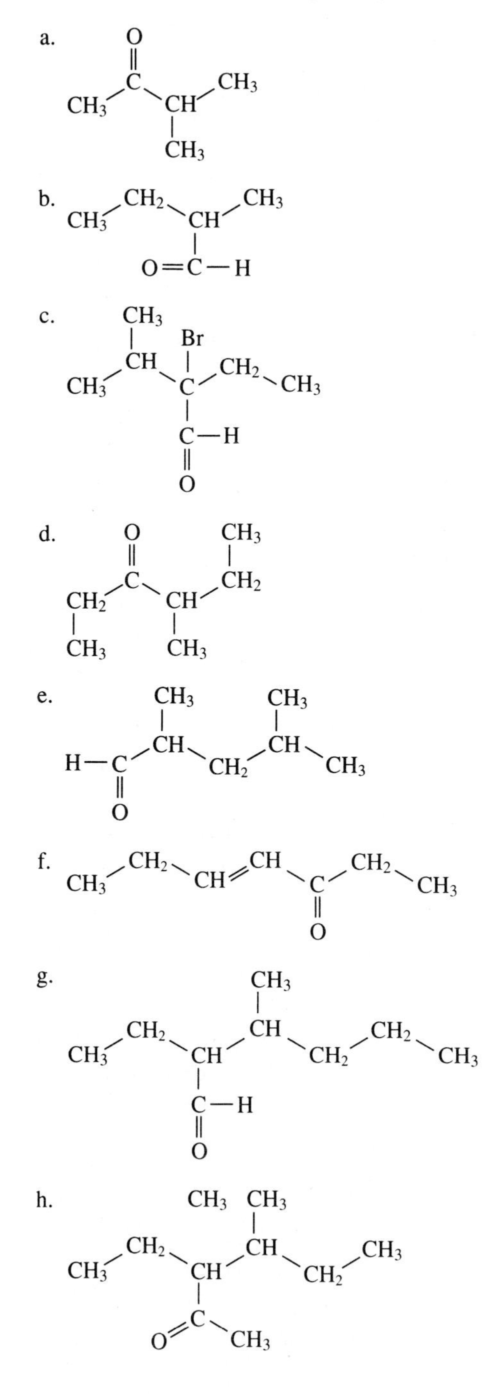

i.

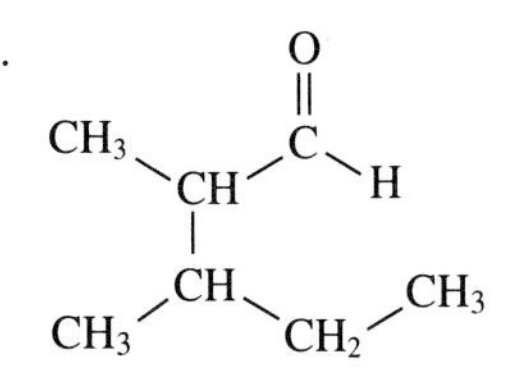

12. Give structures for each of the following compounds:
    a. 5-bromopentanal
    b. 3,4-dimethylcyclohexanone
    c. *p*-chlorobenzaldehyde
    d. 2,2,5,5-tetramethyl-4-heptanone
    e. *m*-nitrobenzaldehyde
    f. 4,4-diiodo-2-heptanone
    g. *trans*-3,4-dimethylcyclohexanone
    h. 2-benzyl-3-heptanone
    i. methyl ethyl ketone
    j. 5-phenylhexanal

Reactions (Objective 3)

Give products for each of the following reactions:

13. *Reductions:*

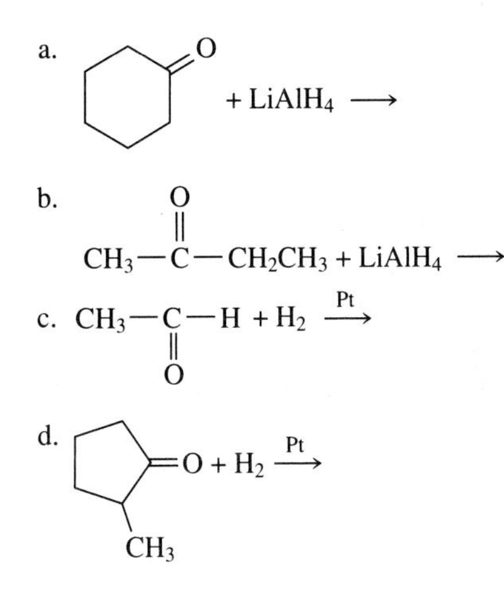

14. *Acetal Formation:*

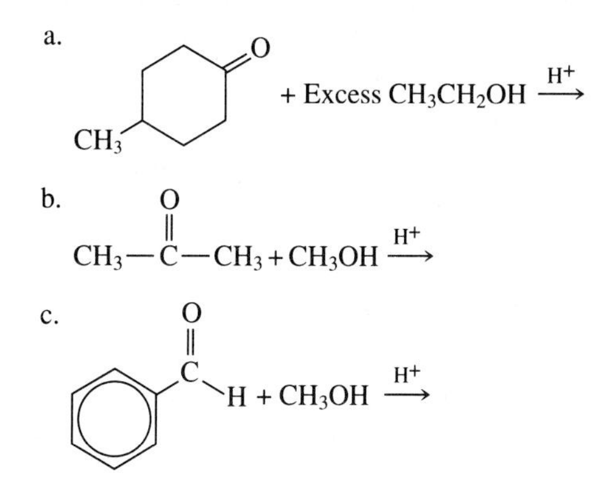

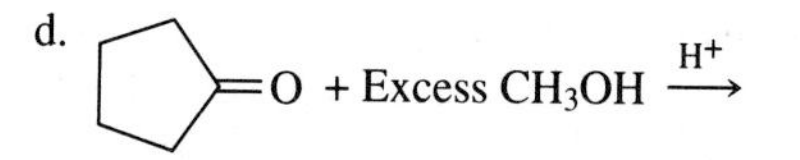

15. *Oxidations:*

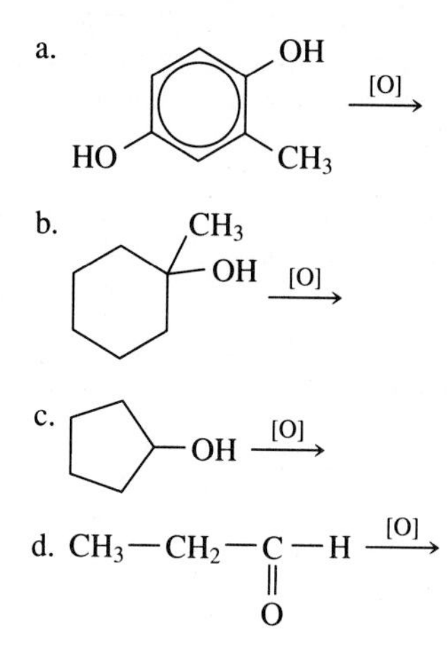

16. *General Survey of Carbonyl Reactions:*

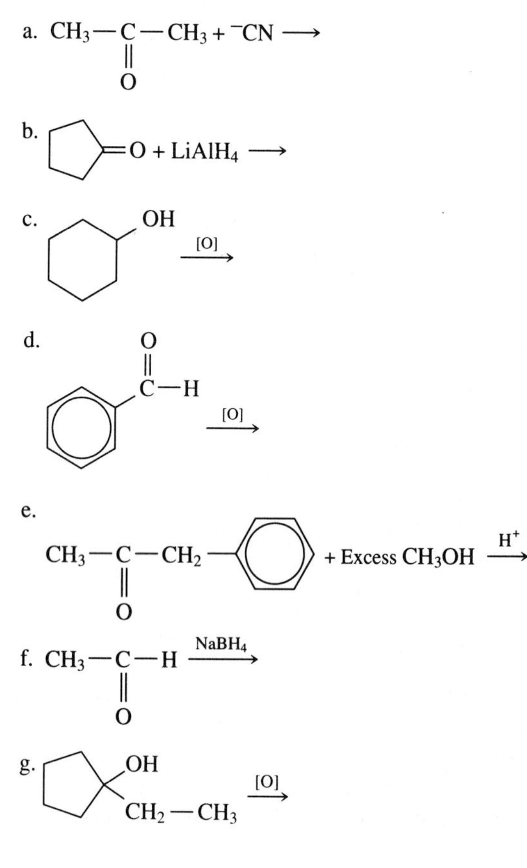

h. $CH_3CH_2CH_2-OH \xrightarrow{[O]}$

i.

OH
Cl Cl
Cl Cl
OH
$\xrightarrow{[O]}$

j.

O
+ $H_2 \xrightarrow{Pt}$

Thought Exercises

17. Using the oxidation model introduced in this chapter, show how the oxidation of methane produces carbon dioxide.
18. Using the same model, show the oxidation of methanol and ethanol.

Aldehyde Nomenclature (Objective 2)

19. From the following line drawings suggest appropriate names:

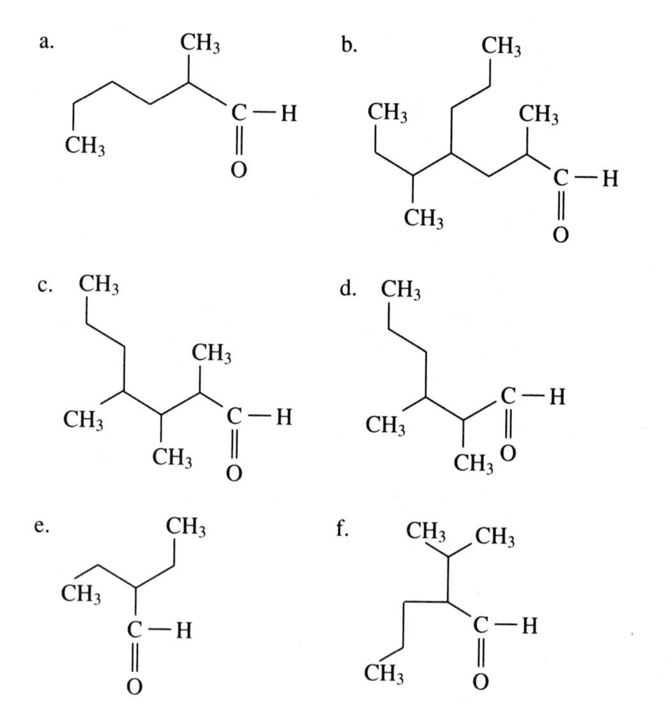

Ketone Nomenclature (Objective 2)

20. From the following line drawings, give appropriate names to the compounds:

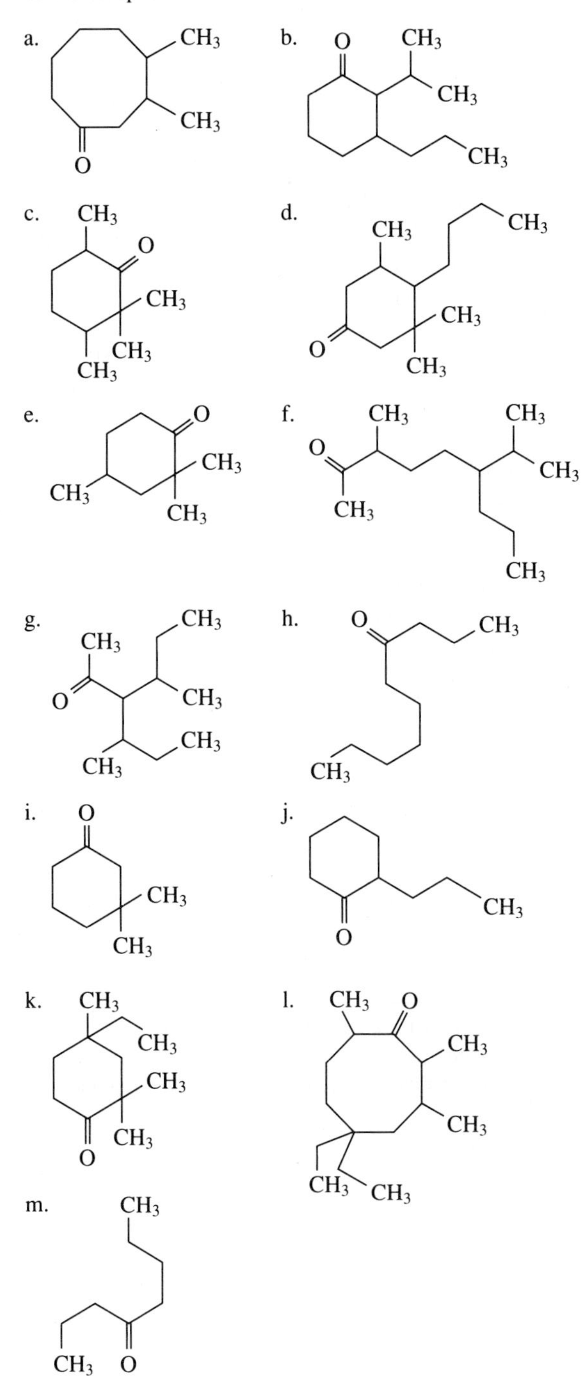

Structure and Properties (Objective 1)

21. Describe the bonding between carbon and oxygen in the carbonyl group.
22. What is the hybridization of carbon in the carbonyl group?
23. Describe dipole attractive forces for carbonyl compounds.
24. How are the electrons polarized on the carbon and oxygen atoms of the carbonyl group?
25. Describe how the carbonyl group resembles an alkene. What is the major difference?
26. How does the bonding of the carbonyl group influence the addition reactions to the carbonyl double bond?

Nomenclature (Objective 2)

27. Give names for the following compounds:

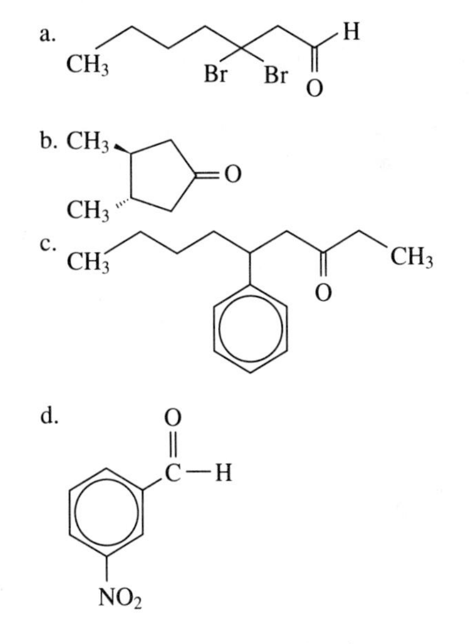

28. Draw structures for the following compounds
    a. *cis*-2,3-dimethylcyclohexanone
    b. 3-isopropylheptanal
    c. 2-phenylcyclobutanone
    d. 3-hexenal

Reactions (Objective 3)

29. Describe why a nucleophile reacts at the carbonyl carbon of a carbonyl group.
30. What is a cyanohydrin?
31. What is a hydrate of a carbonyl compound?
32. A secondary alcohol is formed by the reduction of what?

33. Give products for each of the following reactions:

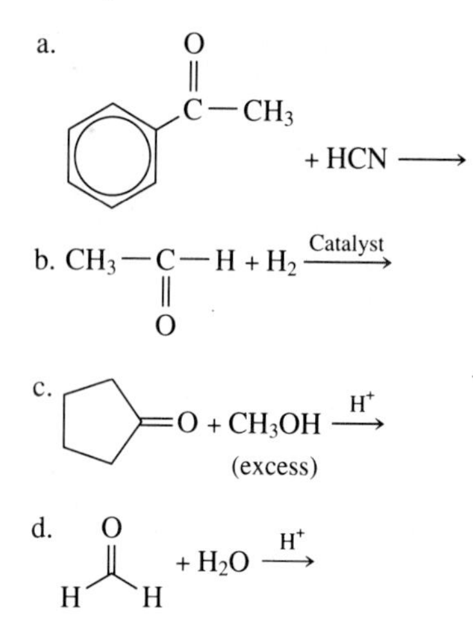

34. In a Tollen's test for aldehydes, what is reduced?
35. What does oxidation of a tertiary alcohol produce?
36. What does oxidation of a hydroquinone produce?
37. Give *general* products from the following general reactions:
    a. An aldehyde + oxidation →
    b. An aldehyde + reduction →
    c. A ketone + oxidation →
38. Give products for the following reactions:

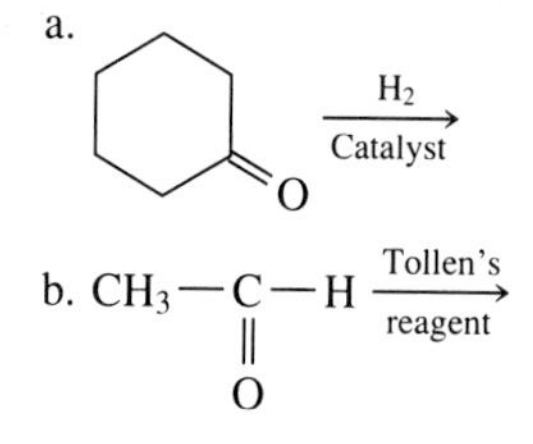

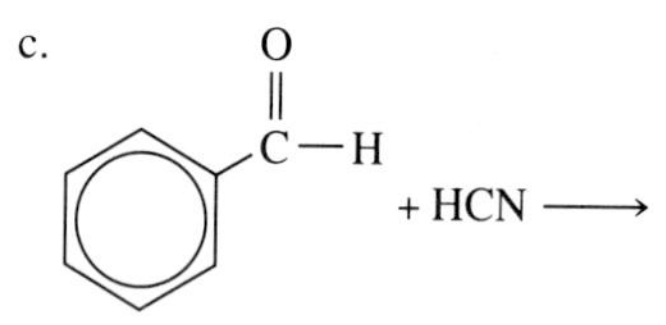

Unclassified Exercises

39. Carbonyl carbon reactivity is best understood by using which representation?

    a. $>C=O$ b. $>\overset{+}{C}-\overset{-}{O}$

    c. $>\overset{-}{C}-\overset{+}{O}$ d. none of these

40. Which of these compounds is *not* easily oxidized?
    a. aldehyde c. secondary alcohol
    b. primary alcohol d. tertiary alcohol
41. The compound

    $$CH_3-CH_2-CH_2-CH_2-\underset{\underset{O}{\|}}{C}-CH_3$$

    is best named:
    a. 5-hexanal c. 1-methylpentanal
    b. 5-hexanone d. 2-hexanone
42. The oxidation of cyclopentanol gives
    a. no reaction c. cyclopentene
    b. cyclopentanone d. cyclopentanal
43. The carbon of a carbonyl carbon is ______ hybridized.
    a. *sp* b. sp^2 c. sp^3 d. not
44. Aldehydes are oxidized to
    a. alcohols c. acetals
    b. ketones d. carboxylic acids
45. The reduction of cyclohexanone with $NaBH_4$ gives
    a. an organic acid
    b. an alcohol
    c. an aldehyde
46. When the alpha proton is removed from a carbonyl compound, the resulting anion is *not*
    a. a nucleophile c. a Lewis acid
    b. a Lewis base
47. Aldehydes have boiling points that are ______ the corresponding primary alcohols.
    a. lower than b. about the same as c. higher than

CARBOXYLIC ACIDS AND THEIR DERIVATIVES

Crystals of aspirin are seen under polarized light.

OUTLINE

7.1 Carboxylic Acids: Structure, Properties, and Nomenclature
7.2 Reactions and Preparations of Carboxylic Acids
7.3 Esters
7.4 Acid Anhydrides
7.5 Important Organic Acids and Acid Derivatives

OBJECTIVES

After completing this chapter, you should be able to

1. Explain the reasons for the acidity of carboxylic acids.
2. Name and identify representative acids and their derivatives.
3. Predict the products of reactions of acids and of their derivatives.
4. Describe the process of saponification.
5. Explain the chemical behavior of soaps and detergents.

7.1 Carboxylic Acids: Structure, Properties, and Nomenclature

Structure

From application of the rules for Lewis electron-dot structures, an organic acid can be characterized by the following general structural representation:

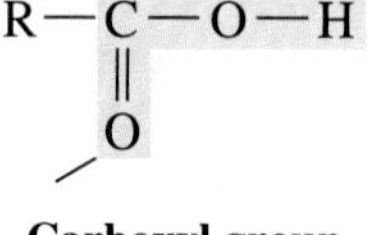

Carboxyl group

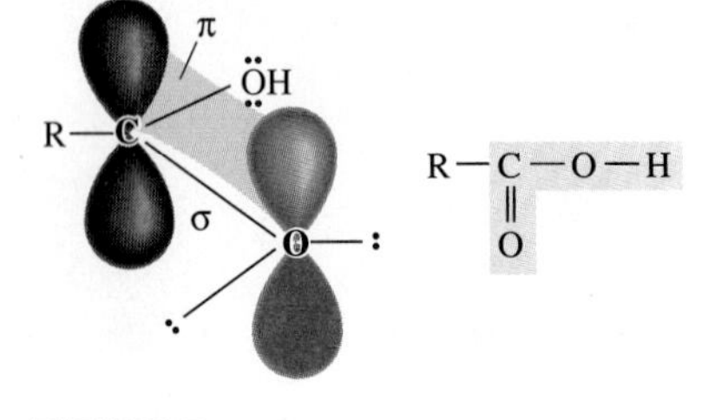

FIGURE 7.1
The Lewis structure of carboxylic acid. Notice the carbonyl portion of the carboxyl group.

The —COOH is the **carboxyl group** (a contraction of the words *carb*onyl and hydr*oxyl*), which is the functional group that characterizes the organic **carboxylic acids**. We see within the carboxyl group a C=O and an —OH group. However, the acid group does *not* react as though it were an alcohol plus a carbonyl—it is a unique functional group. The C=O group, just as in the case of the aldehydes and ketones, has both a sigma and a pi bond (Figure 7.1).

The acid functional group can be designated in several ways, and you should be able to recognize each one.

$$-\underset{\underset{O}{\|}}{C}-O-H \qquad -CO_2H \qquad -COOH$$

Hydrogen Bonding

The electronegative oxygen of the carbonyl participates well in hydrogen bonding with an alcohol.

$$>C=O\cdots H-O-R$$

Because the carboxyl group has two different opportunities for hydrogen bonding, it is observed that carboxylic acids form *dimers*. A dimer is a compound made up of two identical pieces. The bonding in the dimer structure is sufficiently strong so that in some methods of determining molecular weights, organic acids seem to be twice as heavy as they really should be. This is due to the formation of the dimer (Figure 7.2).

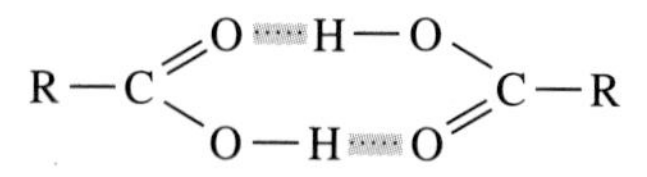

FIGURE 7.2
Strong hydrogen bonding causes the formation of dimers. The carbonyl portion of the carboxyl group of one molecule is hydrogen bonded to the —OH group of the other molecule. Because of the relative positions of the two portions of the functional group, each molecule effectively participates twice in this arrangement, forming a relatively strong attraction of two molecules for one another.

Boiling Points

Because the carboxylic acids are able to form intermolecular dimers and also have the potential for normal, linear hydrogen bonding between adjacent molecules, the boiling points of acids are even higher than that of the alcohols. Figure 7.3 shows the boiling points of the alcohols, aldehydes, and carboxylic acids. Notice that as the carbon chain gets longer, the influence of the functional group decreases, and the carboxylic acid behaves more and more like a simple hydrocarbon. (Compare this with Figure 5.1 where the relationship of the alcohols to the alkanes was shown.)

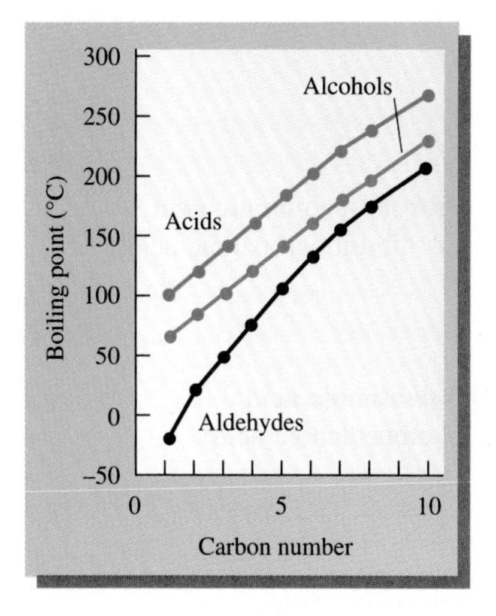

FIGURE 7.3
The stronger hydrogen bonding of acids causes the boiling points to be higher than for the corresponding alcohols. Again notice how the differences become less important as the carbon chain length increases.

Solubility

The solubility of a carboxylic acid is readily understood by considering the hydrogen bonding of the carboxyl group with water (Figure 7.4). Just as the solubility of alcohols in water rapidly decreases as the carbon number increases, the same trends can be seen for the organic acids. As the carbon chains increase in size, the acids become more and more like a hydrocarbon (Figure 7.5).

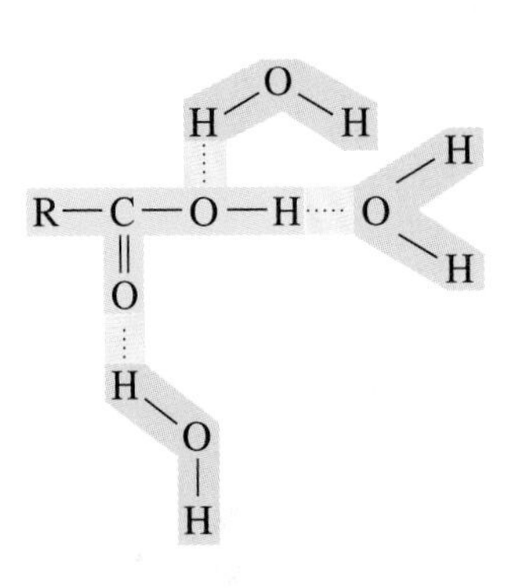

FIGURE 7.4
Extensive hydrogen bonding of the carboxyl group with water gives the acids high water-solubility for the shorter-chain-length acids.

Nomenclature of Carboxylic Acids

General Rules: The Carboxyl Group Is Number One

The nomenclature of carboxylic acids is very similar to that already represented by most of the other functional groups. Here the carboxyl group, —COOH, takes priority, and the carboxyl carbon is assigned the number-one position in any numbering scheme. The *-e* ending of alkane, alkene, and alkyne is replaced by *-oic acid.* Other substituents on the carbon chain are named and numbered as previously discussed. If there is a carbonyl group somewhere in the molecule (other than that contained in the carboxyl group), it is given a number, to show its position, and the designation *oxo*. Another way of naming acids that also contain a ketone carbonyl is to use the prefix *keto* to designate the position and functional group. Often when this is done, Greek symbols are used rather than numbers to show position.

EXAMPLE 7.1

Name these carboxylic acids:

$$\overset{\beta}{\underset{3}{CH_3}}—\overset{\alpha}{\underset{2}{CH}}(Br)—\overset{1}{COOH}$$

Solution:
2-Bromopropanoic acid
(or α-bromopropanoic acid)

$$\overset{4}{CH_3}—\overset{3}{CH}=\overset{2}{C}(CH_3)—\overset{1}{COOH}$$

2-Methyl-2-butenoic acid

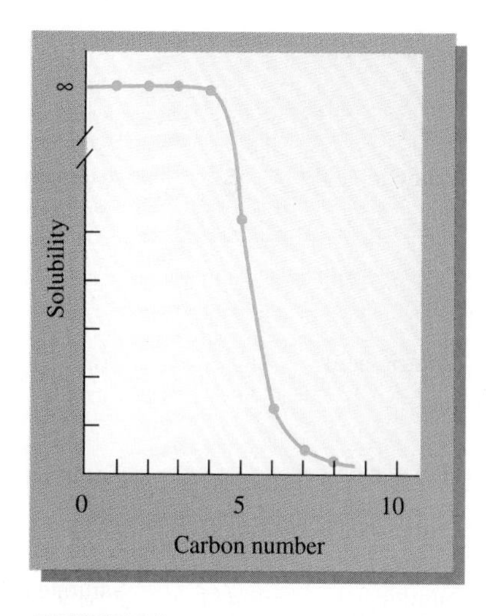

FIGURE 7.5
The solubility of acids decreases as the carbon number increases. Again, the dominating role of the functional group rapidly decreases as the carbon chain length increases.

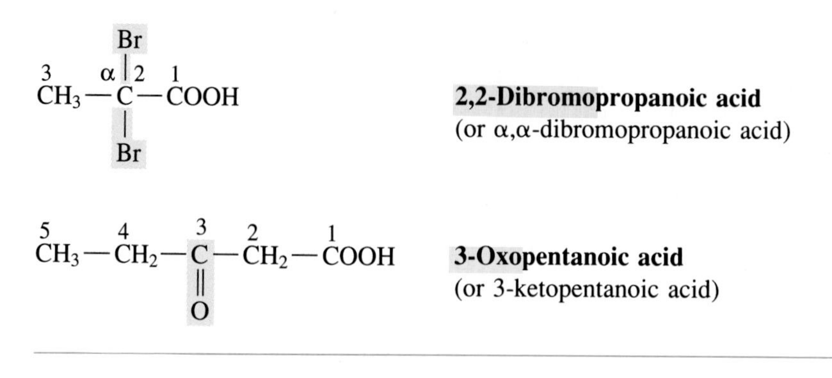

Common Names for Carboxylic Acids

Aromatic carboxylic acids are often named on the basis of the common name, *benzoic acid*, which serves as the parent name. Several other common names are routinely used for organic acids, and several representative examples are shown in Table 7.1.

TABLE 7.1
Representative Naturally Occurring Organic Acids

Name	Description	Structure
Formic acid	Found in ant venom	$H—COOH$
Acetic acid	Found in vinegar (4 to 5%)	$CH_3—COOH$
Benzoic acid		$C_6H_5—COOH$ (benzene ring with COOH)
Salicylic acid	A constituent of aspirin and oil of wintergreen	benzene ring with COOH and ortho OH
Tartaric acid	Common in grapes	$HOOC—CH(OH)—CH(OH)—COOH$
Malic acid	Common in apples	$HOOC—CH(OH)—CH_2—COOH$
Lauric acid (n = 10) Myristic acid (n = 12) Palmitic acid (n = 14) Stearic acid (n = 16)	Saturated fatty acids	$CH_3—(CH_2)_n—COOH$
Oleic acid	Unsaturated fatty acid	$CH_3—(CH_2)_7—CH{=}CH—(CH_2)_7—COOH$ (cis, H on each C)

CHEMISTRY CAPSULE

Aspirin

For over 200 years it has been known that willow bark contains substances useful for reducing pain and fever. In the late 1800s, the sodium salt of salicylic acid was used as a medication; however, this proved to irritate the lining of the stomach. *Salol*, the phenyl ester of salicylic acid, is a milder form of the drug, but the byproduct of its hydrolysis in the intestine is phenol, a very toxic compound. In 1899, *aspirin*, or *acetylsalicylic acid*, was introduced by the Baeyer company of Germany. In many countries aspirin means only the Baeyer brand, but in the United States Baeyer no longer holds rights to the name. Thus, aspirin is the generic term used for all brands of salicylic acid. It remains on the market in the same form today.

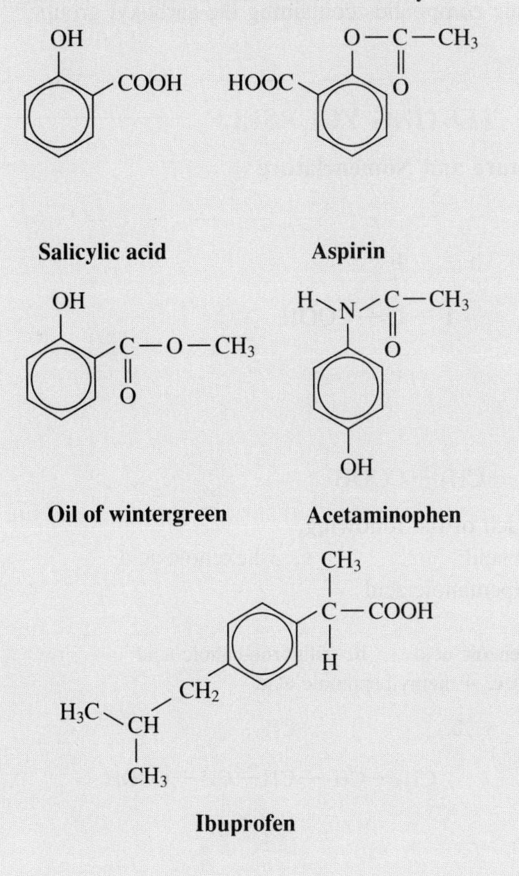

A powerful boost to the sales of aspirin came in late 1987 with the report that one aspirin tablet a day can reduce the risk of heart attack for middle-aged men by almost 50 percent. The important activity that appears to be helpful in this respect is the ability of aspirin to reduce blood clotting. This may have a negative effect for some people because of a slight increase in the probability of a stroke. However, these data are not yet well understood, and much more research needs to be done. The search for compounds that have the medicinal values of aspirin is a big business. Companies are involved in synthesizing new compounds that have the properties of reducing inflammation and fever, without irritating the stomach lining as aspirin does. Two common compounds found in a number of medicines are acetaminophen and ibuprofen. Check the ingredients in common cold medicines and over-the-counter pain relievers to see what compounds are used.

A number of linaments for relieving pain of sore muscles have a strong odor of wintergreen, which is also a derivative of salicylic acid! Structures of important compounds are shown below.

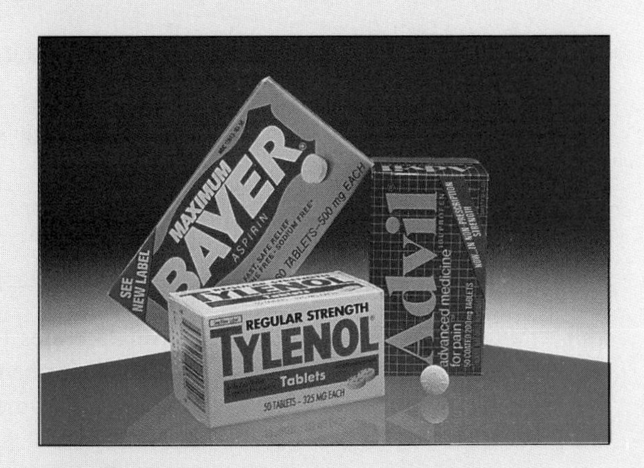

FIGURE 7.A
Common over-the-counter (OTC) aspirin and aspirin-related products.

$HOOC—(CH_2)_n—COOH$

Name	n
Oxalic acid	0
Malonic acid	1
Succinic acid	2
Glutaric acid	3
Adipic acid	4
Pimelic acid	5

Names for Dicarboxylic Acids

There are also a number of *di*carboxylic acids, and many of these have common names. The sentence, "*o*h *m*y, *s*uch *g*ood *a*pple *p*ie" is a useful mnemonic device for remembering the list of dicarboxylic acids that have an increasing number of CH_2 groups attached. Some of the dicarboxylic acids and their derivatives play important roles in biological reactions, and they will be seen again in later chapters.

NEW TERMS

carboxyl group The functional group that characterizes the organic acids:

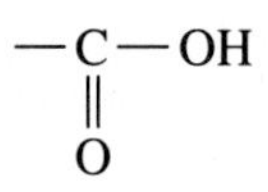

carboxylic acid Organic compounds containing the carboxyl group

TESTING YOURSELF

Carboxylic Acid Structure and Nomenclature

1. Name the following:

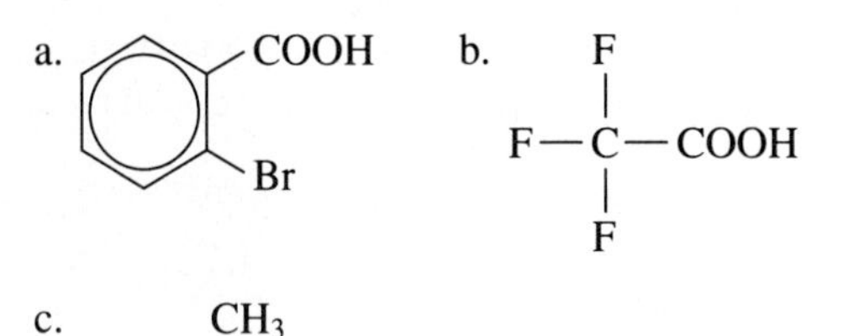

c.
$$\begin{array}{c} CH_3 \\ | \\ CH_3—CH—CH_2—CH_2—COOH \end{array}$$

2. Give structures for each of the following.
   a. 3,5-dinitrobenzoic acid
   b. 2-bromo-3-methylpentanoic acid
   c. 3-hexenoic acid

Answers **1. a.** *o*-bromobenzoic acid **b.** trifluoroethanoic acid (or trifluoroacetic acid) **c.** 4-methylpentanoic acid

2. a. **b.**

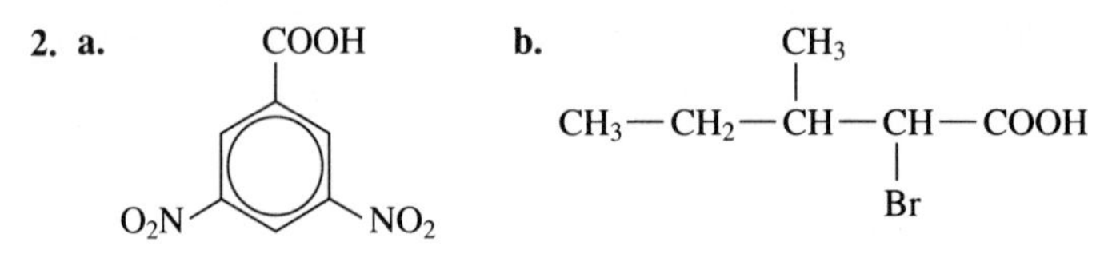

c. $CH_3—CH_2—CH{=}CH—CH_2—COOH$

7.2 Reactions and Preparations of Carboxylic Acids

Organic Acids Are Weaker Than Mineral Acids

The class of organic compounds called *carboxylic acids* are characterized by a special acidity.

Resonance Stability of the Carboxylate Anion

♦ An acid readily loses a proton and forms resonance-stabilized anions.

The same generalization already made for the acidity of phenols and the α-hydrogens on carbonyl compounds can be applied to carboxylic acids. After loss of the proton from the carboxylic ȧcid, a **carboxylate anion** remains. By resonance there is charge delocalization of the carboxylate anion. Further, as with the resonance structures of benzene, two *equivalent* resonance structures can be drawn for the carboxylate anion. These resonance structures are shown in Figure 7.6.

Quantitative Look at the Acidity of Carboxylic Acids

Unlike HCl, which we know to be *completely* ionized or dissociated into H^+ and Cl^-, all of the organic carboxylic acids are weak acids. That is, they do *not* ionize completely. As the data in Table 7.2 show, however, the organic acids are more acidic, that is, they have lower pK_a values than most other common functional groups.

It is one thing to use numbers to explain relative acidities and another to realize what the numbers mean. It turns out that a simple acid–base test can show the ranges of acidity. We need only to compare the strength of base required to

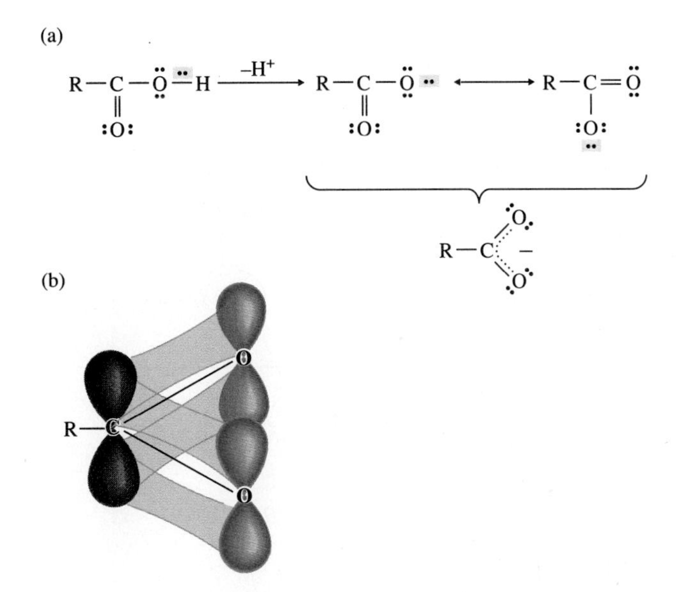

FIGURE 7.6
Structures for the carboxylate anion. (a) Resonance structures are one simple way to examine electron delocalization. (b) An orbital view shows the same features, but it is harder to show chemical behavior with these more complex drawings.

TABLE 7.2
Effects of Functional Groups on Hydrogen Acidity

Functional Group	Approximate K_a	Approximate pK_a
R—H	10^{-50}	50
R—O—H	10^{-18}	18
H—O—H	10^{-16}	16
R—S—H	10^{-12}	12
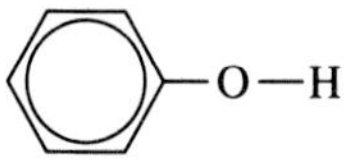 C_6H_5—O—H	10^{-10}	10
R—C(=O)—O—H	10^{-5}	5

(Arrow: Increase in acidity, from top to bottom)

remove the acidic hydrogen. Comparisons of alcohols, phenols, and acids are given in Table 7.3. In this table we can see that a carboxylic acid has a proton that can be removed even by the weakest bases, whereas an alcohol shows no reaction. The phenol exhibits acidity between acids and alcohols.

TABLE 7.3
Tests for Acidity

Base	Functional Group: *Alcohol* R—OH	Functional Group: *Phenol* C_6H_5—OH	Functional Group: *Acid* R—COOH
Na (metal)	R—O⁻	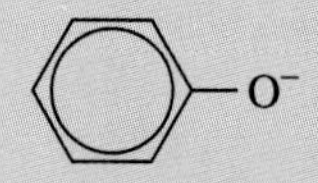 C_6H_5—O⁻	R—COO⁻
NaOH		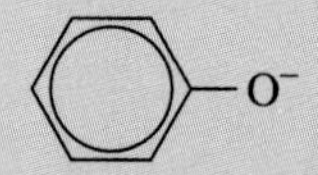 C_6H_5—O⁻	R—COO⁻
$NaHCO_3$			R—COO⁻

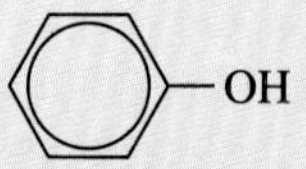

Preparation of Carboxylic Acids by Oxidation

The oxidation of aldehydes to organic acids as a test for aldehydes was already discussed in Chapter 6. An aldehyde may serve as a starting material for the preparation of acids. As in the case of alcohol oxidation, the symbol [O] is employed to denote an oxidizing agent. The usual reagents for oxidizing aldehydes to acids are oxides of chromium or manganese ($KMnO_4$, CrO_3, or $K_2Cr_2O_7$). Recall that primary alcohols may be oxidized to aldehydes. However, it is difficult to stop the oxidation at this point, and it is often observed that continuation of the oxidation process provides a synthetic method for preparing acids from alcohols.

◆ Acids are easily prepared by the oxidation of primary alcohols and aldehydes.

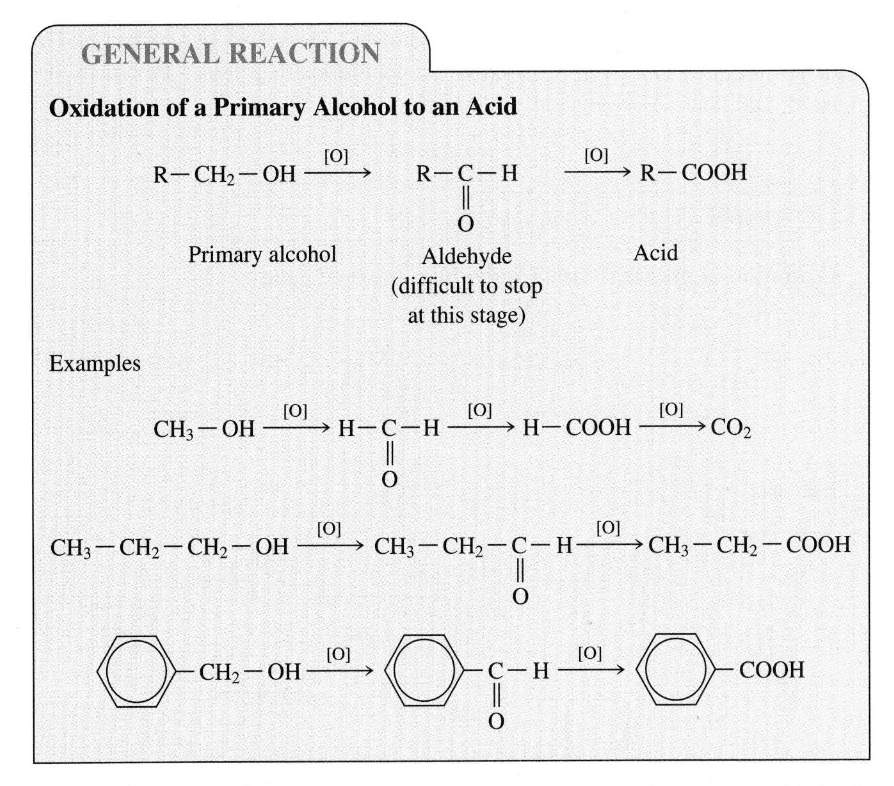

How can we tell what kind of organic compound can undergo easy oxidation? Is there a way to know when oxidation is likely to continue to another compound, as in the case of primary alcohols to aldehydes and then to acids? These questions can be answered by looking at a general scheme of oxidation (Figure 7.7). Let us *assume* that an oxidation involves the replacement of a C—H bond by a C—OH bond. We see that the C—H bonds on methane are replaced, one at a time, by —OH. In the previous chapter we saw that a hydrate of a carbonyl readily lost water to produce the carbonyl. The approach shown here for primary alcohols demonstrates the same thing. Thus, we can see why primary alcohols lead to aldehydes and in turn produce acids. Since ketone has no hydrogens attached to the carbonyl-containing carbon atom, we correctly predict that ketones are resistant to oxidation. We also correctly predict that tertiary alcohols resist chemical oxidation.

$$\mathrm{H{-}\overset{H}{\underset{H}{C}}{-}H \xrightarrow{[O]} H{-}\overset{H}{\underset{H}{C}}{-}OH \xrightarrow{[O]} R{-}\overset{OH}{\underset{H}{C}}{-}OH \xrightarrow{-HOH}}$$

$$\mathrm{(H)(H)C{=}O \xrightarrow{[O]} (HO)(H)C{=}O \xrightarrow{[O]} (HO)(HO)C{=}O \xrightarrow{-HOH} O{=}C{=}O}$$

FIGURE 7.7
A simple model for an oxidation can be considered as replacing a C—H bond with a C—OH bond. Whenever there are two —OH bonds on the same carbon atom, water is removed and a C=O is formed. This model does not represent how the oxidation takes place but simply shows the increasing oxidation states as H is replaced by O.

◆ Alkyl side chains on aromatic rings are oxidized to acids.

In another important oxidation reaction, an alkyl side chain attached to a benzene ring can be converted to a benzoic acid derivative by oxidation. This reaction requires more vigorous oxidation conditions than those previously discussed, and heat (Δ) is generally required.

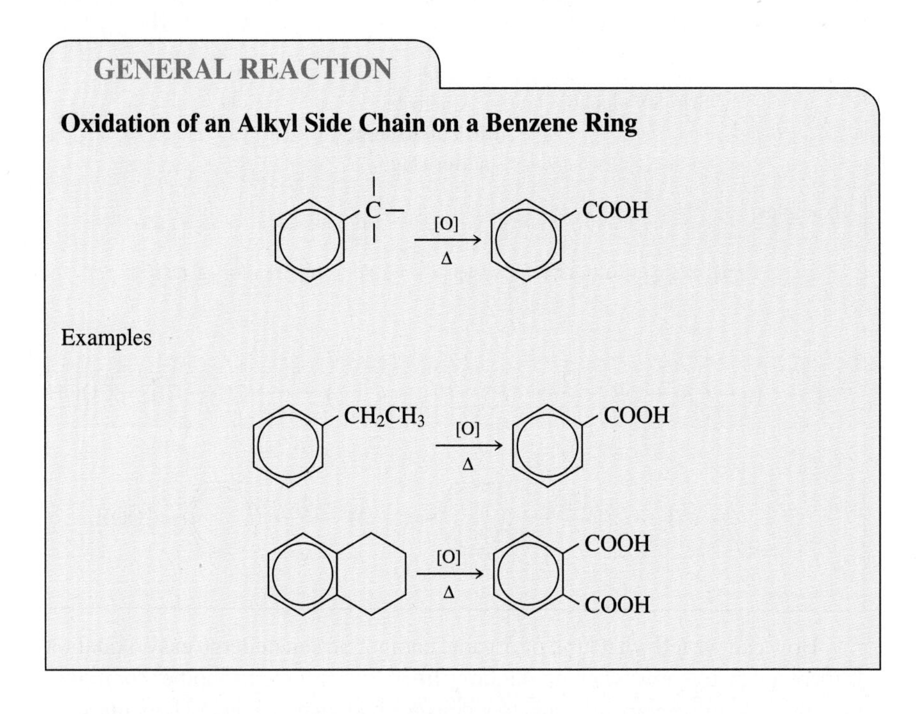

As a result of this special ability of alkylated benzene rings to become oxidized at the side chain, both in the laboratory as well as under biological conditions, alkyl-substituted benzene derivatives are often easily degraded in the environment. These alkylated compounds are also usually less toxic to living organisms than benzene rings without alkyl groups. Recall from Chapters 4 and 5 that the biological oxidation of benzene gives phenol, a poisonous compound. The oxidation of toluene forms benzoic acid, a relatively harmless compound that can be excreted from the body. It is noteworthy that an aromatic ring is not attacked by the usual oxidizing agents.

Decarboxylation of Beta Keto Acids

Organic acids that also contain a carbonyl group are often given the generic name **keto acid**. When you begin your study of biochemistry, you will find two common kinds of acids that also contain a carbonyl group—the *alpha keto* acids and the *beta keto* acids.

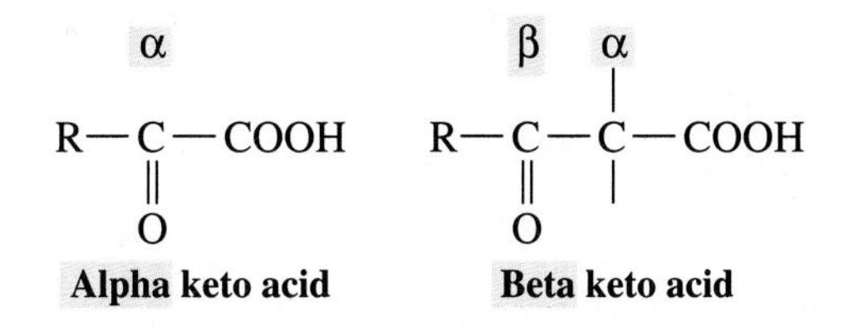

One of the most interesting aspects of the beta keto acid is its ability to undergo easy **decarboxylation**, or loss of carbon dioxide. This decarboxylation provides a useful and simple way to make ketones. In biological systems, the loss of CO_2 is a very important process.

♦ Beta keto acids are easily decarboxylated (lose carbon dioxide).

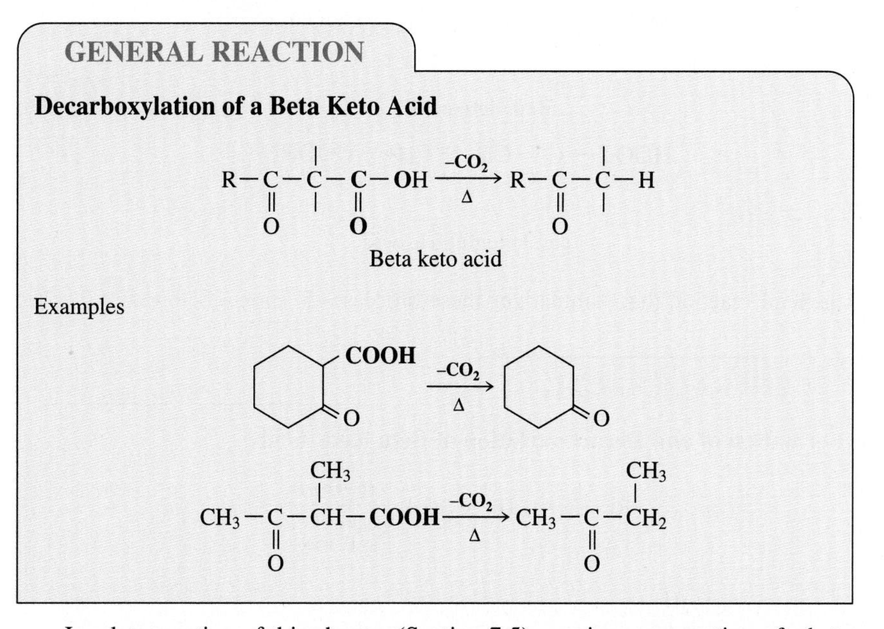

In a later section of this chapter (Section 7.5), a unique preparation of a beta keto acid will be provided. One approach to the beta keto acid involves the oxidation of a beta hydroxy acid, which can be produced by the hydration (adding of water) of a 2,3-unsaturated acid. This method is an important feature in biochemical degradation processes. For example, one of the byproducts of oxidation of carbon compounds in our body is carbon dioxide. We are constantly breathing out the carbon dioxide produced as our metabolism breaks down carbon compounds for energy. One of the sources of carbon dioxide is the *citric acid cycle* (or TCA cycle) to be discussed in a later chapter. (Notice the number of simple organic reactions!)

1. Dehydration
2. Hydration, with reversal of positions for H and OH
3. Oxidation of a secondary alcohol. Notice that the alcohol before step 1 is tertiary and resistant to oxidation.
4. Beta keto decarboxylation

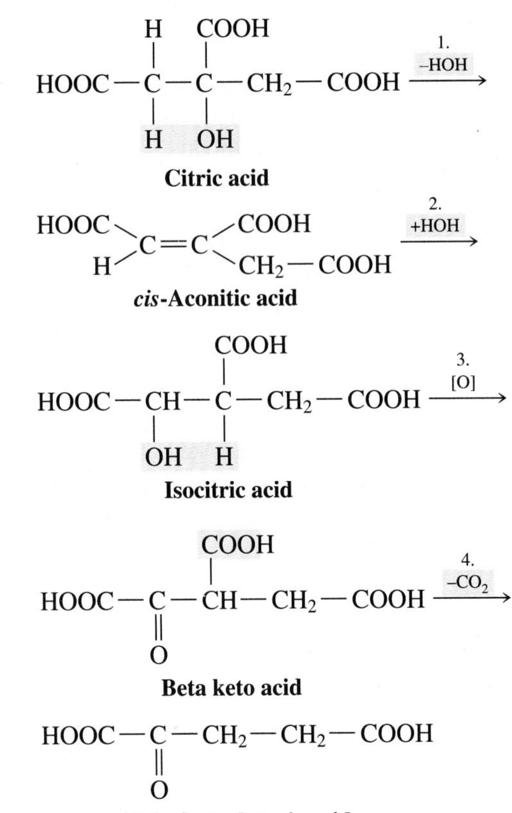

A general reaction that summarizes these processes is shown below.

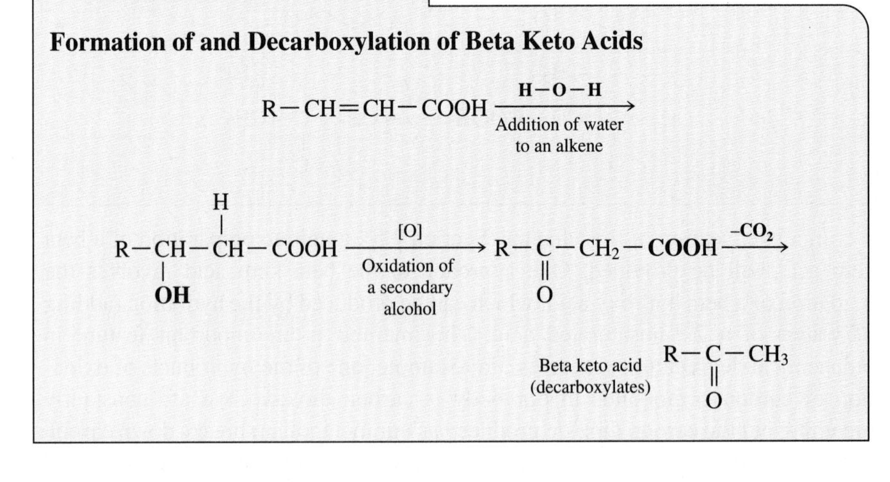

carboxylate anion The resonance-stabilized anion resulting from removal of the acid proton of an organic acid.

keto acids Organic acids that also contain a carbonyl functional group

decarboxylation The loss of CO_2. This reaction takes place very readily for beta keto acids.

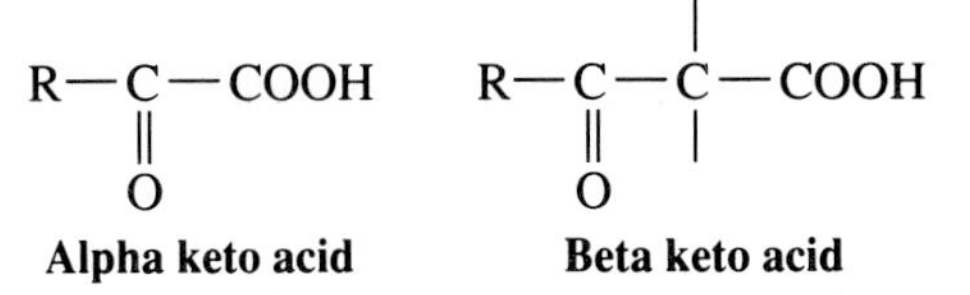

Alpha keto acid **Beta keto acid**

TESTING YOURSELF

Reactions of Carboxylic Acids

1. Place the following compounds in order of *increasing* acidity: phenol, alcohol, water
2. What is the organic acid formed in each of the following reactions?

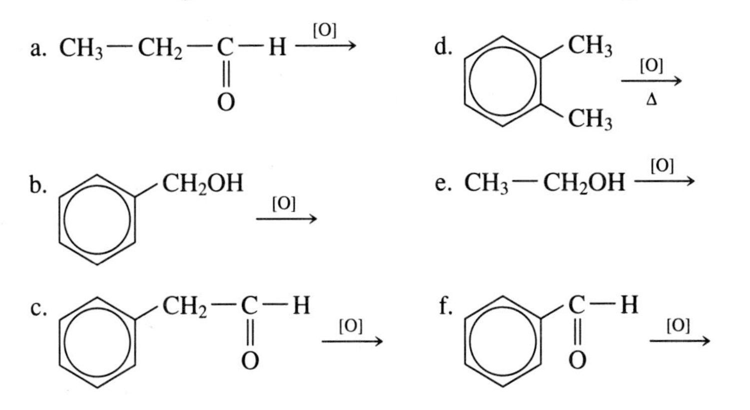

3. Give products for each of the following beta keto acid reactions:

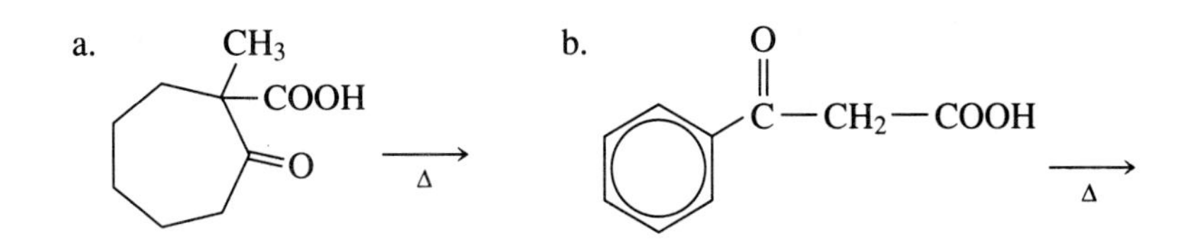

Answers **1.** alcohol, water, phenol
2. a. propanoic acid **b.** benzoic acid **c.** phenylacetic acid (or phenylethanoic acid)

d.

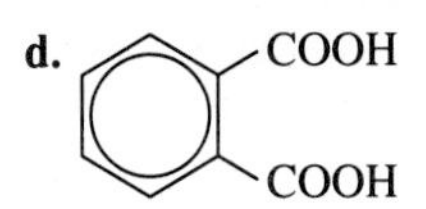

e. acetic acid **f.** benzoic acid

3. a. **b.**

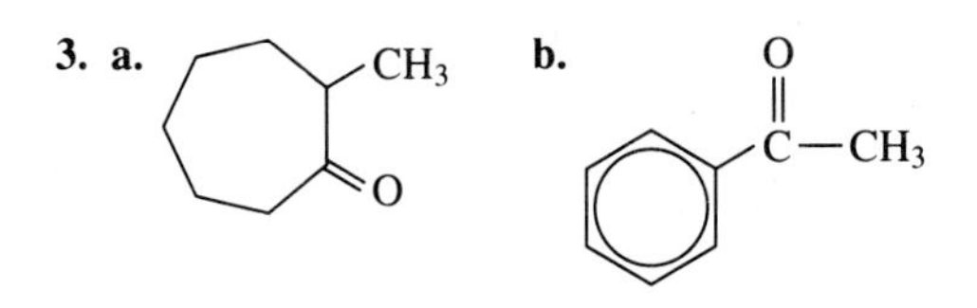

7.3 Esters

Structure and Reactivity

Organic Esters

♦ *Esters* are derivatives of acids and alcohols:

$$R-\underset{\underset{O}{\|}}{C}-OH + R'-OH \longrightarrow R-\underset{\underset{O}{\|}}{C}-O-R'$$

One of the most common derivatives of an organic acid is the **ester**. An ester has the structure

$$R-\underset{\underset{O}{\|}}{C}-O-R'$$

Esters are most commonly formed by reaction of an acid and an alcohol. For example, propanoic acid and methanol react in the presence of an acid catalyst to give an ester:

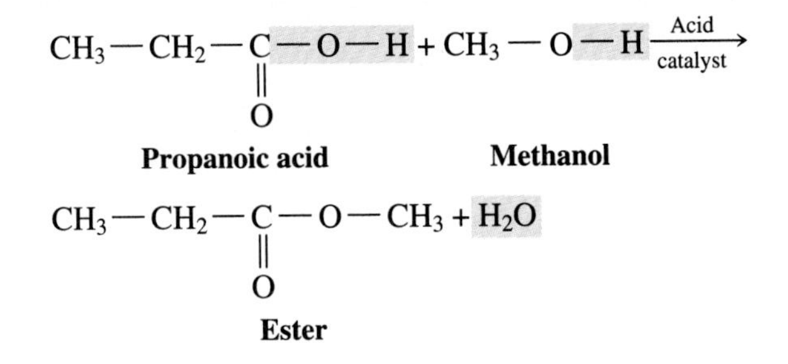

Note that in the formation of an ester, the hydroxyl group of the acid (*not* of the alcohol) is lost as water.

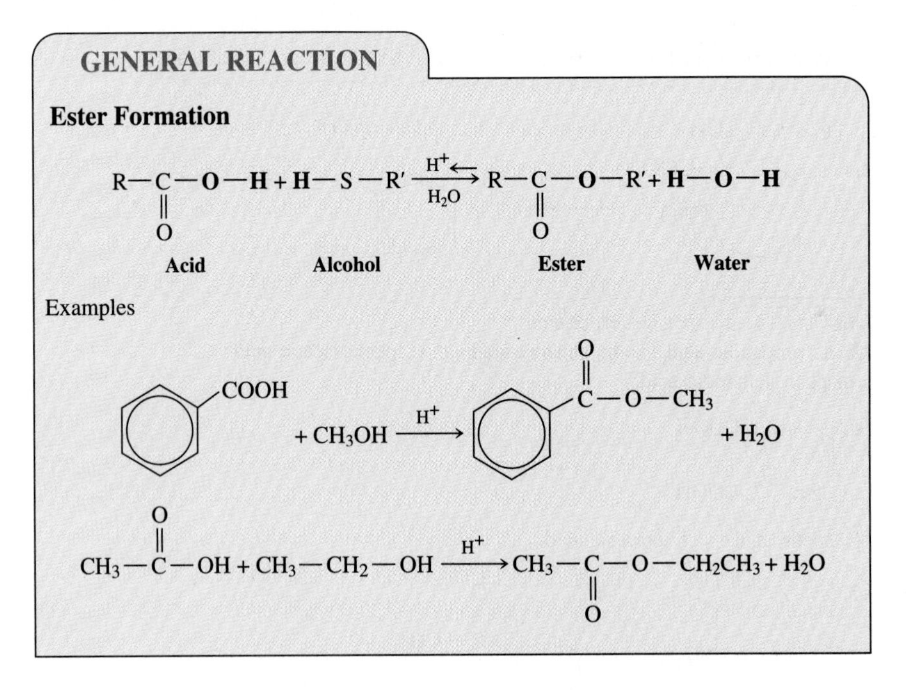

This reaction is an *equilibrium reaction* that is catalyzed by acid. Thus, to obtain good yields of the ester product, the equilibrium must be forced in the direction of the ester. According to Le Chatelier's principle, the easiest way to unbalance the system is to remove the water as it forms.

Conversely, we could correctly predict that if water was in excess, the reaction would proceed backward and the ester would react with the water to form the starting acid and alcohol. This is indeed observed. We will discuss this reaction in a later section of this chapter.

Ester Flavors and Odors

Many esters are used as flavorings and fragrances because a common characteristic of many esters is a "fruity" odor. Several common esters and their characteristic odors are shown in Table 7.4. It should be mentioned that the authentic odors and flavors are generally a complex mixture of many compounds. This is one of the reasons that synthetic fragrances are not quite the same as those provided by nature.

♦ Esters are often used in flavorings.

Thioesters

We will find a number of **thioesters** in our study of biological chemistry. They have the same structure as esters but have sulfur replacing one of the oxygen atoms, just as in a thiol.

$$R{-}\underset{\underset{O}{\|}}{C}{-}S{-}R'$$

It is appropriate to recognize that their chemistry is not unique; we have already seen that thiols are the sulfur analogs of alcohols. Thus, when an organic acid is reacted with a thiol, a thioester results. Notice in the following general reaction

TABLE 7.4
Esters and Their Odors

$R{-}\underset{\underset{O}{\|}}{C}{-}OH$	$R'{-}OH$	$R{-}\underset{\underset{O}{\|}}{C}{-}O{-}R'$
First Side Group (R) (acid)	***Second Side Group (R′) (alcohol)***	***Fruity Odor***
H	Ethyl	Rum
H	Isobutyl	Raspberry
Methyl	Isopentyl	Banana
Methyl	Octyl	Orange
Propyl	Ethyl	Pineapple

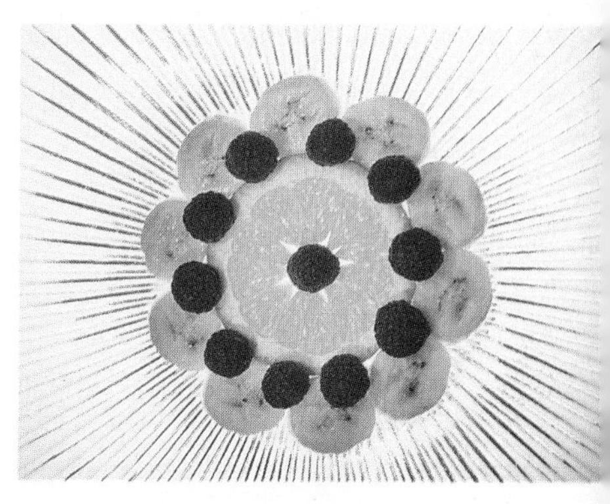

Esters are responsible for many of the pleasant smells of fruits, such as raspberries, bananas, and oranges.

that if the —OH from the acid were not involved in ester formation, it would not be possible to prepare thioesters from thiols.

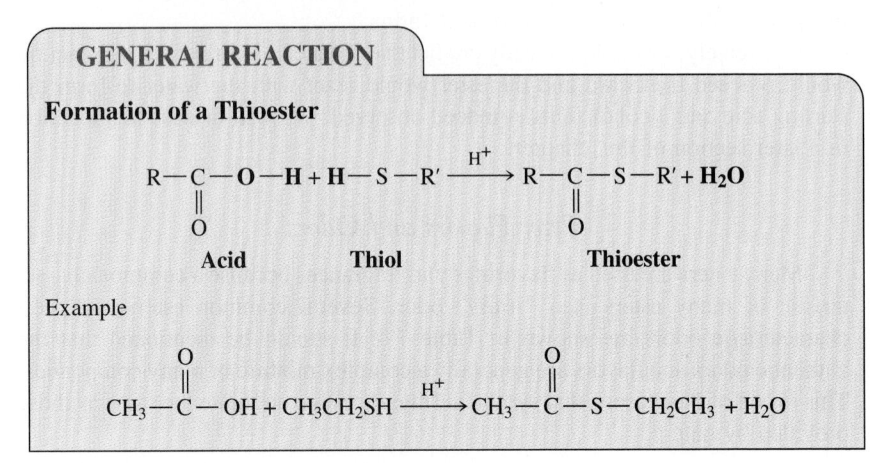

♦ Acetyl group

$CH_3—C(=O)—$

An important thioester found throughout the study of biochemistry is acetyl-coenzyme A, which serves as an agent for transferring the acetyl group

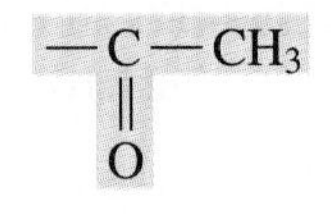

Many biochemical reactions are derived from this acetyl-group transfer.

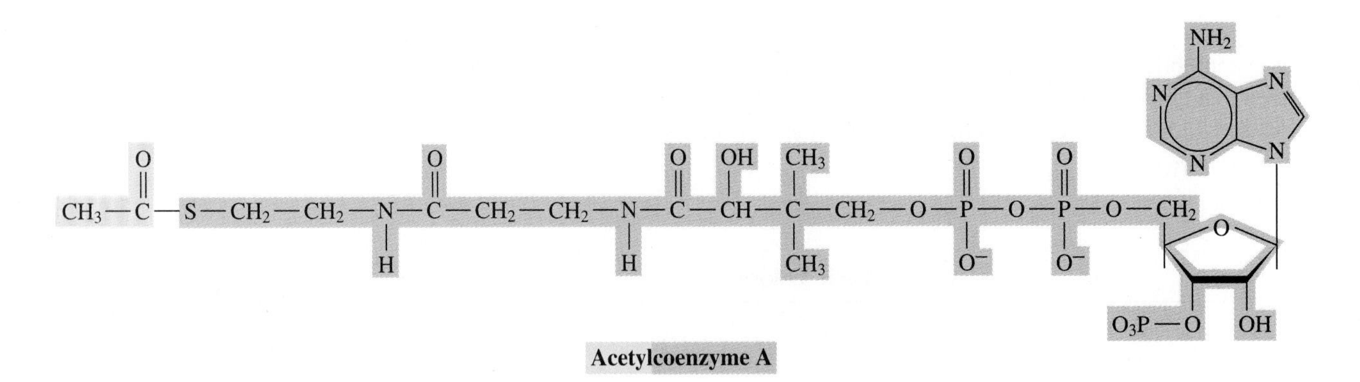

Acetylcoenzyme A

Coenzyme A is often abbreviated CoA. You can see that it is a thiol. Reaction with acetic acid results in the thioester acetylCoA

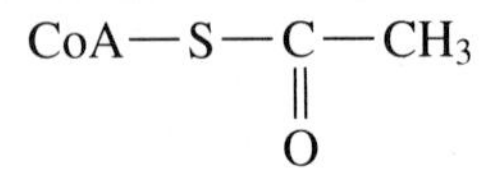

Inorganic Esters

It is also possible to prepare esters of inorganic acids, particularly sulfuric acid (cold), phosphoric acid, and nitric acid.

$$R-O-H + H-O-SO_3H \longrightarrow R-O-SO_3H + H_2O$$

$$R-O-H + H-O-PO_3H \longrightarrow R-O-PO_3H_2 + H_2O$$

$$R-O-H + H-O-NO_2 \longrightarrow R-O-NO_2 + H_2O$$

Phosphate and sulfate esters are widely used as **detergents** (see the section on Detergents). Nitrate esters are often explosive, as best exemplified by *nitroglycerine*, a trinitrate ester of the triol glycerol.

$$\begin{array}{l} CH_2-O-NO_2 \\ | \\ CH-O-NO_2 \\ | \\ CH_2-O-NO_2 \end{array}$$

Nitroglycerine

Alfred Nobel discovered that by mixing nitroglycerine with an inert material (such as diatomaceous earth), the highly explosive and difficult to handle nitroglycerine was rendered relatively safe. This was applied in the early manufacture of dynamite. This was the start of new techniques for preparing "safe" explosives. Nobel endowed the Nobel prizes from the great wealth that he gained from his work.

Nitroglycerine is also used in medicine. Because it has the ability to dilate blood vessels and arteries, it is used to alleviate heart pain (angina).

Nomenclature of the Esters

By examining the structure of an ester we see that there are two R groups, one belonging to the acid and one to the alcohol.

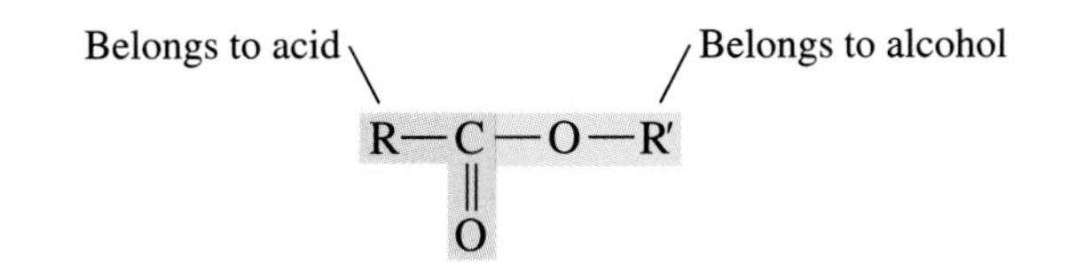

Esters are named by first assigning the name of the alcohol alkyl group (methyl, ethyl, and so on), followed by the acid in which the *-oic acid* suffix is replaced by *-oate*.

EXAMPLE 7.2

Name these esters:

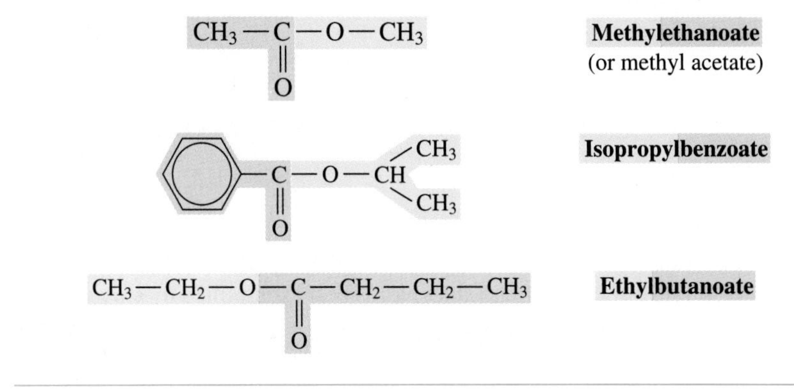

♦ Only the alkyl group of the alcohol is important for the naming of esters.

Saponification: The Basic Hydrolysis of an Ester

The base-catalyzed hydrolysis of any ester to the salt of the acid and the alcohol is called **saponification**. Here is the general reaction.

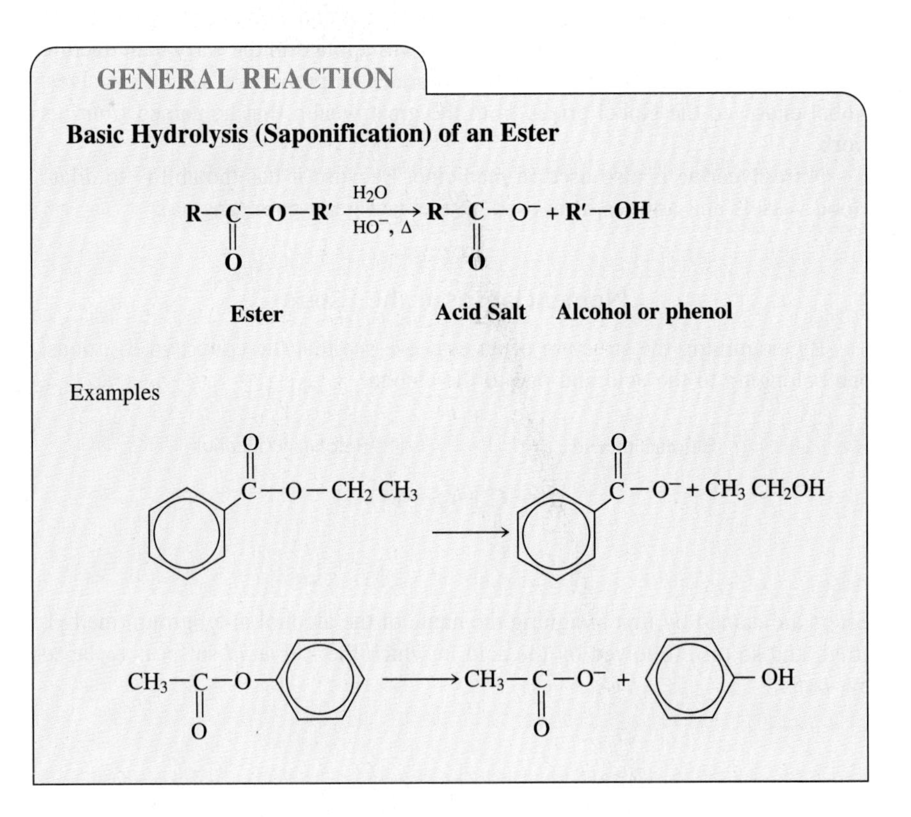

Fat is a complex organic *tri*ester (that is, it has *three* connected esters) of glycerol and long-carbon-chain acids called **fatty acids**. Fat is also called a *triglyceride*. It can be hydrolyzed by lye (NaOH) to form the salt of the fatty acid and glycerol. This salt is called **soap**.

♦ *Soap* is a salt of a long-chain fatty acid.

$$\begin{array}{l} CH_2-O-\overset{\overset{\displaystyle O}{\|}}{C}-(CH_2)_n-CH_3 \\ | \\ CH-O-\overset{\overset{\displaystyle O}{\|}}{C}-(CH_2)_n-CH_3 + 3\,NaOH \longrightarrow \\ | \\ CH_2-O-\overset{\overset{\displaystyle O}{\|}}{C}-(CH_2)_n-CH_3 \end{array}$$

$$\begin{array}{l} CH_2-OH \\ | \\ CH-OH + 3\,CH_3-(CH_2)_n-\overset{\overset{\displaystyle O}{\|}}{C}-O^-Na^+ \\ | \\ CH_2-OH \end{array}$$

Sodium salt of a fatty acid (soap)

Detergents

Soap as One Kind of Detergent

Soaps and detergents are based on molecules having a long hydrocarbon chain and either a polar end, as the salt of a carboxylic acid for a soap, or a different ionic end, as found in many detergents.

Although known by the ancient Romans, it was not until the late 1700s that the process of soap making had developed to the craft that allowed almost everyone to have soap. The soap industry didn't start in the United States until the early 1800s. Today, soap is a very common material in our daily world.

But what is soap? It is the salt of a fatty acid. The constitution of the long-chain alkyl group (the number of carbon atoms is generally in the 12 to 18 range) determines the specific properties of the soap. Similarly, the way a soap feels depends to some extent on the cation involved. For example, potassium salts are often used for the very fine soft soaps associated with shaving creams.

How does soap work? In our earliest discussion of organic chemistry, we noted that water-insolubility was an important feature of hydrocarbons. A particulary important relationship was noted in the solubility of alcohols: the short-chain alcohols are very soluble in water as a result of the ability of the OH group to hydrogen bond with water molecules. As the carbon chain gets larger, the solubility rapidly decreases. The alcohol more resembles a hydrocarbon. A similar relationship is found for organic acids. Thus, we expect the hydrocarbon portion of a fatty acid salt to be insoluble in water. However, the ionic end *is* water-soluble, even more than an alcohol, since it has an ionic group. In a simplified explanation, "dirt," grime, and grease have more hydrocarbon character than water character and thus dissolve in the collective hydrocarbon chains of the soap molecules. The action of soap is very similar to the way biological molecules are

♦ The long hydrocarbon chain dissolves "dirt and grime" while the polar end dissolves in water.

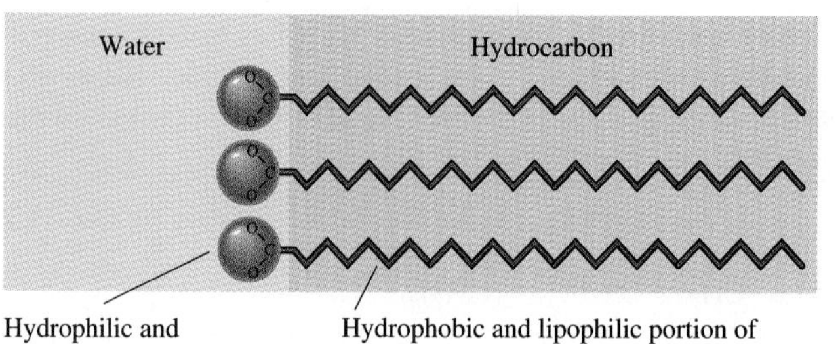

FIGURE 7.8
The chemical and physical behavior of soap is readily understood in terms of the water-solubility of the ionic end of the soap and the hydrocarbon-solubility of grease, dirt, and oil.

carried through cell membranes. There is almost always a hydrocarbon (*hydrophobic*) portion of the molecule and a water-soluble (*hydrophilic*) portion (Figure 7.8).

Micelles and Surface Chemistry

From Figure 7.8, it is apparent that the hydrocarbon "tail" of soap is not soluble in water. These molecules have to find a unique orientation that keeps the hydrocarbon portion out of the water. This is achieved by forming clusters of molecules in which the hydrocarbon tails are in the interior of the cluster and the ionic ends are on the surface of the cluster. This formation is called a **micelle**. Soap in the form of a micelle is able to clean since the oily dirt is collected in the center of the micelle. The micelles stay in solution as a colloid and do not come together to precipitate because of the ion–ion repulsion. Thus, the dirt suspended in the micelles is also easily rinsed away. This process is illustrated in Figure 7.9.

♦ The hydrocarbon part of a soap molecule interacts to form a region of high hydrocarbon density. The ionic ends are on the outside of this region and are in the water phase.

Soap and Hard Water

What are the effects of hard water on soap? The term *hard water* simply means that there is a relatively high concentration of divalent metal cations (Fe^{2+}, Mg^{2+}, and Ca^{2+}) in the water. These react with the soap. To balance the charge, however, *two* soap anions must react with one ion. The resulting product is an insoluble salt that forms the characteristic scum or "bathtub ring" in a sink or tub. It also serves as a water softener by removing the divalent ions from solution.

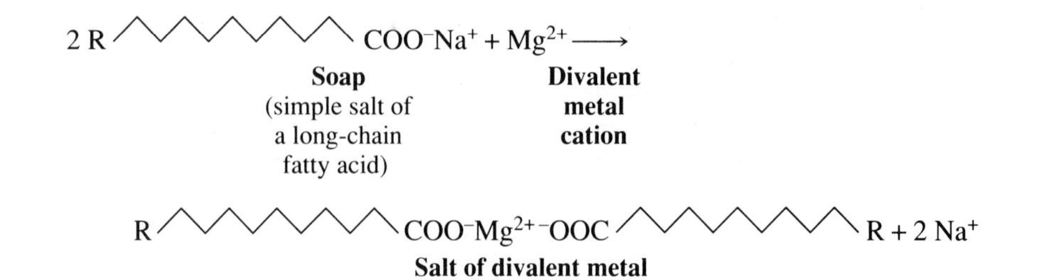

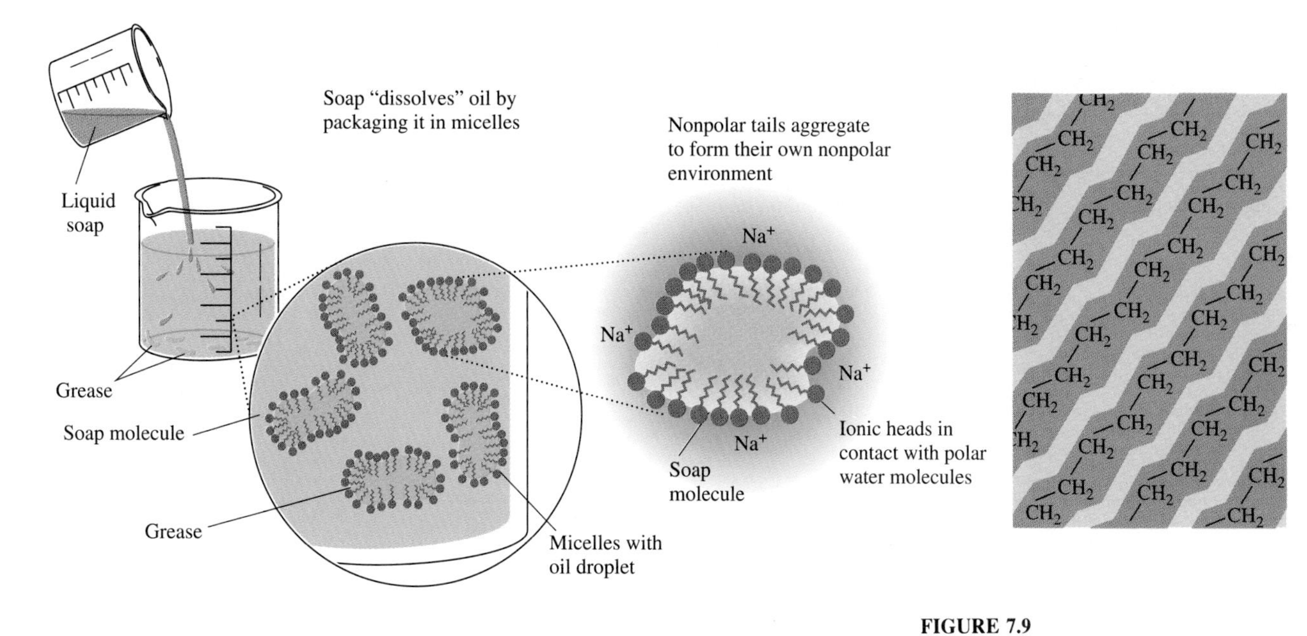

FIGURE 7.9
The chemistry of micelles. This series of magnifications shows what happens to soap in water. The hydrocarbon portion of the soap can dissolve in itself to form a micelle. Grime and grease, being very similar to hydrocarbon, dissolve in the hydrocarbon portion of the soap. We are reminded of the general statement, "like dissolves like."

Other Detergents

There are two ways to get around the problem of hard water: (1) soften the water or (2) use a different type of detergent. In the former method, the divalent cations are exchanged for sodium ions by using an ion-exchange material that has the capacity to "trade" ions. Once the divalent ions in the water are exchanged for sodium ions, the soap works well. The material in the ion-exchange apparatus must be exchanged regularly because after a time, all of the sodium ions are used up. The ion-exchange column is rinsed with a concentrated solution of sodium salts (often sodium chloride) to replace the divalent ions, and the hard water rinse is discarded.

The second way to get around the problem of hard water is to use a detergent (although soap is classified as a detergent, common use makes a distinction). The detergents have a long hydrocarbon portion, just as a soap does. However, the polar ends are different (Table 7.5). Most important, however, is the observation that the polar ends *do not* react with divalent ions in hard water. The charged ends *do not* form insoluble salts with divalent ions and thus remain effective in hard water. Additionally, most common washing detergents have additives to enhance their cleansing power. These are generally inorganic salts that have the ability to complex with the metal ions in hard water, thus removing them from the possibility of reacting with the soap.

NEW TERMS

ester A functional group derived from an acid and an alcohol

$$\mathrm{R{-}C({=}O){-}O{-}R'}$$

TABLE 7.5
Examples of Detergents

Type	Name	Structure
Cationic detergent	Trimethylhexadecylammonium chloride	$CH_3—(CH_2)_{15}—\overset{+}{N}(CH_3)_3Cl^-$
Anionic detergent	Sodium dodecylbenzenesulfonate	$CH_3—(CH_2)_{11}—C_6H_4—SO_3^- Na^+$
Neutral detergent	Pentaerythrityl palmitate	$CH_3—(CH_2)_{14}—C(=O)—O—CH_2—C(CH_2—OH)_3$

thioester A functional group derived from an acid and a thiol

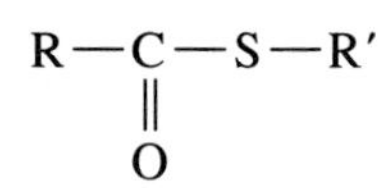

detergent A material that is a surface-active agent having a long hydrocarbon chain and a polar end. If the polar end is a carboxylate anion, the detergent is called a soap. Most other polar groups are often generically classified as detergents.

saponification The basic hydrolysis of an ester to form the salt of the acid and the alcohol

fatty acid A long-carbon-chain carboxylic acid

soap A detergent consisting of the salt of a long-chain fatty acid

micelle An aggregation of hydrocarbon materials having polar ends in the water phase

TESTING YOURSELF

Ester Formation and Nomenclature

1. Give the names of the products of these acid-catalyzed ester reactions:

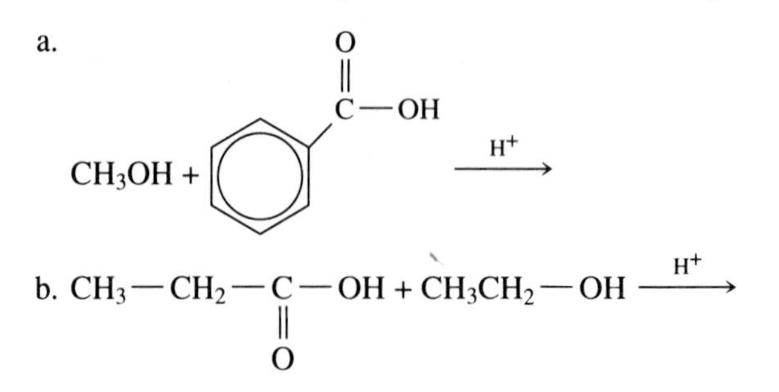

c. $CH_3-\underset{\substack{|\\CH_3}}{CH}-\underset{\substack{\|\\O}}{C}-OH + CH_3-\underset{\substack{|\\CH_3}}{CH}-OH \xrightarrow{H^+}$

2. Give the product for the following reaction:

$$2\ CH_3-(CH_2)_8-COO^-Na^+ + Fe^{2+} \longrightarrow$$

Answers **1. a.** methyl benzoate **b.** ethyl propanoate **c.** isopropyl 2-methylpropanoate

2. $CH_3-(CH_2)_8-\overset{}{\underset{\substack{\|\\O}}{C}}-O^-Fe^{2+-}O-\underset{\substack{\|\\O}}{C}-(CH_2)_8-CH_3 + 2\ Na^+$

7.4 Acid Anhydrides

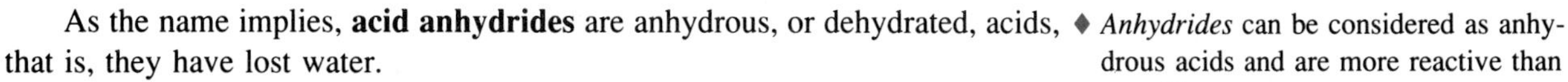

As the name implies, **acid anhydrides** are anhydrous, or dehydrated, acids, that is, they have lost water.

♦ *Anhydrides* can be considered as anhydrous acids and are more reactive than an acid.

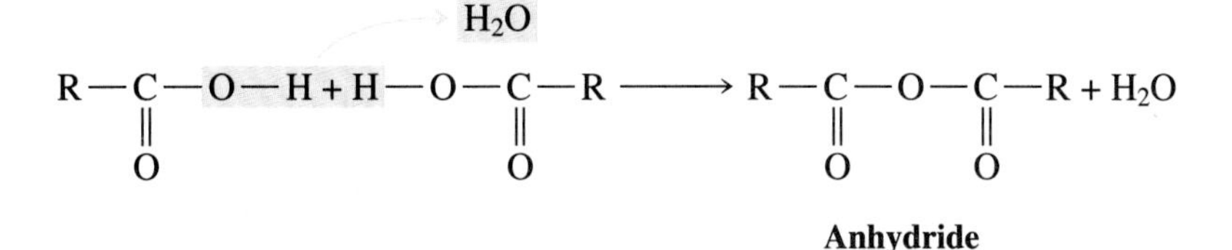

Anhydride

Anhydrides are more reactive than acids. The anhydride can be hydrolyzed by water or esterified with alcohol, but at a slower rate than that observed with an acid chloride.

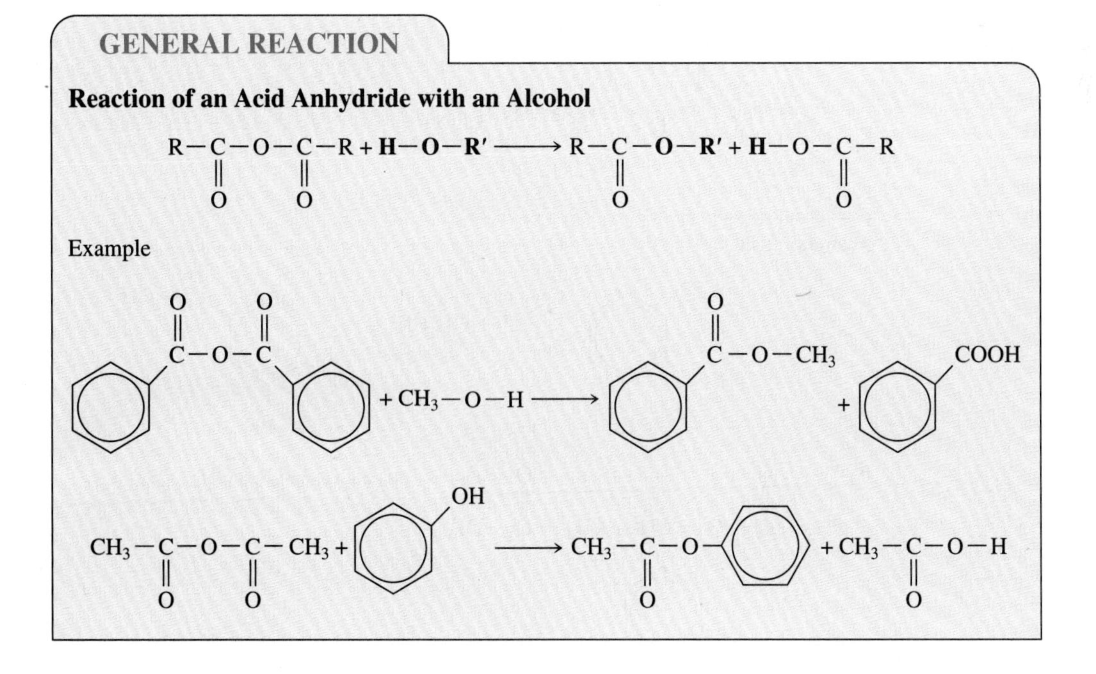

NEW TERM

acid anhydride A derivative of an organic acid that is dehydrated, having the structure

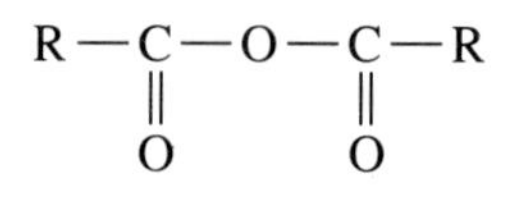

TESTING YOURSELF

Reactions of Acid Anhydrides

1. Give the structures for products from the following reactions:

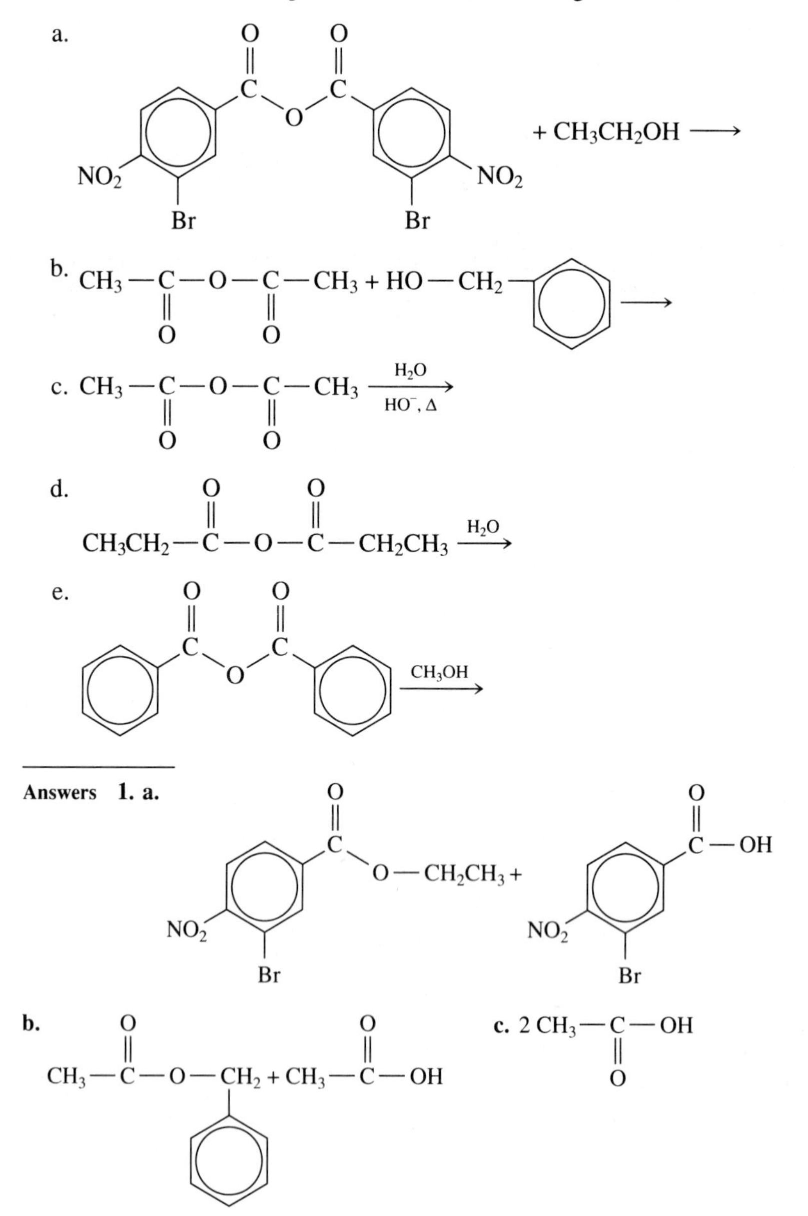

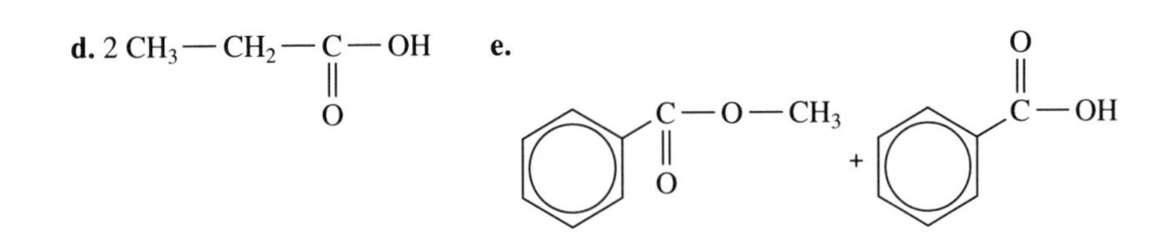

7.5 Important Organic Acids and Acid Derivatives

Salts of benzoic acid and propanoic acid are used as preservatives. Sodium benzoate is used as an antimicrobial in beverages, jams and jellies, and a number of other food items. Sodium and/or calcium propionate is used as a preservative in many bread products, where it reduces or inhibits mold growth.

Sorbic acid and its salts have found use as food additives to inhibit mold and yeast growth. They are often the preservatives found in smoked, pickled, and dried foods.

♦ $CH_3CH{=}CH{-}CH_2{-}COOH$

Sorbic acid

A number of interesting esters are found in everyday life. The esters of *p*-hydroxybenzoic acid (the parabens) are used for inhibiting growth of yeasts and molds, but they are not as useful as antibacterials. The polyester fiber *Dacron* is a polymeric ester of terphthalic acid and ethylene glycol.

Sources of some naturally occurring carboxylic acids.

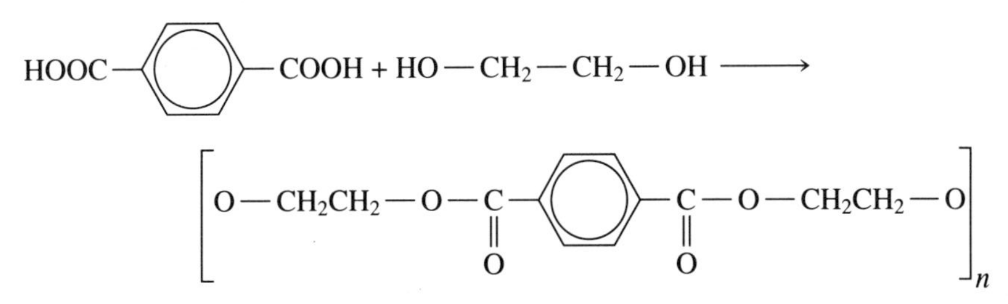

This important polyester has medical applications because it is inert to normal biochemical processes and can be used to repair blood vessels.

Summary

Organic acids are less acidic than the more common mineral acids, but they are more acidic than other organic functional groups. The increased ability of the *carboxyl group* to give up a proton can be attributed to the delocalization of charge on the remaining anion—much the same reasoning as is used to explain phenol acidity. *Carboxylic acids* are prepared by oxidations of alcohols, aldehydes, and alkyl-substituted aromatic systems. *Beta keto acids* undergo very easy loss of carbon dioxide. When an acid and an alcohol are mixed under proper conditions, an *ester* is obtained. *Acid anhydrides* are derivatives of acids and are more reactive than the acids.

Reaction Summary

Oxidation of a Primary Alcohol to an Acid

$$R{-}CH_2{-}OH \xrightarrow{\text{Oxidation}} R{-}\underset{\underset{O}{\|}}{C}{-}H \xrightarrow{[O]} R{-}COOH$$

Oxidation of an Alkyl Side Chain on a Benzene Ring

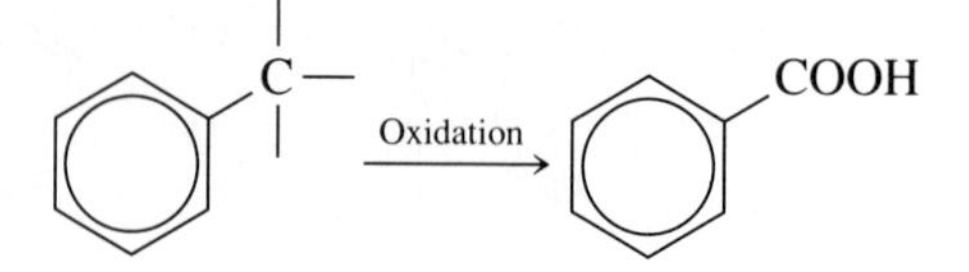

Decarboxylation of a Beta Keto Acid

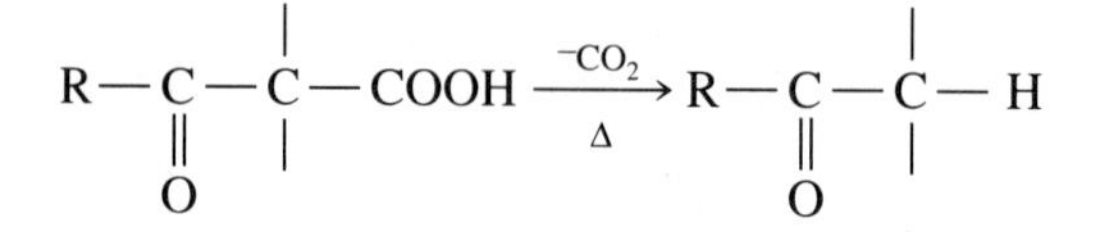

Ester Formation

$$\mathrm{R{-}\overset{\displaystyle O}{\overset{\|}{C}}{-}OH + HOR' \xrightarrow{H^+} R{-}\overset{\displaystyle O}{\overset{\|}{C}}{-}O{-}R' + H{-}O{-}H}$$

Formation of Thioesters

$$\mathrm{R{-}\overset{\displaystyle O}{\overset{\|}{C}}{-}OH + HSR' \xrightarrow{H^+} R{-}\overset{\displaystyle O}{\overset{\|}{C}}{-}S{-}R' + HOH}$$

Basic Hydrolysis (Saponification) of an Ester

$$\mathrm{R{-}\overset{\displaystyle O}{\overset{\|}{C}}{-}O{-}R' \xrightarrow[HO^-,\ \Delta]{H_2O} R{-}\overset{\displaystyle O}{\overset{\|}{C}}{-}O^- + R'OH}$$

Reaction of an Acid Anhydride with an Alcohol

$$\mathrm{R{-}\overset{\displaystyle O}{\overset{\|}{C}}{-}O{-}\overset{\displaystyle O}{\overset{\|}{C}}{-}R + HOR' \longrightarrow R{-}\overset{\displaystyle O}{\overset{\|}{C}}{-}O{-}R' + R{-}\overset{\displaystyle O}{\overset{\|}{C}}{-}OH}$$

Reaction of an Acid Anhydride with Water

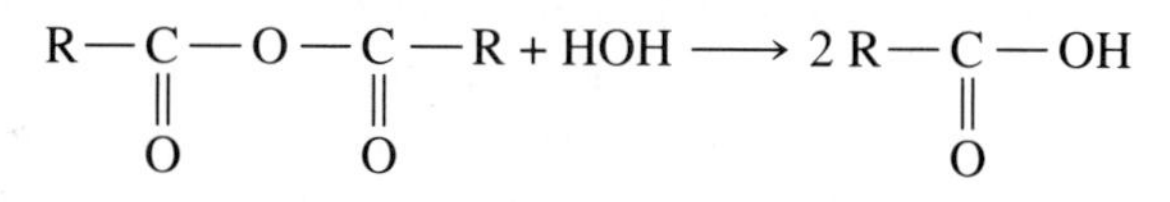

Terms

carboxyl group (7.1)
carboxylic acid (7.1)
carboxylate anion (7.2)
keto acids (7.2)
decarboxylation (7.2)
ester (7.3)
thioester (7.3)
detergent (7.3)
saponification (7.3)
fatty acid (7.3)
soap (7.3)
micelle (7.3)
acid anhydride (7.4)

Exercises

Structure and Physical Properties (Objective 1)

1. Ethanol, ethanal, and ethanoic acid have the same number of carbon atoms, yet their boiling points differ greatly. Please explain.
2. Why is propanoic acid more water-soluble than propanal?
3. By some measurements, acetic acid seems to have a molecular weight two times larger than it really is. Why?

Nomenclature (Objective 2)

4. Name the following:

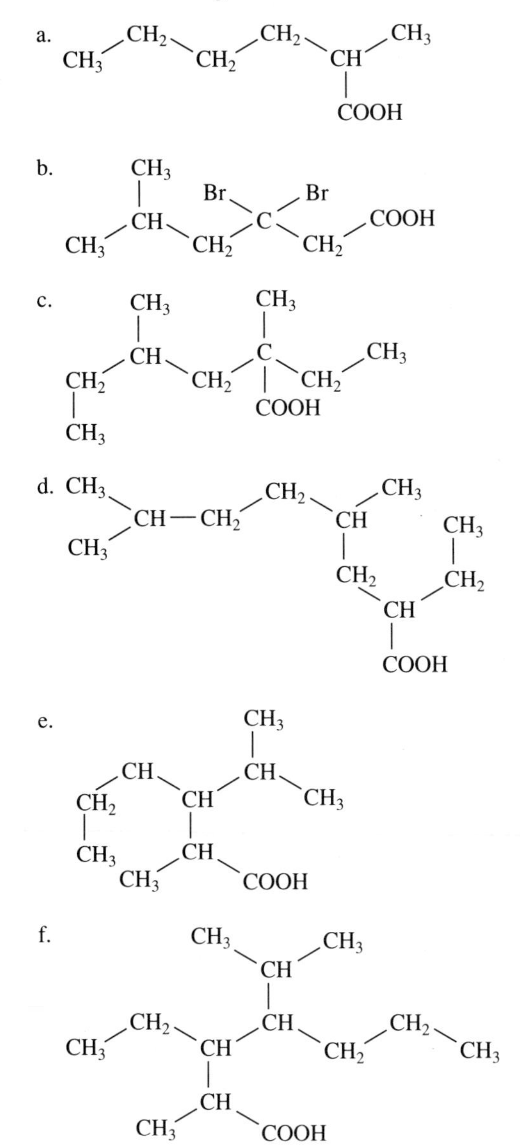

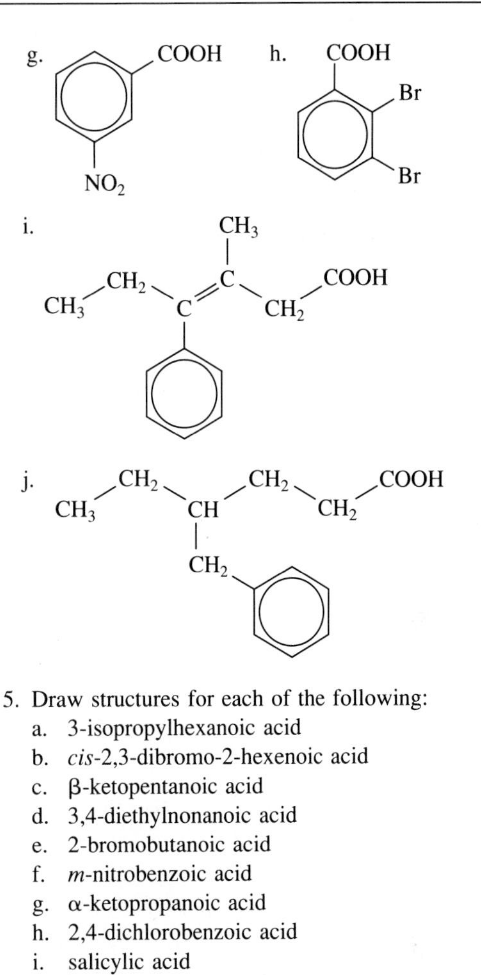

5. Draw structures for each of the following:
   a. 3-isopropylhexanoic acid
   b. *cis*-2,3-dibromo-2-hexenoic acid
   c. β-ketopentanoic acid
   d. 3,4-diethylnonanoic acid
   e. 2-bromobutanoic acid
   f. *m*-nitrobenzoic acid
   g. α-ketopropanoic acid
   h. 2,4-dichlorobenzoic acid
   i. salicylic acid
6. Name the following compounds:

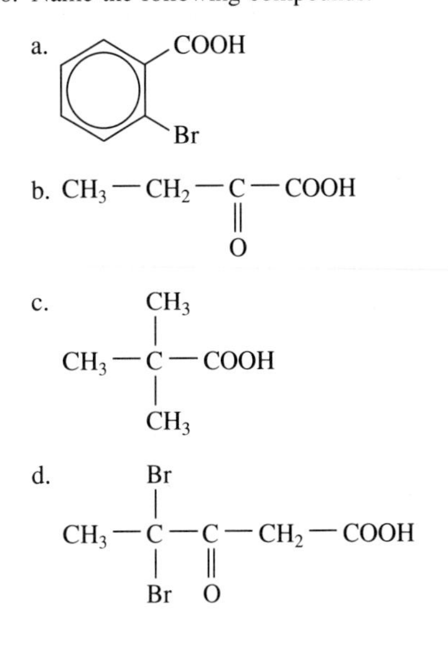

e. $(CH_3)_2CH{-}CH(Cl){-}COOH$

f. $CH_3{-}CH(OH){-}CH_2{-}CH_2{-}COOH$

Reactions of Acids (Objective 3)

7. Give products for each of the following reactions:

   a. $CH_3{-}COOH + NaOH \longrightarrow$

   b. $CH_3{-}CH_2{-}C(=O){-}CH(CH_3){-}COOH \xrightarrow{\Delta}$

   c. $(CH_3)_2CH{-}CH_2{-}CH_2{-}OH \xrightarrow{[O]}$

Derivatives of Organic Acids: Esters
Nomenclature (Objective 2)

8. Name the following esters:

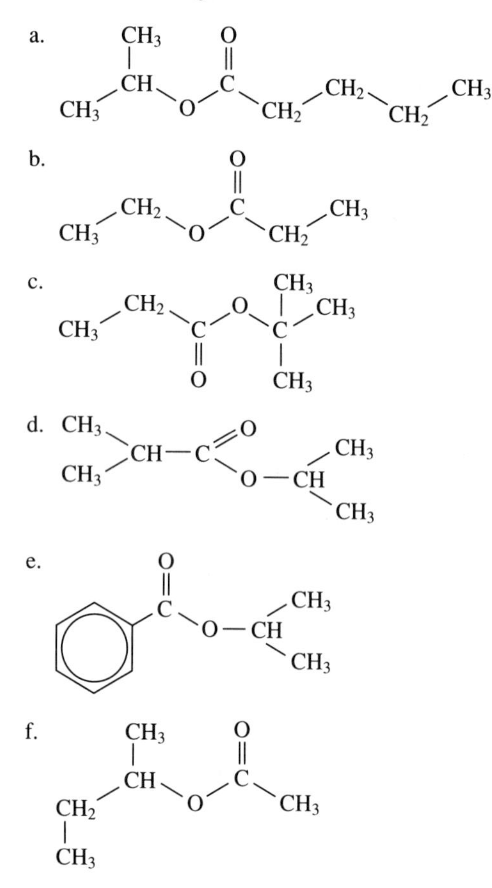

9. Draw structures for the following esters:
   a. methyl benzoate
   b. isopropyl acetate
   c. ethyl hexanoate
   d. (3-methyl)hexyl propanoate
   e. methyl formate
   f. benzyl benzoate

Reactions of Esters (Objective 3)

10. Give products for each of the following reactions:

    a. Benzoic acid + methanol $\xrightarrow{H^+}$

    b. Ethyl acetate $\xrightarrow[\Delta]{H_2O,\ H^+}$

    c. $CH_3{-}CH_2{-}C(=O){-}OH$ + isopropyl alcohol $\xrightarrow{H^+}$

    d. $CH_3{-}C(=O){-}OH + CH_3{-}OH \xrightarrow{H^+}$

11. Give products for each of the following reactions:

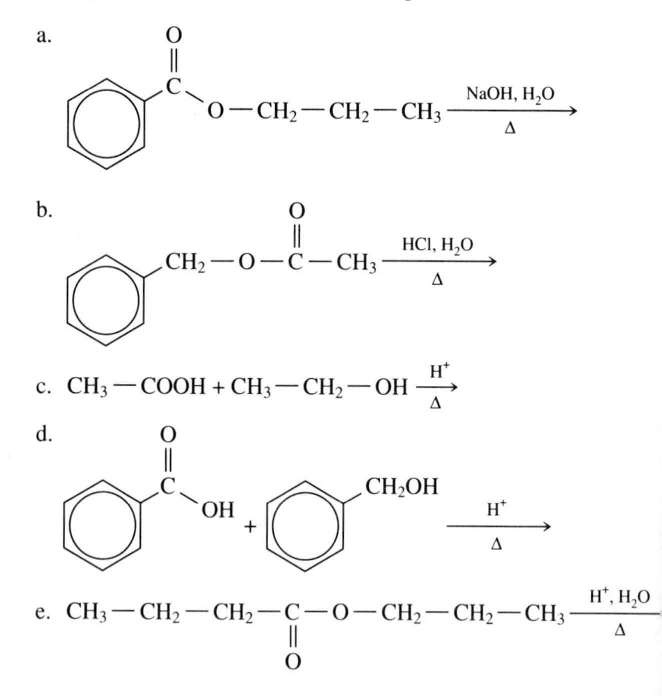

    e. $CH_3{-}CH_2{-}CH_2{-}C(=O){-}O{-}CH_2{-}CH_2{-}CH_3 \xrightarrow[\Delta]{H^+,\ H_2O}$

Acid Anhydrides (Objective 3)

12. Give products for the following reactions:

    a.

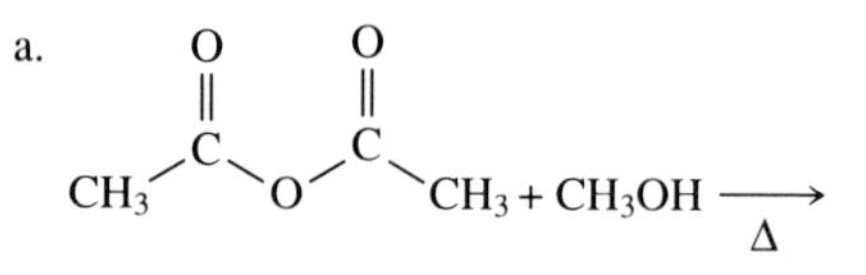

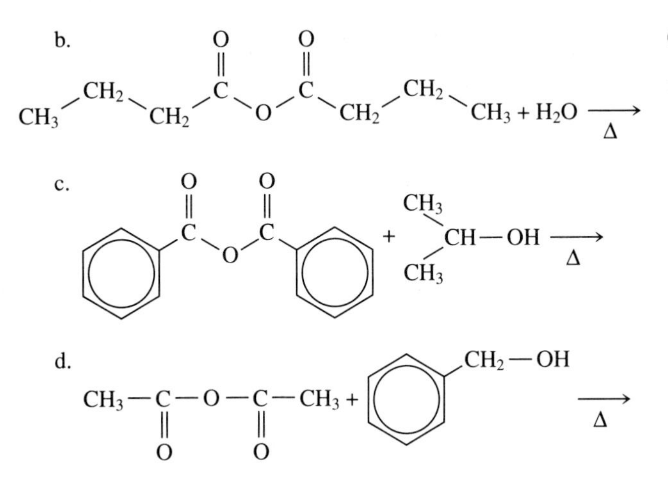

General Nomenclature (Objective 2)

13. Give appropriate names for each of the following compounds:

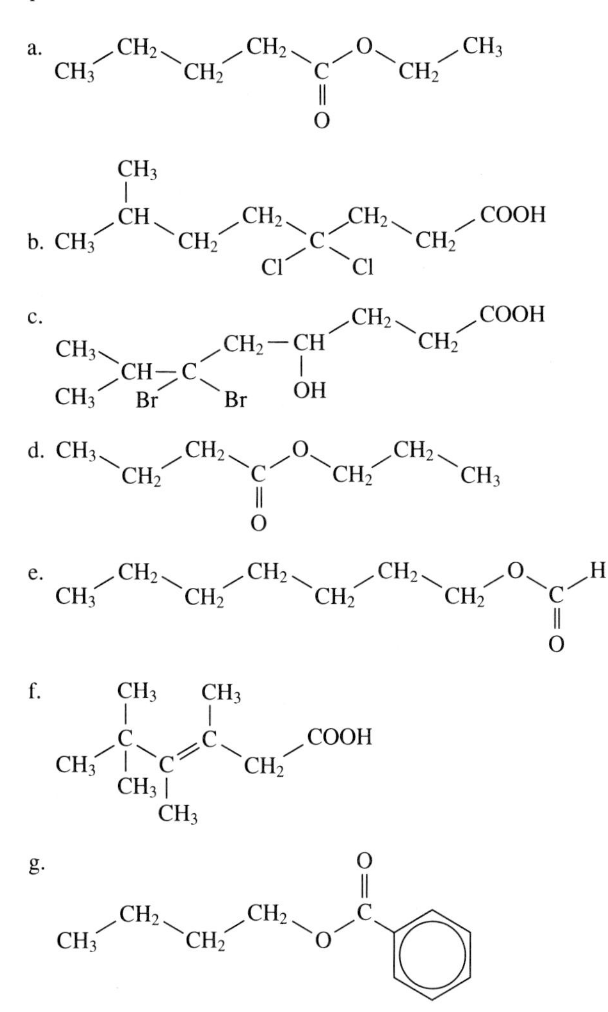

General Reactions (Objective 3)

14. Give products for each of the following reactions:

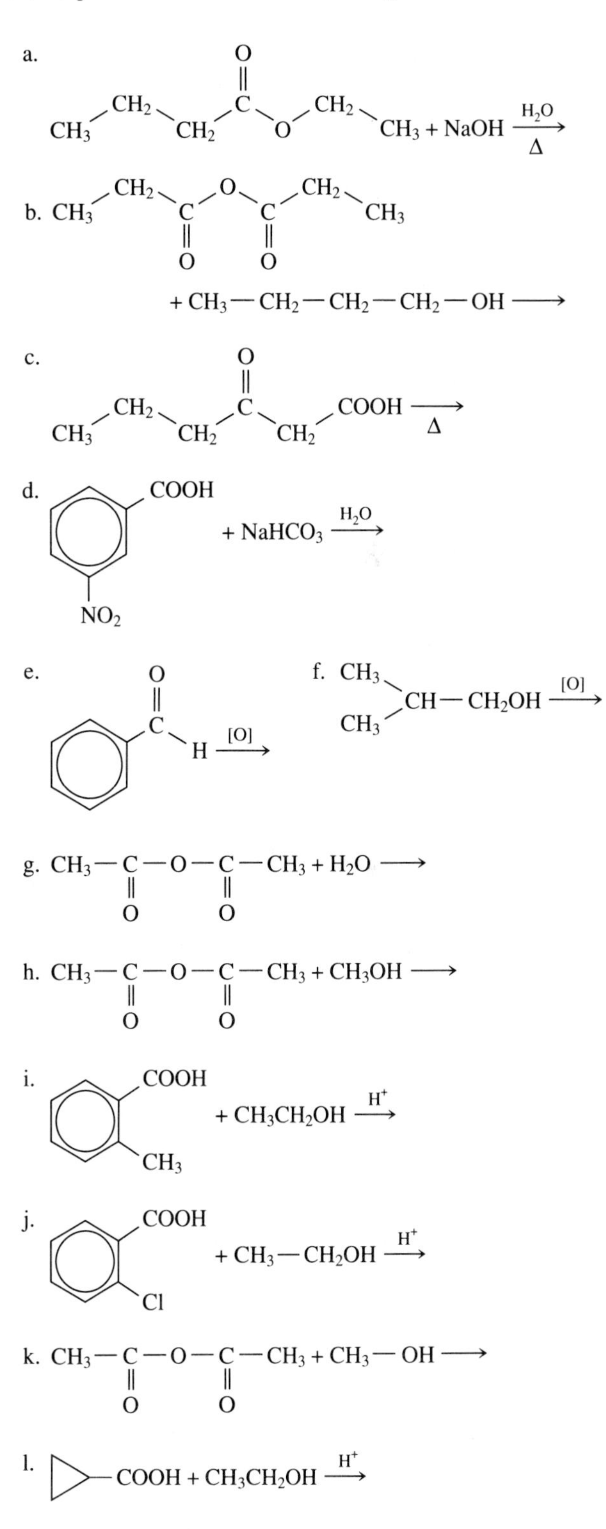

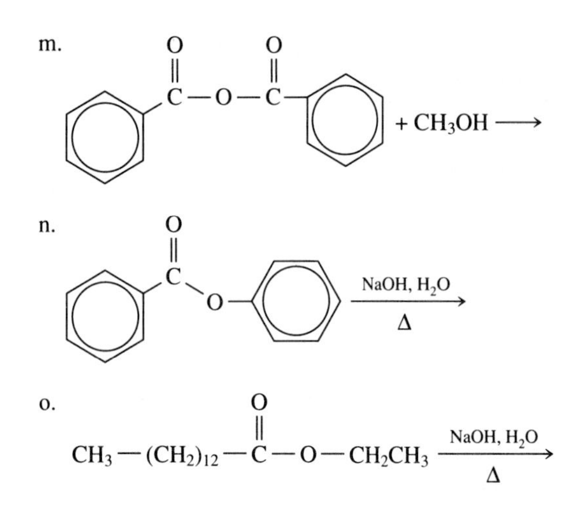

Thought Exercises

15. The preparation of aspirin and oil of wintergreen are both esterification reactions, and they both start from the same starting material—salicylic acid.

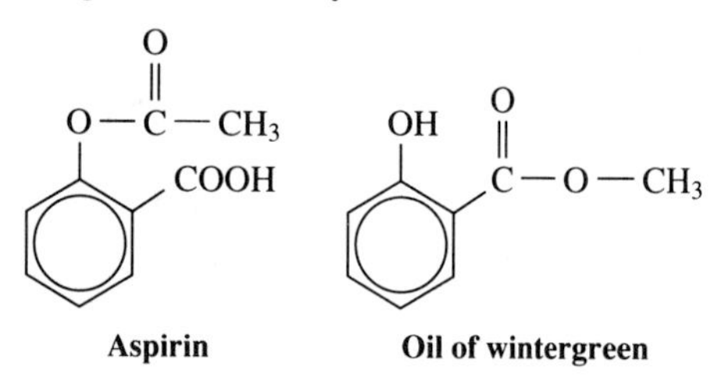

Aspirin **Oil of wintergreen**

How would you prepare these two molecules?

16. In diabetes, a person may exhibit symptoms of ketosis and may even have "acetone breath." If one of the metabolic products is β-ketobutanoic acid, show how it forms acetone.

(Objective 5)

17. Explain why a concentrated detergent might be as useful as turpentine in cleaning a paint brush used for oil paints.

Structure and Physical Properties (Objective 1)

18. What is the carboxyl group?
19. What is the hybridization of the carbon of the carboxyl group?
20. Why do acids exhibit "higher than expected" boiling points?
21. Why are acids of lower carbon number water-soluble?
22. Why are long-chain carboxylic acids *not* water-soluble?

Nomenclature (Objective 2)

23. Give names for each of the following compounds:

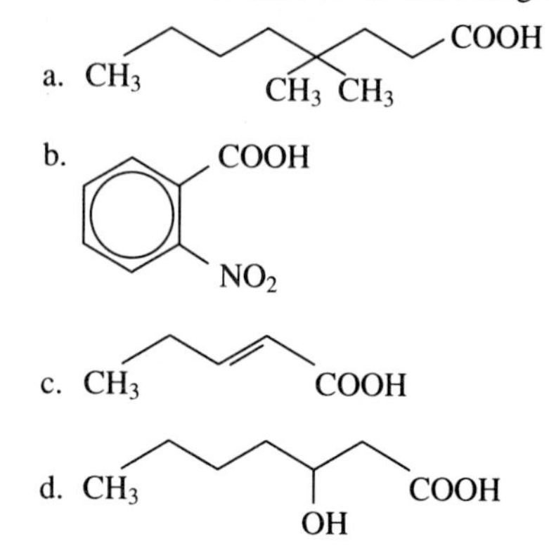

24. Give structures for the following compounds:
    a. *o*-chlorobenzoic acid
    b. 3-isobutyloctanoic acid
    c. *cis*-3-pentenoic acid
    d. 2-hydroxybutanoic acid
    e. 3-oxoheptanoic acid

Reactions (Objective 3)

25. Show the resonance structures for the carboxylate anion formed from reaction of acetic acid with sodium hydroxide.
26. How is an acid prepared from an aldehyde?
27. What is the product formed from heating 3-oxooctanoic acid?
28. What is the product of the reaction of acetic acid with sodium hydroxide?
29. If you mix acetic acid with sodium bicarbonate, a gas is produced. What is the gas? Write an equation to show this reaction.

Esters

30. What is an ester, and how is one prepared?
31. Draw a structure for methyl pentanoate.
32. What is saponification?
33. Describe the action of a detergent in water.
34. Why wouldn't the salt of propanoic acid serve as a soap?

Unclassified Exercises

35.

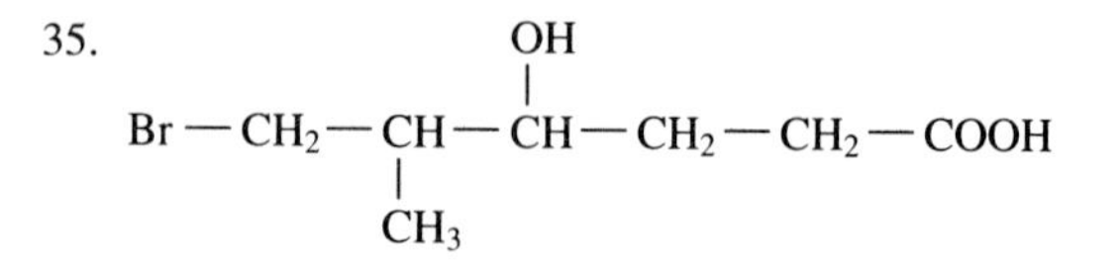

is best named
a. 1-bromo-2-methyl-3-alcoholhexanoic acid
b. 1-carboxyl-4-methyl-5-bromo-3-pentanol
c. 1-bromo-3-hydroxy-2-methylhexanoic acid
d. 6-bromo-4-hydroxy-5-methylhexanoic acid

36. Which of these is most reactive toward water?
a. acid anhydride b. ester c. organic acid

37. A beta keto acid will not
a. be called a 3-keto acid
b. lose carbon dioxide when heated
c. be called a 2-keto acid
d. form a ketone when heated

38. An acid is not formed by
a. oxidation of a primary alcohol
b. ester hydrolysis
c. aldehyde oxidation
d. oxidation of a ketone

39. In an ester

$$R-\underset{\underset{O}{\|}}{C}-O-R'$$

the highlighted oxygen comes from
a. alcohol b. acid c. water

40. The salt of a long-chain organic acid is called a
a. carbanion
b. cationic detergent
c. soap
d. saponification

41. Oxidation of toluene gives
a. benzyl alcohol
b. benzaldehyde
c. benzoic acid
d. quinone

42. What is the product of this reaction?

$$CH_3-\underset{\underset{O}{\|}}{C}-O-\underset{\underset{O}{\|}}{C}-CH_3 + CH_3-CH_2-OH \longrightarrow$$

a. methyl propanoate
b. ethyl propanoate
c. methyl acetate
d. ethyl acetate

Nylon is a polymer made with amines.

AMINES AND AMIDES

OUTLINE

8.1 Amines: Structure, Properties, and Nomenclature
8.2 Reactions of Amines
8.3 Amides
8.4 Special Amines

OBJECTIVES

After completing this chapter, you should be able to

1. Discuss the origins of basicity of amines.
2. Predict the products of amines with acids and carbonyl compounds.
3. Name simple amines.
4. Discuss how amide bonding controls structure.
5. Recognize the amine groups in alkaloids.

TABLE 8.1
Elemental Abundance in the Earth's Crust

Rank (in Abundance)	Element	Atoms per 1000 Atoms on Earth
1	Oxygen	533
3	Hydrogen	151
5	Carbon	30
12	Chlorine	10
14	**Nitrogen**	**9**

Nitrogen is an abundant element on the earth, ranking fourteenth in elemental abundance in the earth's crust (Table 8.1). Thus, it should be no surprise to see that nitrogen is incorporated into a large number of organic compounds. In the biochemistry chapters that follow, nitrogen-containing organic compounds will be seen to have a central role in human physiology. Many biologically important compounds are formed from nitrogen-containing **amines**, which are organic derivatives of ammonia. If one of the hydrogen atoms of ammonia is replaced by an R group, a new compound called a *primary amine* is formed. Replacement of two hydrogens results in a *secondary amine*, and replacement of all three hydrogens gives a *tertiary amine*. The R groups *do not* have to be the same.

♦ Amines are the organic equivalent of ammonia.

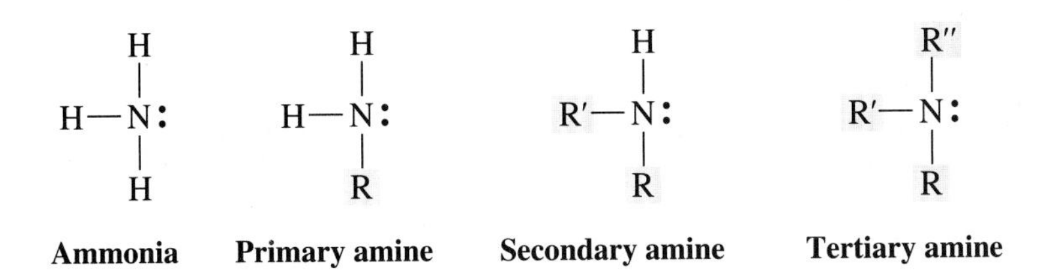

8.1 Amines: Structure, Properties, and Nomenclature

Hybridization and Structure

Ammonia is a weak base as a result of its lone pair of electrons (Lewis base). Application of the rules used for Lewis electron-dot structures gives a structure in which the nitrogen atom of the amine has three bonds and a lone pair of electrons. This requires that amines have sp^3 hybridization. In general, an N—H bond of ammonia can be replaced by an N—R group to make several different kinds of substituted amines. The lone pair of electrons is responsible for the

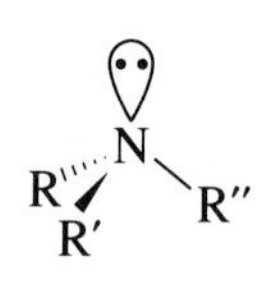

unique chemical behavior and properties of the amine, just as it provides for the chemical behavior of ammonia. A general structure of an amine is given here (left), in which R is an alkyl or H.

A logical question to ask at this point is why the amines are considered to be basic as a result of having lone-pair electrons, but ethers are not, even though they also contain lone pairs of electrons. Recall that an ether *is* a base in the Lewis sense—it is an electron donor. However, amines are better electron donors. This is because nitrogen is less electronegative than oxygen, thus, the electrons on an amine are not held as tightly.

Hydrogen Bonding

Since nitrogen is more electronegative than carbon, polarization of the carbon–nitrogen bond in an amine is expected.

♦ The hydrogen bonding of amines isn't as strong as that of alcohols or acids.

The C—N bond is a polar covalent bond with a dipole such that the carbon has a partial positive charge and the nitrogen a partial negative charge. However, the difference in electronegativity is less than for a C—O or C—Cl bond, and less polarization occurs. Similarly, in the N—H bond, the H acts as a partially positive atom. This is similar to the H—O bond of water (but to a lesser extent), so that amines can also hydrogen-bond with one another and with other polar molecules.

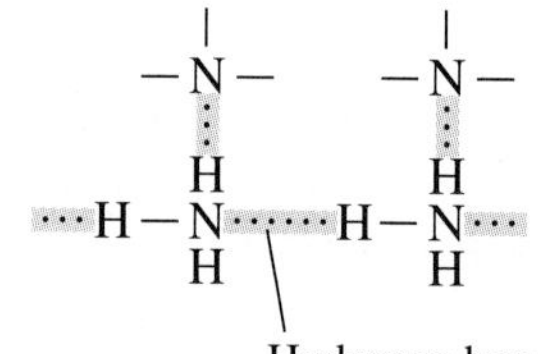

Just as the alcohols had fewer hydrogens for hydrogen bonding than water, the amines decrease in their ability to hydrogen-bond as they become more substituted. Since the difference in electronegativity between N and H is less than the difference between O and H in water, hydrogen bonding is less important for the amines. This is illustrated by comparison of boiling points for a series of amines with boiling points of alcohols (Figure 8.1).

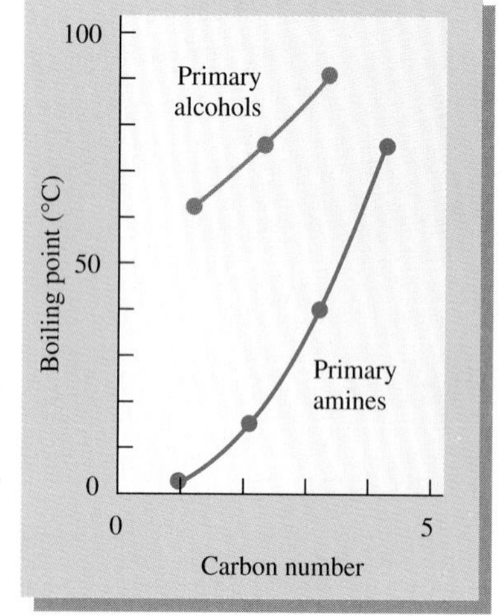

FIGURE 8.1
The boiling points of amines are considerably lower than the boiling points of the corresponding alcohols. This is a result of decreased hydrogen bonding.

Nomenclature

Amines are generally named by identifying all of the alkyl groups on the nitrogen, followed by the suffix *-amine*.

EXAMPLE 8.1

Name these amines.

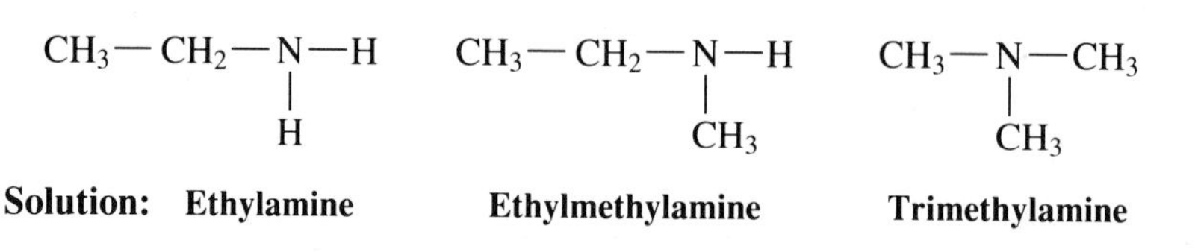

Solution: **Ethylamine** **Ethylmethylamine** **Trimethylamine**

A ring system can also have a nitrogen atom as part of the ring. In general, any ring that contains some atom other than carbon is called a **heterocycle**. Here are three examples of heterocycles:

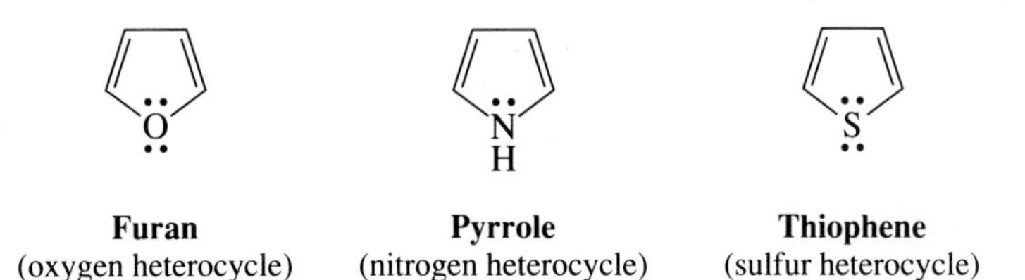

Furan (oxygen heterocycle) **Pyrrole** (nitrogen heterocycle) **Thiophene** (sulfur heterocycle)

Note that the one in the middle (pyrrole) is an amine because it contains nitrogen. Many nitrogen-containing heterocycles have common names, such as those in Table 8.2.

Many amines are characterized by strong, fishy odors; in fact, some have very bad odors. Consider the names of two common diamines, *cadaverine* and *putrescine*; these are produced by microorganisms in rotting meat. Note how these names suggest the odor associated with the amine!

$$H_2N-CH_2-CH_2-CH_2-CH_2-NH_2$$

Putrescine

$$H_2N-CH_2-CH_2-CH_2-CH_2-CH_2-NH_2$$

Cadaverine

One type of derivative of the common amine has the nitrogen atom substituted by an alkyl group. These compounds are called *N*-alkyl derivatives and have the prefix *N*- in their names, which indicates the alkyl group is attached to nitrogen. Thus, *N*-methylhexylamine can be differentiated from 2-methylhexylamine. In this nomenclature, the parent amine will have the longest carbon chain.

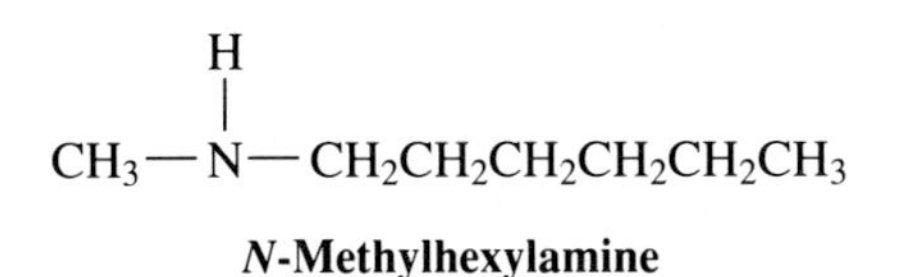

***N*-Methylhexylamine**

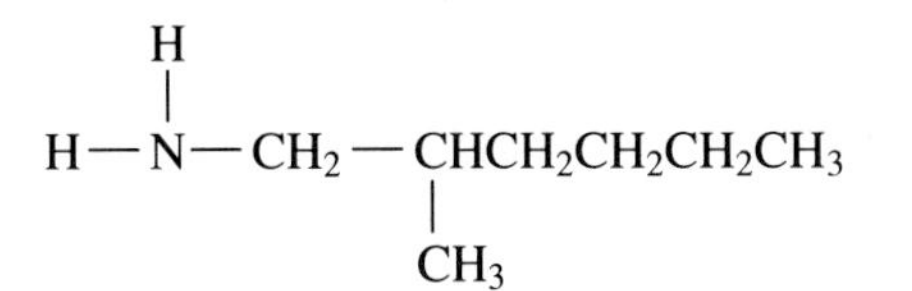

2-Methylhexylamine

TABLE 8.2

Heterocyclic Amines Having Common Names

Name	Structure
Aniline	
Pyridine	
Pyrrolidine	
Piperidine	
Pyrrole	
Nicotinic acid	
Purine	
Pyrimidine	
Imidazole	

EXAMPLE 8.2

Name these *N*-alkyl derivatives of amines:

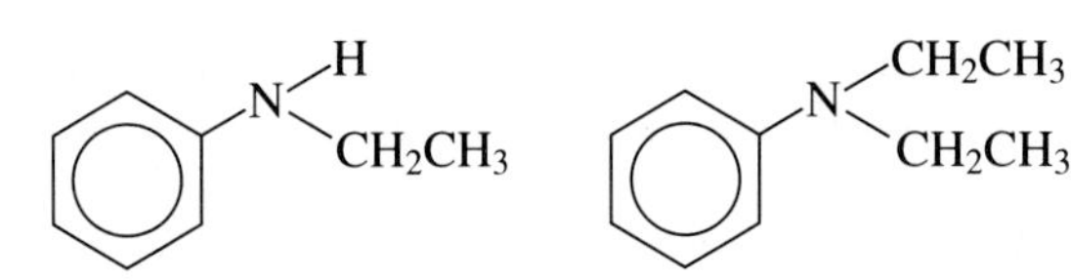

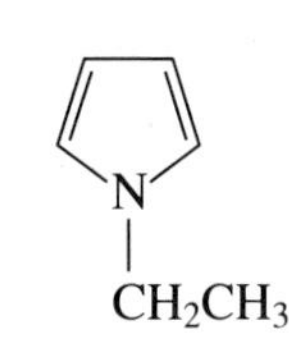

Solution: ***N*-Ethylaniline** ***N,N*-Diethylaniline** ***N*-Ethylpyrrole**

NEW TERMS

amine An organic functional group containing a nitrogen

heterocycle A cyclic system having some atom other than carbon as a structural component of the ring. Some heterocycles are amines.

TESTING YOURSELF

Amine Structure and Nomenclature

1. Name the following amines.

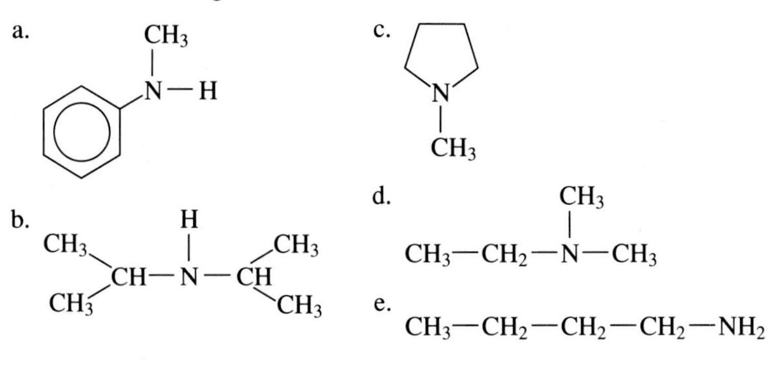

Answers **1. a.** *N*-methylaniline **b.** diisopropylamine **c.** *N*-methylpyrrolidine **d.** *N,N*-dimethylethylamine or dimethylethylamine **e.** butylamine

8.2 Reactions of Amines

Basicity

The lone pair of electrons on an amine gives it the characteristic basic properties. An amine reacts with a proton from an acid according to the general reaction at the top of page 223.

Pyridine is a basic aromatic amine. There are six electrons in the pi system, giving pyridine the same kind of stability demonstrated earlier for benzene. However, since nitrogen needs only three bonds, we see that the bonding electrons are used; thus, a lone pair of electrons is left on nitrogen. This lone pair gives pyridine its observed basicity. A similar argument can be made to account for the basicity of pyrimidine and purine.

The amines pyrrole and pyrrolidine look quite similar in structure, but there is a basic difference. Whereas pyrrolidine is a typical amine base, pyrrole is relatively neutral. These differences in basicity are easily explained by examining the lone-pair electrons (Figure 8.2). In pyrrolidine (as in other typical bases), the lone pair of electrons occupies an sp^3 orbital and is available to react with a proton. In pyrrole the electrons are in a *p* orbital and are used to complete the delocalized electron network. These electrons are *not* available to react with a proton.

♦ Delocalization of electrons from the nitrogen lone pair into the ring is responsible for pyrrole and aniline being weaker bases than other amines.

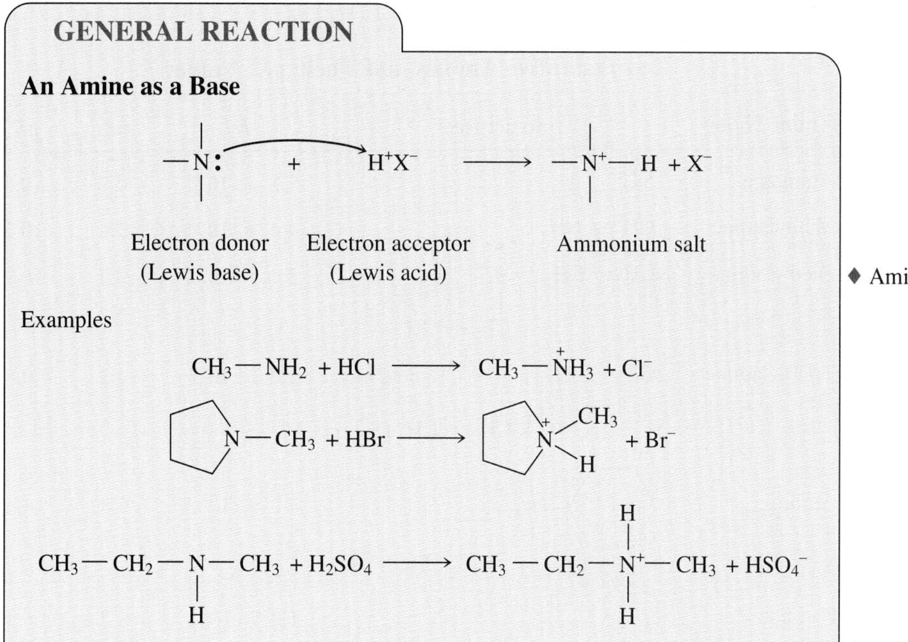

♦ Amines are Lewis bases.

From the examples of amines in Table 8.3, we notice that the range of basicity does not vary too much with structure. A striking exception to this is the relative nonbasic nature of aniline. Why is aniline so weakly basic? Since basicity is related to the ability to donate electrons, the aromatic system electron delocalization essentially removes the electron lone pair from the nitrogen. If the electrons are *not* available on nitrogen, they cannot impart basic character to that site. This idea is well demonstrated by a resonance analysis (Figure 8.3). Note that this rationale is identical to that invoked to explain the pronounced acidity of phenols.

Reactions as a Nucleophile

Reaction with an Alkyl Halide

In the previous section, we saw that the electrons on nitrogen made amines basic. An amine can effectively use its electron lone pair for purposes other than

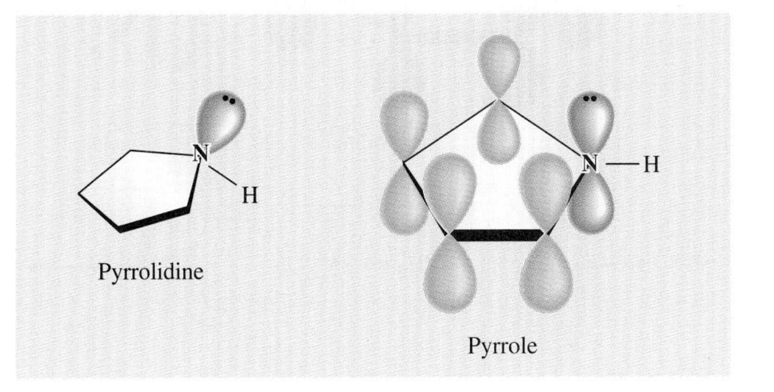

FIGURE 8.2
Reasons for the difference in basicity of pyrrolidine and pyrrole are found in the observation that the lone-pair electrons of pyrrole are delocalized into the pi system of the ring. These electrons then are not available to act as a base. In pyrrolidine the lone pair of electrons is localized on nitrogen and is available to act as a base.

TABLE 8.3

Representative Amines and Their pK_b Values

Amine Name	Structure	K_b	pK_b	pK_a
Ammonia	NH_3	1.8×10^{-5}	4.7	9.3
Methylamine	$CH_3—NH_2$	4.4×10^{-4}	3.4	10.6
Diethylamine	$CH_3—CH_2—NH(—CH_2—CH_3)$	3.1×10^{-4}	3.3	10.7
Triethylamine	$CH_3—CH_2—N(—CH_2—CH_3)—CH_2—CH_3$	1.0×10^{-3}	4.2	9.8
Aniline	$C_6H_5—NH_2$	4.2×10^{-10}	9.4	4.6
Pyridine	C_5H_5N	1.8×10^{-9}	8.7	5.3

pulling off hydrogens. Some of the more important chemical properties of amines center on their ability to act as nucleophiles. Recall the reaction of a nucleophile with an alkyl halide:

$$Nu: + —\overset{|}{\underset{|}{C}}—X \longrightarrow Nu—\overset{|}{\underset{|}{C}}— + X^-$$

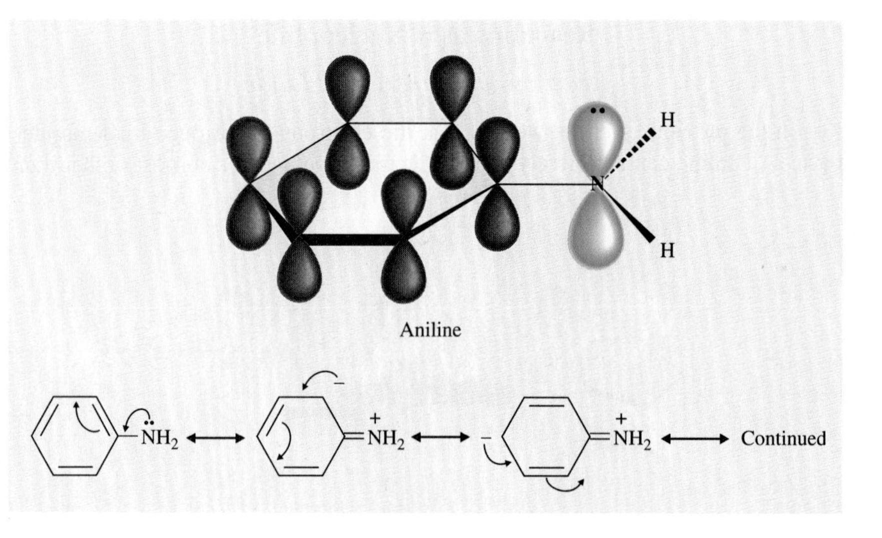

FIGURE 8.3
The electron delocalization of the nitrogen electrons into the ring is the reason why aniline is not very basic. This delocalization is very similar to that already discussed for phenol.

The lone pair of electrons on an amine can participate in the same kind of reaction. Looking at this reaction from another viewpoint, we can also say that alkyl halides can *alkylate* the nitrogen atom of an amine.

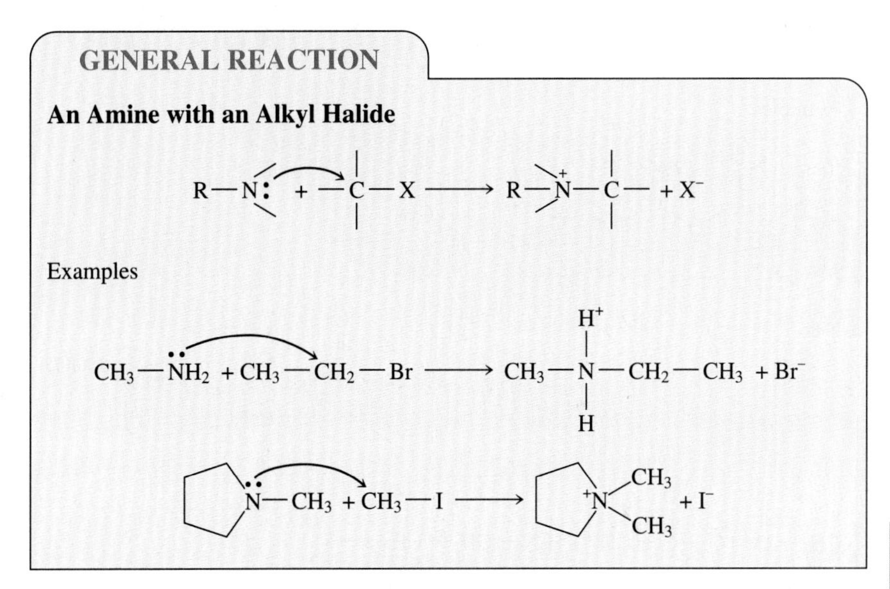

If the alkylation of pyridine is carried out with a long-chain alkyl halide (such as *cetyl chloride*, $C_{16}H_{33}Cl$), a detergent is obtained (in this case, cetylpyridinium chloride.)

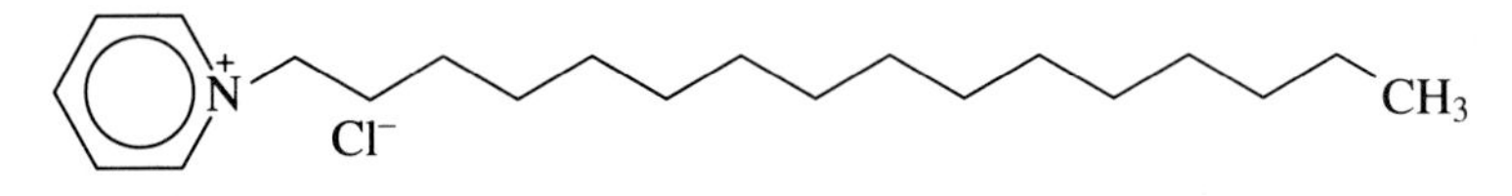

Cetylpyridinium chloride

Many mouthwashes have cetylpyridinium chloride as an active ingredient.

Notice that this compound has a long hydrocarbon portion and an ionic end. This particular compound is the active ingredient in many mouthwashes. Recall how your mouth bubbles when you gargle. You are gargling with a detergent!

The alkylation reaction is biologically significant because a number of alkylating agents (compounds that readily donate an alkyl group) are considered to be the source of genetic mutations. Compounds such as some alkyl halides and dimethylsulfate ($CH_3—O—SO_2—O—CH_3$) alkylate the nitrogens of some of the purines and pyrimidines that are common to DNA.

Reaction with an Acid Anhydride

◆ Amines react with acid anhydrides to form amides.

When an acid anhydride is reacted with an amine, an amide is produced with the byproduct:

$$\mathrm{H{-}O{-}\underset{\underset{\displaystyle O}{\|}}{C}{-}R}$$

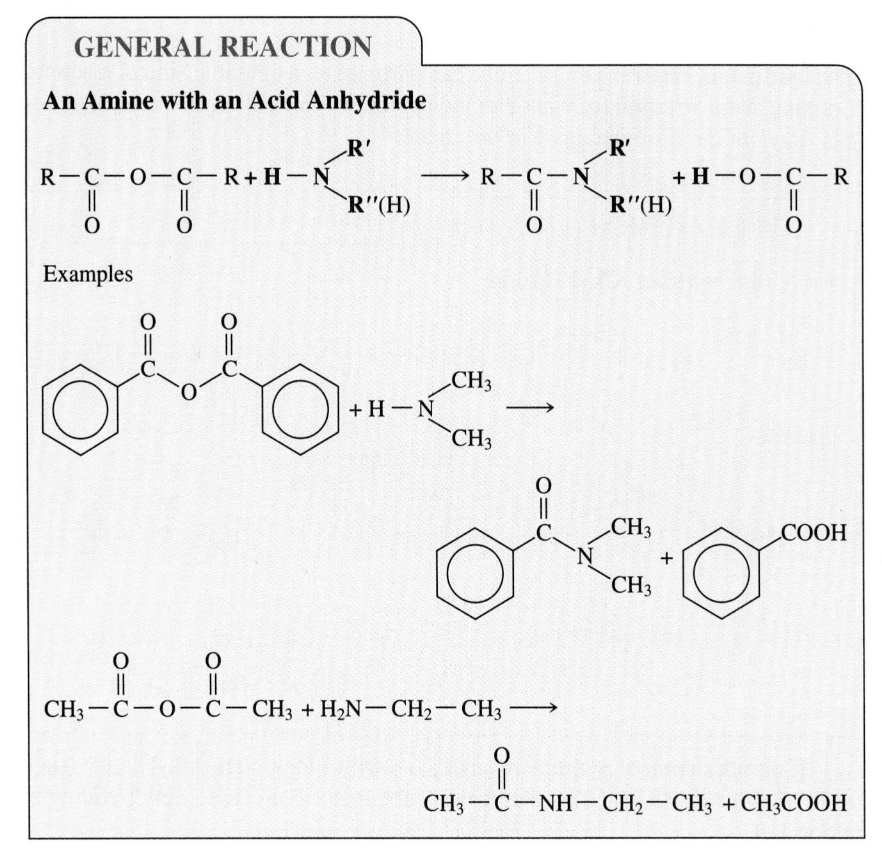

Reactions with Carbonyl Compounds

♦ Amines react with carbonyl compounds to form *imines*.

We have already shown that the amines have a lone pair of electrons that makes them both basic *and* nucleophilic. These electrons can be used for nucleophilic attack at the carbonyl group of aldehydes and ketones. The product of the reaction of a primary amine with a carbonyl group produces an **imine**, which has a double bond between the carbon and nitrogen.

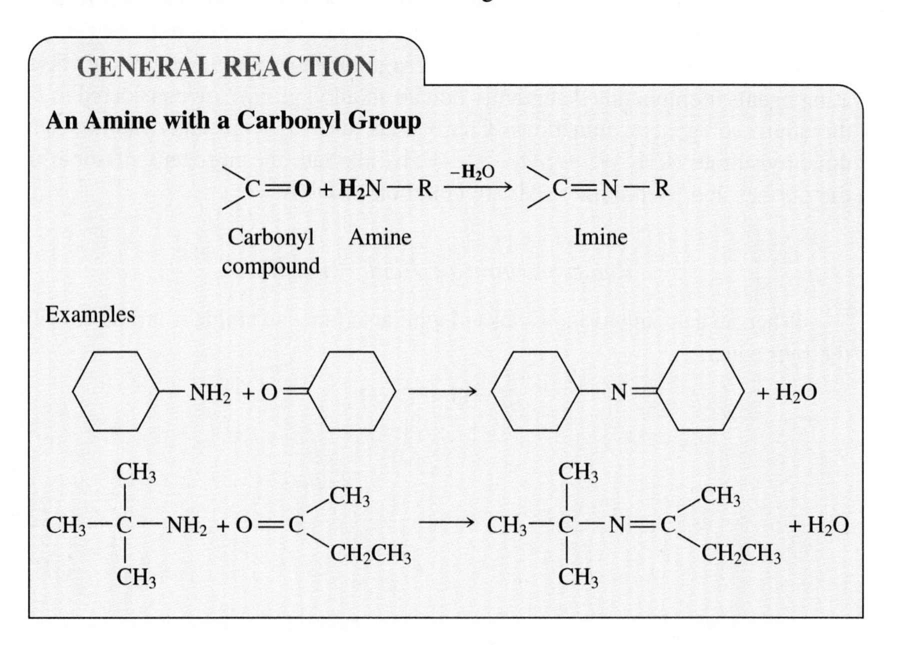

In biological systems the formation of an imine (also called a *Schiff base*) is one of the steps for making amines.

$$\rangle C{=}O + H_2N{-}R \xrightarrow{-H_2O} \rangle C{=}N{-}R \xrightarrow{[H]} \rangle CH{-}NH{-}R$$

Imine

This reaction is significant in biological systems as a way of converting keto acids to *amino acids* (transamination). Unique to this reaction is the realization that a keto acid and amino acid form a different keto acid and amino acid.

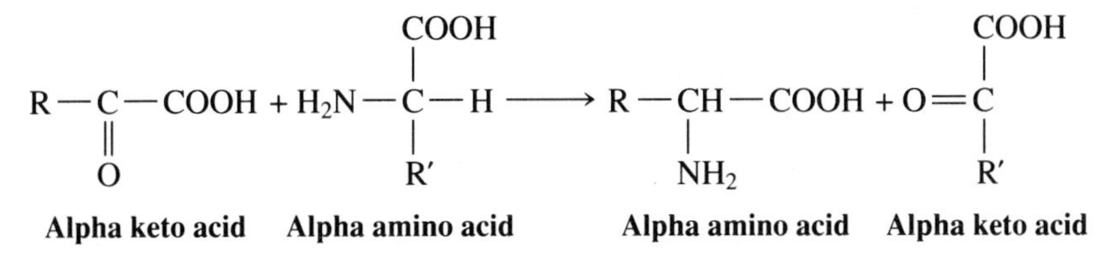

Hydrazine is the simplest diamine, $H_2N{-}NH_2$. In terms of their reactions, hydrazines should be considered similar to primary amines, except the R is another NH_2. We now expect that a primary amine reacts with a carbonyl to give an imine. We should expect a similar reaction with hydrazine; this product is called a *hydrazone*. Substituted hydrazines react to give the corresponding substituted hydrazones. In particular, 2,4-dinitrophenylhydrazine is useful for making unique derivatives of carbonyl compounds. Most of the hydrazones are easily purified and characterized, and thereby serve as derivatives useful for identifying a particular aldehyde or ketone.

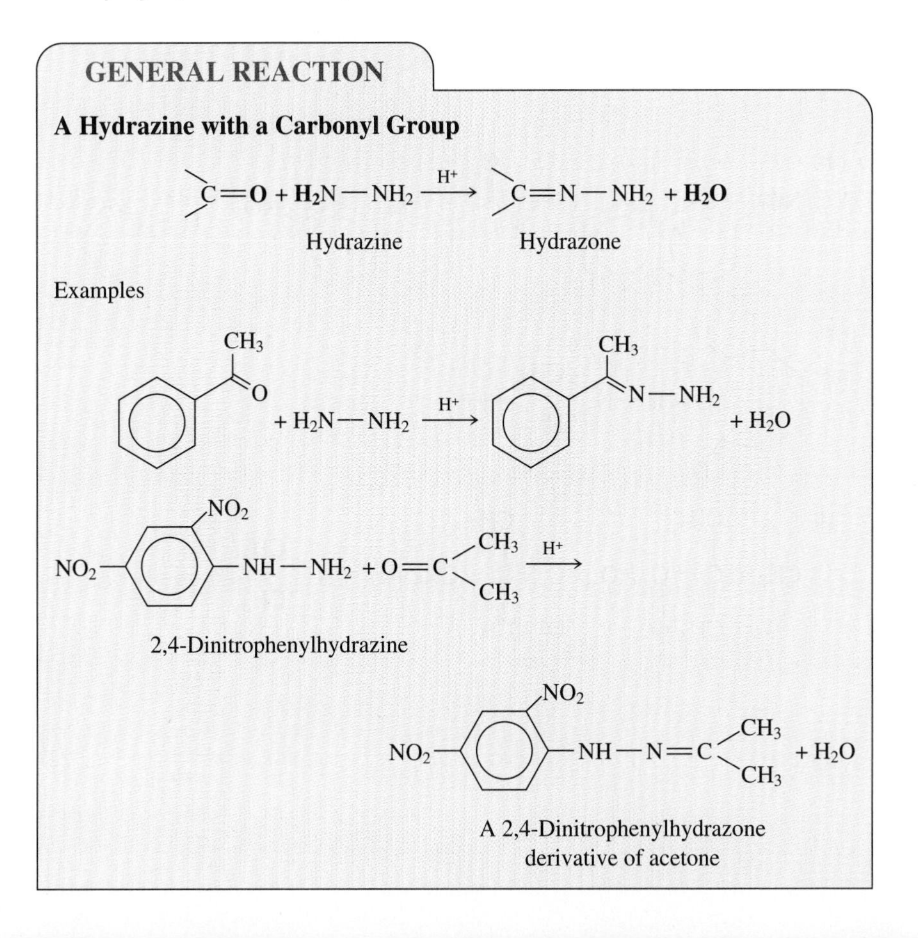

Many prepared meats contain nitrites as a preservative.

Oxidation of Secondary Amines to Nitrosamines

Secondary amines are oxidized by nitrous acid to nitrosamines:

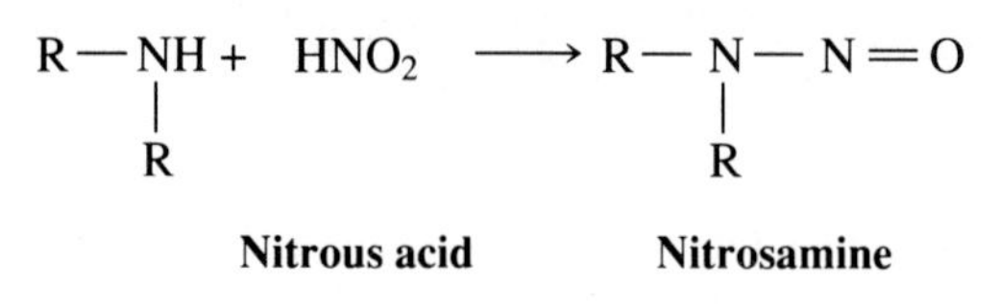

Nitrous acid **Nitrosamine**

There is some concern that nitrites added as preservatives to certain foods, such as packaged meats, hotdogs, and bacon, are harmful to humans. In the presence of stomach acid, these nitrites form nitrous acid that can then convert amines to nitrosamines. These nitrosamines are known carcinogens. Whether the levels are high enough to be a problem is not yet established.

NEW TERM

imine A compound formed from the reaction of a carbonyl-containing compound and a primary amine

TESTING YOURSELF

Amine Reactions

1. Give the amide product of these reactions of an organic acid derivative with an amine:

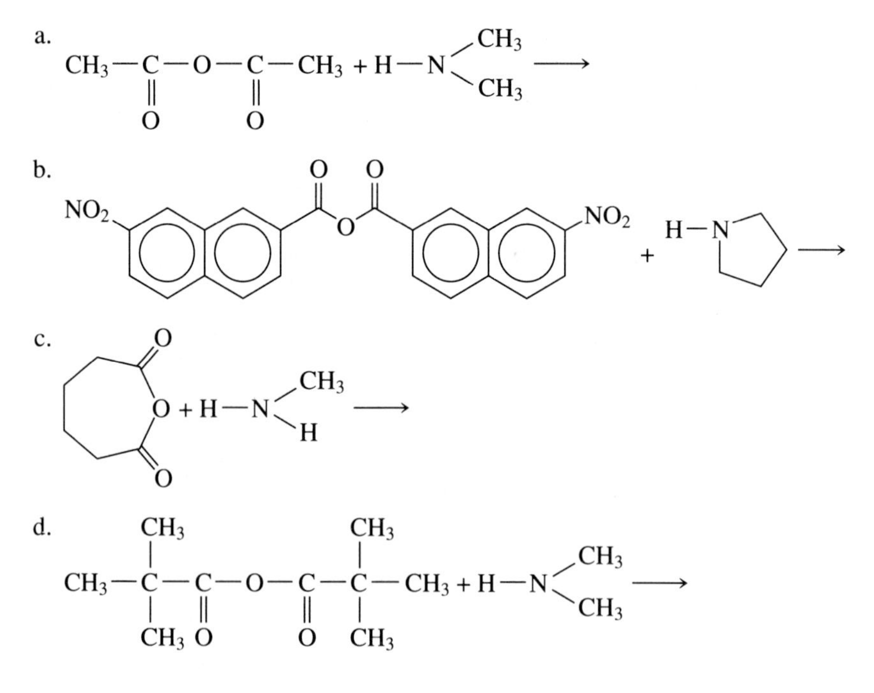

2. Give products for the following amine reactions.

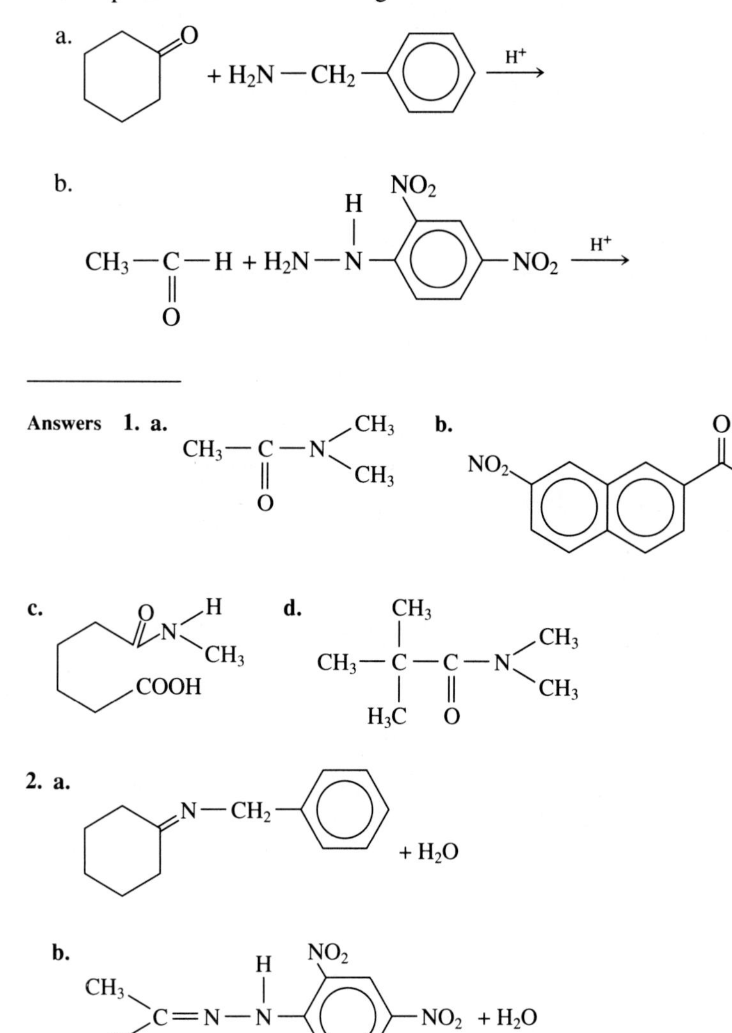

8.3 Amides

Structure

An **amide** is a nitrogen-containing compound with this general structure

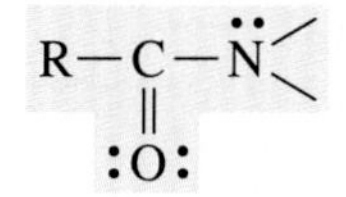

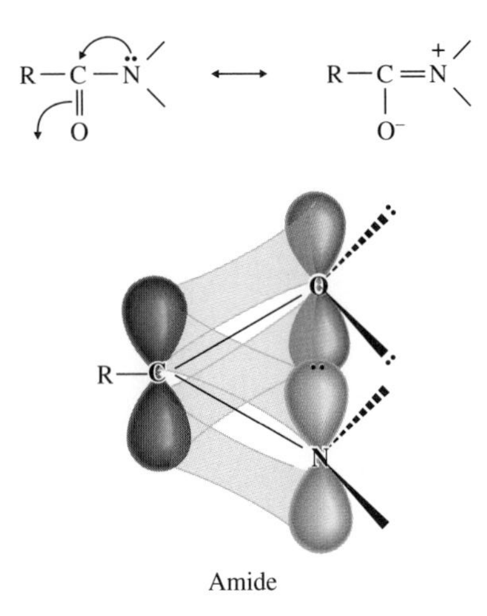

FIGURE 8.4
The resonance structures of an amide can account for the lack of basicity of an amide relative to an amine. The resonance structure having the "charge on oxygen" contributes greatly to the "real" structure of an amide—particularly the observed lack of rotation about the C—N bond.

Amides are related to acids and amines in much the same way that esters are related to acids and alcohols. Note the similarity between these two reactions:

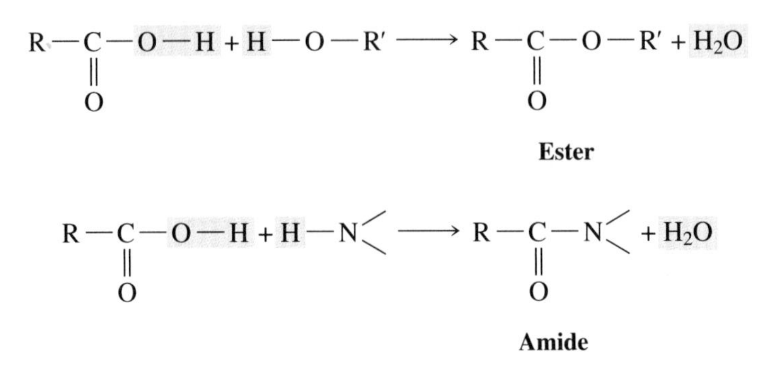

Amides are planar around the nitrogen atom, thus the nitrogen is sp^2 hybridized. As a result of an amide's resonance structure having the nitrogen lone pair delocalized to the carbon (Figure 8.4), amides are not very basic. Also, this causes the amide to be planar around the amide functional group. The planarity of the amide bond plays an important role in the structure of proteins. Because the resonance structure can be drawn to show a positive charge on nitrogen, it is possible to conclude that an N—H bond of an amide is more acidic than an N—H bond of an amine.

Reactions

Amides can be hydrolyzed by reactions similar to the hydrolysis of esters; however, they are generally much more resistant to hydrolysis than esters. This has important biological consequences because the stability of the amide functional group contributes to protein stability.

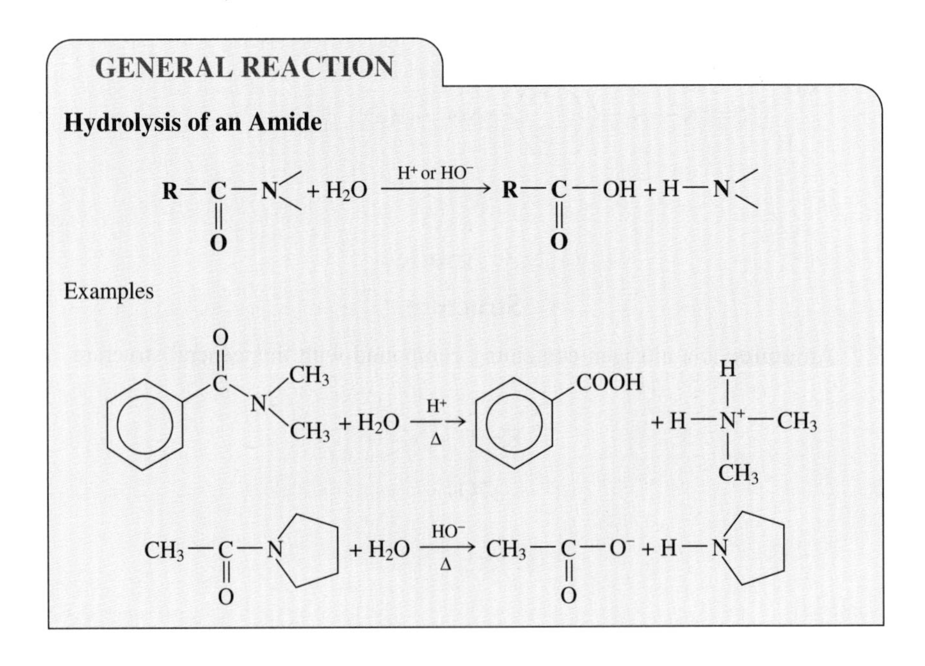

In basic hydrolysis, the products are the free amine and the salt of the acid, whereas in acidic hydrolysis the products are the salt of the amine and the free acid.

Amides as Polymers

We have mentioned polymers in preceding chapters. The amide bond is often involved in natural polymeric materials (proteins) and also in synthetic polymers. One of the best known synthetic polymers, *nylon*, is composed of amide linkages. Nylon is made commercially by heating a derivative of adipic acid with 1,6-diaminohexane.

Nylon is a synthetic polymer formed by connected chains of amides.

$$\left[\begin{array}{l}\overset{\text{O}}{\overset{\|}{\text{C}}}-CH_2-CH_2-CH_2-CH_2-\overset{\text{O}}{\overset{\|}{\text{C}}}-\underset{\text{H}}{\underset{|}{\text{N}}}-CH_2-CH_2-CH_2-CH_2-CH_2-CH_2-\underset{\text{H}}{\underset{|}{\text{N}}}-\end{array}\right]_n$$

The strength of nylon results from hydrogen bonding between chains, much in the same way as we shall see in the chemical behavior of proteins.

$$\left[\underset{\text{O}}{\underset{\|}{\text{C}}}-CH_2-CH_2-CH_2-CH_2-\overset{\text{O}}{\overset{\|}{\text{C}}}-\underset{\text{H}}{\underset{|}{\text{N}}}-CH_2-CH_2-CH_2-CH_2-CH_2-CH_2-\underset{\text{H}}{\underset{|}{\text{N}}}-\right]_n$$

Hydrogen bond → H ⋮ O

$$\left[\underset{\text{O}}{\underset{\|}{\text{C}}}-CH_2-CH_2-CH_2-CH_2-\overset{\text{O}}{\overset{\|}{\text{C}}}-\underset{\text{H}}{\underset{|}{\text{N}}}-CH_2-CH_2-CH_2-CH_2-CH_2-CH_2-\underset{\text{H}}{\underset{|}{\text{N}}}-\right]_n$$

NEW TERM

amide A nitrogen-containing compound derived from an acid and an amine with the general structure

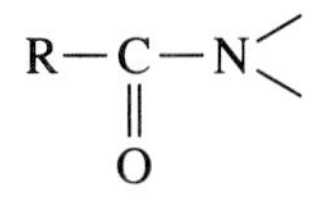

$$R-\underset{\text{O}}{\underset{\|}{\text{C}}}-N\!\!<$$

Nylon is formed at the interface where 1,6-diaminohexane (in the lower water layer) and a derivative of adipic acid (in the upper hexane layer) react. The Nylon can be drawn out and wound on a stirring rod.

TESTING YOURSELF

Amide Structure and Reactivity

1. Which of the two resonance structures in Figure 8.4 (the one on the left or right) is the more important contributor to the overall structure and reactivity of the amide?
2. Show products from the following hydrolysis reactions of amides:

a.

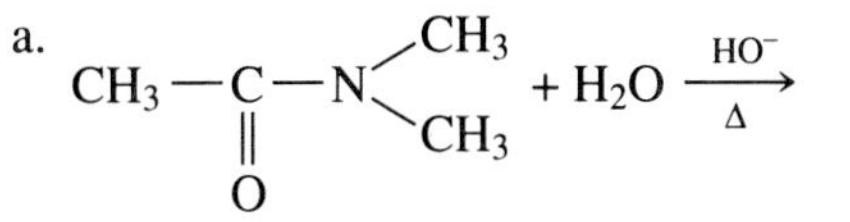

$$CH_3-\underset{\text{O}}{\underset{\|}{\text{C}}}-N\begin{matrix}CH_3\\CH_3\end{matrix} + H_2O \xrightarrow[\Delta]{HO^-}$$

b.

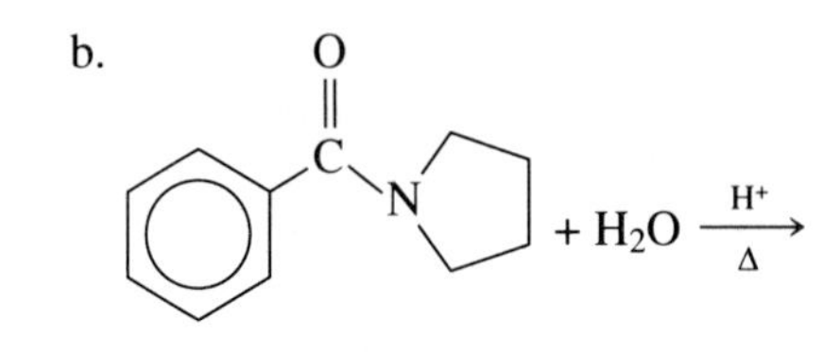

Answers **1.** The one on the left is more important because it is neutral.

2. a. **b.**

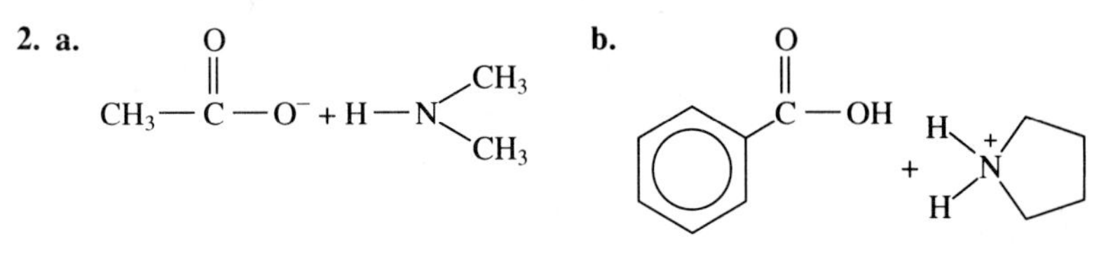

8.4 Special Amines

Amino Acids

As the name implies, these **amino acids** have *both* an amino and a carboxyl group. The chemistry of amino acids will be discussed in detail in Chapter 12. They are introduced at this point to show that they are special amines. Most of the natural amino acids are alpha amino acids.

$$\begin{array}{c} R-CH-COOH \\ \quad | \\ \quad NH_2 \end{array}$$

An alpha amino acid

Alkaloids

♦ *Alkaloids* are examples of biologically active natural basic products.

Many of the best-known natural products can be classified as **alkaloids,** which at one time were defined as nitrogen-containing basic compounds obtained from plants. There is some debate as to whether this definition is too limited because a number of basic natural products are now known from animal sources. Nevertheless, compounds in this very diverse family have been used as drugs (quinine for malaria), pest poisons (strychnine), analgesics (morphine), and stimulants (caffeine). In the following pages, we discuss many examples of this important group of special amines.

Quinine, obtained from the bark of the cinchona tree, has been used as an effective remedy for malaria for hundreds of years. Long before the structure of quinine was known, quinine bark was used to relieve symptoms of malaria. Who first discovered this use will never be known, but folk medicines have long been an important source in finding active medicinal ingredients. Malaria is the most widespread infectious disease in the world and is caused by a protozoan (*Plasmodium* sp.) transmitted through mosquito bite. Quinine effectively destroys the parasite. Today there are many synthetic analogs of quinine.

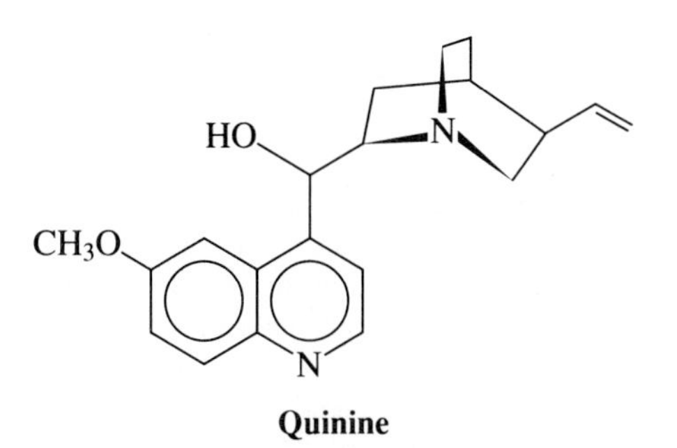

Quinine

Strychnine has long been known as a deadly killer. It is a complex alkaloid with an extremely bitter taste. It was once common to "tonics" that helped to increase appetites; other, less toxic compounds are now used for this purpose. It is a common ingredient in rodent poisons.

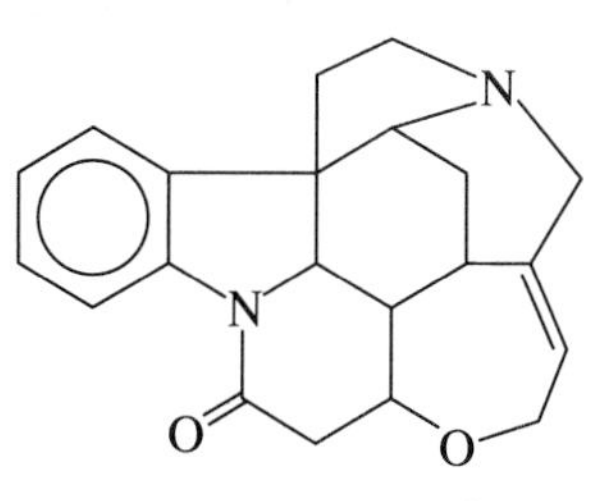

Strychnine

Morphine, one of the major alkaloids of the oriental poppy plant, is one of the most effective pain killers known. It is also very addictive. Until the early 1900s, opium (the crude latex that contains morphine and codeine) was found as a common ingredient in many commercial medicines. Indeed, it is estimated that by 1900 there were about 250,000 legally addicted people in the United States.

Another natural opiate is *codeine*, which is used for cough medicines. It can also be addictive, and its use is only by prescription. Another derivative of morphine is diacetylmorphine or *heroin*, which is much more addictive and is a major problem as a street drug. Chemists have maintained an active interest in preparing compounds that have the pain-killing ability of morphine without its addictive effects. One result of this work is meperidine, best known by the trade name Demerol. *Methadone* is used as a substitute for heroin in drug-treatment programs.

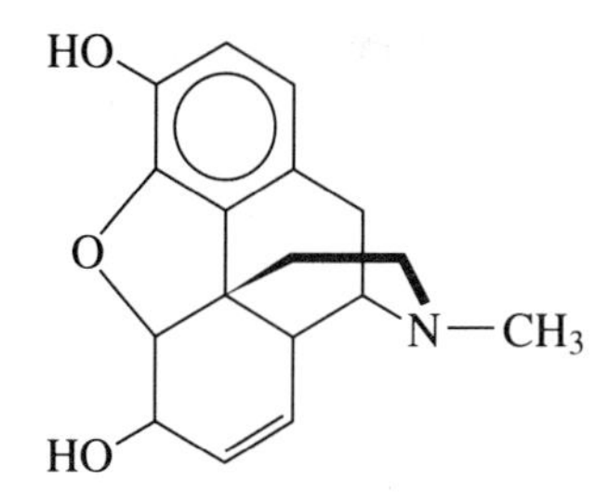

Morphine

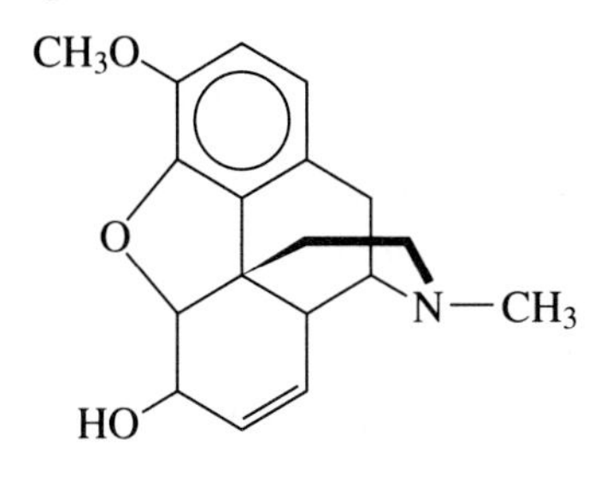

Codeine

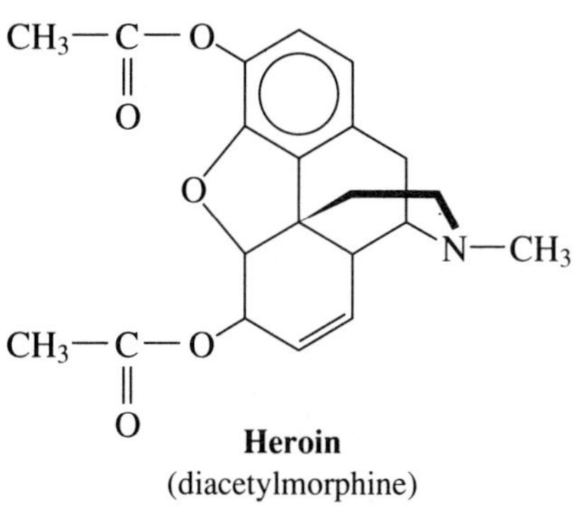

Heroin
(diacetylmorphine)

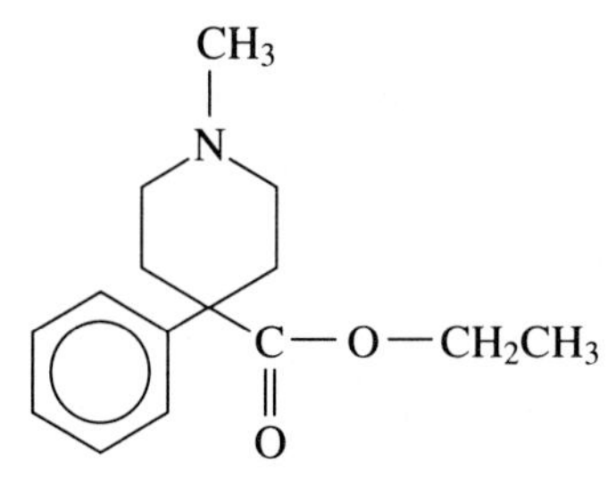

Demerol
(meperidine)

Methadone

CHEMISTRY CAPSULE

Ergotism

The fungus *Claviceps purpurea* parasitizes growing kernels of a number of cereal grains. As the fungus takes over the plant, it produces a number of alkaloids. These are called ergot alkaloids. Ergotism (a generic term describing the effects of ingesting ergot alkaloids) often causes gangrene by causing a decrease in blood circulation in the extremities. Another symptom is epileptic-like convulsions known as St. Anthony's fire. Thousands died of ergot poisoning during the Middle Ages. Infected cereal grains, used by local bake shops, were the reason for many outbreaks of ergotism. In A.D. 994, more than 40,000 people died from a single outbreak.

Hallucinations as a result of ingestion of these alkaloids were common. Ergotism has been used to explain the witchcraft thought to be practiced by a number of young women in New Salem in the late 1690s.

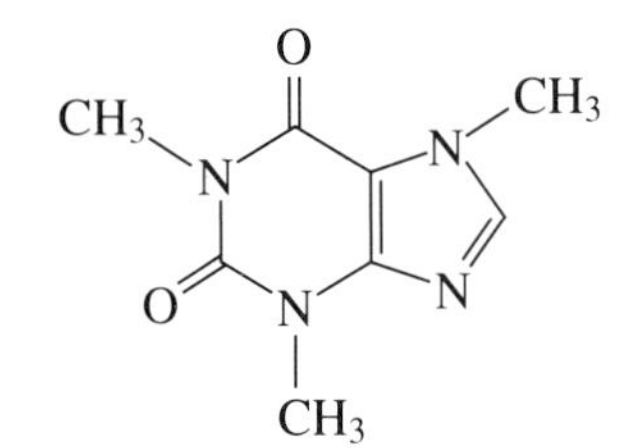

Caffeine

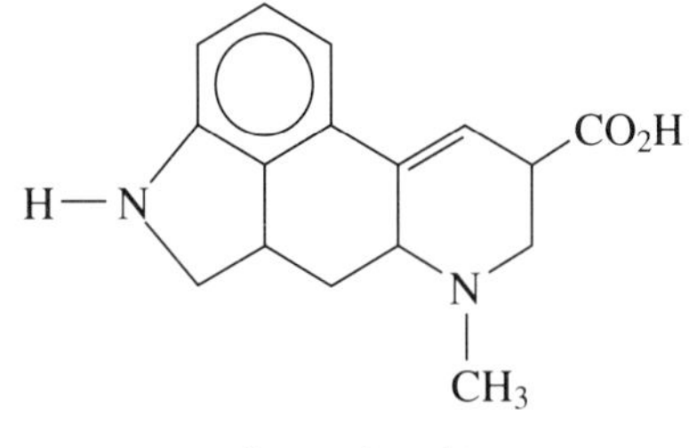

Lysergic acid

Caffeine is a stimulant found in coffee, tea, and many cola drinks. Millions of people begin their days with coffee to "get started." Caffeine affects the central nervous system by stimulating the cerebral cortex, resulting in more rapid and clearer thinking and improved coordination. It is also used to prevent some migraine headaches by acting as a vasoconstrictor.

Lysergic acid is an alkaloid produced on cereal grains that are infected with a certain fungus called *ergot*. Many derivatives of this alkaloid are produced by this fungus. *LSD* ("acid") is a synthetic derivative of lysergic acid and is a potent hallucinogen (a drug that changes perception). As little as 20 μg (micrograms) can result in noticeable effects. Although the questions of long-term health problems as a result of taking this drug are not fully resolved, there is no question that the possibility of a "bad trip" is very real. Under the influence of LSD, serious mental feelings of impending death, insanity, and paralysis are common. Also, users of LSD have reported "flashbacks"—recurrences of the feelings associated with taking the drug even when the drug is not used. These effects have been known to cause serious mental stress.

Phencyclidine (PCP, or "angel dust") is one of the common street drugs often distributed as LSD. Its effects are quite different, though, more closely resembling alcoholic intoxication. However, feelings of schizophrenia are common, and flashbacks can recur unpredictably for months. Significant numbers of deaths have been reported due to use of PCP. It has also had some use in veterinary medicine as an analgesic and anesthetic.

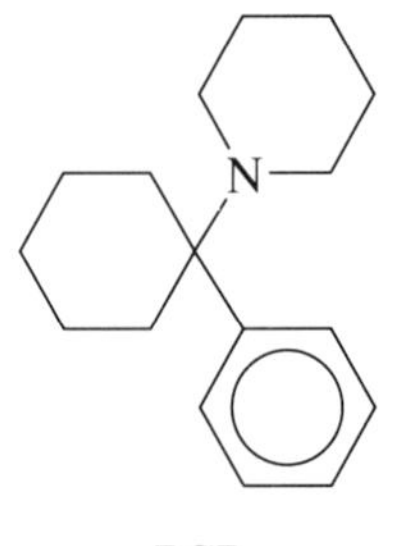

PCP

CHEMISTRY CAPSULE

Crack

"Crack" is a solid form of cocaine that became the major drug problem in the 1980s. Heating this substance and inhaling the vapors is said to cause such an intense response that many users become addicted after their first exposure. Unlike many drugs, the effects of crack are very short-lived, and to maintain the feelings obtained, a person has to have more and more in ever-increasing amounts. A recent essay on the drug sums it up in a most graphic way:

*These are the two Americas. No other line you can draw is as trenchant as this. On one side, people of normal human appetites for food and sex and creature comforts; on the other, those who crave only the roar and crackle of their own neurons, whipped in a frenzy of synthetic euphoria.**

---

* *Newsweek*, Nov. 28, 1988, p. 64.

Atropine and cocaine are structurally similar but have markedly different activities. *Atropine* (from the plant Deadly Nightshade: *Atropa belladonna*) is very toxic, but dilute solutions of this alkaloid cause dilation of the eye pupil and find some use in eye examinations. Atropine has also been used as an antidote for opium poisoning.

Cocaine is the active compound in the coca shrub. The leaves of the plant have been chewed by the natives of Peru as a stimulant. Since it was known to have a numbing effect on the lips and gums, there was early interest in its use as an anesthetic. However, the addictive nature of this drug and its resulting adverse side effects have limited its medical applications. Synthetic cocaine analogs, such as procaine (Novocain) and lidocaine (Xylocaine), have now received much more acceptance.

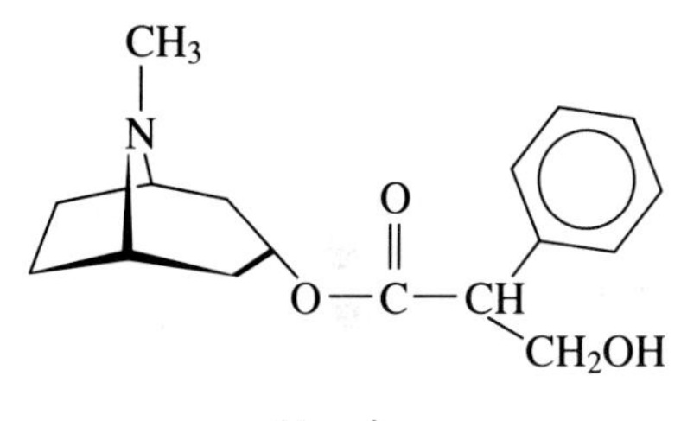

Atropine

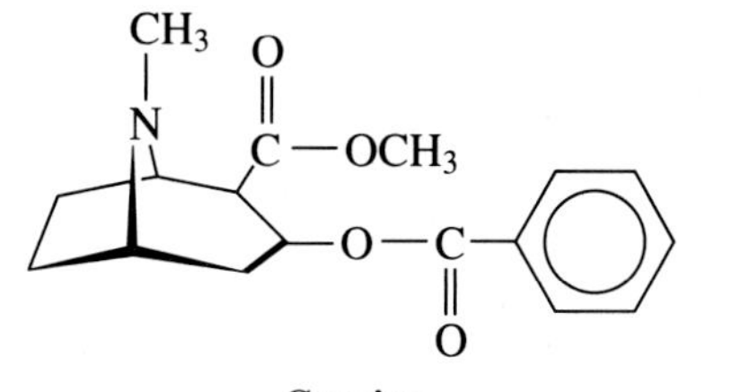

Cocaine

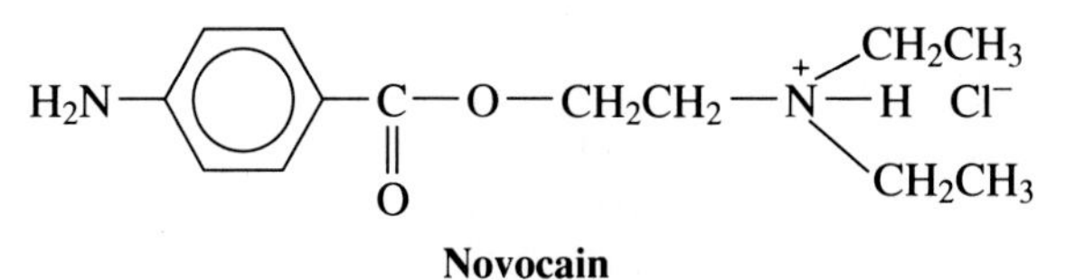

Novocain
(procaine)

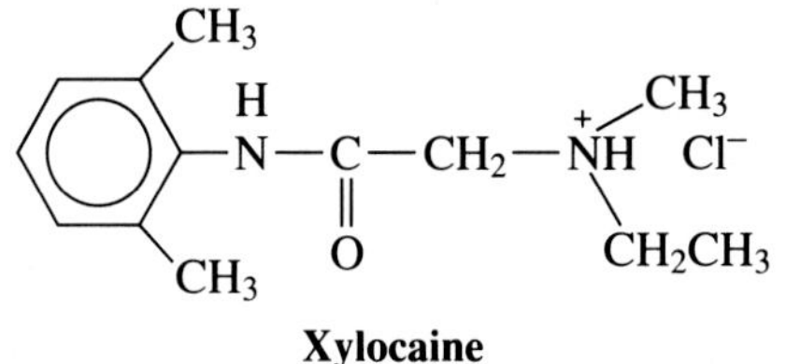

Xylocaine
(lidocaine)

♦ Amine groups are often converted to salts to increase water-solubility.

One of the major alkaloids found in tobacco, *nicotine*, has recently been described by the surgeon general as an addictive drug.

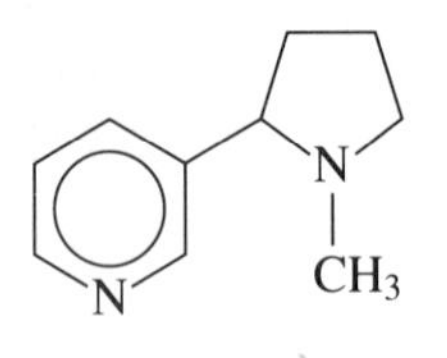

Nicotine

The most common alkaloid is *nicotine*. Nicotine is the major alkaloid constituent of tobacco, a violent toxin. Solutions of nicotine are used as an insecticide. In small doses, nicotine can act as a stimulant and it is addictive. In larger amounts it can cause convulsions, nausea, and death.

Many other diverse structural types are associated with the class of natural products called the alkaloids. Additional examples of interesting alkaloids are given in Table 8.4.

Amines Related to Epinephrine

Amines related to epinephrine are all based on a similar skeleton:

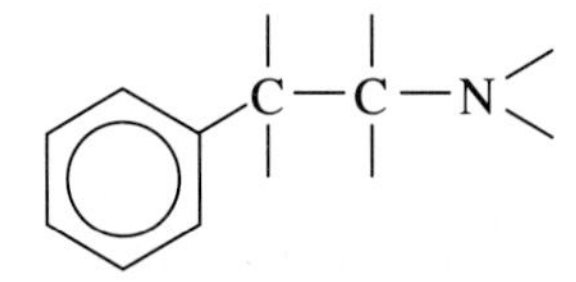

Epinephrine (also known as *adrenaline*) is a hormone produced in the adrenal gland. It is the "fight-or-flight" hormone that prepares the body for instant action. The norepinephrine (*nor* means "loss of methyl") is the hormone that controls blood pressure.

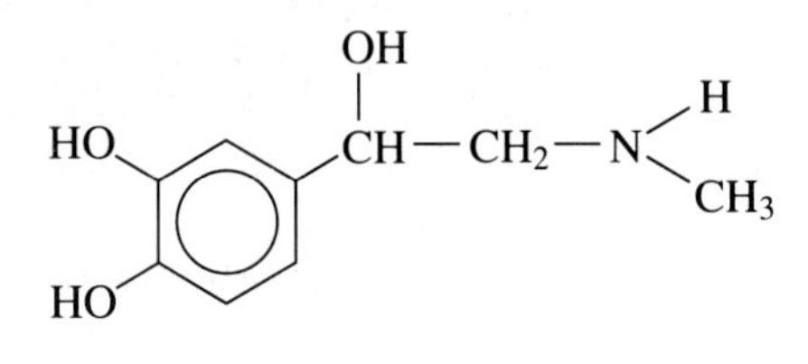

Epinephrine

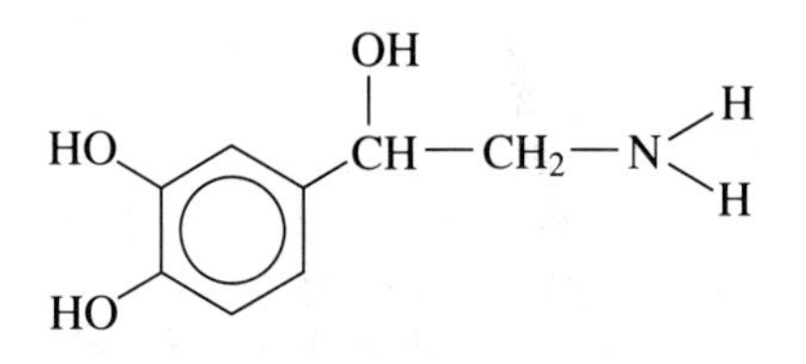

Norepinephrine

A number of compounds have been prepared that mimic some of the activity of these hormones. These molecules have powerful physiological activities and are illicit drugs. The parent amphetamine ("an upper"), methamphetamine ("speed"), and methoxyamphetamine ("STP") are examples of these hormone mimics.

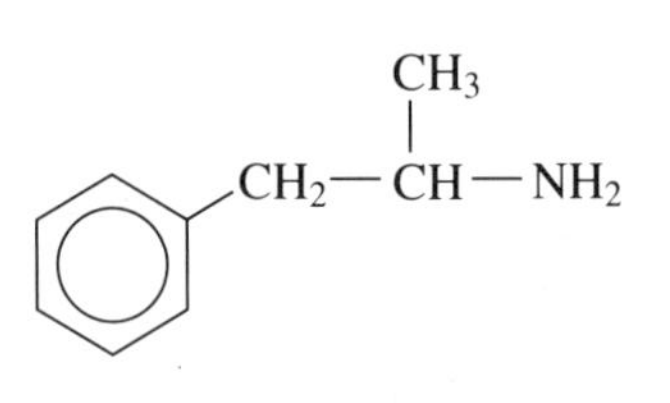

Amphetamine

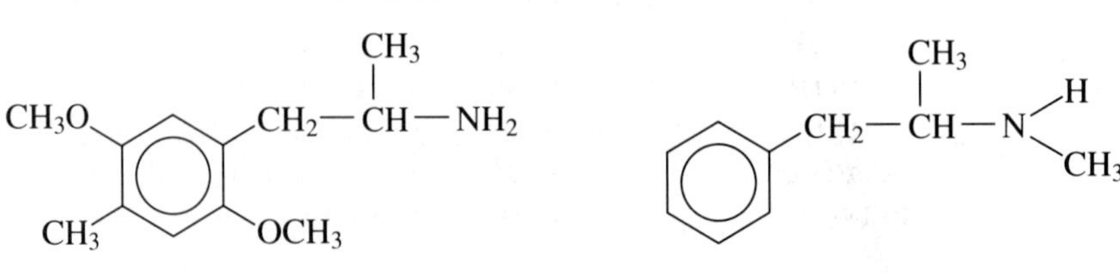

Methoxyamphetamine

Methamphetamine

TABLE 8.4

Examples of Important Alkaloids

Name	Description	Structure
Emetine	Active ingredient of ipecac, used as an emetic (to induce vomiting in cases of poisoning)	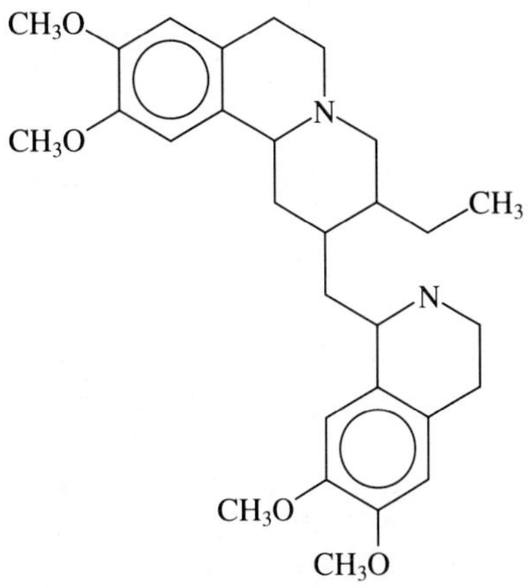 CH_3O, CH_3O, N, CH_3, N, CH_3O, CH_3O
Ergotamine	Used in treatment of migraine headaches	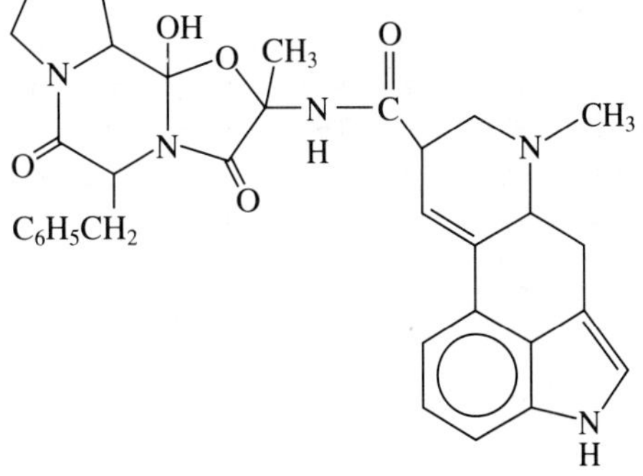 N, OH, O, CH_3, O, N—C, H, N, CH_3, O, O, $C_6H_5CH_2$, N H
Tubocurarine	Active ingredient in curare (South American arrow poison); also used in surgery as a muscle relaxant	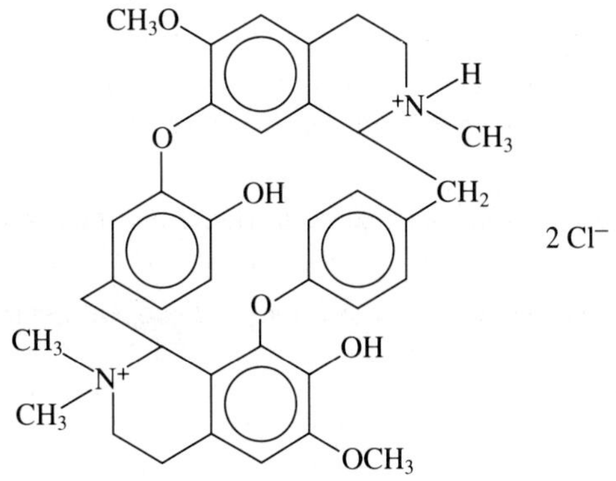 CH_3O, ^+N, H, CH_3, O, OH, CH_2, $2\ Cl^-$, O, CH_3, N^+, CH_3, OH, OCH_3

(continued)

TABLE 8.4 *(continued)*

Name	Description	Structure
Psilocybin	Hallucinogen; active ingredient of the mushroom *Psilocybe mexicana*	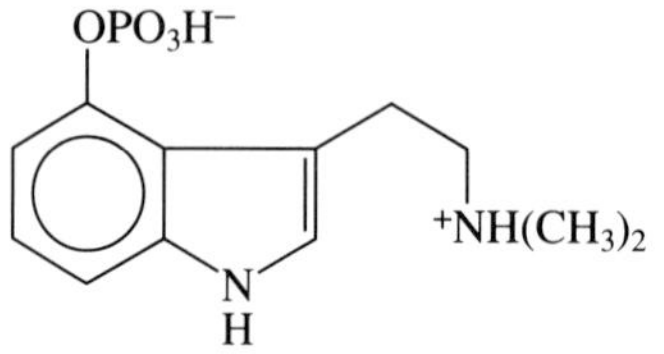
Dopamine	A neural transmitter; a lack of this compound results in the spastic motions and tremors associated with Parkinson's disease	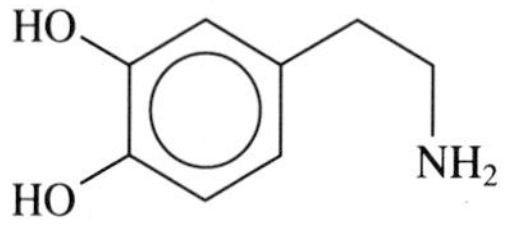
Serotonin	A vasoconstrictor present in platelets; also thought to be involved in neural mechanisms of sleep and sensory perception.	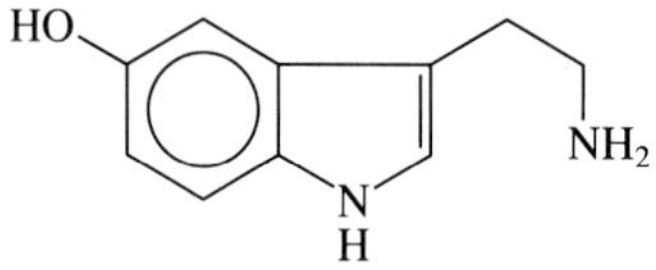
Reserpine	Used to treat hypertension and as a tranquilizer for the emotionally disturbed.	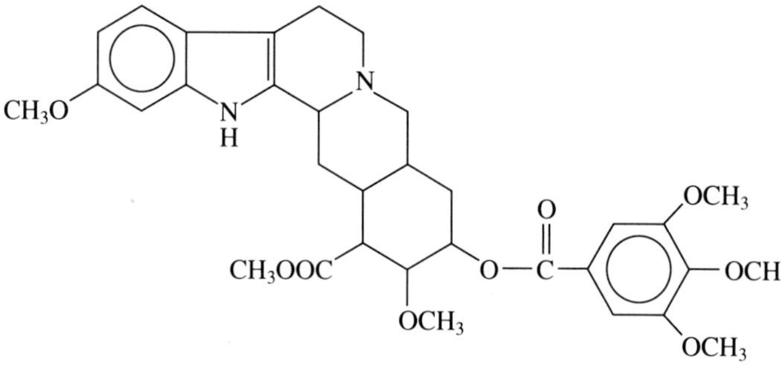

Important Heterocyclic Compounds Used in Medicine

We have mentioned that heterocyclic molecules are very important. Many are structurally complex. However, there are a large number of important drugs that are composed of these complex heterocyclic molecules. In Table 8.5 a few of the representative heterocyclic molecules are shown.

CHEMISTRY CAPSULE

Barbiturates

Barbiturates are a group of synthetic molecules that have important application in medicine as sedatives and hypnotics. The basic arrangement of atoms is derived from the reaction of a malonic acid ester with urea. A number of drugs are related to this general structure. Representative examples are listed in the table below.

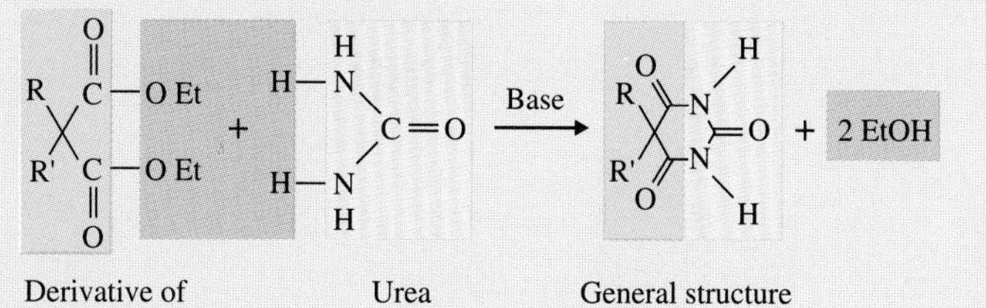

FIGURE 8.A
Barbiturates are formed by condensing malonic acid derivatives with urea.

Name	Drug* and Description	R-Group	R′-Group
Barbital	Veranol, a hypnotic and sedative	Ethyl	Ethyl
Phenobarbital	Luminal, an anticonvulsant, hypnotic, and sedative	Ethyl	Phenyl
Secobarbital	Seconal, a hypnotic	$CH_2=CH-CH_2-$	$CH_3CH_2CH_2CHCH_3$
Butalbital	Sandloptal, a sedative	$CH_2=CH-CH_2-$	$CH_3-CH-CH_2-$ $\vert$ CH_3
Amobarbital	Amytal, a sedative	Ethyl	$CH_3-CHCH_2CH_2-$ $\vert$ CH_3
Pentobarbital	Nembutal, a hypnotic and sedative	Ethyl	$CH_3CH_2CH_2CHCH_3$ $\vert$

* Names in this column are proprietary names.

The sedative properties of these drugs vary, ranging from the long-lasting sedative properties of barbital and phenobarbital to the intermediate sedative properties of secobarbital and pentobarbital. One of the most potent, fast-acting drugs is sodium pentothal. This molecule differs slightly in structure from the other barbiturates by the inclusion of a *thiocarbonyl*.

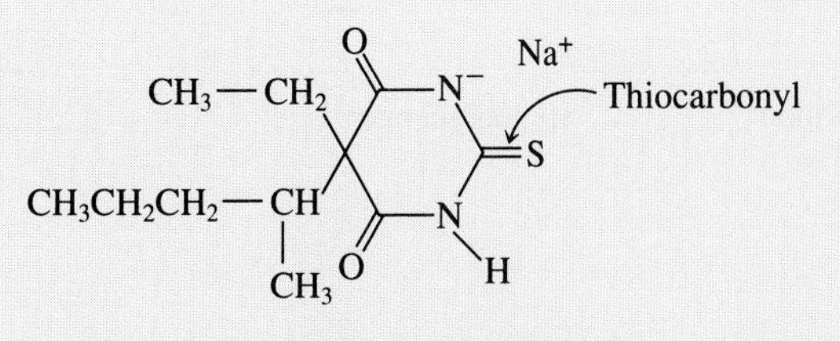

Sodium pentothal

Barbiturates are addictive drugs. Because people can build a tolerance to these drugs, a dosage-to-effect ratio is difficult to maintain. The continual need to increase the dose to obtain a given response can lead to serious overdoses and even death. In combination with alcohol, the barbiturates are very dangerous compounds.

TABLE 8.5

Examples of Important Heterocyclic Amines

Name	Description	Structure
Diphenhydramine (Benadryl)	Antihistamine	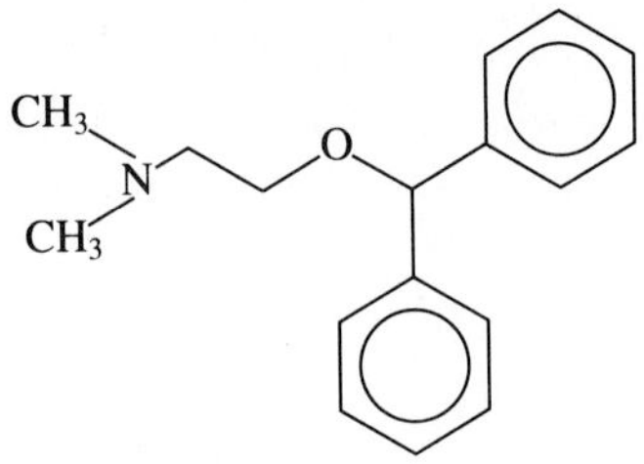
α-Propoxyphene (Darvon)	Analgesic	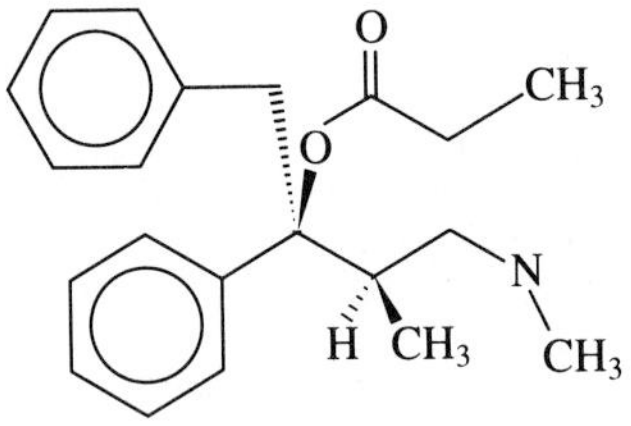
Methyldopa (Medopren)	Antihypertensive	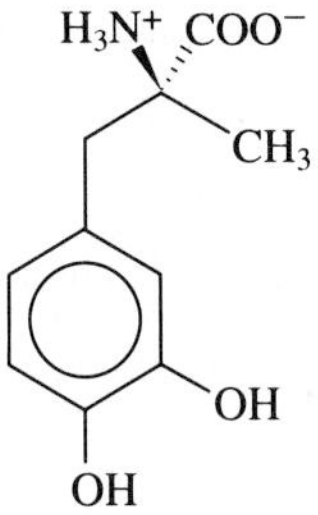
Phenylephrine (Neo-Synephrine)	Adrenergic, mydriatic, and decongestant	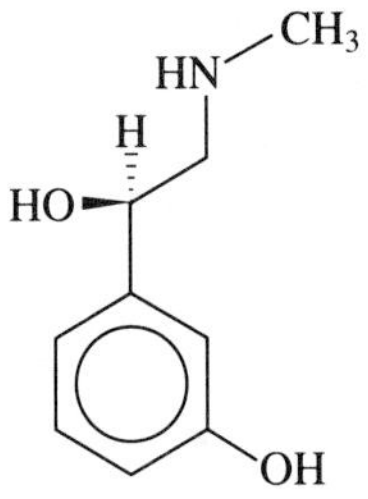
Tolbutamide (Orinase)	Hypoglycemic	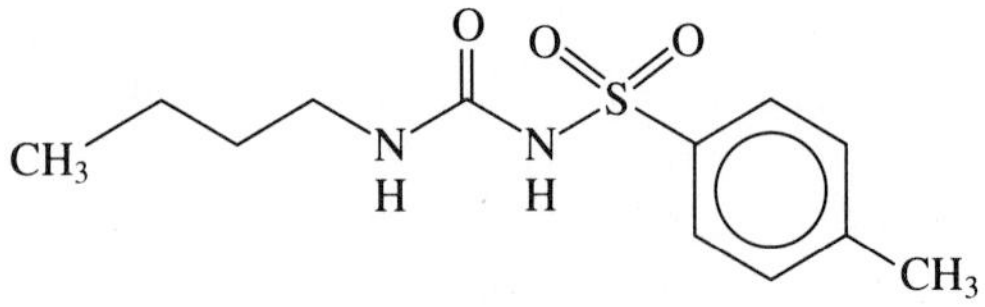
Meprobamate (Miltown)	Tranquilizer and anticonvulsant	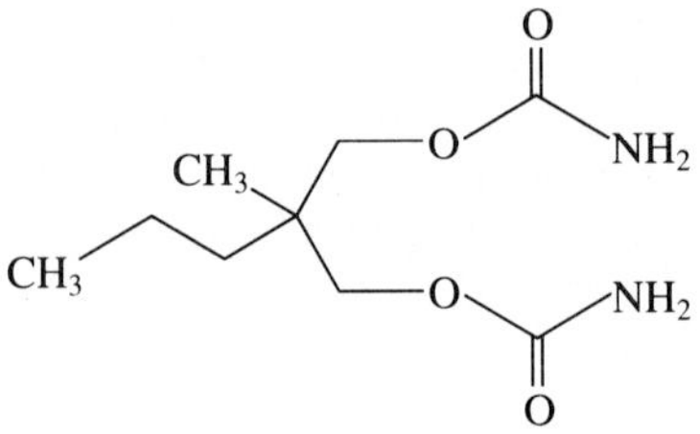

TABLE 8.5 (*continued*)

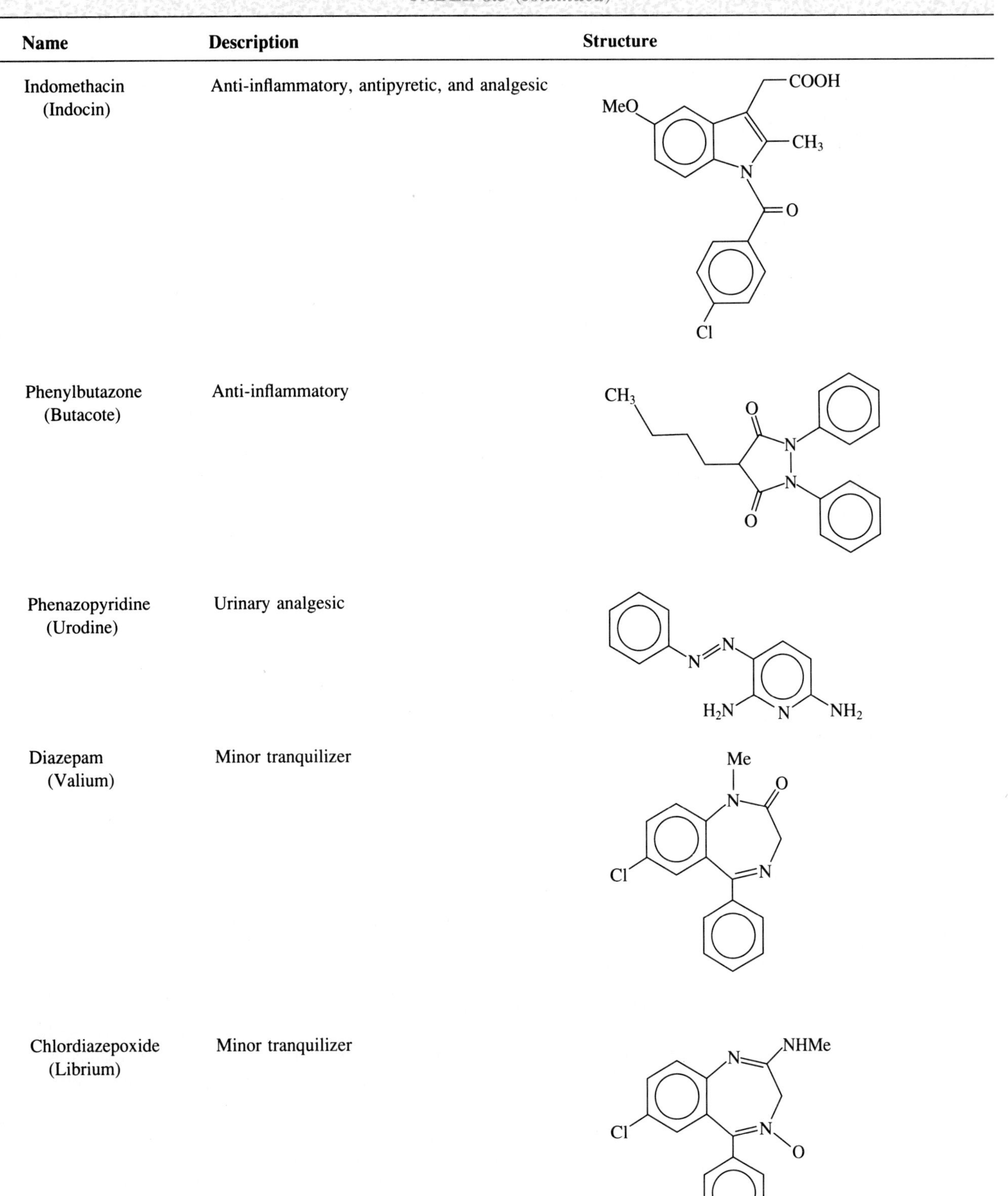

Name	Description	Structure
Indomethacin (Indocin)	Anti-inflammatory, antipyretic, and analgesic	
Phenylbutazone (Butacote)	Anti-inflammatory	
Phenazopyridine (Urodine)	Urinary analgesic	
Diazepam (Valium)	Minor tranquilizer	
Chlordiazepoxide (Librium)	Minor tranquilizer	

(continued)

TABLE 8.5 (*continued*)

Name	Description	Structure
Penicillins	If R = $PhCH_2$; Benzylpenicillin = penicillin G (usual fermentation product) If R = $PhCH_2O$: Phenoxymethyl-penicillin = penicillin V If R 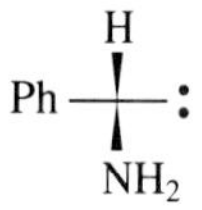 = Ph—C(H)(NH₂)— : Ampicillin	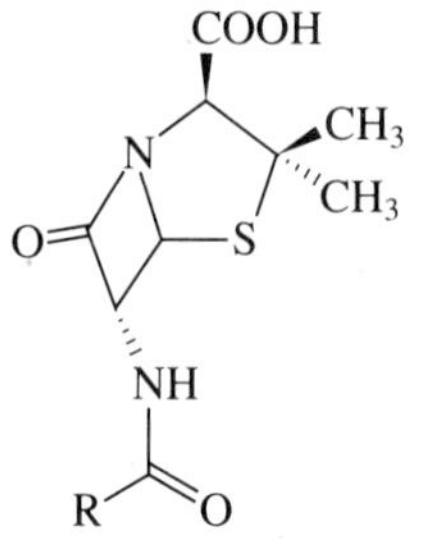 COOH, N, CH_3, CH_3, O, S, NH, R, O

NEW TERMS

amino acid Important molecules containing both an amine functional group and a carboxylic acid functional group. Most of the important natural amino acids are alpha amino acids.

alkaloid A naturally occurring, basic, nitrogen-containing compound

TESTING YOURSELF

Special Amines

1. Based on your understanding of solubility, why would you predict that amino acids are water-soluble?
2. You find a rare plant in an isolated region of the world and learn that the local people use this plant as a medicine. If you think an alkaloid is responsible for the biological activity, how would you get the alkaloids out of the plant?

Answers **1.** Since amines are basic and acids are acidic, you expect these two functional groups to undergo acid–base chemistry. An acid and a base react to form a salt. Many salts, being ionic, are water-soluble. **2.** Since alkaloids are basic, grind the plant up and extract with acid. The basic alkaloids form water-soluble salts.

Summary

The *amines* are an important functional group. As a result of the electron lone pair on nitrogen, amines are basic. The lone pair also makes it possible for an amine to act as a nucleophile. Amines undergo a variety of reactions as a consequence of the nitrogen lone pair. Reactions include acid–base chemistry, nucleophilic substitution with organohalogen compounds, reaction with carbonyl compounds to form *imines*, reactions with hydrazine to form hydrazones, and reactions with acid derivatives to form amides.

Amides are important derivatives of amines, and the amide bond is central to the structure and chemical behavior of proteins. The amide bond is also an important structural feature of nylon.

Alkaloids constitute a diverse and interesting class of basic natural products. Many of these have dramatic pharmacological activity.

Reaction Summary

An Amine as a Base Reacts with Acids

$$-\overset{|}{\underset{|}{N}} + HX \longrightarrow -\overset{|}{\underset{|}{N^+}}-H + X^-$$

An Amine with an Alkyl Halide

$$R-N\langle + -\overset{|}{\underset{|}{C}}-X \longrightarrow R-N^+\langle -\overset{|}{\underset{|}{C}}- + X^-$$

An Amine with an Acid Anhydride

$$R-\underset{\underset{O}{\|}}{C}-O-\underset{\underset{O}{\|}}{C}-R + H-N\begin{matrix} R' \\ R''(H) \end{matrix} \longrightarrow R-\underset{\underset{O}{\|}}{C}-N\begin{matrix} R' \\ R''(H) \end{matrix} + R-\underset{\underset{O}{\|}}{C}-O-H$$

An Amine with a Carbonyl Group

$$\rangle C=O + H_2N-R \xrightarrow{-H_2O} \rangle C=N-R$$

A Hydrazine with a Carbonyl Compound

$$\rangle C=O + H_2N-NH_2 \xrightarrow{H^+} \rangle C=N-NH_2 + H_2O$$

Hydrolysis of an Amide

$$R-\underset{\underset{O}{\|}}{C}-N\langle + H_2O \xrightarrow{H^+ \text{ or } HO^-} R-\underset{\underset{O}{\|}}{C}-OH + H-N\langle$$

Terms

amine (introduction)	amide (8.3)
heterocycle (8.1)	amino acid (8.4)
imine (8.2)	alkaloid (8.4)

Exercises

Structure and Physical Properties (Objective 1)

1. Why is methylamine a gas and methanol a liquid at room temperature?
2. Which would you expect to be more soluble in water, and why?
   a. *N*-propylamine
   b. ethylmethylamine
   c. trimethylamine

Nomenclature (Objective 2)

3. Give names for the following amines:

a. 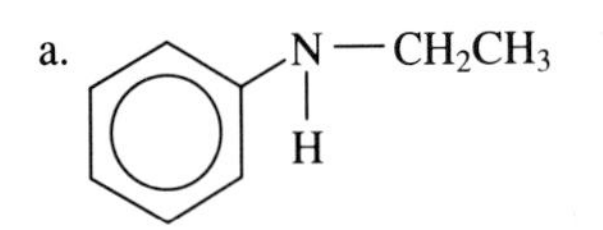

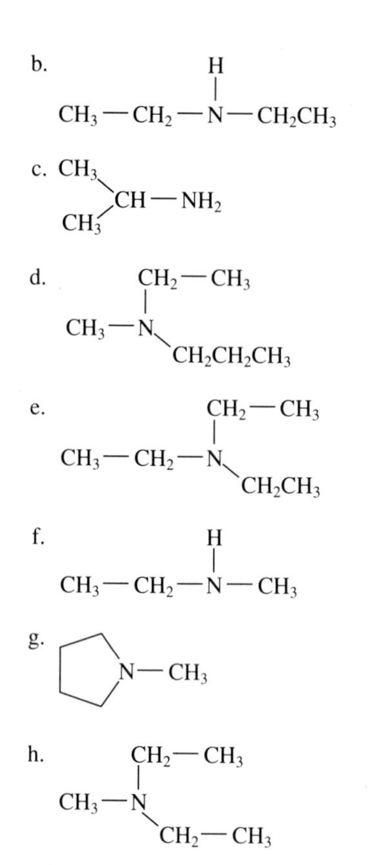

4. Give structures for the following named compounds:
   a. diisopropylamine
   b. isobutylmethylamine
   c. benzylamine
   d. triethylamine
   e. *N*-methylaniline
   f. pyrrole
   g. *N,N*-diethylaniline
   h. 3-(*N*-methylamino)hexane
   i. 1,6-diaminohexane
   j. *o*-bromoaniline

Reactions (Objective 3)

5. Give products for each of the following reactions:

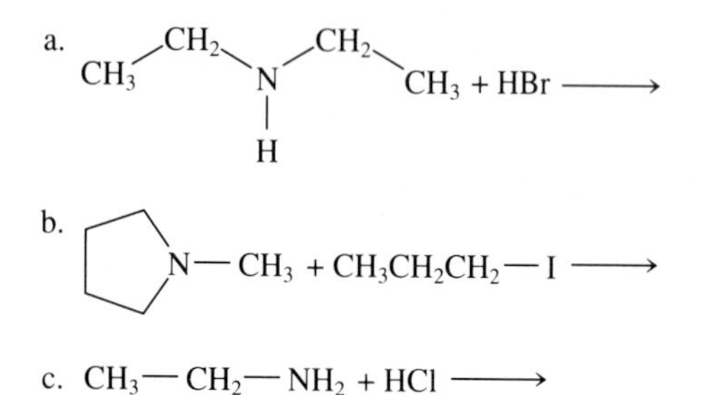

d. (quinolizidine, N) $+ HBr \longrightarrow$

e. $CH_3—N(H)—CH_3 + CH_3—Br \longrightarrow$

f. $CH_3—NH_2 + Cl—CH_2—C_6H_5 \longrightarrow$

g. (pyrrolidine)N—$CH_2—C_6H_5 + Br—CH_2—CH_2—CH_3 \longrightarrow$

h. $CH_3CH_2I + C_6H_5—CH_2—NH_2 \longrightarrow$

i. (piperidine)N—$CH_3 + HCl \longrightarrow$

j. (isoindoline)N—$CH_3 + CH_3Br \longrightarrow$

k. $C_6H_5—C(=O)—H + H_2N—N(H)—C_6H_3(NO_2)—NO_2 \xrightarrow{H^+}$

l. $CH_3—C(=O)—O—C(=O)—CH_3 + H—N$(pyrrolidine) $\longrightarrow$

6. Give products for each of the following reactions:

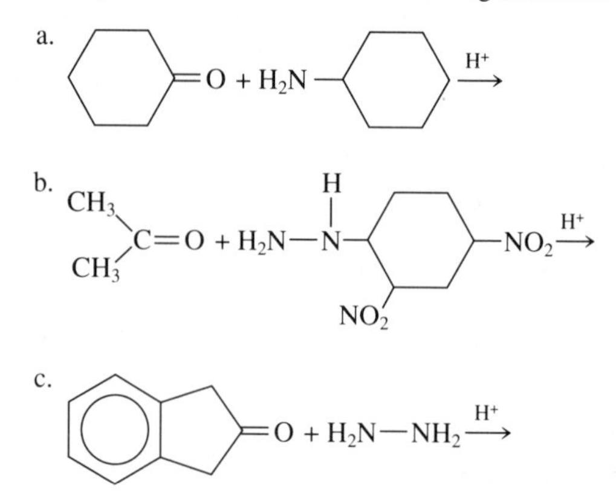

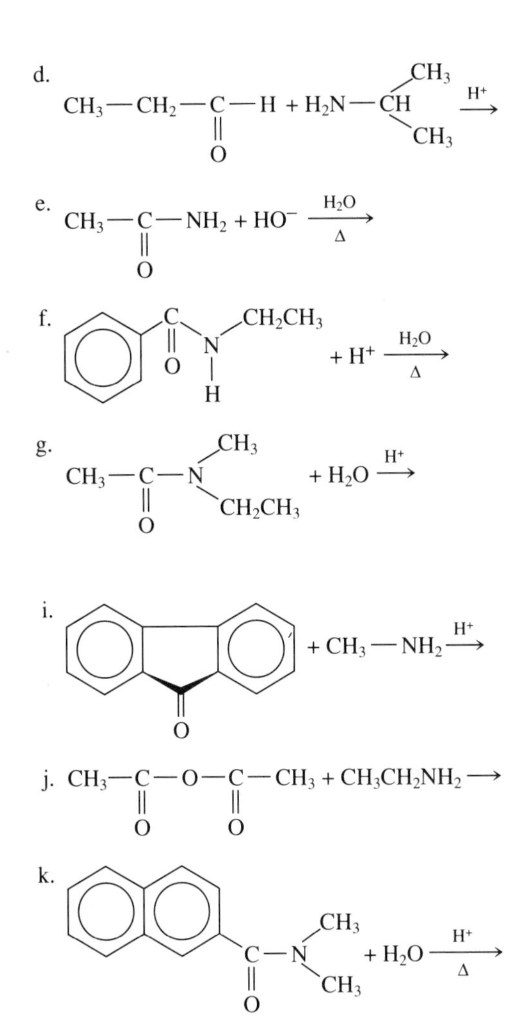

Structure and Properties (Objective 1)

7. Why is methylamine considered a Lewis base?
8. Give a Lewis electron-dot structure for methylamine.
9. Why does methylamine have a lower boiling point than methanol?
10. What are the characteristics that make a molecule a Lewis base?
11. Explain why aniline is considerably less basic than cyclohexylamine.
12. Why would an alcohol have a higher boiling point than an amine with the same number of carbon atoms?

Special Amines

13. Give the general structure for an alpha amino acid.
14. What is an alkaloid?

Amides

15. Why is an amide less basic than an amine?
16. Give products for the following reactions of amides:

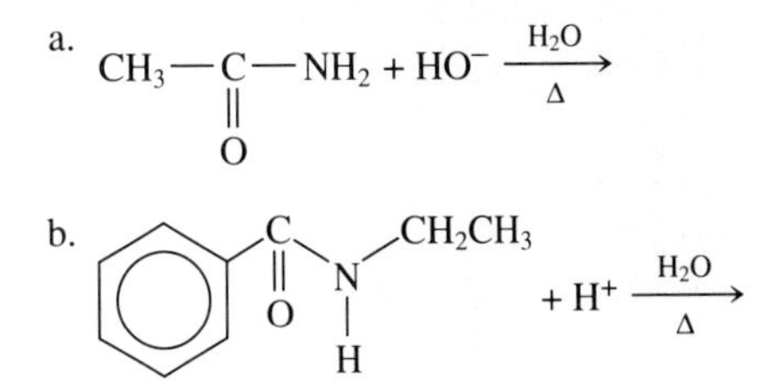

Unclassified Exercises

17. The best name for

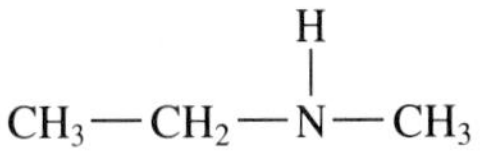

$$CH_3{-}CH_2{-}\overset{\displaystyle H}{\overset{|}{N}}{-}CH_3$$

is
a. ethylhydromethylamine
b. 2-nitrobutane
c. *N*-methylethylamine
d. 2-aminobutane

18. The hybridization of nitrogen in an amine is
a. *sp* b. sp^2 c. sp^3 d. sp^4

19. In the reaction: $CH_3{-}NH_2 + CH_3{-}Cl \longrightarrow$: the amine acts as
a. an electrophile c. an alkylating agent
b. a nucleophile d. a base

20. Amines are generally classified as
a. weak acids c. strong acids
b. weak bases d. strong bases

21. The reaction of an acid anhydride with methylamine yields
a. an amide b. an acid salt c. an imine d. no reaction

22. A hydrazone results from the reaction of hydrazine with
a. an acid b. an aldehyde c. an alcohol d. a phenol

23. An amine exhibits ______ hydrogen bonding than an alcohol.
a. more b. the same c. less

24. Reduction of an imine gives an
a. acid b. amide c. amine d. alcohol

25. The reaction of

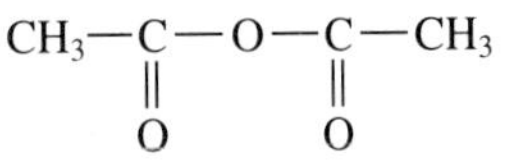

$$CH_3{-}\underset{\displaystyle O}{\underset{\|}{C}}{-}O{-}\underset{\displaystyle O}{\underset{\|}{C}}{-}CH_3$$

with $CH_3{-}NH_2$ gives
a. an acid + amine c. an amide + acid
b. an amide + amine d. no reaction

26. A secondary amine is
a. methylamine c. ethylmethylpropylamine
b. ethylmethylamine d. dimethylethylamine

INTRODUCTION TO BIOCHEMISTRY

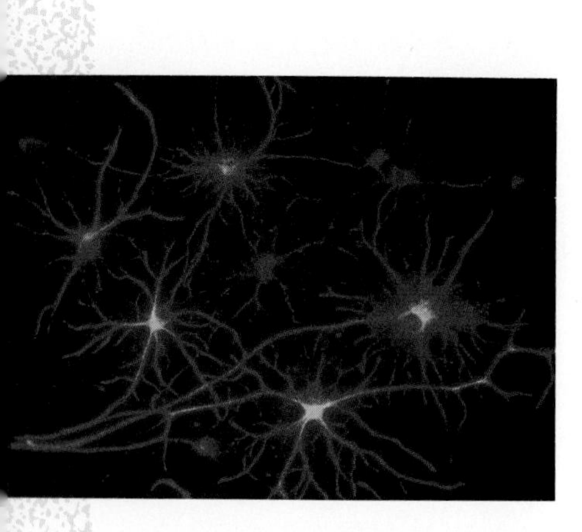

Cells like these astrocytes are the basic unit of life. Most of the reactions of the body occur within cells.

OUTLINE

9.1 Introduction to Cells
9.2 Procaryotic Cells
9.3 Eucaryotic Cells
9.4 Biomolecules
9.5 General Principles of Nutrition

OBJECTIVES

After completing this chapter, you should be able to

1. Describe a cell and explain its role in organisms.
2. Name the organelles of a procaryotic cell and describe their function.
3. Name the organelles of a eucaryotic cell and describe their function.
4. Describe the role chirality plays in the interaction of biomolecules.
5. Define the term *nutrition* and name the major groups of nutrients.

9.1 Introduction to Cells

From the simplest organism to the most complex, one structural feature remains constant: all living organisms contain one or more cells. The **cell** is the basic structural unit of life, and every cell, given the proper environment, appears to be capable of living in the absence of other cells. Most of the nutrients absorbed by the human body are used by cells, and much of the chemistry that occurs within the body takes place within cells. Cells contain a wide variety of substances dissolved or dispersed in water, but they also contain specialized structures called **organelles**. Organelles carry out many cellular functions. Some knowledge of cell structure will provide a framework for understanding biochemistry.

◆ The cell is the smallest unit of a living organism that can live independently.

Cell Size

With the exception of some egg cells, most cells are rather small, having dimensions that range from a micrometer to tens of micrometers. There appears to be some lower and upper limits to cell size. Typical cell volume is about 10^{-9} mL (10^{-6} μL). Why are cells not smaller? The most likely reason is the requirement for some minimum number of compounds and some minimum number of copies of these compounds that must be present to sustain life. But why then are cells not larger? Movement of cell contents depends primarily upon *diffusion* of the reactants through the cell medium. Diffusion must be fast enough to maintain the proper concentration of the reactants in the various parts of the cell. As a cell gets larger in size, diffusion cannot supply the various reactants fast enough.

◆ Cell size is limited to a certain range by several physical and molecular factors.

Aside from this restriction imposed by diffusion rates, a second factor restricting cell size is the movement of materials into and out of the cell. Nutrients must be able to enter the cell, and waste products must be able to leave. This transport is dependent upon the surface area of the cell. But as cells grow larger, surface area increases more slowly than volume. Consider a cube as an example: the cube's surface area is related to its length squared; the cube's volume is its length cubed. If a cell is too large, the surface area is too small to provide adequate transport of nutrients and wastes.

What is the best size for a cell? There is no single correct answer, but a cell must be large enough to hold the needed cell constituents, yet small enough to allow for ready cellular diffusion within the cell and to have a favorable surface-area-to-volume ratio for movement of materials into the cell.

Cells in Multicellular Organisms

The cell is the basic structural unit of life. Cells of different organisms show far more similarity than differences. If cells from multicellular and single-celled organisms are compared, both cell types lie within the same size range, both have many of the same molecules, and both carry out many of the same reactions. The major differences between multicellular organisms and single-celled ones is the number of cells and the specialization of cells. Multicellular organisms are not simply collections of independent cells.

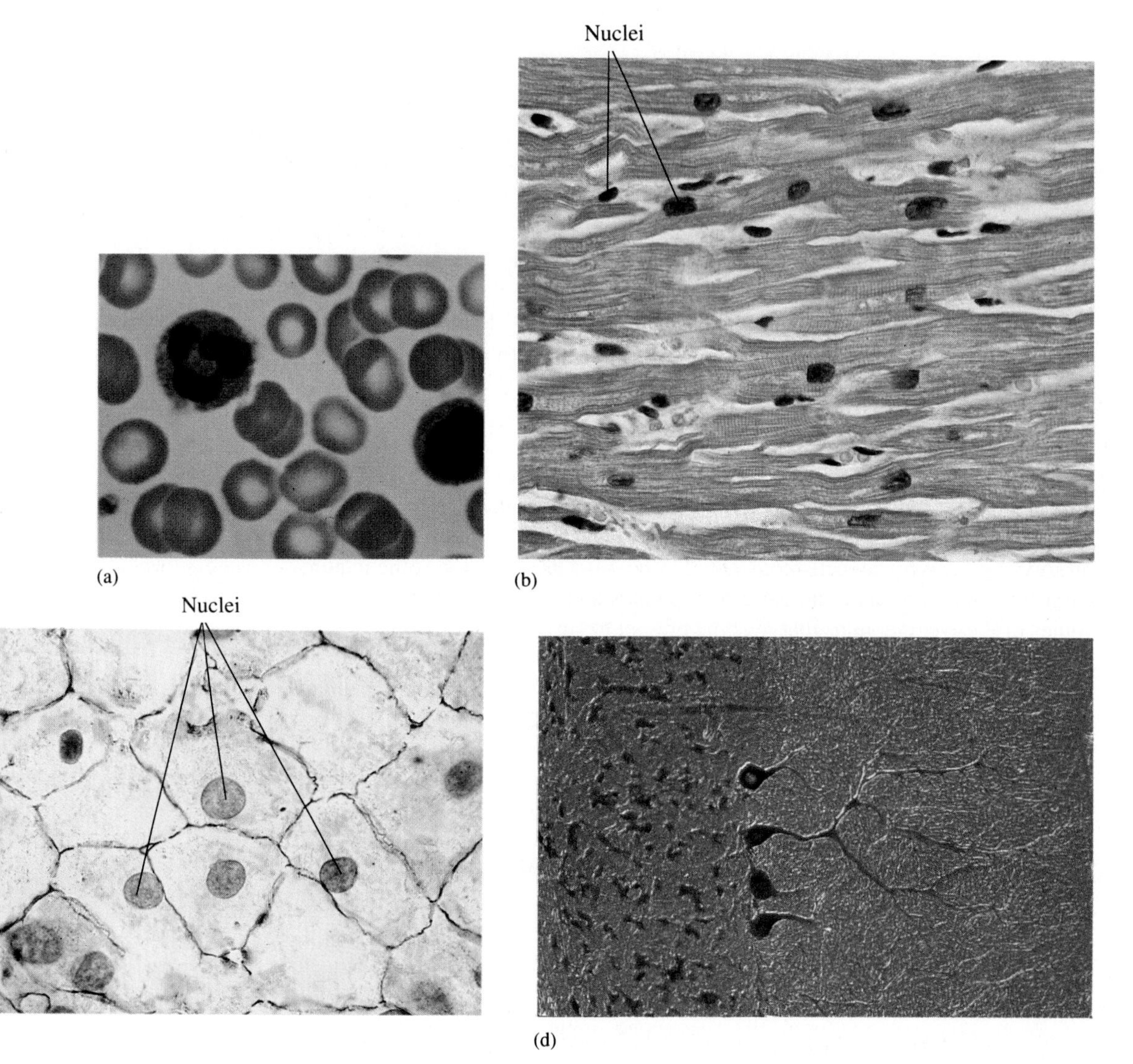

FIGURE 9.1
The cell is the basic unit of life. These are some representative cells of the human body. (a) Blood cells. (b) Cardiac muscle cells. (c) Simple squamous epithelium cells. (d) Cells of nerve tissue of the cerebellum.

There are many different kinds of cells in the human body (Figure 9.1), and these cells are arranged into various tissues, organs, and organ systems. There is significant interaction between the cells of the body. Many of these interactions are molecular, thus biochemistry provides insight not only into the workings of cells, but also into how the body functions as a whole.

NEW TERMS

cell The basic structural unit of life; the smallest part of a living organism that can carry out life processes by itself

organelle A structure within a cell that carries out one or more functions of the cell

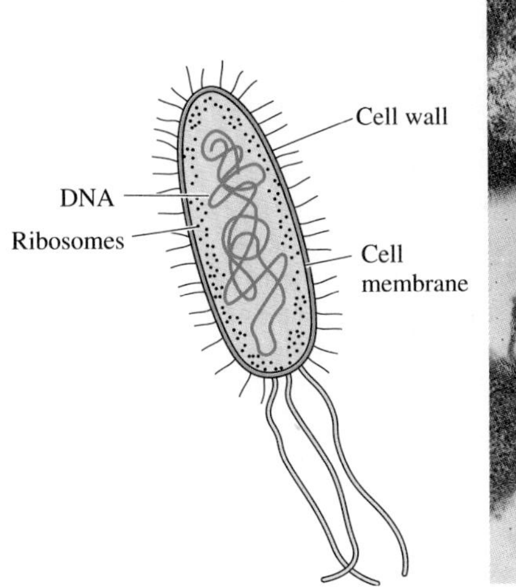

(a)

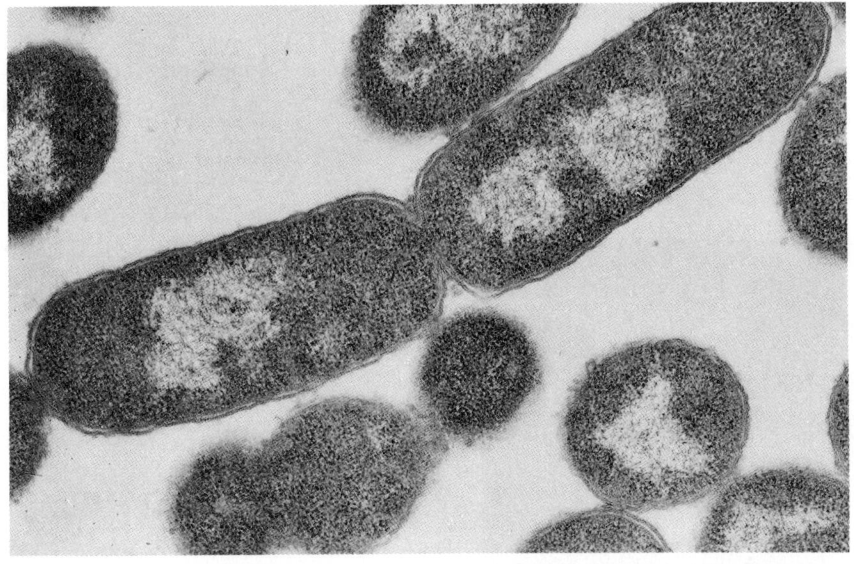

(b)

FIGURE 9.2
The bacterial cell is a typical procaryotic cell. The procaryotic cell is the simplest cell type. (a) Schematic drawing of a procaryotic cell. (b) Electron micrograph of bacterial cells.

9.2 Procaryotic Cells

Bacteria are **procaryotes**, simple one-celled organisms whose cells distinctly differ from the cells of higher organisms, the *eucaryotes*. The procaryotic cell does not contain a true nucleus and is both smaller and structurally simpler than the eucaryotic cell. The structural differences between bacterial cells and animal cells provide a basis for the effectiveness of a number of antimicrobial drugs. These drugs kill or retard growth in bacteria (procaryotic cells) but not in eucaryotic cells because they affect a particular bacterial cell structure or component that is uniquely different from its counterpart in the animal cell. Several of these differences will be pointed out during the discussion of bacterial cell structure.

◆ Antibiotics used in medicine are effective because they adversely affect bacteria with little or no damage to human cells.

The major structural components of the procaryotic cell include the cell membrane, the cytosol, the cell wall, and in some species, an outer layer beyond the cell wall (Figure 9.2). The **cell membrane** separates the contents of the cell from the external environment. The cell membrane has a large surface area, but it is only a few molecules thick. It is not a strong, rigid structure, nor does it just act as a passive barrier between the internal and external environment. The cell membrane is responsible for controlling the entry of nutrients and the departure of waste products. Many transport systems are located in the cell membrane.

In bacteria, the **cytosol** consists of the water, ions, and molecules that are separated from the exterior environment by the cell membrane. This is the internal environment of the cell. The cytosol makes up the bulk of the bacterial cell volume. With an electron microscope, two structures can be seen in the cytosol of procaryotes. One of them is a huge circular molecule called DNA (deoxyribonucleic acid). In bacteria, the DNA is sometimes called the *nucleoid*, or *nuclear region*. Also visible in the cytosol are **ribosomes**. These organelles are the site of protein synthesis, and numerous ribosomes are present in each cell. Procaryotic

TABLE 9.1
Some Common Antibiotics and Their Mode of Action

Antibiotic	Mode of Action in Bacteria
Cephalosporins	Disrupt cell-wall synthesis
Chloramphenicol	Disrupts protein synthesis
Erythromycin	Disrupts protein synthesis
Novobiocin	Disrupts DNA synthesis
Penicillins	Disrupt cell-wall synthesis
Streptomycin	Disrupts protein synthesis

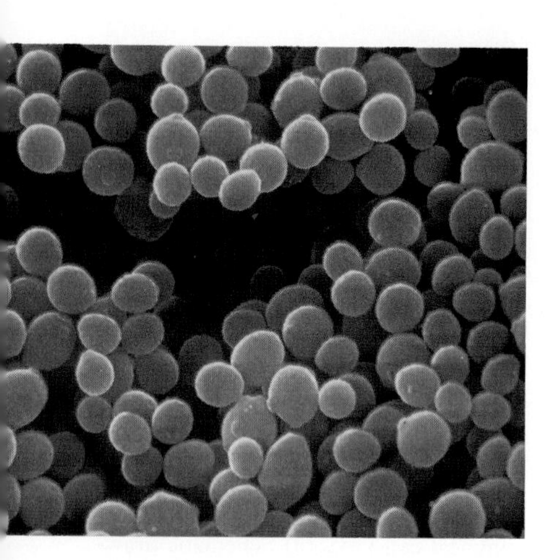

FIGURE 9.3
These gram-positive staphylococci are quite vulnerable to penicillin because penicillin disrupts normal cell wall synthesis.

♦ Cell walls protect bacteria from bursting in hypotonic solutions.

♦ Penicillin was the first antibiotic used in humans. Penicillin and its derivatives are still highly useful and effective.

DNA and ribosomes are structurally different from their eucaryotic counterparts. These differences can be exploited in treatment of bacterial infections. The antibiotic *streptomycin*, for example, affects bacterial ribosomes, resulting in disruption of bacterial protein synthesis (Table 9.1). The ribosomes in the cells of the invaded host organism are not affected.

The bacterial **cell wall** lies outside the cell membrane. It is actually one gigantic molecule that forms by bonding numerous smaller molecules together. This structure provides mechanical protection to the cell. If a bacterial cell were placed in a solution hypotonic to it, water would move into the cell via osmosis. The relatively weak cell membrane has little strength, so the membrane would expand. In the absence of a cell wall, it would expand indefinitely, then burst. The cell wall prevents cellular expansion that would rupture the cell membrane. Animal cells lack cell walls. A cell wall is not needed because the composition of the body fluids bathing the cells is maintained within a narrow range of concentrations.

The bacterial cell wall is another feature that can be exploited by antibiotics. Penicillins and cephalosporins are antibiotics that prevent the normal formation of cell walls in some bacterial species (Figure 9.3). Without normal cell walls, the bacteria perish. Except for allergic reactions in some people, penicillins are one of the safest of all antimicrobials because nothing in the host organism appears to be affected by them.

Some bacterial species possess an outer layer beyond the cell wall. This layer is composed of a material called *lipopolysaccharide*, which contains both lipids (nonpolar biomolecules) and carbohydrates. This outer layer is *pyrogenic* in mammals, that is, it causes fever. Even if the bacterium is dead, the lipopolysaccharide can cause adverse reactions in humans. The next time you are around sterile medical equipment such as syringes, note that the writing on the wrapper indicates that the contents are both sterile and nonpyrogenic. This certifies that no living microorganisms are present, and no pyrogens are present that could cause fever in the patient.

NEW TERMS

procaryote The simplest and smallest cell type, characterized by the absence of a true nucleus and other membranous organelles. Bacteria and blue-green algae are procaryotic organisms.

cell membrane Sheet-like structure that separates the interior of the cell from the external environment

cytosol The fluid interior of the cell, containing the ions and molecules that support life. Organelles are suspended in it

ribosome Small organelle that is the site of protein synthesis

cell wall Rigid structure that surrounds bacterial and plant cells. It protects the cell from rupturing in hypotonic solutions.

TESTING YOURSELF

Cells and Procaryotes

1. Cells fall within a certain size range. (a) Why are they not smaller than this? (b) Why are they not larger?
2. What is the function of the cell membrane of a cell?
3. Compare what would happen to a bacterial cell and an animal cell in a hypotonic solution.

Answers **1. a.** Cells must contain a minimum number of different kinds of biomolecules, and a minimum number of copies of each kind. **b.** Diffusion cannot provide nutrients fast enough if cells are too large. **2.** The cell membrane separates the cell from the external environment and regulates the passage of substances into and out of the cell. **3.** The animal cell would expand and probably burst due to osmotic pressure. The bacterial cell would not be harmed because the cell wall would prevent expansion of the cell membrane.

9.3 Eucaryotic Cells

The cells of plants and animals are **eucaryotic** cells. They are generally larger than procaryotic cells and have more structural complexity. All eucaryotes possess cytosol and a cell membrane. The cytosol contains ribosomes, but the ribosomes of eucaryotes are larger and differ somewhat in structure from the ribosomes of procaryotes. Some cells, like those of plants and algae, also possess cell walls and outer layers, but animal cells generally lack these features. Four major organelles are found in eucaryotes that are not found in procaryotes.

♦ Eucaryotic cells are larger and more complex than procaryotic cells.

1. Eucaryotes possess a true *nucleus*, the organelle in which DNA is located. In bacteria, the DNA is not located in a separate organelle but is simply suspended in the cytosol.
2. Eucaryotic cells also possess an *endoplasmic reticulum*, a complex of membranes that penetrates throughout the cytosol and is involved in synthesis of some biomolecules.

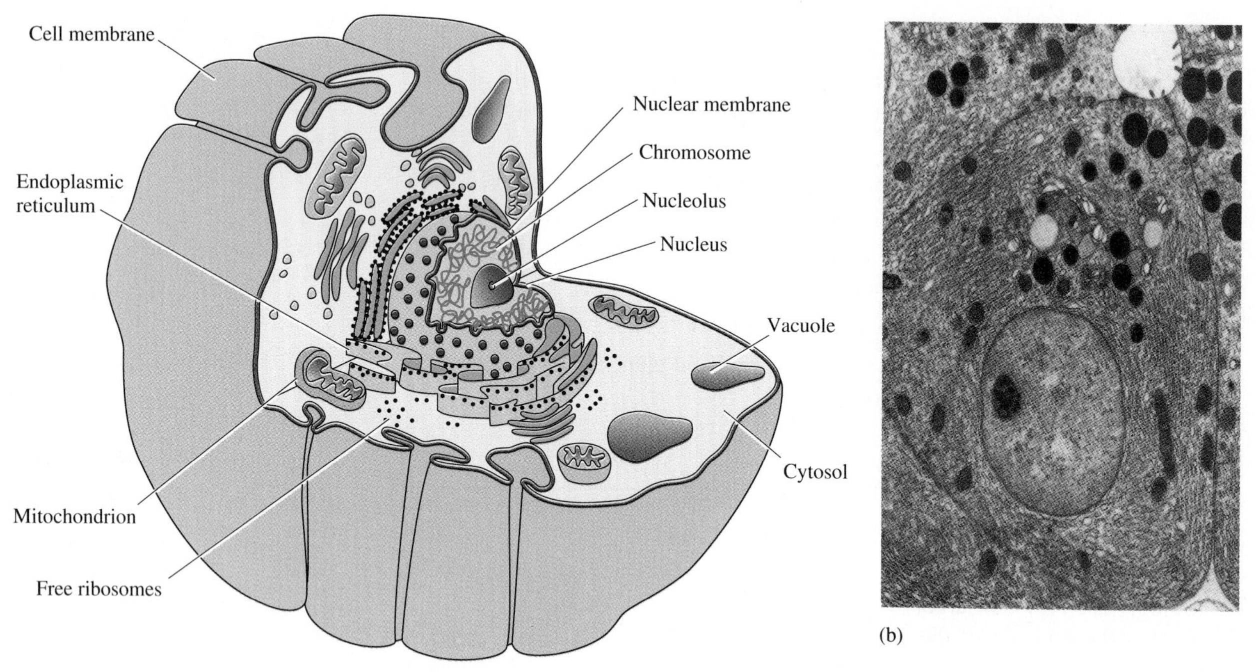

FIGURE 9.4
(a) This generalized eucaryotic cell illustrates the structures found in an animal cell. Eucaryotic cells are larger and more complex than procaryotic cells. (b) Electron micrograph of typical mammalian cell.

3. Eucaryotes possess *mitochondria*, organelles that are the site of aerobic energy production. Most of the useful energy that is produced in animal cells is produced in the mitochondria.
4. Most plants and algae possess *chloroplasts*, organelles that are the site of photosynthesis. When light is available, the chloroplast is the major site of energy production in the plant.

Figure 9.4 illustrates the general features of a eucaryotic cell.

The **nucleus** is the portion of the cell where information needed for cell growth, division, and maintenance is located. The nucleus is generally visible in cells prepared for viewing with a microscope. It is here that most of the DNA of the eucaryote is found. The nuclear DNA of eucaryotes is present as several pieces instead of the single molecule found in procaryotes. Each of these DNA molecules is associated with proteins in larger structures called *chromosomes*. Chromosomes are complex structures that contain genetic material. Also located in the nucleus is the *nucleolus*, which has a major role in the synthesis of ribosomes.

The **endoplasmic reticulum** is a complex network of membranes that permeates and divides the cytosol into various compartments (Figure 9.4). It plays a role in the synthesis and transport of a number of molecules and structures that are used inside and outside the cell. Much remains to be learned about this organelle.

The **mitochondrion** (plural, mitochondria) is a bacteria-sized organelle where much of the energy production of the cell takes place (Figure 9.5). Mitochondria are generally located where energy demand is highest in the cell. This organelle has two membranes: a relatively smooth outer membrane that separates the mitochondrion from the cytosol, and a highly folded inner membrane

(Figure 9.5). The folds of the inner membrane, which are called *cristae*, greatly increase the surface area of this membrane. Many of the proteins needed for aerobic energy production are located in this membrane. They carry electrons and are involved in the synthesis of ATP. The core of the mitochondrion is called the *matrix*. The matrix contains another group of enzymes that are involved in energy production, the enzymes of the tricarboxylic acid cycle. Curiously, mitochondria possess DNA and ribosomes that are similar to those of bacteria. Some antibiotics that affect bacteria are toxic to mitochondria.

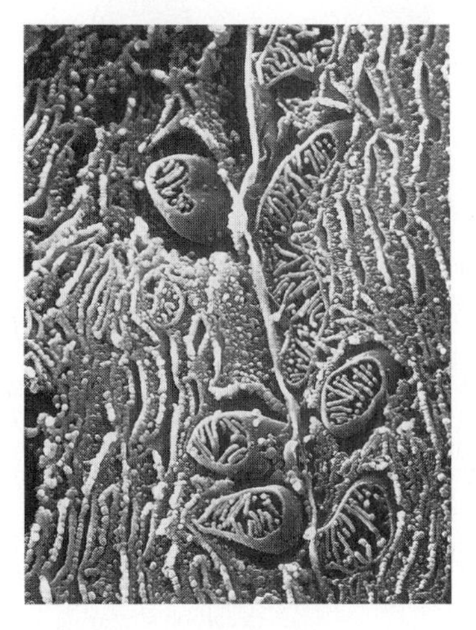

FIGURE 9.5
Mitochondria surrounded by endoplasmic reticulum. In eucaryotes, mitochondria are the sites of aerobic energy production. Structural similarities between mitochondria and bacteria have led many scientists to believe mitochondria evolved from ancestors of bacteria.

NEW TERMS

eucaryote Organism with cells that possess a true nucleus and a number of other membranous organelles that are not present in the procaryotic cell

nucleus Eucaryotic organelle that contains most of the DNA of the cell and directs cell division, growth, and maintenance of the cell

endoplasmic reticulum The membranous organelle within the eucaryotic cell where some of the cellular synthesis and transport of materials occurs

mitochondrion The organelle of eucaryotes where oxygen-requiring energy production occurs

TESTING YOURSELF

Eucaryotic Cells

1. Compare the structure of a procaryotic cell with that of a eucaryotic cell.
2. What role does the mitochondrion play in a cell?
3. Compare the location and organization of DNA in procaryotes and eucaryotes.

Answers **1.** Eucaryotes are larger and more complex with various membranous organelles and a true nucleus. Procaryotes are smaller and simpler, lack a nucleus, and have only a cell membrane that is membranous. **2.** Mitochondria are the sites of aerobic energy production and thus provide most of the useful energy in a eucaryotic cell. **3.** Most eucaryotic DNA is in the nucleus and is complexed with proteins to form highly organized structures called chromosomes. The DNA of procaryotes is naked and is not contained in a nucleus.

9.4 Biomolecules

The compounds found in living organisms do not differ fundamentally from the substances you have studied in the organic chemistry chapters of this book. They possess no vital force nor other mystic properties. Many of them are, however, considerably more complex than the relatively simple organic molecules you have studied so far.

Size and Functional Groups of Biomolecules

♦ The complexity of many biomolecules is due to their relatively large size, the presence of multiple functional groups, and their distinctive three-dimensional shapes.

Many biomolecules are quite large. Proteins typically have molecular weights in the tens to hundreds of thousands; DNA molecules often have molecular weights in the millions. Even some of the smaller biomolecules such as simple

(a)

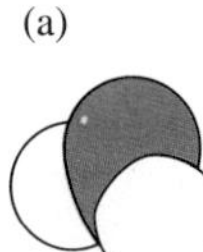
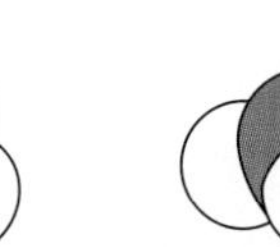
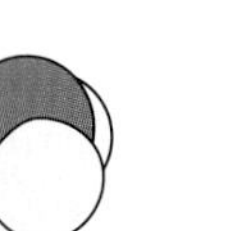
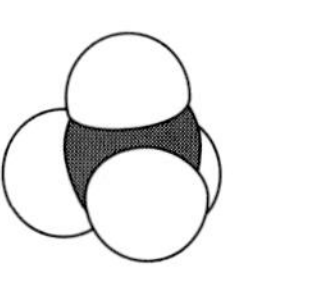
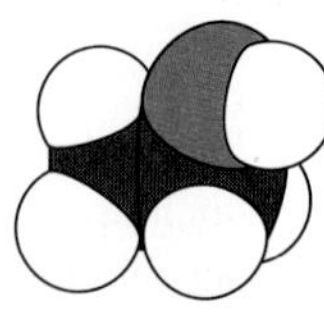

Water Ammonia Methane Ethanol

β-D-Glucose
(a carbohydrate)

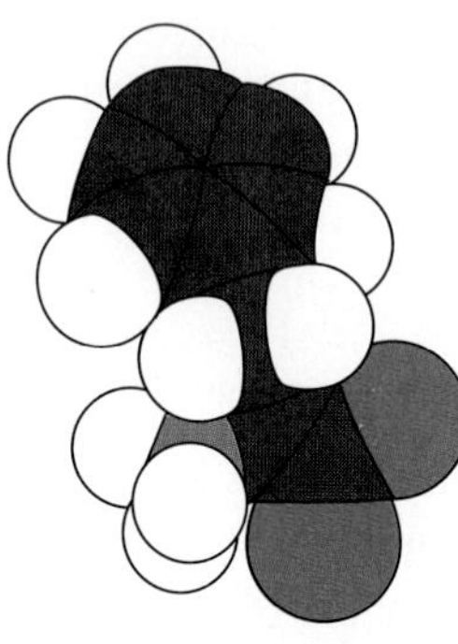

Phenylalanine
(an amino acid)

(b)

Water

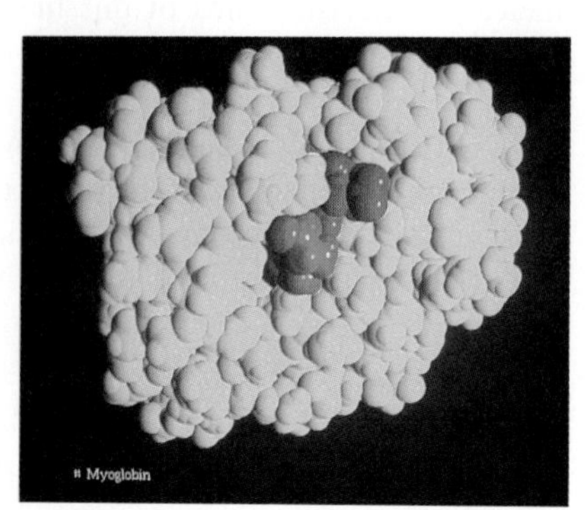

FIGURE 9.6
(a) These space-filling models illustrate that biomolecules are generally larger and more complex than the compounds you studied in general and organic chemistry. Several simple molecules are shown for comparison along the top. (b) A water molecule is shown next to the protein myoglobin to show relative size.

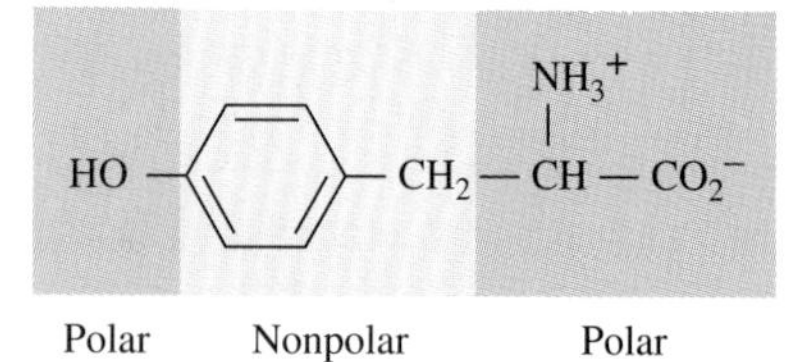

Polar Nonpolar Polar

FIGURE 9.7
This is the amino acid tyrosine. Although this is a small biomolecule, it has several functional groups and three regions of distinctly different polarity.

sugars and fats have molecular weights in the hundreds. To help visualize their relative sizes, Figure 9.6 shows space-filling models of several important biomolecules. Models of the more familiar compounds water, ammonia, methane, and ethanol are included for comparison.

Many biomolecules possess two or more functional groups within the molecule, thus they are a mosaic with respect to chemical and physical properties. Some parts of these molecules are nonpolar, whereas other regions are polar. Some regions react rapidly with certain other molecules, but other regions are inert. The size and multiple functionality of biomolecules allows them to interact with other biomolecules in important and unique ways. For example, look at Figure 9.7, which illustrates the amino acid tyrosine. Note that even in this small biomolecule, three different functional groups are present and two polar regions are separated by a nonpolar region within the molecule.

Stereoisomerism in Biomolecules

♦ Stereoisomers differ only in the three-dimensional arrangement of atoms in space.

Many biomolecules show a type of isomerism known as *stereoisomerism*. As with other isomers, **stereoisomers** have the same composition. Stereoisomers also have the same atoms attached to the same atoms, but they differ in the way the atoms are oriented in space. You have already seen examples of stereoisomerism in cis–trans isomerism in alkenes and ring compounds. These isomers have the same composition, and they have the same atoms attached to the same atoms, but the atoms have different orientations in space. (Make a pair of cis–trans isomers and examine them for these features.) The stereoisomerism seen in many biomolecules meets these same conditions, but it involves a different structural relationship.

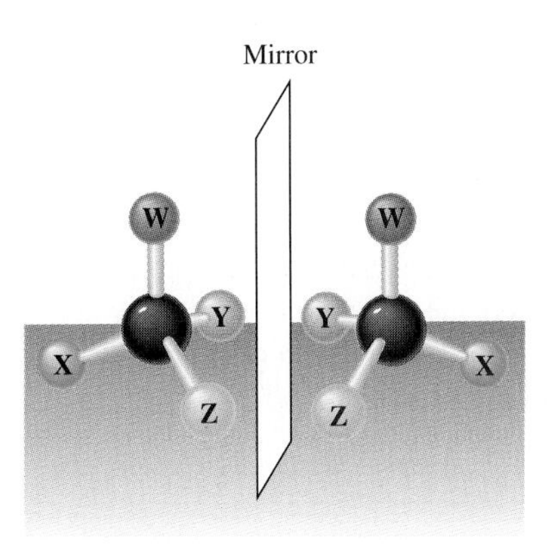

FIGURE 9.8
These two molecules are stereoisomers. They have the same atoms attached to the same atoms, yet they are not identical structures because their orientation in space is different.

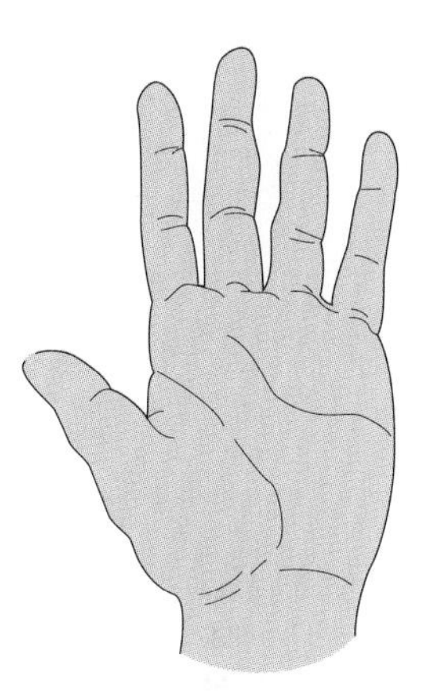

This structural relationship is illustrated with the two models shown in Figure 9.8. Each molecule has the chemical formula CWXYZ, and each has the atoms W, X, Y, and Z attached to the carbon (C) atom. But these two molecules are *not* simply copies of the same molecule. They are uniquely different. They are nonsuperimposable mirror images of each other.

These stereoisomers are related in the same way your hands are related to each other. Are your two hands identical? Look at the hand drawn in the margin. Is it a left hand or a right hand? You have no trouble distinguishing the hand in the margin even though your hands are very similar. If they were identical, you could not have identified the hand correctly except by chance. Each hand has five fingers, a palm, and a back; they have the same composition. The components are attached to the same places on each hand. Identical composition, same attachment, yet your hands are different. Try one more test with your hands. Try to superimpose one of them on the other, keeping both palms up. Superimposition simply means that the two forms can be mentally merged into one as the objects are brought together (Figure 9.9).

A pair of hands display **chirality**, that is, they are *chiral* objects. (The word "chiral" is derived from the Greek word for *hand.*) The pairs of molecules shown in Figure 9.9 are also chiral objects, since they cannot be superimposed upon

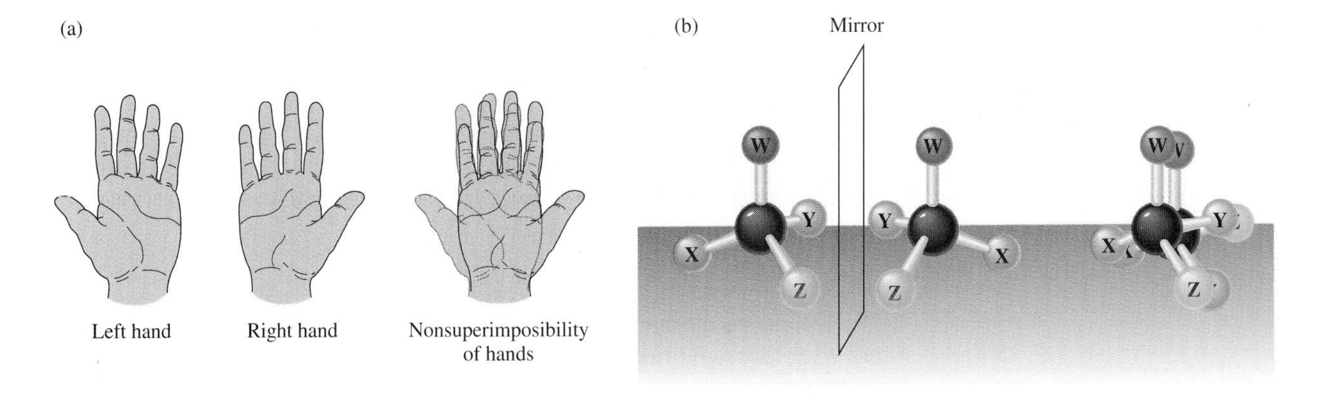

FIGURE 9.9
(a) Hands are mirror images, but they are not superimposable upon each other. (b) These two molecules are said to be chiral objects because they, like hands, are nonsuperimposable mirror images.

each other. A pair of molecules are chiral if they are mirror images of each other and are not superimposable. Each molecule in the pair is an **enantiomer** of the other.

◆ A chiral carbon has four different substituents bonded to it.

Why are these molecules chiral? It is due to the presence of a tetrahedral atom with four different substituents. This atom, usually a carbon atom, is called a **chiral center**, once called an *asymmetric carbon*. Whenever an atom has four different substituents, there are always two ways the atoms can be arranged about that atom. There is always the possibility of two isomers. In most biomolecules, however, only one of this possible pair actually exists.

EXAMPLE 9.1

Which of the following objects are chiral: a tennis racquet, a baseball glove, a football, a right ear.

Solution: To determine whether an object is chiral, try to superimpose the object on its mirror image. If it cannot be superimposed, it is chiral. Most tennis rackets and footballs (disregard any markings on these objects) superimpose on their mirror images; they are not chiral. A baseball glove does not superimpose (that is why there are right- and left-handed gloves), nor does the right ear; thus these objects are chiral.

A pair of enantiomers are identical in most physical and chemical properties, but they do differ in two properties. First, they differ in their interaction with plane polarized light and are thus sometimes called *optical isomers*. Second, they differ in their interaction with other chiral compounds. This second difference is of great biological importance; the first difference is useful in studying these compounds.

Enantiomers interact differently with plane-polarized light. This interaction can be seen with an instrument called a polarimeter (Figure 9.10). Light that has passed through a polarizing filter is aligned in one plane and is said to be plane-polarized (Figure 9.10a). One enantiomer of a pair rotates this plane-polarized light clockwise and is designated as the (+) isomer. The other enantiomer rotates the light counterclockwise by the same amount (Figure 9.10b), and is designated as the (−) isomer. Matter that is not chiral does not rotate the light in either direction. In addition, an equal mixture of two enantiomers, a **racemic mixture**, shows no rotation of polarized light. This is because the enantiomers rotate the light in equal but opposite directions, canceling each other. Thus, there is no net rotation of the light.

Some molecules possess two or more chiral centers. Since the presence of a chiral center results in a pair of stereoisomers called enantiomers, how many possible stereoisomers result from two or more chiral centers? This can be determined with this simple equation:

$$\text{Number of stereoisomers} = 2^n$$

The symbol n is the number of chiral centers. A compound with two chiral centers has 2^2, or four stereoisomers. Consider the sugar D-erythrose (you will learn more about sugar structures and D- L- designations later).

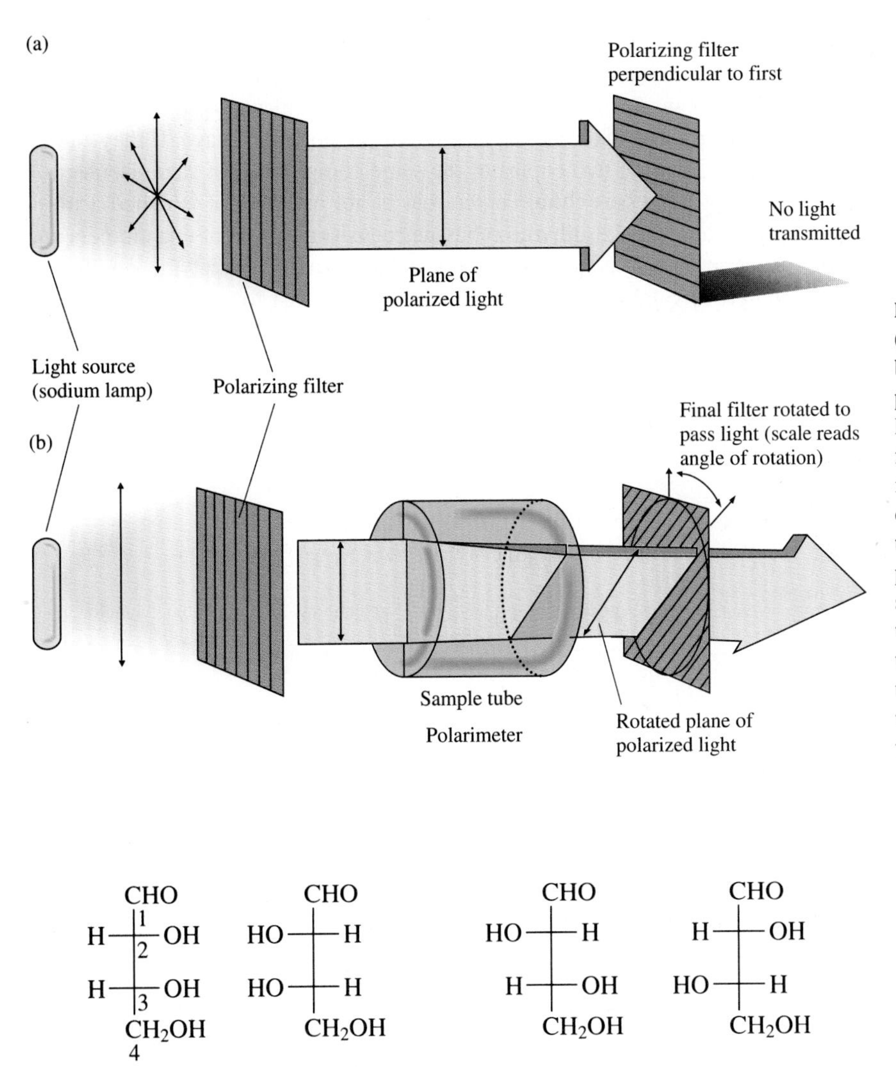

FIGURE 9.10
(a) Plane-polarized light is produced by passing a beam of light through a polarizing filter. This plane-polarized light can pass through any polarizing filter with the same alignment, but it is not transmitted by a polarizing filter of another alignment. (b) A polarimeter measures the rotation of plane-polarized light as it passes through a sample. A sample of an enantiomer rotates light; the observer must rotate the eyepiece containing the final filter to observe the light. The amount of rotation, in degrees, is read directly from the eyepiece.

Notice that the carbon atoms numbered 2 and 3 are chiral centers. Because there are two chiral centers, there should be four stereoisomers, one of which is D-erythrose. D-Erythrose and the stereoisomer that is its mirror image are a pair of enantiomers. The other two are also a pair of mirror images and are therefore enantiomers. The two pairs of enantiomers are **diastereomers,** which are stereoisomers that are not enantiomers. Consider another set of stereoisomers, the 2,3-dihydroxybutanes.

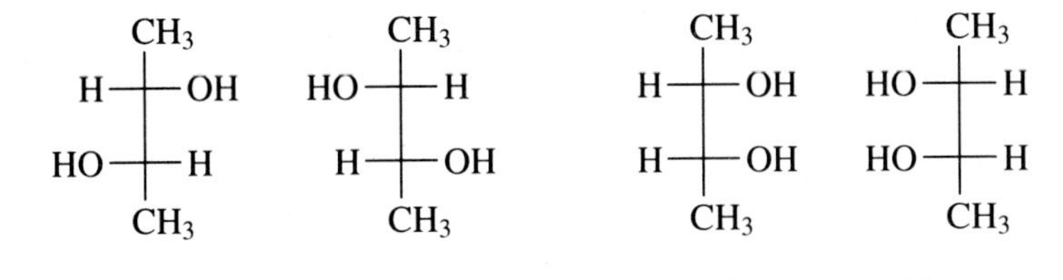

Nonsuperimposable
(A pair of enantiomers)

Superimposable
(A meso compound)

One set is a pair of enantiomers, but what appears to be another pair is really two images of the same compound. There exists a plane of symmetry in the molecule (you can cut it in half to yield two identical parts), thus one of the images can be rotated 180° and superimposed on the other image. If they can be superimposed, they are two copies of the same compound, not stereoisomers. A compound that possesses chiral centers but is not itself chiral is called a **meso compound**. Thus, the 2,3-dihydroxybutanes consist of a pair of enantiomers and one meso compound. Note that the formula given to predict the number of stereoisomers is valid only if no meso compounds exist for the set of isomers.

EXAMPLE 9.2

Determine the number of chiral centers and stereoisomers for each of these molecules:

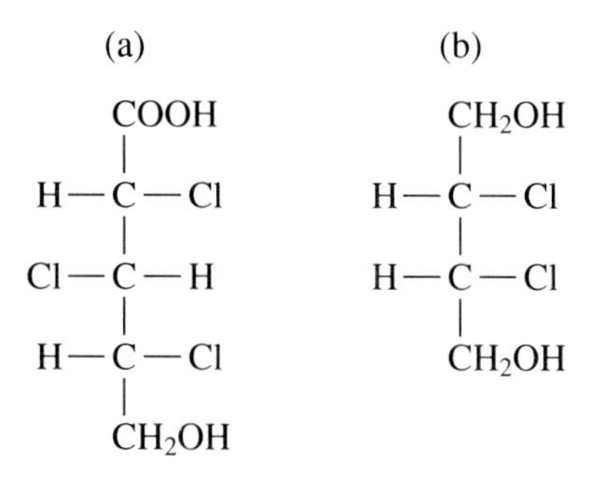

Solution: In a compound, any carbon atom with four different substituents is a chiral center. Thus, (a) has three chiral centers and (b) has two. The number of stereoisomers can be determined by raising 2 to the nth power and subtracting one for each meso compound that can exist. Molecule (a) has three chiral centers, and 2^3 is 8. This compound does not have any symmetry, thus, no meso compound can exist. Molecule (a) is one of eight possible stereoisomers. Molecule (b) has two chiral centers, and 2^2 yields 4 possible stereoisomers. But molecule (b) is symmetrical, thus two of the possible stereoisomers are identical; a meso compound exists. Subtract one from four to account for the meso compound. Molecule (b) is one of three stereoisomers.

♦ One reason enzymes bind some small biomolecules with high specificity is because the enzymes and the small molecules are both chiral.

Many biomolecules are chiral. Many of the interactions that occur between biomolecules can occur only because of this chirality. Consider for a moment the binding of a small molecule to the surface of a hypothetical macromolecule such as a protein. In Figure 9.11 a pair of enantiomers are shown colliding with the surface of the same protein. The surface has a unique shape due to its chirality, and because of this, one of the enantiomers binds to the surface but the other does not.

Your sense of smell is based upon this type of interaction. Consider the two enantiomers shown here.

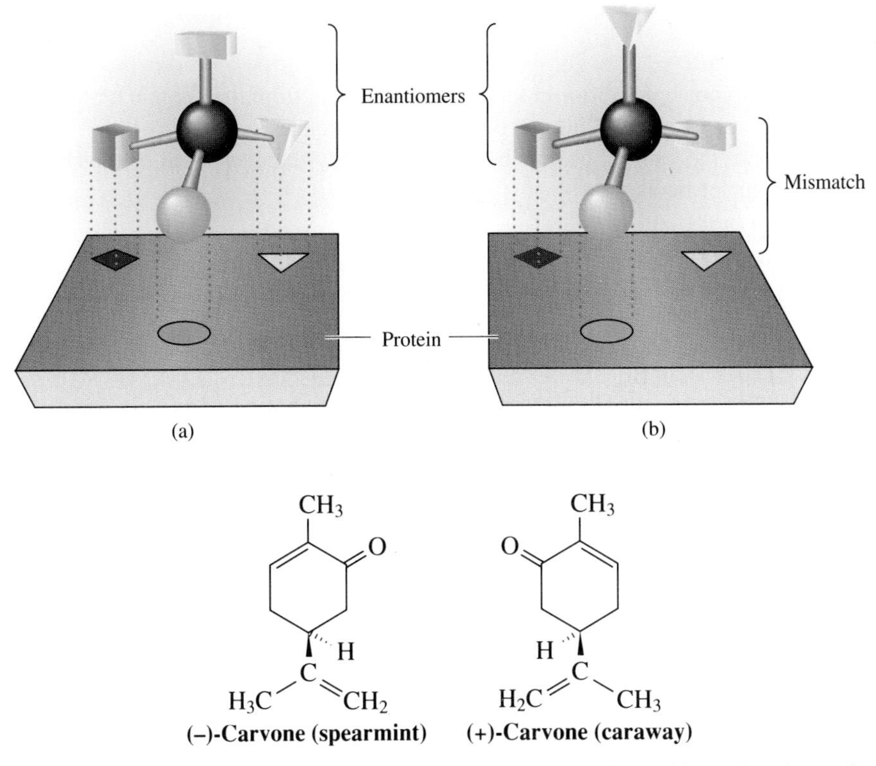

FIGURE 9.11
The matching of two enantiomers to the surface of a protein. (a) Three groups of this enantiomer fit perfectly into the three sites on the surface of the protein, thus, this enantiomer binds with the protein. (b) The groups are oriented in space such that only two groups fit into the sites on the protein suface. This enantiomer does not bind with the protein.

Your nose possesses receptors for odor, proteins that bind specific molecules and initiate the perception of smell for that compound. The compound (+)-carvone smells like caraway, and (−)-carvone smells like spearmint. One binds to one receptor, and the other to a different one. Thus, they have distinctly different smells. Many of the reactions of the body show equal selectivity. Often one enantiomer is biologically active and the other is totally inactive.

NEW TERMS

stereoisomers Isomers that differ only in the arrangement of atoms in space

chirality An object is chiral if it cannot be superimposed on its mirror image. This word is derived from the Greek word for hand because hands are chiral.

enantiomers A pair of isomers that are mirror images of each other and are not superimposable; a pair of chiral molecules

chiral center A tetrahedral atom with four different substituents

racemic mixture A mixture that contains equal amounts of a pair of enantiomers and thus does not rotate plane-polarized light

diastereomers These are stereoisomers that are not enantiomers, that is, they are not mirror images of each other

meso compound A compound with chiral centers that is nonchiral because of molecular symmetry

TESTING YOURSELF

Biomolecules

1. Most biomolecules differ in three important ways from the organic molecules you studied earlier. What are these three differences?
2. Compare the pairs of objects below and identify them as chiral or nonchiral.
   a. a pair of boots
   b. two basketballs
   c. two baseball bats
   d.

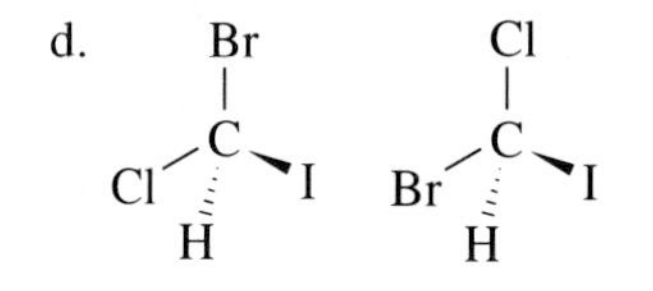

Answers **1.** Biomolecules are generally larger, often multifunctional, and are usually chiral. **2. a.** chiral **b.** nonchiral **c.** nonchiral **d.** chiral

9.5 General Principles of Nutrition

Biochemistry is linked to nutrition by biomolecules. These molecules are found in living organisms, but they are also the molecules found in food. **Nutrition** is that branch of science that studies diets and food, and nutrition is playing an ever-increasing role in health care. This section provides a brief introduction to nutrition.

♦ Nutrients are the substances needed by the body to sustain life.

Food is anything which, when eaten, provides substances that the body needs to grow or maintain itself. These substances are referred to as **nutrients**. Food consists of biomolecules. You will find that many of the nutritional terms you have heard on TV commercials or read about on food packaging labels are general classes of biomolecules.

♦ The Calorie of nutrition is 1000 times larger than the "small" calorie used by chemists.

Food provides matter and energy for the body. Through a large number of chemical reactions that are collectively called **metabolism**, foods are converted to energy and molecules needed by the body. Food contains energy in the form of potential energy stored in the chemical bonds of molecules. The total amount of potentially available energy in food is measured in **Calories**, a unit of energy used in nutrition that is equal to the kilocalorie of the chemist. (One Calorie equals 1000 "small" calories.) In the body, the bonds of food molecules are broken and other bonds are formed when wastes are produced. The potential energy in the bonds of waste products is less than that in the bonds of food molecules. This difference in energy is available to the body; it is useful energy that a chemist calls *free energy*. Part of it is always lost as heat, but the rest is conserved by the body as the potential energy in certain chemical bonds. This energy is used as needed by the body to carry out the numerous tasks that require energy (Figure 9.12).

Food also provides matter needed by the body. During growth, the mass of an organism increases. Since matter cannot be created nor destroyed, the matter must be added to the body from an external source—food. In addition, during normal life maintenance processes, molecules of the body are broken down and excreted as wastes. This matter must be replaced, and again food is the source.

Foods contain nutrients, but the nutritional content of foods varies greatly. Similarly, the needs of the body for various nutrients are different. Nutrients can be classified into six major classes: (1) water, (2) carbohydrates, (3) lipids, (4) proteins, (5) vitamins, and (6) minerals (Table 9.2). These groups collectively provide all of the energy and matter needed by the body. Some classes, such as carbohydrates and lipids, are most important as sources of energy, whereas other classes, such as vitamins and minerals, are important as sources of specific compounds or elements. Water serves yet another and very vital role. It is the solvent in which all the other molecules of the body are either dissolved, dispersed, or suspended. The nutrition of the most important biomolecules is discussed at the end of the appropriate chapters that follow.

FIGURE 9.12
Potential energy stored in chemical bonds is used for a variety of purposes. Motion results from the mechanical energy produced by muscles. That mechanical energy is derived from energy stored in chemical bonds in adenosine triphosphate (ATP) (see Chapter 15).

NEW TERMS

nutrition The branch of science that deals with the composition, preparation, consumption, and processing of food

nutrients Substances required by the body that are provided by food

Calorie The unit of energy used in nutrition. It is equal to 1 kcal or 1000 "small" calories.

TABLE 9.2
Nutrients Required by Humans

Class	Examples and Some Sources	Principal Function
Carbohydrates	Sugars, starches (complex carbohydrates), and fiber from grains, fruits, and vegetables	Provide energy and fiber
Lipids	Fats and oils from vegetable oils, nuts, and nonlean meats	Provide energy and an essential fatty acid
Protein	Proteins are widespread in animal and plant tissues; meats are an excellent source of protein	Provide amino acids for protein synthesis and energy
Vitamins	Water-soluble vitamins: niacin, riboflavin, thiamine, pyridoxal, pantothenic acid, folic acid, biotin, cobalamin, ascorbic acid; fat-soluble vitamins: A, D, E, K	Serve as coenzymes and have a variety of other roles
Minerals	Calcium, chromium, cobalt, copper, iron, magnesium, manganese, molybdenum, potassium, selenium, sodium, zinc, chlorine, fluorine, iodine, nitrogen, phosphorus, sulfur	Many roles—calcium and phosphorus found in bones and teeth; sodium, potassium, and chloride are electrolytes, and so on
Water	Unpolluted, uncontaminated fresh water	Solvent for biomolecules and the reactions of metabolism

TESTING YOURSELF

Nutrition

1. a. Convert 7.2 Cal into kilocalories and calories.
   b. If this energy were used to increase the temperature of a liter of water (mass = 1000 g), what temperature increase would be seen?
2. List the six major classes of nutrients.

Answers **1. a.** 7.2 kcal, 7200 cal **b.** 7.2°C **2.** carbohydrates, lipids, proteins, vitamins, minerals, and water

Summary

The *cell* is the basic structural unit of life. The two basic types of cells are the *procaryote* and the *eucaryote*. The pyrocaryotic cell is smaller and simpler and is typical of bacteria. Procaryotes typically have a *cell membrane, cytosol, cell wall*, and perhaps an outer layer. Eucaryotic cells are larger, more complex cells found in plants and animals. In addition to a cell membrane, cytosol, and perhaps cell walls, eucaryotes contain a *nucleus, mitochondria*, and *endoplasmic reticulum*. The biomolecules in cells possess the characteristics of other organic compounds but are generally larger, multifunctional, and *chiral*. *Nutrition* is the science that deals with foods and the nutrients they contain. *Nutrients* are substances required by the body for normal growth and maintenance and include water, carbohydrates, lipids, proteins, vitamins, and minerals. Within the cells of the body, the nutrients are processed by the chemical reactions that make up metabolism, to produce the energy and substances needed by the body.

Terms

cell (9.1)
organelle (9.1)
procaryote (9.2)
cell membrane (9.2)
cytosol (9.2)
ribosome (9.2)
cell wall (9.2)
eucaryote (9.3)
nucleus (9.3)
endoplasmic reticulum (9.3)
mitochondrion (9.3)
stereoisomers (9.4)
chirality (9.4)
enantiomers (9.4)
chiral center (9.4)
racemic mixture (9.4)
diastereomers (9.4)
meso compound (9.4)
nutrition (9.5)
nutrients (9.5)
Calorie (9.5)

Exercises

The Cell (Objective 1)

1. The cell is called the basic structural unit of life. Why is it given this distinction?
2. Calculate the surface-area-to-volume ratio of a cube that is 1 cm on each side.
3. Cells have a large surface-area-to-volume ratio. What is gained by this large ratio?
4. Calculate the volume in cubic micrometers of a typical bacterium and a eucaryotic cell. Consider the bacterium to be a cylinder 10 μm in height and 2 μm in diameter. Consider the cell to be a cube with 100-μm sides. (*Hint:* The volume of a cylinder is $\pi r^2 h$.)

5. Compare the surface-area-to-volume ratio of the bacterium and eucaryotic cell in exercise 4.
6. a. Are all of the cells of the body identical?
   b. What are tissues?
   c. What are organs?
7. What physical process is responsible for the movement of substances into and out of cells?
8. Define the following terms:
   a. nucleus
   b. organelle
   c. procaryote
   d. eucaryote
   e. cytosol

Procaryotes (Objective 2)

9. The cell membrane is more than a passive barrier to movement of substances into and out of the cell. Why is this?
10. What synonyms are used to describe DNA in procaryotes?
11. What is the physiological role of ribosomes?
12. Where are ribosomes located in the procaryotic cell?
13. What is the mode of action of these common antibiotics?
    a. penicillin
    b. erythromycin
    c. chloramphenicol
14. Streptomycin disrupts protein synthesis. It is given to humans as an antibiotic to treat certain infections. Why are the human host cells not killed?
15. What is the composition of a lipopolysaccharide?
16. What effect does lipopolysaccharide have on humans?
17. What is the meaning of the word "nonpyrogenic"?
18. Bacterial cells possess a cell wall that is lacking in animal cells. Why do animal cells lack a cell wall?
19. Does the cell wall protect a bacterial cell in hypertonic solutions? Why or why not?
20. Compare the fate of a bacterial cell and an animal cell in these solutions:
    a. hypotonic b. isotonic c. hypertonic

Eucaryotes (Objective 3)

21. What broad classes of organisms have eucaryotic cells?
22. What is the physiological role of these organelles?
    a. nucleolus c. endoplasmic reticulum
    b. mitochondrion d. chloroplast
23. Make a small table showing which of the organelles in exercise 22 are found in animals, plants, and bacteria.
24. Where is the nucleolus found in eucaryotes? Where is it found in procaryotes?
25. What macromolecule is found in the cytosol of bacteria but is found in the nucleus of eucaryotic cells?
26. Which of the following organelles is found in both procaryotic and eucaryotic cells?
    a. nucleus
    b. ribosome
    c. mitochondrion
    d. endoplasmic reticulum
27. The endoplasmic reticulum serves what role?
28. What organelles and macromolecules are found in the nucleus?

Biomolecules (Objective 4)

29. Define the following terms:
    a. racemic mixture
    b. chirality
    c. chiral center
    d. meso compound
    e. diastereomers
30. Identify the polar and nonpolar regions in a molecule of lysine (a common amino acid).

$$^{+}NH_3CH_2CH_2CH_2CH_2\underset{\displaystyle |\atop {}^{+}NH_3}{C}HCOO^-$$

31. Two molecules that are nonsuperimposable mirror images of each other are called ______.
32. Which of the following pairs of objects are superimposable?
    a. a pair of gloves c. your ears
    b. two tennis rackets d. two footballs
33. Which of the pairs of objects in Exercise 32 are chiral?
34. A pair of enantiomers are identical in all but two properties. What are these properties?
35. You have isolated two liquid samples that are pure compounds. They have the same composition. What simple tests could you use to determine whether they are the same compound, enantiomers, or some other type of isomer?
36. Use a molecular model kit to make (−)- and (+)-carvone. Note that the ring can have several conformations since it is not aromatic. Are these models identical? Are they mirror images? Are they superimposable?
37. Identify the chiral center in the carvones in Exercise 36.
38. Plane-polarized light is affected differently by what kind of molecules?
    a. identical molecules
    b. chiral molecules
    c. macromolecules
    d. all kinds of biomolecules
39. Two compounds have the same composition and also have the same atoms attached to the same atoms, although with different orientations in space. These compounds are
    a. identical c. positional isomers
    b. structural isomers d. stereoisomers

40. Mark with an asterisk any chiral centers in these compounds:

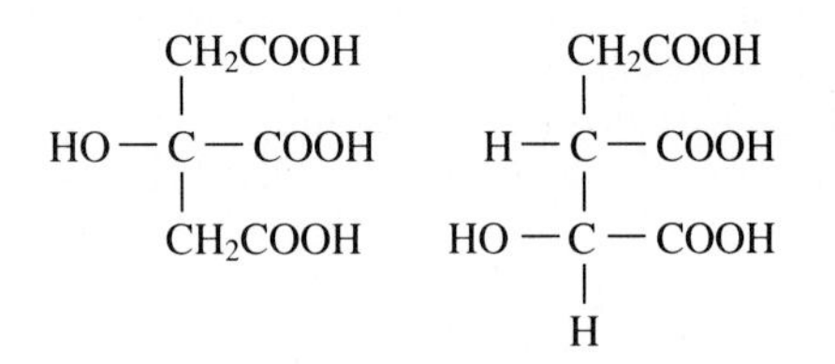

a. Citric acid **b. Isocitric acid**

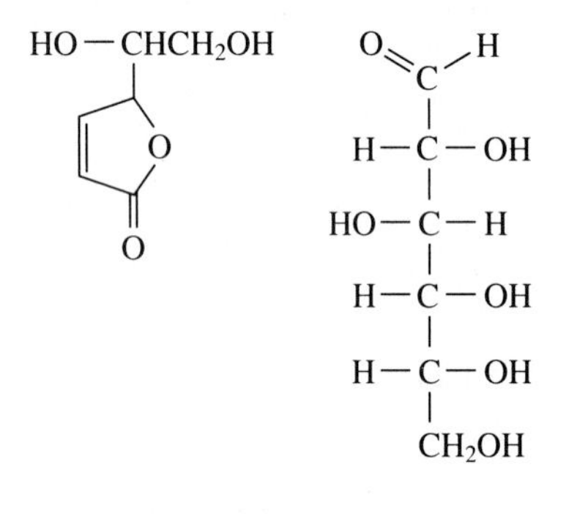

c. Ascorbic acid (vitamin C) **d. Glucose**

41. How many stereoisomers can exist for each molecule in Exercise 40?

42. Mark with an asterisk any chiral centers in these amino acids:

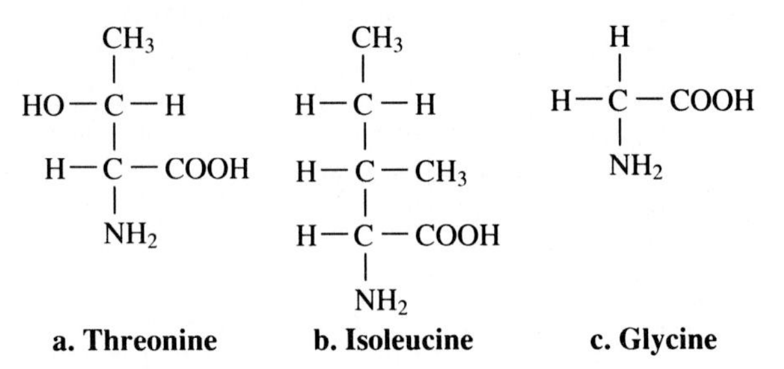

a. Threonine **b. Isoleucine** **c. Glycine**

43. How many stereoisomers can exist for each molecule in Exercise 42?
44. A student finds that two models of a 2,3-dihydroxybutane superimpose. She correctly identifies the compound as a meso compound. She then shows that two models of butane superimpose. Explain why butane is or is not a meso compound.

Nutrition (Objective 5)

45. Define the term *nutrient.*
46. Give examples of substances that are classified as
    a. carbohydrates
    b. lipids
    c. minerals
    d. vitamins
47. Molecules from food are broken down to yield smaller molecules and energy. What happens to this energy?
48. Food provides energy. How is energy stored in food?
49. Water is a nutrient. What is its biological role?
50. Define the following terms:
    a. food
    b. nutrition
    c. Calorie

CARBOHYDRATES

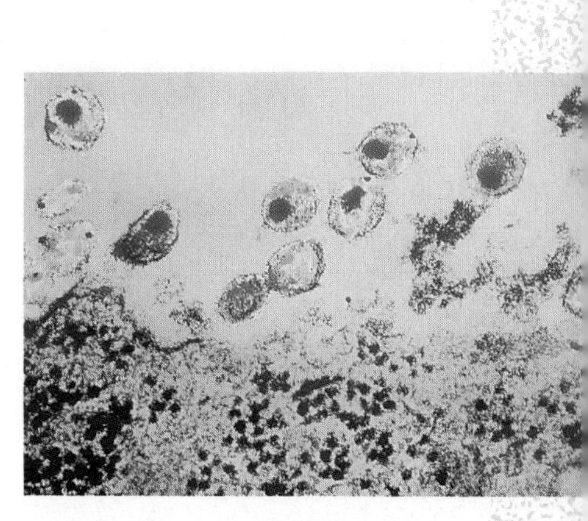

Carbohydrates are found on the outside surface of cells and appear to be involved in cell recognition. Viruses like this human immunodeficiency virus (HIV) use surface molecules to locate host cells in an organism.

OUTLINE

10.1 Classification of Carbohydrates
10.2 Monosaccharides
10.3 Properties and Reactions of Sugars
10.4 Oligosaccharides
10.5 Polysaccharides

OBJECTIVES

After completing this chapter, you should be able to

1. Identify and classify some important monosaccharides.
2. Describe the glycosidic bond.
3. Identify some of the common disaccharides and polysaccharides.
4. Recognize some of the chemical properties of carbohydrates and relate them to the chemistry of alcohols and carbonyl compounds.
5. Describe the major role of carbohydrates in energy metabolism and structure.

Carbohydrates are widespread and occur in many forms.

We correctly perceive paper, cotton, crab shells, nectar, and blood as quite different substances. Yet all of these materials are related by composition—they all contain glucose. The glucose in these substances is not in the same form. Blood has glucose (blood sugar) in solution. Glucose provides a significant amount of the body's energy needs. Nectar contains sucrose, common table sugar, which in turn contains glucose. Paper and cotton contain cellulose, a macromolecule containing only glucose. Crab shells are formed from chitin, a macromolecule that is essentially a modified cellulose. Glucose is a widely distributed and abundant compound. Glucose is a carbohydrate.

Carbohydrates are polyhydroxy aldehydes, ketones, or derivatives of these compounds. Although *poly-* normally means many, as used here it means two or more groups.

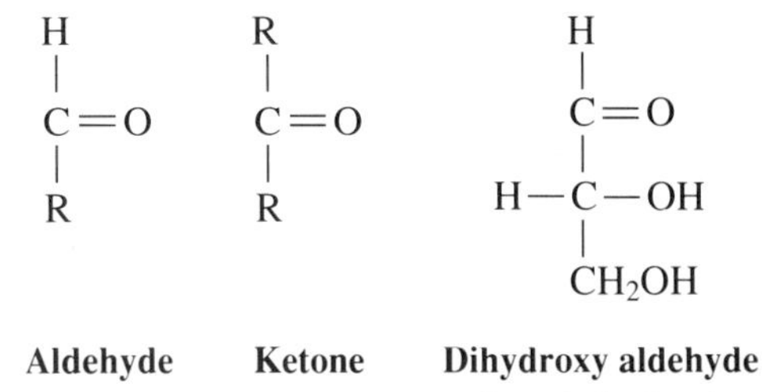

Aldehyde **Ketone** **Dihydroxy aldehyde** (a carbohydrate)

This chapter introduces the major carbohydrate classes—the monosaccharides, oligosaccharides, and polysaccharides. The structure, nomenclature, and some of the chemical properties of these organic compounds will be presented. In later chapters some of their roles in energy storage and cellular and extracellular structure will be examined.

10.1 Classification of Carbohydrates

The name *carbohydrate* comes from the early observation that some members of this class had formulas that could be written $C_n(H_2O)_n$. They appeared to be "hydrates of carbon." A simple test that seems to confirm this notion is to treat a sugar with a strong mineral acid such as concentrated sulfuric acid. When this acid is added to sugar, the white sugar is converted to a blackish residue that is nearly pure carbon and water is liberated as steam (Figure 10.1).

$$C_n(H_2O)_n \xrightarrow[\text{(dehydrating agent)}]{H_2SO_4} n\,C + n\,H_2O$$

Early chemists named these compounds for this property, although today it is clear that this definition is not really accurate nor does it reflect the general chemical properties of these compounds.

♦ The suffix *-ose* identifies many compounds as carbohydrates.

The names of many carbohydrates contain the suffix *-ose*. Gluc*ose*, sucr*ose*, and cellul*ose* are clearly identified as carbohydrates. Some carbohydrates do not follow this simple rule of nomenclature—starch, glycogen, and chitin are examples. Carbohydrates can be named by IUPAC nomenclature, but generally common names are used because they are simpler and are universally understood by chemists and biologists alike. The examples cited above are common names.

FIGURE 10.1
The reaction of sucrose, table sugar, with concentrated sulfuric acid yields a carbon residue and water.

Carbohydrates vary in size, with the larger ones containing two or more basic carbohydrate units. This basic unit is called a *saccharide*, derived from the Latin word for sugar. One classification of carbohydrates is based on the number of these units found in the molecule. The smallest carbohydrates, the **monosaccharides**, or simple sugars, are the basic units found in other carbohydrates. These compounds have the general formula $C_nH_{2n}O_n$. The monosaccharides *cannot* be converted to smaller carbohydrates by hydrolysis in dilute acid. Glucose is a monosaccharide.

◆ Carbohydrates are classified by the number of saccharide units found in the molecule

$$\underset{\textbf{(Monosaccharide)}}{\text{Glucose}} + H_2O \xrightarrow{\text{dil. } H^+} \text{No reaction}$$

Oligosaccharides are carbohydrates that have two or more monosaccharides covalently linked together. Mild hydrolysis breaks these covalent bonds to yield the individual monosaccharides. As previously mentioned, sucrose is the sugar found in nectar. Mild hydrolysis of sucrose yields two monosaccharides, glucose and fructose.

$$\underset{\textbf{(Disaccharide)}}{\text{Sucrose}} + H_2O \xrightarrow{\text{dil. } H^+} \underset{\textbf{(Monosaccharides)}}{\text{Glucose} + \text{Fructose}}$$

The prefix *oligo-* means several and will be used in this text to mean two through ten. The oligosaccharides are designated by the actual number of monosaccharides present in the molecule. For example, the *di*saccharide lactose contains *two* monosaccharides, and the *tetra*saccharide maltotetrose contains *four* monosaccharides. Note that the prefixes *di-*, *tri-*, *tetra-*, and so on represent the same numerical value you learned earlier. Most of the oligosaccharides discussed in this text are disaccharides.

Cellulose, starch, and chitin are examples of very large carbohydrates. These substances are classified as **polysaccharides**. They are polymers of monosaccharides that yield many monosaccharide molecules upon hydrolysis.

$$\underset{\textbf{(Polysaccharide)}}{\text{Cellulose}} + n\,H_2O \xrightarrow{\text{dil. } H^+} n + 1 \underset{\textbf{(Monosaccharide)}}{\text{Glucose}}$$

A polysaccharide contains more than ten monosaccharide units.

EXAMPLE 10.1

Classify these compounds as mono-, oligo-, or polysaccharides:

(a) When 0.25 mmol (millimoles) of a sugar from insects is hydrolyzed in weak acid, the product is 0.51 mmol of the monosaccharide glucose.

(b) When 0.013 mmol of a carbohydrate is hydrolyzed in weak acid, over 300 mmol of glucose are formed.

(c) A sugar isolated from honey does not react in weak acid.

Solution: Classification of these carbohydrates is based on the number of saccharide units found in the carbohydrate. This number can be determined by hydrolysis of the carbohydrate. Divide the number of moles of product by the moles of reactant. In (a), 0.51 mmol/0.25 mmol = 2.0. There are two glucose molecules per sugar molecule. This sugar is an oligosaccharide, a disaccharide. In (b), $>$300 mmol/0.013 mmol = $>$23,000 glucose molecules per carbohydrate molecule. Molecule (b) is a polysaccharide. Since (c) cannot be hydrolyzed by weak acid, it is probably a monosaccharide.

NEW TERMS

carbohydrate A class of compounds consisting of polyhydroxy aldehydes and ketones and derivatives of these compounds

monosaccharides Simple sugars, such as glucose; the smallest compounds that are carbohydrates. These are the basic units from which larger carbohydrates are made.

oligosaccharides Carbohydrates that contain two to ten monosaccharides, such as table sugar and milk sugar (disaccharides)

polysaccharides Carbohydrates that contain many monosaccharide units, such as starch and cellulose.

TESTING YOURSELF

Classification of Carbohydrates

1. How many monosaccharide units are present in a trisaccharide?
2. What name is given to a carbohydrate that contains many monosaccharide units?

Answers **1.** three **2.** polysaccharide

TABLE 10.1
Some Representative Sugars

Common Name	Molecular Formula	Structure	Classifications
Glyceraldehyde	$C_3H_6O_3$	O ‖ H—C | H—C—OH | CH_2OH	Triose, aldose, aldotriose
Erythrose	$C_4H_8O_4$	O ‖ H—C | H—C—OH | H—C—OH | CH_2OH	Tetrose, aldose, aldotetrose
Ribose	$C_5H_{10}O_5$	O ‖ H—C | H—C—OH | H—C—OH | H—C—OH | CH_2OH	Pentose, aldose, aldopentose
Fructose	$C_6H_{12}O_6$	CH_2OH | C=O | HO—C—H | H—C—OH | H—C—OH | CH_2OH	Hexose, ketose, ketohexose

10.2 Monosaccharides

Monosaccharides are sugars that are found free in nature or as components of other compounds. Monosaccharides vary by the number of carbon atoms that are present in the chain and by the functional groups that are present. The number of carbon atoms is indicated by a prefix such as *tri-* (three carbons), *tetr-* (four carbons), *pent-* (five carbons), *hex-* (six carbons), or *hept-* (seven carbons). This numerical prefix is combined with the suffix *-ose*. Thus, a *triose* is a sugar with three carbon atoms, and a *pentose* contains five carbon atoms. Examples of representative sugars are presented in Table 10.1.

◆ Classification by carbon number refers to the number of carbon atoms in the monosaccharide.

♦ All monosaccharides contain hydroxy groups. Classification by functional group refers to the presence of an aldehyde group or a ketone group.

The functional group present in the sugar may also be used for classification. For example, sugars having an *ald*ehyde group as part of their structure are classified as **aldoses**. Other sugars have a *ket*one carbonyl group and are called **ketoses**. A sugar can be classified by both functional group and chain length. The *aldo* or *keto* prefix is placed before the designation for chain length. Thus, in the examples found in Table 10.1 there is an *aldotriose*, an *aldotetrose*, an *aldopentose*, and a *ketohexose*.

Monosaccharides, like many biological molecules, possess one or more chiral centers. The presence of chiral centers means stereoisomers may exist. Consider the aldotriose glyceraldehyde, which is the smallest monosaccharide with a chiral center. There are two enantiomers for this monosaccharide, and each of them must be uniquely named. Biochemists distinguish between the two using the prefixes D- or L- to designate the stereochemistry.

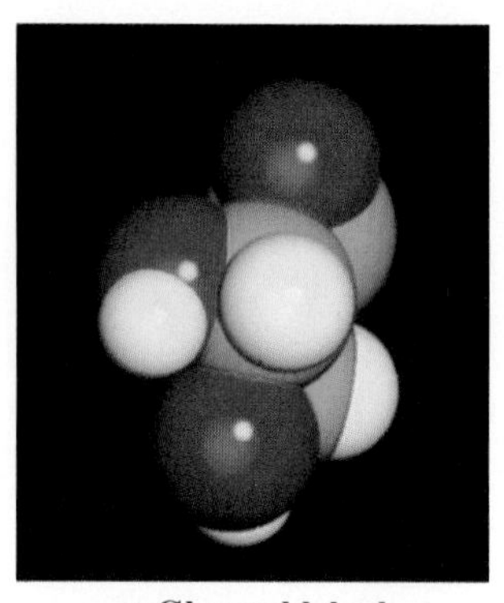

L-Glyceraldehyde

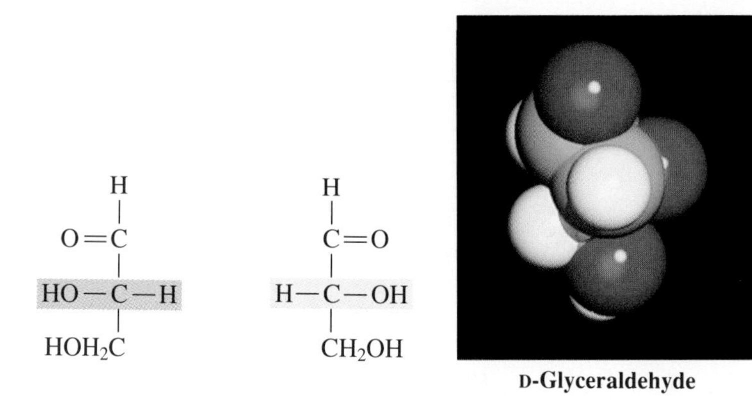

D-Glyceraldehyde

When glyceraldehyde is drawn with the carbonyl group at the top (as drawn here), the —OH group on the middle carbon of the left molecule is on the left side of the molecule. This enantiomer is designated L-glyceraldehyde. The other enantiomer is designated D-glyceraldehyde. The letters are derived from *levo-* (left) and *dextro-* (right). The stereochemistry of the glyceraldehydes is used to assign all other carbohydrates as D- or L-. For example, D-glucose and L-glucose have the following stereochemistry:

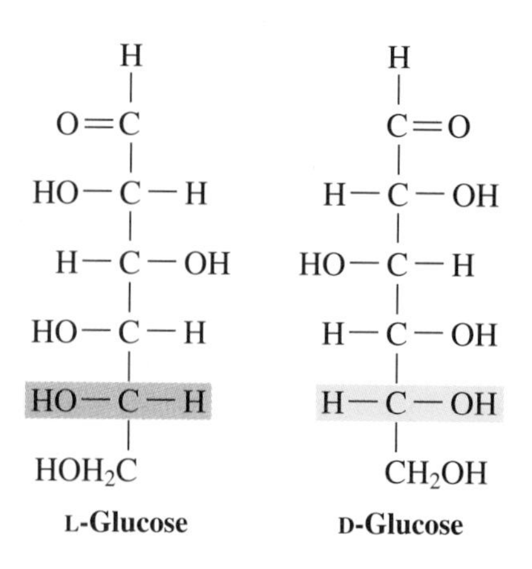

L-Glucose D-Glucose

Note that L-glucose is the mirror image, the enantiomer, of D-glucose. Also note that the hydroxyl group (highlighted) most distant from the carbonyl group has the orientation designated by D- and L-. Since there may be more than one chiral center in the molecule, one particular chiral center must be used to determine whether the molecule is in the D- or L- family. The chiral center most distant from the carbonyl group is used for the determination. Most of the naturally occurring sugars belong to the D- family.

♦ The D- and L-isomers of a compound are enantiomers.

♦ Most naturally occurring carbohydrates belong to the D-stereochemical family.

EXAMPLE 10.2

Examine this carbohydrate and classify it with respect to functional group, number of carbon atoms, and stereochemical family.

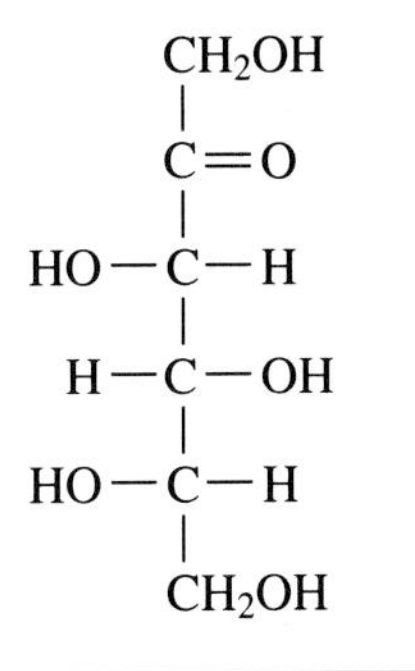

Solution: This molecule has a keto group in it; thus, the sugar is a ketose. There are six carbon atoms, so it is a hexose. These terms can be combined to ketohexose. The hydroxyl group on the chiral carbon most distant from the keto group points to the left. Thus, this is an L- sugar.

Hemiacetals and hemiketals are formed when an alcohol reacts with an aldehyde or ketone, respectively.

$$\text{R—C(=O)—H} + \text{HOR}' \overset{H^+}{\rightleftharpoons} \text{R—C(OH)(H)—OR}'$$

Aldehyde **Alcohol** **Hemiacetal**

$$\text{O=C(R)(R}') + \text{HOR}'' \overset{H^+}{\rightleftharpoons} \text{HO—C(R)(R}')\text{—OR}''$$

Ketone **Alcohol** **Hemiketal**

Because sugars have both hydroxyl groups and an aldehyde or ketone group, intramolecular hemiacetals or hemiketals are certainly possible and, in some sugars, quite likely. Generally, pentoses and hexoses exist as either a five- or six-membered ring. A sugar with fewer than five carbon atoms cannot form a stable ring because the ring would have three or four atoms. Just as for cyclopropane and cyclobutane, these are unstable.

♦ The cyclic forms of sugars are hemiacetals or hemiketals.

♦ Cyclic sugars typically have five- or six-membered rings.

Ribose forms a five-membered ring in the following reaction:

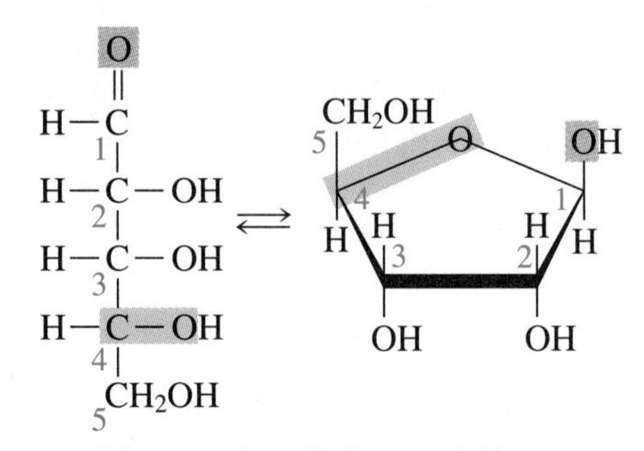

Linear and cyclic forms of ribose

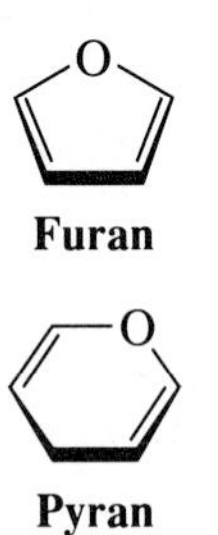

Monosaccharides with five-membered rings are called **furanoses**, because the cyclic ether *furan* has the same ring structure with four carbon atoms and a single oxygen atom. Glucose forms a six-membered ring in the following reaction. A monosaccharide with a six-membered ring is a **pyranose**, because the cyclic ether *pyran* contains five carbon atoms and an oxygen atom.

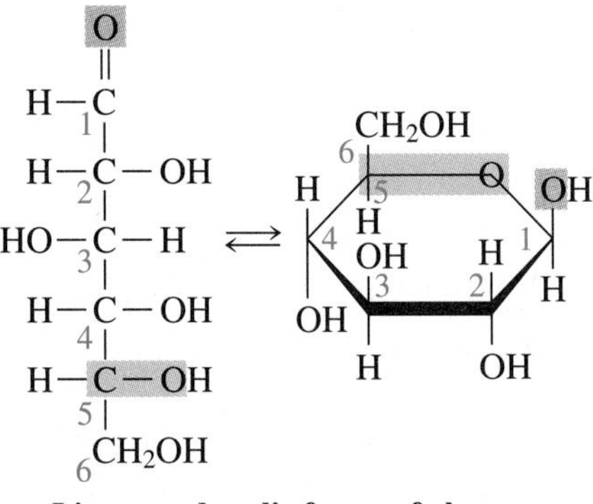

Linear and cyclic forms of glucose

Pentoses and hexoses exist as cyclic forms, but the introduction of a ring into the sugar results in the formation of two cyclic isomers. Examine Figure 10.2. Glucose forms two different pyranoses. In both cases this occurs through a reversible reaction between the carbonyl group (C-1) and the C-5 hydroxyl group to yield a cyclic hemiacetal. The carbon atoms of the open chain and ring forms of glucose are labeled in Figure 10.2 to assist you in visualizing the formation of the ring. Notice that the reaction yields two different products that differ with respect to the orientation of the —OH group on carbon atom 1. Because the carbonyl group is planar, the alcohol can attack either side of the planar carbonyl carbon, yielding two isomers. In sugar chemistry, the carbon atom that was the carbonyl carbon before ring formation is called the **anomeric carbon** and the two products formed are called **anomers**. In the ring form the anomeric carbon has an —OH that can be drawn either above the ring (up), or below the ring (down). If the anomeric —OH is drawn "up" (on the same side as the —CH_2OH), then the anomer is said to be a beta (β) anomer; the other orientation is "down" and is called an alpha (α) anomer. Glucose exists as two different ring forms or

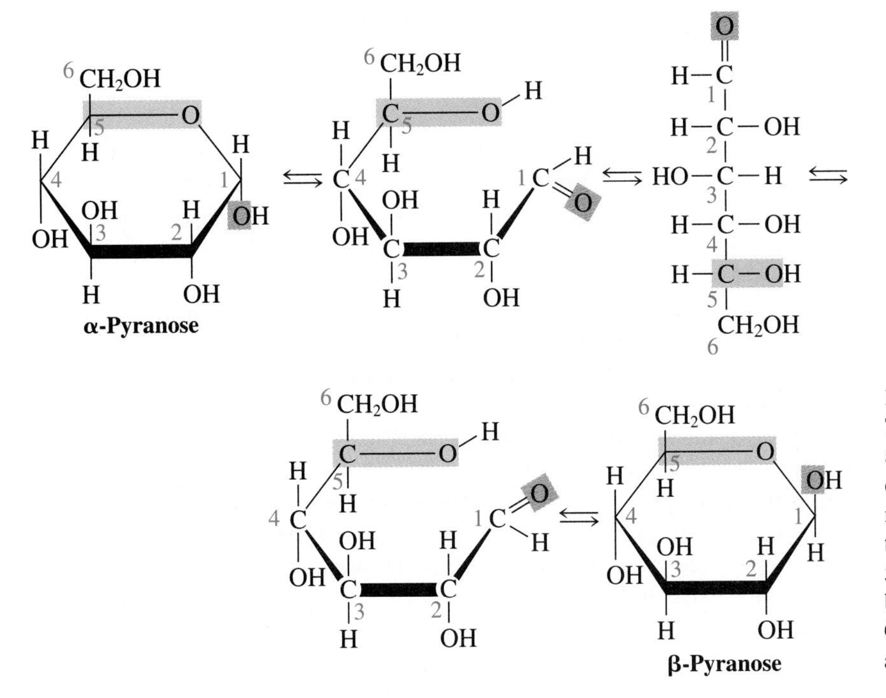

FIGURE 10.2
The formation of α-D-glucopyranose and β-D-glucopyranose from the open-chain form of glucose. Two possible ring forms of glucose exist because the hydroxyl group on carbon number 5 can add to either side of the carbonyl group of carbon number 1. α-D-Glucopyranose and β-D-glucopyranose are anomers.

anomers, both of which are glucopyranoses. Although they are distinguished formally as α-D-glucopyranose and β-D-glucopyranose, they are more commonly called α-D-glucose and β-D-glucose.

♦ α-D-Glucose and β-D-glucose are anomers.

The structures of several common monosaccharides have been shown to you during this introduction. Let's consider these sugars and some others in more detail.

The two smallest monosaccharides, *glyceraldehyde* and *dihydroxyacetone*, each contain three carbon atoms. These sugars are not common in the free form, but their derivatives are important in some reactions of glycolysis.

H
|
C=O
|
H—C—OH
|
CH_2OH

Glyceraldehyde

CH_2OH
|
C=O
|
CH_2OH

Dihydroxyacetone

Ribose and *2-deoxyribose* are five-carbon sugars. They are components of several biomolecules, including nucleotides and nucleic acids, that are important in heredity and cell activity. Note that these sugars and the three that follow are shown in both the straight-chain forms and cyclic forms.

♦ Ribose and 2-deoxyribose are components of RNA and DNA, respectively.

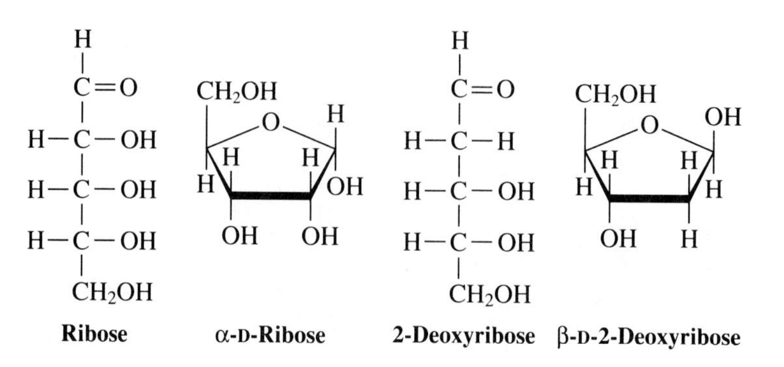

Ribose **α-D-Ribose** **2-Deoxyribose** **β-D-2-Deoxyribose**

Glucose is one of several important monosaccharides that contains six carbon atoms. It is found as a component of many other carbohydrates, but its role as blood sugar makes it of great importance. Blood circulates glucose throughout the body, making it available to all cells. The normal range for blood glucose concentration is about 60 to 100 mg/100 mL. If the concentration is consistently above this normal range, the condition known as *hyperglycemia* exists. This is seen in the disease diabetes mellitus. If concentrations below the normal range persistently exist, then the person has *hypoglycemia* (low blood sugar). The regulation of blood sugar concentration is discussed in more detail later.

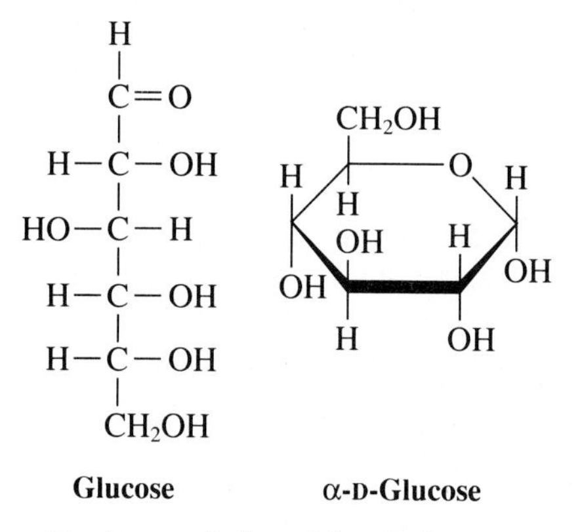

Glucose **α-D-Glucose**

Fructose, along with glucose, is found free in honey. Derivatives of fructose are important in energy metabolism. Fructose is now used as a commercial sweetener in the form of high-fructose syrups. On a weight basis, fructose is 1.7 times sweeter than table sugar (sucrose), therefore less fructose yields the same degree of sweetness as a given amount of table sugar. For example, 0.6 g of fructose is as sweet as 1 g of sucrose. Since less is needed for any desired sweet taste, fewer calories are added to the food.

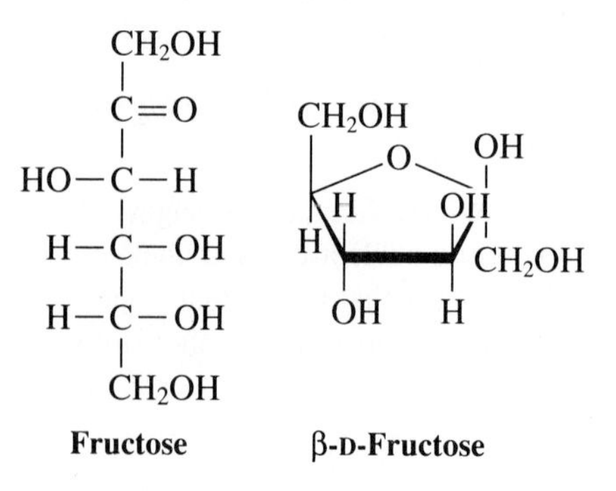

Fructose **β-D-Fructose**

CHEMISTRY CAPSULE

Artificial Sweeteners

Many people are overweight and seek to correct this by reducing their caloric intake. Since a significant amount of energy is derived from dietary carbohydrate, many of them reduce their intake of carbohydrate. This can be accomplished by reducing the total intake of food or by reducing or eliminating specific carbohydrate-rich foods from the diet. Many have turned to artificial sweeteners to satisfy their "sweet tooth" while reducing dietary intake of refined sugar and other sweeteners. Today and in the recent past, several artificial sweeteners have been used, including saccharin, cyclamates, and aspartame. The relative sweetness of equal masses of some of the more common natural and artificial sweeteners are listed in the accompanying table.

Sweeteners	Sweetness Relative to Sucrose
Glucose (blood sugar)	0.5
Fructose (a sugar in honey)	1.7
Sucrose (table sugar)	1.0
Cyclamate	30
Saccharin	450
Aspartame	180

Saccharin was discovered in 1879, and it is the oldest known low-calorie sweetener. It was in common use until the early 1960s, when it was supplemented by *cyclamates*. A major reason for the switch to cyclamates was the "aftertaste" associated with saccharin. However, although cyclamates were the preferred sweetener during this time, saccharin was still used in some cyclamate preparations because it is so much sweeter. Both of these sweeteners are nonnutritive, that is, the body does not use them for energy or matter.

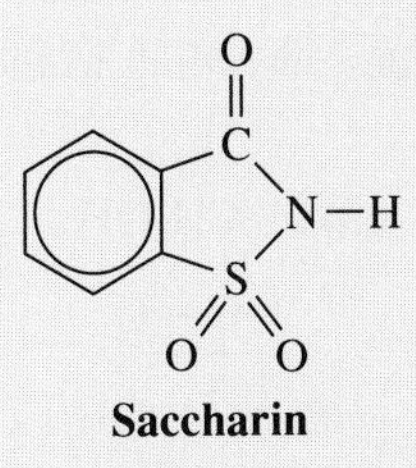

Saccharin

Concern over the safety of cyclamates caused their removal from use. Furthermore, since 1978, a warning appears with every product containing saccharin, which states that saccharin causes cancer in laboratory animals. In 1981, *aspartame* was approved as a low-calorie sweetener. Aspartame lacks the aftertaste of saccharin, and in exhaustive testing, no apparent health concerns were found for this compound. Testing for adverse effects of aspartame continues. Aspartame, like sugars, is a nutritive sweetener, that is, it can be used by the body. Although aspartame has about the same caloric value per gram as sucrose, it is 180 times sweeter and so much less can be used.

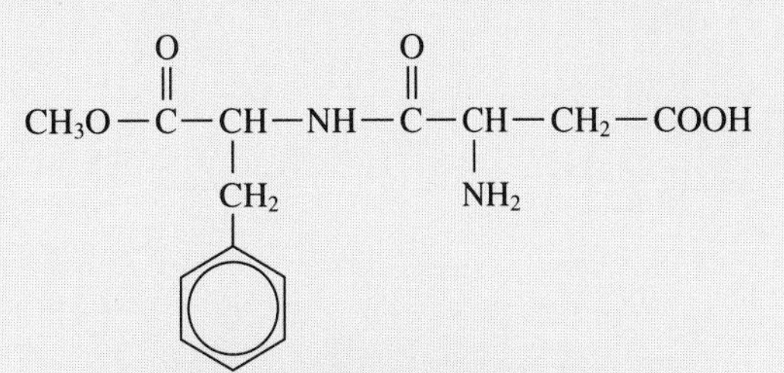

Aspartame

FIGURE 10.A
Aspartame tastes sweet like many sugars, but it is not a sugar; it is a peptide (Chapter 12).

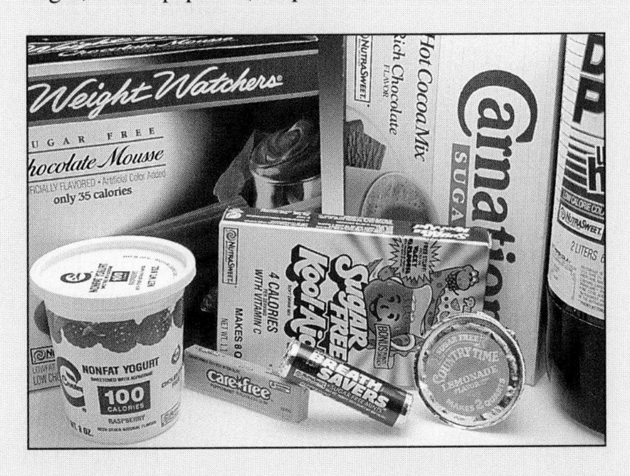

FIGURE 10.B
Aspartame is present in many of the artificially sweetened foods and beverages. Several promising low and noncaloric sweeteners may soon be competing for this market.

Galactose is a third important monosaccharide with six carbon atoms. It is a component of lactose (milk sugar), and it forms part of a variety of larger molecules used by the body. Several genetic diseases involve impaired galactose metabolism.

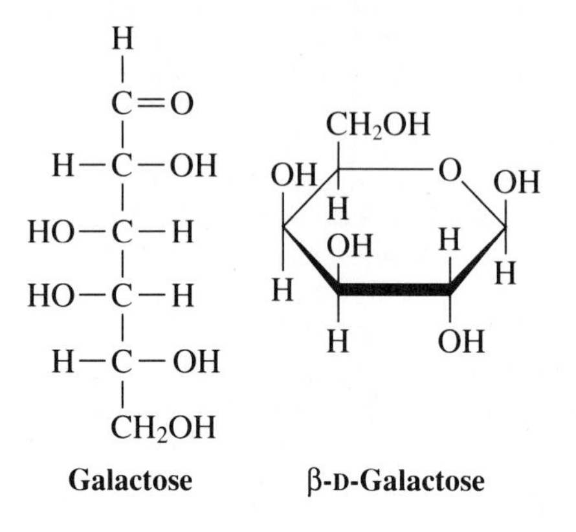

Galactose **β-D-Galactose**

NEW TERMS

aldose A monosaccharide containing an aldehyde group. The prefix *ald-* indicates the presence of an aldehyde group, and the *-ose* suffix signifies a carbohydrate.

ketose A monosaccharide containing a ketone group. Note the prefix *keto-* and the suffix *-ose*.

furanose The five-membered ring form of monosaccharides. The ring resembles the cyclic ether furan.

pyranose The six-membered ring form of monosaccharides. The ring resembles the cyclic ether pyran.

anomeric carbon The carbon atom in the cyclic form of sugars that had been the carbonyl carbon of the open-chain form.

anomers The pair of isomers that forms when a monosaccharide forms a ring. The oxygen attached to the anomeric carbon can have either a *beta* (β) ("up") or an *alpha* (α) ("down") orientation.

TESTING YOURSELF

Monosaccharides

1. Describe the following monosaccharides according to their functional group and number of carbon atoms.

a.
```
      H
      |
      C=O
      |
   H—C—OH
      |
      CH2OH
```

b.
```
      CH2OH
      |
      C=O
      |
   H—C—OH
      |
      CH2OH
```

c.
```
      CH2OH
      |
      C=O
      |
   H—C—OH
      |
   H—C—OH
      |
      CH2OH
```

2. Classify glucose according to its carbonyl functional group and the number of carbon atoms it contains.
3. Identify the rings found in deoxyribose, fructose, and galactose as furans or pyrans.
4. Name the ring forms of deoxyribose, fructose, and galactose.
5. An aldohexose such as glucose contains four chiral centers. How many stereoisomers exist?

Answers **1. a.** aldotriose **b.** ketotetrose **c.** ketopentose **2.** aldohexose **3.** Deoxyribose and fructose contain a furan ring, and galactose contains a pyran ring. **4.** deoxyribofuranose, fructofuranose, and galactopyranose **5.** $2^n = 2^4 = 16$

10.3 Properties and Reactions of Sugars

Solubility

Sugars are water-soluble. This solubility is easily explained because sugars form hydrogen bonds between water and the atoms of their hydroxyl groups. Carbohydrates contain two or more hydroxyl groups that hydrogen-bond with water. This accounts for the high water-solubility of the smaller carbohydrates. Because of their large size, the polymeric carbohydrates are not water-soluble. They form colloidal suspensions or are insoluble.

♦ The hydroxyl groups in sugars greatly influence their water-solubility.

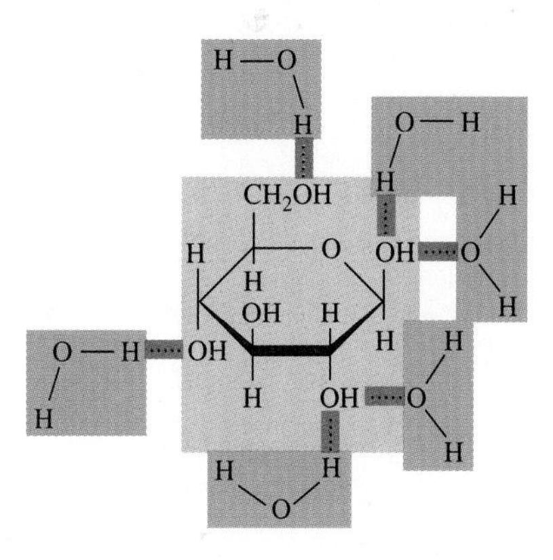

Optical Properties of Sugars

Sugars contain one or more chiral centers. Compounds with chiral centers are optically active because they rotate plane-polarized light. One of the enantiomers rotates light clockwise, and the other rotates light counterclockwise by the same amount. The clockwise rotation of light, dextrorotatory, is designated by a (+) symbol, and counterclockwise rotation, levorotatory, is specified by a (−). D-Glyceraldehyde and D-glucose rotate light to the right (+), and L-glyceraldehyde and L-glucose rotate the light to the left (−). It is simply chance that these two D-monosaccharides both rotate light to the right. There is *no* necessary relationship between family and the rotation of light. For example, D-fructose rotates plane-polarized light to the left (−). The D- and L- family classification simply relates the arrangement of atoms about a single chiral center in the molecule to the arrangement of atoms about the chiral center in glyceraldehyde. It is not related to optical activity.

♦ The direction of rotation of plane-polarized light cannot be predicted by stereochemical family.

Mutarotation

Much of the chemical behavior of a sugar can be related to the form of the carbonyl group (aldehyde or ketone) and the equilibrium between the carbonyl (open-chain) form and the cyclic (hemiacetal or hemiketal) forms. Proof of this interconversion is provided by the phenomenon of **mutarotation**, which is a change in the optical activity of sugar solutions.

When one of the anomers of glucose, α-D-glucopyranose, is dissolved in water, the solution initially has a specific rotation of +112°, as measured by a polarimeter. A similar solution of the other anomer, β-D-glucopyranose, gives a specific rotation of +15°. After either pure anomer is dissolved in water, the specific rotation changes. Eventually, each sugar solution exhibits the same rotation of +52°.

What has happened? The change in specific rotation (mutarotation) can be explained by a dynamic equilibrium between the cyclic hemiacetal forms and the open-chain carbonyl compound (look at Figure 10.2 again). The aldehyde group and the alcohol group of the open-chain structure can form a cyclic hemiacetal, and the hemiacetal can form the open-chain carbonyl compound. Since the reaction is reversible, the hemiacetal products eventually isomerize, since each can return to the open-chain carbonyl compound and then re-form either hemiacetal. The alpha anomer and the beta anomer undergo interconversion. The pure alpha anomer rotates light +112°, and the pure beta form rotates light +15°. Since they can interconvert, either one eventually forms an equilibrium mixture of the two anomers.

♦ Anomers readily convert to each other through a common intermediate, the free aldehyde or ketone.

Oxidation–Reduction of Sugars

The aldehyde group of aldoses is easily oxidized to a carboxyl group, and ketoses can be isomerized to aldoses. Thus, all monosaccharides are reducing agents, that is, they give up electrons as they are oxidized. For this reason, the monosaccharides are classified as **reducing sugars**. Furthermore, the carbonyl groups of both ketoses and aldoses can be reduced. Several derivatives of sugars found in living systems are oxidized or reduced sugars.

The reducing properties of monosaccharides can be used to identify carbohydrates and to determine how much is present. Several convenient analyses for sugars are available, including Benedict's and Fehling's tests. The basis for these tests is the reduction of colored metal ions to other products. Although glucose and many other simple sugars exist in ring forms, they still undergo reactions characteristic of the aldehyde group because they are in equilibrium with the open-chain form. These sugars can be analyzed by observing the oxidation of the aldehyde group to a carboxylic acid. Since every oxidation requires a reduction, progress of the reaction is observed by following color changes that occur as metal ions are reduced to other products. This process is illustrated in Figure 10.3.

Benedict's and Fehling's tests for reducing sugars use the reduction of Cu^{2+} to Cu^{+} as the basis of the tests. Solutions of Cu^{2+} ions are blue, whereas the product Cu_2O is brick-red. Another useful reaction is the reduction of Ag^{+} to free Ag^{0}. If this reaction is carried out in the presence of clean glass, the free silver will deposit on the glass to form a mirror.

EXAMPLE 10.3

Why is this sugar a reducing sugar?

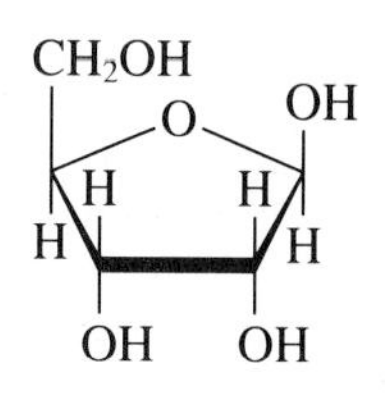

Solution: This is the cyclic form of D-ribose, which is in equilibrium with the open-chain form. The open-chain form has an aldehyde group, and this can reduce certain metal ions.

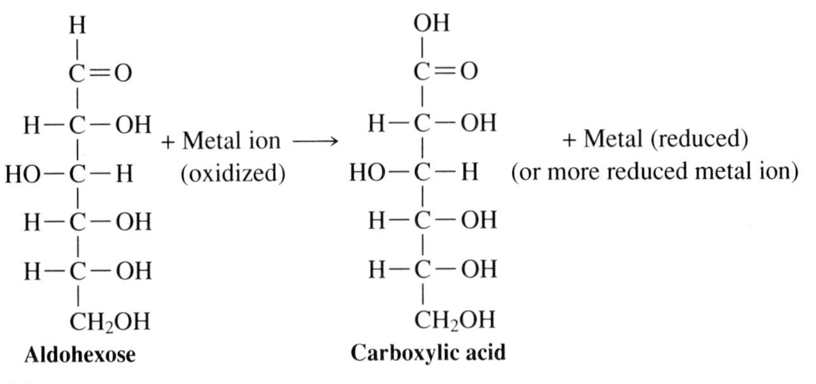

(a)

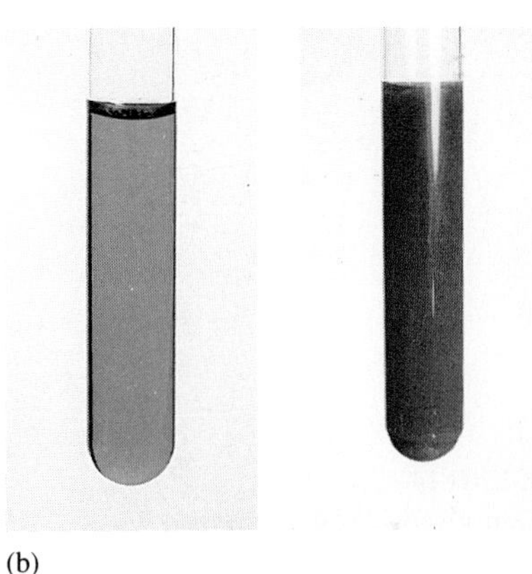

(b)

FIGURE 10.3
(a) The oxidation of an aldohexose to a carboxylic acid by a metal ion. The ions copper(II) and silver(I) are commonly used for this reaction. (b) Benedict's solution is used to show that glucose has reduced Cu(II) ions (blue) to Cu(I) ions (red). Under the conditions of this experiment, Cu(II) is soluble, Cu(I) is rather insoluble.

Glycosidic Bonds

The anomeric hydroxyl group of a cyclic sugar is the hydroxyl group of a hemiacetal or hemiketal. This hydroxyl group is somewhat more reactive than the other —OH groups of a sugar molecule. Hemiacetals and hemiketals react with alcohols to form acetals and ketals, respectively.

$$\text{R(H)C(OH)(OR}') + \text{R}''\text{OH} \overset{\text{H}^+}{\rightleftharpoons} \text{R(H)C(OR}'')(\text{OR}') + \text{H}_2\text{O}$$

Hemiacetal **Alcohol** **Acetal**

When the hemiacetal of a sugar reacts with an alcohol, the product is a carbohydrate equivalent of an acetal (ketal) called a **glycoside**. The covalent bond between the hemiacetal or hemiketal and the alcohol is called a **glycosidic bond** (Figure 10.4). The glycoside has either an alpha or a beta glycosidic bond, depending on the orientation of the sugar's anomeric —OH group when the glycosidic bond formed. Once the glycosidic bond has formed, the sugar is locked into the anomeric form it had when the glycosidic bond formed. The ring does not open as long as the glycosidic bond is present.

♦ Glycosidic bonds link the cyclic forms of sugars to alcohols.

The name of the sugar of the glycoside is used to name the glycoside. For example, if the sugar of a glycoside is glucose, the compound is named as a glucoside. Fructose is the sugar in a fructoside, and galactose the sugar in a galactoside. The stereochemistry of the sugar is included, as is the name of the alcohol.

NEW TERMS

mutarotation The change in specific rotation seen in sugar solutions that results from equilibrium between anomeric forms

reducing sugars Sugars that have a free carbonyl group that can be oxidized. They are thus reducing agents.

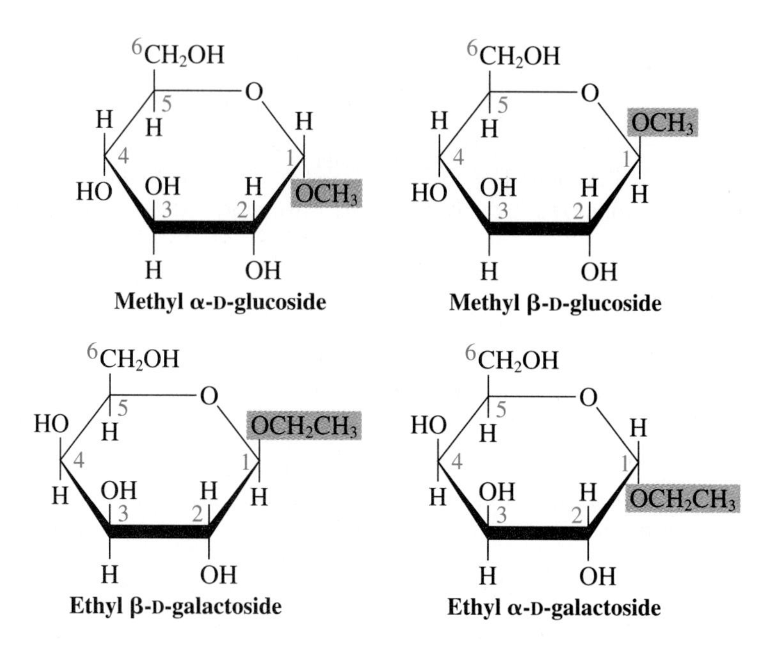

FIGURE 10.4
These glycosides were formed when glucose and galactose reacted with methanol and ethanol. The name indicates which alcohol (red) and sugar formed the glycoside. Note that two possible isomers are formed for each sugar and alcohol because the sugar can exist as either of two anomers.

glycoside Compound formed when a cyclic sugar is bonded to an alcohol through a glycosidic bond

glycosidic bond Bond between the anomeric carbon of a cyclic sugar and the —OH group of another sugar or an alcohol. This bond links sugars together in oligosaccharides and polysaccharides.

TESTING YOURSELF

Properties of Sugars and Glycosidic Bonds

1. a. Classify glucose, lactose, and sucrose as reducing or nonreducing sugars. (Refer to Section 10.4 for the structure of lactose and sucrose.)
   b. Why is each reducing or nonreducing?
2. Do the following glycosides have alpha or beta linkage?

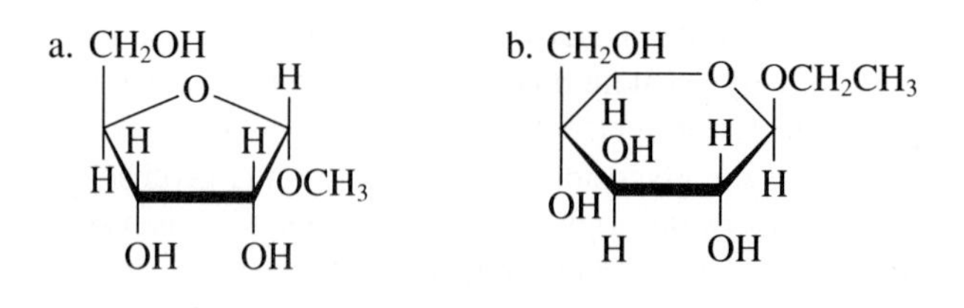

Answers **1. a.** Glucose and lactose are reducing sugars; sucrose is nonreducing. **b.** Glucose and lactose contain a free anomeric carbon that can yield a free carbonyl group. Sucrose has both anomeric carbon atoms tied up in a glycosidic bond. **2. a.** alpha **b.** beta

10.4 Oligosaccharides

Acetals and ketals form when hemiacetals or hemiketals react with an alcohol. Cyclic sugars are hemiacetals and hemiketals with additional hydroxyl groups present. These hydroxyl groups are available to react with the anomeric hydroxyl group of another sugar molecule. For example, the hydroxyl group on carbon 4 of glucose could react with the anomeric hydroxyl of another glucose.

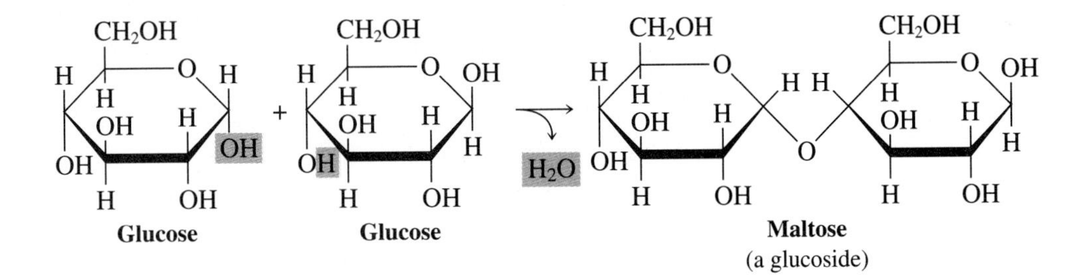

Oligosaccharides are sugars consisting of two to several monosaccharides linked by glycosidic bonds. In biochemistry, the positions within a larger molecule that are occupied by smaller units or building blocks are referred to as **residues**. The residues in an oligosaccharide or polysaccharide are occupied by monosaccharides.

By far the most common oligosaccharides are *disaccharides*. There are several common disaccharides that you should know. Lactose, which is milk sugar, is an important source of energy for the young of mammals. It makes up about 5 percent of cow's milk, but human milk contains somewhat more. Lactose consists of the monosaccharides galactose and glucose linked by a β 1→4 glycosidic bond (Figure 10.5). As shown in the figure, the oxygen on the anomeric carbon of galactose is in the beta position and is bonded to the number 4 carbon of glucose. The designation β 1→4 is used to identify which atoms of the residues are bonded together and indicates the orientation of the anomeric hydroxyl group when the glycosidic bond was formed. The anomeric carbon atom of the glucose residue in lactose does not participate in a glycosidic bond. Like free glucose, this glucose residue exists in an equilibrium between the beta anomer, the alpha anomer, and the open-chain form. Since the open-chain form has a free carbonyl group that can reduce metal ions, lactose is a reducing sugar.

◆ The most important biological oligosaccharides are the disaccharides.

◆ Lactose is milk sugar.

Sucrose is a disaccharide found in plants, where it is used for transportation of sugar residues. In addition, sucrose is found in the nectar of flowers, where it serves as an inducement to insects and other animals, which in turn pollinate the flowers. Sucrose is a disaccharide of glucose and fructose linked by an α,β 1→2 glycosidic bond (Figure 10.5b). Examine the structure of fructose in the figure to be sure you understand what the designation α,β 1→2 means. Note that in sucrose, the anomeric hydroxyl group of both fructose and glucose are involved in the glycosidic bond. This is stated in the nomenclature of the bond, and since neither has a free anomeric hydroxyl group, neither can exist as the free carbonyl compound. Unlike the aldoses and lactose, sucrose is not a reducing sugar.

◆ Sucrose has other common names besides table sugar. Sucrose obtained from cane is called cane sugar; from beets, it is beet sugar.

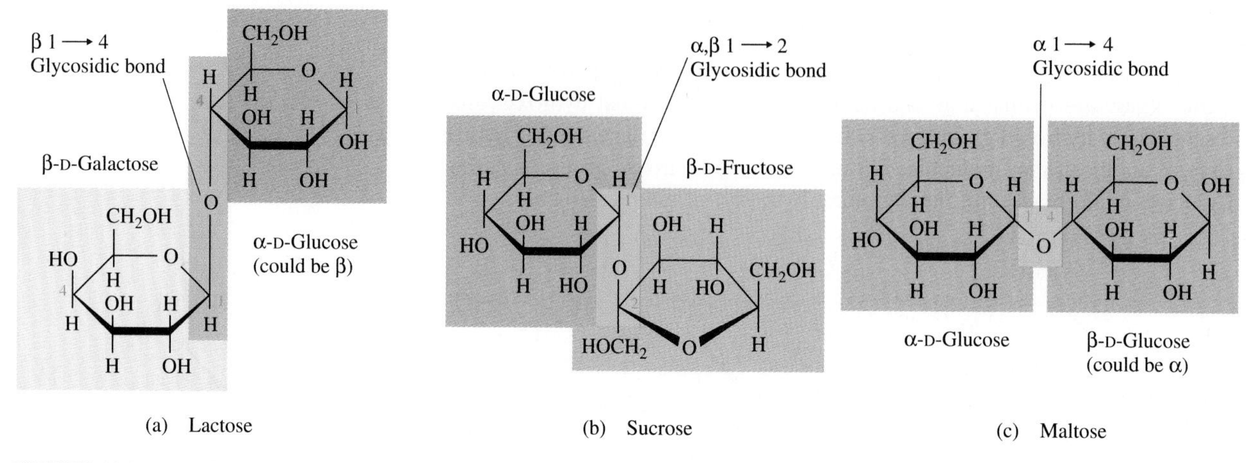

FIGURE 10.5
Three important disaccharides that are common components of foods and partially digested starch are lactose, sucrose, and maltose. Take a moment to compare their monosaccharides and glycosidic bonds (blue).

Maltose is a disaccharide made up of two residues of glucose linked by an α 1→4 glycosidic bond (Figure 10.5c). Maltose is obtained from the malt of germinating grain and is present in the digestive tract during the digestion of starch. In germinating grain or in the digestive tract, maltose is a degradation product of polysaccharides. Malt, which is obtained from germinating barley, is used to make beer. The sugars in malt, including maltose, serve as the energy source for yeast during the fermentation process.

Although there are many oligosaccharides besides the disaccharides that play a role in the normal structure and function of the body, only a few will be mentioned in this text. For example, carbohydrates are bonded to many of the proteins and lipids of the body.

NEW TERM

residue A position within an oligomer or polymer; the residues of an oligosaccharide are occupied by monosaccharides.

TESTING YOURSELF

Oligosaccharides

1. Name the simple sugars in (a) lactose, (b) maltose, and (c) sucrose.
2. Explain the designation β 1→4 used to describe the glycosidic bond of lactose.

Answers **1. a.** galactose and glucose **b.** glucose **c.** fructose and glucose **2.** The anomeric —OH of galactose is in the beta position and is bonded to an —OH group on carbon number 4 of glucose.

10.5 Polysaccharides

♦ Polysaccharides are polymers of monosaccharides.

When many monosaccharide units are linked together by glycosidic bonds, the resulting macromolecules are called *polysaccharides*. These are polymers of simple sugars, and some are very abundant in nature. Starch, glycogen, cellulose, and chitin are common examples of polysaccharides. Although each of these polysaccharides is a polymer of glucose or a glucose derivative, they have important differences in properties that result from differences in structure. Among these differences is the suitability of the polysaccharide as a nutrient.

Starch and Glycogen: Storage Polysaccharides

Starch and **glycogen** are similar molecules that contain D-glucose molecules linked by α 1→4 glycosidic bonds and, in some cases, α 1→6 bonds (Figure 10.6). Starches are storage polysaccharides found in plants and are the **complex carbohydrates** of the human diet. There are two different types of polymers found in starches, *amylose* and *amylopectin*. These polymers differ in size and shape. Amylose is a linear polymer, that is, it has all of its glucose residues connected together into one long chain by α 1→4 glycosidic bonds (Figure 10.6a). Amylopectin is larger than amylose and is branched. Some of the glucose residues have both α 1→4 and α 1→6 glycosidic bonds, and a branch occurs at each of these residues. The difference in size, from several hundred glucose residues for amylose to several thousand for amylopectin, accounts for the difference in their behavior in water. Neither is truly soluble in water because both are too big to form true solutions. They form colloidal dispersions. Amylose forms these more readily in hot water because it is smaller. In water, the α 1→4 bonds in starch molecules are relatively exposed to water molecules. Thus, these bonds can be cleaved readily by hydrolysis.

Glycogen, like amylopectin, is also highly branched (Figure 10.6b). The branching in glycogen has physiological significance. In liver and muscle cells, a specific, highly regulated enzyme named glycogen phosphorylase releases individual phosphorylated glucose molecules at appropriate times from the ends

FIGURE 10.6
A comparison of amylose and glycogen. (a) Amylose is a linear polymer composed only of glucose residues linked by α 1→ 4 bonds.
(b) Glycogen (and the similar substance amylopectin) consists of glucose residues linked by α 1→ 4 and α 1→ 6 glycosidic bonds. Each α 1→ 6 bond introduces another branch point that provides an additional terminal-sugar residue. These terminal sugars in glycogen can be quickly cleaved, yielding a glucose derivative for quick energy production. Residues are shown as glucose molecules or as small circles.

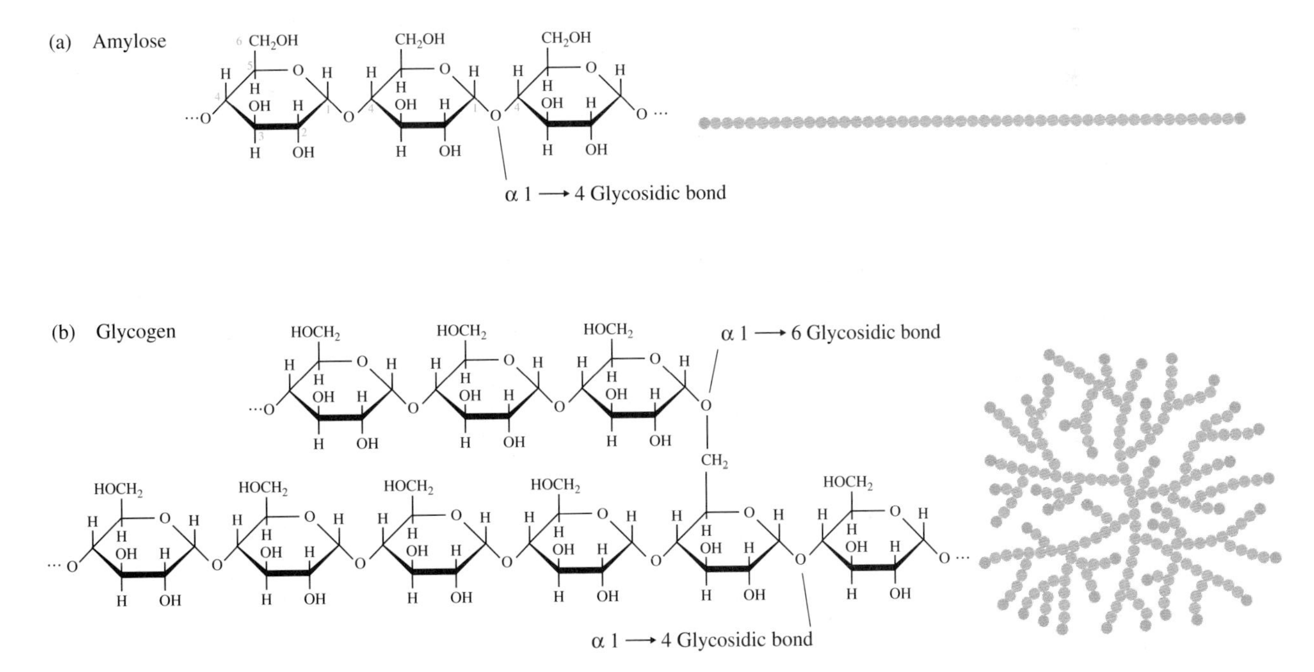

CHEMISTRY CAPSULE

Carbohydrates and Nutrition

The carbohydrates can be classified into three categories as nutrients—sugars, complex carbohydrates, and fiber. Dietary sugars are water-soluble monosaccharides and disaccharides. Sucrose, more commonly known as table sugar, is perhaps the most common dietary sugar, but glucose (blood or grape sugar), fructose, and lactose (milk sugar) are other common sugars of the diet. During digestion, oligosaccharides are broken down to simple sugars.

Complex carbohydrates and fiber are polysaccharides. The complex carbohydrates, which are principally starches, are broken down by digestion to simple sugars that are used by the body. Fiber consists primarily of polysaccharides and complexes of polysaccharides that cannot be digested. There has been speculation that the higher incidence of cancer of the lower intestinal tract for people of "developed" countries might be related to the relatively low fiber content of their diet. Studies of this nature have resulted in modification by some people of their diets and have led to additional studies that may ultimately clarify the role of fiber in diet.

Simple sugars in the diet and those produced by digestion are a major source of energy for the body. From each gram of digestible dietary carbohydrate 4 Cal (4 kcal) are available. Through a series of chemical reactions, glucose is broken down to carbon dioxide and water. The overall process is highly exothermic, and much of the released energy is conserved by the body for various tasks. Simple sugars are also used by the body to make a variety of biomolecules. Sugars are found as components of all of the major classes of biomolecules in the body. Some lipids, some proteins, and all of the nucleotides and nucleic acids contain sugars.

FIGURE 10.C
Carbohydrates are widespread, but fruits, grain products, and some vegetables are good sources of sugars and complex carbohydrates.

There are many dietary sources of carbohydrates. Generally foods from plants are relatively higher in carbohydrates than foods derived from animals. Roughly 40 to 50 percent of the calories consumed in the U.S. diet are derived from carbohydrates. In other cultures the percentage is somewhat higher. At the present time, nutritionists believe a larger proportion of dietary energy should come from dietary carbohydrate and that an increase in complex carbohydrates and natural sugars should be accompanied by a decrease in consumption of processed sugars such as table sugar.

of the chains of glycogen. These molecules can be used by the cell for energy or other needs. Since glycogen is highly branched, there are many ends to bind phosphorylase molecules. Thus, at any given moment, many phosphorylated glucose molecules can be released. In contrast, a linear polysaccharide like amylose has only one or two ends to act as sites where an enzyme could release individual molecules from the polymer. Clearly the highly branched nature of glycogen facilitates its rapid breakdown when glucose residues are needed.

♦ The highly branched structure of glycogen enhances rapid mobilization of energy reserves.

Cellulose and Chitin: Structural Polysaccharides

Cellulose and **chitin** are structural polymers with glycosidic bonds that are β 1→4. This is in contrast to starches and glycogen (used for energy storage), which have α 1→4 bonds. In general, α 1→4 glycosidic bonds are more easily

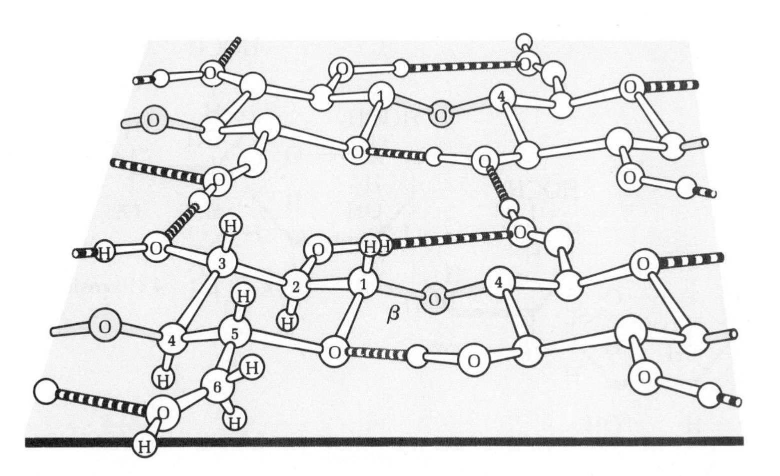

FIGURE 10.7
Hydrogen bonding between cellulose molecules forms sheet-like structures. These sheets in turn are hydrogen-bonded to sheets located above and below. Because of this extensive hydrogen bonding, the β 1→4 bonds in cellulose are not exposed, and hydrolysis is slow.

cleaved than β 1→4 bonds because they are more exposed. Figure 10.7 illustrates a small piece of a cellulose fiber, which consists of numerous cellulose molecules that are hydrogen-bonded to each other. Note the extensive hydrogen bonding that extends up, down, and sideways. As a consequence, few of the β 1→4 bonds are exposed, and the rate of cellulose breakdown is slower than starches.

This difference in rates of breakdown between starches and cellulose makes sense. Structural materials should be relatively resistant to breakdown; energy-storage forms should be converted readily to smaller, more easily used energy forms. Cellulose is a polymer of glucose, but in chitin the —OH group of carbon number 2 in each glucose molecule is replaced by an amide group (Figure 10.8). Chitin is the polysaccharide found in the exoskeletons of crabs, shrimp, insects, and other arthropods. Cellulose is the principal structural polysaccharide of plants, and because plants are so abundant, it is estimated that for every person on earth, the plant world synthesizes about 100 lb of cellulose per day! This biomass has great potential as a source of energy and chemicals.

♦ Cellulose and chitin are structural polysaccharides that help give rigidity to plants and arthropod exoskeletons, respectively.

Wood is primarily cellulose fibers embedded in a highly polymerized substance called lignin. This complex of cellulose and lignin is a strong, enduring substance that is ideal for many structural uses. Steel-reinforced concrete mimics the structure of natural wood because the steel rods are analogous to the cellulose fibers and the concrete is like the lignin. Similarly, fiberglass fishing rods consist of fiberglass embedded in resin.

Starch (α 1→4) and cellulose (β 1→4) differ only in the orientation of the glycosidic linkage between the glucose residues, yet humans efficiently digest starch in food but cannot utilize cellulose. In general, mammals and other higher animals do not possess the necessary enzymes to catalyze the cleavage of cellulose to glucose. However, grazing animals, such as sheep and cattle (ungulates), possess intestinal microflora that make these enzymes. Cellulose hydrolysis is catalyzed by microbial enzymes, and the products are thus available to the microflora and the host animal. Thus, cattle and related organisms can use many plants for food that are not suitable for other animals. Dietary cellulose, however, is not a completely inert, wasted portion of the diet for humans and other animals. Cellulose is a major part of the indigestible, tasteless **fiber** present in our diet as roughage. There appear to be some benefits of a high-fiber diet.

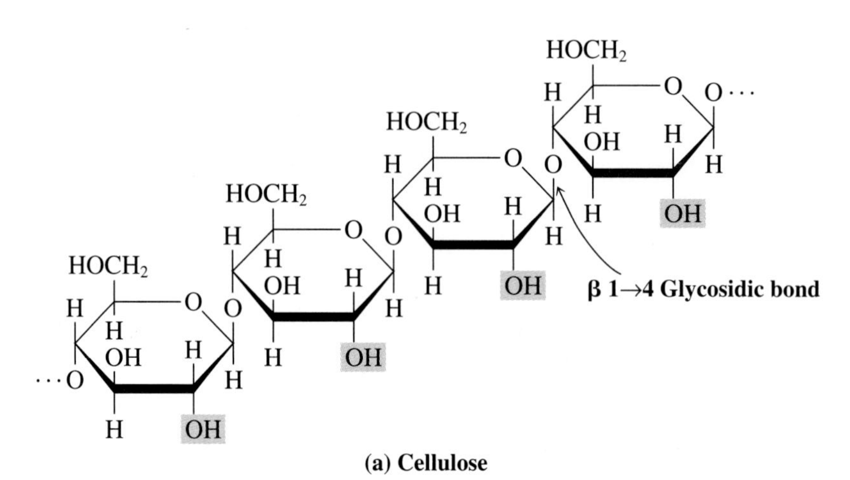

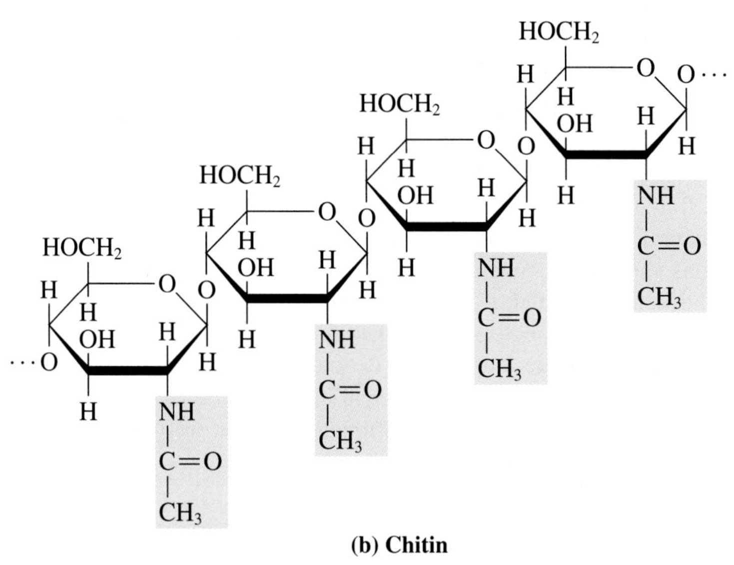

FIGURE 10.8
A comparison of (a) cellulose and (b) chitin. These structural polysaccharides possess beta glycosidic bonds that are not readily broken. Chitin has a different sugar than cellulose; cellulose contains glucose, whereas chitin has *N*-acetylglucosamine residues (see highlighted portions for structural difference).

Other Polysaccharides: Hyaluronic Acid and Heparin

The polysaccharides discussed so far have had repeating units consisting of a single sugar or a single-sugar derivative. Some important polysaccharides have a more complex repeating unit. **Hyaluronic acid** is found in the connective tissue of higher animals. It is part of the viscous material around bone joints that absorbs shock and acts as a lubricant for the bone surfaces. In Figure 10.9 the unique glycosidic bonding and the *two*-sugar repeating units of hyaluronic acid are shown.

Heparin is a polysaccharide produced by some cells of the circulatory system. It is a powerful blood anticoagulant because it inhibits blood clotting. Heparin is used in a number of medical applications where the chance of blood clotting must be reduced or eliminated. Heparin is common to most animals and is an almost linear polysaccharide. It also has a unique two-sugar repeating unit and is characterized by a large number of sulfate groups (Figure 10.10).

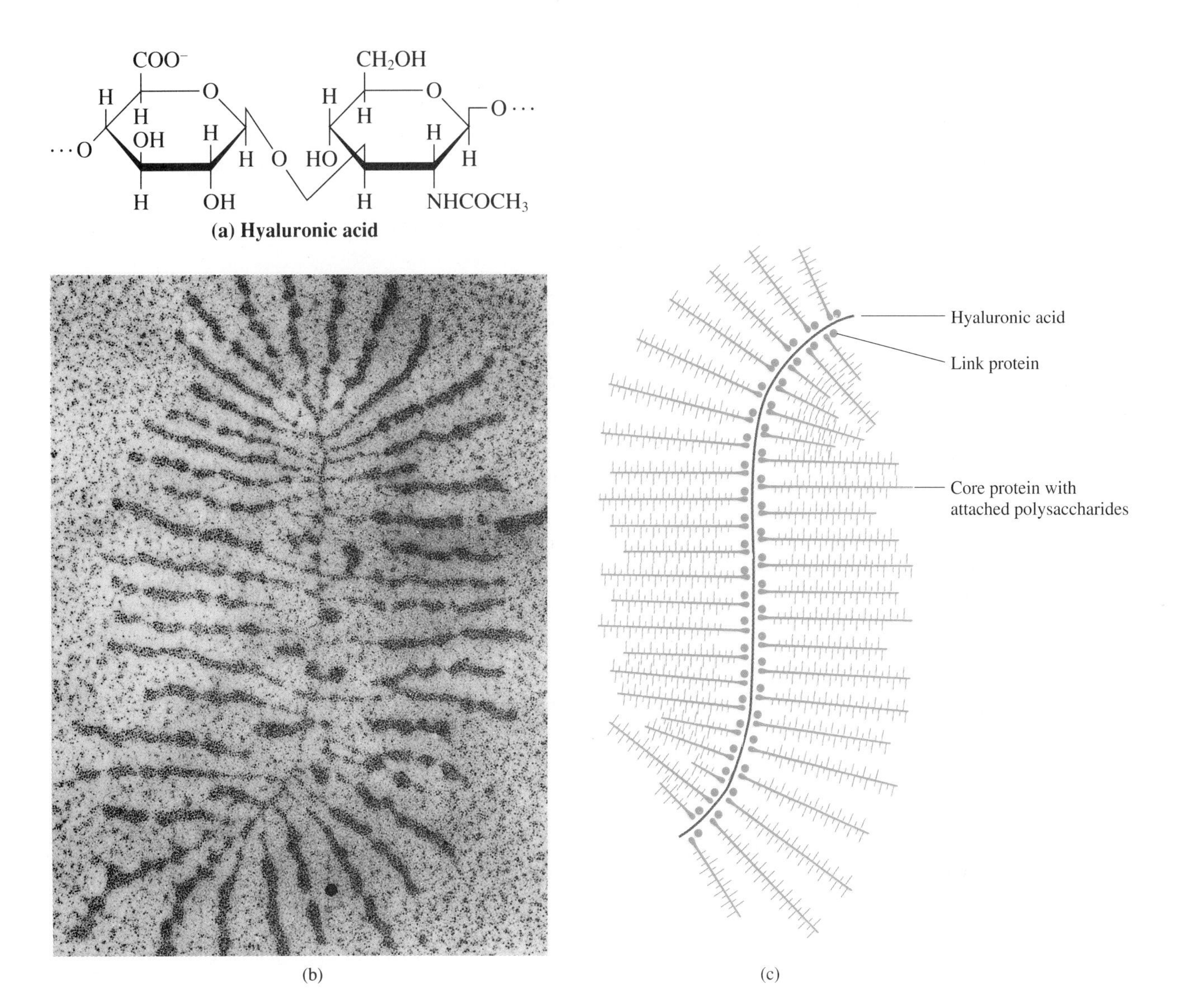

FIGURE 10.9
Hyaluronic acid is a polysaccharide found in connective tissues. (a) This disaccharide is the repeating unit found in hyaluronic acid. (b) An electron micrograph and (c) interpretative drawing of developing cartilage. Polysaccharides (blue) found in connective tissue are normally complexed with proteins (blue) to form proteoglycans. Hyaluronic acid makes up the central backbone of this aggregate (purple).

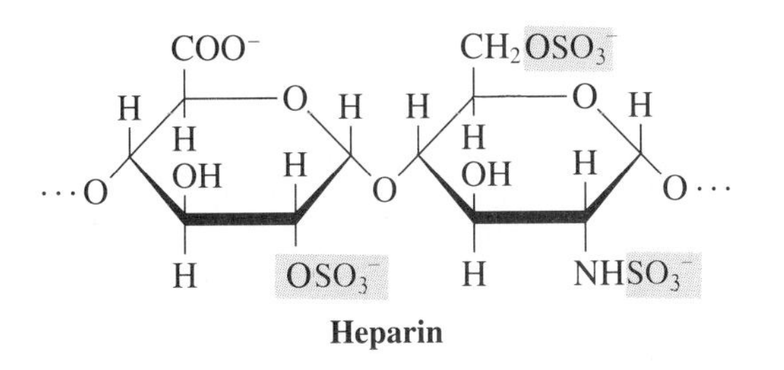

FIGURE 10.10
The structure of a short segment of heparin. Note the repeating unit of heparin and the sulfate groups (yellow) that are present.

NEW TERMS

starch A readily digestible plant polysaccharide made of glucose residues bonded by α 1→4 and some α 1→6 glycosidic bonds

glycogen The animal equivalent of starch; the polysaccharide that is the storage form of glucose in animals

complex carbohydrates The digestible polysaccharides of the diet; mostly starches

cellulose A plant polysaccharide made up of β 1→4 linked glucose; the structural material in plants

chitin A structural polysaccharide found in arthropods, similar to cellulose but has an amide in place of an —OH on carbon number 2 of the glucose residues

fiber The indigestible carbohydrates of the human diet; mostly polysaccharides such as cellulose and complexes of polysaccharides

hyaluronic acid A complex polysaccharide found in connective tissue and in bone

heparin A polysaccharide that functions as an anticoagulant

Summary

Carbohydrates are polyhydroxy aldehydes and ketones and derivatives of these compounds. The nomenclature of *monosaccharides* (simple sugars) is closely related to the structure of the sugar. Thus, carbon number, carbonyl type, ring size, and stereochemistry may be included in typical names for monosaccharides. Monosaccharides, particularly glucose, are linked together by *glycosidic bonds* to form *oligosaccharides* and polymers called *polysaccharides*. Most carbohydrates, such as *starch* and *glycogen*, serve as energy reserves or as structural elements in or between cells.

Terms

carbohydrate (10.1)
monosaccharides (10.1)
oligosaccharides (10.1)
polysaccharides (10.1)
aldose (10.2)
ketose (10.2)
furanose (10.2)
pyranose (10.2)
anomeric carbon (10.2)
anomers (10.2)
mutarotation (10.3)
reducing sugars (10.3)
glycoside (10.3)
glycosidic bond (10.3)
residue (10.4)
starch (10.5)
glycogen (10.5)
complex carbohydrates (10.5)
cellulose (10.5)
chitin (10.5)
fiber (10.5)
hyaluronic acid (10.5)
heparin (10.5)

Exercises

Monosaccharides (Objective 1)

1. Define the term *monosaccharide*.
2. Identify each of the following sugars by functional group, carbon length, and D- or L- family:

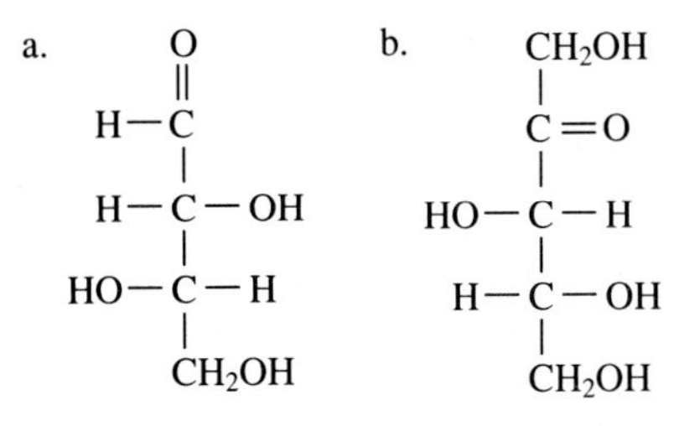

3. What information is contained in the term *ketopentose*?
4. How can an L- sugar be recognized?
5. Identify each of the simple sugars shown here:

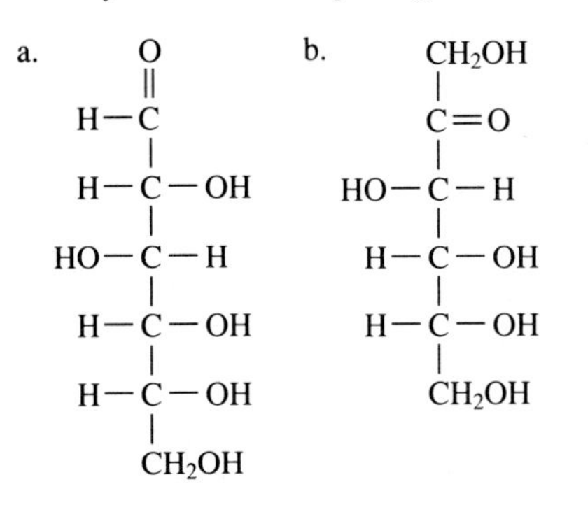

6. Draw the structure of the simplest ketose and aldose.
7. Name the smallest aldose and ketose.
8. What functional groups are found in carbohydrates?
9. Identify the functional groups on the carbon atoms of glucose and fructose (use the open-chain forms).
10. Draw the open-chain structure of:
    a. an aldopentose
    b. a ketohexose
    c. an aldotetrose in the L-stereochemical family
11. If you have a drawing of an open-chain D-aldohexose, what minimal structural change(s) could you make to convert it into an L-aldohexose?
12. Which is a monosaccharide?
    a. sucrose b. galactose c. maltose d. cellulose
13. Which of the following sugars is a monosaccharide and a furanose?
    a. glucose b. galactose c. fructose d. sucrose
14. What are the similarities and differences of glucose and fructose?
15. What are the similarities and differences of ribose and 2-deoxyribose?
16. Which carbon atom of an aldopentose is the anomeric carbon atom in the cyclic hemiacetal? Which carbon atom is the anomeric carbon atom in a ketohexose?
17. One of your classmates says α-D-glucopyranose and β-D-glucopyranose are enantiomers, another says D-glucose and L-glucose are enantiomers. Which is correct? Justify your choice.
18. Examine the structures below:
    a. Indicate which of them is a cyclic hemiacetal.
    b. Circle the anomeric carbon atom in each hemiacetal.

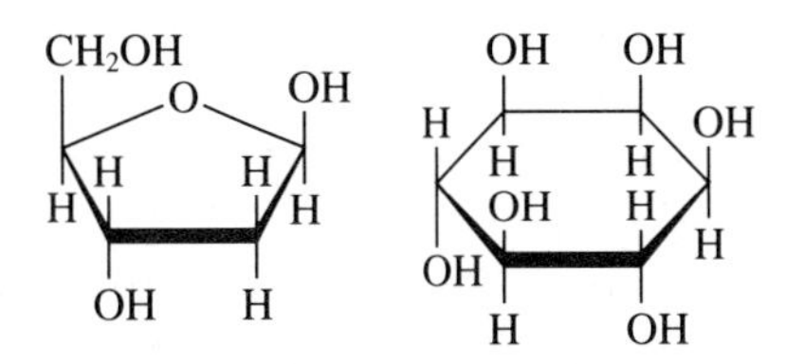

19. How is an alpha anomer recognized?
20. Describe the furanose and pyranose forms of carbohydrates.
21. Show how hemiacetal formation and the cyclic forms of sugars are related.
22. A carbohydrate is shown to be an aldotetrose with two chiral centers. It is shown to be in the D- stereochemical family. Draw the possible structure or structures for this compound.
23. How many chiral centers are present in D-fructose? How many stereoisomers exist for ketohexose?

Reactions and Glycosides (Objectives 2, 3, and 4)

24. Why are monosaccharides and disaccharides water-soluble?
25. D-Ribose is levorotatory but D-glucose is dextrorotatory. How can this be?
26. Mutarotation is a term related to
    a. interconversion of anomers
    b. relationship of D- to L- families
    c. hydrolysis of sucrose
    d. number of simple sugars in a carbohydrate
27. Describe the process of mutarotation.
28. A reducing sugar will
    a. react with Benedict's reagent
    b. have fewer calories
    c. always be a ketose
    d. none of these
29. Describe the chemistry involved in the designation "reducing sugar."
30. Sucrose is not a reducing sugar. Why?
31. Fructose is not an aldose, yet it is classified as a reducing sugar. How can this be?

32. Identify each of the following glycosidic linkages as alpha or beta:

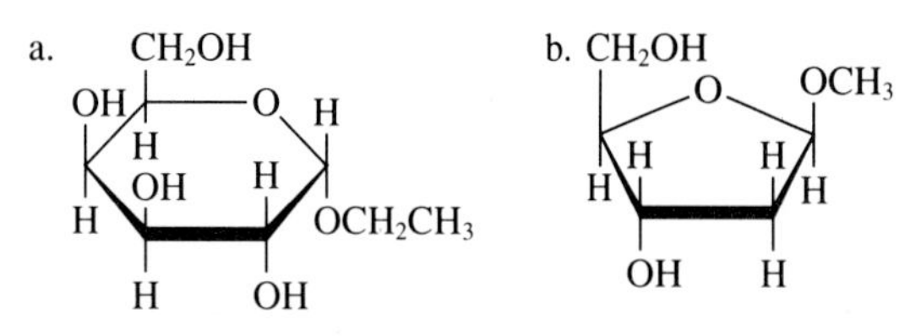

33. Examine the structures of ethyl β-D-galactoside and ethyl α-D-galactoside in Figure 10.4. Are these substances enantiomers? Why or why not?
34. Would methyl α-D-glucoside (Figure 10.4) show mutarotation? Why or why not?
35. Identify each of these disaccharides:

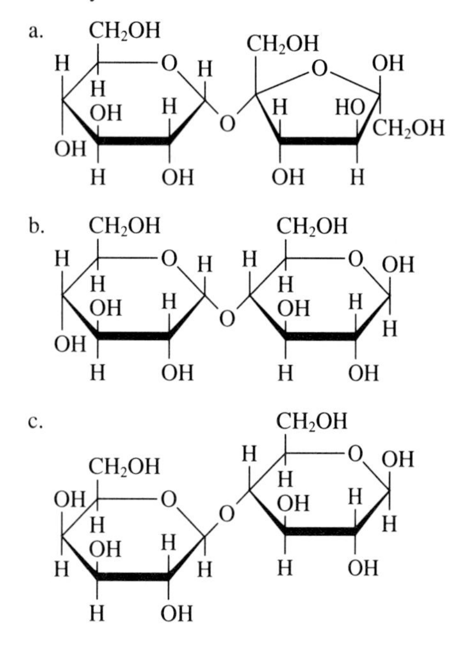

36. Common table sugar is
    a. glucose b. fructose c. sucrose d. maltose
37. Identify the aldose and ketose found in honey.
38. Draw the structure of a disaccharide discussed in this chapter that is a reducing sugar and yields only glucose upon hydrolysis. Draw the structure of another disaccharide that meets these conditions.
39. The sugar shown here is found in some mushrooms. How does this disaccharide differ from most other disaccharides you have seen? Is it a reducing sugar?

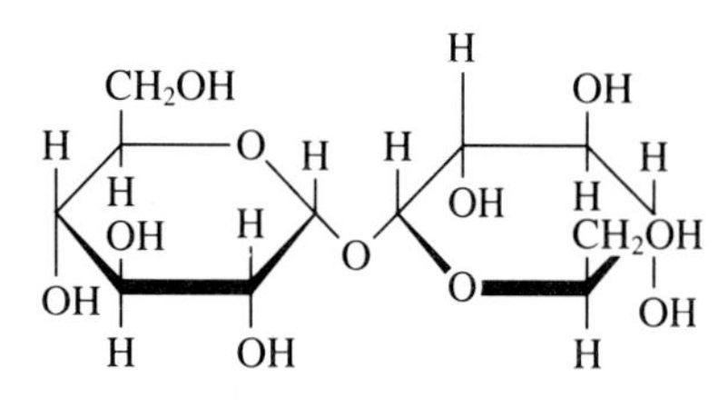

40. Hydrolysis of disaccharides yields two simple sugars. What is (are) the simple sugar(s) obtained from the hydrolysis of the following?
    a. sucrose b. maltose c. lactose
41. What glycosidic linkages are found in the disaccharides in exercise 40?
42. Draw the structure of a disaccharide formed from two 2-D-deoxyribose molecules linked by an α 1→3 glycosidic bond.
43. Consider the molecule you have drawn in exercise 42. Is it a reducing sugar?
44. Draw one of the two structures of maltose.
    a. Circle any hemiacetal that is present.
    b. Circle any acetal that is present.
    c. What other name can be used to identify the acetal group (include stereochemistry)?

Polysaccharides (Objectives 3 and 5)

45. What determines whether a carbohydrate is classified as a polysaccharide or an oligosaccharide?
46. What polysaccharides yield only D-glucose when hydrolyzed?
47. What is the structural difference between amylose and cellulose?
48. What is the structural difference between amylose and amylopectin?
49. Describe the similarities and differences between cellulose and chitin.
50. What is the physiological role for glycogen?
51. Explain why glycogen can be broken down rapidly in liver or muscle cells.
52. Why can cattle digest cellulose and humans cannot?
53. Where is hyaluronic acid found and what is its role?
54. What is heparin and what is its physiological role?
55. A starch sample is estimated to have an average molecular weight of 140,000. How many glucose residues are in an average starch molecule of this sample? (*Hint:* first calculate the molecular weight of glucose, then subtract the molecular weight of water from it because a water molecule is lost during glycosidic bond formation.)
56. Cellobiose is a disaccharide obtained by partial hydrolysis of cellulose. What is the composition of this disaccharide? What glycosidic bond is present?

Unclassified Exercises

57. Match the term on the left with the description on the right:

A. amylose	a. ketose
B. glycogen	b. table sugar
C. fructose	c. a polymer of glucose with beta glycosidic linkages
D. sucrose	d. a polymer of glucose stored in the body
E. triose	e. a three-carbon sugar
F. cellulose	f. a form of starch

58. For each of the following words, give an explanation, definition, or example that demonstrates your understanding of the word:
    a. carbohydrate
    b. pentose
    c. aldose
    d. ketose
    e. ketohexose
    f. L-family
    g. furanose
    h. anomeric carbon
    i. anomers
    j. glycosidic linkage
    k. glycoside
    l. monosaccharide
    m. disaccharide
    n. polysaccharide
    o. glycogen
    p. cellulose
    q. amylose
    r. reducing sugar
    s. aspartame
    t. lactose
    u. mutarotation

59. Draw structures for each of the following:
    a. D-aldopentose
    b. L-glyceraldehyde
    c. α-D-glucopyranose
    d. β-D-galactopyranose
    e. the repeating unit for starch and glycogen
    f. the repeating unit for cellulose
    g. the isomers of "ketotetrose"

60. Which of the following carbohydrates are classified as reducing sugars?
    a. galactose
    b. fructose
    c. starch
    d. glycogen
    e. sucrose
    f. lactose

61. Compare the appearance of a dilute glucose solution and a more concentrated glucose solution when they are mixed with Benedict's solution.

62. Examine the formulas of ribose and glucose and use them to explain the origin of the term *carbohydrate*.

63. If 0.03 mol of a carbohydrate yield 0.12 mol of glucose upon hydrolysis, what kind of carbohydrate is it?

64. A children's breakfast cereal contains 11 g of sugar and 15 g of complex carbohydrate per 1-oz serving. (a) Determine the number of calories provided by these carbohydrates. (b) What percentage is provided in the form of complex carbohydrates?

65. Many distance runners practice "carbo loading" (eating carbohydrate-rich foods) for one to three days preceding a race. What do they hope to accomplish?

CHAPTER 11

LIPIDS

Bilayers are the main structural feature of this soap bubble and biological membranes.

OUTLINE

11.1 Classification of Lipids
11.2 Fatty Acids
11.3 Triacylglycerols
11.4 Saponifiable Lipids of Membranes
11.5 Nonsaponifiable Lipids
11.6 Liposomes and Membranes

OBJECTIVES

After completing this chapter, you should be able to

1. Describe the general structures of the various lipids.
2. Classify lipids as saponifiable or nonsaponifiable.
3. Identify the structure and role of fatty acids.
4. Describe the structural differences between fats and oils.
5. Describe the function of lipids in membranes.
6. Identify the common structure of steroids.

Unlike carbohydrates, the compounds known as lipids are not composed of similar, chemically related structural units. Instead, lipids contain a variety of functional groups and structural features. **Lipids** are the biological molecules that are insoluble in water and soluble in organic solvents. As the hydrocarbon component (the alkyl group) of an organic compound increases in size, the relative contribution of a functional group to the physical properties of the molecule decreases. For example, low-molecular-weight acids and alcohols are readily soluble in water because the polar —COOH or —OH groups hydrogen-bond with water. High-molecular-weight acids and alcohols still have the polar functional groups, but they are water-insoluble because the large nonpolar alkyl group does not bond to water. Alkyl groups are *hydrophobic* ("water hating") and therefore they are water-insoluble. As the size of an alkyl group increases, the water-solubility of the compound decreases. Lipids are molecules that have a large hydrocarbon component and are largely or entirely nonpolar.

♦ Lipids are soluble in nonpolar organic solvents but insoluble in water.

Because lipids are defined in terms of their solubility properties, it should not be surprising to learn that lipids vary greatly in structure and function. For example, fats and oils are involved in energy storage and utilization, but phospholipids are structural elements of membranes and other macromolecular complexes. Some lipids are hormones, some play protective roles, and others aid in digestion and nutrient transport. Lipids are an important and diverse group of biomolecules.

11.1 Classification of Lipids

Lipids are often classified as either saponifiable or nonsaponifiable. Saponification occurs in two steps. In the first step hydroxide ion of a strong base, such as sodium hydroxide, catalyzes the hydrolysis of an ester bond. The carboxylic acid then reacts with the hydroxide ion of the base to yield the salt of the acid and water. The net reaction is:

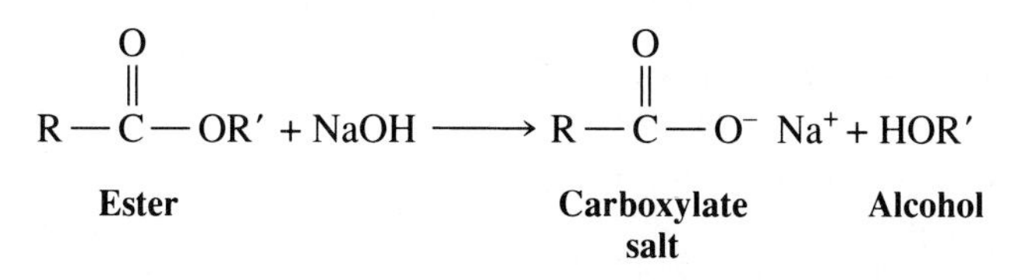

Saponifiable lipids are hydrolyzed by base to give the salts of large carboxylic acids (*fatty acids*) and an alcohol. Saponifiable lipids include the triacylglycerols, glycolipids, sphingolipids, some waxes, and a variety of phosphate-containing lipids called phospholipids. **Nonsaponifiable lipids** are those that are *not* hydrolyzed by base and include the steroids, prostaglandins, leukotrienes, some waxes, and terpenes.

♦ Saponifiable lipids are hydrolyzed in base to yield fatty acid salts and alcohols.

NEW TERMS

lipid A class of biomolecules characterized by insolubility in water and solubility in organic solvents

saponifiable lipid A lipid that is hydrolyzed by base to one or more fatty acids and one or more other molecules

nonsaponifiable lipid A lipid that does not contain fatty acids and therefore cannot be saponified by base

TESTING YOURSELF

Lipid Classification

1. An organic compound has been isolated from animal tissue with chloroform and methanol. It does not dissolve in water. To what class of biomolecules does this compound belong?
2. A water-insoluble compound is heated in the presence of aqueous base. Several fatty acids and the alcohol glycerol were obtained. How should this lipid be classified?

Answers **1.** lipid **2.** saponifiable lipid

11.2 Fatty Acids

♦ Fatty acids are large, unbranched carboxylic acids.

The carboxylic acids found in the saponifiable lipids are rather large carboxylic acids. These acids are called **fatty acids** and typically have 10 to 24 carbon atoms in the carbon chain.

$$CH_3CH_2CH_2CH_2CH_2CH_2CH_2CH_2CH_2CH_2CH_2CH_2CH_2COOH \equiv CH_3(CH_2)_{12}COOH$$

Myristic acid
(a fatty acid)

Fatty acids nearly always contain an even number of carbon atoms and they are rarely branched. Fatty acids may have zero, one, or more carbon–carbon double bonds. These structural features are a consequence of their synthesis in the body, as we shall learn later in our discussion of metabolism. Although fatty acids are a principal component of saponifiable lipids, they are rarely found free in cells or living organisms. In fact, free fatty acids are somewhat toxic to cells. Soaps, which are the salts of fatty acids, have some antibacterial action on the surface of the skin.

Classification

Fatty acids are classified as saturated fatty acids or unsaturated fatty acids. **Saturated fatty acids** contain no carbon–carbon double or triple bonds. The term *saturated* means these fatty acids do not accept any more hydrogen, that is, they are saturated with hydrogen. Table 11.1 shows the names and structures of some common fatty acids. Fatty acids can be named by the IUPAC rules for carboxylic acids, but more frequently, common names are used. For example, the unsaturated fatty acid linolenic acid has the IUPAC name *cis,cis,cis*-9,12,15-octadecatrienoic acid. Systemic names are descriptive, but they are often too long for common usage.

TABLE 11.1

Some of the Common Fatty Acids Found in Animals and Plants

Number of Carbon Atoms and Number of Double Bonds	Common Name	Structure
Saturated fatty acids		
10:0	Capric acid	$CH_3(CH_2)_8COOH$
12:0	Lauric acid	$CH_3(CH_2)_{10}COOH$
14:0	Myristic acid	$CH_3(CH_2)_{12}COOH$
16:0	Palmitic acid	$CH_3(CH_2)_{14}COOH$
18:0	Stearic acid	$CH_3(CH_2)_{16}COOH$
20:0	Arachidic acid	$CH_3(CH_2)_{18}COOH$
Unsaturated fatty acids		
16:1	Palmitoleic acid	$CH_3(CH_2)_5-\overset{H}{\overset{\vert}{C}}=\overset{H}{\overset{\vert}{C}}-(CH_2)_7COOH$
18:1	Oleic acid	$CH_3(CH_2)_7-\overset{H}{\overset{\vert}{C}}=\overset{H}{\overset{\vert}{C}}-(CH_2)_7COOH$
18:2	Linoleic acid	$CH_3(CH_2)_4-\overset{H}{\overset{\vert}{C}}=\overset{H}{\overset{\vert}{C}}-CH_2-\overset{H}{\overset{\vert}{C}}=\overset{H}{\overset{\vert}{C}}-(CH_2)_7COOH$
18:3	Linolenic acid	$CH_3CH_2-\overset{H}{\overset{\vert}{C}}=\overset{H}{\overset{\vert}{C}}-CH_2-\overset{H}{\overset{\vert}{C}}=\overset{H}{\overset{\vert}{C}}-CH_2-\overset{H}{\overset{\vert}{C}}=\overset{H}{\overset{\vert}{C}}-(CH_2)_7COOH$

Some of the fatty acids were named for the source from which they were first isolated. Palmitic acid is the saturated fatty acid typical of palm oil. Stearic acid is the principal fatty acid obtained from beef tallow. A convenient shorthand notation for fatty acids consists of two numbers separated by a colon, the first indicating the number of carbon atoms and the second the number of carbon–carbon double bonds in the molecule. For example, palmitic acid is designated as 16:0 and stearic acid as 18:0.

Unsaturated fatty acids contain one or more carbon–carbon double bonds. If the acid contains just one double bond, it is called a *monounsaturated* fatty acid. If two or more double bonds are present, it is a *polyunsaturated* fatty acid. Table 11.1 lists the more common unsaturated fatty acids found in foods. Since palmitoleic acid (16:1) and oleic acid (18:1) contain one carbon–carbon double bond, they are monounsaturated. Linoleic (18:2) and linolenic (18:3) acids are polyunsaturated fatty acids. In organic chemistry you learned that the orientation and position of the double bond as well as the number of double bonds should be expressed in a name because several isomers of each compound are possible. In nature, the isomers shown in Table 11.1 are by far the most common. The fatty acids in plants and animals nearly always have the cis configuration. In monounsaturated fatty acids, the double bond is usually between carbon atoms 9

♦ The carbon–carbon double bonds of most unsaturated fatty acids are cis.

and 10. In linoleic acid the double bonds are between carbon atoms 9 and 10 and between carbons 12 and 13. Linolenic acid is similar to linoleic acid, but has a third carbon–carbon double bond between carbons 15 and 16.

Physical Properties

The solubility properties of fatty acids depend upon pH. Fatty acids are carboxylic acids with pK_a values between 4 and 5. They can exist in a protonated form as a carboxylic acid or in an unprotonated carboxylate form, which is the conjugate base of the carboxylic acid. These two forms have different solubility properties.

$$CH_3(CH_2)_{14}\,COOH \rightleftharpoons CH_3(CH_2)_{14}\,COO^- + H^+$$

Carboxylic acid	**Carboxylate**
(palmitic acid)	(palmitate)
Predominates in acidic solutions	Predominates in neutral and basic solutions

Carboxylic acids have a polar head group, the —COOH group, but the large nonpolar alkyl group makes them insoluble in water and soluble in nonpolar organic solvents. The salts of fatty acids are in the carboxylate form, which is ionic. Ionic compounds are generally insoluble in nonpolar organic solvents, but the interactions of water molecules with the ionic group are sufficiently strong that these salts appear to "dissolve or suspend in water." True solutions are not formed, however; micelles form instead (see Figure 11.13).

Fatty acids show increasing melting points and boiling points with increasing size. This increase is also apparent in the other classes of organic compounds that you have studied. The melting points of fatty acids and of the compounds that contain them are also influenced by the presence of cis carbon–carbon double bonds. The cis double bonds in the unsaturated fatty acids impart a definite structure on the acid by, in effect, introducing "kinks" into the molecule (Figure 11.1). These kinks or bends in the molecule reduce the ability of the molecules to "stack" easily in an orderly fashion in the solid state. The net effect is that compounds containing cis double bonds remain liquids at lower temperatures than comparable compounds that lack cis double bonds. Biomolecules that contain appreciable amounts of unsaturated fatty acids are liquids rather than solids at room temperature.

♦ Compounds containing cis double bonds have lower melting points than the corresponding saturated compound.

Chemical Properties

The chemical properties of fatty acids are those predicted by the presence of a carboxyl group and carbon–carbon double bonds. The carboxyl group of fatty acids can undergo the same reactions as any carboxyl group, but in biological tissues, the most common derivative is an ester. Ester bonds are very common between fatty acids and a variety of alcohols. The ester bond is seen repeatedly in the sections on saponifiable lipids that follow. Examples of amide bonds will also be seen.

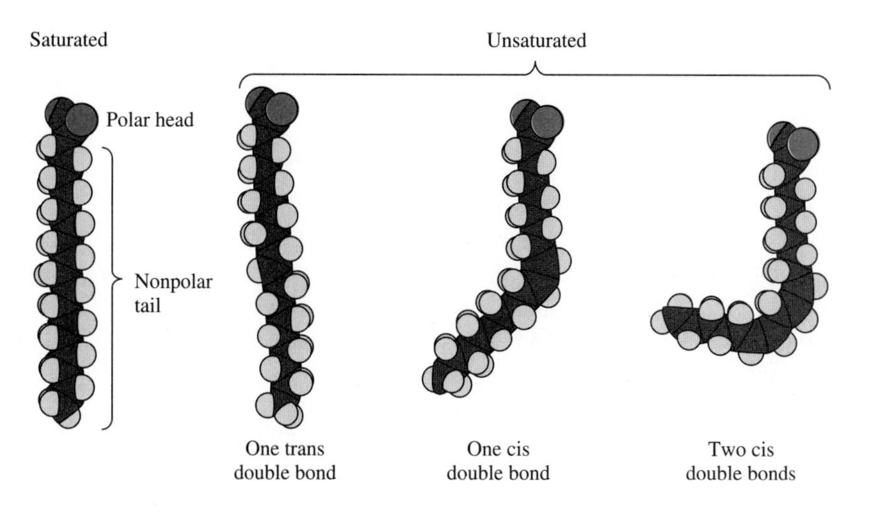

FIGURE 11.1
These space-filling models illustrate the structure of some saturated (far left) and unsaturated fatty acids. Cis double bonds introduce a "kink" into the molecule that greatly affects the structure.

Unsaturated fatty acids are common in most organisms. The carbon–carbon double bonds of these compounds undergo the typical addition reactions of carbon–carbon double bonds, three of which will be discussed. In the first reaction, hydration, water adds to these double bonds to introduce a hydroxyl group into the molecule. This is important in the metabolic reactions where fatty acids are broken down to produce energy (Chapter 15).

♦ Carbon–carbon double bonds in fatty acids undergo hydration, hydrogenation, halogenation, and other reactions typical of this functional group.

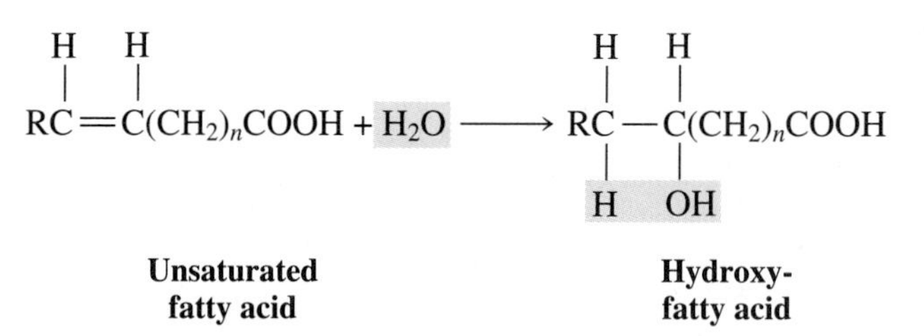

In a second reaction, hydrogenation, hydrogen adds to these double bonds:

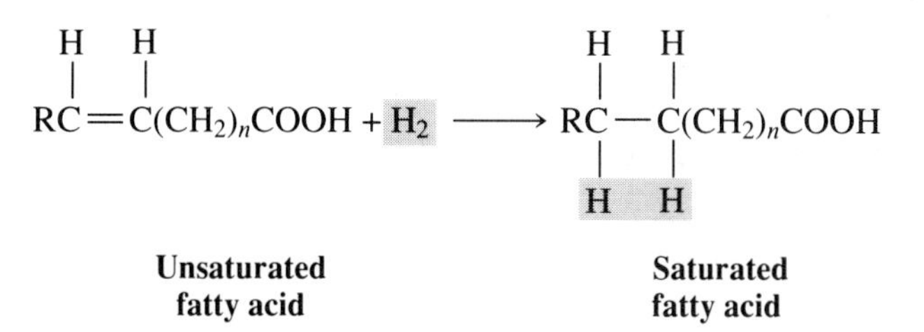

A number of important reactions of the body involve addition of hydrogen to these bonds, but this reaction is also important in the food industry. When you get a chance, read the label on a number of food packages. Look for the term *partially hydrogenated*, which is used to describe some oils. This term indicates that some of the carbon–carbon double bonds found in the fatty acids of these oils have had hydrogen added to them. A third addition reaction of importance is halogenation, the addition of a halogen to the double bond.

Hydrogenation converts vegetable oils to solid fats.

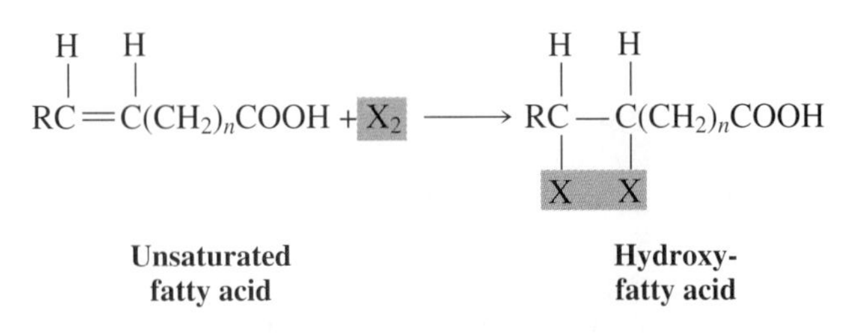

The halogens add to carbon–carbon double bonds of fatty acids, and this reaction can be used to measure unsaturation in lipids.

One other reaction of unsaturated fatty acids is important to an understanding of the properties of fats and oils. Carbon–carbon double bonds can be oxidized by oxygen. This reaction cleaves the double bond, yielding products that contain the aldehyde group or carboxylic acid group. If the product molecules are small enough, they will be volatile.

$$CH_3(CH_2)_nCH{=}CH(CH_2)_nCOOH \xrightarrow{\text{Oxidation}}$$

Fatty acid

$$CH_3(CH_2)_nCHO + CHO(CH_2)_nCOOH$$

Volatile aldehydes and acids

EXAMPLE 11.1

A solution containing 1.55 mmol of linoleic acid is placed in an atmosphere of hydrogen gas (with platinum as a catalyst). How many moles of hydrogen react with the linoleic acid? What is the structure of the product?

Solution: Linoleic acid contains two carbon–carbon double bonds per molecule, therefore there are 2 mol of carbon–carbon double bonds per mole of linoleic acid. Each of these double bonds reacts with 1 mol of H_2 to yield a product that is the saturated fatty acid of the same chain length. The balanced equation is:

$$CH_3(CH_2)_4CH{=}CHCH_2CH{=}CH(CH_2)_7COOH + 2\,H_2 \xrightarrow{\text{Pt}} CH_3(CH_2)_{16}COOH$$

Use the balanced equation and dimensional analysis to determine the amount of hydrogen consumed:

$$1.55 \times 10^{-3} \text{ mol linoleic acid} \times 2 \text{ mol } H_2/\text{mol linoleic acid} = 3.10 \times 10^{-3} \text{ mol } H_2$$

The product is the fatty acid with 18 carbon atoms and no carbon–carbon double bonds (18:0). This is stearic acid.

NEW TERMS

fatty acids Carboxylic acids from biological sources that generally contain 10 or more carbon atoms

saturated fatty acid A fatty acid that has no C=C double bonds

unsaturated fatty acid A fatty acid that has one or more C=C double bonds

TESTING YOURSELF

Fatty Acids

1. Name the following fatty acids.
   a. The saturated fatty acid with 18 carbon atoms.
   b. The unsaturated fatty acid with 16 carbon atoms and one double bond.
2. The double bonds of most fatty acids have what configuration?
3. a. What is the appropriate name for the predominate form of oleic acid at pH 7?
   b. Draw its structure at pH 7.

Answers **1. a.** stearic acid **b.** palmitoleic acid **2.** cis **3. a.** oleate **b.** $CH_3(CH_2)_7CH=CH(CH_2)_7COO^-$

11.3 Triacylglycerols

Triacylglycerols are nonpolar, saponifiable lipids that serve as energy reserves. These lipids are triesters formed from three fatty acids esterified to the three hydroxyl groups of glycerol (Figure 11.2). All triacylglycerols contain glycerol, but they may contain different fatty acids. Some contain three copies of the same fatty acid, but more commonly two or three different fatty acids are present in a triacylglycerol. During certain reactions in the body, one or two fatty acids may be bonded to a glycerol. These molecules are called *monoacylglycerols* and *diacylglycerols*, respectively. These compounds, when present at all, are not generally found in high concentrations in living systems.

♦ Triacylglycerols are triesters of three fatty acids and glycerol.

Triacylglycerols are known by several synonyms. Triglyceride is an older name that means the same thing: three fatty acids esterified to glycerol. You may see the term *triglyceride* used frequently, but it is slowly being replaced by triacylglycerol. The terms *fats* and *oils* are common terms used to describe

FIGURE 11.2
The structure of glycerol and a triacylglycerol. (a) Glycerol is a trihydroxy alcohol that is a component of many lipids. (b) Triacylglycerols have three fatty acids bound to glycerol through ester bonds. Mono- and diacylglycerols are much less common than triacylglycerols.

(a) Glycerol

CH_2-O-H
$CH-O-H$
CH_2-O-H

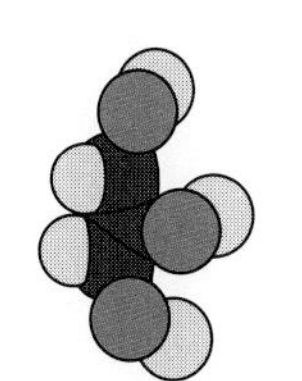

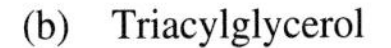

(b) Triacylglycerol

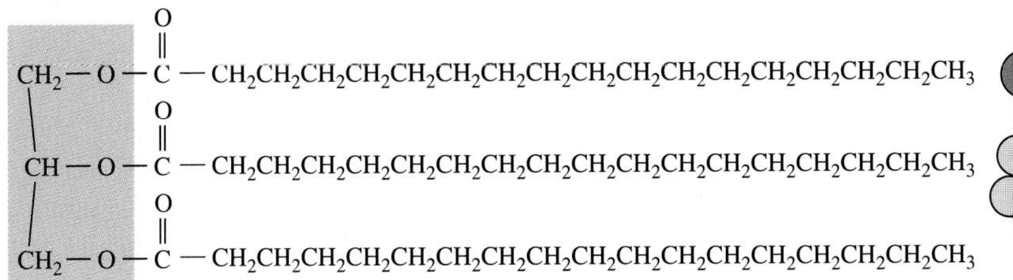

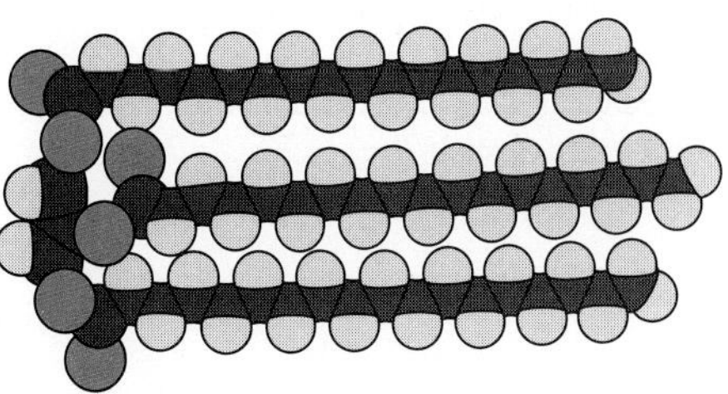

♦ Fats and oils differ primarily in their melting point.

triacylglycerols. **Fats** are triacylglycerols that are solid at room temperature. Lard from pork and tallow from beef are fats. The triacylglycerols of most animals are fats. **Oils** are triacylglycerols that are liquids at room temperature. Corn oil, cotton seed oil, and sunflower oil are examples. Many plant triacylglycerols are oils. The word "oil" is also used for other liquids like crude oil (petroleum), mineral or paraffin oil, and motor oil. However, these materials are hydrocarbons, not triacylglycerols.

Physical and Chemical Properties

If fats and oils are both triacylglycerols, why are some triacylglycerols solids at room temperature (fats) and others are liquids (oils)? The basis for this difference is the degree of unsaturation found in the fatty acids in the triacylglycerols. Fats contain fatty acids that are predominantly saturated. The fatty acid molecules fit together relatively well and thus remain in the solid phase until the temperature is higher. Oils contain a larger percentage of fatty acids that are more highly unsaturated. These fatty acids do not fit together as well because of the "kinks" resulting from the double bonds and thus tend to form a liquid phase at room temperature. The physical state of a triacylglycerol provides some clues to the degree of unsaturation of the fatty acids in it.

A convenient test for estimating the amount of unsaturation in a sample of triacylglycerols is the *iodine test*. Iodine as ICI (iodine monochloride) readily adds to carbon–carbon double bonds.

$$...CH_2CH_2CH{=}CHCH_2CH_2... + ICl \longrightarrow$$

$$...CH_2CH_2CH\ I\ CH\ Cl\ CH_2CH_2...$$

♦ A high iodine number indicates a high degree of unsaturation in the oil.

The more double bonds present in the sample, the more iodine required. The **iodine number** of a fat or oil is defined as the grams of iodine required to react with all of the carbon–carbon bonds in 100 g of the sample. A high iodine number indicates a high degree of unsaturation. A low number means the triacylglycerol is relatively saturated. Lard has an iodine number of about 65 to 70, whereas corn oil has an iodine number of 125 to 130. Iodine numbers vary slightly from sample to sample because environmental factors and diet influence the fatty acid composition of plants and animals. The degree of unsaturation in fats and oils has significant implications in health and nutrition.

♦ Added antioxidants, such as BHA or BHT, or partial hydrogenation of the oil reduces the chance of rancidity and thus extends shelf life.

The presence of carbon–carbon double bonds makes triacylglycerols susceptible to oxidation. They can readily become **rancid**, a state in which fats and oils smell and taste bad due to the presence of volatile organic acids and aldehydes that have formed via oxidation of carbon–carbon double bonds. Fats and especially oils may become rancid with time after exposure to air. A small amount of oxidative cleavage makes an oil or fat unsuitable for eating. The food industry can do two things to reduce the chance of a food becoming rancid. They can add antioxidants such as BHA (butylated hydroxyanisole) or BHT (butylated hydroxytoluene) that react with the oxidizing species to prevent their reaction with double bonds, or they can partially hydrogenate an oil to reduce the number of double bonds present. This hydrogenation of some of the double bonds does indeed increase shelf life and extend the useful life of a cooking oil, but the degree of saturation in dietary oils and fats is correlated to coronary disease. Extended shelf life may be gained at the expense of good nutrition.

Biological Roles

Many animals and plants store triacylglycerols as energy reserves. Triacylglycerols are excellent molecules for energy storage because they are high in energy and are water-insoluble. In general, the less oxygen a molecule contains, the more energy it contains. Thus, on a per gram basis, fats and oils have more energy than sugars, starches, and protein. These compounds contain numerous oxygen atoms in the hydroxyl groups, and thus the carbon atoms of the molecule are more oxidized than the carbon atoms of the long hydrocarbon chain of fats and oils. The carbon atoms of amino acids in proteins are also more oxidized than the carbon atoms of fats and oils. As a rough guide, fats and oils have about 9 kcal (Cal) of energy per gram of triacylglycerol. Dry carbohydrates and proteins have about 4 kcal of energy per gram. Thus, just on a mass basis, fats and oils are more than twice as efficient for energy storage. Furthermore, fats and oils are stored "dry" in the fat cells of the body, but glycogen is hydrated by numerous water molecules. Thus, 1 g of hydrated glycogen in the body contains far less energy than 1 g of anhydrous body fat. Most people have a significant amount of body fat. The average North American has about 10 to 30 percent body fat. The energy stored in fat can provide usable energy for several months.

◆ Triacylglycerols are the most efficient molecular storage form for energy.

EXAMPLE 11.2

A 12-oz serving of chocolate shake from a fast food establishment contains 10 g of protein, 72 g of carbohydrate, and 9 g of fat. Determine the number of Calories (kcal) present in the shake, and determine the percentage of the calories that come from fat.

Solution: To obtain the number of Calories, use dimensional analysis to determine the number of Calories contributed by each type of biomolecule, then sum the Calories:

$$10 \text{ g protein} \times 4 \text{ Cal/g protein} + 72 \text{ g carbohydrate} \times 4 \text{ Cal/g carbohydrate} + 9 \text{ g fat} \times 9 \text{ Cal/g fat} = 40 \text{ Cal} + 288 \text{ Cal} + 81 \text{ Cal} = 409 \text{ Cal}$$

(Apply the rules of significant figures to get 400 Cal). The shake is reported to have 403 Cal/serving.

To determine the percentage of the energy that is contributed by fat, divide the energy from fat by the total energy, then multiply by 100:

$$\frac{81 \text{ Cal}}{409 \text{ Cal}} \times 100 = 19.8\%$$

(Corrected for number of significant figures yields 20%)

Triacylglycerols are broken down in the body during metabolism and most of the glycerol and fatty acids are used for energy production. However, some of the fatty acids are used as building blocks for other molecules. One of these fatty acids is the polyunsaturated fatty acid linoleic acid. This is an *essential fatty acid* because the body cannot synthesize it, yet it is needed to synthesize other necessary molecules in the body. There is some evidence that a second polyunsaturated

◆ Linoleic acid and perhaps linolenic acid are essential fatty acids.

CHEMISTRY CAPSULE

Sperm Whales and Lipids

An interesting use of a lipid is found in the sperm whale. Well known for being a rich source of oil, the sperm whale was extensively hunted in the past. The head of the sperm whale contains a large amount (sometimes as much as 4 tons!) of a mixture of oils called *spermaceti oil.* This mixture tends to solidify as it is cooled. Most solids have greater densities than liquids; thus, as the whale feeds at great depths in the cold oceans, the oil freezes, the amount of freezing depending on the temperature. At the colder temperatures, the density of the whale increases because the mass remains constant, but the volume of the whale decreases as the oil solidifies. The buoyancy of the whale decreases, and it is able to stay submerged without exerting a great amount of energy. The whale is able to control its density and thus its buoyancy by diverting either cold sea water or warm blood through the chambers in its head, thus either freezing or melting the spermaceti oil.

fatty acid, linolenic acid, may also be essential. The human body cannot introduce carbon–carbon double bonds into the proper places in monounsaturated fatty acids to make linoleic or linolenic acids. Many oils are good sources of the polyunsaturated fatty acids as are some marine fish.

Large amounts of stored fat are common in mammals that hibernate or live in cold climates (Figure 11.3). This fat storage serves two useful purposes: (1) it provides an energy source during hibernation or fast, and (2) it acts as a source of insulation against cold. Animals such as seals use fat as insulation against the cold polar waters, and as an energy source during the breeding and pup-rearing seasons. Hibernating animals, such as rodents and grizzly bears, use it as both insulation and a food source during hibernation. Subcutaneous fat in humans may serve some insulating role. Slender people may feel cold during conditions in which heavier people are comfortable, yet when it is hot, the better insulated individual is uncomfortable. Fatty tissue also serves a protective role: fatty layers around the internal organs of vertebrates serve as a cushion to protect the organs from injury.

FIGURE 11.3
This polar bear relies on stored fat for energy and insulation. Some marine mammals have several inches of blubber for insulation against cold arctic waters.

NEW TERMS

triacylglycerols Triesters of glycerol and three fatty acids

fats Triacylglycerols that are solids at room temperature

oils Triacylglycerols that are liquids at room temperature

iodine number An index used to indicate the degree of unsaturation present in a fat or oil

rancid Term used to indicate that a fat or oil is foul-smelling and bad-tasting due to the presence of volatile acids and aldehydes

TESTING YOURSELF

Triacylglycerols

1. A sample of triacylglycerol is found to have an iodine number of 139. Is this sample an oil or fat?

CHEMISTRY CAPSULE

Waxes

Waxes are a heterogeneous group of waxy solids found in nature. Some are hydrocarbons, others are alcohols or ketones, and some are esters and thus saponifiable. These waxes, unlike triacylglycerols, are not used for energy storage. Instead they serve structural roles. Waxes are found as protective coatings of the skins of fruits and leaves and on the exterior surface of the exoskeleton of insects. Waterfowl use wax from preen glands to coat feathers for water-proofing. Repelling water is the role of all of these waxes. Birds use waxes to keep feathers dry, but plants and insects use waxes to prevent water from leaving the organism via transpiration. Bees also use waxes to make combs. These wax cells are marvelous vessels for storing honey and pollen and also serve as growth chambers for the developing young.

Saponifiable waxes are esters of a long-chain fatty acid and a long-chain alcohol.

$$CH_3(CH_2)_{14}\overset{\overset{\large O}{\|}}{C}O(CH_2)_{29}CH_3$$

Myricyl palmitate
(a typical ester in beeswax)

FIGURE 11.A
Waxes have a wide range of consumer and industrial uses.

The melting points of waxes vary according to their structures in much the same way as fats and oils; thus waxes from different plants and animals have different physical properties. Some of these waxes have past or current commercial uses, in cosmetics, candles, candies, and floor and automobile waxes.

2. How many Calories (kilocalories) are found in 1 oz of margarine? Assume the margarine is pure triacylglycerol.
3. If a mole of triacylglycerol is hydrolyzed, what are the products?

Answers **1.** Corn oil has an iodine number of 125 to 130, and since this sample has a larger iodine number, it is probably an oil. **2.** 1 oz $\times$ 28.4 g/oz $\times$ 9 kcal/g $\times$ 1 Cal/kcal = 256 Cal. (The actual value is slightly smaller because margarines are not pure oil or fat.) **3.** Hydrolysis cleaves the ester bonds, yielding 1 mol of glycerol and 3 mol of fatty acids.

11.4 Saponifiable Lipids of Membranes

Some saponifiable lipids are structural components of cell membranes. Some are also found in the circulatory system, where they play a role in normal transport of other lipids. These structural lipids are often referred to as *polar lipids* because one end of the molecule contains one or more polar functional groups, but the other end of the molecule is nonpolar. Molecules that possess a highly polar end and a nonpolar end are **amphipathic molecules**. The carboxylate form of a carboxylic acid is amphipathic.

♦ Amphipathic molecules possess polar and nonpolar regions.

$$\underbrace{CH_3CH_2CH_2CH_2CH_2CH_2CH_2CH_2CH_2CH_2CH_2CH_2CH_2CH_2CH_2}_{\text{Nonpolar}}\underset{\text{Polar}}{COO^-}$$

Some of these lipids are saponifiable, others are not. The saponifiable lipids of membranes and the circulatory system are the topic of this section.

The most abundant class of lipids in the membranes are the lipids that contain phosphorus, the *phospholipids*. Many of these phospholipids are **phosphoacylglycerols**. This important class of compounds is related structurally to the triacylglycerols. Instead of three fatty acids bonded to glycerol through ester bonds, the phosphoacylglycerols have two fatty acids esterified to the first and second hydroxyl groups of glycerol and a phosphoric acid esterified to the third hydroxyl group of the glycerol molecule (Figure 11.4a). Molecules with this structure are called *phosphatidic acids*, but normally an additional molecule is bonded to the phosphate group.

If the compound choline is bonded to the phosphate, then the compound is a phosphatidylcholine, the most abundant phospholipid. It is also known as lecithin (Figure 11.4b). If ethanolamine is bonded to the phosphate group, the phospholipid is called phosphatidylethanolamine (Figure 11.4c). These compounds are also known as cephalins. Other derivatives of phosphatidic acids, including phosphatidylserine and phosphatidylinositol, are also found in cells. All of these compounds are polar lipids because they contain a polar head group, the phosphate and its attached compound, and a nonpolar region commonly referred to as the "tail," which is made up of the two fatty acids (Figure 11.4d).

The **sphingolipids** are another group of saponifiable lipids found in membranes. These lipids contain the base *sphingosine*, a fatty acid, and one or more other molecules (Figure 11.5a). The fatty acid is bonded to sphingosine through an amide bond to the amino group. The other molecules are polar molecules such as sugars or phosphate and choline. These polar groups are bonded to the sphingosine through a hydroxyl group. The sphingolipids are not really similar to phosphoacylglycerols in composition, but they do resemble each other in structure. Sphingolipids are amphipathic because each has a polar end and a nonpolar end consisting of two long hydrocarbon chains. This similarity in structure is why they appear to have similar roles in membranes.

There are two types of sphingolipids, *sphingomyelin* (Figure 11.5b), which contains phosphate and choline, and *cerebroside*, which contains sugars. In your studies you may encounter the term *glycolipid*. This term is used to describe a lipid that has a carbohydrate as a component. Phosphatidylinositol and the cerebrosides are examples of glycolipids.

NEW TERMS

amphipathic molecules Molecules with both polar and nonpolar regions

phosphoacylglycerols Amphipathic molecules similar to triacylglycerols in structure, but with a substituted phosphoric acid in place of the fatty acid on the third hydroxyl group of glycerol

sphingolipids A group of polar membrane lipids characterized by the presence of sphingosine

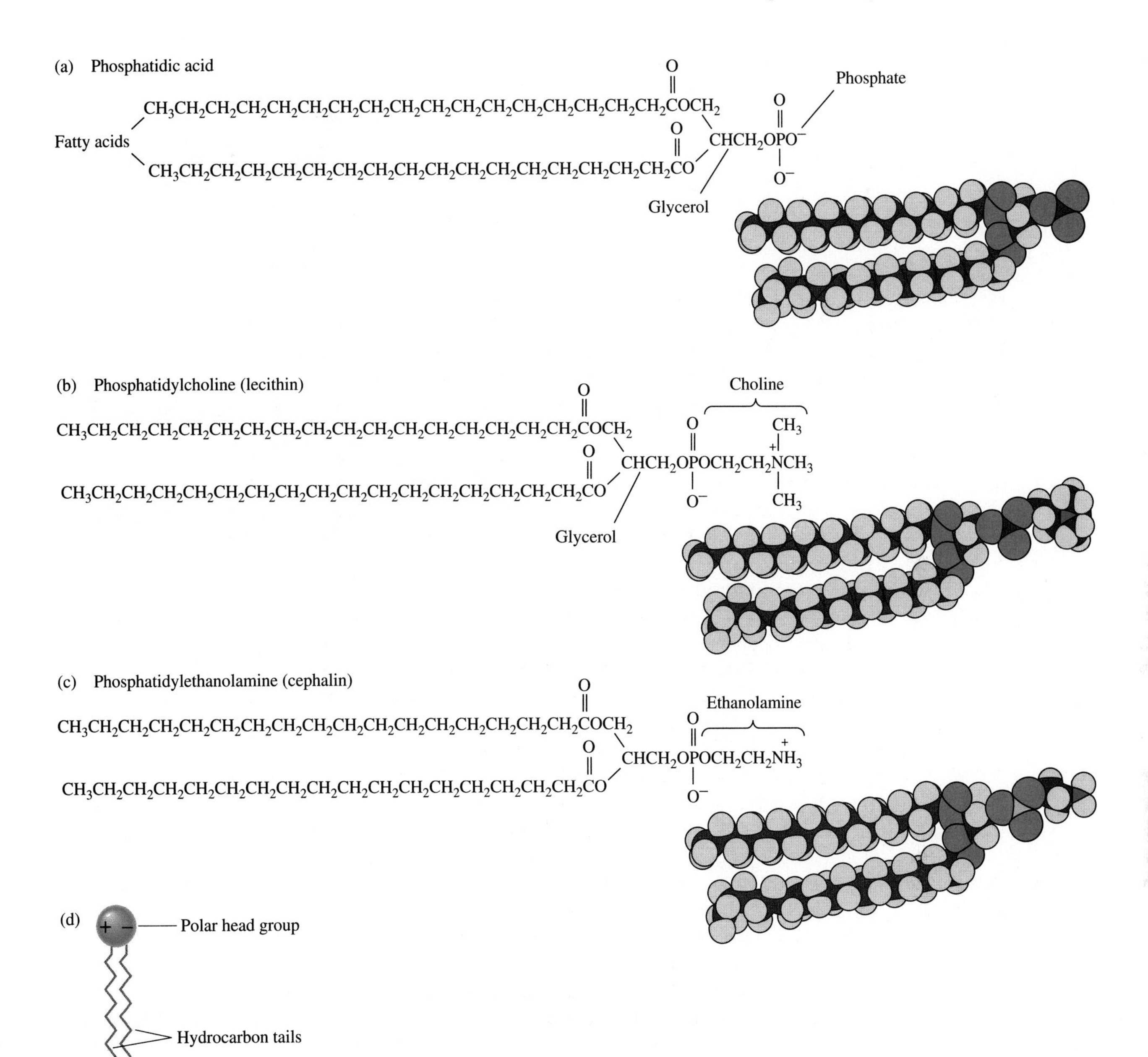

FIGURE 11.4
Phosphoacylglycerols are polar lipids containing phosphatidic acids. (a) Phosphatidic acid contains glycerol, two fatty acids, and a phosphate group. These lipids are the basic unit of phosphoacylglycerols. (b) Phosphatidylcholine (lecithin) is the most common phosphoacylglycerol. It contains a phosphatidic acid and a molecule of choline. Its space-filling model is also shown. (c) Phosphatidylethanolamine (cephalin) is another common phosphoacylglycerol. It contains a phosphatidic acid and ethanolamine. (d) Most polar lipids can be represented by a polar head group with two hydrocarbon tails.

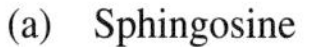

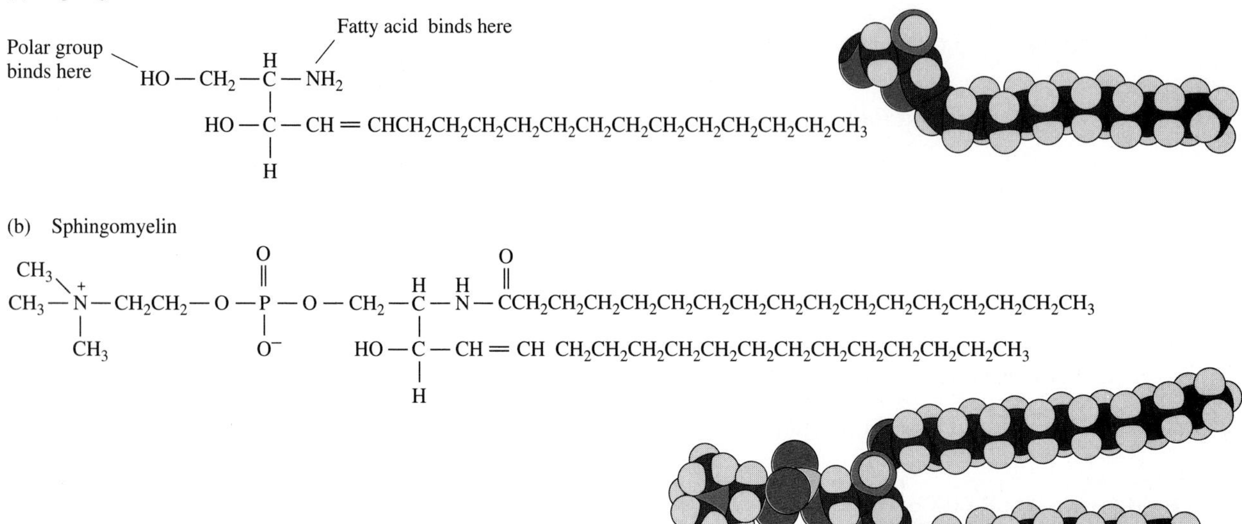

FIGURE 11.5
Sphingosine and sphingomyelin. (a) Sphingosine is found in all sphingolipids. In sphingolipids, a fatty acid is bonded to the amino group through an amide bond, and a polar group is attached to the primary alcohol. (b) Sphingomyelin is a typical sphingolipid. The polar group for this compound is a phosphate group and a molecule of choline. Compare the space-filling model of sphingomyelin with the space-filling model of phosphatidylcholine in Figure 11.4(b).

TESTING YOURSELF

Saponifiable Lipids of Membranes

1. Why is a phospholipid more likely to be slightly water-soluble than a triacylglycerol?
2. Hydrolysis of glycolipids yields what classes of compounds?
3. What polarity characteristics must a molecule possess if it is amphipathic?

Answers **1.** The polar head group of the phospholipid is water-soluble and increases the overall solubility compared with triacylglycerols. **2.** Carbohydrates and lipids **3.** The molecule must have both polar and nonpolar ends.

11.5 Nonsaponifiable Lipids

The nonsaponifiable lipids do not possess fatty acids bonded to some other molecules. Base hydrolysis, (saponification), has no apparent effect on their structure. This section discusses the steroids, prostaglandins, and leukotrienes.

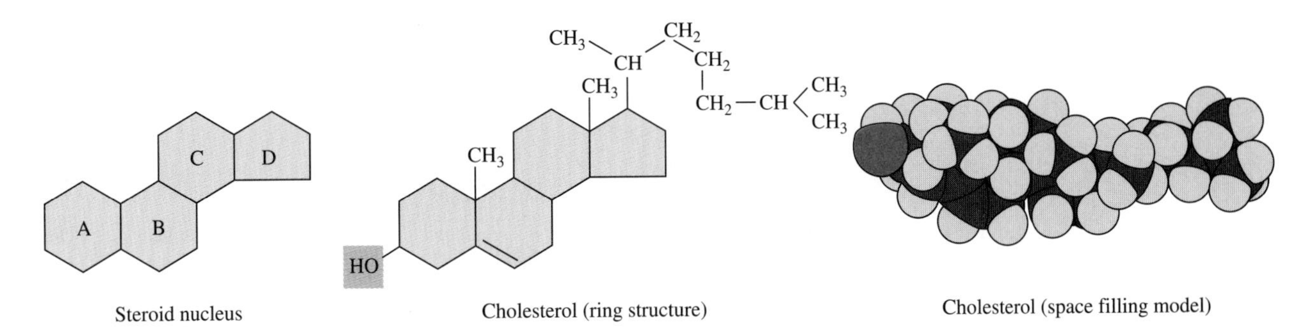

FIGURE 11.6
All steroids possess the steroid nucleus (yellow). Cholesterol, the most abundant steroid, is shown here as both a flat ring structure and as the more realistic space-filling model. The hydroxyl group (red) adds polarity to the molecule.

Steroids

The **steroids** are those compounds that contain a fused-ring system commonly known as the *steroid nucleus* (Figure 11.6). Although the steroids are related by structure and synthesis, they possess a great variety of functions. Regardless of the function, however, you can always recognize a steroid by its steroid nucleus.

♦ All steroids possess a four-membered ring system called the steroid nucleus.

The most abundant of the steroids in animals is **cholesterol** (Figure 11.6). Cholesterol is a component of all membranes in animals, and like other membrane lipids, it is an amphipathic compound. The steroid nucleus is nonpolar, but the hydroxyl group can hydrogen-bond with water or other polar molecules. Many of the other steroids in the body are synthesized from cholesterol.

♦ Most steroids are synthesized from cholesterol, which is the most abundant steroid in animals.

Cholesterol is obtained from the diet and is synthesized primarily by the liver. The circulatory system transports cholesterol to the rest of the body. In some people, deposits containing cholesterol build up on the interior surface of arteries. This condition is known as *atherosclerosis*, the accumulation of fatty substances and growth of abnormal muscle cells in the artery wall. These deposits reduce the diameter of the arteries supplying the heart and other vital organs. This reduced diameter means these vessels are more easily blocked, perhaps by a small clot of blood that would normally pass through. If no blood reaches portions of the heart, that portion of the heart is deprived of oxygen and nutrients and may die. Surgical techniques can be used to reduce constrictions in arteries, but this is only useful before the artery is totally clogged and while the heart tissue is still alive. Although the specific role of cholesterol in this disease process is not understood, it can be shown that a relationship exists between blood cholesterol levels and atherosclerosis. In general, the more elevated are blood cholesterol levels, the higher is the probability for heart attack.

Another group of steroids found in the body are the *steroid hormones*. Although never abundant, these compounds have a great influence over body function and activity. Some are involved in salt and water regulation, others in the body's response to stress, still others in sexual function and the development and maintenance of secondary sexual characteristics.

♦ The steroid hormones, although present only in trace quantities, exert powerful effects on the body.

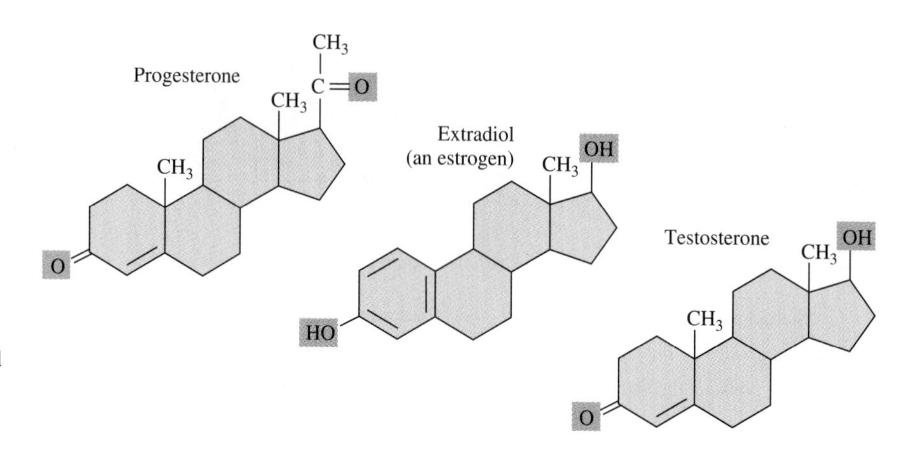

FIGURE 11.7
The common human sex hormones. Note the steroid nucleus (yellow) and the polar oxygen-containing groups (red).

Sex hormones are steroids that determine the secondary sexual characteristics of males and females. Although both sexes carry some of each of the hormone types, a careful balance must be maintained to provide normal growth and development. The male hormones, testosterone and androsterone, are quite similar in structure to the female hormone progesterone. Other common female hormones, the estrogens, possess an aromatic ring in the fused-ring system (Figure 11.7).

In the human female the concentrations of estrogen and progesterone change during the menstrual cycle. Prior to ovulation, there is a high estrogen level, but after ovulation there is an increase in progesterone. This increase in progesterone helps to prepare the uterine lining for implantation of a fertilized egg. If this does not occur, there is a "sloughing off" of the blood cells that had been stored for the implantation, resulting in menstruation. A generalized sequence of this monthly cycle is outlined in Figure 11.8.

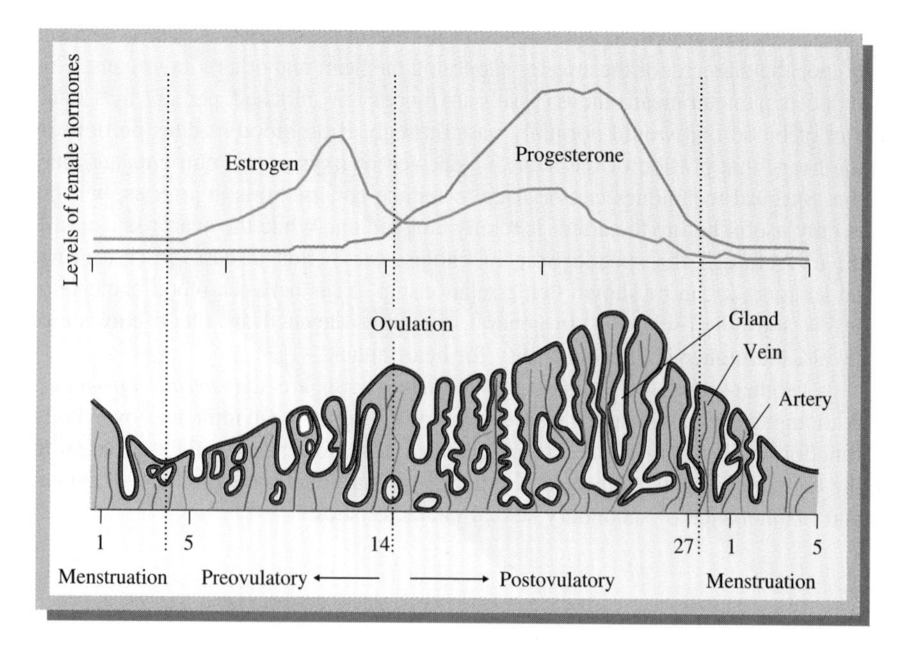

FIGURE 11.8
The menstrual cycle. The concentrations of estrogen and progesterone vary throughout the cycle (top). These hormones cause significant physiological changes in their target tissues. Changes that occur in the uterine wall during the menstrual cycle are shown along the bottom.

Correlations of Lipids and Health

There are several correlations between dietary lipids and good health. The most obvious one is the relationship between good health and obesity. People who are significantly overweight are at greater risk for several health problems, including cardiovascular disease. To reduce weight, caloric intake must be less than the caloric needs of the body. The body must then use stored fat to make up the energy deficiency. Because fats and oils are the most energy-rich nutrients, reducing their amounts in the diet is an excellent way to lose weight and ultimately bring energy use and energy intake into balance. However, it is neither easy nor desirable to eliminate fat from the diet. A small amount of the polyunsaturated fatty acids should be eaten regularly. In addition, fat contributes significantly to the taste and texture of foods. A nearly fat-free diet is difficult for most people to maintain.

A second correlation between dietary lipids and health is related to the degree of saturation in dietary fats and oils. Fats are more saturated than oils. There appears to be a correlation between the amounts of saturated fats and polyunsaturated fats in the diet and cardiovascular disease. In general, if the ratio of polyunsaturated fats to saturated fats is greater than 1, then the risk of heart disease is reduced. If the ratio is below 1, the risk is increased. Since foods from animal sources are higher in saturated fats, and plants are higher in highly unsaturated fats, a reduction in foods from animal sources and an increase in consumption of fruits and vegetables normally increases the ratio of these fats in the diet.

FIGURE 11.B
Cholesterol is found in foods from animal sources but is absent in foods of plant origin.

There is also a correlation between blood cholesterol levels and cardiovascular disease. Humans obtain cholesterol in the diet and through synthesis. Normally the amount of cholesterol eaten and synthesized is balanced with the amount that is lost from the body or used to synthesize other compounds. Cholesterol and its derivatives play essential roles in the body, thus it is neither necessary nor desirable to rid our systems of cholesterol. Although it can be shown that a correlation exists between blood cholesterol concentrations and heart disease, a clear link between dietary cholesterol and the development of heart disease in humans is not yet established. It has been shown that high levels of dietary cholesterol do increase blood cholesterol levels by, in essence, overwhelming the cellular transport system that removes cholesterol from blood. But whether cholesterol-free diets or reductions below normal dietary amounts of cholesterol will reduce the risk of heart disease in the general population is not yet clear.

Many factors appear to be correlated to cardiovascular disease—smoking, exercise, obesity, the amount of dietary fats and oils, the degree of saturation in dietary fats and oils, and blood cholesterol levels are all known factors that affect the cardiovascular system. A health-conscious individual may wish to consider all of these factors.

TABLE 11.A
Amount of Cholesterol Found in Some Common Foods

Food	Serving	Amount of Cholesterol (mg)
Egg	1	250
Tuna (canned in oil, drained)	184 g	116
Beef, pork, or turkey (dark meat)	84 g	67
Chicken or turkey (light meat)	84 g	55
Halibut	84 g	55
Salmon	40 g	55
Butter	1 tbsp	35
Hot dog	1	34
Whole milk	1 cup	34
Cheddar cheese	28 g	28
Ice cream	0.5 cup	27
Skim milk	1 cup	5

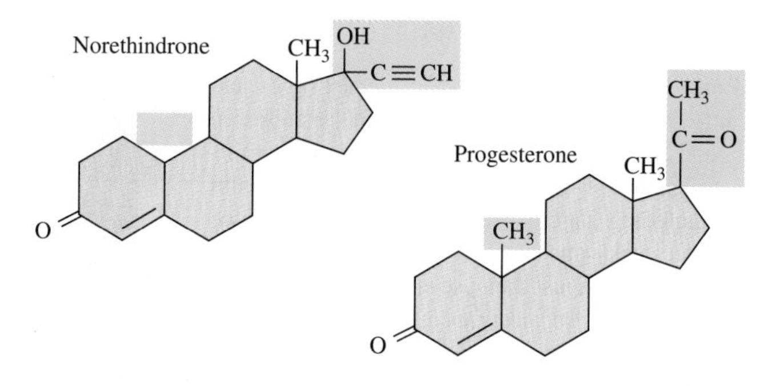

FIGURE 11.9
A comparison between norethindrone and progesterone with the differences highlighted in blue for emphasis.

An interesting application of structural modification of biomolecules can be found in the steroid analogs used as oral contraceptives. A specific example is *norethindrone*, which closely resembles progesterone (Figure 11.9). By the slight modification in structure shown in the figure, a natural product that cannot be taken orally because of rapid biological degradation is replaced by a synthetic material that has biological activity and is stable enough to be taken orally. This synthetic steroid mimics the action of progesterone and prevents further ovulation. Synthetic estrogens are also used in some contraceptive formulations.

Another important group of steroid hormones are produced by the cortex of the adrenal gland and are known as the *adrenocorticoid* hormones. These steroids play a central role in the regulation of some aspects of metabolism. Structurally, they are classified according to their degree of oxygenation. Several alcohol and carbonyl groups within a molecule are not uncommon. The most common member of this class is *hydrocortisone* (also known as *cortisol*), a powerful anti-inflammatory agent (Figure 11.10). Hydrocortisone and some of its active derivatives, such as cortisone and prednisolone, are used to treat a variety of skin inflammations, arthritis, and asthma.

Although hydrocortisone and its derivatives have highly beneficial properties, prolonged therapeutic use can result in problems, including excessive breakdown of protein (reflected in muscle deterioration, weakness, and excessive excretion of nitrogen-containing waste products), sodum retention, calcium excretion, and increased susceptibility to infection.

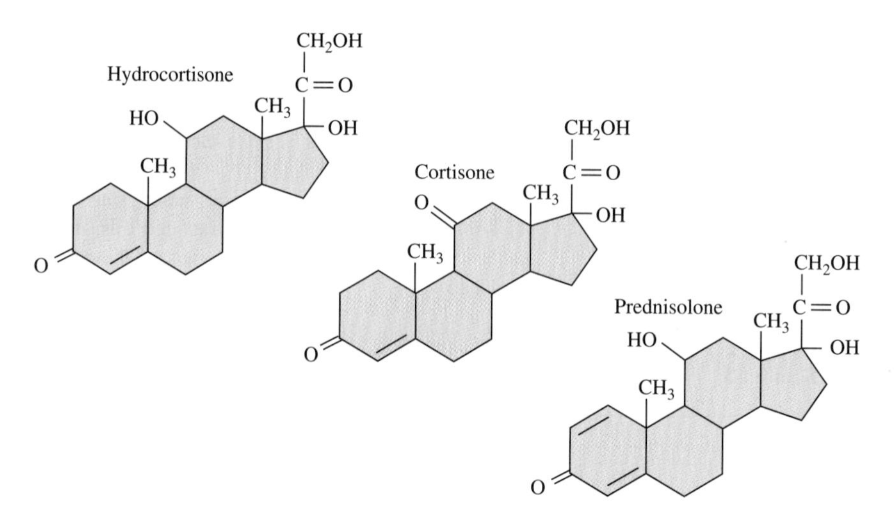

FIGURE 11.10
Hydrocortisone, cortisone, and the synthetic analog prednisolone.

CHEMISTRY CAPSULE

Anabolic Steroids in Athletics

Humans thrive on competition. Athletics provides an acceptable outlet for the competitive drive. But the internal need to excel or external pressure to succeed puts athletes in an ethical dilemma. Should they win at all cost or play the game by the rules? All serious competitors train hard. Many eat balanced diets, conscientiously obtain enough rest, and abstain from tobacco and other substances that reduce performance. Some take vitamin and mineral supplements in an attempt to enhance their performance. But which actions to enhance performance are acceptable and which are not?

Some athletes have overstepped the boundary between acceptable and unacceptable by using performance-enhancing drugs. Within this group of drugs are the *anabolic steroids*. These compounds are natural steroid hormones or, more commonly, synthetic analogs of these steroid hormones. In biochemistry, anabolic refers to reactions that build or synthesize. Anabolic hormones stimulate reactions related to synthesis and growth.

The male sex hormone testosterone is an anabolic hormone. It stimulates muscle development. Thus, at puberty, males develop increased muscle mass, which is a secondary sex characteristic of males. Some athletes use anabolic steroids to stimulate muscle growth. This provides additional strength that is beneficial in many sports, including weight lifting, sprinting, and football. But is this action to enhance performance acceptable? Most people think that it is not. It provides an unfair edge. Furthermore, there are a number of known or suspected side effects that are potentially quite harmful: liver tumors, aggressive behavior, testicular atrophy in males, and the development of masculine traits in females.

Bile salts are a group of steroids that play a role in digestion. These compounds are synthesized from cholesterol in the liver and then stored in the gallbladder. The bile salts are transported to the small intestine during the process of digestion, where they aid in the processes of digestion and absorption of dietary lipids. The specifics of lipid digestion are discussed in Chapter 15, but for now note that these compounds have polar groups projecting from one face of the steroid ring system, whereas the other face is nonpolar. These compounds are amphipathic, and their role in digestion requires this property. Two typical bile salts, *taurocholate* and *glycocholate*, are shown in Figure 11.11.

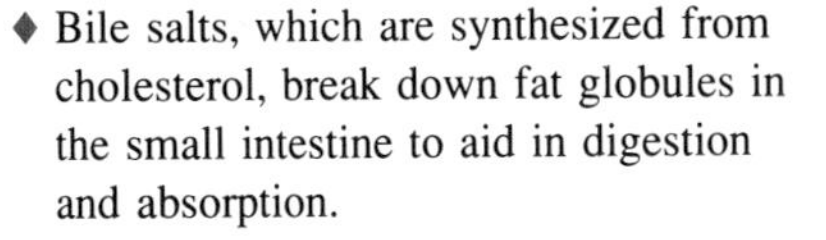
◆ Bile salts, which are synthesized from cholesterol, break down fat globules in the small intestine to aid in digestion and absorption.

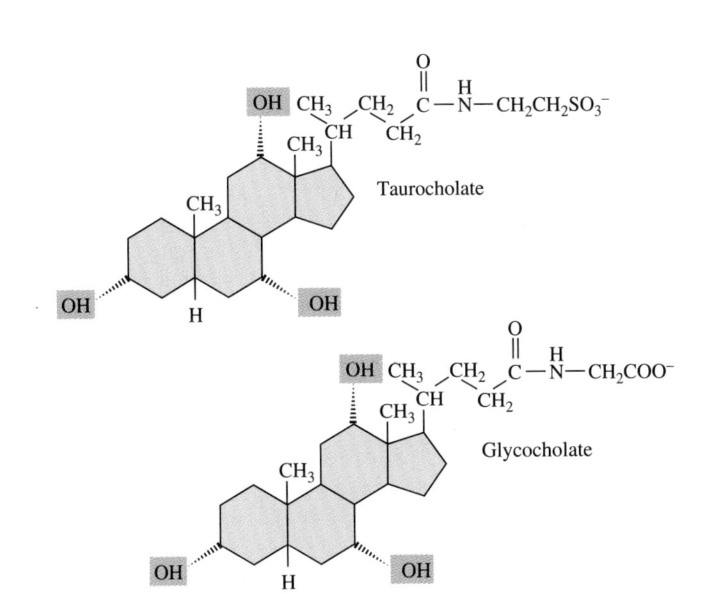

FIGURE 11.11
The common bile salts of mammals. Note that the polar hydroxyl groups (red) are all located on the same face of the steroid nucleus.

Prostaglandins and Leukotrienes

Prostaglandins and leukotrienes are two more classes of nonsaponifiable lipids that play a role in regulation of body function. These compounds are never abundant in the body, but they are very active in trace amounts. These lipids are synthesized from three polyunsaturated fatty acids—linoleic acid, linolenic acid, and arachidonic acid. The **prostaglandins** were first isolated from secretions of the male reproductive tract and were thought to originate in the prostate gland. It is now apparent that they are widespread in both sexes. They are present in extremely small amounts in many tissues of the body and have pronounced regulatory properties. Prostaglandins have been shown to exhibit powerful biological activity in many body functions, including blood pressure regulation, operation and functioning of the lungs, reproductive physiology, and smooth-muscle action, including uterine contraction. They also appear to participate in the physiology of fever and pain.

◆ Prostaglandins and leukotrienes are regulatory molecules.

Aspirin is the world's most common analgesic and antipyretic. Although aspirin has been used for over 100 years, the details of how it controls pain and reduces inflammation were not understood until recently. It now appears that aspirin interferes with the body's ability to synthesize prostaglandins. This implies that the prostaglandins somehow function in the perception of pain. Laboratory syntheses of many prostaglandins now provide enough material for wide-scale biological testing and evaluation. The structures of two common prostaglandins are shown in Figures 11.12(a) and (b). Arachidonic acid, the unsaturated fatty acid from which prostaglandins are synthesized, is shown for comparison.

One of the most recently discovered groups of lipid compounds are the **leukotrienes**. These compounds are also derived from arachidonic acid and appear to play a role in allergic and inflammation responses. Some of these

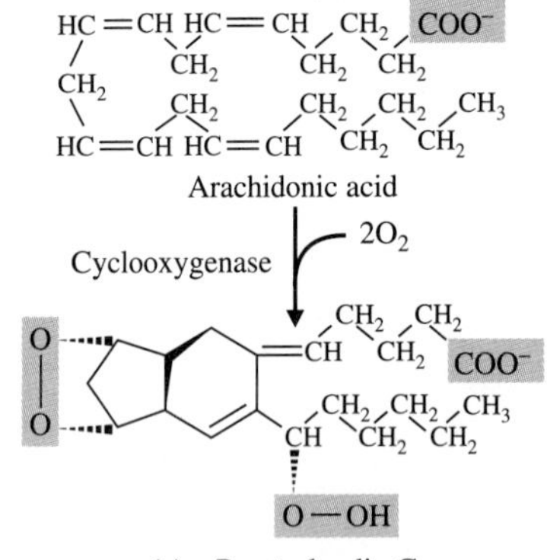

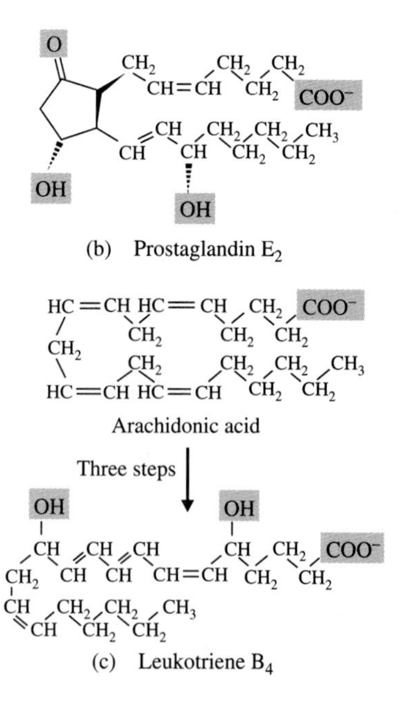

FIGURE 11.12
Prostaglandins and leukotrienes are made from arachidonic acid and other polyunsaturated fatty acids.
(a) Prostaglandin G_2 is made from arachidonic acid in a reaction catalyzed by cyclooxygenase (which is a part of the enzyme prostaglandin synthase). Aspirin inhibits cyclooxygenase activity. Because prostaglandins enhance inflammation, aspirin is an anti-inflammatory agent. (b) Prostaglandin E_2 stimulates contraction of uterine muscles. (c) Leukotriene B_4 is made from arachidonic acid.

compounds were isolated from leukocytes, and the name reflects this origin as well as the unsaturated nature of the compounds. The structure of a common member of this group is shown in Figure 11.12(c).

NEW TERMS

steroids Compounds possessing the steroid nucleus

cholesterol The most abundant steroid in animals. An important membrane lipid with significant health implications

prostaglandins Regulatory lipids derived from unsaturated fatty acids. Prostaglandins often have a cyclic portion based on cyclopentane

leukotrienes Lipids derived from arachidonic acid that are implicated in allergic responses

TESTING YOURSELF

Nonsaponifiable Lipids

1. Steroids, which are a large and diverse group, share one structural feature. Name and draw this feature.
2. Estrogens and progesterone have what general physiological role?
3. The body uses arachidonic acid to synthesize what two important classes of regulatory molecules?

Answers **1.** the steroid nucleus

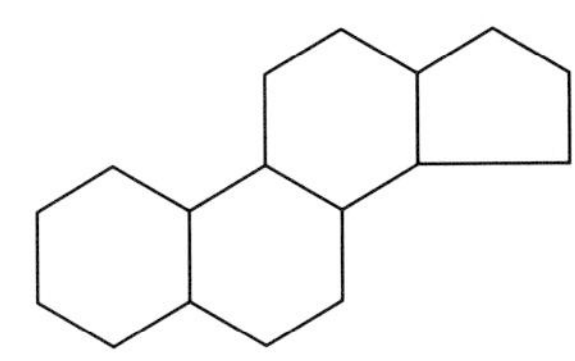

2. female sex hormones that control reproductive activity and influence secondary sex characteristics **3.** prostaglandins and leukotrienes

11.6 Liposomes and Membranes

Micelles, Bilayers, and Liposomes

Micelles are aggregates of amphipathic molecules that assume a spherical shape in water. Soaps and many other amphipathic molecules form micelles under certain conditions. In micelles, the long hydrophobic tails of the molecules are clustered together in the core of the micelle, with water excluded from the core (Figure 11.13). Micelles provide two distinctly different environments: (1) a nonpolar or hydrophobic environment in the core of the micelle, and (2) a polar, aqueous environment on the surface of the micelle.

The structure of a micelle can be explained in several ways, but the most accurate picture includes hydrogen bonding between water molecules that effectively squeezes out the nonpolar part of the amphipathic molecule. A water

Polar (aqueous) surface

Nonpolar (hydrophobic) core

FIGURE 11.13
The structure of a micelle.

(a)

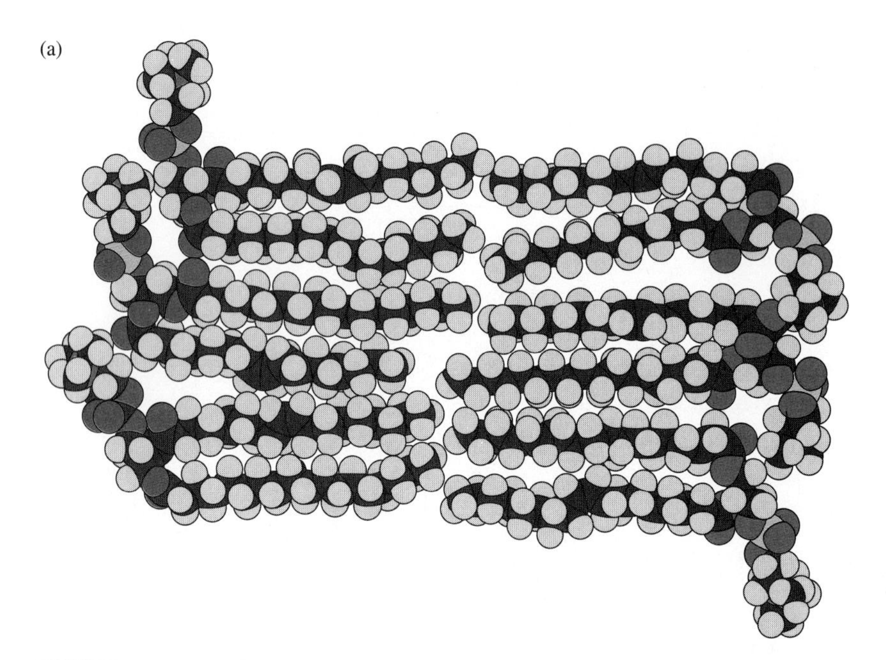

(b)

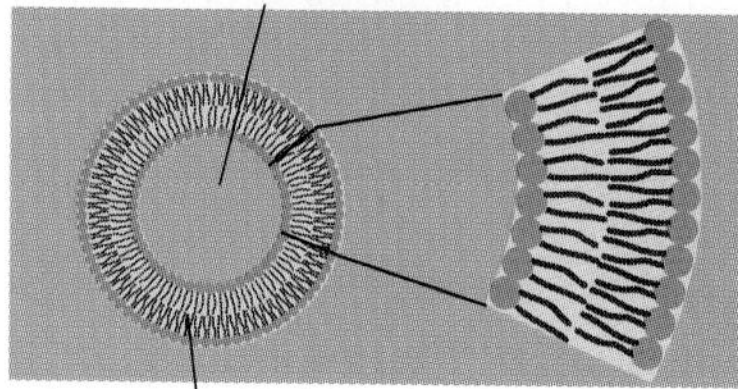

FIGURE 11.14
The structure of bilayers and liposomes. (a) Bilayers consist of polar lipids that have their nonpolar regions aggregated together and their polar head groups facing the aqueous solution. (b) Liposomes are spherical structures that are, in essence, bilayers folded around a core of polar solvent. Researchers are developing drug-filled liposomes as a more efficient drug-delivery system.

molecule normally hydrogen bonds to other water molecules. When a nonpolar entity is present between two water molecules, the strong hydrogen bond between them is broken and replaced by weaker London forces between the water molecules and the nonpolar species. In terms of energy, this is unfavorable because it is less stable. The tendency is thus for the two water molecules to bond together, effectively squeezing the nonpolar molecule out. If several nonpolar entities are in an aqueous solution, they end up clustered together. This is not because they have a particularly strong attraction for each other, but rather because water molecules prefer to bond to other water molecules. Nonpolar molecules or nonpolar parts of amphipathic molecules are squeezed into nonaqueous cavities in a water solution.

The clustering of nonpolar species in an aqueous solution is referred to as *hydrophobic interactions*. These interactions play a very important role in biological membranes and protein structure. Remember, the term hydrophobic interactions implies that water-hating, nonpolar molecules prefer to be together, but the reality is that they have been squeezed together by polar water molecules.

Another structure that some amphipathic molecules can form when placed into an aqueous solution is a **bilayer** (Figure 11.14a). The amphipathic molecules in a bilayer are arranged with the nonpolar tails projecting into the core of the bilayer and the polar head groups positioned on the exterior surface of the bilayer. Soap bubbles are bilayers of soap molecules with a film of water on either side. Bilayers can close back upon themselves to form a continuous bilayer surrounding a core of water. These structures are called **liposomes** (Figure 11.14b).

♦ Micelles, bilayers, and liposomes are aggregates of amphipathic molecules that have a hydrophobic core and a polar surface.

Liposomes can be prepared as relatively stable structures. They show promise in the pharmaceutical industry as "casings" for drugs. The polar surface of the liposome keeps them suspended in blood and other body fluids, and the core carries a solution or suspension of the drug. In the body, the liposomes are slowly broken down, resulting in a gradual, time-released delivery of the drug in appropriate concentrations.

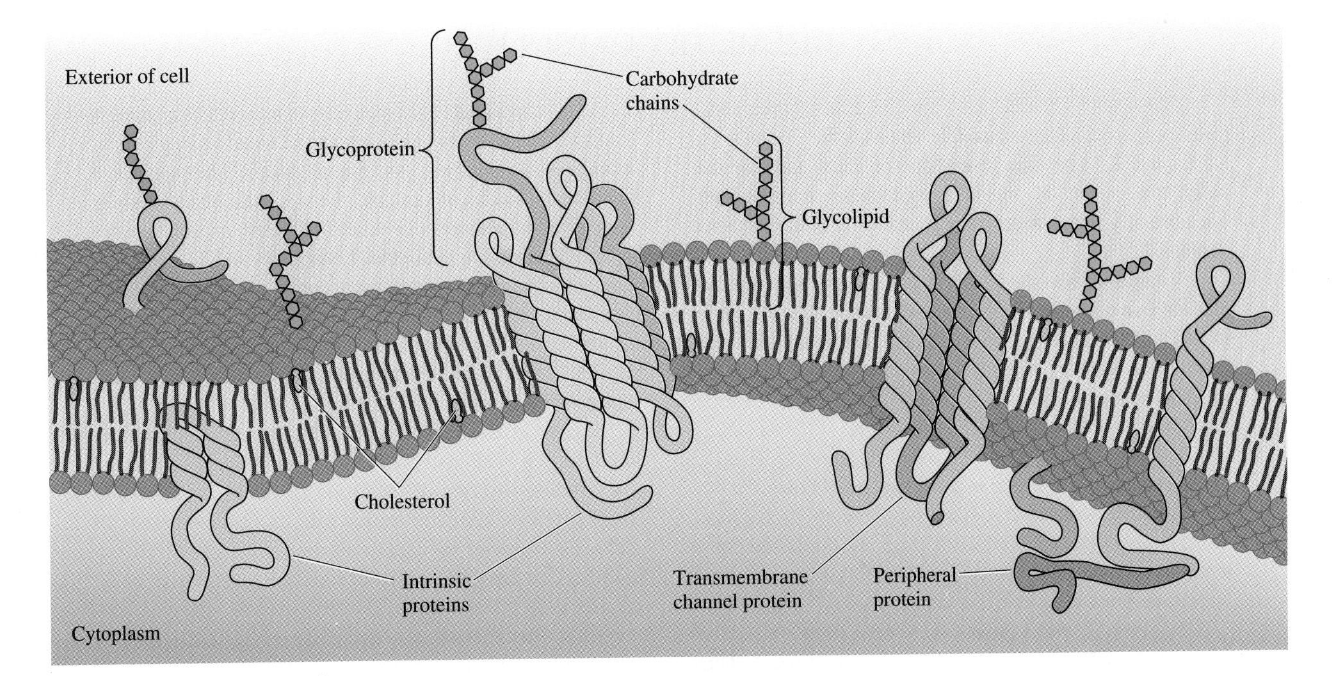

FIGURE 11.15
The fluid mosaic model of biological membranes. Note the proteins that are imbedded in the lipid matrix. The lipid matrix is made up of polar lipids, primarily phosphoacylglycerols, with some cholesterol, glycolipids, and others. Because covalent bonds do not exist between the components of the membrane, they are free to diffuse throughout the membrane. The membrane is thus "fluid."

Membranes

Micelles, bilayers, and liposomes have served another important role in biology—as models for developing an understanding of biological membranes (Figure 11.15). **Membranes** are bilayer-like structures that separate a cell from the external environment and divide the cell into numerous compartments. The most abundant components of membranes are amphipathic molecules, such as phospholipids, glycolipids, and cholesterol. Like micelles and bilayers, membranes have a nonpolar core that is hydrophobic and surfaces that are polar. But polar parts of these molecules include positive and negative charges that attract each other. This attraction significantly stabilizes membranes.

♦ Membranes possess amphipathic lipids and proteins.

Another major component of membranes are proteins. These macromolecules are discussed in detail in Chapter 12, but for now picture them as large molecules that have polar and sometimes nonpolar regions on their surface. Membrane proteins are imbedded in the lipids of membranes. The nonpolar portions of the protein are in contact with the hydrophobic interior of the membrane, and the polar portions of the protein face out toward the water. Again, hydrophobic interactions hold the nonpolar portion of the protein in the core, whereas various polar interactions keep the polar part facing the water.

The presently accepted model for cellular membranes is called the *fluid mosaic model* because the proteins form a mosaic in the membrane (Figure 11.15).

CHEMISTRY CAPSULE

Fat-soluble vitamins are lipids because of their solubility properties. The fat-soluble vitamins are vitamins A, D, E, and K. They are stored in the fatty parts of the body; thus a diet lacking in any of these vitamins does not normally cause a deficiency until this pool has been depleted.

Vitamin A was the first human vitamin to be discovered. In two of its forms, retinol and retinal, it serves a role in detection of light in the eye. A deficiency may manifest itself in a form of night blindness. It is also involved in maintenance of normal skin and other surface tissues, growth, and perhaps membrane function.

Toxicity from vitamin A overdose can occur because excess vitamin A is stored in body fat and is not excreted. Most overdoses involve dietary supplements or medications; rarely does diet provide toxic quantities. (Large daily amounts of liver could cause it, but who among you eats large quantities of liver every day?)

Vitamin D, which exists in several forms, is active in the body as dihydroxycholecalciferol (calcitriol). This compound is involved in several ways in calcium uptake and utilization. Proper bone mineralization does not occur unless adequate amounts are present. The disease rickets may result from serious deficiency of vitamin D in children.

Vitamin D is sometimes called the "sunshine vitamin" because ultraviolet light converts some steroids to this vitamin. These reactions occur in the skin when a person is exposed to sunlight. As long as people are exposed to adequate amounts of sunlight, there is little or no need for supplementary sources of vitamin D. Few natural sources of this vitamin are available. Liver, fatty fishes, egg yolk, and butter provide some. A more common source is supplementary vitamin D added to dairy products and some other foods. This is the major source of dietary vitamin D in the United States. Strict vegetarians should have no deficiency unless they lack access to sunlight.

Vitamin E is a collection of compounds known as *tocopherols*. The tocopherols are antioxidants—that is, they reduce the rate of oxidation of double bonds—and they appear to function in this role in the membranes of cells. Deficiencies in humans are rare. In animals, several deficiencies have been seen or induced by deficient diets. These symptoms include sterility, muscle atrophy, and nervous disorders. Seeds and oils are good sources of vitamin E.

Vitamin K is found in leafy vegetables. In animals, it has a role in the synthesis of some proteins required for normal blood clotting. Without vitamin K, these proteins are not adequately synthesized, and normal clotting may not occur.

FIGURE 11.C
Spinach, sweet potatoes, liver, and carrots provide significant amounts of vitamin A.

FIGURE 11.D
Cereals and grain products are good sources of Vitamin E.

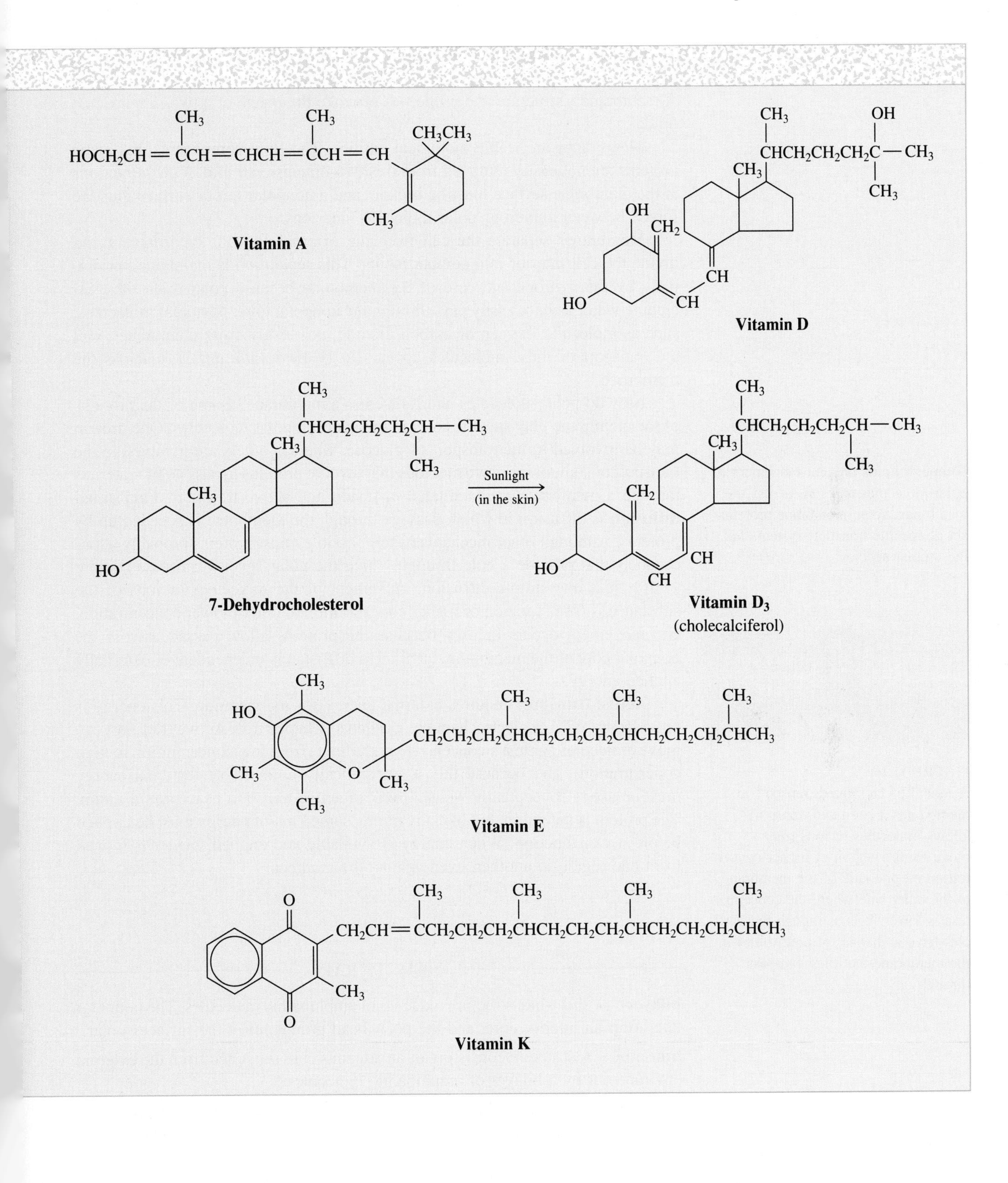

Vitamin A
Vitamin D
Sunlight
(in the skin)
7-Dehydrocholesterol
Vitamin D3
(cholecalciferol)
Vitamin E
Vitamin K

Protein molecules and lipid molecules are free to diffuse around the surface of the membrane because only weak, noncovalent bonds hold them in the membrane. A membrane's structure and properties provoke the image of a two-dimensional fluid.

Membranes are stable biological entities. The lipid components and some proteins are not easily removed from this two-dimensional film. Ionic attractions at the membrane surface must be broken, and a nonpolar tail or surface must be forced between numerous polar molecules and ions.

Membranes separate the cell from the external aqueous environment and divide the cell interior into compartments. This separation is maintained principally by the hydrophobic core of the membrane because polar molecules, including water, cannot easily pass through the nonpolar core. Nonpolar molecules, such as molecular oxygen or carbon dioxide, do readily cross membranes, and the transport of these molecules appears to be by *simple diffusion* across the membrane.

How do polar molecules and ions cross a membrane? Some of the proteins of the membrane play specific roles in transporting polar molecules. One protein may be involved in the transport of glucose, whereas others are involved in the transport of sodium ion. Two methods that involve proteins to pass polar materials through a membrane are facilitated diffusion and active transport. **Facilitated diffusion** is diffusion in which passage through the membrane is speeded up by proteins. Although other mechanisms are possible, most proteins probably act as channels that provide a hole through which the polar species can pass (Figure 11.16). Just like simple diffusion, movement of the molecules or ions of the substance is from a region of higher concentration to one of lower concentration. But movement occurs quickly because the proteins allow passage across the nonpolar core of the membrane. Facilitated diffusion is independent of externally applied energy.

♦ Although lipids prevent movement of polar molecules and ions across the membrane, some membrane proteins act as specific transport systems for these polar species.

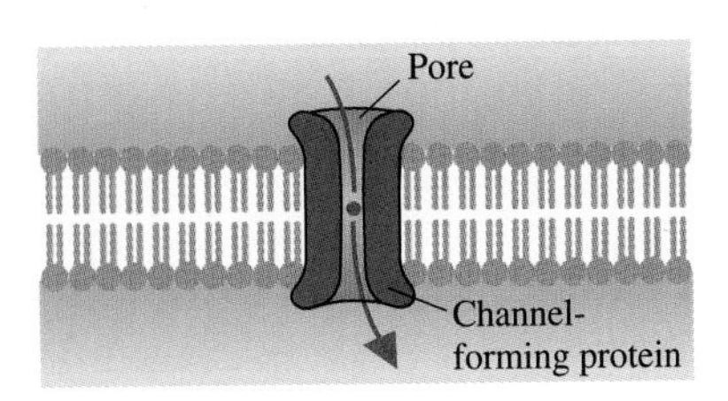

FIGURE 11.16
A model of facilitated transport in membranes. Facilitated transport allows molecules or ions (red) to move from a region of higher concentration on one side of the membrane to the other side where the concentration is lower. It is as though the particles have a channel or pore (blue) in the membrane that they can pass through.

Active transport requires external energy for its operation. This energy is provided by ATP or electrochemical gradients (Chapter 15). Active transport can move molecules against a concentration gradient (from low concentration to high concentration), and because this is energetically unfavorable, external energy must be used. Conceptually we can think of active transport as a form of pump. The protein is the pump, but like any pump, some kind of energy must be applied before it can function. When energy is available and applied, matter is moved from one region to another, even against the gradient.

NEW TERMS

bilayer A sheet-like structure made up of amphipathic molecules. The nonpolar tails form an interior core, and the polar head groups make the surfaces polar.

liposome A structure consisting of an aqueous core separated from the external environment by a bilayer of amphipathic molecules

membranes The sheet-like structures in cells that separate the cell interior from the external environment and divide the cell interior into compartments

facilitated diffusion The transport of substances across the membrane of a cell in which proteins speed up the movement. The substances can only flow down their gradient.

active transport The energy-requiring transport of substances across the membrane of a cell.

TESTING YOURSELF

Membranes

1. What are the two principal classes of compounds found in membranes?
2. Consider a protein in a membrane. Describe the interaction (bonding) of the molecule with the molecules surrounding it.
3. What name is given to the energy-requiring process that passes material through a membrane?

Answers **1.** lipids and proteins **2.** The protein has hydrophobic interactions with the nonpolar tails of the lipids in the interior of the membrane and polar interactions with the polar head groups of the lipids and the components of the aqueous environment surrounding the membrane. **3.** active transport

Summary

Lipids are water-insoluble biomolecules that can be extracted from tissues by nonpolar solvents. Lipids can be classified according to their hydrolysis properties. *Saponifiable lipids* are hydrolyzed in aqueous base, whereas *nonsaponifiable lipids* are not. *Fatty acids* are a component of all saponifiable lipids. These carboxylic acids are long, unbranched molecules containing zero to several cis carbon–carbon double bonds. *Triacylglycerols* (fats and oils) and phospholipids are common saponifiable lipids involved in energy storage and structure, respectively. *Prostaglandins, leukotrienes*, and *steroids* are common nonsaponifiable lipids. All three have regulatory roles, but steroids have structural roles as well.

Terms

lipid (introduction)
saponifiable lipid (11.1)
nonsaponifiable lipid (11.1)
fatty acids (11.2)
saturated fatty acid (11.2)
unsaturated fatty acid (11.2)
triacylglycerols (11.3)
fats (11.3)
oils (11.3)
iodine number (11.3)
rancid (11.3)
amphipathic molecules (11.4)
phosphoacylglycerols (11.4)
sphingolipids (11.4)
steroids (11.5)
cholesterol (11.5)
prostaglandins (11.5)
leukotrienes (11.5)
bilayer (11.6)
liposome (11.6)
membranes (11.6)
facilitated diffusion (11.6)
active transport (11.6)

Exercises

Classification (Objectives 1 and 2)

1. All of the compounds discussed in this chapter are lipids. What do they have in common?
2. What is the difference between a saponifiable and a nonsaponifiable lipid?
3. List the common saponifiable lipids introduced in this chapter.
4. List the nonsaponifiable lipids discussed in this chapter.
5. Which of the following solvents may be suitable for extracting (dissolving) lipids from biological tissues?
   a. chloroform
   b. ethanol
   c. saturated NaCl (aq.)
   d. toluene
   e. water
6. Is petroleum a lipid?

Fatty Acids (Objective 3)

7. Fatty acids are carboxylic acids. What distinguishes them from other carboxylic acids?
8. Fatty acids are sometimes classified as saturated, monounsaturated, or polyunsaturated. What do these terms mean?
9. What effect does a cis double bond have on the shape of a fatty acid molecule? What effect does this have on melting points of fatty acids and compounds that contain them?
10. Consider linolenic acid (18:3). All three double bonds are cis in this molecule. How many isomers exist if both cis and trans isomers are considered for each carbon–carbon double bond?
11. Which of the following groups are likely to be found in a fatty acid?
    a. $C{=}C$ c. $-CH(CH_3)_2$ e. $-OH$
    b. $C{\equiv}C$ d. $-COOH$
12. Name and write the structures of the common unsaturated fatty acids that contain eighteen carbon atoms. Show the correct stereochemistry.
13. Write the equations for the reaction of palmitoleic acid with the following:
    a. water b. hydrogen c. iodine
14. One of your classmates has drawn the structures of some fatty acids he thinks will be in a lipid. Included is this compound:

$$(CH_3)_2CH(CH_2)_{13}COOH$$

    Explain why you believe this compound would or would not be present in common lipids.
15. Write a balanced reaction for the hydrogenation of the following:
    a. oleic acid b. palmitic acid c. linolenic acid
16. Write a balanced reaction for the iodination of the three acids in exercise 15.

Triacylglycerols (Objective 4)

17. Draw the structure of a triacylglycerol that contains palmitoleic, stearic, and linoleic acids.
18. A food package label lists an oil as "partially hydrogenated." What does this mean?
19. An oil is said to have an iodine number of 110. Define the term *iodine number*, and describe what it indicates.
20. Fats and oils are classified as triacylglycerols. What structural features do they share?
21. What structural features are likely to make a triacylglycerol an oil?
22. The short-chain fatty acids have bad odors. Why do some fats and oils become "rancid" when exposed to light, air, and water?
23. What can be done to reduce the chances of rancidity in fats and oils found in food?
24. There are two main factors that make fats more efficient for energy storage than glycogen. What are these factors?
25. A 50-kg woman with 20% body fat begins a fast. Her daily activity consumes 2000 Cal per day during the fast. If we assume that all of the energy used comes from metabolism of fat (not strictly true) and that all 20% of the body fat is available (also not strictly true), how long will her fat sustain her energy needs?
26. Hydrogenation of a vegetable oil gives
    a. soap
    b. waxes
    c. spermaceti
    d. vegetable fat
27. How many different triacylglycerol molecules can be made from one glycerol molecule, one molecule of oleic acid, and two molecules of palmitic acid?
28. How many of the molecules formed in exercise 27 could be optically active?
29. Enzymes are generally specific for substrate and product. For example, usually only one of two possible enantiomers act as a reactant or is produced by the reaction. Given this information, how many of the possible optically active molecules in exercise 27 would actually be formed?

Saponifiable Lipids of Membranes (Objective 5)

30. Lecithin is a phospholipid that contains
    a. sugar
    b. ethanolamine
    c. choline
    d. sphingosine
31. What is an amphipathic molecule? What cell structure contain this type of molecule?
32. The basic unit of the phosphoacylglycerols is phosphatidic acid. Draw its structure.
33. Phospholipids are found in membranes, but triacylglycerols are not. Why?

34. What is the structural difference between a cephalin and a lecithin?
35. What is a sphingolipid?
36. Draw the structure of sphingomyelin.
37. One mmol of a phosphatidylcholine was hydrolyzed to yield 1 mmol each of oleic acid, stearic acid, phosphate, glycerol, and choline. Draw a structure for this compound. How many other structures are possible?

Nonsaponifiable Lipids (Objectives 5 and 6)

38. What steroid is found in cell membranes?
39. Draw the structure of cholesterol. What structural features does it have in common with other steroids?
40. What is atherosclerosis, and how is blood cholesterol related to it?
41. Name the two major female sex hormones and describe their biological function.
42. There has been an increased awareness of athletes taking anabolic steroids (those resembling testosterone). Why would an athlete want to take these?
43. How do the estrogens vary in structure from most other steroids?
44. What is the best known corticosteroid? What is its biological role?
45. Name the two common bile salts. What is their biological role?
46. What is a prostaglandin, and what role does it play in our well-being?
47. How does aspirin reduce inflammation and pain?
48. What synthetic relationship exists between prostaglandins and leukotrienes? Where in the body are leukotrienes found?
49. Vitamin D is synthesized from what precursor molecule?
50. The precursor to steroids in the human body is
    a. cholesterol c. cortisone
    b. testosterone d. bile acids
51. The steroid responsible for maintenance of the uterine lining is
    a. progesterone c. estrogen
    b. testosterone d. cholic acid
52. Which of the following is a male sex hormone?
    a. progesterone c. estrone
    b. androsterone d. cortisone

Liposomes and Membranes (Objective 5)

53. Show the arrangement of amphipathic molecules in a micelle.
54. Compare the polarity of a micelle's core to the polarity of its surface.
55. *Hydrophobic interactions* is a term used to describe the interaction of nonpolar molecules or pieces of molecules when they are in an aqueous environment. Use examples of bonding between these molecules and bonding between water molecules to describe hydrophobic interactions.
56. Draw a short piece of lipid bilayer. Label the parts of the molecules with respect to polarity.
57. Draw a liposome. What potential use might the pharmaceutical industry have for liposomes?
58. What two types of biological molecules are found in biological membranes?
59. Describe the fluid mosaic model for biological membranes.
60. Compare active transport, facilitated diffusion, and simple diffusion.
61. Which are essential in membrane formation?
    a. waxes c. phospholipids
    b. triacylglycerols d. terpenes
62. What biomolecules are found in cell membranes?

Unclassified Exercises

63. Match the term on the left with the description on the right.
    A. triacylglycerols a. female sex hormone
    B. phospholipid b. derived from arachidonic acid
    C. prostaglandin c. fats and oils
    D. progesterone d. constituent of membrane
64. For each of the following words or phrases, provide a definition, description, or example to show that you understand it:
    a. lipid i. prostaglandin
    b. triacylglycerol j. steroid
    c. unsaturated fatty acid k. estrogen
    d. fat l. testosterone
    e. phosphoacylglycerol m. bile salts
    f. facilitated transport n. cholesterol
    g. phosphatidylcholine o. hydrocortisone
    h. glycolipid
65. Much of the chemical behavior of compounds can be directly related to the structures of the compounds. For each of the following pairs of compounds, describe the similarities and differences.
    a. oil/fat c. estrogen/progesterone
    b. fat/phospholipid d. prostaglandin/leukotriene
66. What are the names of the essential fatty acids?
67. What correlations exist between lipids and cardiovascular disease?

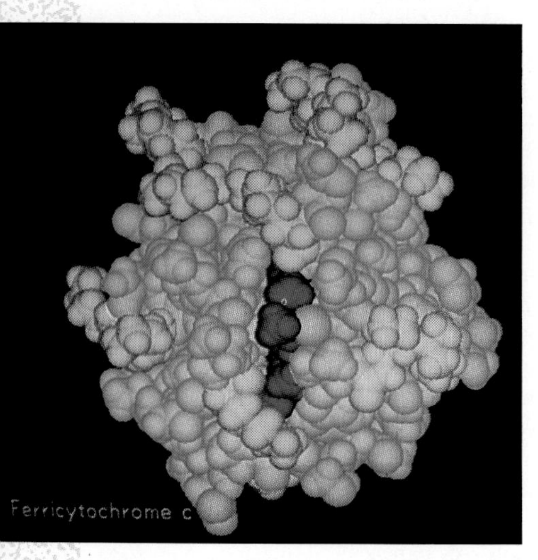

This spider's web is made of protein. Proteins serve many diverse roles in organisms.

CHAPTER 12

AMINO ACIDS, PEPTIDES, AND PROTEINS

OUTLINE

12.1 Protein Function

12.2 Alpha Amino Acids

12.3 Peptide Bonds and Peptides

12.4 Protein Structure

12.5 Properties and Classification of Proteins

OBJECTIVES

After completing this chapter, you should be able to

1. Write the structure of a generalized amino acid.
2. Define the electronic properties of amino acids, peptides, and proteins.
3. Describe the peptide bond.
4. Describe the four levels of protein structure.
5. Explain the importance of amino acid sequence in protein structure and function.
6. Describe some of the properties of proteins.
7. Describe some common functions of proteins.

12.1 Protein Function

Protein is the single most abundant class of biomolecules in cells; protein makes up one half of the dry mass of a cell. The name *protein* is derived from the Greek word *proteios*, which means "of the first rank or importance." Proteins play a fundamental role in life. The structure of a cell is determined by the proteins it contains. In addition, what a cell does or can do is determined by the proteins it contains. Since organisms are made of cells, the capabilities and shape of an organism are determined by proteins. Understanding protein function is a major goal of biochemistry.

Proteins have a wide variety of functions. One important group of proteins, the *enzymes*, serve as catalysts for the multitude of reactions that occur in the body. Enzymes are discussed in more detail in Chapter 14. Proteins are also involved in transport of materials in the blood. Hemoglobin helps transport oxygen and carbon dioxide, and lipoproteins transport lipids that would otherwise be insoluble. Various membrane proteins are involved in movement of substances across membranes via active transport or facilitated diffusion.

♦ Proteins have many important roles in the body.

Proteins serve a wide variety of other critical roles in living organisms. Some proteins have protective roles in the body. Antibodies are proteins that bind to specific foreign substances, including viruses and bacteria. The antibody–particle complex can then be more easily destroyed by various mechanisms. Lysozyme is an enzyme that cleaves or lyses the cell wall of some bacterial species, thus providing a protective role. Some proteins play a role in movement, such as actin and myosin, which are two contractile proteins in muscle. Some proteins serve as storage proteins. Myoglobin stores oxygen in muscle, and casein is a protein in milk that stores amino acids for the nursing young. Some proteins have structural roles. Collagen is the most abundant protein in animals and is a component of skin, bone, teeth, ligaments, tendons, and other extracellular connective structures. Some proteins play a role in regulation of body functions. Insulin, glucagon, and growth hormone help control cellular and body activities. Several examples of these assorted functions are presented in Table 12.1.

The function of a protein is determined by its structure. Therefore, an understanding of protein structure must precede an understanding of how a protein works. Proteins are polymers of amino acids, and cells make proteins by linking amino acids together by amide (peptide) bonds. The protein assumes a specific shape that is determined by the amino acids in that protein. This means the specific shape and properties of each protein are determined by the specific amino acids found in the protein.

NEW TERM

proteins Polymers of amino acids that have important roles in living systems

12.2 Alpha Amino Acids

Much of this chapter describes the structural properties of proteins, peptides, and amino acids. The common component of these biomolecules is the amino acid. Peptides and proteins are oligomers and polymers of amino acids, just

♦ Amino acids are the building blocks of proteins.

TABLE 12.1

Common Proteins and Their Functions

Name	Function
Enzymes	
Amylase	Catalyzes hydrolysis of bonds in starch
Carbonic anhydrase	Catalyzes hydration of carbon dioxide in red blood cells
Structural proteins	
α-Keratin	Found in skin and hair
Collagen	Fibrous connective tissue
Contractile proteins	
Myosin	Thick muscle filaments
Actin	Thin muscle filaments
Transport proteins	
Hemoglobin	Oxygen transport in blood
Serum albumin	Fatty acid transport in blood
Cytochrome c	Transport of electrons
Protective proteins	
Antibodies	React with foreign particles
Fibrinogen and thrombin	Important for blood clotting
Regulatory proteins	
Insulin	Involved in regulation of metabolism
Growth hormone	Involved in regulation of metabolism
Storage proteins	
Myoglobin	Stores oxygen in muscle
Ovalbumin	Protein of egg white. Provides amino acids to developing young

as oligosaccharides and polysaccharides are made from monosaccharides. An understanding of protein structure first requires an understanding of amino acids.

Structure of the Amino Acids

◆ The α-carbon of an amino acid has four substituents: a hydrogen, an amino group, a carboxyl group, and a side chain.

Amino acids contain both an amino group and a carboxylic acid group. Both functional groups are attached to the same carbon atom, which is designated in common nomenclature as the α-carbon (alpha carbon) because it is the first carbon of the main chain. Thus, the amino group is said to be an α-amino group because it is bonded to the α-carbon. Figure 12.1 shows a generalized α-amino acid. The α-carbon atom has four substituents. R in this figure represents an organic group or side chain, such as an alkyl group. There are 20 amino acids commonly found in proteins, each with a different side chain. The fourth bond to the α-carbon is always to a hydrogen atom.

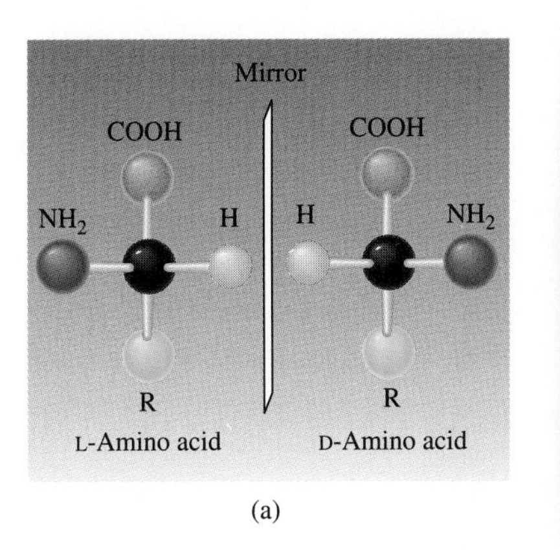

(a)

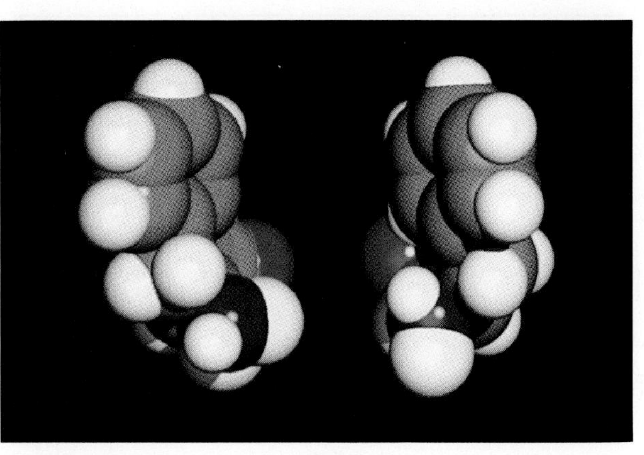

(b)

FIGURE 12.1
The generalized α-amino acid possesses a chiral center, and thus exists as a pair of enantiomers. (a) Both enantiomers are shown here aligned as the mirror images of each other. The amino acids in proteins are L-amino acids. (b) The two forms of the amino acid phenylalanine are shown here in space-filling models.

The α-carbon atom has four substituents, and unless R represents a hydrogen atom (as it would for the amino acid glycine), all four substituents are different. This makes the α-carbon a chiral center, and a pair of enantiomers (mirror image isomers) exist. Figure 12.1 shows both of these enantiomers, the D- and L-forms. These designations were used earlier to compare the stereochemistry of monosaccharides to glyceraldehyde. The terms mean the same thing here. When the molecules are arranged with the carboxyl group at the top of the illustration and the side chain down, the L-amino acid has the amino group facing to the left, and the D-amino acid has the amino group pointing to the right. The amino acids found in proteins are L-amino acids. D-Amino acids are uncommon, but they are found in some bacterial cell walls.

♦ The amino acids of the body are L-amino acids.

With the exception of proline, the α-amino acids are identical except for the side chain (Table 12.2). The structure of these side chains distinguishes amino acids from one another physically and chemically. There are several ways to classify the amino acids. The system used in this book is based on the polarity of the side chain. Since the structure of the amino acids in a protein ultimately determines the structure of that protein, an understanding of amino acid structure must be developed to appreciate their contribution to protein structure.

The *nonpolar amino acids* contain a side chain that is nonpolar. These side chains may be alkyl groups, aromatic rings, or other nonpolar groups. These amino acids play an important role in determining and maintaining protein structure. *Proline* is nonpolar and has a ring of five atoms that includes the α-carbon atom and the α-nitrogen atom of the amino group. One end of the side chain is bonded to the α-carbon atom, and the other end is bonded to the nitrogen atom of the α-amino group. This ring structure makes proline unique among the amino acids, and it thus has a special role in protein structure.

♦ The nonpolar amino acids all possess a nonpolar *side chain.*

TABLE 12.2

The 20 Common Amino Acids

Amino Acid Name	Abbreviations	Structure	Isoelectric point* pI
Nonpolar R Groups			
Alanine	Ala or A	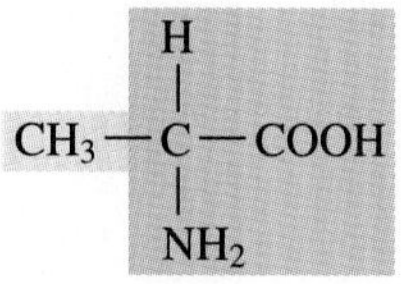 $CH_3-CH(NH_2)-COOH$	6.00
Valine	Val or V	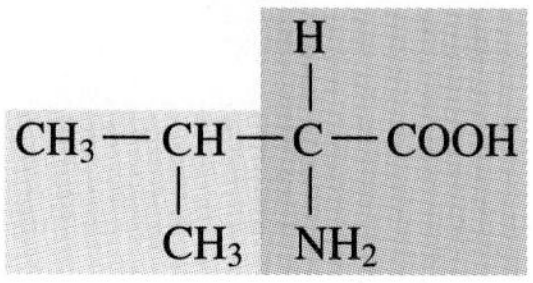 $CH_3-CH(CH_3)-CH(NH_2)-COOH$	5.96
Leucine	Leu or L	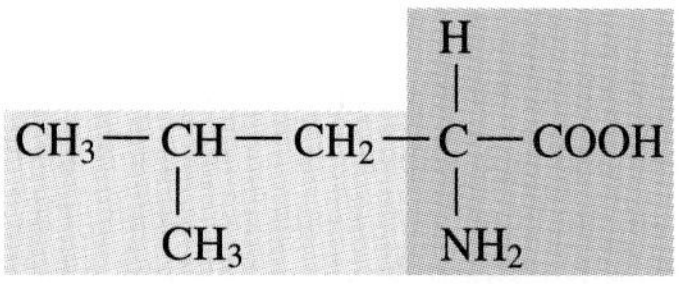 $CH_3-CH(CH_3)-CH_2-CH(NH_2)-COOH$	5.98
Isoleucine	Ile or I	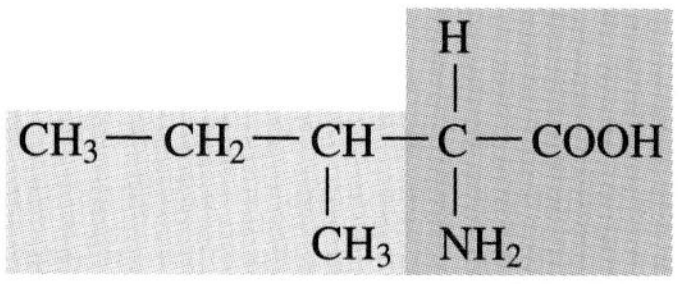 $CH_3-CH_2-CH(CH_3)-CH(NH_2)-COOH$	6.02
Phenylalanine	Phe or F	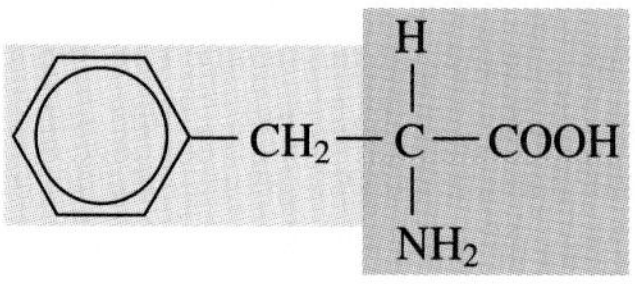 $C_6H_5-CH_2-CH(NH_2)-COOH$	5.48
Tryptophan	Trp or W	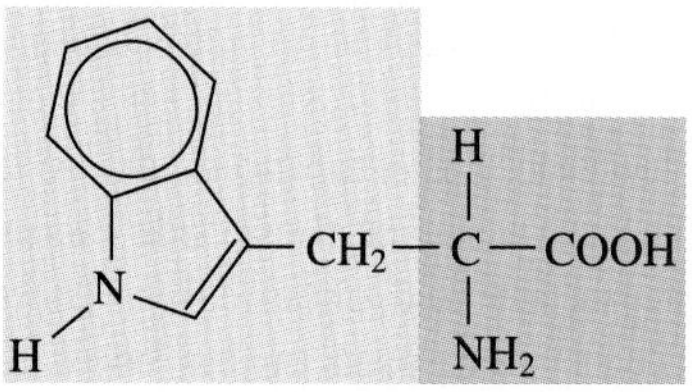 indolyl(N–H)$-CH_2-CH(NH_2)-COOH$	5.89
Methionine	Met or M	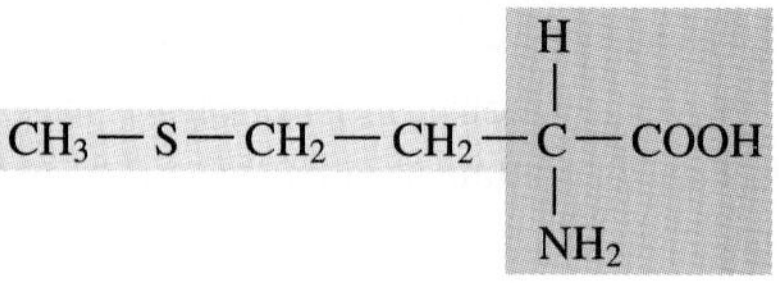 $CH_3-S-CH_2-CH_2-CH(NH_2)-COOH$	5.74
Proline	Pro or P	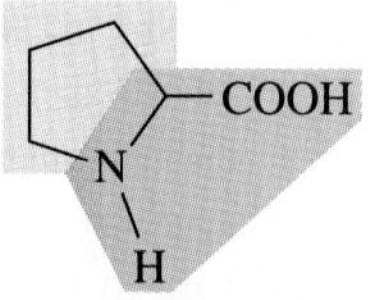 pyrrolidine ring (N–H) $-COOH$	6.30
Glycine	Gly or G	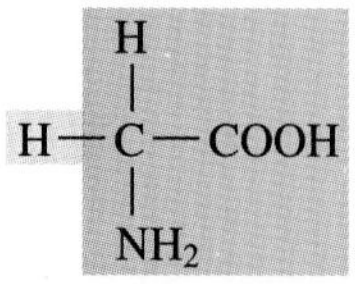 $H-CH(NH_2)-COOH$	5.97

Table 12.2 (*continued*)

Amino Acid Name	Abbreviations	Structure	Isoelectric point* pI
Polar but Neutral R Groups			
Serine	Ser or S	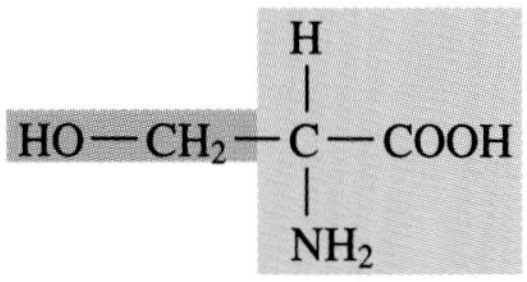 $HO-CH_2-CH(NH_2)-COOH$	5.68
Threonine	Thr or T	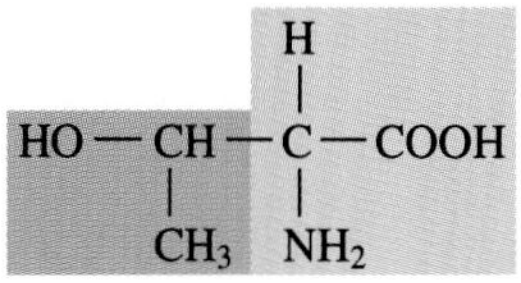 $HO-CH(CH_3)-CH(NH_2)-COOH$	5.60
Tyrosine	Tyr or Y	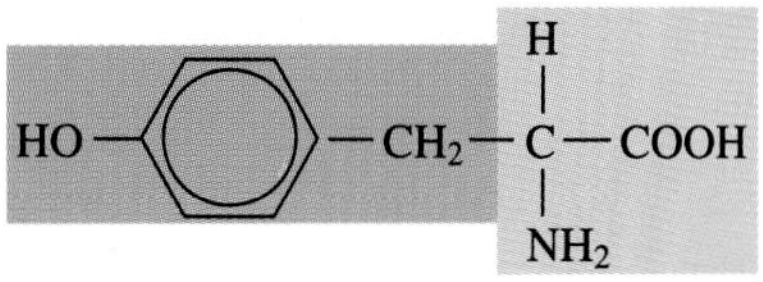 $HO-C_6H_4-CH_2-CH(NH_2)-COOH$	5.66
Cysteine	Cys or C	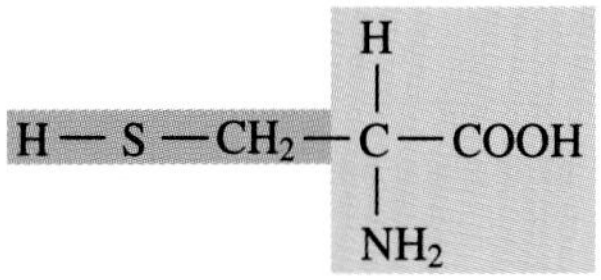 $H-S-CH_2-CH(NH_2)-COOH$	5.07
Asparagine	Asn or N	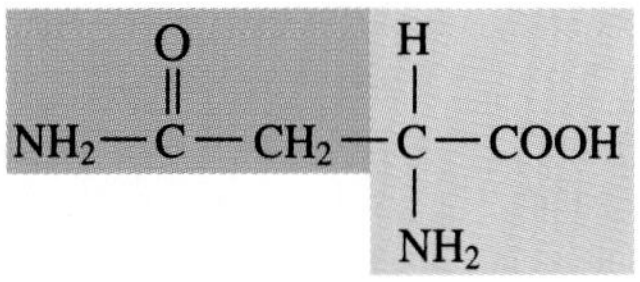 $NH_2-C(=O)-CH_2-CH(NH_2)-COOH$	5.41
Glutamine	Gln or Q	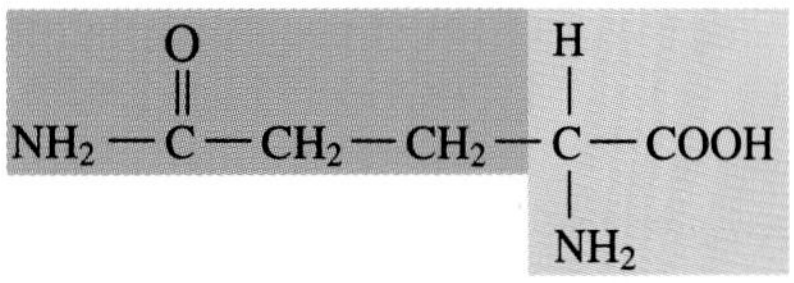 $NH_2-C(=O)-CH_2-CH_2-CH(NH_2)-COOH$	5.65
Acidic R Groups			
Glutamic acid	Glu or E	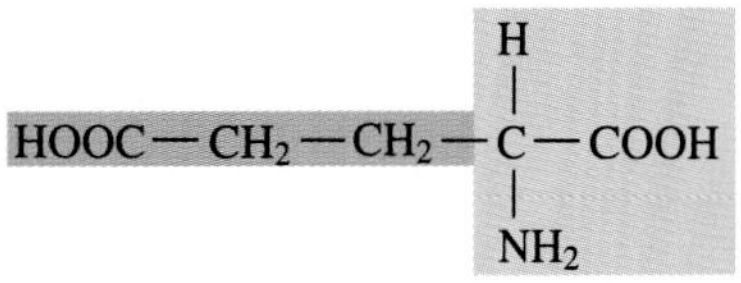 $HOOC-CH_2-CH_2-CH(NH_2)-COOH$	3.22
Aspartic acid	Asp or D	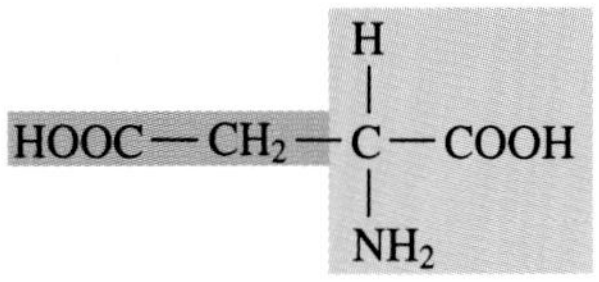 $HOOC-CH_2-CH(NH_2)-COOH$	2.77

(*continued*)

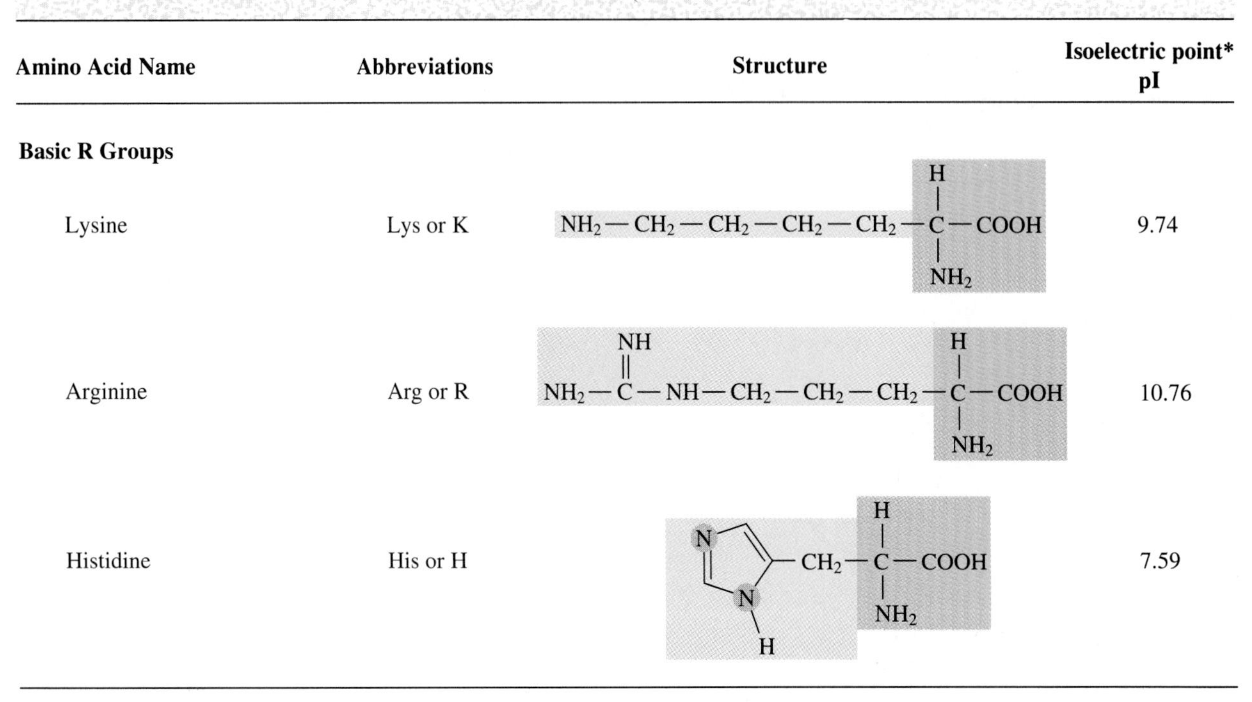

Table 12.2 (*continued*)

Amino Acid Name	Abbreviations	Structure	Isoelectric point* pI
Basic R Groups			
Lysine	Lys or K	$NH_2-CH_2-CH_2-CH_2-CH_2-CH(NH_2)-COOH$	9.74
Arginine	Arg or R	$NH_2-C(=NH)-NH-CH_2-CH_2-CH_2-CH(NH_2)-COOH$	10.76
Histidine	His or H	(imidazole ring, N, N–H)$-CH_2-CH(NH_2)-COOH$	7.59

* Isoelectric point is defined in the section titled "Acid–Base Properties of Amino Acids."

The rest of the amino acids are considered *polar* because the side chain contains one or more polar groups. These amino acids also play vital roles in protein structure, but some of them may play special roles in the function of the protein. The polar amino acids are placed into three different subgroups. The first subgroup of the polar amino acids is known as the *neutral polar amino acids.* The side chains of these amino acids are polar, but not ionic. The *acidic amino acids* make up the second group of polar amino acids. These amino acids contain a carboxyl group in the side chain; this second carboxyl group is *in addition* to the one attached to the α-carbon atom. The third group of polar amino acids is known as the *basic amino acids*. Each contains one or more nitrogen atoms that have an unshared pair of electrons; they are Lewis bases.

◆ The side chain of the neutral polar amino acids is uncharged but polar.

◆ The side chain of the acidic amino acids contains a carboxyl group.

◆ The side chains of the three basic amino acids have a nitrogen atom that can accept a hydrogen ion.

Acid–Base Properties of Amino Acids

The α-amino and α-carboxyl groups of amino acids act as acids and bases. Acids and bases react in neutralization reactions, and the α-amino group and carboxyl group of an amino acid are not exceptions. These groups react in an intramolecular acid–base reaction. The carboxyl group donates a hydrogen ion to the amino group.

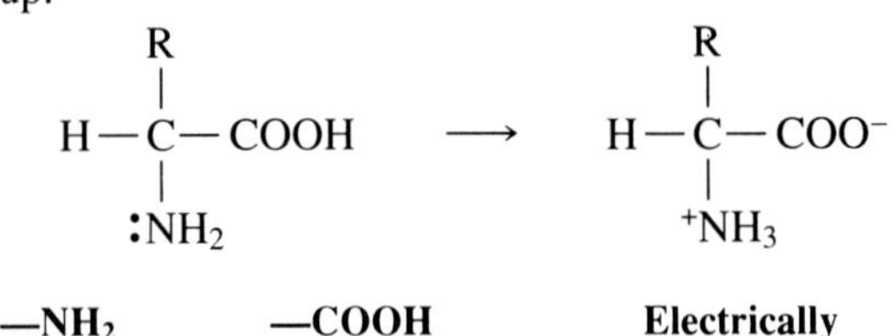

$$H-C(R)(:NH_2)-COOH \longrightarrow H-C(R)(^{+}NH_3)-COO^-$$

—NH₂ Amino group (basic) **—COOH Carboxyl group** (acidic) **Electrically neutral** (zwitterion)

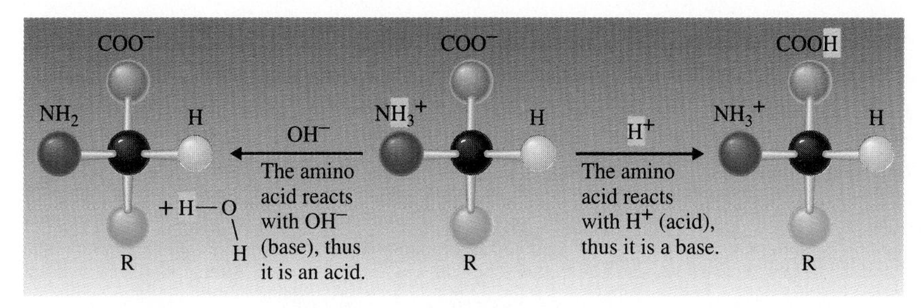

FIGURE 12.2
Amino acids are amphoteric molecules because they can act as both acids and bases. The protonated amino group acts as an acid, and the carboxylate group acts as a base.

The product of this reaction is a dipolar ion that is called a **zwitterion**. Dipolar ions are electrically neutral because the opposite charges cancel. A consequence of the ionic nature of the zwitterions is the water-solubility and high melting points of the α-amino acids. These are just the properties you would expect of ionic compounds.

♦ Amino acids can exist as dipolar ions called *zwitterions*.

A second acid–base property is seen in amino acids. They are **amphoteric**; that is, they can function as *either* acids or bases. In the presence of a base, the amino acid behaves as an acid because it gives up a proton from the ammonium salt portion of the molecule. Likewise, in an acidic solution, the amino acid functions as a base because the carboxylate group accepts a hydrogen ion (Figure 12.2). Although it is important to recognize this dipolar ion character of amino acids, in this text the amino acids are represented in either the dipolar or nonionic form.

In aqueous solutions around pH 6 to 7, many of the amino acids are neutral dipolar ions. At other pH values, these amino acids are charged. In an acidic solution a hydrogen ion is gained by the molecule, and the electric charge of the amino acid is changed from zero to plus one.

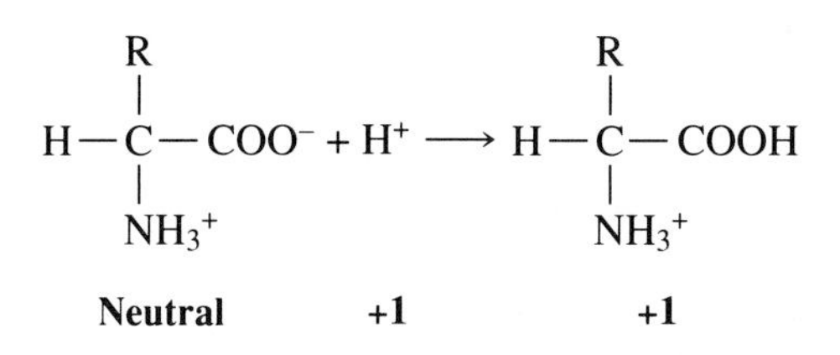

Similarly, in basic solutions hydroxide ion removes a hydrogen ion from the neutral molecule changing the charge on the amino acid from neutral to minus one.

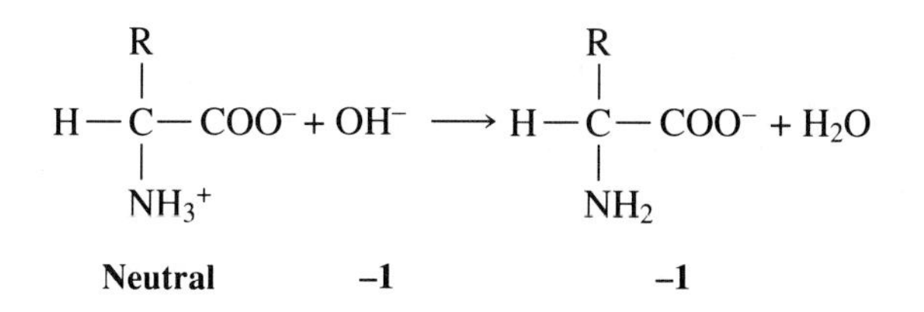

Amino acids are neutral at only one pH value. If the pH is higher or lower, then the amino acid has a negative or positive charge. The pH where an amino acid is neutral, where it has no net charge, is called the **isoelectric point**, or pI. Table 12.2 contains the pI values of the amino acids.

♦ The charge on an amino acid varies with pH.

When amino acids are present in peptides and proteins, the α-amino and carboxyl groups are not free but instead are involved in covalent bonds. In

CHEMISTRY CAPSULE

Proteins and Amino Acids in Nutrition

Amino acids are needed by the body to synthesize proteins. Some amino acids can be synthesized by animals, others cannot. Those that cannot be synthesized are called essential amino acids. The essential amino acids for humans are:

histidine	isoleucine
leucine	lysine
methionine	phenylalanine
threonine	tryptophan
valine	

These amino acids must be included in the diet to ensure proper growth or function because adequate protein synthesis cannot occur unless all 20 amino acids are present simultaneously in adequate amounts. Two of the nonessential amino acids are synthesized from essential amino acids. Tyrosine is made from phenylalanine, and the sulfur of cysteine comes from methionine. Thus, the diet must provide extra methionine and phenylalanine if tyrosine or cysteine are present in inadequate quantities.

In a nutritional sense, proteins are the sources of amino acids. Dietary proteins are broken down during digestion to amino acids, which are then absorbed. They are used by cells to make proteins or other compounds. If the amino acids are in excess of body needs for protein synthesis, they are broken down to yield energy. A gram of protein yields 4 Cal (4 kcal) of energy, about the same as carbohydrate and much less than 1 g of fat.

FIGURE 12.A
Proteins are found in most foods, but meats, dairy products, nuts, grains, and other seeds are good sources.

Although many foods contain protein, meat and dairy products are among the best sources; other good sources include foods derived from some seeds and nuts. Generally, foods from animal sources are richer in protein and contain a better blend of essential amino acids than foods from plants. Most vegetables and fruits contain relatively little protein.

contrast, the side chain is rarely involved in a covalent bond. It is the side chain of an amino acid that contributes most to the acid–base and other properties of a peptide or protein. A brief review of the side chains of the amino acids in Table 12.2 reveals that a variety of functional groups and polarities exist. In Section 12.4, the properties of the amino acid side chains are used to explain protein structure.

EXAMPLE 12.1

Alanine is a nonpolar amino acid and serine is a polar one. Yet these two amino acids have similar solubilities in water. Why?

Solution: The polarities of the amino acids are described in terms of the properties of the side chain: alanine has a nonpolar side chain; serine has a polar one. But both amino acids possess a carboxyl group and an α-amino group, either one or both of which may be ionized. The ionized group has greater influence on solubility than the nonpolar side chains of some amino acids, thus the amino acids have similar solubility properties.

NEW TERMS

amino acids A class of biological compounds whose members possess both an amino group and a carboxylic acid group. Amino acids are the building blocks of peptides and proteins.

zwitterion A dipolar ionic form of an amino acid that is formed by donation of a H^+ from the carboxyl group to the α-amino group. Because both charges are present, the net charge is neutral.

amphoteric molecule A molecule that functions as a base in the presence of an acid and as an acid in the presence of a base

isoelectric point (pI) The pH at which an amphoteric molecule, such as an amino acid or protein, has no net charge

TESTING YOURSELF

Amino Acids

1. Identify each of the following amino acids as neutral, acidic, or basic.

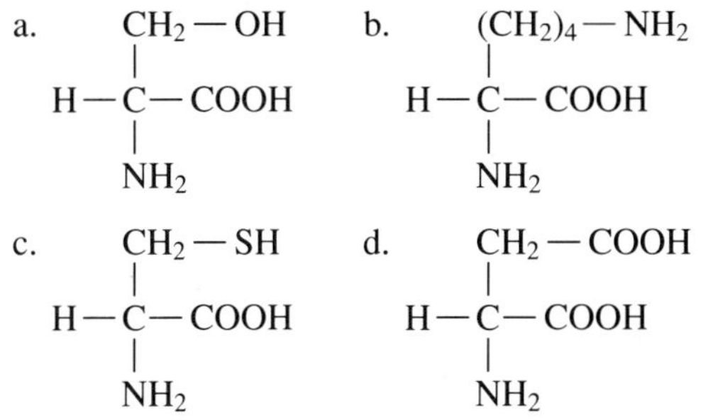

a. $H-C(CH_2-OH)(NH_2)-COOH$

b. $H-C((CH_2)_4-NH_2)(NH_2)-COOH$

c. $H-C(CH_2-SH)(NH_2)-COOH$

d. $H-C(CH_2-COOH)(NH_2)-COOH$

2. Give a product for each of the following reactions.

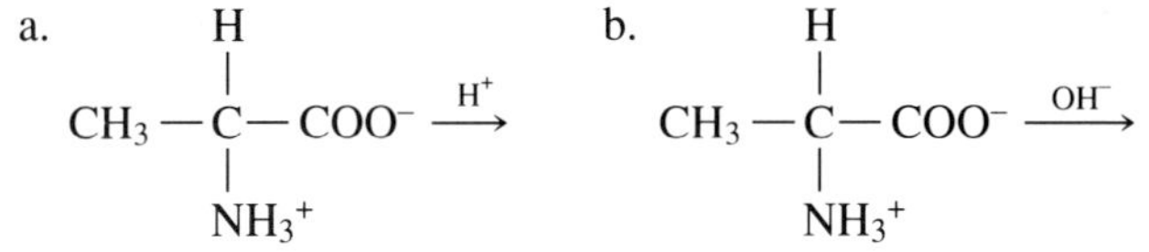

a. $CH_3-C(H)(NH_3^+)-COO^- \xrightarrow{H^+}$

b. $CH_3-C(H)(NH_3^+)-COO^- \xrightarrow{OH^-}$

Answers **1. a.** neutral **b.** basic **c.** neutral **d.** acidic

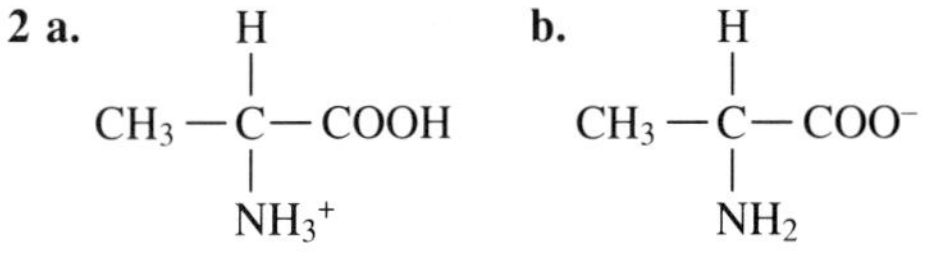

2 a. $CH_3-C(H)(NH_3^+)-COOH$ **b.** $CH_3-C(H)(NH_2)-COO^-$

12.3 Peptide Bonds and Peptides

The Peptide Bond

Although amino groups do not readily react with carboxylic acids to form amides, the formation of an amide can be summarized by this equation.

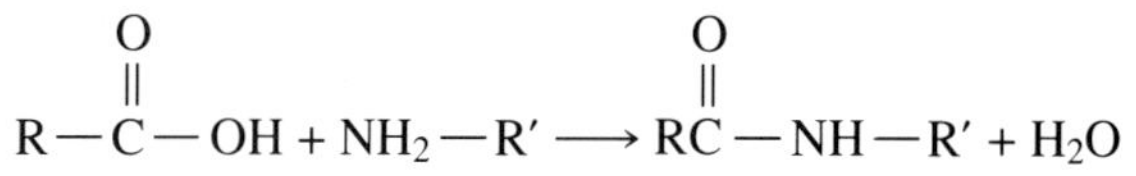

$$R-\overset{O}{\overset{\|}{C}}-OH + NH_2-R' \longrightarrow R\overset{O}{\overset{\|}{C}}-NH-R' + H_2O$$

In organic chemistry, the bond formed between the carbon atom of the carbonyl group of a carboxylic acid and the nitrogen atom of an amino group is called an amide bond. When this bond involves amino acids, it is called a peptide bond. **Peptide bonds** connect amino acids together in peptides and proteins.

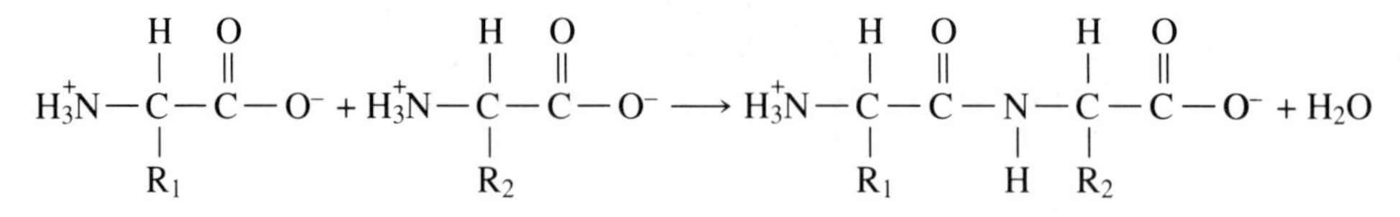

The reaction that forms an amide or peptide is a readily reversible reaction. This reverse reaction is a hydrolysis reaction, that is, a cleavage by water.

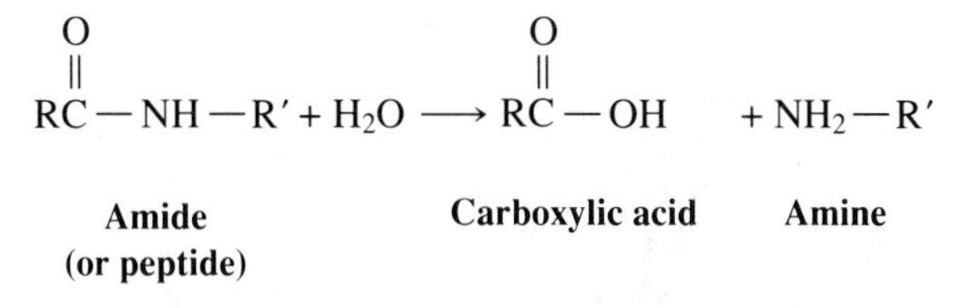

Both of these reactions occur in the body. The body uses energy to form peptide bonds to build the proteins and peptides needed to carry out the normal functions of the body. Proteins and peptides are broken down by hydrolysis; dietary proteins are cleaved in the digestive tract, and cellular proteins and peptides are hydrolyzed and replaced routinely during the normal course of cellular activity.

The peptide bond has some structural features that are not readily apparent in formulas used to represent it. In the preceding dipeptide reaction, the peptide bond was shown as a single bond between the carbon atom of the carbonyl group and the nitrogen atom, but this bond actually has some double bond character due to delocalization of electrons. A better representation of this bond would be a rigid, plate-like structure (Figure 12.3). Because of its double-bond character, the peptide bond lacks the free rotation typical of a single bond. This results in the two atoms of the carbonyl group and the nitrogen atom with its hydrogen atom lying in the same plane. The atoms of a peptide bond act as a single rigid unit in a peptide or protein.

When a peptide bond is formed between two amino acids, the amino group of one and the carboxyl group of the other participate in the reaction. That leaves one of the amino acids with a free carboxyl group and the other with a free amino group. Each of the amino acids can form two peptide bonds, one with its amino

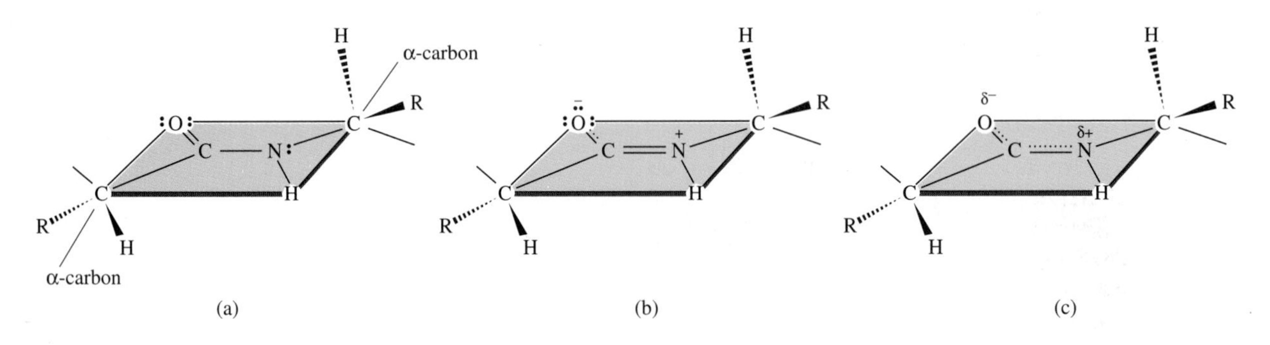

FIGURE 12.3
The structure of the peptide bond. (a) The peptide bond is typically represented as an amide bond with a carbonyl group and a C—N single bond. (b) Because the atoms of the peptide bond have unshared electron pairs and pi electrons, another resonance form for this bond can be drawn. (c) The forms in (a) and (b) can be mentally combined to illustrate the partial double bond character of the C—N bond and the C—O bond. Since free rotation does not occur around double bonds (or partial double bonds), these atoms all lie in a plane.

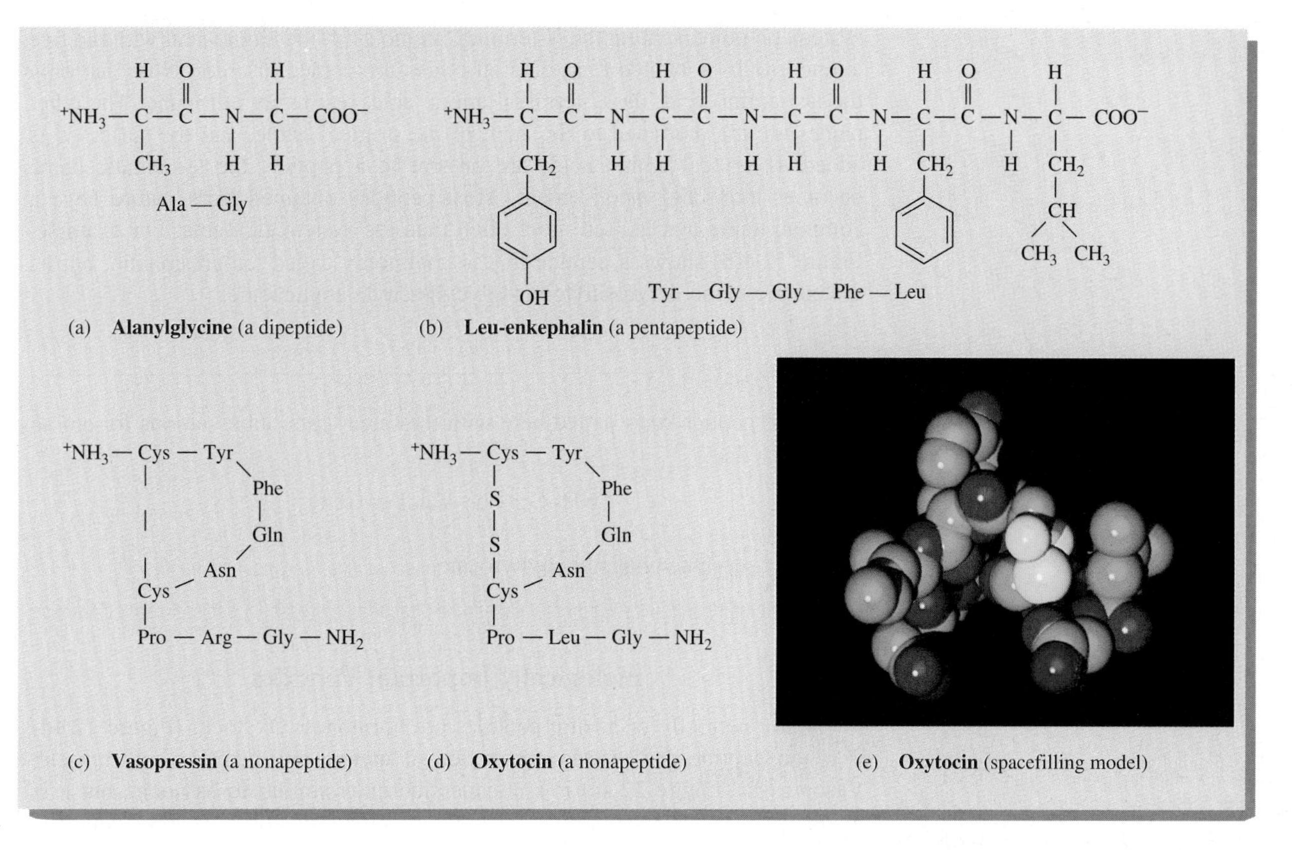

FIGURE 12.4
Some common peptides. The entire structures of alanylglycine (a) and leu-enkephalin (b) are shown; abbreviations are normally used for larger peptides. Amino acid composition and covalent bonding are shown for the other peptides. Note the structural similarity between (c) vasopressin and (d) oxytocin, even though they cause quite different physiological responses. Gly—NH_2 in these two peptides indicates that the carboxyl group of glycine is bonded to ammonia via an amide bond. A space-filling model of oxytocin is shown in (e) for comparison.

group and one with its carboxyl group. Because amino acids are bifunctional, several to many amino acids can be linked together to form oligomers and polymers called *peptides* or even larger polymers called *proteins*.

Structure and Nomenclature of Peptides

Peptides are oligomers and polymers of amino acids. There is no fixed number of amino acids that determines an upper limit for classifying a compound as a peptide. Instead, a compound composed of amino acids with a molecular weight less than 5000 is often classified as a peptide. The average molecular weight of an amino acid is somewhat over 100, so on the average, peptides have fewer than 50 amino acids.

A peptide containing two amino acids is called a *dipeptide*. One with three is a tripeptide and one with four is a tetrapeptide. The prefixes used for the number of amino acids are the same as in organic nomenclature and in the nomenclature of the carbohydrates. Oligopeptides contain several amino acids. For our purposes, an oligopeptide has ten or fewer amino acids, **polypeptides** contain more than ten amino acids. The boundary between oligopeptides and polypeptides is arbitrary.

Peptides are sometimes named by a systematic nomenclature. For example, the dipeptide in Figure 12.4(a) is called alanylglycine, because it contains alanine which has a free amino group and glycine with a free carboxyl group. In the systematic nomenclature the *N*-terminal amino acid (the amino acid with the free amino group) is named first, then all others are named in order. Note that only the last amino acid, the *C*-terminal amino acid, retains its full name. The other names are all shortened to the stem of the original name, and the suffix *-yl* is added. If several amino acids are present in a peptide, the systematic name becomes large and inconvenient. Most peptides obtained from nature have a common name that is used more often than the systematic name. For example, Figure 12.4(b) shows a peptide that is commonly called leu-enkephalin, but its systematic name is tyrosinylglycylglycylphenylalanylleucine.

EXAMPLE 12.2

Name the peptide represented here with the three-letter abbreviations for amino acids:

$$NH_2\text{-Ser-Phe-Val-Tyr-}COO^-$$

Solution: serylphenylalanylvalyltyrosine

Biologically Important Peptides

Many naturally occurring peptides are hormones. Oxytocin (Figure 12.4d) is a nonapeptide that causes contraction of uterine and other smooth muscle. Vasopressin (Figure 12.4c) is a nonapeptide quite similar to oxytocin, but it is involved in water retention and blood pressure regulation. Several peptides, including the endorphins and enkephalins (Figure 12.4b), are neuropeptides that seem to be involved in pain and pleasure sensation. Morphine and heroin appear to mimic some of the effects of endorphins and enkephalins. (The term *endorphin* is a contraction of *end*ogenous m*orphin*e.) These peptides may also be the agent responsible for the "runner's high" that follows intense aerobic exercise. Another example is glutathione, which is a tripeptide that is not a hormone. It appears to serve a role as a reducing agent in cells. The artificial sweetener aspartame is a modified dipeptide that is many times sweeter than sucrose.

NEW TERMS

peptide bond The amide bond between the amino group of an amino acid and the carboxylic acid group of another amino acid

peptide A compound consisting of amino acids linked by peptide bonds. Often the number of amino acids is indicated by prefixes such as *di-*, *tri-*, or *oligo-*.

polypeptide A macromolecule containing many (ten or more) amino acids. This term is sometimes used to mean the chain in a protein.

TESTING YOURSELF

Peptide Bond and Peptides

1. The peptide bond is essentially what bond of organic chemistry?
2. Examine oxytocin in Figure 12.4d. Which amino acid is *N*-terminal? Also

name the C-terminal amino acid of leu-enkephalin (Figure 12.4b).

3. You have isolated a molecule from a human neuron that is made from numerous amino acids connected by peptide bonds. What criterion would you use to determine if it is a peptide or protein?

Answers **1.** the amide bond **2.** cysteine; leucine **3.** Molecular weight. If the compound has a molecular weight greater than 5000, then it is a protein; if less than 5000, it is a peptide.

12.4 Protein Structure

Proteins are macromolecules containing one or more polypeptides plus, in some cases, one or more other components. These polymers of amino acids have molecular weights greater than 5000. Proteins and the larger peptides are polypeptides and are distinguished arbitrarily by size. Most proteins are very large; many have molecular weights greater than 50,000, and some are more than one million. The amino acids of proteins are linked by peptide bonds. To distinguish each amino acid residue from another in a peptide or polypeptide, each specific residue is numbered. The N-terminal amino acid occupies the first residue in the protein, the next one is the second residue, and the C-terminal amino acid is the last residue in the protein.

◆ A molecular weight of 5000 is the arbitrary boundary between large peptides and proteins.

If a few reasonable assumptions are made, you can get an idea of how many possible proteins can be formed from amino acids. Consider a protein with 100 residues. This is a rather small protein. Assume that any of the 20 different amino acids could be in any of the 100 residues. Each of these combinations is a unique protein. The number of possible combinations that the amino acids can have is equal to 20 times itself 100 times or 20^{100}. This is about 1×10^{130}! And remember, this number was calculated using only 100 amino acid residues. Any number of amino acids from about 50 up to many hundreds could actually be present in a protein. Compare this gigantic number of possible combinations to the estimated number of atoms in the entire universe, which is about 10^{80}, give or take a few orders of magnitude. The number of possible proteins that could be made from the 20 amino acids is truly astronomical. In reality, the number of actual proteins that are found in the living organisms on earth is much smaller. It is estimated that the number of different protein structures present in all organisms is greater than one trillion (10^{12}).

EXAMPLE 12.3

The peptide oxytocin contains nine amino acids. How many possible nonapeptides can be made from the 20 amino acids?

Solution: Since any of the 20 amino acids can occupy any of the residues; the answer is $(20)_1 \times (20)_2 \times \cdots \times (20)_9 = (20)^9 = 5.12 \times 10^{11}$ nonapeptides.

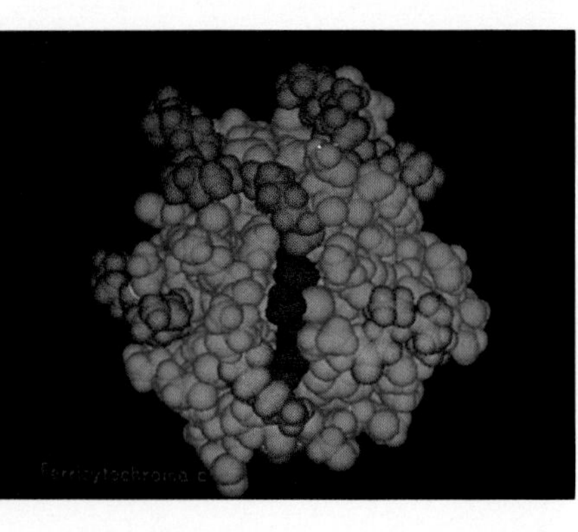

FIGURE 12.5
This computer-drawn image of cytochrome c shows the complex three-dimensional nature of proteins. Each protein has its own shape.

Protein Conformation

Each amino acid sequence in a protein is unique for that protein. Obviously, the number of possible amino acid sequences for proteins is very large, but what about the orientation in space of the atoms of any one of these proteins? Free rotation about single bonds allows for an infinite number of conformations for a molecule of even modest size or complexity. There are large numbers of single bonds in a protein, so an infinite number of possible conformations exist for each protein. Yet biochemists are quite sure that all of the copies of any given functional protein are in one or, at most, a very few conformations (Figure 12.5). Why does each have only one conformation? Because it is the most stable and lowest energy conformation. There are forces that operate within a protein that cause it to be in that most stable conformation.

The normal conformation of a protein is called its **native conformation**. In this conformation the protein has its native function. If a protein's conformation is changed, it loses its native function or activity. This process is called **denaturation**, and it illustrates that a protein must have its native conformation to be functional. Because protein structure is critical to protein function, much of this chapter deals with protein structure. Protein structure is complex, and to facilitate understanding, protein chemists have described protein structure at four different levels called *primary, secondary, tertiary*, and *quaternary* structure.

Primary Structure

♦ The amino acid sequence of a protein chain is the primary structure of the protein.

The **primary structure** of a protein is the amino acid sequence of that protein. The primary structure is known when it has been determined what amino acids are present in the protein, how many copies of each amino acid are present, and which amino acids are bonded to each other. The amino acid sequences of many proteins are now known, and each unique protein appears to have its own combination of amino acids in their own unique sequence. Each protein has its own unique primary structure. The amino acid sequence of the small protein myoglobin is shown in Figure 12.6.

♦ Primary structure determines the shape of a protein.

Ultimately, the amino acid sequence of a protein, the primary structure, determines the other levels of structure. Since various amino acids are present in different places within each protein, their interactions with water and with each other differ from protein to protein. This results in unique combinations of interactions within each protein, which in turn produces a unique conformation for the protein. *Each protein has its own unique shape because it has its own unique amino acid sequence.*

A single, unique conformation for a molecule should not really be surprising to you. Recall from Chapter 12 that the small hydrocarbon ethane has an infinite number of possible conformations, but only one conformation, the staggered one, is the most stable. It is the bonding of atoms within ethane that ultimately determines its most stable conformation. The hydrogens of ethane get in each other's way in other conformations, making them less stable. Although more factors than hindrance are involved in protein structure, the same basic argument holds. A protein has some conformation that is more stable, and that particular conformation depends on the atoms in the protein and how they are bonded to each other.

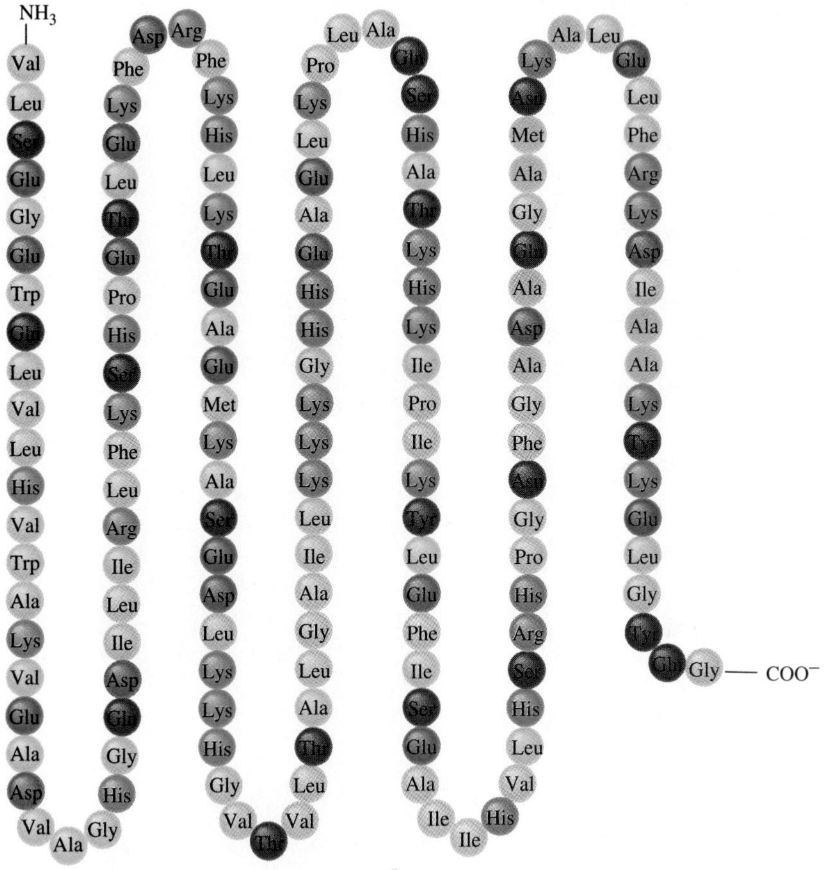

NH_3

Val Leu Ser Glu Gly Glu Trp Gln Leu Val Leu His Val Trp Ala Lys Val Glu Ala Asp Val Ala Gly His Gly Gln Asp Ile Leu Ile Arg Leu Phe Lys Ser His Pro Glu Thr Leu Glu Lys Phe Asp Arg Phe Lys His Leu Lys Thr Glu Ala Glu Met Lys Ala Ser Glu Asp Leu Lys Lys His Gly Val Thr Val Leu Thr Ala Leu Gly Ala Ile Leu Lys Lys Lys Gly His His Glu Ala Glu Leu Lys Pro Leu Ala Gln Ser His Ala Thr Lys His Lys Ile Pro Ile Lys Tyr Leu Glu Phe Ile Ser Glu Ala Ile Ile His Val Leu His Ser Arg His Pro Gly Asp Phe Gly Ala Asp Ala Gln Gly Ala Met Asp Lys Ala Leu Glu Leu Phe Arg Lys Asp Ile Ala Ala Lys Tyr Lys Glu Leu Gly Tyr Gln Gly — COO^-

FIGURE 12.6
The primary structure of myoglobin from a sperm whale. This protein has these 153 amino acids in this specific sequence.

Secondary Structure

The **secondary structure** of a protein involves the arrangement of the atoms of the polypeptide chain or "backbone" of the protein. This chain is the continuous line of atoms that includes the atoms of the peptide bonds plus the α-carbon atom of the amino acids. The atoms that are not part of this chain are the hydrogen atom attached to the α-carbon and the atoms of the side group. Figure 12.7 shows the backbone of a part of a polypeptide chain. There are three different kinds of bonds found in the polypeptide backbone. Two of these three bonds are single bonds that allow for free rotation, that is, twisting of the chain. But the third bond, the peptide bond, forms a rigid, plate-like structure that does not allow free

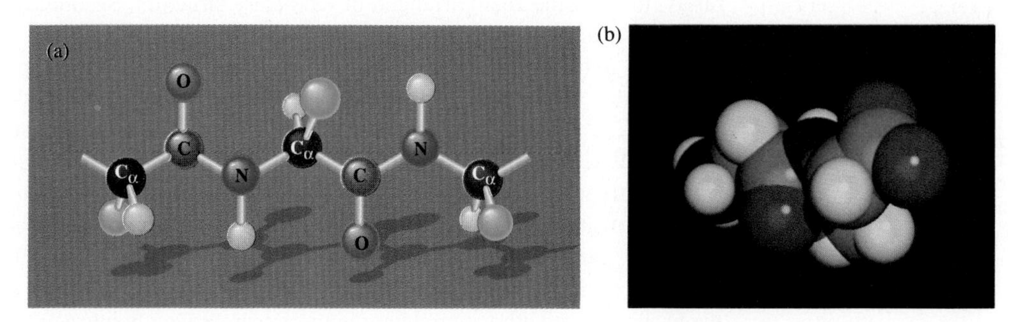

FIGURE 12.7
(a) The atoms of part of a polypeptide chain are shown here. Secondary protein structure is defined as the orientation of these atoms with respect to each other. (b) This is a space-filling model of a peptide bond between glycine and alanine.

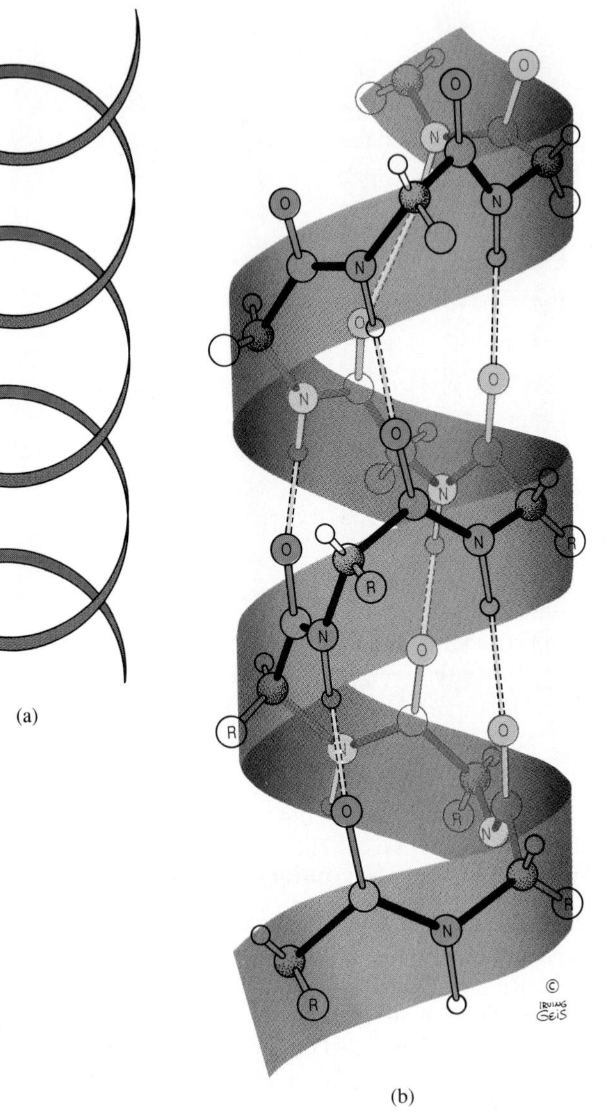

FIGURE 12.8
The arrangement of the polypeptide chain of a protein into a spiral is referred to as an α-helix. (a) Coiled springs and slinkies are common objects with spiral or helical structures. (b) Hydrogen bonds (dashed lines) between atoms of the polypeptide chain maintain the α-helix of the protein.

rotation. Because of these restrictions, the atoms of the chain can be arranged in only a few ways. Many proteins have portions of the peptide chain or, more rarely, most of the chain arranged in secondary structure. Secondary structure is classified into two categories: (1) α-helix or (2) β-pleated sheet.

In an **α-helix**, the atoms of the backbone are arranged into a spiral shape, a helix. Coiled springs and the toys called slinkies are common examples of helices (Figure 12.8a). The atoms of a peptide bond form a rigid plate-like structure, and in an α-helix, these plates are arranged in an orderly, repeating fashion that places them in this spiral arrangement. They are arranged in this way because hydrogen bonds form between a hydrogen atom in an N—H group on one peptide bond and the oxygen atom of a carbonyl group in another loop (Figure 12.8b). Each hydrogen bond is a weak dipolar interaction, but the α-helix brings the atoms of the backbone into a position where numerous hydrogen bonds can form between atoms of different coils.

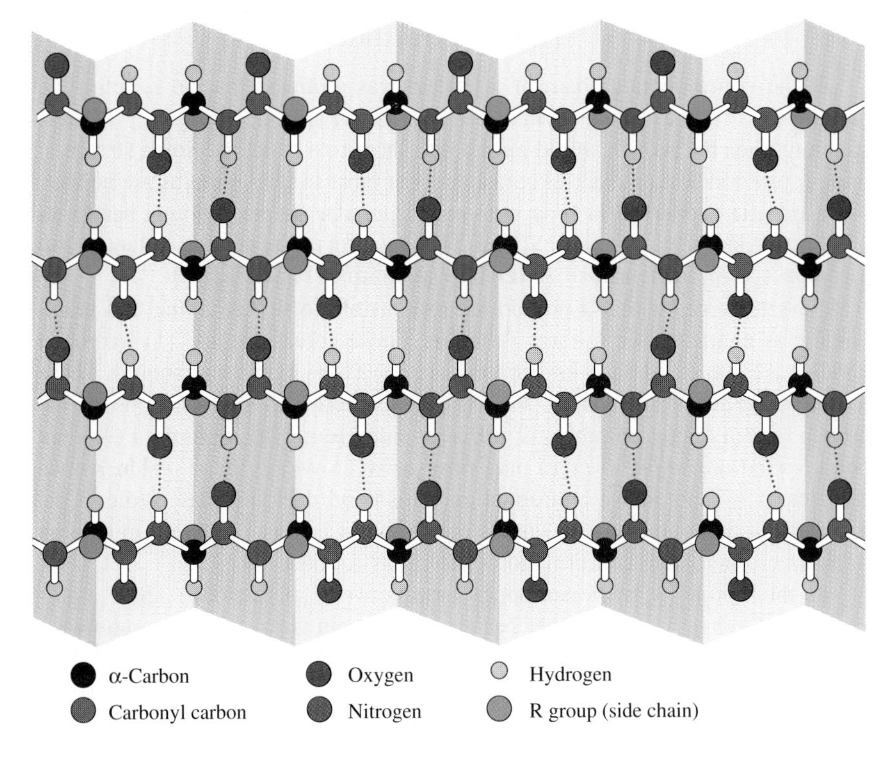

FIGURE 12.9
A representation of a β-pleated sheet in a protein. In this type of secondary protein structure, polypeptide chains are held in place by hydrogen bonding (dots) between atoms of different chains.

It appears that as long as no other forces are present to prevent an α-helix from forming, the numerous hydrogen bonds stabilize a protein into an α-helix. Some proteins, such as the hair and skin protein α-keratin, have virtually the entire length of the polypeptide chain arranged in an α-helix. Other proteins have only portions of the chain in this conformation (for example, see myoglobin in Figure 12.10). What prevents the chain from assuming an α-helix conformation? Whenever a proline is present, its unique shape introduces a "kink" in the chain. Just as a cis bond in the fatty acid portion of a triacylglycerol prevents orderly clustering of these lipid molecules, a proline bends the chain and prevents the α-helix conformation. The α-helix can also be prevented by certain side groups. If the size or shape of a side group places a portion of it between atoms that would hydrogen-bond, then that large bulky R group acts as a physical barrier that prevents stable formation of hydrogen bonds in that region. Thus, the α-helix does not form. Similarly, side chains with similar charge would repel each other, therefore adjacent or nearby charged side groups may prevent the formation of a stable α-helix.

The second type of secondary structure found in proteins is called the **β-pleated sheet** (Figure 12.9). This structure also results from hydrogen bonding between atoms of the peptide bond, but the bond forms between atoms that are in different polypeptide chains. β-Sheets are found in a number of proteins. The structure of fibroin, the protein of silk, is mostly β-pleated sheets.

♦ Examples of secondary protein structure are α-helix and β-pleated sheet.

If the atoms of a polypeptide backbone are arranged into an α-helix or a β-pleated sheet, then the term *secondary structure* applies, but most proteins are not entirely helical or sheet-like. Higher levels of structure are needed to adequately explain the conformations of these proteins.

Tertiary Structure

If the entire peptide chain of a protein was arranged into an α-helix, the protein would have a long, cylindrical, rod-like shape. If it had only β-sheet structure, then the protein would have a flat, sheet-like structure. Some structural proteins are rod-like or sheet-like in shape, but most soluble proteins are neither. They are often spherical or have a compact globular shape, although nearly all other possible shapes have been found. This bending of the peptide chain to form a compact shape is described as **tertiary structure** (Figure 12.10).

Several forces within a protein are responsible for its assuming and maintaining its characteristic tertiary structure. Those forces include (1) hydrogen bonding, (2) ionic and polar interactions, and (3) hydrophobic interactions. These forces result from interactions between atoms from different portions of the protein, usually between atoms of different side groups. In addition, a covalent bond (called a *disulfide bridge*) may form between two cysteine residues once the protein is in its native conformation. This bond does not play a role in the protein's assumption of native conformation, but it is very important in helping to maintain the native conformation once it has formed.

♦ Disulfide bridges are covalent bonds between cysteine residues that help maintain the tertiary structure of a protein.

In most proteins, hydrogen bonds form between some atoms. These atoms have been brought close together because the protein is folded in a certain way, even though they are atoms in amino acids that may be many residues apart (Figure 12.11). The hydrogen bond may form between atoms of two different side chains or an atom of a side chain and an atom of the backbone. In either case, the formation of hydrogen bonds stabilizes the protein and helps hold the protein in that shape. Again, any one hydrogen bond is weak, but *many* may be present in a globular protein.

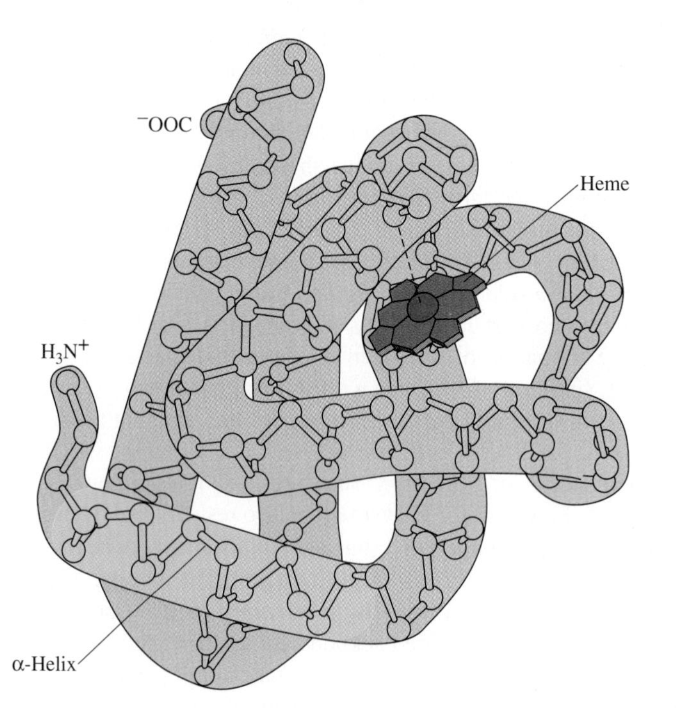

FIGURE 12.10
The tertiary structure of myoglobin. Much of the length of this protein exists as regions of α-helix, one of which is highlighted. Take a moment to identify other regions of α-helix. Heme is the prosthetic group of myoglobin.

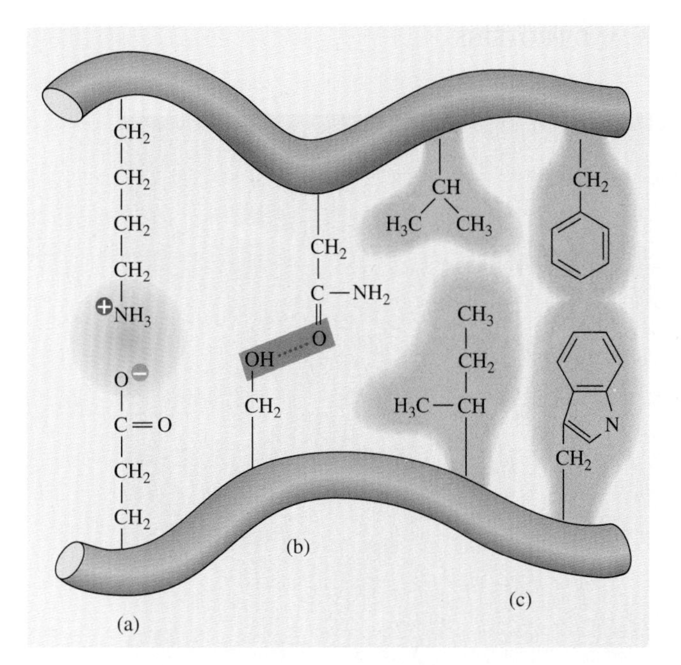

FIGURE 12.11
Forces responsible for tertiary structure of a protein include: (a) ionic and polar bonding (green), (b) hydrogen bonding (red), and (c) hydrophobic interactions (yellow).

Ionic and other polar interactions result whenever a full or partial positive charge is close to a full or partial negative charge. In a compactly folded protein, multiple ionic and polar interactions may be present. Ionic interactions are quite strong, whereas other polar interactions are weaker, but again, these interactions collectively contribute to help make the protein conformation stable (Figure 12.11).

Hydrophobic interactions also contribute to protein shape and stability. Hydrophobic side groups are found within the core of many proteins. In the core, few or no water molecules are found. Most or all of the amino acid side chains within the core of a protein are nonpolar or hydrophobic. This clustering of nonpolar side chains is much like the clustering of the hydrophobic tails of soap molecules in micelles or the clustering of the nonpolar tails of membrane lipids in the core of a membrane. Water-heating nonpolar molecules or parts of molecules are "squeezed out" of water; they prefer to be together in a hydrophobic environment. Much of the stability of many globular proteins seems to be due to these hydrophobic interactions (Figure 12.11).

Earlier you were introduced to a covalent bond that forms between two thiols. This bond, a **disulfide bridge** or bond, forms by oxidation. Once formed, the bond can be broken by a reduction reaction.

$$\underset{\textbf{Thiol}}{R-SH} + \underset{\textbf{Thiol}}{HS-R} \underset{\text{Reduction}}{\overset{\text{Oxidation}}{\rightleftharpoons}} \underset{\textbf{Disulfide}}{R-S-S-R}$$

Once a newly synthesized protein assumes its native conformation, one or more pairs of cysteine side chains (the sulfhydryl groups) may be brought next to each other. These pairs oxidize to form disulfide bonds (Figure 12.12). These covalent bonds are stronger than the individual bonds and interactions previously discussed, and they significantly stabilize the protein in this conformation.

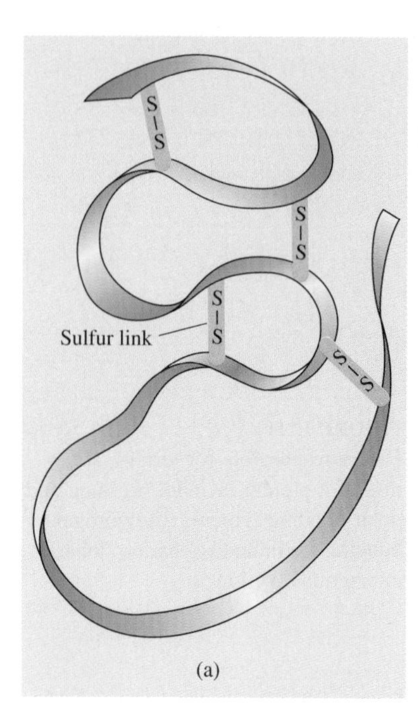

(b)

FIGURE 12.12
Disulfide bonds are covalent bonds in proteins that help maintain tertiary structure. (a) Schematic drawing of a hypothetical protein with the bonds highlighted. (b) A three-dimensional representation of the enzyme ribonuclease with four disulfide bonds.

EXAMPLE 12.4

Consider the amino acids serine, lysine, and tryptophan. What types of interactions might the side chains of these amino acids be involved in within a protein?

Solution: Serine might be involved in hydrogen bonding. Lysine could be involved in an ionic interaction or hydrogen bonding. Tryptophan is nonpolar and might be involved in hydrophobic interactions.

Quaternary Structure

♦ The number and arrangement of polypeptide chains in a protein is called its quaternary structure.

Some proteins possess more than one polypeptide chain, and the term **quaternary structure** is used to describe this level of structure. These proteins are called oligomeric proteins, which means they are oligomers of polypeptide chains. For example, the oxygen-transport protein hemoglobin has four *subunits*, two identical subunits designated as α (alpha) and two other identical subunits designated as β (beta). The four subunits in hemoglobin have a very specific arrangement designated $\alpha_2\beta_2$.

What forces hold these subunits together in this particular way? The same forces that are responsible for tertiary structure are responsible for quaternary structure—hydrophobic interactions, hydrogen bonds, and ionic and polar interactions. It may be helpful to picture the two interacting faces of a dimeric protein as being complementary to each other (Figure 12.13). A portion of the face of one of the subunits may have a hydrophobic surface that matches with a hydrophobic surface of the other. One subunit may possess a region with positively charged groups that is attracted to negative groups on the corresponding region of the

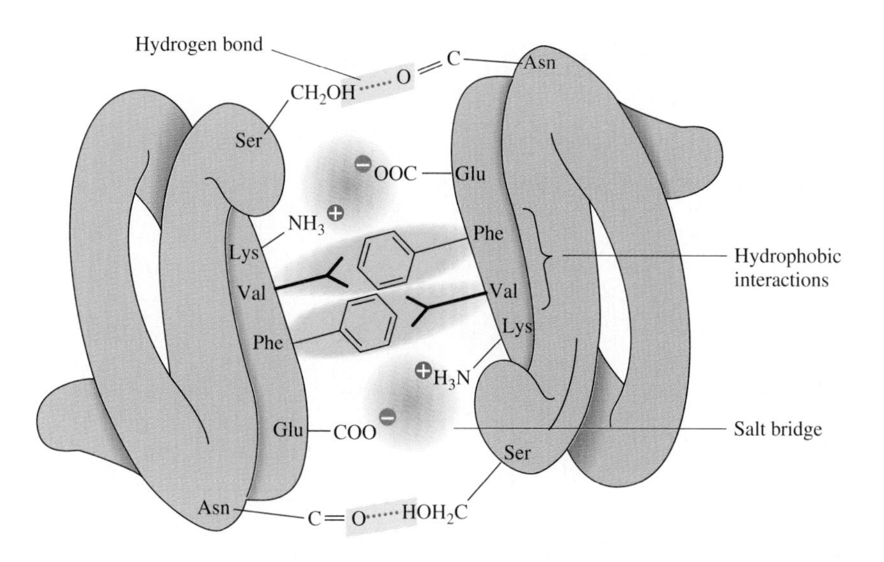

FIGURE 12.13
Subunit interaction in a hypothetical dimeric protein. The two faces bind to each other through ionic and polar bonding (green), hydrogen bonding (red), and hydrophobic interactions (yellow). The forces responsible for quaternary protein structure are essentially the same as those responsible for tertiary structure.

other subunit. One subunit has just the right groups in just the right orientation to form hydrogen bonds to groups on the face of the other subunit. Collectively, these forces hold the subunits together in a specific orientation.

Other Components in Proteins

So far we have presented a picture of a protein as one or more polypeptide chains arranged into a specific shape. Proteins possessing only amino acids connected by peptide bonds are called **simple proteins**. Other proteins, called **conjugated proteins**, possess one or more additional components called prosthetic groups. A **prosthetic group** is a non–amino acid component of a protein. For example, the prosthetic group in hemoglobin is *heme*. Hemoglobin contains four copies of heme, one bound to each of the four polypeptides of hemoglobin. A second example is the digestive enzyme carboxypeptidase, which contains zinc ion (Zn^{2+}) as its prosthetic group. Conjugated proteins lose their native activity if the prosthetic group is removed.

♦ The prosthetic group of a conjugated protein is the part that is not an amino acid.

Macromolecular Complexes

Proteins have a unique shape because certain forces and interactions occur that cause the protein to assume this most stable conformation. Are these same forces responsible for other molecular complexes and organelles? Most definitely. Hydrophobic interactions, hydrogen bonding, and ionic interactions all play a part in holding many different macromolecular complexes together. For example, ribosomes (organelles involved in protein synthesis) are macromolecular complexes consisting of many different proteins and a few large molecules of ribonucleic acid (RNA). Each of the proteins and RNAs of a ribosome are synthesized separately, yet they bind together to form the subunits of a ribosome. This process is called *self-assembly*. Self-assembly occurs because the components are more stable when they are bound to other components rather than being free in solution. Self-assembly is a characteristic of all macromolecular complexes.

CHEMISTRY CAPSULE

Sickle Cell Anemia

A look at the human genetic disease *sickle cell anemia* illustrates the importance of protein structure for proper function. Portions of the human population that originated in the malarial belt of the world possess a type of hemoglobin called hemoglobin S. In the United States, some people of African descent and a smaller proportion of the people of Mediterranean descent possess this type of hemoglobin.

Hemoglobin S differs from normal hemoglobin by one amino acid in the β-chain. In normal hemoglobin, the sixth residue is a glutamic acid, whereas in hemoglobin S it is a valine. Both hemoglobins bind oxygen properly, but hemoglobin S binds to other hemoglobin S molecules in a way that normal hemoglobin does not. Hemoglobin is tightly packed inside erythrocytes (or red blood cells), and normal hemoglobin molecules do not bind to each other within the erythrocyte. But hemoglobin S molecules can bind to one another if many of them contain no oxygen. When the hemoglobin S molecules bind to each other, they form long filaments. These filaments change the shapes of the red blood cells into sickle-shaped cells (see figure).

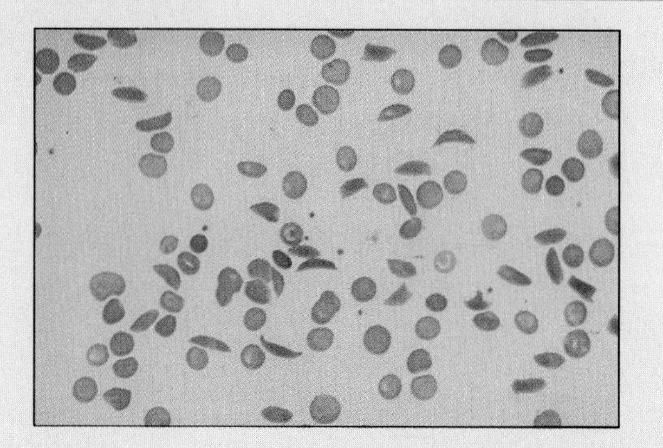

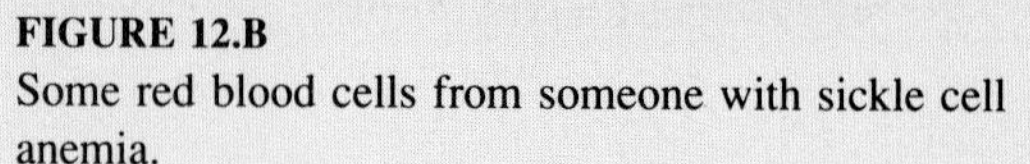

FIGURE 12.B
Some red blood cells from someone with sickle cell anemia.

These sickle cells are fragile, easily broken, and do not pass as readily through capillaries. Normal blood flow to tissues may be impaired, and anemia (iron deficiency) may result from breakage of the sickled erythrocytes. A single amino acid difference in this protein causes a distinct difference in the solubility behavior of the protein. This affects cell shape and stability, which can lead to impaired cardiovascular function.

NEW TERMS

native conformation The normal shape or conformation that a protein has in its biological setting performing its normal biological activity

denaturation Process that results in the loss of a protein's native conformation, and therefore its activity

primary structure The amino acid sequence in a protein or peptide

secondary structure The helical or pleated structure of a protein that is due to hydrogen bonding between atoms of the polypeptide backbone

α-helix Secondary structure of a protein in which the polypeptide backbone is arranged into a regular spiral shape

β-pleated sheet Secondary structure of a protein in which the polypeptide chain or chains are arranged into a sheet-like structure

tertiary structure The compact, three-dimensional shape of globular proteins

disulfide bridge A covalent bond between two sulfur atoms

quaternary structure The structure of an oligomeric protein resulting from specific interactions among the subunits

simple protein A protein containing only amino acids

conjugated protein A protein that contains one or more prosthetic groups in addition to amino acids. (The groups may be organic or inorganic.)

prosthetic group A group in a protein that is not an amino acid and that plays a major role in the activity of conjugated proteins

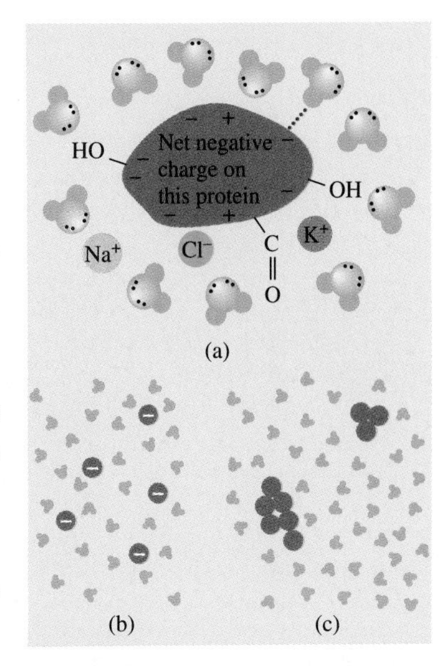

FIGURE 12.14
An illustration of protein–solvent and protein–protein interactions. (a) The polar and ionic groups on a protein surface interact with water and solutes through noncovalent bonds. The protein is solvated and remains in solution; it has a net negative charge at this pH. (b) The protein molecules do not bind to each other because their net negative charges effectively repel each other. The proteins remain in solution. (c) If the pH of the solution changes, the net charge on the protein changes. At the pI, the protein has no net charge, and the protein molecules no longer repel each other. They aggregate and precipitate from solution. A step in cheese making involves precipitating milk protein at pH 4.7.

TESTING YOURSELF

Protein Structure

1. What is the primary structure of a protein?
2. What forces are responsible for maintaining tertiary structure in a protein?
3. A protein is broken down in the laboratory to yield amino acids and an organic molecule called pyridoxal. How would you classify this protein? What is pyridoxal in this case?

Answers **1.** the amino acid sequence in the protein **2.** hydrogen bonding, ionic and other polar interactions, hydrophobic interactions, and disulfide bridges **3.** This is a conjugated protein. It is the prosthetic group of this protein.

12.5 Properties and Classification of Proteins

Properties of Proteins

Proteins are amphoteric molecules, that is, they can act as acids or bases. The amphoteric nature of proteins is due to the presence of acidic and basic side chains in some of the amino acids found in the protein. At the neutral pH typical of a cell, the carboxyl groups in side chains are deprotonated and exist as carboxylate ions, —COO^-. The basic groups on lysine and arginine are protonated; for example, the amino group of lysine is —NH_3^+. These ionic groups are generally located on the surface of the protein, so the surface is a mosaic of charges, some positive and some negative. If the solution were to become more acidic, the carboxyl groups would accept some of the hydrogen ions.

$$—COO^- + H^+ \longrightarrow —COOH$$

A negatively charged carboxylate ion is changed to a neutral carboxyl group. In addition, if any histidine residues are present, these would also accept a hydrogen ion to become positively charged. Under acidic conditions, the charge on a protein becomes more positive. In basic solutions, the protonated side chains of arginine and lysine give up hydrogen ions. These positively charged groups become neutral, and the net charge on the protein becomes more negative. The net charge on a protein changes with the pH of the solution.

◆ As the pH of a solution changes, the charge on a protein changes.

What effects do changes in pH have on the properties of a protein? Since pH affects the charge of a protein, then any property that depends on the charge of the protein is influenced by pH. Three properties that are influenced are solubility, conformation, and mobility in an electric field.

Consider a protein in an aqueous solution. Each of the protein molecules interacts with the water molecules and solutes that may be present (Figure 12.14). Water and water-soluble solutes interact with polar and ionic side chains located on the surface of the protein. These polar interactions between the protein and the solvent keep the protein in solution. As long as the interaction of the protein molecules with the solvent are stronger than the interactions between protein molecules, the protein molecules remain in solution.

Why don't protein molecules bind to each other, since their surfaces are polar and therefore should be able to form polar interactions? Each of the protein molecules has some net charge that is a reflection of the ionizable side groups found in the protein (Figure 12.14a). A protein may have more acidic side groups than basic ones, therefore the net charge on the protein is negative because at normal cellular pH, the carboxyl groups have lost their hydrogen ion and have a negative charge. Since the protein molecules have a net negative charge, they tend to repel each other. They remain independent in the solvent rather than aggregating into a complex too large to remain in solution (Figure 12.14b). If the solution becomes more acidic, some of the carboxylate ions will be protonated, which will decrease the negative charge on the protein. At a certain pH value, enough of the carboxylate ions are protonated that the net charge on the protein is at zero. Just as amino acids are neutral at their isoelectric pH, so are proteins. At the isoelectric point, the protein molecules are neutral and no longer repel each other. The protein molecules tend to aggregate and precipitate from solution (Figure 12.14c). In general, proteins are least soluble at their isoelectric point.

Changes in pH can denature a protein. Denaturation occurs whenever a protein changes from its normal or native conformation to some other conformation. Recall that the structure of a protein may depend on specific ionic interactions. If the pH changes, certain ionizable side chains may accept or lose hydrogen ions. Ions may form or disappear, specific stabilizing ionic interactions may be lost, and other destabilizing ionic interactions may occur. Generally, proteins are denatured if the pH of the solution goes beyond a certain range.

♦ Several factors disrupt normal bonding and interactions in proteins and thus lead to denaturation.

The pH of a solution is not the only factor that can affect the native conformation. Heat, solvents, and heavy metals can also change it. Heating a solution increases the thermal energy of the molecules in the solution. As the protein molecules gain thermal energy, the bonds in the molecule vibrate and rotate more strongly. Eventually enough heat is gained that the hydrogen bonds and other interactions maintaining the protein conformation are no longer strong enough to hold the protein in this conformation. The protein assumes other random shapes, and native function and properties are lost. For example, when you fry an egg, the soluble proteins in the egg white become denatured. These proteins are highly insoluble in the denatured form, thus a white solid forms from the clear, viscous solution that is the egg white. Proteins also denature in solvents that do not resemble the cytosol or blood. Many salt solutions and organic solvents denature proteins. Similarly, heavy-metal ions cleave disulfide bonds, which may result in denaturation.

EXAMPLE 12.5

The evils of alcohol abuse could perhaps be illustrated by treating egg white with alcohol. The egg white turns from clear to white and looks like it has been cooked. What is occurring, and why?

Solution: The proteins in the egg white are being denatured because the solvent they are in has been changed from an aqueous one to one consisting of water and alcohol.

CHEMISTRY CAPSULE

Protein–Calorie Malnutrition

Inadequate protein in the diet is most harmful to children. If protein synthesis cannot occur adequately during development, temporary or permanent physical or mental impairment may result. *Protein–calorie malnutrition (PCM)* is a disease seen primarily in children of developing nations. This disease appears to be caused by inadequate amounts of protein in the diet, but the total amount of energy available in the diet is also often inadequate. *Kwashiorkor* is a form of the disease in which the child has edema, an enlarged liver, and peeling skin. This form of the disease may be more common when caloric intake is close to adequate but too little dietary protein is present. Marasmus is another form of the disease in which the body appears more wasted away, but edema tends to be absent. Diets too low in both calories and protein may lead to this form of PCM. Often victims of this disease show some symptoms of both forms (Figure 12.C).

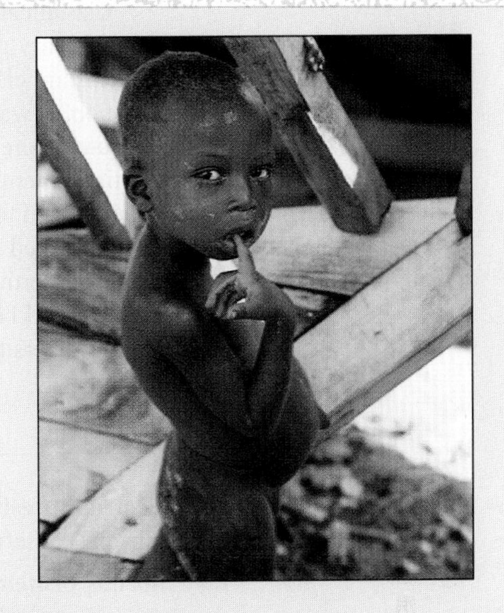

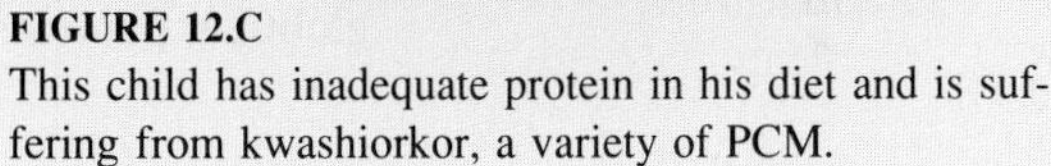

FIGURE 12.C
This child has inadequate protein in his diet and is suffering from kwashiorkor, a variety of PCM.

Mixtures of proteins are separated in many medical and research applications. Because proteins vary in size and charge, they can be separated by **electrophoresis**. This technique, which separates particles located in an electric field, is very useful for analyzing proteins, peptides, and amino acids. Figure 12.15 illustrates a common medical application, the analysis of serum proteins.

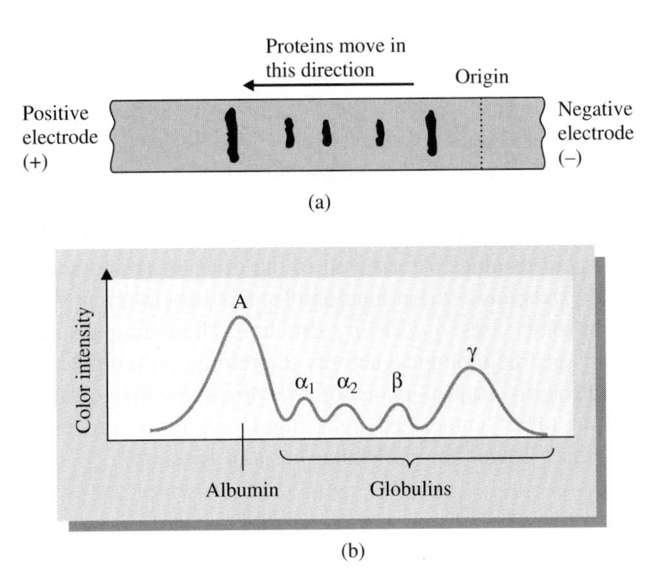

FIGURE 12.15
These blood proteins were separated by electrophoresis. (a) Serum was placed onto an electrophoresis strip as a band at the origin. The strip was placed into an electric field (designated by the positive and negative electrodes). At the pH value of this experiment, blood proteins have a net negative charge and are attracted to the positive electrode. Because the proteins have different sizes and different net charges, they migrate at different rates. (b) After separation, the strip is stained with a protein-specific dye. The intensity of the stain indicates the relative amount of each protein.

Classification of Proteins

Classification of proteins by one simple system is not possible because proteins vary greatly in size, shape, function, and properties. As a consequence, proteins are classified by a variety of systems. They are sometimes classified by function, as was shown in Table 12.1. They can be classified by shape or solubility. **Fibrous proteins** are fiber-like and are generally insoluble, although some contractile proteins of muscle and other cells are exceptions. **Globular proteins** are spheroid and tend to be soluble. They can be classified by composition as simple or conjugated proteins. Proteins are indeed numerous, complex, and varied in many properties. In Chapter 14 the enzymes, which are proteins with catalytic activity, will be studied in more detail. Even within this single class of proteins, a great diversity in function and properties is seen.

NEW TERMS

electrophoresis Technique that separates charged particles in an electric field; especially useful for analysis of amino acids, peptides, or proteins

fibrous proteins Fiber-like proteins that are usually insoluble

globular proteins Spheroid, generally soluble proteins

TESTING YOURSELF

Proteins

1. Why are proteins least soluble at their pI?
2. What factors can denature proteins?
3. Compare the solubility of fibrous and globular proteins.

Answers **1.** Proteins have no net charge at the isoelectric point, thus they do not repel each other. **2.** Heat, pH, change in solvent, and some heavy metals. **3.** Most fibrous proteins are insoluble, whereas many of the globular proteins are soluble.

Summary

Amino acids are bifunctional molecules that are the building blocks of *proteins* and *peptides*. All amino acids contain an α-hydrogen, α-amino group, and α-carboxyl group, but each of them has a unique side chain. Many of the properties of amino acids are determined or influenced by the side group. Amino acids can be linked together by *peptide bonds* to form peptides or proteins characterized by unique amino acid residues. As a result of the order of amino acids in a protein chain (*primary structure*), these macromolecules have very precise *secondary*, *tertiary*, and in some cases, *quaternary* structure. These unique structural characteristics impart specific biological activity to proteins, including catalytic activity to enzymes. Proteins control many of the processes that occur within cells. Proteins are, in turn, synthesized by a highly regulated process discussed in Chapter 13, which covers the molecular basis of heredity.

Terms

proteins (12.1)	secondary structure (12.4)
amino acids (12.2)	α-helix (12.4)
zwitterion (12.2)	β-pleated sheet (12.4)
amphoteric molecule (12.2)	tertiary structure (12.4)
isoelectric point (pI) (12.2)	disulfide bridge (12.4)
peptide bond (12.3)	quaternary structure (12.4)
peptide (12.3)	simple protein (12.4)
polypeptide (12.3)	conjugated protein (12.4)
native conformation (12.4)	prosthetic group (12.4)
denaturation (12.4)	electrophoresis (12.5)
primary structure (12.4)	fibrous proteins (12.5)
	globular proteins (12.5)

Exercises

Protein Function (Objective 7)

1. List the general functions of proteins discussed in this chapter.
2. Describe the specific function of these proteins.
   a. enzymes
   b. antibodies
   c. blood lipoproteins
   d. hemoglobin
   e. myoglobin
   f. lysozyme
   g. casein
   h. insulin
   i. collagen
3. Which of the following proteins serves a defensive role?
   a. casein
   b. myoglobin
   c. lysozyme
   d. cytochrome c

Amino Acids (Objectives 1 and 2)

4. Draw the structure of an α-amino acid. Why are these compounds called α-amino acids?
5. Almost all of the common amino acids
   a. are in the D-family
   b. are in the L-family
   c. are split equally between both families
6. Complete the structural skeleton below to obtain the structure of L-valine.

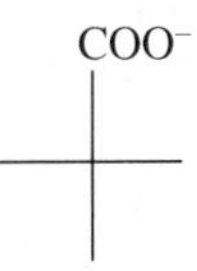

7. Examine the structure of threonine. How many chiral centers are present in this molecule? Which is compared with the glyceraldehydes to determine stereochemical family?
8. The amino acids of the body are related to which glyceraldehyde? Draw the structure of alanine, showing the correct stereochemistry.
9. List the nonpolar amino acids.
10. How does proline differ from the other amino acids?
11. Draw the structure of a generalized α-amino acid in the zwitterion form.
12. A zwitterion is *not*
    a. soluble in hydrocarbons
    b. soluble in water
    c. electrically neutral
    d. polar
13. What is the difference between a basic and an acidic amino acid? Draw the structures of one basic and one acidic amino acid in both their protonated and unprotonated forms.
14. What property must a compound possess to be classed as amphoteric?
15. Is there a pH value where an α-amino acid exists in a form that is un-ionized? Why or why not?
16. Draw the structure of lysine showing the correct charge on the ionizable groups at pH 1, pH 7, and pH 13.
17. Draw a molecule of aspartic acid showing the correct charge on the ionizable groups at pH 1, pH 7, and pH 13.

Peptide Bond and Peptides (Objective 3)

18. What is a peptide bond?
19. Why is there no free rotation about a peptide bond?
20. There are two different dipeptides containing glycine and alanine. Draw the structure of each. How do they differ?
21. Draw the structure of the tripeptide glutamylphenylalanylleucine.
22. Draw the tripeptide Ala-Gly-Leu, with alanine as the *N*-terminal amino acid. *Hint:* Refer to the list of amino acids.

23. Name this tripeptide:

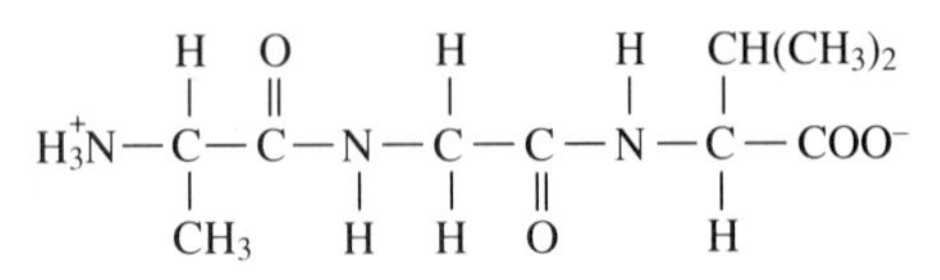

24. Which of the following is a peptide?
    a. hemoglobin
    b. casein
    c. myoglobin
    d. vasopressin
25. Make a table of the biologically active peptides discussed in this chapter. Include the names, structures, and biological role or effect.
26. Compare the number of tripeptides that could be made from all 20 amino acids to the number that could be made from the three basic amino acids. (Show your calculations.)

Protein Structure (Objectives 4 and 5)

27. Describe what is meant by primary, secondary, tertiary, and quaternary structure of proteins.
28. Describe the forces that are responsible for the maintenance of the protein structures in Exercise 27.
29. The α-helix of a protein is an example of
    a. primary structure
    b. secondary structure
    c. tertiary structure
    d. quaternary structure
30. The shape of globular proteins is an example of
    a. primary structure
    b. secondary structure
    c. tertiary structure
    d. quaternary structure
31. A β-pleated sheet is an example of
    a. primary structure
    b. secondary structure
    c. tertiary structure
    d. quaternary structure
32. Hemoglobin is an example of
    a. primary structure
    b. secondary structure
    c. tertiary structure
    d. quaternary structure
33. The clustering of the side chain of leucine with the side chain of valine is referred to as
    a. disulfide bonding
    b. hydrogen bonding
    c. hydrophobic interactions
    d. ionic bonding
34. The bonding of the hydroxyl hydrogen of serine to the side chain carbonyl oxygen of aspartate is referred to as
    a. disulfide bonding
    b. hydrogen bonding
    c. hydrophobic interaction
    d. ionic bonding
35. Draw a small region of polypeptide in an α-helix. What forces maintain the chain in this conformation?
36. What is a *disulfide bridge* in protein structure?
37. The structure of the enzyme RNA polymerase is sometimes shown as $\alpha_2\beta\beta'\sigma$. What do you think this symbolism means?
38. What does *self-assembly* mean?
39. Compare the composition of a simple protein and a conjugated protein.
40. Consider a globular protein that is water-soluble. Which of these amino acids probably are on the exterior surface and which are more likely to be buried within the core of the protein?
    a. alanine
    b. glutamate
    c. leucine
    d. lysine
    e. threonine
    f. valine
    g. phenylalanine
41. Why are the amino acid residues in Exercise 40 arranged this way?

Use the amino acids listed in Exercise 40 to answer Exercises 42 through 45.

42. Which of these amino acids could form ionic interactions when in a protein?
43. Which could form hydrogen bonds?
44. Which could form hydrophobic interactions?
45. How many pentapeptides can you make from these amino acids? (You may use any of them as often as you wish.)

Properties of Proteins (Objective 6)

46. What is denaturation of a protein?
47. Which amino acids would contribute a positive charge to a protein in solution? Which would contribute a negative charge?
48. A protein has no net charge at pH 6.13. What charge does it have at pH 7? At pH 9? At pH 5.5?
49. At which pH does the protein in Exercise 48 have the lowest solubility?
50. Explain what is meant by the term *isoelectric point*.
51. Examine Figure 12.15. Predict the migration of these proteins if the pH of the solution were made more acidic.
52. Changes in pH most affect what type of bonding in proteins?
    a. hydrophobic interactions
    b. ionic interactions
    c. hydrogen bonding
    d. disulfide bridges
53. Proteins that have been isolated from a biological source are typically kept in buffered solutions at cold temperatures. Why?

Unclassified Exercises

54. Explain the molecular basis of sickle cell anemia.
55. Two protein samples have been shown to have the same amino acid composition. Does it follow that they must have the same structure?
56. An ounce of a common breakfast cereal contains 3 g of protein, 1 g of fat, and 23 g of carbohydrate. Estimate the caloric content of 1 oz of this cereal. (If necessary, look up the caloric content of these food groups in Chapters 10, 11, and 12.)

57. An essential amino acid
    a. is incorporated into all proteins
    b. is found in all food
    c. cannot be synthesized by the human body
58. List the essential amino acids.
59. A diet for experimental rats is prepared containing the minimal amounts of the essential amino acids needed for normal growth in this species. The diet is lacking in the amino acid cysteine. Describe the growth of these rats, and explain the basis for the observed growth.
60. What names are given for the forms of protein–calorie malnutrition? What nutritional conditions contribute to these two forms?
61. A list of some words and phrases introduced in this chapter is presented below. For each, provide a description, definition, or example to show that you understand the word or phrase.
    a. α-amino acid
    b. L-amino acid
    c. zwitterion
    d. amphoteric
    e. isoelectric point
    f. dipeptide
    g. peptide linkage
    h. protein
    i. primary structure
    j. secondary structure
    k. tertiary structure
    l. hydrophobic interaction
    m. disulfide bridge
    n. denaturation

MOLECULAR BASIS OF HEREDITY

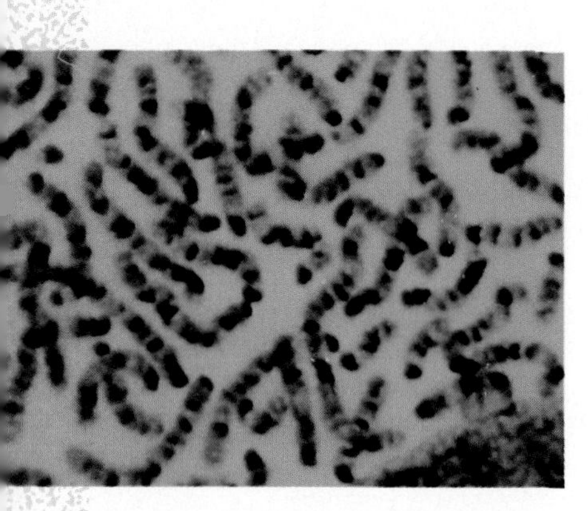

Nucleic acids, like the DNA in the chromosomes shown here, are the carriers of the information needed for reproduction and other vital functions.

OUTLINE

13.1 Search for the Molecular Basis of Heredity
13.2 Nucleotides
13.3 Structure and Replication of DNA
13.4 RNA
13.5 Transcription
13.6 Translation
13.7 Mutagenesis
13.8 Regulation of Gene Expression
13.9 Genetic Engineering

OBJECTIVES

After completing this chapter, you should be able to

1. Recognize a nucleotide, and identify those that are important in DNA and RNA.
2. Describe the structure of DNA, and explain why specific bases are paired.
3. Explain the process of DNA replication.
4. Describe the structure of RNA, and explain why it is complementary to a strand of DNA.
5. Identify the roles of DNA, mRNA, and tRNA in the synthesis of proteins.
6. Define the term *codon*, and explain the role of codons in protein synthesis.
7. Describe the molecular basis for the introduction of errors into genetic material.
8. Explain how induction and repression regulate protein synthesis in bacteria.
9. Recognize the overall strategy of genetic engineering for introducing a gene into an organism.

FIGURE 13.1
Individuals always have offspring that are recognizable as members of that species, yet considerable differences are found within a species. This chapter presents the molecular basis for this consistency and variability.

13.1 Search for the Molecular Basis of Heredity

Living organisms are unique among the materials of the earth because they have the ability to replicate themselves. "Like begets like" is a biblical phrase that shows this was known to ancient peoples. Yet variability is present within the constancy—cats beget cats, but kittens differ from each other. We see this in humans, too; although sets of identical twins appear very similar or nearly identical, other people are distinctly different in appearance (Figure 13.1). What is the molecular basis for these similarities and differences? Genetics and molecular biology are disciplines that study how traits are passed from one generation to another. Although the study of the biological processes that control heredity began well over a century ago with the pioneering experiments of Gregor Mendel (1822–1884), the molecular basis of heredity was not established for nearly another century.

The search for the molecular basis of heredity has been long and fascinating. It began when scientists studying cells and genetics became convinced that the cell nucleus contained the molecules that controlled heredity. Biochemical studies

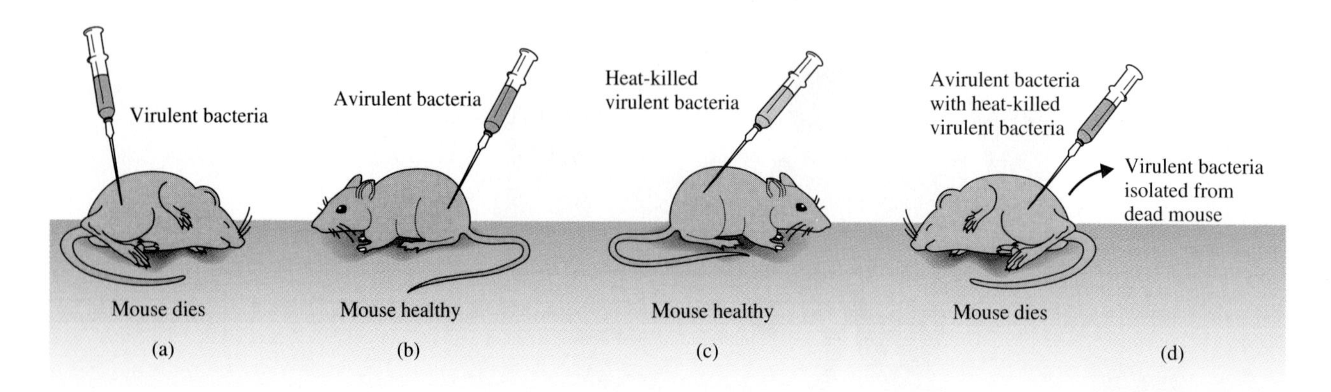

FIGURE 13.2
Experiment on bacterial transformation. (a) Virulent *Streptococcus pneumoniae* kill mice, (b) but the avirulent strain does not. (c) Heat-killed virulent strain cannot kill mice, (d) but avirulent strain mixed or grown with heat-killed virulent strain does kill mice. The avirulent strain becomes virulent because it takes up molecules from the killed virulent bacteria. These molecules transform the avirulent strain into a disease-causing strain.

of the nucleus revealed numerous small molecules, many proteins, and two classes of large acidic molecules called **deoxyribonucleic acid (DNA)** and **ribonucleic acid (RNA)**. At first, it was not known which, if any, of these molecules were the molecules of heredity.

EXAMPLE 13.1

Name the kittens in Figure 13.1.

Solution: (Just kidding, but the fourth one from the left in the top row should be named Igor.)

Then, in the late 1920s, some experiments by Fred Griffith provided clues that ultimately led to the determination of the molecular basis of heredity. These experiments and later ones showed that non-disease-producing (*avirulent*) strains of a bacterium could be converted into a disease-producing (*virulent*) strain simply by growing the avirulent strain in the culture medium where a disease-producing strain had been grown (Figure 13.2). Apparently, molecules in the medium that came from the virulent strain were able to change the avirulent strain into a virulent one. Most important, the transformation was permanent; virulence remained through all future generations of the strain. The heredity of the strain had been altered.

A reasonable hypothesis for these observations was that molecules containing genetic information were released into the culture medium by the virulent strain. The cells of the avirulent strain then took up these molecules and made them a permanent part of the cell. These experiments indicated that molecules are involved in heredity, but they did not establish the identity of the molecules involved.

In the early 1940s, Oswald Avery and his colleagues at the Rockefeller Institute used these bacteria to determine the molecules of heredity. They took virulent cells and removed several kinds of molecules from them. Avirulent cells were mixed with each of these fractions and allowed to grow. These cells were then tested for virulence (Figure 13.3). Only the cells grown in the presence of the DNA fraction became virulent; DNA was the molecule of heredity. What remained was to determine how this huge molecule could carry information.

♦ DNA is the molecule that carries genetic information.

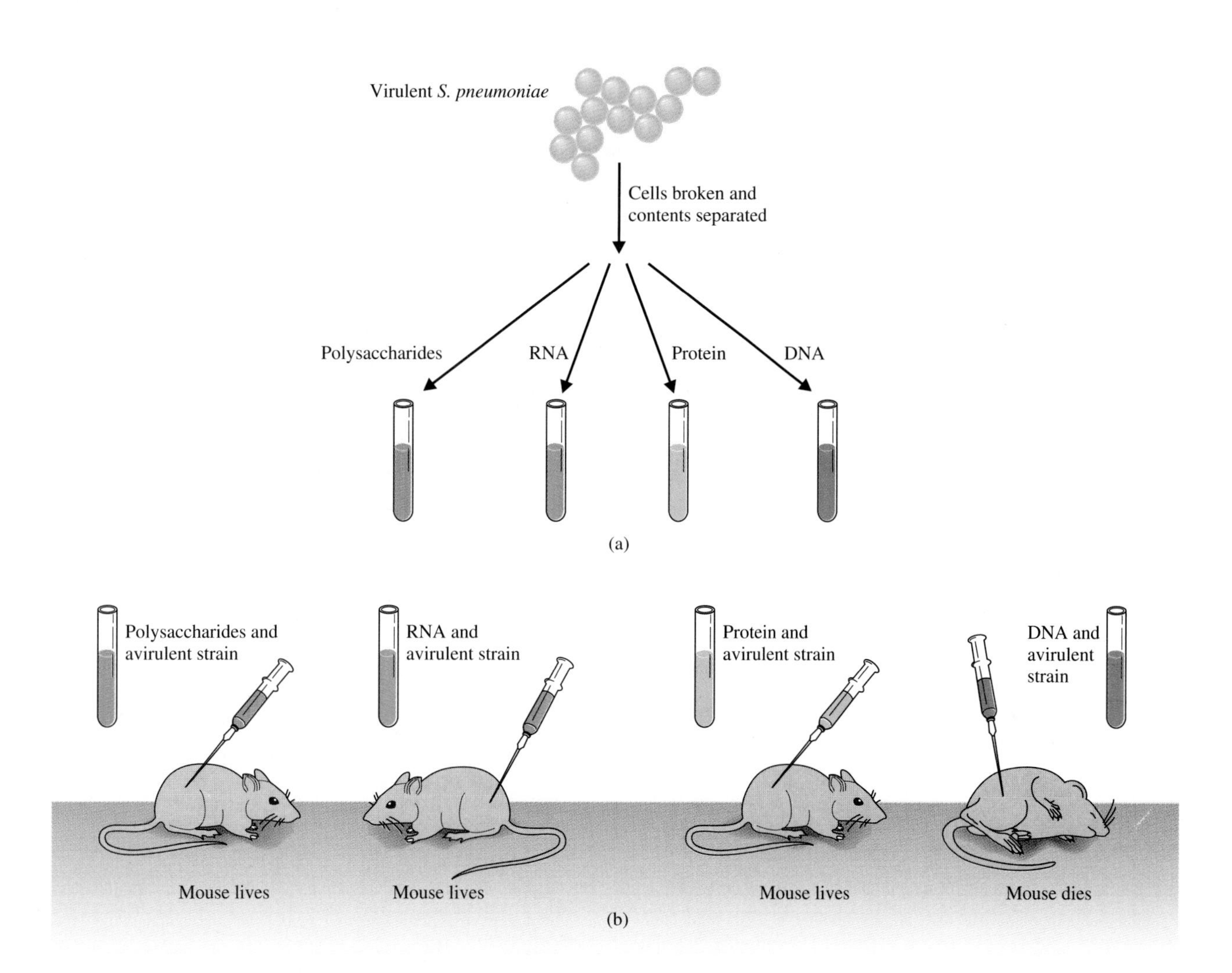

FIGURE 13.3
Avery and his colleagues showed that DNA transforms avirulent bacteria into a virulent strain. (a) Virulent *Streptococcus pneumoniae* were killed, broken open, and the cellular contents separated from each other. (b) Avirulent *S. pneumoniae* were transformed into a virulent strain only by the fraction containing DNA.

NEW TERMS

deoxyribonucleic acid (DNA) A very large, acidic macromolecule found in the cell nucleus. This molecule is the carrier of genetic information.

ribonucleic acid (RNA) Several kinds of acidic macromolecules found in the nucleus and other parts of the cell.

13.2 Nucleotides

Deoxyribonucleic acid, or DNA, is the molecule of heredity, and it is made up of nucleotides. DNA is a polynucleotide, that is, a polymer of nucleotides. **Nucleotides** contain a sugar, a nitrogenous base, and one or more phosphate groups. The **nitrogenous base** is a heterocyclic base containing nitrogen and is typically bonded to a carbon on one side of the sugar. The phosphate is typically bonded to a carbon on the other side of the sugar. Very roughly, a nucleotide can be represented as

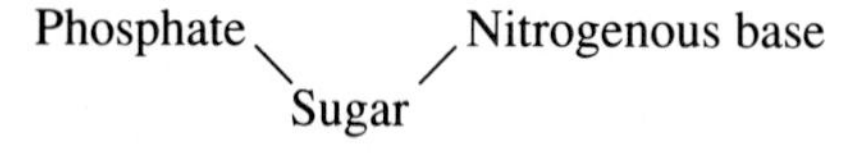

Several different nitrogenous bases are found in nucleotides, but only two sugars are common. Generally one to three phosphates are present in nucleotides. Nucleotides are the building blocks of DNA and the other nucleic acid, RNA. They are also present in a number of coenzymes, such as ATP and NAD. Because nucleotides are common in biochemistry, a brief look at their structure and nomenclature is in order.

♦ Adenine and guanine are purines; cytosine, thymine, and uracil are pyrimidines.

Although a number of different nitrogenous bases can be found in nucleotides, five are of immediate interest. These five bases are *adenine, guanine, cytosine, thymine*, and *uracil*. These bases are heterocyclic compounds, but adenine and guanine have the fused-ring system found in the compound purine (Figure 13.4a),

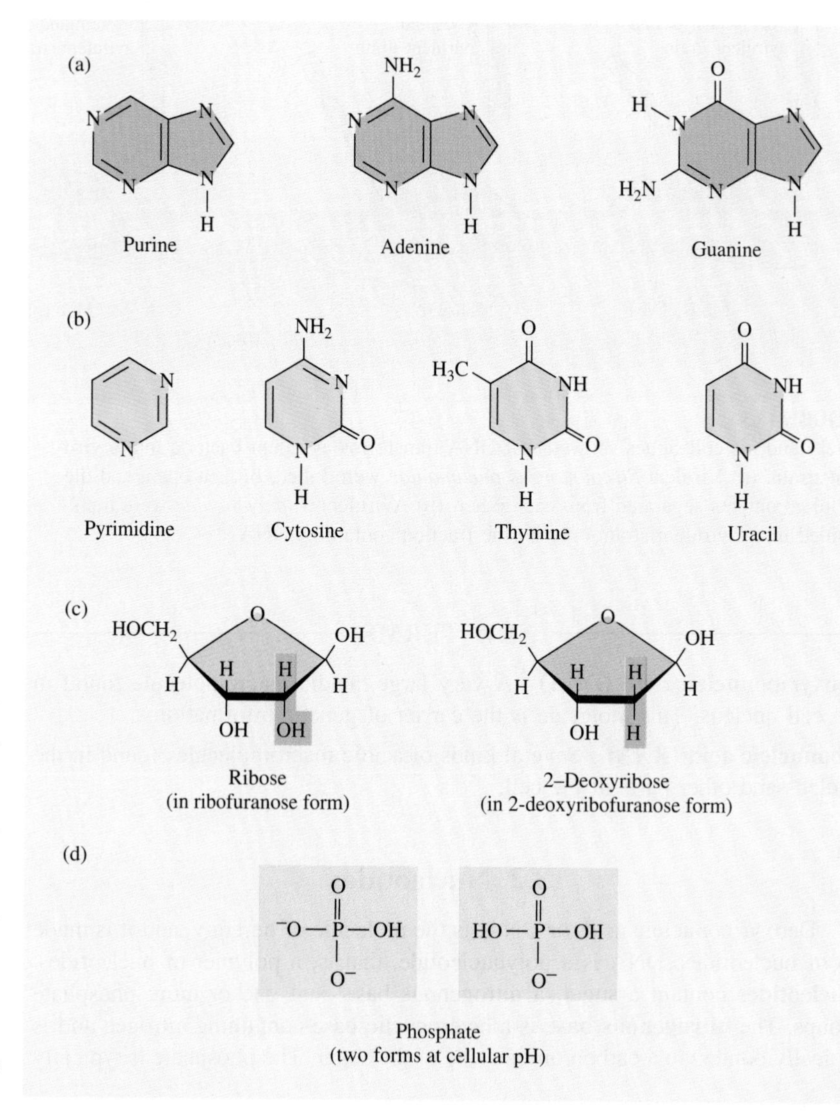

FIGURE 13.4
The nitrogenous bases, sugars, and phosphate found in common nucleotides. (a) The nitrogenous bases adenine and guanine are shown with purine. The similar ring structure identifies these compounds as purines. (b) The nitrogenous bases cytosine, thymine, and uracil are shown with pyrimidine. Because they contain the same ring system, they are referred to as pyrimidines. (c) Ribose and 2-deoxyribose are the common sugars found in nucleotides. The structural difference between these sugars is highlighted. (d) Phosphate is the common name given to the forms of phosphoric acid found at cellular pH.

and cytosine, thymine, and uracil resemble the compound pyrimidine (Figure 13.4b). Because of these resemblances, adenine and guanine are called *purines* and the other three bases are called *pyrimidines*. Take a moment to study the ring systems and the functional groups attached to the rings because these features distinguish the purines and pyrimidines from each other.

The sugars most commonly found in the nucleotides are ribose and deoxyribose (Figure 13.4c). Each is an aldopentose that exists as a furanose in living systems. These two sugars differ only in the substituents on carbon number two (C-2). Ribose has one hydrogen atom and one hydroxyl group as substituents, whereas deoxyribose lacks the hydroxyl group and thus has two hydrogen atoms as substituents. The prefix *deoxy-* signifies the absence or loss of an oxygen from the molecule. In nucleotides, a bond exists between a specific nitrogen atom of a nitrogenous base and C-1 of the furanose ring. This bond is highlighted in Figure 13.5.

Nucleotides also contain phosphate, which exists as an anion or dianion at cellular pH (see Figure 13.4d). When phosphate is in a nucleotide, it typically has a single negative charge.

Compounds containing only a sugar and a nitrogenous base are called **nucleosides**. Although these compounds are rarely found in biological samples, they are important because the nomenclature of nucleotides is based on nucleoside nomenclature. The name of a nucleoside is derived from the nitrogenous base it contains. For example, a nucleoside containing adenine bonded to deoxyribose is called deoxyadenosine. Note as a general rule that

$$\text{Base} + \text{Sugar} = \text{Nucleoside}$$

$$\text{Base} + \text{Sugar} + \text{One or more phosphates} = \text{Nucleotide}$$

Table 13.1 summarizes the nomenclature of the nucleosides and nucleotides.

FIGURE 13.5
The structures of the deoxyribonucleotides. The highlighted bond between the base and the sugar is a glycosidic bond, and the bond between the sugar and the phosphate is an ester bond.

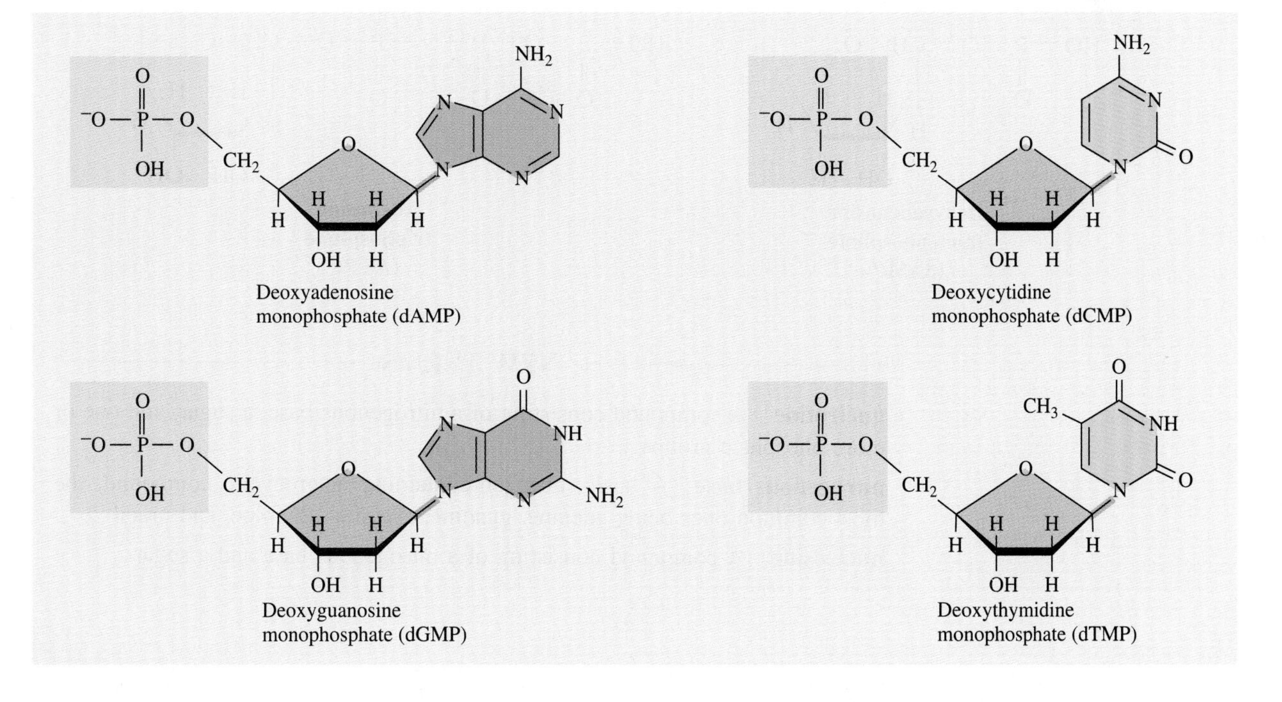

TABLE 13.1

Nomenclature of the Nucleosides and Nucleotides of DNA and RNA

Symbol	Base	Sugar	Nucleoside	Nucleotide
A	Adenine	Deoxyribose	Deoxyadenosine	Deoxyadenosine monophosphate (dAMP)
G	Guanine	Deoxyribose	Deoxyguanosine	Deoxyguanosine monophosphate (dGMP)
C	Cytosine	Deoxyribose	Deoxycytidine	Deoxycytidine monophosphate (dCMP)
T	Thymine	Deoxyribose	Deoxythymidine	Deoxythymidine monophosphate (dTMP)
A	Adenine	Ribose	Adenosine	Adenosine monophosphate (AMP)
G	Guanine	Ribose	Guanosine	Guanosine monophosphate (GMP)
C	Cytosine	Ribose	Cytidine	Cytidine monophosphate (CMP)
U	Uracil	Ribose	Uridine	Uridine monophosphate (UMP)

Nucleotides are nucleosides with one or more phosphates added through an ester bond between the phosphate and the hydroxyl group on C-5 of the sugar. An ester bond involving phosphate is often called a *phosphoester bond.*

A nucleotide is named from the nucleoside it contains. For instance, deoxyadenosine monophosphate is the nucleotide containing deoxyadenosine and one phosphate (Table 13.1). Adenosine triphosphate, ATP, is the nucleotide containing the nucleoside adenosine and three phosphates.

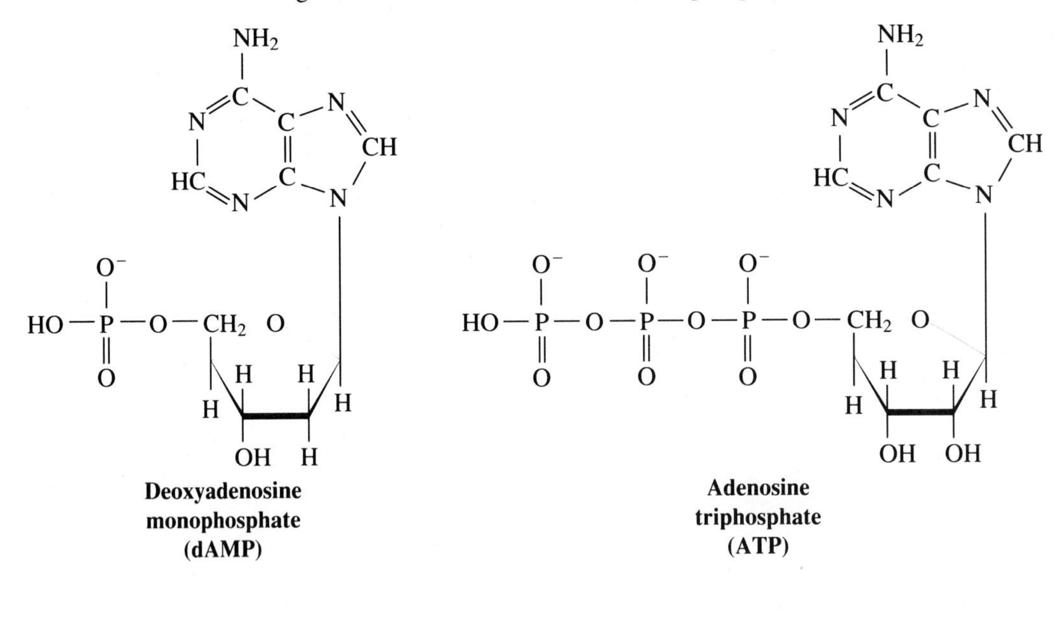

Deoxyadenosine monophosphate (dAMP)

Adenosine triphosphate (ATP)

NEW TERMS

nucleotide A compound consisting of a nitrogenous base, a sugar, and one or more phosphate groups

nitrogenous base A basic, nitrogen-containing heterocyclic compound, the most common ones being adenine, guanine, cytosine, thymine, and uracil

nucleoside A compound consisting of a nitrogenous base and a sugar

TESTING YOURSELF

Nucleotides

1. Name the nitrogenous bases that are (a) purines and (b) pyrimidines.
2. Name the compound containing ribose, one phosphate, and uracil.

Answers **1. a.** adenine and guanine **b.** cytosine, thymine, and uracil
2. uridine monophosphate (UMP)

13.3 Structure and Replication of DNA

Earlier it was stated that DNA is a polymer of nucleotides. It is perhaps better to say that it is a polymer of deoxyribonucleotides, for DNA is an abbreviation of *deoxyribo*nucleic acid. The name clearly identifies deoxyribose as the sugar found in this polymer. DNA is a polymer of nucleotides, a nucleic acid, but how are the nucleotides arranged in DNA? What is the structure of DNA? Often the structure of a biomolecule provides important clues to its function, and much of DNA's function was predicted once its structure became known.

Determination of DNA Structure: The Double Helix

The determination of DNA structure relied on several lines of evidence. One involved the base composition of DNA. Four bases are found in the deoxynucleotides of DNA; the purines adenine and guanine and the pyrimidines cytosine and thymine. In a sample of DNA, the amount of thymine found is nearly equal to the amount of adenine, and the amount of cytosine is nearly equal to the amount of guanine. This means that the amount of purines equals the amount of pyrimidines. Except for some viruses, this is true for the DNA of all species (Table 13.2). Such similarity could not occur by chance in all species. The structure of DNA must provide an explanation for the existence of these equalities.

A second line of evidence involved the size and shape of DNA. It was known for a long time that DNA was a very long molecule, but in the early 1950s Maurice Wilkins and Rosalind Franklin determined some other dimensions for the molecule and showed that it had repeating units at specific intervals along its length. In 1953, James Watson and Francis Crick proposed a structure for DNA

TABLE 13.2

Percentage Composition of the Bases of DNA in Several Species

Species	% A	% T	% G	% C	% Purines	% Pyrimidines
Escherichia coli	24.7	23.6	26.0	25.7	50.7	49.3
Yeast	31.3	32.9	18.7	17.1	50.0	50.0
Salmon	29.7	29.1	20.8	20.4	50.5	49.5
Human	30.9	29.4	19.9	19.8	50.8	49.2

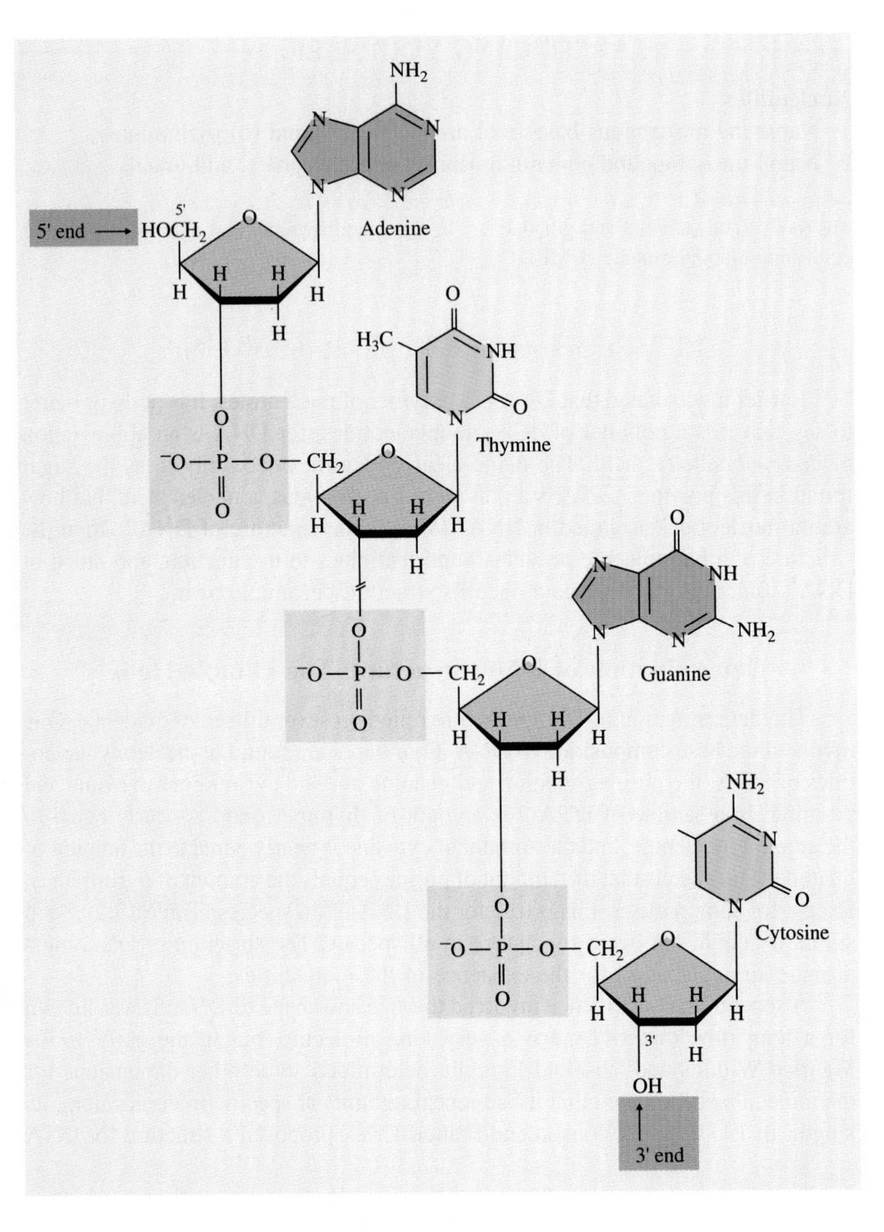

FIGURE 13.6
The bonding of nucleotides to each other to form a strand of DNA. The deoxyriboses (red) and phosphates (yellow) alternate to form a backbone, and the nitrogenous bases (blue and green) all lie on the same side of the strand. At one end of the strand, there is a hydroxyl group on the number 5 carbon of ribose that is not bound to another nucleotide. This end of the molecule is called the 5′ end. The other end of the strand has a free hydroxyl group on carbon number 3 of a sugar, thus it is the 3′ end.

that accounted for the data of Wilkins and Franklin, the equality of bases, and all other known structural data for DNA. They proposed that DNA is a *double helix*, two strands wrapped around each other in a helical fashion. The specific structure they proposed also suggested how DNA could split to provide the exact genetic material for two cells.

Before an understanding of the double helix can be reached, the structure of a single strand of DNA should be examined. The deoxyribonucleotides of a strand are covalently linked to each other through ester bonds between the phosphates and the hydroxyl groups of the sugars (Figure 13.6). Just as proteins are polymers of amino acids bonded through peptide bonds, DNA is a polymer of deoxynucleotides bonded through phosphate ester bonds. If you mentally extend this linear polymer of deoxynucleotides in both directions, a large ribbon-like molecule is formed.

The double helix proposed by Watson and Crick has two strands of DNA present in each molecule. The two strands are arranged with the bases of one strand close to the bases of the other. The bases form the core of the molecule with the sugar–phosphate backbones wound around the core in a helical fashion. The DNA molecule is indeed a double helix (Figure 13.7).

Hydrogen Bonding in DNA

What forces or bonding holds the two strands together? The strands are held together by hydrogen bonds between specific bases and by hydrophobic interactions that are not specific. The force of each hydrogen bond is individually weak, typically only a few kilocalories per mole, but the thousands upon thousands of hydrogen bonds present in a DNA molecule collectively hold the two strands together. Furthermore, hydrophobic interactions involving the bases add more stability. The strands are held tightly together by these interactions, but during certain biological processes they are temporarily separated.

♦ DNA is a double-stranded molecule with the strands held together by hydrophobic interactions and specific hydrogen bonding between complementary bases.

In DNA, adenine (A) pairs with thymine (T) through two hydrogen bonds, and guanine (G) pairs with cytosine (C) through three hydrogen bonds.

♦ In DNA, adenine is complementary to thymine, and guanine is complementary to cytosine.

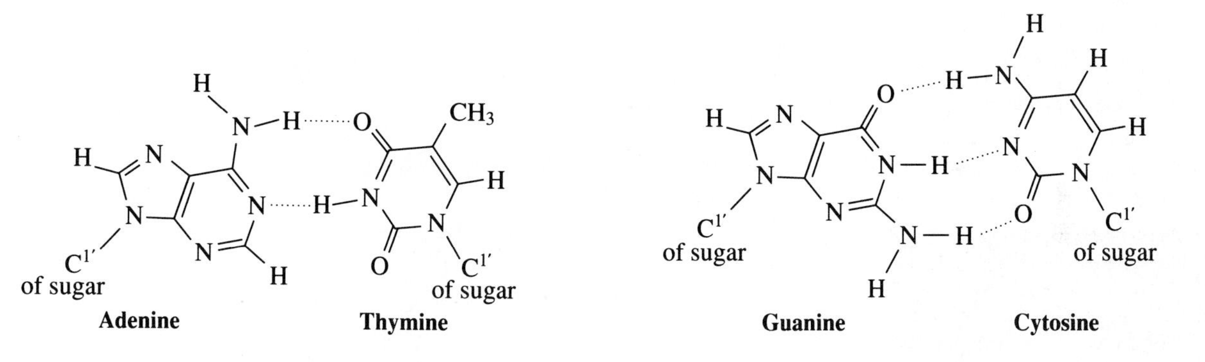

Bases that pair with each other in nucleic acids are called **complementary bases.** This accounts for the equality of bases that is noted in Table 13.2, but why is this specific combination of bases seen rather than other ones? This base pairing occurs because of size and shape factors. The core of the double helix is only large enough for one purine and one pyrimidine to fit. Two purines are too large to fit into the core without distorting the shape of the molecule, thus A does not pair with G. Two pyrimidines would not be close enough to each other for hydrogen bonds to form effectively, so C does not pair with T. Furthermore, the shape of A is ideal for hydrogen bonding to T, but in the core of a DNA molecule, A has neither the correct functional groups nor the orientation to form stable hydrogen bonds with C. Similarly, G can hydrogen-bond to C but not T (Figure 13.7). Thus, DNA is a double helix held together by hydrogen bonds between specific base pairs.

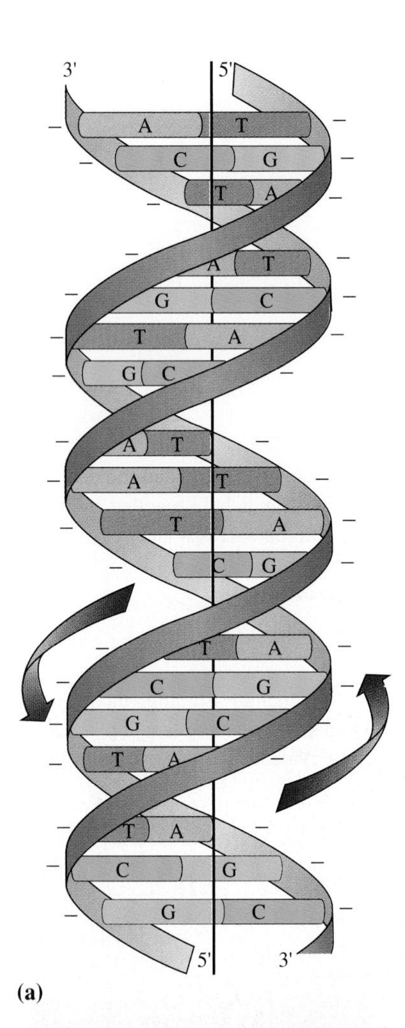
3'
5'
A
T
C
G
T
A
A
T
G
C
T
A
G
C
A
T
A
T
T
A
C
G
T
A
C
G
G
C
T
A
T
A
C
G
G
C
5'
3'

(a)

(c)

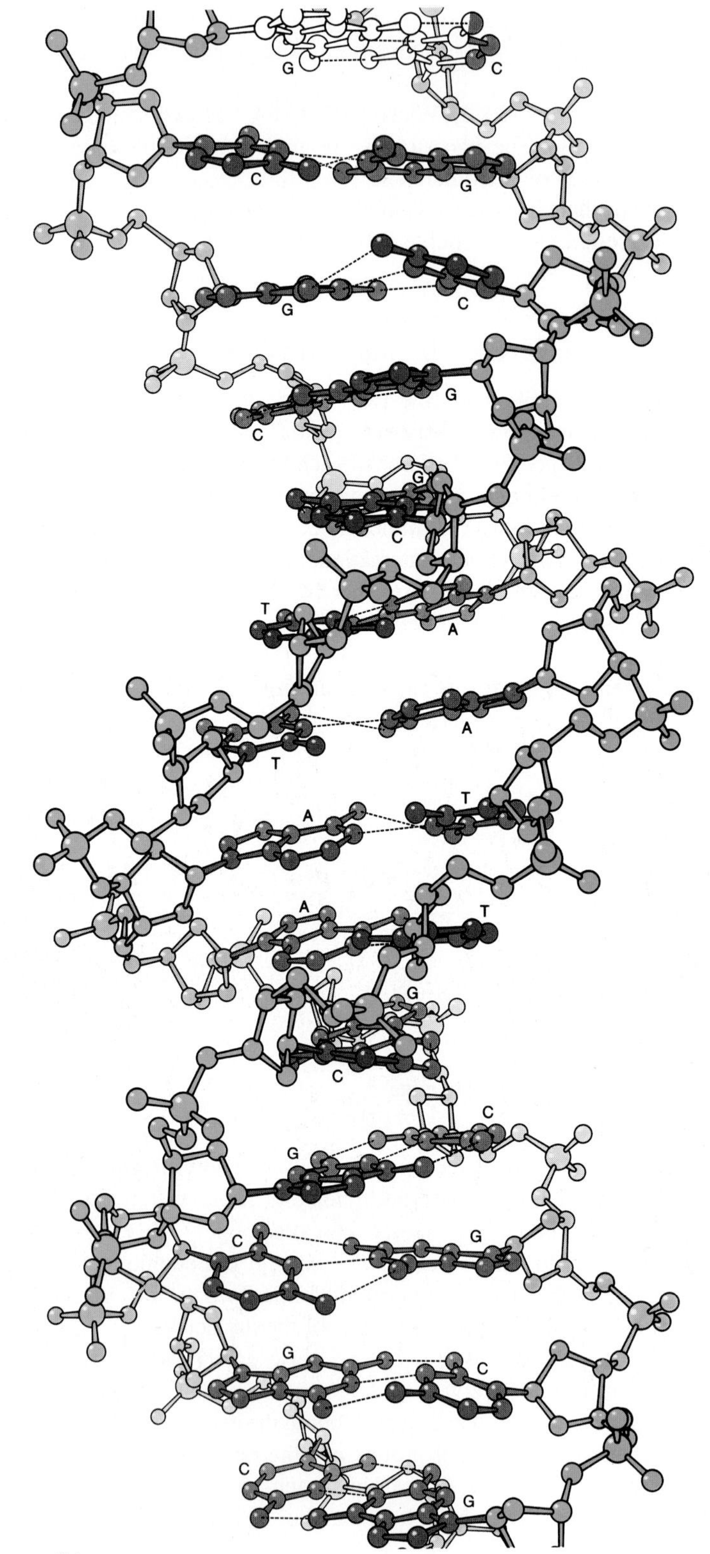
G
C
C
G
G
C
C
G
G
C
T
A
T
A
A
T
A
T
G
C
G
C
C
G
G
C
C
G

(b)

FIGURE 13.7 ***(facing page)***
Three representations of the structure of a short piece of DNA. (a) In this ribbon model, the two strands of a DNA molecule are held together by hydrogen bonds between bases. Two hydrogen bonds form between adenine (A) and thymine (T), and three hydrogen bonds form between guanine (G) and cytosine (C). (b) This illustration shows the specific hydrogen bonds between groups on pairs of nitrogenous bases. (c) This is a space-filling model of DNA. This model provides the best overview of DNA, but specific interactions between bases are difficult to see.

Higher Levels of DNA Structure

As described so far, DNA is a long, linear molecule with two strands wrapped around each other. But DNA is not found in this form in living cells. In a typical bacterium, the length of the DNA molecule is greater than 1 mm, yet the entire length of the bacterial cell is only about 10 μm (one one-hundredth as long as the DNA). This long piece of DNA is not just shoved into the volume of the cell like thread into a thimble, but instead the DNA is arranged into a more compact, orderly form. In bacteria, the ends of the DNA molecule are covalently attached to each other to form circular DNA (Figure 13.8a). The circular DNA molecule

◆ Bacterial DNA is circular and supercoiled.

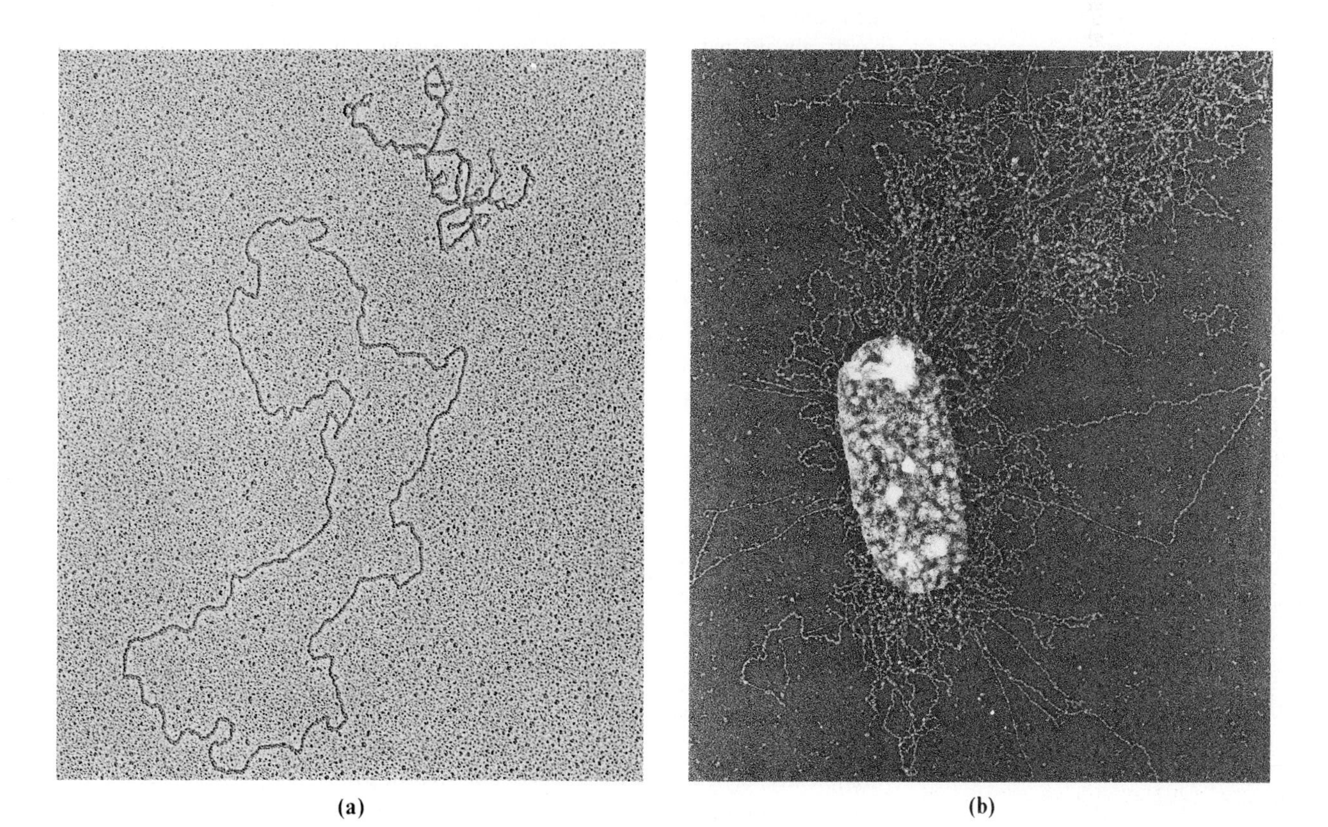

FIGURE 13.8
Electromicrographs of DNA. (a) Bacterial DNA is circular (below) and supercoiled (above). (b) DNA is spilling from a ruptured bacterial cell. DNA is much longer than the cell that contains it.

is supercoiled, that is, the molecule is wrapped around itself to yield a more compact and orderly structure (Figure 13.8a). This arrangement fits into the cell better and allows for the many functions and changes typical of DNA in living cells.

♦ Eucaryotic DNA is complexed with proteins to form chromosomes.

The cells of higher organisms contain several to many long, linear DNA molecules. The DNA molecules of a typical mammalian cell have a combined length of over 1 m, yet the DNA is contained in a nucleus that is much less than 1 mm in diameter. The DNA of these higher cells is arranged into **chromosomes**. The DNA in chromosomes is complexed with a large number of proteins to yield a more compact structure (Figure 13.9). Much of the protein consists of *histones*, which are small positively charged proteins that bind through electrostatic interactions to DNA and, in effect, neutralize the negative charge associated with the phosphates. Protein–DNA interactions make the structures more compact, yet allow for all of the normal functions of DNA.

Replication of DNA

Role of Replication

DNA stores genetic information. This information is used to control all of the activities of a cell and, ultimately, those of a multicellular organism. DNA is nearly always passed on exactly to cells arising from cell division. Thus, with rare exceptions, the activities and capabilities of daughter cells are identical to the cell from which they arose.

♦ DNA replication is semiconservative because each daughter cell gets one of the original strands.

What structural features of DNA allow genetic information to be passed to daughter cells? The double-stranded nature of the molecule and the complementary nature of the bases allow for it. The two strands of DNA are *complementary*. This means that the specific composition and sequence of one strand is related to the other because a specific base of one strand is always paired through hydrogen bonding to a specific base on the other strand. If the two strands are separated and a new strand is made for each of the original strands, then each new strand is complementary to the old strand it is now paired with (Figure 13.10). This process is called **semiconservative replication** because one strand of the original DNA molecule is conserved intact in each of the daughter DNA molecules. Through replication, two new DNA molecules are made that are identical in composition and sequence to the DNA molecule from which they were made. Each daughter cell receives one of the two molecules of DNA that were formed by replication and thus has the same genetic information possessed by the original cell.

Process of Replication

Although the concept of replication is straightforward, the cellular and molecular mechanics of the process are complex. Two factors contribute greatly to this complexity: (1) the helical nature and size of DNA and its association with proteins complicate the separation of the strands, and (2) the requirement that a

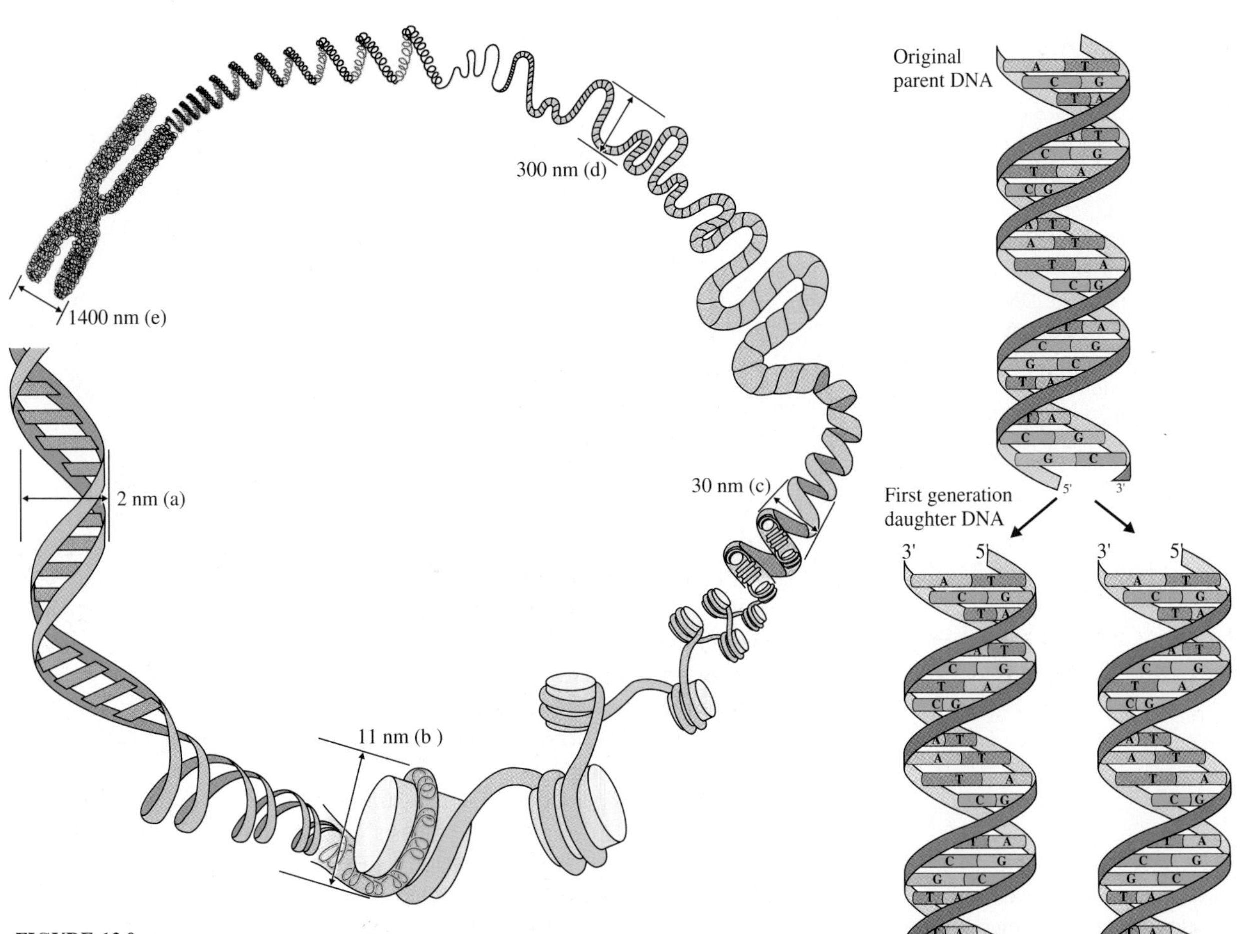

FIGURE 13.9
The arrangement of DNA in chromosomes. (a) DNA is a double helix of two strands approximately 2 nm in diameter (blue). (b) In chromosomes, the DNA molecule is wrapped around an aggregate of proteins called histones (yellow) to form nucleosomes that have a diameter of about 11 nm. (c) The nucleosomes are in turn wound about each other to form orderly clusters, 30 nm in diameter (brown). (d) The clusters are part of various size loops that may be 300 nm across. (e) During certain phases of the cell cycle, chromosomes are visible with a microscope (1400 nm).

FIGURE 13.10
Semiconservative replication of DNA. Each of the original strands of the parental DNA molecule is paired with a new strand in the daughter molecules. Note that the bases of the new strands are complimentary to the bases of the original strands.

specific base must pair with another specific base means that the process must be very accurate. What follows is an overview of replication; no attempt is made here to provide all of the specific details.

Replication begins at a specific site called the *origin of replication.* In bacteria, there is only one origin per DNA molecule, but higher organisms have multiple origins per chromosome. Multiple origins are necessary to speed up

♦ The increased size and complexity of eucaryotic DNA requires multiple origins of replication.

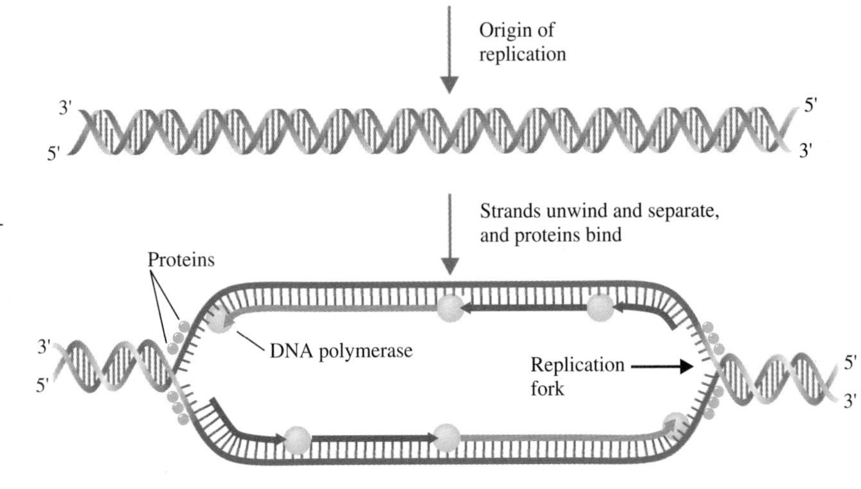

FIGURE 13.11
The origins of replication and the replication forks of DNA. Proteins aid in the separation of the DNA strands at the origins of replication. The separation yields two replication forks, which are regions where DNA polymerase can bind to the DNA strand and synthesize DNA complementary to the strand.

replication because the chromosomes of higher organisms are much larger and more complex than the bacterial DNA molecule. At the origins of replication, several proteins unravel the DNA molecule, separate the two strands, and maintain them as separated strands (Figure 13.11). This separation is necessary because the enzymes responsible for the synthesis of the new strands must be able to bind to the individual DNA strand and because the nitrogenous bases of the strand must not be bonded to the bases of the other strand.

Unwinding and separation of the two strands of DNA yield two *replication forks* (Figure 13.11). At each fork, synthesis of both strands occurs simultaneously. Although several enzymes and proteins are involved in replication, the principal enzymes responsible for the synthesis of the new strands are the *DNA polymerases*. DNA polymerase functions in the following way. DNA polymerase binds noncovalently to one of the parent DNA strands. A deoxyribonucleotide triphosphate then binds to the complex (Figure 13.12a). There are four deoxyribonucleotide triphosphates in cells: dATP, dGTP, dCTP, and dTTP. The nucleotide that binds is the one whose base is complementary to the base on the DNA strand at the point where the DNA polymerase is bound. This specific binding occurs because only that base can form hydrogen bonds to the base on the strand. In Figure 13.12(a) note that the first nucleotide has adenine, which is complementary to thymine in the strand of DNA. The other three deoxyribonucleotides cannot form these hydrogen bonds with thymine, so they cannot bind to the complex. Hydrogen bonding between the base of a nucleotide from solution and a base on the strand of DNA correctly places the incoming base needed to synthesize the new strand of DNA.

◆ Enzymes are discussed in detail in Chapter 14. For this overview, consider them simply as efficient catalysts.

DNA polymerase can bind two nucleotides (Figure 13.12b). A second nucleotide from solution binds to the complex. Again the nucleotide that binds is the one whose base is complementary to the base on the strand of DNA. In Figure 13.12(b) the second nucleotide has cytosine, which is the complementary base to guanine on the DNA strand. At this point, the two nucleotides are positioned next to each other, each hydrogen bonded to its complementary base on the parent strand. The DNA polymerase now catalyzes the formation of a phosphoester

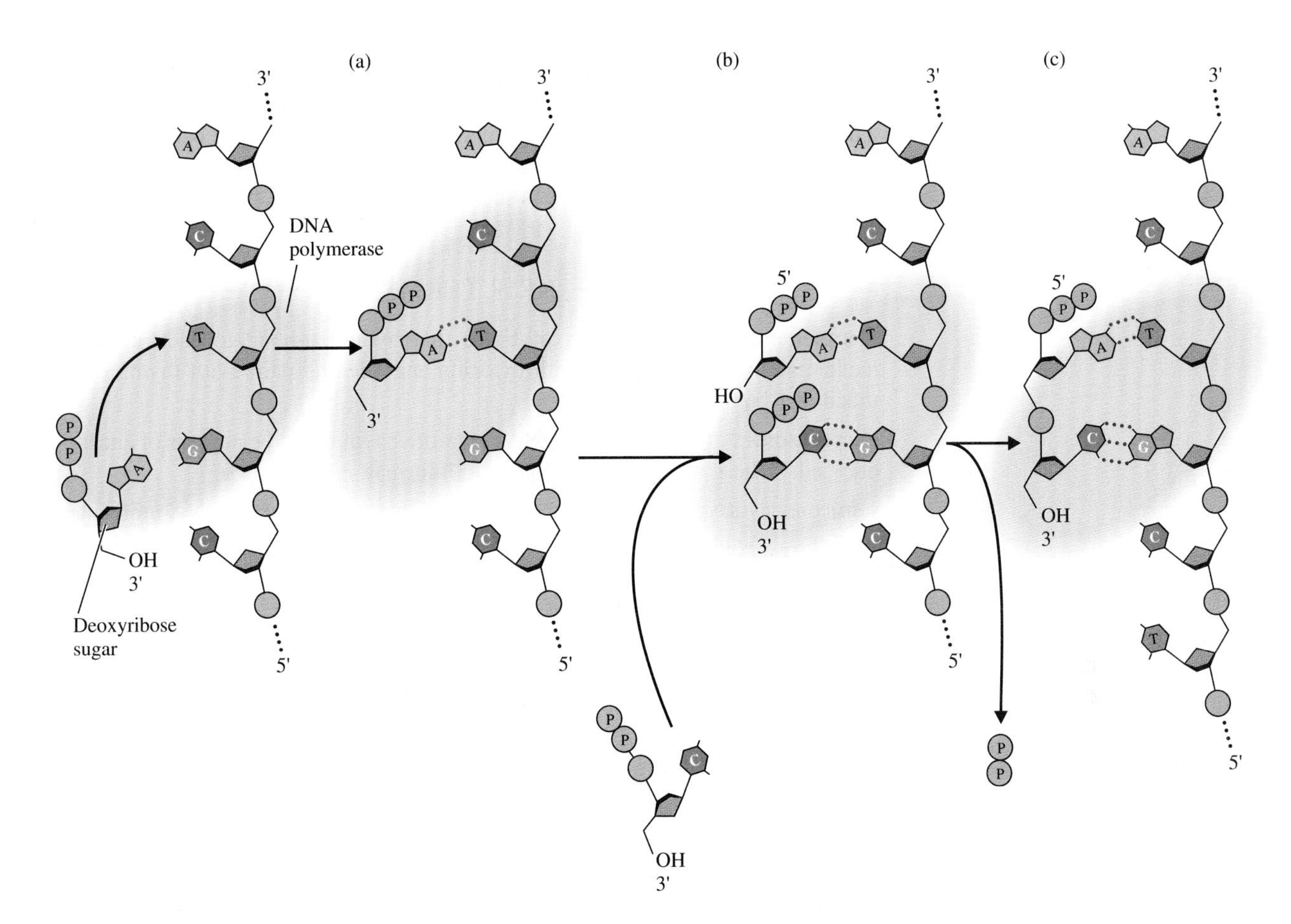

FIGURE 13.12
The role of DNA polymerase in replication and the formation of phosphoester bonds. (a) The first nucleotide of the daughter strand binds to the DNA polymerase and DNA strand through hydrogen bonds to the complementary base in the strand of DNA. (b) A second nucleotide binds to the DNA–DNA polymerase complex. (c) DNA polymerase catalyzes the formation of a phosphoester bond between the nucleotides.

bond between these two nucleotides (Figure 13.12c). The nucleotides are now covalently attached to each other.

The DNA polymerase moves down one more base on the DNA strand, another nucleotide binds (it is complementary to the next base on the strand), and the DNA polymerase catalyzes the formation of an ester bond between this new nucleotide and the previous one. One by one, nucleotides are added to the end of a growing strand of DNA, with the base of each new nucleotide complementary to the base on the parent strand.

DNA polymerase can only move toward the 3′ end of DNA, thus DNA synthesis is said to be 5′ → 3′ (see Figure 13.6 caption). Since DNA synthesis is occurring on both strands simultaneously, one of the new strands, the leading strand, is synthesized as a long continuous piece and the other, the lagging strand, is synthesized as a series of shorter pieces called *Okazaki fragments*. A ligase connects these pieces into a long, continuous piece.

During replication, a new strand of DNA is synthesized on both parent strands of DNA at both replication forks of each origin of replication. When the ends of these newly synthesized strands come together, they are covalently bonded together by the ligase. Eventually, all of the pieces have been linked together, and a new intact complementary strand is complexed with each of the parent strands, as was shown in Figure 13.10. If no errors have occurred during replication, each of the daughter DNA molecules is identical to the parent DNA molecules.

NEW TERMS

complementary bases Pairs of bases that hydrogen-bond to each other in nucleic acids: guanine pairs with cytosine and adenine pairs with thymine (or uracil in RNA).

chromosomes Complexes of DNA and proteins found in the nucleus of eucaryotic cells. These structures carry genetic information.

semiconservative replication The process that produces two DNA molecules from one. The process is semiconservative because each daughter DNA molecule receives one of the strands from the parent molecule.

TESTING YOURSELF

DNA

1. What bond links nucleotides in DNA?
2. A short segment of a DNA strand has the sequence AACTGGC. What is the sequence of the complementary strand?
3. In eucaryotes, DNA is complexed with proteins to form what structures?
4. What enzyme is responsible for the synthesis of the complementary strand of DNA during replication?

Answers **1.** phosphoester bond **2.** TTGACCG **3.** chromosomes **4.** DNA polymerase

13.4 RNA

Besides DNA, cells contain another kind of nucleic acid called ribonucleic acid, or RNA. RNA is also a polymer of nucleotides, but there are several important structural differences between RNA and DNA. First, the sugar found in RNA is ribose, whereas DNA contains deoxyribose. Second, RNA contains the base *uracil* instead of the thymine found in DNA. Third, RNA is a single-stranded molecule, and DNA is double-stranded. In addition, DNA is a much larger molecule than RNA. The structural differences between DNA and RNA are summarized in Table 13.3.

♦ There are three classes of RNA: messenger, transfer, and ribosomal.

There are three main classes of RNA found in cells. **Messenger RNA (mRNA)** carries the genetic information stored in DNA to the rest of the cell. **Ribosomal RNA (rRNA)** is RNA found in ribosomes, the organelles responsible for protein synthesis. **Transfer RNA (tRNA)** is a smaller form of RNA that carries amino acids to the ribosome and helps ensure that the correct amino acids are used at the right time during protein synthesis. Each of these molecules is

TABLE 13.3
Structural Differences Between DNA and RNA

Nucleic Acid	Sugar	Base	Strands	Size
DNA	Deoxyribose	Thymine	Two	Largest molecule in cell
RNA	Ribose	Uracil	One	Macromolecules, but much smaller than DNA

synthesized by a process called *transcription*. The specific structure and role of each of these RNA types will be discussed more fully in Sections 13.5 and 13.6 on transcription and translation.

NEW TERMS

messenger RNA (mRNA) RNA that carries genetic information from the nucleus to the rest of the cell. The information is used to direct protein synthesis.

ribosomal RNA (rRNA) RNA molecules found in ribosomes, the site of protein synthesis.

transfer RNA (tRNA) RNA molecules that transfer the correct amino acids into the protein that is being synthesized by ribosomes.

TESTING YOURSELF

RNA

1. What are the three classes of RNA?
2. List four differences between DNA and RNA.

Answers **1.** messenger RNA, transfer RNA and ribosomal RNA. **2.** DNA has the sugar deoxyribose, RNA has ribose. DNA is double-stranded, RNA is single-stranded. DNA has the base thymine, RNA has the base uracil. DNA is the largest molecule in the cell, RNA is much smaller.

13.5 Transcription

Role of Transcription

Information stored in DNA ultimately controls the activities of cells. These activities are primarily chemical reactions that are catalyzed and controlled by enzymes. Enzymes, and all the other proteins required for life, are synthesized using information contained in DNA. The process called *transcription* synthesizes a molecule of messenger RNA using DNA as a template or guide. This mRNA then carries the information needed to synthesize the protein to the ribosomes. Ribosomes, through the process called *translation*, use the information in the mRNA to synthesize the protein (see Section 13.6). To a large extent, a cell controls its molecular activities by controlling which proteins are present and by controlling how much of each protein is present at any given time.

♦ Transcription synthesizes an RNA molecule that is complementary to a part of a DNA strand.

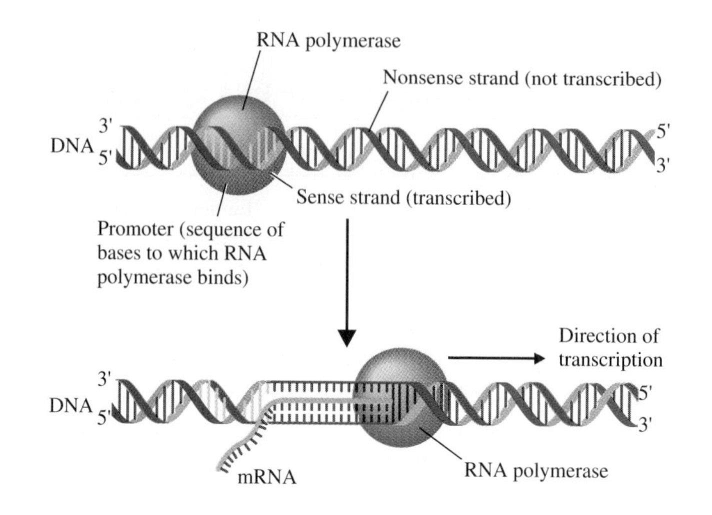

FIGURE 13.13
Some specific steps of transcription. (a) RNA polymerase (purple) binds to the sense strand of DNA at a sequence of bases called the promoter (yellow). (b) The RNA polymerase catalyzes the sequential connection of ribonucleotides that are complementary to the bases of the sense strand of DNA. The newly synthesized RNA molecule (green) is thus complementary to a portion of a DNA strand.

Process of Transcription

♦ Transcription begins at promoters, which are specific sites on the sense strand of DNA.

Transcription is a process similar to replication. Replication yields a strand of DNA that is complementary to the parent strand of DNA, and transcription yields an RNA molecule that is complementary to a portion of a strand of DNA. Transcription begins with the binding of the enzyme *RNA polymerase* to one of the strands of DNA at one of many specific sites on the DNA molecule. These sites are called *promoters*, sequences of bases recognized by RNA polymerase as binding sites. Once bound, the RNA polymerase begins to synthesize a strand of RNA that is complementary to the strand of DNA. This strand of DNA is referred to as the *sense strand*, because the information that is needed to synthesize the protein is present in this strand. The other strand is called the *nonsense strand.* Remember, the strands of DNA are complementary, not identical; only one can carry the exact information needed for the synthesis of a polypeptide.

Ribonucleotides from solution bind to the RNA polymerase–sense strand complex. The specific ribonucleotide that binds depends upon the base on the sense strand in the complex. Like DNA polymerase, RNA polymerase binds nucleotides from solution that are complementary to the bases of the DNA strand, then it covalently links them. The specifics of RNA synthesis are similar to those shown for DNA polymerase in Figure 13.12, but ribonucleotides bind rather than deoxyribonucleotides. In Figure 13.13(b), RNA polymerase is moving along a sense strand, building an RNA molecule by the sequential addition of ribonucleotides to the end of the growing RNA molecule. Each new base is complementary to the base on the sense strand. The RNA polymerase molecule continually moves down one more base, then another nucleotide binds, and a phosphoester bond forms between this base and the previous one. This continues until the RNA polymerase comes to a sequence of bases on the sense strand that signals the termination of transcription. The complex dissociates to give the newly synthesized RNA, the RNA polymerase, and the DNA molecule.

It is worthwhile to compare the accuracy of replication and transcription. Transcription need not be as accurate as replication. Replication yields two daughter DNA molecules that contain all of the genetic information of the organism. If one of them has been synthesized incorrectly, the genetic information of that daughter cell and all the descendants of that cell will be in error. If an error is made in transcription, then some of the copies of a particular protein may be incorrect. Since other correct copies of the protein are also present, the function of that protein is fulfilled. The cell wastes some material and energy making the defective protein, but that is far less serious than a genetic error.

The RNA formed by transcription is called the **primary transcript**, but the transcript must usually be altered before it can function in the cell. One exception is the transcript in bacteria; this transcript becomes a functional mRNA as soon as transcription is complete. Messenger RNA in higher organisms is formed from transcripts by the action of specific enzymes in the nucleus. These enzymes remove the unneeded segments called *introns* and splice together the remaining segments called *exons*. This splicing plus some other processing converts the transcript into mRNA that diffuses from the nucleus into the rest of the cell. Although some base pairing can occur in mRNA, it is generally regarded as a long linear molecule randomly oriented in solution.

♦ In eucaryotes, the primary transcript must be processed to obtain a functional mRNA.

Ribosomal RNA is synthesized from specific transcripts in a slightly different way. A specific transcript is cleaved by enzymes in the nucleus into smaller pieces of RNA. These smaller pieces are the rRNA molecules that join with specific proteins to form functional ribosome subunits. The RNA in a ribosome has a specific shape and is bound to certain proteins in the ribosome.

Transfer RNA is made from specific transcripts that are cleaved by enzymes to yield RNA of the proper size. Other enzymes catalyze covalent changes to some of the nucleotides in the RNA. There are many different tRNA molecules in a cell, each of which undergoes different processing to become a functional tRNA. All of the tRNAs have the same overall shapes, but each is uniquely different. In two dimensions, tRNAs are often represented as cloverleaf-shaped molecules with complementary bases hydrogen-bonded to other bases of the same strand. In three dimensions the tRNAs look more L- or club-shaped (Figure 13.14).

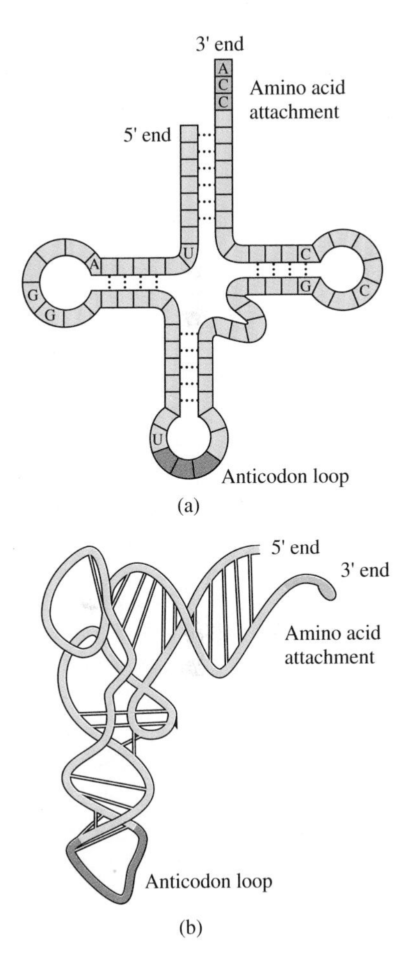

FIGURE 13.14
(a) Two- and (b) three-dimensional representations of tRNA. The anticodon (in red) of a tRNA molecule pairs with the codon of mRNA during translation. The 3′ end of the molecule (blue) is the attachment site for the amino acid.

NEW TERMS

transcription The process that synthesizes RNA molecules using a DNA molecule as a template

primary transcript The first product of transcription. In higher cells it is modified to yield functional RNA molecules.

TESTING YOURSELF

Transcription

1. What enzyme is responsible for RNA synthesis?
2. What bases in a strand of DNA are complementary to this segment of RNA?

—A—U—G—C—

—?—?—?—?—

3. Transcription begins where on the sense strand of DNA?

Answers **1.** RNA polymerase **2.** —T—A—C—G— **3.** at a promoter

13.6 Translation

♦ The process called protein synthesis literally translates the information stored in the base sequence of mRNA into a sequence of amino acids in a polypeptide chain.

Proteins are synthesized by **translation**, a process that uses information stored in mRNA as instructions for protein synthesis. Because mRNA is synthesized by transcription from DNA, the information comes ultimately from DNA. The term **gene** is often used for that part of a DNA molecule that contains the information needed for the synthesis of a polypeptide or protein. The mRNA transcribed from one gene has the information needed for the synthesis of one particular polypeptide. If a protein has only one kind of polypeptide, then only one mRNA is needed for the synthesis of that protein. If two or more different peptides (subunits) are present in the protein, then two or more kinds of mRNA are needed to synthesize the polypeptides of that protein.

The Genetic Code

Since mRNA molecules differ from one another only in the sequence of the purine and pyrimidine bases, how does the information stored in a specific set of mRNA molecules convey the correct information for the synthesis of that specific protein? An analogy is helpful. There are 26 letters in the English alphabet. From various combinations of these 26 letters we can make words and sentences that express our thoughts. Analogously, the information stored in each mRNA molecule differs from that in all other mRNA molecules because the bases of each mRNA molecule are arranged in a unique order. An "alphabet" of only four "letters"—adenine, guanine, cytosine, and uracil—provides all the information needed to synthesize all of the proteins of all living organisms.

♦ Three consecutive bases in mRNA make up a codon.

The information contained in mRNA is in three base sequences called **codons**. All of the 64 codons have meaning, none are nonsense (Figure 13.15). Most codons specify an amino acid, although a few codons serve as punctuation for the process of translation. For example, this nine-base sequence in an RNA molecule, AAUGCUGGA, corresponds to three codons: AAU, GCU, and GGA. They code for the amino acids asparagine, alanine, and glycine. Table 13.4 shows that many of the amino acids have two or more codons.

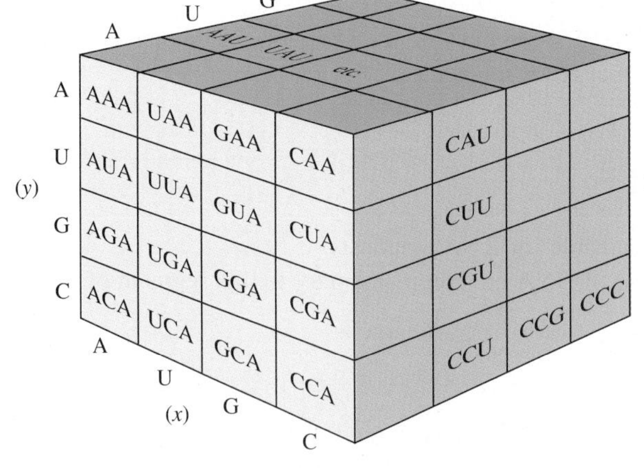

FIGURE 13.15
The genetic code. There are 64 combinations of four bases arranged into sequences of three. These three-base sequences are called *codons*, and codons correspond to a particular amino acid or serve as a stop signal during translation.

Messenger RNA molecules are longer than the message they contain, that is, there are bases at either end of the molecule that are not used as parts of codons. The message for the amino acid sequence of the protein begins somewhere inside one of the physical ends of the mRNA, and the end of the message occurs before the other end of the mRNA molecule. Certain codons are required to start and stop the process of translation. The stop codons are highlighted (blue) in Table 13.4. The codon for start, AUG (highlighted in red), is also the codon for the amino acid methionine. How can one codon have two meanings? Wherever AUG occurs in a message, a methionine is inserted into the protein. But if the AUG codon is preceded by some specific bases in a particular order, then the process of translation will begin with that codon. AUG can thus serve two roles in translation.

EXAMPLE 13.2

What amino acids or physiological roles correspond to these codons: GAA, CGC, UAG, AUG?

Solution: Table 13.4 contains information about codon assignments. GAA corresponds to glutamate, CGC to arginine, UAG is a stop codon, and AUG corresponds to methionine and is the start codon.

TABLE 13.4
The Genetic Code

Codon	Amino Acid	Codon	Amino Acid	Codon	Amino Acid	Codon	Amino Acid
UUU	Phe	UCU	Ser	UAU	Tyr	UGU	Cys
UUC	Phe	UCC	Ser	UAC	Tyr	UGC	Cys
UUA	Leu	UCA	Ser	UAA	Stop	UGA	Stop
UUG	Leu	UCG	Ser	UAG	Stop	UGG	Trp
CUU	Leu	CCU	Pro	CAU	His	CGU	Arg
CUC	Leu	CCC	Pro	CAC	His	CGC	Arg
CUA	Leu	CCA	Pro	CAA	Gln	CGA	Arg
CUG	Leu	CCG	Pro	CAG	Gln	CGG	Arg
AUU	Ile	ACU	Thr	AAU	Asn	AGU	Ser
AUC	Ile	ACC	Thr	AAC	Asn	AGC	Ser
AUA	Ile	ACA	Thr	AAA	Lys	AGA	Arg
AUG	Met	ACG	Thr	AAG	Lys	AGG	Arg
GUU	Val	GCU	Ala	GAU	Asp	GGU	Gly
GUC	Val	GCC	Ala	GAC	Asp	GGC	Gly
GUA	Val	GCA	Ala	GAA	Glu	GGA	Gly
GUG	Val	GCG	Ala	GAG	Glu	GGG	Gly

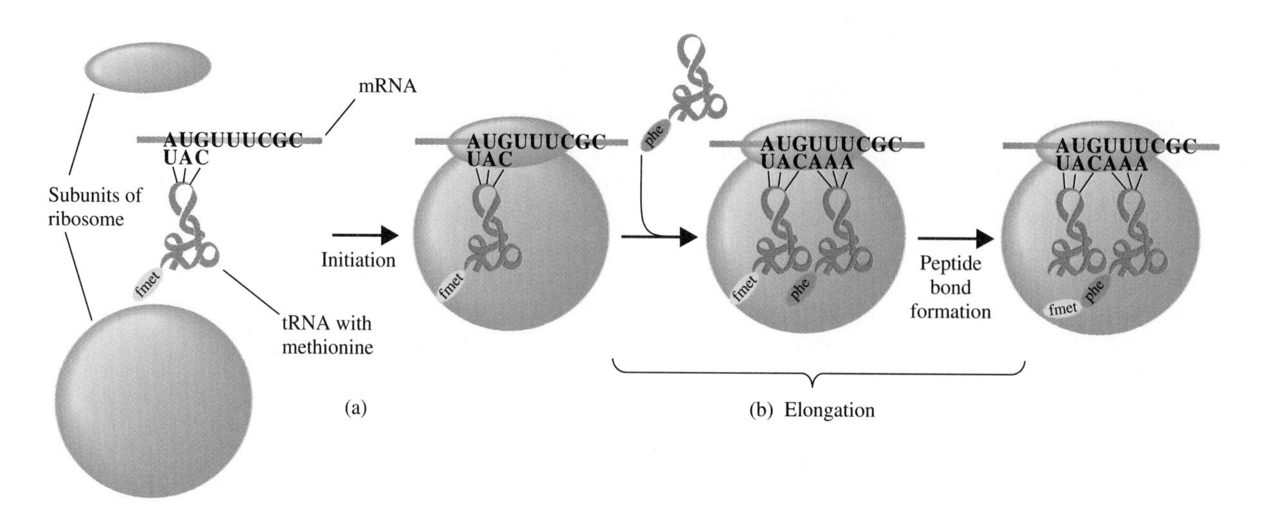

FIGURE 13.16
The process of translation. (a) Initiation. Translation is initiated by the formation of a complex of mRNA, a ribosome, and a tRNA bound to methionine. (b) Elongation. A second tRNA with its amino acid (Phe) binds, then a ribosomal enzyme forms a peptide bond between the amino acids, leaving a free tRNA and a tRNA with a dipeptide. (c) Through translocation, the ribosome and tRNA bearing the peptide move along the mRNA by one codon. The next tRNA with its amino acid then binds, and peptide bond formation and translocation repeat. (d) Termination of translation. When a stop codon (green) is present in the ribosome–mRNA complex, an enzyme cleaves the bond between the tRNA and the polypeptide. The complex then dissociates.

Process of Translation

♦ The molecular interactions that occur during translation are among the most complex yet studied.

Translation involves numerous molecules, including mRNA, ribosomes, many tRNAs with amino acids bound to them, and a variety of proteins. Translation requires energy that is provided by GTP and ATP. The process can be summarized by a series of steps. The steps described here are a simplified version of the actual process.

In the first step of translation, the *initiation* step, one of the subunits of a ribosome binds to the mRNA at the start codon, AUG. Remember, AUG serves as the start codon only if certain sets of bases precede it. A tRNA that has methionine covalently bound to it then binds by hydrogen bonding between three of its bases and the three bases of the codon AUG. These three bases on the tRNA are called an **anticodon** because they are complementary to the bases of the codon. Only this tRNA of the many kinds of tRNA in the cell binds, because only this tRNA has the proper anticodon. Now the other subunit of the ribosome binds to form a complex of a ribosome, mRNA, and the proper tRNA bound to the start codon of the mRNA. Figure 13.16(a) summarizes this first step.

In the second step of translation, a second tRNA with its amino acid binds to the complex (Figure 13.16b). The tRNA that binds is the one whose anticodon is complementary to the second codon of the mRNA. This tRNA has the amino

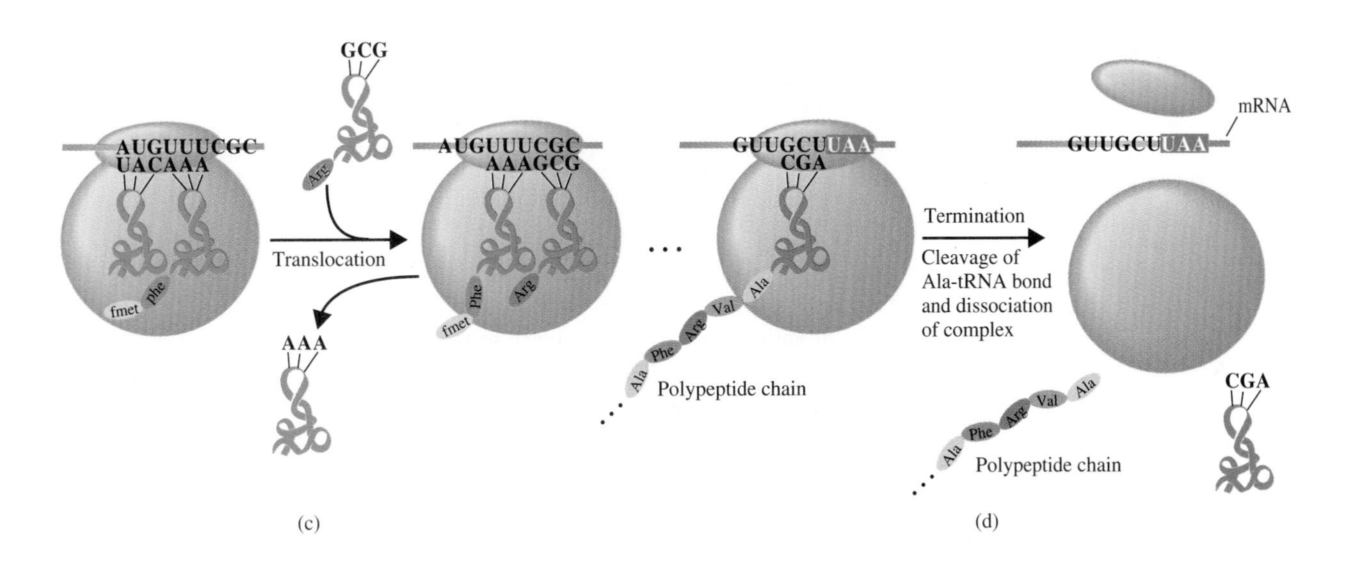

(c)

(d)

acid bound to it that is coded for by the codon on the mRNA. Note in Figure 13.16 that codon AUG is next to codon UUU, and at the other end of the tRNA, the amino acid methionine is next to the amino acid phenylalanine. The tRNA molecules serve as adapters or guides that ensure that the correct amino acids, corresponding to specific codons, are inserted into the growing polypeptide chain. A protein in the ribosome now catalyzes the formation of a peptide bond between the two amino acids, leaving an empty tRNA and a tRNA that now has a peptide bound to it (Figure 13.16b).

Next the process of *translocation* occurs, and the ribosome moves one codon along the mRNA. The empty tRNA falls from the complex, and the peptidyl–tRNA remains with its codon (Figure 13.16c). Another tRNA with its amino acid will bind to the next codon, a peptide bond will form, and translocation will occur again. These events continue as long as the next codon is a codon for an amino acid.

When the next codon is a stop codon, termination ocurs. There are no tRNAs for the stop codons. When a stop codon enters the ribosome–mRNA complex, a protein in the ribosome cleaves the polypeptide from the tRNA (Figure 13.16d). The freed polypeptide diffuses from the complex, and the two subunits of the ribosome fall from the mRNA. Many times the polypeptide is modified slightly following translation. Often one or more polypeptide bonds are cleaved, reducing the size of the polypeptide. If the polypeptide requires a prosthetic group or if it is a part of an oligomeric protein, then it will bind to these other parts to become a functional protein.

♦ Most polypeptides are modified slightly or extensively after translation.

NEW TERMS

translation A synonym for protein synthesis. Information stored in mRNA is used to direct the synthesis of protein.

gene That portion of a DNA molecule that codes for a specific transcript. Since most transcripts become mRNA molecules, a gene can be considered the information in a DNA molecule that codes for one polypeptide chain.

codon The three-base sequence in mRNA that determines what amino acid is inserted into the polypeptide chain. Ultimately the sequence of codons in mRNA determines the primary sequence of proteins.

anticodon The three-base sequence in tRNA that is complementary to a codon on mRNA

TESTING YOURSELF

Translation

1. What anticodon is complementary to the codon AGC?
2. Use Table 13.4, the genetic code, to determine which amino acid corresponds to the codon CGU.
3. What molecules ensure that the correct amino acid is inserted for a codon on mRNA?
4. What name is given to the process in which a ribosome moves stepwise along a mRNA molecule?

Answers **1.** UCG **2.** arginine **3.** tRNA molecules **4.** translocation

13.7 Mutagenesis

Mutations are permanent changes in genetic material. They arise whenever the structure of DNA is permanently altered. Since the sequence of bases in DNA determines the information that is stored, any change in this sequence changes the information contained in DNA. These changes usually show up as proteins with altered structure and function. The process that leads to the formation of mutations is called *mutagenesis*, and there are several ways a mutation can arise. Mutations can arise spontaneously, that is, without external agents.

Spontaneous Mutations

♦ The small but measurable error rate in replication is one source of mutations.

One form of spontaneous mutation involves replication. Replication is a very accurate process, but it is not perfect. If an error occurs, the information contained in the DNA is altered and a spontaneous mutation has occurred. It is estimated that during replication an incorrect base is added to the growing DNA strand less than one time per one billion base additions. Most enzymes are not this accurate, so why is DNA polymerase an exception? It is believed that the inherent accuracy for base addition by DNA polymerases is about one error per 100,000 base insertions, but these polymerases possess an additional capability not seen in most other enzymes—DNA polymerases have "proofreading" ability.

After the phosphoester bond is formed between the most recent nucleotide and the previous one, the DNA polymerase checks again (proofreads) to see that the newest base is really base paired correctly to the complementary base on the parent strand. It can sense correct geometric alignment, and if two bases are not hydrogen-bonded correctly they will not have the correct shape. If the wrong base has been inserted, then the DNA polymerase cleaves the phosphoester bond and the nucleotide diffuses into solution. Another nucleotide, presumably the correct one, then diffuses in and binds to the complex, and the process is repeated. It is estimated that the error rate of this proofreading operation is also around one error per 100,000 base insertions. The overall error rate is simply the mathematical product of the error rate of each: $(1/100{,}000) \times (1/100{,}000) = 1/10^{10}$. This is one error per ten billion base additions.

What happens to a base pair in a newly synthesized DNA molecule if the bases are not complementary? Cells possess a set of enzymes that repair DNA. These DNA repair enzymes have a variety of specific functions. The base pair that is not complementary does not hydrogen-bond properly, so the strands are not held together at that point. Repair enzymes cut out the incorrect base, insert the proper base, and form phosphoester bonds between this base and the neighboring nucleotides. This operation restores the correct base sequence to the daughter strand. The repair enzymes are generally able to distinguish between the parent and daughter strands. Normally some of the bases of a strand of DNA are modified slightly by addition of methyl groups or other changes. A mature strand of DNA (the parent strand) has these modifications on it, but a newly synthesized strand does not. Immediately following replication, the two strands are physically different and can be distinguished by the repair enzymes. Normally the incorrect nucleotide is removed from the daughter strand.

If the base is removed from the parent strand and replaced with the base that is complementary to the base in the daughter strand (which happens occasionally), then the DNA molecule is *not* identical to the DNA molecule from which it was synthesized. It differs by one base pair (Figure 13.17a). A permanent genetic

Normal gene

DNA template strand AAA TAG CGG TCC···
mRNA UUU AUC GCC AGG···
Normal protein chain Phe Ile Ala Arg ···

(a) Substitution of G by A on DNA
Template AAA TAG CAG TCC···
mRNA UUU AUC GUC AGG···
Mutant protein Phe Ile Val Arg ···

(b) Insertion of a C base in DNA
Template AAA TAG CCG GTC C···
mRNA UUU AUC GGC CAG G···
Mutant protein Phe Ile Gly Gln ···

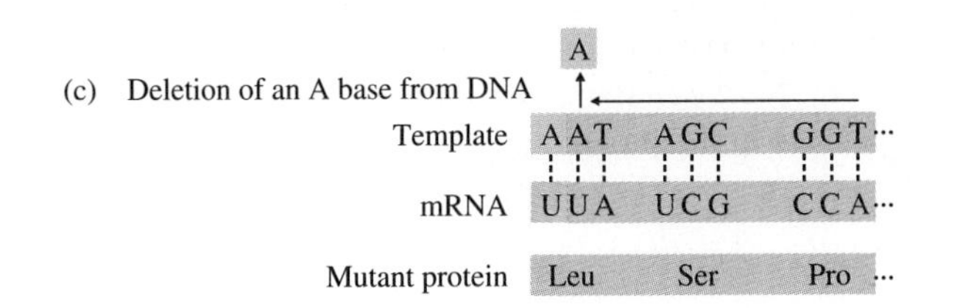

FIGURE 13.17
Introduced changes in DNA structure. (a) Substitution of one base for another in DNA results in one altered codon in mRNA. This means a different amino acid may be inserted into the protein. (b and c) Frameshift mutations result when a base is inserted or deleted from a DNA strand. This results in altered codons from that corresponding point in mRNA to the end of the gene. Many amino acids will be altered from that point in the protein to the end of the protein.

change involving a replacement of one base for another in DNA is called **substitution**. The change of one base means that one codon has been changed. A change in a codon could result in a whole spectrum of possible consequences. If the codon has been altered to a different codon but for the same amino acid, then no observable change will occur. The protein produced from this DNA molecule is identical to the protein produced from the parent DNA molecule. Another possibility is that the new codon may be for a different but similar amino acid. This may have little or no effect on the structure of some proteins, so even though it is different, the protein may still be functional. The codon may also be for a very different amino acid, or perhaps the codon is now a stop codon. The synthesized protein has the wrong structure in the first case and an incomplete structure in the second case. In either case, the protein is inactive. Much of the time a mutation results in a defective protein rather than a functional one.

Mutagens Induce Changes in DNA

♦ Mutagens are agents that increase the mutation rate in cells.

Mutations can also be induced by a variety of agents called **mutagens**. An agent is a mutagen if an organism shows a higher rate of mutation in the presence of the agent than in its absence. For example, several forms of electromagnetic radiation cause mutations. Ultraviolet (UV) light and ionizing radiation such as x-rays alter the structure of bases in DNA. Again, repair mechanisms normally repair the damage, but this process is not perfect. If an error is made, a mutation arises. Each time an individual is exposed to sunlight, there is a chance that UV light may cause a change in the DNA of skin cells. The greater the intensity of the sunlight, or the longer the duration of exposure, the greater the chance of a mutation.

There is a reasonable correlation between mutagenesis and *carcinogenesis*, the generation of cancer. The greater the exposure to UV light, the greater the chance of skin cancer, perhaps because one or more mutations in skin cells have resulted in a change in that cell from a normal growth pattern to the growth pattern of a cancerous cell.

A number of substances, such as nitrous acid and 2-aminopurine, cause mutations. Some of these mutagens chemically modify bases in DNA, and as before, the repair may not be accurate. The result is a substitution. Some other mutagens become wedged between base pairs in a DNA molecule. Some of these molecules contain three fused rings that roughly resemble a purine–pyrimidine base pair. If this DNA molecule were replicated, a base could be inserted where it should not be, or less frequently, a base may be left out. Again, if this is not corrected, a mutation will result. These mutations are called *insertion* or *deletion* mutations. Collectively they are called **frameshift mutations**, and both change the codon at the point of base insertion or deletion and all of the codons that follow (Figure 13.17b). Nearly all frameshift mutations result in defective proteins since numerous residues in the protein are affected.

EXAMPLE 13.3

If mutations changed DNA to yield the following codons in mRNA, what would be the effects on the structure of the resulting protein?

$$\text{GAC} \longrightarrow \text{GAA}; \qquad \text{UAC} \longrightarrow \text{UAA}$$

Solution: Table 13.4 provides the information to interpret these changes. The change GAC → GAA results in the substitution of glutamate for aspartate. Although mutations are usually harmful, because these are both acidic amino acids, the effect on the protein could be small. The change UAC → UAA substitutes a stop codon for tyrosine. This means that protein synthesis of this polypeptide chain now stops at this codon. The protein will most probably be nonfunctional.

Mutations are the Source of Genetic Variability

Mutations can occur in two types of cells—the somatic (or body) cells and the germ (or sex) cells. The consequences of mutation are quite different in these two cell types. The DNA found in the sperm cells of males and in the eggs (or ova) of females is called *germinal DNA*. The DNA found in all other cells is called *somatic DNA*. An error in somatic DNA results in a mutation of a cell in a specific part of the body. That cell may divide to yield a clone of cells with the same genetic change. The result may eventually be a cancerous tumor, but the results are confined to that individual. The offspring from this individual do not carry the same error because somatic cells are not passed on to offspring.

If a change occurs in the germinal DNA of sperm or ova, it will be transmitted to the offspring as a permanent genetic change. Exposure to mutagens can result in alteration of the germinal DNA, thus directly affecting the offspring. Or instead, the offspring may become a carrier of this genetic trait, which can then be passed on to subsequent offspring.

These genetic changes or errors are the ultimate source of genetic variation seen in organisms. There is no unique sequence of DNA that is the DNA of a species. Instead, the individuals of a species contain very similar, but nonidentical DNA. This is why each kitten of a litter is unique, why each human is an individual distinctly different from all other humans. Unfortunately, some of these genetic differences are the source of genetic diseases or deficiencies that are passed from generation to generation. Some genetic diseases and their causes are listed in Table 13.5.

◆ Mutations are the ultimate source of genetic variation.

Sometimes a greater number of people in specific populations are more prone to certain disorders than others. For example, Tay-Sachs disease, a disorder affecting lipid metabolism, is more common in Eastern European Jewish populations and their descendants than in other populations. About 10 percent of the African-American population carry the sickle cell trait, meaning they are carriers but do not suffer the full effects of the disorder. About 0.4 percent are afflicted with sickle-cell anemia. Sickle-cell anemia is a well-documented genetic disorder that was discussed in Chapter 12.

TABLE 13.5
Some Human Genetic Diseases

Disorder	Cause
Phenylketonuria (PKU)	An enzyme for converting phenylalanine to tyrosine is defective. The result is that phenylalanine is converted to phenylpyruvate and other phenylketones. The accumulation of these compounds in the developing young causes neurological damage.
Galactosemia	The enzyme needed for changing galactose to glucose is missing. The buildup of galactose is very serious in infants. Mental disorders as well as cataracts of the eye lens result from the buildup of galactose in the body.
Albinism	An enzyme needed for the conversion of tyrosine to a diphenolic compound is missing. This diphenolic compound is needed for formation of normal pigmentation.
Fabry's disease	An enzyme needed for removal of a galactose from glycolipids is defective. This results in accumulation of glycolipid in a number of tissues. Death is often a result of kidney or cardiac failure due to this buildup.

NEW TERMS

mutation A permanent change in the base sequence of DNA that changes genetic information

substitution A mutation in which a single base has substituted for another in a DNA strand; a single codon is affected

mutagens Agents that cause mutations

frameshift mutation A mutation resulting from insertion or deletion of a base from a DNA strand. The codon at that point plus all that follow are altered.

TESTING YOURSELF

Mutations

Show how substitution, insertion, and deletion mutations can affect protein synthesis. Consider this hypothetical base sequence in mRNA:

· · · C—G—A—C—U—G—G—A—U—A—A—U—C—C · · ·
1 2 3 4 5 6 7 8 9 10 11 12 13 14

1. What is the corresponding amino acid sequence?
2. Replace ^{4}C (C at number four position) with G. What is the new amino acid sequence?
3. Between ^{3}A and ^{4}C, insert a U. What is the amino acid sequence?
4. Delete ^{5}U. What is the amino acid sequence?

Answers **1.** Arg—Leu—Asp—Asn **2.** Arg—Val—Asp—Asn
3. Arg—Ser—Gly—chain termination **4.** Arg—Arg—Ile—Ile

13.8 Regulation of Gene Expression

Need for Regulation of Cellular Activity

Information stored in DNA as genes, through the processes of transcription and translation, is used to synthesize the variety of proteins that carry out the numerous and necessary functions of a cell. The activities of all cells of an organism, in turn, contribute to the well-being of the organism. All of the somatic cells of a person have essentially the same genetic composition because they have the same DNA. They are derived from a single fertilized cell that marked the beginning of life for that individual. Yet the cells of the body are not identical. Cells of the heart have a different shape and function than liver cells. Neurons do not resemble heart or liver cells in either structure or function. Furthermore, any given cell may vary slightly in capability and function with time. Consider the changes that occur as a fertilized egg develops into an embryo, then a fetus, an infant, a child, and finally an adult.

The structure and function of a cell are a reflection of the proteins that make up the cell. Each cell type is different because it has a different set of proteins. Some proteins, like DNA polymerase, are found in virtually all cells. Other proteins are found in only one cell type; for instance, hemoglobin is found only in red blood cells and the cells that develop into them. Why do different cells have different proteins when they contain the same genes?

Regulation in Bacteria

Cells possess the ability to synthesize specific proteins in specific amounts. Cells regulate the amount and timing of protein synthesis; only some of the genes of the cell are used at any given time. An understanding of the regulation of protein synthesis in higher organisms is just beginning. Because regulation in bacteria is better understood, two examples of regulation in bacteria will be presented here. Some of these principles extend, with modification, to higher organisms.

♦ Cells differ because they contain differing amounts of proteins. This results from regulation of when and how much protein is synthesized.

Induction

Many bacteria possess the ability to use several different organic compounds as energy sources and as sources of carbon atoms. For example, the intestinal bacterium *Escherichia coli* grows on glucose, lactose, and a variety of other organic compounds. But if you look at a population of these bacteria that have been grown on glucose, they lack the enzymes needed to consume lactose. Yet if some of these bacteria are placed into a solution that contains only lactose, they will use the lactose and grow well in this solution. Furthermore, if we look for the enzymes needed to consume lactose, they are now present in the bacteria. The presence of lactose *induced* the appearance of these enzymes in the bacteria. This process is called **induction**.

Induction works by altering the rate of formation of mRNA for a specific set of proteins. If the mRNA is not formed, no proteins will be made. This is because mRNA in bacteria exists for only a short time; it is broken down to nucleotides by specific enzymes. To continually synthesize proteins, mRNA must be constantly formed by transcription. If transcription is stopped or blocked, no more molecules of that protein will be made.

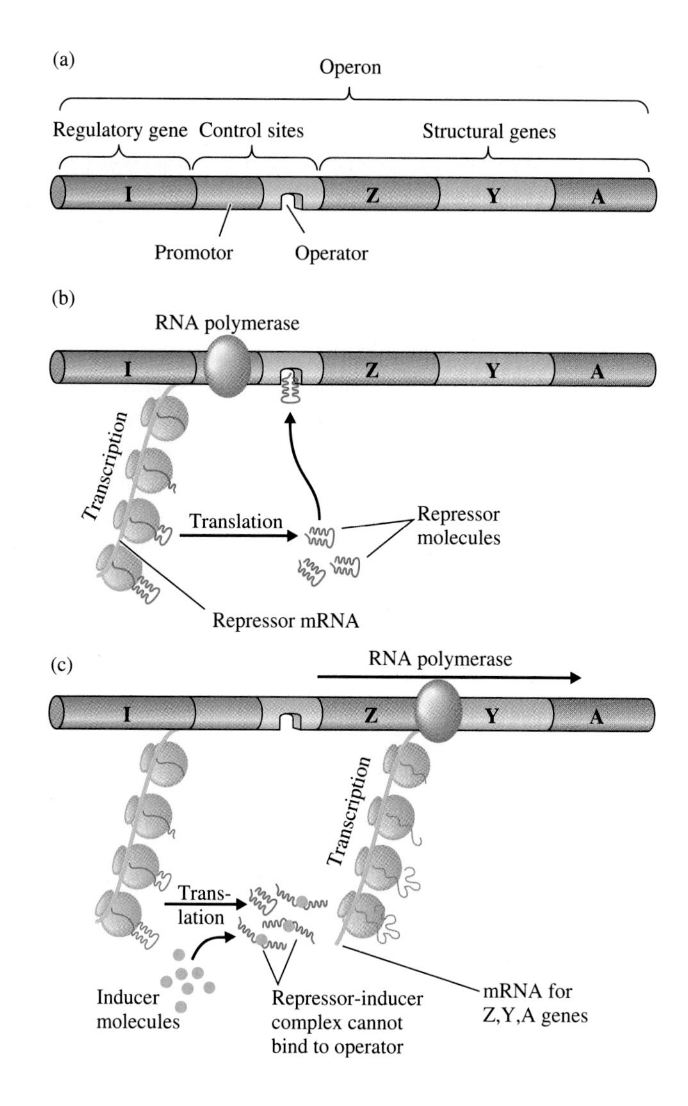

FIGURE 13.18
In some bacteria, lactose in the culture medium induces the appearance of the enzymes that break down lactose. (a) The arrangement of the genes in the lactose operon. (b) When lactose is absent, the repressor protein binds to the operator. RNA polymerase cannot transcribe the structural genes because the repressor prevents the polymerase from moving from the promotor to the structural genes. (c) When the inducer (lactose) is present, it binds to the repressor and alters its shape. The complexed repressor can no longer bind to the operator. RNA polymerase can now transcribe RNA. In this figure we have shown that the repressor blocks movement of RNA polymerase. In some other examples it blocks the binding of RNA polymerase to the promoter.

Induction by lactose works in the following way. The genes for the proteins that are needed to use lactose are adjacent to each other on the DNA molecule. In Figure 13.18(a) these genes are called structural genes and are labeled Z, Y, and A, symbols chosen for these genes when they were first discovered. In this figure, a promoter is near the Z gene. Remember, promoters are the sites where RNA polymerase binds to begin transcription. If synthesis of the lactose-utilizing proteins were not regulated, RNA polymerase would freely bind here, and transcription would occur to yield mRNA that contains information for the translation of all three genes. But the synthesis of these proteins is regulated. Next to the promoter is a region that is called the operator. The operator is involved in controlling whether transcription does or does not occur. The operator and promoter are labeled control sites in Figure 13.18. The final piece of DNA that is involved in the regulation of lactose use is the regulatory gene. This gene directs the synthesis of a repressor protein. A regulatory gene, a promoter, an operator, and the structural genes are a unit that is called an **operon**. This particular operon is the lactose operon, which provides for the regulated synthesis of lactose-utilizing enzymes.

In the bacterial cell is a *repressor* protein called the lactose repressor. This protein's function is to bind to the operator of the lactose operon (Figure 13.18b). There are other repressors that bind to the operators of the other operons. When the repressor is bound to the operator, RNA polymerase *cannot* move past the promoter, thus transcription cannot occur. The presence of the repressor prevents transcription because it physically blocks the movement of RNA polymerase. In the absence of lactose, the repressor remains on the operator.

If some lactose diffuses into the cell, some of it binds to the repressor molecule. The binding of lactose to the repressor changes the conformation of the repressor. The ability of a protein to bind to a molecule depends on the shape of both the protein and the molecule. If the shape of either changes, the ability to bind changes. In this new conformation the repressor no longer binds to the operator, so the repressor–lactose complex diffuses from the DNA. Now RNA polymerase molecules are free to go past the promoter (Figure 13.18c), thus transcription and subsequently translation occurs. The enzymes needed for utilization of lactose are synthesized.

Repression

Bacteria possess a second way to regulate protein synthesis. **Repression** is the *turning off* of protein synthesis when the proteins are not needed by the cell. As an example, consider amino acid synthesis in bacteria. Bacteria can produce all 20 of the amino acids used in protein synthesis. If they are given a diet lacking all 20, they synthesize all of them. If one of the amino acids, say, tryptophan, is provided to the bacteria, they stop synthesizing the enzymes needed for the synthesis of tryptophan. The presence of tryptophan in the diet repressed the synthesis of the tryptophan-synthesizing enzymes. Repression by tryptophan also involves an operon with an operator, a promoter, the genes for the proteins, and a repressor. The main difference is in the binding properties of the repressor. The repressor, by itself, *does not* bind to the operator, but if tryptophan is bound to it, the repressor–tryptophan complex does bind to the operator.

Consider the initial amount of tryptophan in the cell to be low because there is none in the solution surrounding the bacterium. When the concentration of tryptophan is low, it does not bind to the repressor (Figure 13.19). If tryptophan is not bound to the repressor, then it will not bind to the operator. If the repressor–tryptophan complex is not bound to the operator, then RNA polymerase can go past the operator and transcription can occur. The enzymes needed for tryptophan are made by translation of the mRNA that has been transcribed.

But if the concentration of tryptophan is high, perhaps because it was present in the environment, then tryptophan binds to the repressor. The repressor with bound tryptophan has the correct conformation needed to bind to the operator. Once the repressor–tryptophan complex is bound to the operator, RNA polymerase can no longer pass the operator, transcription no longer occurs, and the proteins are not synthesized. Tryptophan represses the synthesis of the enzymes needed for its synthesis.

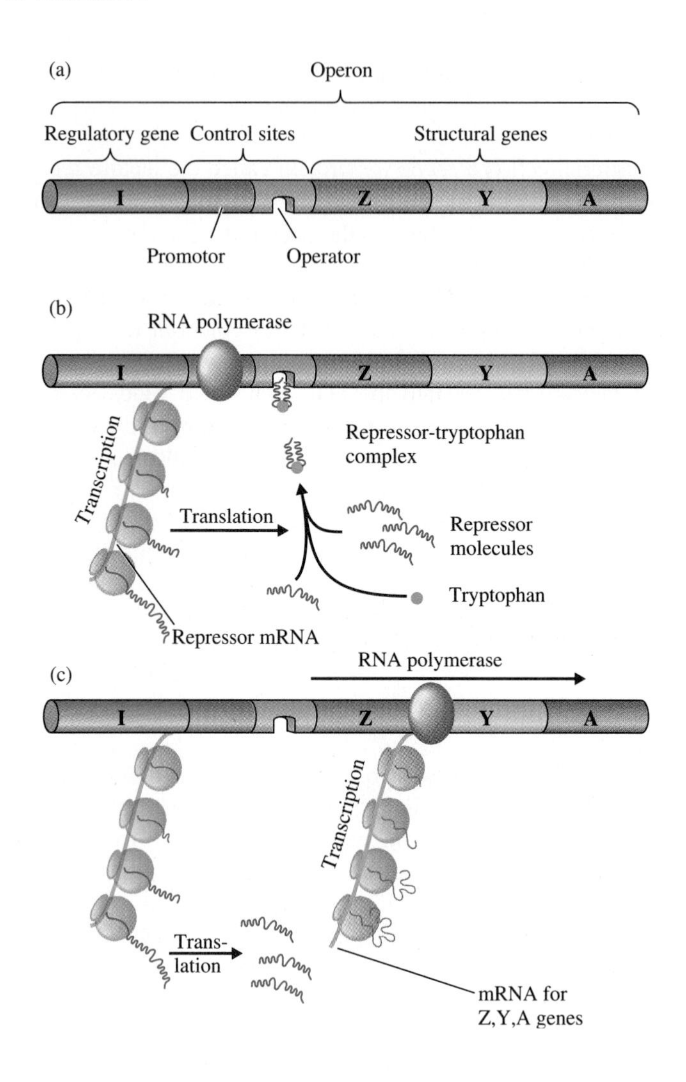

FIGURE 13.19
Tryptophan repression in bacteria. (a) The arrangement of genes in the tryptophan operon. (b) When tryptophan is present, it binds to the repressor protein and alters the shape of the repressor. When tryptophan is bound, the repressor binds to the operator. This prevents RNA polymerase from transcribing the structural genes. (c) When tryptophan is absent in the cell, the repressor cannot bind to the operator. RNA polymerase is free to transcribe the structural genes. As with induction, the complex could either prevent movement or binding of RNA polymerase.

Regulation in Higher Organisms

♦ Specific interactions between proteins and DNA are involved in regulation in eucaryotic cells.

Regulation in higher organisms is not well understood. Specific proteins may bind to DNA and influence the transcription of genes, somewhat like induction turns on bacterial genes and repression turns them off. Operons have not been found in higher organisms, but specific interactions between proteins and some portions of DNA molecules do occur. More research is needed to determine the details of regulation in humans and other higher organisms.

NEW TERMS

induction Regulation of gene expression involving the turning on of genes by the presence of a compound (inducer)

operon A set of genes in bacteria that work in concert. Genes within the operon are either turned on or off collectively.

repression Regulation of gene expression involving the turning off of genes by the presence of a compound (repressor)

TESTING YOURSELF

Regulation

1. What is the operator in an operon?
2. How does the regulatory gene contribute to regulation of lactose use?
3. Why does RNA polymerase not transcribe the structural genes constantly?

Answers **1.** the operator is the piece of DNA where the repressor binds. **2.** This gene is transcribed to yield mRNA that is translated to yield the lactose repressor protein. **3.** The repressor–operator complex physically blocks the movement of RNA polymerase to the structural genes.

13.9 Genetic Engineering

Mutations are permanent changes in the genetic material of an organism. Today techniques are available that can be used to permanently alter the DNA of organisms in very specific ways. It is now possible to add whole genes into an organism. This scientific and technical area has several names, including **genetic engineering**, *biotechnology*, and *bioengineering*.

Process of Genetic Engineering: An Example

We will consider only one example—the production of synthetic human insulin using genetic engineering techniques. Insulin is needed for treatment of patients with the genetic disease diabetes mellitus. In the past, these people were given insulins obtained from the pancreas of slaughtered animals. These insulins did not always function identically to human insulin, and medical complications resulted. It is now possible to introduce a gene for human insulin into bacteria. These genetically altered bacteria are then grown in large numbers and are able to synthesize usable quantities of human insulin. The insulin is harvested, not unlike how antibiotics are harvested from cultured microorganisms, and then processed and provided to the medical community.

Although the strategy for obtaining quantities of an unavailable human protein is straightforward, the actual process is much harder. The first complication encountered in this strategy is the phenomenon of split genes in higher organisms. The human gene for our desired protein has introns scattered throughout the gene. This is no problem for a higher cell; the introns are cut out of the transcript and the transcript is otherwise modified to a functional mRNA. But bacteria do not have the enzymes needed to process a transcript to a functional mRNA. They do not need them because the transcript in bacteria *is* a functional mRNA. Human DNA cannot just be fragmented, and pieces of it inserted into the bacterial DNA, because in the bacterium, the genes would not yield a functional mRNA for the protein.

Two alternatives can be used to get around this problem of split genes. First, the gene could be chemically synthesized using sophisticated techniques derived from simpler organic syntheses. If the amino acid sequence of the protein is

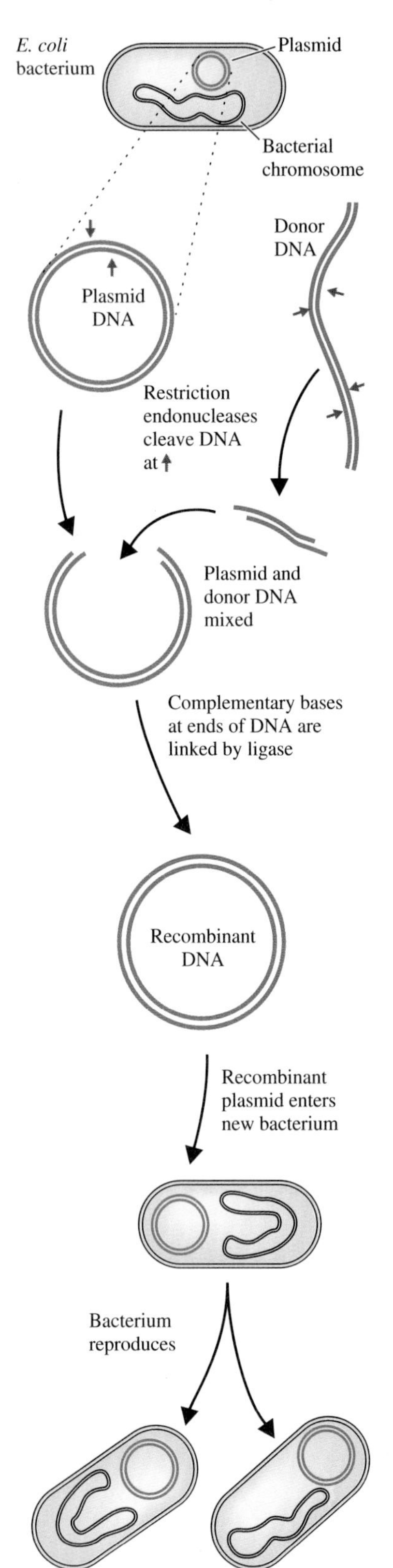

FIGURE 13.20
The insertion of a eucaryotic gene into a bacterium. A plasmid is isolated from bacteria, and the desired eucaryotic DNA is prepared. The plasmid DNA is cleaved with a restriction enzyme. These pieces are mixed, bind together through the complementary bases at the end of the pieces, and are covalently linked by the enzyme ligase. This incorporates the donor DNA into the bacterial plasmid. This plasmid is then inserted into bacteria, which reproduce. Often genes of the donor DNA are expressed in the bacteria.

known, the codons needed for the protein can be determined. A start and stop codon can be added, and perhaps a section at the beginning can be added to act as a promoter. To each end of the synthesized gene is added a short piece of several bases. These pieces are used to connect the gene to bacterial DNA (Figure 13.20).

The second method that could be used involves human mRNA for the desired protein. The mRNA is isolated, then the enzyme *reverse transcriptase* is used to make a single strand of DNA that is complementary to the mRNA. This single strand of DNA is made into double-stranded DNA by DNA polymerase to yield a DNA molecule that codes for the protein. It is not the natural gene since it has no introns. Instead, it is a copy of the expressed portion of the natural gene. It is again necessary to make modifications to the ends of the gene so it can be inserted into bacterial DNA.

The next step is to insert the gene into a smaller piece of bacterial DNA called a **plasmid**. Plasmids, like bacterial chromosomes, are circular DNA molecules, but plasmids are much smaller. The plasmid DNA is cleaved by one of several enzymes that are called *restriction endonucleases*. These enzymes are used because they cleave DNA, leaving a short piece of single-stranded DNA at each end. These short pieces are complementary to the short pieces that were added to the genes. When the copies of the gene are mixed with the cleaved plasmid DNA, the genes are inserted into some of the plasmids and held by hydrogen bonding between complementary bases. An enzyme is added to form covalent bonds between the plasmid DNA and the DNA of the inserted gene. The gene is now an integral part of the plasmid. The plasmid can then be inserted into the bacterium. If the process has been successful, the gene will be transcribed to yield mRNA that is in turn translated into functional protein.

The Future of Genetic Engineering

Bioengineering is already providing us with altered and sometimes useful biological products. It may also be used someday to alter the genes of human beings. Already numerous alterations have been made with microorganisms, and some experiments with higher species have produced organisms with portions of their body derived from one species and other parts from a second species. One example of such an organism is part chicken and part quail (Figure 13.21). Although the genetic makeup of microorganisms are readily changed now, it is not yet feasible to alter the genes of higher organisms in a similar manner. There are indications that such manipulations will be available in the future. Perhaps someday human germ cells will be genetically altered by adding a functional gene to replace a nonfunctional gene. A couple who had previously chosen to not have children because of the risk of a genetic disease, will be able to have children free of the disease. Perhaps someday people with genetic diseases will be cured by inserting a functional gene or set of genes into the individual. A phenylketonuric might metabolize phenylalanine properly, or a diabetic could produce or use adequate amounts of insulin.

As the technology progresses, we can imagine that virtually all genetic traits could be altered at the will or whim of society. The potential for altering humanity is most impressive. When and how these changes come about will be determined ultimately by you, by society, not by bioengineers. All of us should be informed citizens prepared to deal with issues from a wide variety of subjects and areas. Technology opens many doors; society chooses which doors we pass through.

♦ Biotechnology brings not only the promise of great benefits, but the challenge of choice.

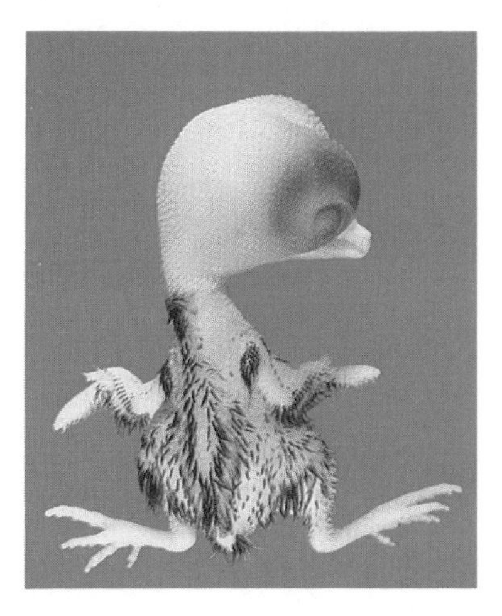

FIGURE 13.21
This chick has the head of a chicken and the body of a quail. Early in development, embryonic cells of these species were mixed and joined into one embryo. The embryo developed into a mosaic, part of which is quail-like and part of which is chicken-like.

NEW TERMS

genetic engineering The branch of technology that manipulates genetic information to produce products or organisms with permanently altered abilities

plasmid Small pieces of DNA found in some bacteria; a common vehicle for introducing new genes into bacteria

TESTING YOURSELF

Genetic Engineering

1. What type of enzyme is used to cut genes out of DNA and to cut plasmids?
2. Why are these enzymes used?
3. What role does the enzyme reverse transcriptase play in the insertion of a human gene into a bacterium?

Answers **1.** Restriction endonucleases. **2.** They cut DNA, leaving single-stranded ends, sticky ends, which permit pieces of DNA to be rejoined. **3.** This enzyme is used to make single-stranded DNA that is complementary to human mRNA.

Summary

Deoxyribonucleic acid, or *DNA*, is a polymer of nucleotides whose sequence of purine and pyrimidine bases provides information to living organisms, much like the alphabet is used to provide information for society. The bases adenine, guanine, cytosine, and thymine are combined with deoxyribose and phosphate to form the *nucleotides* that make up DNA. DNA is duplicated with high fidelity by the process of *replication*, and the process of *transcription* produces another nucleic acid, *ribonucleic acid*, or *RNA*, the carrier of information from DNA to the cell. This information is needed to synthesize, via *translation*, the proteins that carry out the functions of the cell. The activities of the cell are highly regulated, in part by controlling the rates of transcription. The genetic composition of organisms can now be altered through the process of *genetic engineering*. These potential genetic alterations promise great changes for the future.

Terms

deoxyribonucleic acid (DNA) (13.1)
ribonucleic acid (RNA) (13.1)
nucleotide (13.2)
nitrogenous base (13.2)
nucleoside (13.2)
complementary bases (13.3)
chromosomes (13.3)
semiconservative replication (13.3)
messenger RNA (mRNA) (13.4)
ribosomal RNA (rRNA) (13.4)
transfer RNA (tRNA) (13.4)
transcription (13.5)
transcript (13.5)
translation (13.6)
gene (13.6)
codon (13.6)
anticodon (13.6)
mutation (13.7)
substitution (13.7)
mutagens (13.7)
frameshift mutation (13.7)
induction (13.8)
operon (13.8)
repression (13.8)
genetic engineering (13.9)
plasmid (13.9)

Exercises

Nucleotides (Objective 1)

1. What sugars are found in nucleic acids?
2. Which nitrogenous bases are purines? Pyrimidines?
3. What is the difference between ribose and deoxyribose?
4. What is the difference between a nucleoside and a nucleotide?
5. Draw the structure of deoxyadenosine monophosphate and identify the three components of this nucleotide.
6. Draw the nucleoside consisting of uracil and ribose.
7. Name the nucleoside in Exercise 6.
8. If three phosphates were attached to the nucleoside in Exercise 6, what would it be named?
9. To what class of compounds would the compound in Exercise 8 belong?

DNA and Replication (Objectives 2 and 3)

10. The backbone of DNA is *not* composed of
    a. sugars b. bases c. phosphoric acid
11. Describe the base pairing of DNA.
12. What two types of bonding hold the two strands of DNA together?
13. Nucleotides are too big to draw easily, thus, a single-letter code can be used to indicate the base in a nucleotide or nucleic acid. What base sequence is implied by this set of letters: ATTGC? Which base is at the 3′ end of the chain?
14. What structural feature of DNA accounts for the 1:1 ratio observed between adenine and thymine or guanine and cytosine?
15. If the base sequence of a portion of one strand of DNA is known to be . . . AGGCTACGTAAGC . . . , what is the base sequence in the other strand?
16. What bonds attach the nucleotides in a strand of DNA?
17. What enzyme is responsible for the insertion of the correct nucleotide into the growing DNA strand during replication?
18. If guanine is the next base to be inserted into a growing DNA stand, what nucleotide would occupy the active site of the enzyme?
19. What is the difference between the leading strand and lagging strand in replication?
20. During replication, what enzyme ultimately links the newly synthesized pieces of DNA together?
21. Describe the process of DNA replication.
22. DNA is found in what structures in the nucleus?
23. Replication is described as semiconservative. What does this mean?

RNA and Transcription (Objective 4)

24. What are the three major types of RNA? Describe their functions.
25. What sugar is found in RNA?
26. Compare the nitrogenous bases in DNA and RNA.
27. Compare the size of DNA and RNA.
28. If samples of RNA are hydrolyzed to nucleotides, the ratios of any two of the bases are rarely one to one. Why?
29. Transcription begins at sites on DNA where RNA polymerase binds. What are these sites called?
30. What is meant by the terms *sense* and *nonsense strands*?
31. What is the primary transcript?
32. What is an intron?
33. What is an exon?

Genetic Code and Translation (Objectives 5 and 6)

34. What is a gene?
35. Why are three bases needed to determine the genetic code?
36. If a segment of DNA has the base sequence

—A—C—G—G—T—A—C—T—G—

what will be the corresponding sequence on the mRNA? What amino acids would be coded?
37. Define the terms *codon* and *anticodon*.
38. In what molecules are codons and anticodons found?
39. Why isn't the triplet T—A—C a codon for protein synthesis?
40. A classmate asks you which amino acid corresponds to the codon TTT. What is your response?
41. Write a sequence of bases in mRNA that code for the tripeptide phenylalanylproplylleucine.
42. Is the base sequence in Exercise 41 the only base sequence that codes for this tripeptide? How many base sequences are possible for it?
43. The peptide hormone *oxytocin* is formed by the pituitary gland by cutting it from a larger peptide. The amino acid sequence for oxytocin is

$$\mathrm{HOOC{-}\underset{\underset{S\text{ - - - - - - - - - - - - - - - - - - -}}{|}}{Cys}{-}Tyr{-}Ile{-}Gln{-}Asn{-}\underset{\underset{S}{|}}{Cys}{-}Pro{-}Leu{-}Gly{-}NH_2}$$

(Note the disulfide bridge between the two Cys residues.) Write codons that code for oxytocin.
44. A portion of the amino acid sequence for normal hemoglobin is

—Val—His—Leu—Thr—Pro—Glu—Glu—Lys—

The corresponding amino acid sequence for sickle cell hemoglobin (S) is

—Val—His—Leu—Thr—Pro—Val—Glu—Lys—

Write the codon change that has occurred.
45. What organelles are the site of protein synthesis?
46. What nucleic acid brings amino acids to the site of protein synthesis?
47. How can the single codon AUG be used to initiate translation and also serve as the codon for methionine within a growing polypeptide chain?
48. What occurs during the translocation step of translation?
49. What changes can occur in a protein after it is synthesized by translation?

Mutagenesis (Objective 7)

50. The synthesis of ______ is most directly affected by errors in the DNA code.
a. proteins b. lipids c. carbohydrates d. vitamins
51. DNA polymerases are said to have "proofreading" ability. What does this mean?
52. What is a mutagen?
53. Describe how substitution of bases can yield a mutation.
54. What is a frameshift mutation?
55. There is a major effort to determine the human genome. Would you expect the results to yield a single unique sequence for human DNA? Why or why not?

Regulation and Genetic Engineering (Objectives 8 and 9)

56. What name is given to the turning off of a set of genes in bacteria?
57. A graduate student wants to examine a protein involved in the synthesis of the amino acid tryptophan in a bacterium. She provides the bacterium with a complete medium including all 20 amino acids. After substantial growth has taken place, she isolates the proteins from the bacteria, but is unable to isolate any of the particular protein that she is interested in. What went wrong?
58. What is an operon?
59. A bacterium grown in a medium with lactose as the only energy source uses the sugar for energy. A bacterium grown on a medium containing glucose as the only energy source cannot immediately use lactose for energy but adapts in time to use it. Explain this phenomenon.
60. If a gene is to be introduced into a bacterium, what is often used to carry the gene?
61. How are restriction endonucleases used in genetic engineering?
62. What is the name of the enzyme that can synthesize DNA using RNA as a template?

Unclassified Exercises

63. Briefly describe how a gene on DNA becomes expressed to yield protein.
64. Where in cells are DNA and RNA found?
65. Although DNA carries the "information" or message for all processes in an organism, most of the attention is directed toward the role of DNA in protein synthesis. Discuss why this is reasonable.
66. Compare the number of possible trinucleotides from the four bases in DNA to the number of possible tripeptides from the 20 amino acids.
67. Synthetic DNA could be made for a gene if the amino acid sequence of its protein were known. Would the base sequence of the synthetic gene resemble the actual base sequence of the gene? Why or why not?
68. You have isolated a nucleic acid from a virus and have found the bases adenine, guanine, uracil, and cytosine. Which nucleic acid is present in this virus?
69. You have isolated a nucleic acid from a second virus. The bases are adenine, guanine, thymine, and cytosine, but the percentage of adenine does not equal that of thymine, nor does the percentage of guanine equal that of cytosine. Explain these curious observations.

ENZYMES

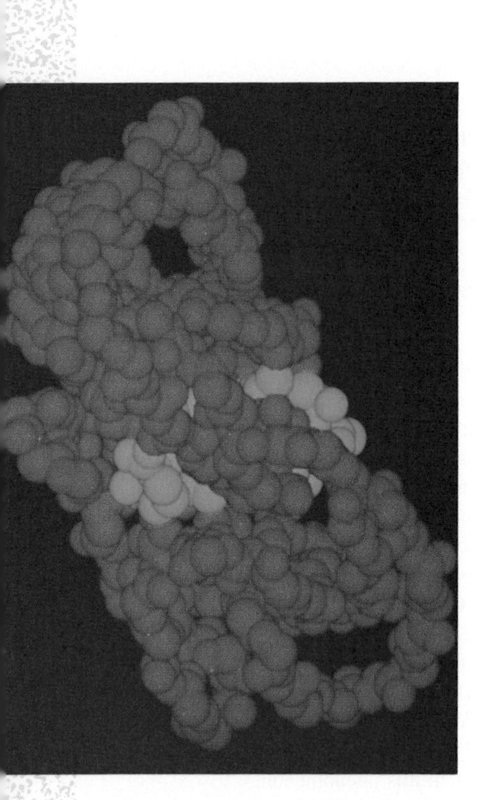

Computer modeling helps biochemists understand how enzymes work. This is a model of an experimental drug (green) inhibiting the AIDS virus by blocking replication of the HIV-1 enzyme (red).

OUTLINE

14.1 The Roles of Enzymes in Metabolism
14.2 Enzymes
14.3 Enzyme Specificity and Activity
14.4 Rates of Enzyme-Catalyzed Reactions
14.5 Regulation of Enzyme Activity

OBJECTIVES

After completing this chapter, you should be able to

1. Define the terms *metabolism, catabolism*, and *anabolism*.
2. Describe the composition and classification of enzymes.
3. Discuss how enzymes work and why they are specific.
4. List factors that influence the rates of enzymic reactions.
5. Discuss how enzyme activity is regulated.

14.1 The Roles of Enzymes in Metabolism

All of the chemical reactions that take place within the body are collectively called **metabolism**. Some of these reactions produce energy, which is used by the body in a variety of ways. These reactions make up **catabolism**, the subject of Chapter 15. Another set of reactions builds larger, more complex molecules by using energy to connect smaller molecules or pieces of molecules together. These reactions constitute **anabolism**, the topic of Chapter 16.

♦ The chemical reactions of the body make up metabolism.

The body's chemical reactions do not occur in an uncontrolled random fashion, nor do they occur completely independent of each other. Many of the reactions are linked sequentially; the product of one reaction is the reactant of another. A sequence of reactions within the body is called a **metabolic pathway**. A metabolic pathway provides a series of small chemical alterations that changes a substance into something quite different. In the pathway illustrated to the right, reactant A is converted to product B in the first reaction (R_1) of the pathway. B, in turn, is the reactant for the second reaction. For the overall process, substance E is formed from A by a sequence of four chemical reactions. Some of the metabolic pathways of the body have more than a dozen steps.

♦ 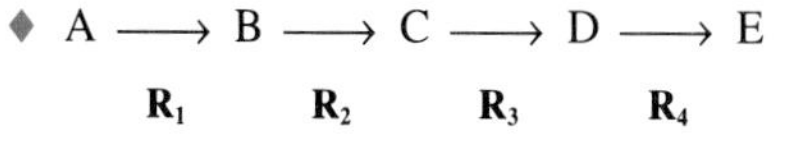

$$A \xrightarrow[R_1]{} B \xrightarrow[R_2]{} C \xrightarrow[R_3]{} D \xrightarrow[R_4]{} E$$

The chemical reactions of a metabolic pathway are catalyzed and controlled by enzymes. These reactions occur under the mild conditions that exist within a cell. For example, through many individual steps, cellular glucose is converted to carbon dioxide, water, and energy. Several dozen reactions occur during this transformation. Glucose is also a component of cellulose in wood, and when glucose burns in a campfire, carbon dioxide, water, and energy are the products. The reactions of the body occur at 37°C, but combustion occurs at temperatures that would destroy living tissues. Enzymes are special-purpose catalysts that permit the occurrence of low-temperature metabolic reactions that are essential for life.

NEW TERMS

metabolism The sum of all chemical reactions in the body that collectively sustain life

catabolism The energy-yielding part of metabolism that breaks down larger, more complex molecules into simpler ones

anabolism The part of metabolism that uses energy to build larger, move complex molecules from simpler ones

metabolic pathway A series of chemical reactions in the body that converts one substance to another distinctly different substance

14.2 Enzymes

Definition of Enzymes

The reactions of the body are catalyzed by **enzymes**. Enzymes are proteins, although recently a few RNA molecules have also been shown to possess some catalytic activity. Catalysts are defined as substances that speed up chemical reactions without themselves being changed during the reaction. Enzymes speed

◆ Catalysis by enzymes allows metabolic reactions to occur at a reasonable rate in the body.

up reactions by lowering the energy of activation of the reaction (see Figure 14.7). A catalyzed reaction requires less time to reach equilibrium and occurs at lower temperatures and under milder reaction conditions than an uncatalyzed reaction. Enzymes are some of the most efficient catalysts known. Many of the reactions in the laboratory require high temperatures, high concentrations of reagents, and a variety of solvents and reagents that would destroy a living cell or not be available to it. Enzymes permit the body's reactions to occur rapidly and efficiently at body temperature with relatively low reagent concentrations in water, the solvent of the cell. For example, one molecule of carbonic anhydrase, an enzyme of red blood cells, can catalyze the hydration of over 600,000 carbon dioxide molecules in 1 s.

Enzymes are more than just efficient catalysts, however. They are often specific for certain substances (see Section 14.3). Cells contain many hundreds of different compounds, yet only one or a few of these compounds are affected by a particular enzyme. Furthermore, the amount and activity of some enzymes varies with the conditions that exist within the cell. These changes in enzyme activity are one important way cells help regulate the body's chemical reactions and activities (see Section 14.5).

Enzyme Nomenclature

◆ The suffix *-ase* in a substance name identifies it as an enzyme.

◆ A substrate is the reactant in an enzyme-catalyzed reaction.

Although most organic molecules are named by composition and structure, proteins are too big and complex to be named this way. Enzyme nomenclature is based on the reactions catalyzed by the enzyme and attempts to provide some information about the function of the enzyme. First, the name identifies the substance as an enzyme with the suffix **-ase**. Less commonly, an enzyme is identified by an older suffix *-in*. Ure*ase*, RN*ase*, and hexokin*ase* are clearly enzymes, as are trypsi*n*, chymotryps*in*, and peps*in*. Second, the name may describe the reaction catalyzed by the enzyme. For example, an *oxid*ase catalyzes an oxidation reaction, and an *isomer*ase catalyzes an isomerization of the reacting molecule to an isomer. Third, the name may provide information about the reacting molecules, the **substrate** or substrates of the reaction. *Glucose oxid*ase catalyzes the oxidation of glucose, and *hexokin*ase catalyzes reactions involving hexoses.

Many of the common enzyme names contain one or two of the desired features of modern enzyme nomenclature but rarely all three. A more formal nomenclature has been developed by the International Enzyme Commission. This system classifies all enzymes with respect to reaction type and substrates. All enzymes are uniquely identified by a name and number code. Common enzyme names are in general use, but a formal name is usually given at least once in a research paper to identify the enzyme unambiguously. Table 14.1 lists the six major classes of enzymes.

Enzyme Composition

Enzymes are proteins that possess catalytic activity. Some enzymes require no other molecule or ion to be active. These enzymes contain only one or more polypeptide chains and are thus simple proteins. Other enzymes are not catalysts unless one or more inorganic ions or organic molecules are present. The ions and

TABLE 14.1
Major Enzyme Classes

Class of Enzyme	Reaction Type
Oxidoreductase	Oxidation–reduction
Transferase	Transfer one or more atoms from one substance to another
Hydrolase	Hydrolytic cleavage or reverse
Lyase	Cleavage or reverse, but not oxidation–reduction or hydrolysis
Isomerase	Intramolecular rearrangements
Ligase	Energy-requiring bond formation

organic molecules required for the activity of these enzymes are called **cofactors.** If the cofactor is an organic compound, it is typically referred to as a **coenzyme.** Some coenzymes are derived from vitamins.

♦ Some enzymes require cofactors for catalytic activity.

An enzyme with its cofactor is called a *holoenzyme.* An enzyme lacking its cofactor is called an *apoenzyme.* If a cofactor is permanently attached to an enzyme, it is called the *prosthetic group* of the enzyme. In other cases, the coenzymes may be attached to the enzyme only part of the time in a manner similar to the substrate.

Most enzymes have very different compositions and structures, but one group of enzymes are very similar. These are the **isozymes** (isoenzymes). Consider an example, lactate dehydrogenase (LDH). This enzyme is a tetrameric protein, but two similar subunits called H and M exist. These designations are related to the organs where the subunit is most abundant; H subunits predominate in heart muscle and M predominates in striated muscle. Although each type of subunit may be more abundant in a specific tissue, each is present, and tetramers form from any combination of the two. Thus, five tetramers designated H_4, H_3M, H_2M_2, HM_3, M_4 exist, and each tetramer catalyzes the same reversible reaction:

$$\text{pyruvate} + \text{NADH} \xrightleftharpoons{\text{LDH}} \text{lactate} + \text{NAD}^+$$

So why do different forms exist? Heart LDH (H_4) appears to be a better catalyst at oxidizing lactic acid, but the other forms may favor reduction of pyruvate. This is appropriate for each tissue's role. Energy production in the heart is principally aerobic (Chapter 15), and under these conditions lactate is consumed. In muscle, the tissues are on occasion anaerobic, and under these conditions rapid consumption of pyruvate occurs. Thus, each organ possesses an LDH that is best suited for the metabolic activity found in that organ.

NEW TERMS

enzyme A protein that has specific catalytic activity

-ase A suffix used to indicate that the substance is an enzyme

substrate The reacting molecule that binds to an enzyme. The enzyme catalyzes its conversion to product

CHEMISTRY CAPSULE

Minerals and Vitamins in Metabolism

Macronutrients, including carbohydrates, proteins, and lipids, provide energy and building blocks for the body. Minerals and vitamins serve a different role—they provide a source of cofactors for catalytic activity. For example, the digestive enzyme carboxypeptidase requires zinc ion, and many enzymes require a coenzyme called NAD^+ (nicotinamide adenine dinucleotide). The minerals in our diet provide inorganic cofactors that enzymes require, and many of the vitamins are used to make a variety of coenzymes needed by the body. The table below provides the names and functions of a few key coenzymes along with the vitamins from which they are formed. Many coenzymes possess nucleotides or nitrogenous bases. In the representative coenzymes shown here, the bases (blue), ribose sugars (orange), and phosphates (yellow) are highlighted.

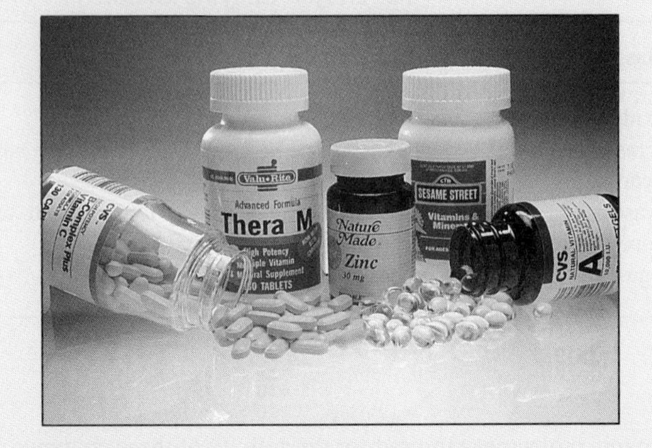

FIGURE 14.A
Vitamins are organic molecules that cannot be made by the body. They are, however, readily available in most foods and as vitamin supplements.

Table 14.A

Coenzymes, Their Functions, and Vitamins from Which They Are Derived

Coenzyme	Function	Vitamin
Flavin adenine dinucleotide (FAD)	Redox reactions	Riboflavin
Nicotinamide adenine dinucleotide (NAD^+)	Redox reactions	Niacin
Coenzyme A	Carrier of acyl groups	Pantothenic acid
Thiamine pyrophosphate (and other forms)	Forms covalent intermediates during catalysis	Thiamine
Pyridoxal phosphate (and other forms)	Forms covalent intermediates during catalysis	B_6
Tetrahydrofolate (and other forms)	Transfers groups containing one carbon atom	Folic acid
Cobalamines	Several reaction types	B_{12}
Biotin	Carboxylations	Biotin

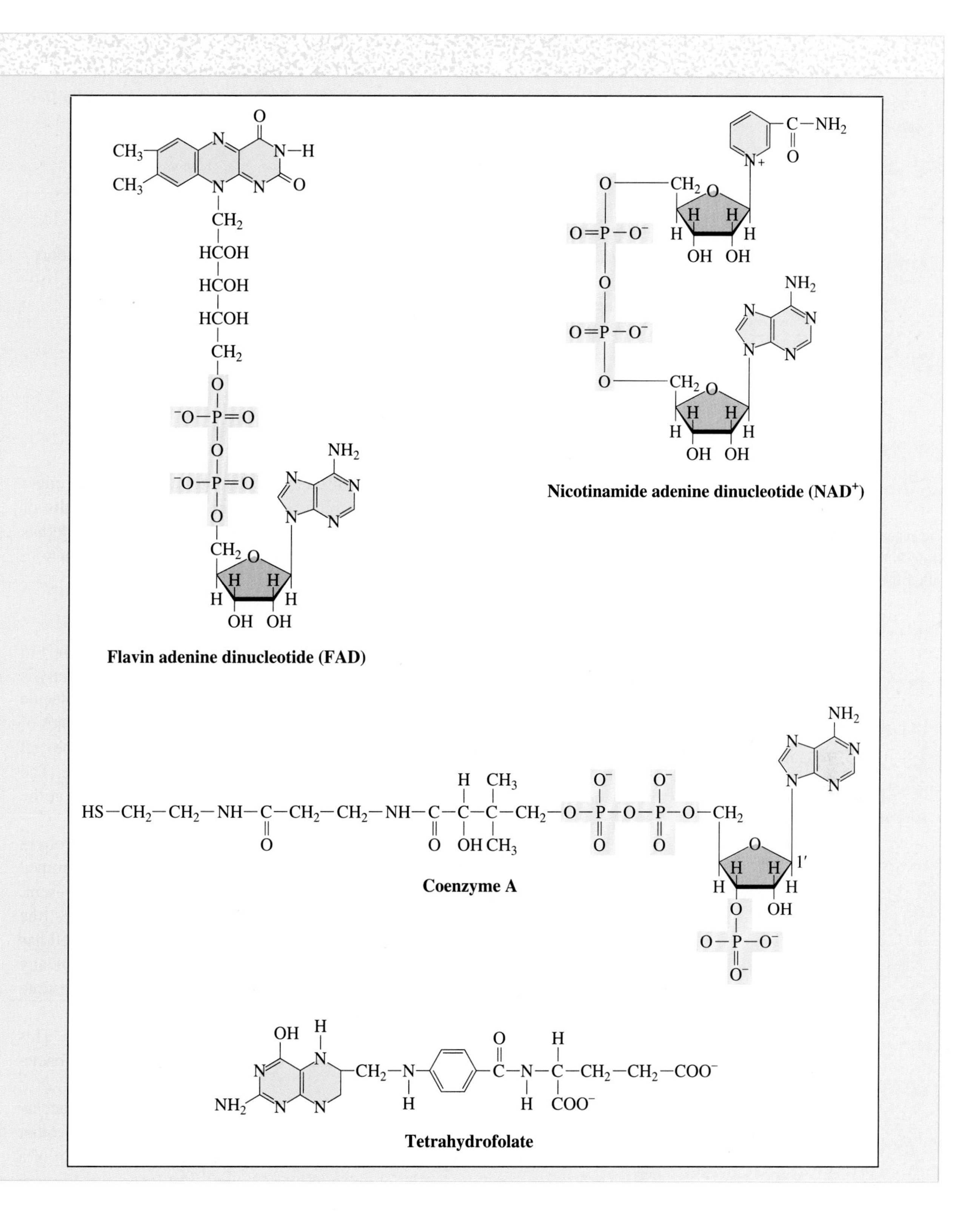

Flavin adenine dinucleotide (FAD)
Nicotinamide adenine dinucleotide (NAD+)
Coenzyme A
Tetrahydrofolate

cofactor The non-amino-acid portion of some enzymes

coenzyme The name given to organic cofactors

isozymes Structurally similar enzymes with similar catalytic activity and different metabolic roles

TESTING YOURSELF

Enzymes

1. What name is given to that portion of metabolism that yields useful energy?
2. Use enzyme nomenclature to predict the function of triose phosphate isomerase.
3. What name is given to the non-amino-acid part of an enzyme?

Answers **1.** catabolism **2.** The enzyme isomerizes triose phosphates. **3.** cofactor

14.3 Enzyme Specificity and Activity

♦ The place on an enzyme where substrate binds and catalysis occurs is called the *active site*.

Enzymes catalyze the conversion of specific substrate molecules to product molecules. The substrate or substrates bind noncovalently to the **active site** of the enzyme, where they react to yield a product or products. The active site has two distinct functions: the binding of substrate and the catalysis of the reaction.

Enzyme Specificity

Why does the active site bind only one or a few of the many compounds in the cell? Why do enzymes show specificity for substrates? Enzyme specificity is determined by fit, that is, proper alignment of the enzyme and the small molecule (see Figure 9.11), and specific noncovalent bonding between the active site of the enzyme and portions of the substrate molecules. Other molecules in the cell do not fit nor form similar bonds, thus they do not bind to the active site. The size and shape of the active site and the specific chemical groups located at the active site contribute to binding specificity.

The active sites of enzymes are on the surfaces and in pockets and grooves of the enzymes (Figure 14.1). The size and shape of the active site are determined by the amino acid residues at the active site and the prosthetic group, if present. Because each enzyme has different amino acids with different orientations, the active site of each enzyme is different. Similarly, each compound in the cell has its own size and shape, so only some of them fit into the active site of any particular enzyme. The size and shape of an active site influences substrate binding.

Consider the enzyme lysozyme shown in Figures 14.1(b) and 14.2. This enzyme catalyzes the hydrolysis of glycosidic bonds in polysaccharides of bacterial cell walls.

The long cleft of this enzyme's active site accommodates the long polysaccharide molecule quite nicely (Figure 14.2), but if you think about it, many other molecules are also linear like a polysaccharide. A polypeptide or the chain of a fatty acid of a triacylglycerol might also fit into the cleft. But lysozyme does not catalyze the hydrolysis of these molecules; it is specific for its substrate.

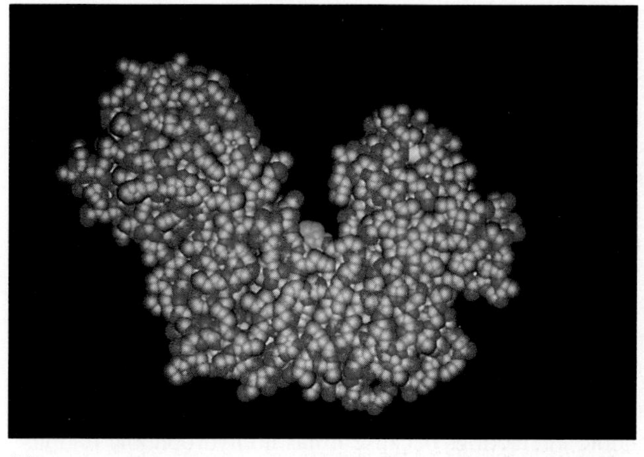

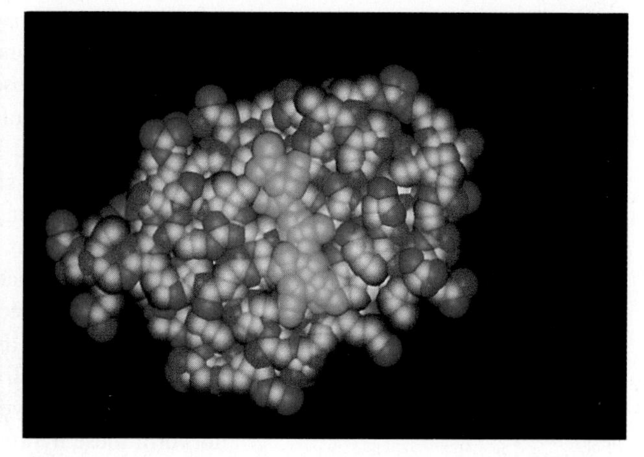

(a) (b)

FIGURE 14.1
The active site of two enzymes. (a) The active site of hexokinase is a pocket that can accommodate the substrate hexose (green) and ATP. (b) The active site of lysozyme is a long cleft that can accommodate the polysaccharide chain (green) that is its substrate.

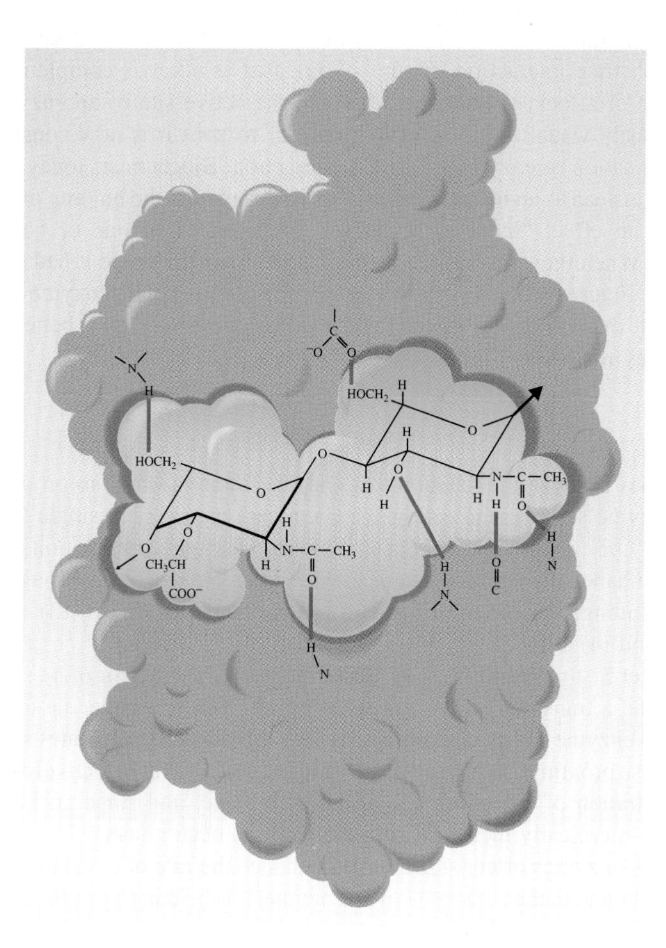

FIGURE 14.2
Binding of a polysaccharide of a bacterial cell wall to the active site of lysozyme. The substrate and active site are complementary in shape, with the long substrate molecule (green) fitting into a cleft in the enzyme. Noncovalent interactions (red), primarily hydrogen bonds, form between groups on the substrate and enzyme to hold the substrate in the active site.

The size and shape of an active site contribute to proper substrate binding, but additional factors are necessary to explain specificity of binding. Look at Figure 14.2. This close view of binding between lysozyme and its substrate shows specific hydrogen bonds between them. Not only are the sizes and shapes of the substrate and active site complementary, but the orientation of groups on the substrate and active site permit hydrogen bonding between them. Fatty acids and polypeptides may fit, but they do not bond to the active site.

The interactions that occur between an enzyme and its substrate are the same interactions that are responsible for tertiary structure in proteins: hydrogen bonding, hydrophobic interactions, and ionic and polar interactions. A lipid binds to the active site of an enzyme because the sizes and shapes are complementary and because the active site of the enzyme is lined with nonpolar amino acid residues that form hydrophobic interactions with the lipid. A polysaccharide could not form these hydrophobic interactions because it has no hydrophobic regions. The digestive enzyme trypsin cleaves peptide bonds in proteins at lysine or arginine residues. The active site of trypsin can accommodate the size and shape of the polypeptide chain and these amino acid residues, but in addition, the active site has a negative charge that forms a strong ionic bond to the positive charge of either lysine or arginine. No other amino acid residue in a polypeptide has the correct shape and charge to bind to the active site. No other peptide bond is normally cleaved by trypsin.

♦ A substrate binds to an enzyme active site because they are complementary in size, shape, and potential for binding to each other.

This description of the active site corresponds to the **lock-and-key model** of enzyme–substrate interaction (Figure 14.3a). Just as a key is complementary to the keyhole of a lock, the substrate fits into the active site of an enzyme. This model is easily visualized, but for a variety of reasons it is now considered an oversimplification of enzyme–substrate interaction. Biochemists today view this interaction instead as an **induced-fit model**. In this model, the binding of substrate to enzyme in effect "induces" a change in the shape of one or both of the molecules. When they are bound together, neither has the shape it had when free in solution (Figure 14.3b). Either model can be used to explain enzyme–substrate binding, but the induced-fit model helps explain enzyme activity better than the lock-and-key model.

♦ The induced fit model and lock-and-key model both explain substrate–enzyme binding.

Enzyme Activity

Enzymes catalyze the reaction that converts their substrate to product. How can the enzyme's active site catalyze a chemical reaction? No single factor can explain enzyme catalysis; any given enzyme may use any combination of several factors to enhance the reaction rate. Enzymes catalyze reactions through several factors including; (1) proximity effects, (2) orientation effects, (3) acid–base catalysis, and (4) strain.

Consider first proximity effects. Some enzymes have two or more substrates. For example, a transferase transfers some group from one molecule to another. One way an enzyme can enhance catalysis is by binding both substrates simultaneously in close proximity at the active site (Figure 14.4a). This effectively increases the concentration of the substrates at the active site, and since an increase in concentration typically increases rate, the reaction occurs faster.

Second, enzymes orient substrates to enhance the rate of a reaction. When a substrate is bound to the active site of an enzyme, it is held in a specific orientation

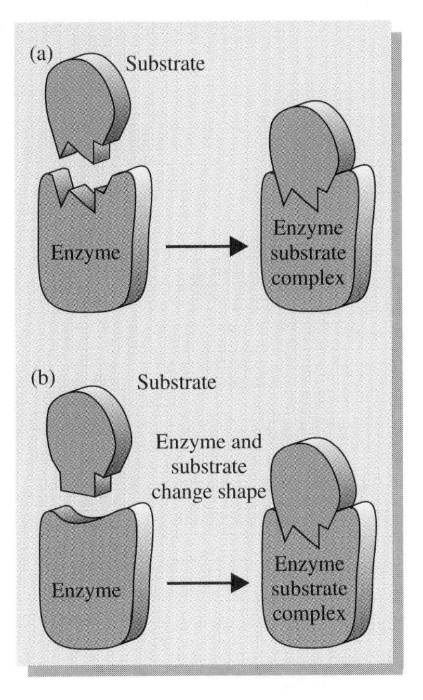

FIGURE 14.3
Binding of substrate to enzyme.
(a) The lock-and-key model describes binding sites on enzymes as complementary to the substrate; the substrate fits into a preexisting site. (b) The induced-fit model states that the binding of the substrate to the enzyme causes changes in shape that make the substrate and binding site complementary.

with respect to another substrate molecule or to amino acid side chains in the active site (Figure 14.4b). The atoms of the substrate are in the proper orientation for fast reaction with the other substrate or the group on the enzyme.

Third, enzymes can act as acids and bases during catalysis. This speeds up the rate of a reaction just as acids and bases catalyze many reactions in general and organic chemistry. Recall that certain amino acid side chains may be acidic or basic (Chapter 12)—they can donate or accept hydrogen ions. They can contribute to catalysis in an enzyme by acting as acids or bases (Figure 14.5). The acidic and basic groups at an active site have one enormous advantage over acids and bases in solution—they are properly positioned to quickly give up or accept protons. This orientation allows acid–base catalysis to proceed rapidly and efficiently.

A fourth factor influencing the rate of enzyme reactions is substrate strain. In the induced-fit model of substrate–enzyme binding, the enzyme or substrate (or both) undergoes conformational changes when they bind. When a substrate is free in solution, it is generally in the lowest energy (or most stable) conformation. But when it binds to the enzyme, the substrate is forced into a higher energy conformation. This strained conformation of the substrate is more reactive than a free substrate molecule because it more closely resembles the transition state of the reaction, the high-energy form between substrate and product (Figure 14.6). Because it is higher in energy, it reacts more quickly. Upon binding, many enzymes turn ordinary substrate molecules into strained, reactive species.

The strain induced in the substrate upon binding to an enzyme in effect lowers the energy of activation of a reaction. This energy is the difference in energy between the substrate and the highest energy form as the substrate changes to product (Figure 14.7). Binding of the substrate to an enzyme results in a change in the substrate to a form that more nearly has the same energy as the highest energy form. Because they are so similar, even the small amounts of energy available at the temperatures within a cell are enough to get the substrate beyond this energy barrier to form the product.

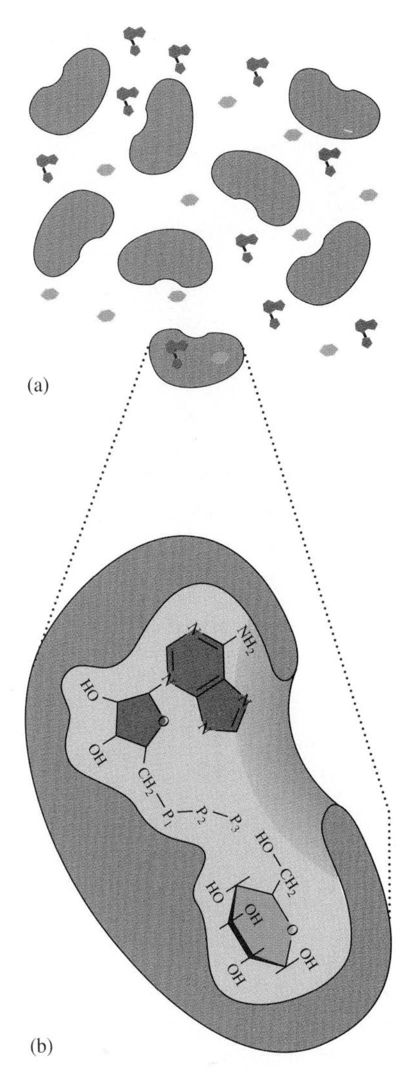

FIGURE 14.4
The binding of substrates to an enzyme can result in two effects.
(a) Proximity effect. Because these two hypothetical substrates bind to the enzyme simultaneously, they are concentrated at the active site. The chance of them encountering and reacting with each other at the active site is much higher than in solution. (b) Orientation effect. ATP and a hexose are used to illustrate orientation in a hypothetical enzyme. The third phosphate of ATP is well aligned for transfer to the 6′ hydroxyl group of hexose.

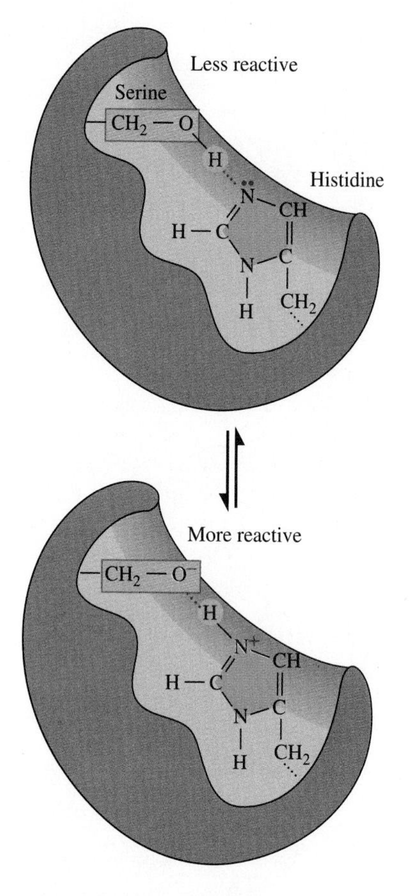

FIGURE 14.5
Some of the side chains of amino acids in an enzyme can act as acids and bases that speed up reactions. In this example, histidine, a base shown in blue at the top of the diagram, accepts a hydrogen ion (yellow) from serine, a weak acid shown in red. The serine on the bottom is in the anionic form, and is much more reactive. It reacts more rapidly with substrate than does the —OH form of serine.

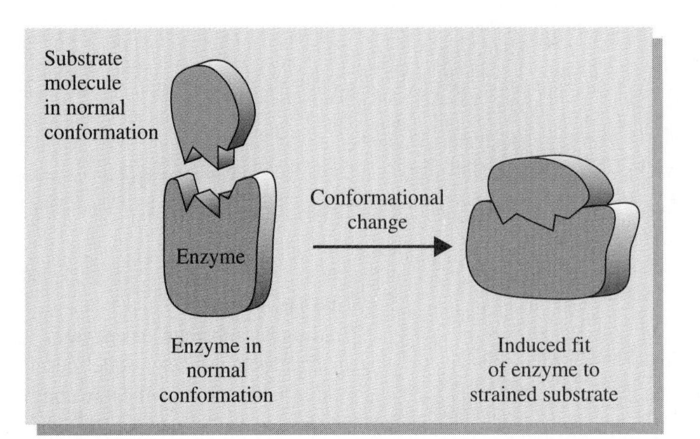

FIGURE 14.6
According to the induced-fit hypothesis, the binding of substrate to enzyme results in conformational changes in these molecules. The substrate in the strained conformation reacts more easily to yield a product than does the unstrained substrate molecule.

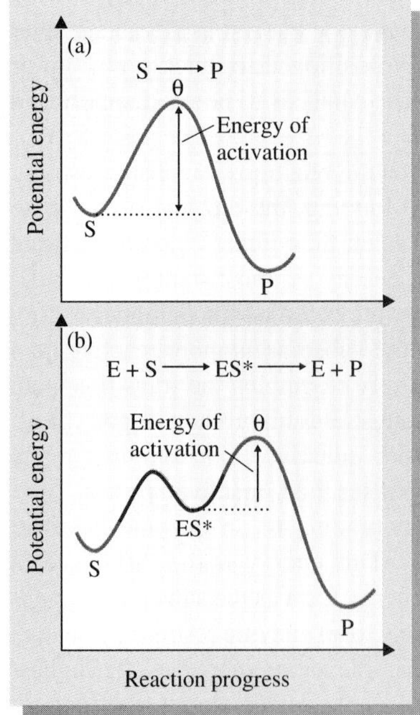

FIGURE 14.7
The energy of activation for a reaction in (a) the absence and (b) the presence of an enzyme that induces strain in the substrate. The energy of activation for enzyme-catalyzed reactions is always lower than for the uncatalyzed reaction, but the difference is not always due to strain in the substrate. (a) As S is converted to P, it passes through a high energy state θ. The difference between this state and S is the energy of activation. (b) When S binds to E, conformational changes convert it to a higher-energy, less-stable form, S*. The difference in energy between the highest-energy form θ and S* is less than the difference shown in part (a).

NEW TERMS

active site The site on an enzyme that binds substrate and catalyzes the reaction to yield product

lock-and-key model Model for substrate–enzyme interaction that states that the two molecules are complementary to each other before binding and fit together like a lock and key

induced-fit model Model for substrate–enzyme interaction that states that the binding of substrate to enzyme causes a change in the shapes of one or both of the molecules. When bound to each other, the two are complementary.

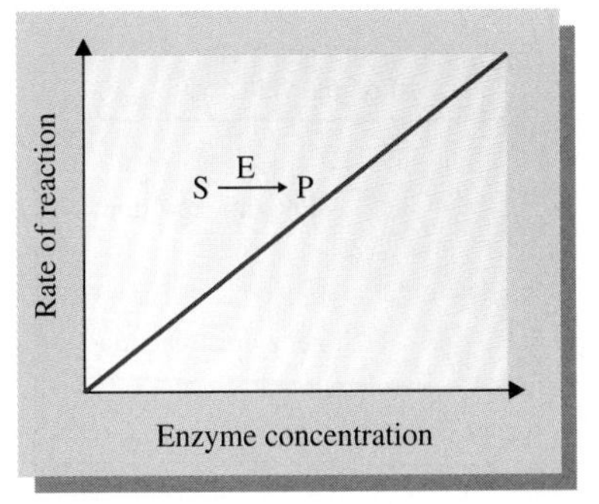

FIGURE 14.8
The effect of increasing concentrations of enzyme on the rate of a reaction. If the enzyme concentration doubles, the probability of substrate binding to an enzyme molecule doubles and the rate of formation of product doubles.

TESTING YOURSELF

Enzyme Activity

1. What noncovalent interactions bind substrates to enzymes?
2. List four ways enzymes catalyze reactions.

Answers **1.** hydrogen bonding, ionic interactions, and hydrophobic interactions **2.** proximity effects, orientation effects, acid–base catalysis, and strain

14.4 Rates of Enzyme-Catalyzed Reactions

Reaction Velocities

Enzyme-catalyzed reactions are always much faster than uncatalyzed reactions occurring under otherwise identical conditions. Whether enzyme-catalyzed or not, **reaction rates**, or **velocities**, are typically expressed in terms of the amount of reactant (substrate) consumed per unit time or the amount of product formed per unit time. The velocity v, of enzyme-catalyzed reactions is affected by a variety of factors, including the concentration of the enzyme. Consider some enzyme, designated as E, with its substrate S. The rate at which S is converted to product P is directly dependent on the amount of enzyme present (Figure 14.8). As the number of molecules (concentration) of the enzyme increases, the rate of the reaction increases. For example, if 4 molecules of E convert 100 molecules of S to product P in 1 s, 8 molecules of E would convert 200 molecules of S. The reaction rate doubles. If all other conditions are kept the same, the rate of the reaction is directly proportional to the amount of enzyme. Living organisms can increase the rate of some reactions by simply making more of the enzyme needed for the reaction (see Section 14.5).

◆ For enzyme-catalyzed reactions, biochemists traditionally, but incorrectly, use the term *velocity* for rate.

The rate of enzyme-catalyzed reactions also varies with the concentration of the substrate. But the rate varies with substrate concentration in a way that is substantially different from "ordinary" chemical reactions. In many nonenzymic reactions, the rate is directly proportional to the concentration of the reactant. Doubling the concentration of the reactant doubles the rate of the reaction (Figure 14.9). Doubling the concentration of S in an enzyme-catalyzed reaction appears to double the rate if the concentrations of S are rather small, but doubling relatively

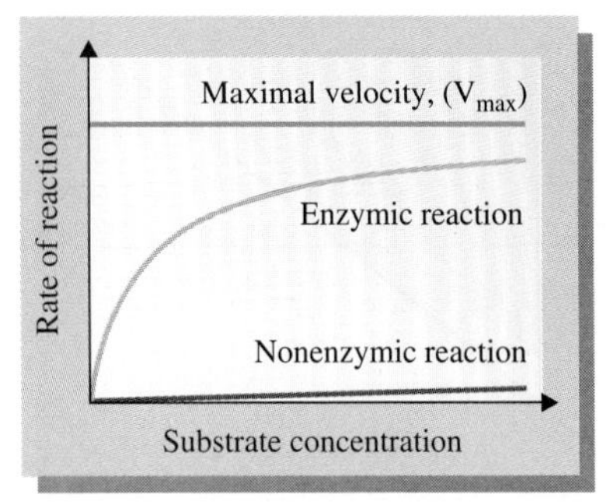

FIGURE 14.9
The effects of substrate concentration on the rate of an uncatalyzed reaction (purple) and an enzyme-catalyzed reaction (yellow). (The reaction rates are not drawn to scale; the actual rate of the enzyme-catalyzed reaction is much larger.) With increasing substrate concentration, the velocity of an enzyme reaction approaches a maximal velocity, V_{max}, (green). As substrate concentration increases, the enzyme becomes more nearly saturated with substrate.

large concentrations of S actually yields only a slight increase in reaction rate (Figure 14.9). Rather than the usual linear graph, a hyperbolic graph is observed. The velocity of the reaction with ever-increasing substrate concentration approaches a maximal velocity designated V_{max}.

The relationship between rate and substrate concentration is explained by the formation of an enzyme–substrate complex. In turn, this complex may break down to yield free E and S or product P and free E.

$$E + S \rightleftharpoons ES \longrightarrow E + P$$

At low concentrations of S, a substrate molecule is likely to collide with an unoccupied E molecule to form ES. Doubling S increases the probability of collision by 2. But at higher concentrations of S, many of the enzyme molecules already have S bound to them. Many of the collisions of S with enzyme molecules involve unproductive collisions with ES rather than with E. Because the number of enzyme molecules is normally small, at high concentrations of S most of the E molecules already have S bound to them; they exist as ES, not E. Doubling the concentration of S increases the amount of ES and decreases the amount of E. Ever-increasing concentrations of S yield ever smaller increases in rate, which produces the hyperbolic graph shown in Figure 14.9.

♦ When all enzyme molecules have substrate bound to them, increasing the concentration of substrate has no effect on the reaction rate.

Effects of Temperature on Enzyme-Catalyzed Reactions

The effects of temperature on an enzyme-catalyzed reaction are similar to the effects on other reactions (up to a point). The rates for most chemical reactions increase regularly with increasing temperature (Figure 14.10a). This increase in

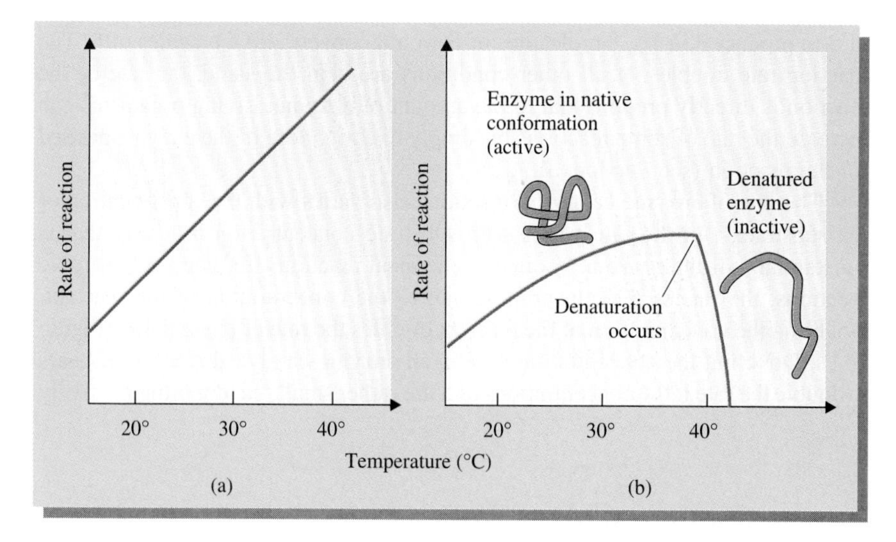

FIGURE 14.10
The effects of temperature on the rate of (a) uncatalyzed and (b) enzyme-catalyzed reactions. (Rates not to scale.) (a) Near room temperature or body temperature, a 10°C rise in temperature roughly doubles the rate of a reaction. (b) The rates of enzyme-catalyzed reactions also increase with temperature until the enzyme denatures. Above this temperature, the rate drops.

CHEMISTRY CAPSULE

Effects of Temperature on Body Function

Humans are homeothermic organisms, that is, we maintain a more or less constant internal temperature. Occasionally, environmental stress or disease causes body temperature to deviate significantly from this normal range. Consider the effects of reduced body temperature. As temperature decreases, life-sustaining biochemical reactions within the body proceed more slowly. For example, normal brain function is dependent on normal metabolic rates in the cells of the central nervous system. As body core temperature decreases, as it does in *hypothermia*, brain chemistry and function slow down. Disorientation, unconsciousness, and death may follow. Conversely, if core temperature increases beyond the normal range, as it can during prolonged high fever or during excessive exposure to very hot weather, some proteins in the body denature. Loss of certain critical enzymes in the central nervous system results in dysfunction and possibly death.

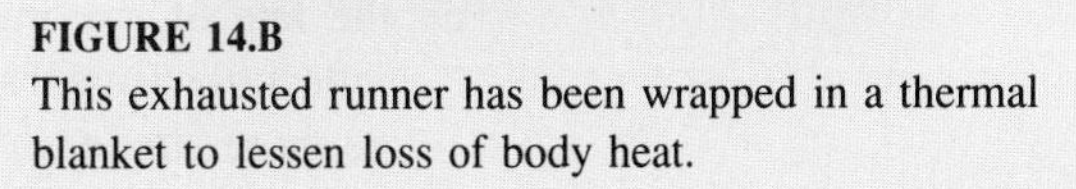

FIGURE 14.B
This exhausted runner has been wrapped in a thermal blanket to lessen loss of body heat.

the rate of a reaction reflects the higher speed and increased energy of molecules. Collisions between reactants are more frequent, and a larger proportion of these collisions possess enough energy to break old bonds and make new ones. The rate of an enzyme-catalyzed reaction also increases with temperature, but the stability of the enzyme is also an important factor. At lower temperatures, enzymes are stable and the rate of the reaction increases with temperature. But there is some upper temperature limit beyond which the protein is unstable, that is, it is *denatured*; it loses its normal shape and therefore its normal function as a catalyst, and the reaction rate drops drastically (Figure 14.10b).

Thermophilic bacteria live in these hot springs. These bacteria have proteins that remain in their native conformation well above 75°C.

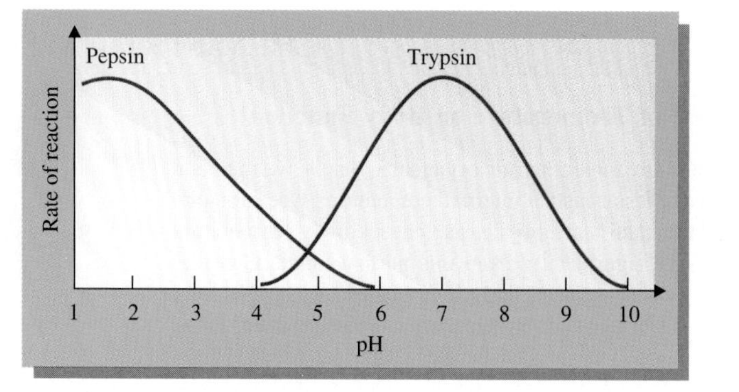

FIGURE 14.11
The effects of pH on enzyme activity. Pepsin, a digestive enzyme of the stomach, is most active under acidic conditions like those found in the stomach. Trypsin, a digestive enzyme of the small intestine, is most active at neutral to slightly alkaline pH, the pH range of the small intestine.

Effects of pH on Enzyme-Catalyzed Reactions

Changes in hydronium ion concentration (that is, changes in pH) also alter the rate of enzyme-catalyzed reactions (Figure 14.11). The effect may simply reflect the denaturation of the protein above or below a stable pH range, or it may be more subtle. In our discussion of enzyme activity in Section 14.3, the acid–base properties of some residues were listed as factors that contribute to catalysis in some enzymes. These groups accept and donate protons during the catalytic cycle. If the pH is too high or too low, then these groups will be unprotonated or protonated too much of the time. The enzyme will function much less efficiently. It is only within a certain range of pH that the catalytic groups of the enzyme are protonated an appropriate amount of time.

♦ Changing the pH results in the protonation or deprotonation of side chains in certain amino acids in an enzyme.

Changes in pH can also alter the substrate. As the pH changes, a group in the substrate may be protonated or deprotonated. The altered form of the substrate may not bind to the enzyme or may react more slowly.

Inhibition of Enzymes

The rate of enzyme-catalyzed reactions can be decreased or inhibited by a group of substances called *inhibitors*. Inhibitors bind to enzymes and alter the ability of the enzyme to carry out catalysis. The inhibition may be permanent, that is *irreversible*, or it may be *reversible* if activity returns when the inhibitor is removed.

♦ Inhibitors bind to enzymes and reduce their activity.

Irreversible inhibition occurs whenever an **irreversible inhibitor**, represented by molecule I, binds permanently to an enzyme, E, and the resulting association has little or no activity.

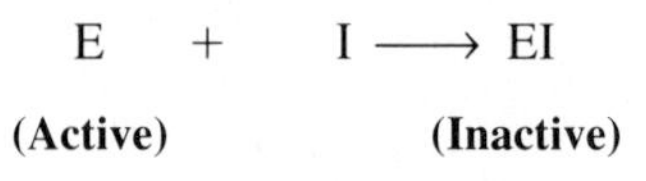

$$\underset{\textbf{(Active)}}{\mathrm{E}} + \mathrm{I} \longrightarrow \underset{\textbf{(Inactive)}}{\mathrm{EI}}$$

The association between the inhibitor and the enzyme might be a covalent bond or a very strong noncovalent interaction, but once it is formed, the complex remains. The enzyme molecule still exists, but without its catalytic activity it makes no contribution to the cell. The apparent concentration of the enzyme has been reduced.

A group of organophosphates provides an example of irreversible inhibition. This group of compounds includes the nerve gases and certain organophosphate

TABLE 14.2
Nerve Gases and Insecticides that Inactivate Acetylcholinesterase

Compound	Structure	Use	Toxicity
Tabun	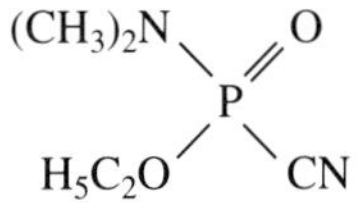 $(CH_3)_2N$ — P(=O) — CN, with H_5C_2O on P	Nerve gas	Highly toxic
Sarin	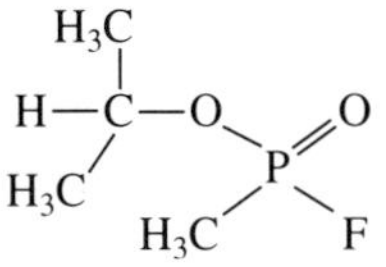 H_3C, H_3C — $H-C-O-$P(=O)(F)(H_3C)	Nerve gas	Highly toxic
Malathion	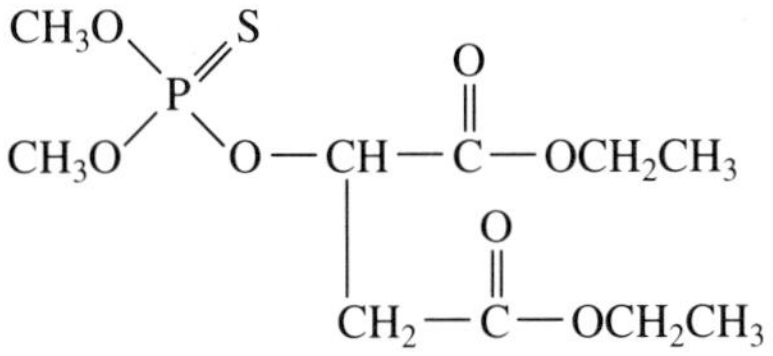 $(CH_3O)_2$P(=S)$-O-CH-C(=O)-OCH_2CH_3$ with $CH_2-C(=O)-OCH_2CH_3$ on CH	Insecticide	Toxic but can be handled safely with proper precautions
Parathion	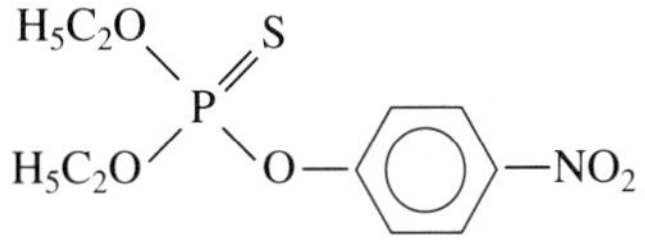 $(H_5C_2O)_2$P(=S)$-O-C_6H_4-NO_2$	Insecticide	Toxic but can be handled safely with proper precautions

insecticides, such as malathion (Table 14.2). They share a common structural feature—each of them possesses a phosphate group that reacts readily with the hydroxyl group of a serine in an enzyme called *acetylcholinesterase* (Figure 14.12). This reaction yields a stable phosphoester bond. Once formed, the complex remains. Because this serine is part of the active site, and is therefore essential for the catalytic activity of this enzyme, the enzyme molecule loses activity once the inhibitor is irreversibly bound.

In some of the junctions of the nervous system, and at neuromuscular junctions, nerve impulses cause the release of acetylcholine into the synapse. In the synapse, acetylcholine stimulates an electric impulse in a second nerve or in a muscle. Normally acetylcholinesterase hydrolyzes acetylcholine in the synaptic junctions, preparing the synapse to respond to future nerve impulses. If acetylcholine accumulates, however, as it will if enough acetylcholinesterase molecules are inhibited, the muscle or nerve will show increased activity even though the normal stimulus is absent. The result is abnormal nerve transmission and muscle contraction, loss of normal function, and often death.

A second type of inhibition is caused by **reversible inhibitors**. Like irreversible inhibitors, these substances also bind to an enzyme molecule and reduce or eliminate enzymic activity. But these substances bind reversibly to the enzyme; they remain associated with the enzyme for a time, then diffuse away. Since a permanent association is not formed, the inhibitor can be removed and activity restored.

$$\underset{\textbf{(Active)}}{E} + I \rightleftharpoons \underset{\textbf{(Inactive)}}{EI}$$

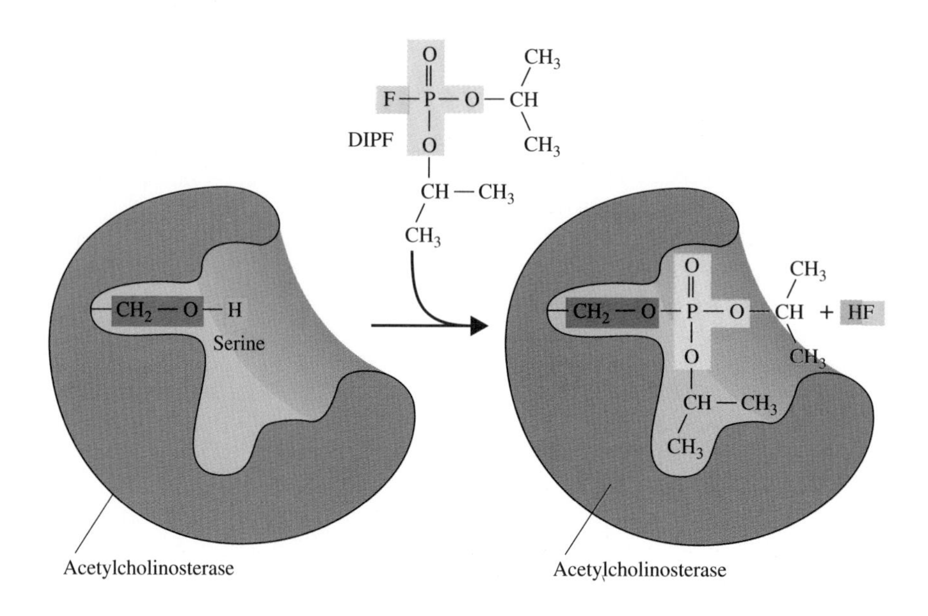

FIGURE 14.12
The reaction of diisopropylphosphofluoridate (DIPF) with the catalytic serine of acetylcholinesterase. The bond between the oxygen of serine (red) and the phosphate (yellow) of DIPF is covalent and strong. Once formed, it remains. Without this free serine, the enzyme molecule is inactive.

♦ Competitive inhibitors compete with substrate for the active site of an enzyme.

Reversible inhibitors may bond to the active site of the enzyme or elsewhere on the enzyme surface. If binding occurs at the active site, the substance is a **competitive inhibitor**, so called because it competes with the substrate for the active site (Figure 14.13). While I is bound to the active site, S cannot bind. For that period, that enzyme molecule is catalytically inactive. A reversible inhibitor reduces the catalytic activity of an enzyme.

The structures of competitive inhibitors resemble the substrate structures. These inhibitors can form many or all of the noncovalent bonds that a substrate forms with the enzyme molecule. But competitive inhibitors, unlike substrate, cannot react to yield a product. Competitive inhibitors bind to the active site, but no reaction occurs.

Sometimes competitive inhibitors of enzymes can be used to advantage. Consider the antitumor agent, methotrexate. This compound is structurally similar to dihydrofolate, a coenzyme involved in the synthesis of thymine (dTMP).

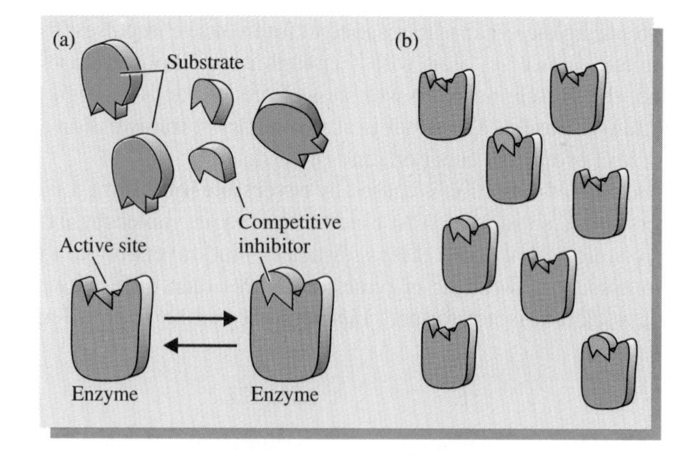

FIGURE 14.13
The reversible binding of a competitive inhibitor to the active site of an enzyme. (a) A competitive inhibitor binds at the active site and prevents binding of substrate. While it is bound, the enzyme is inactive. (b) A competitive inhibitor effectively reduces the concentration of active enzyme by temporarily inactivating some of the enzyme molecules.

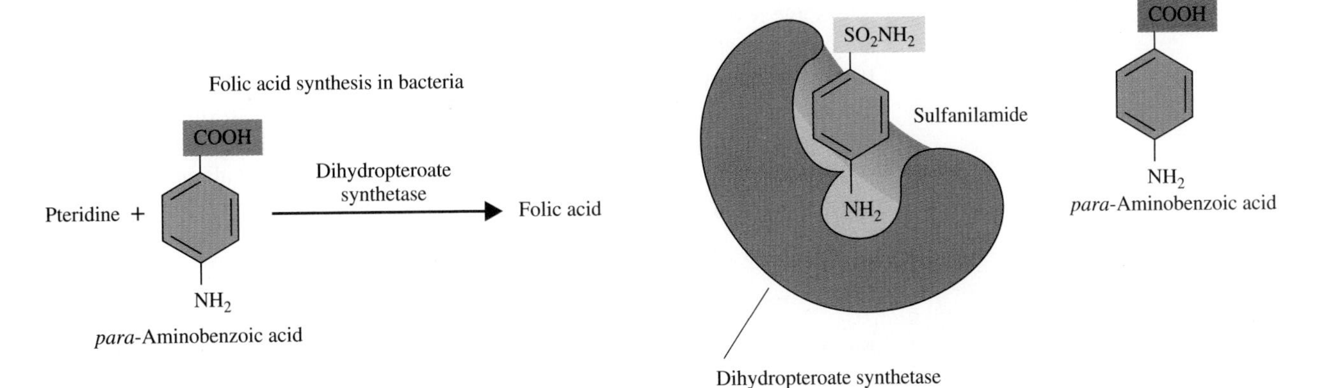

FIGURE 14.14
The effects of sulfanilamide on folic acid synthesis. Sulfanilamide competes with *p*-aminobenzoic acid for the active site of this enzyme. This antimicrobial drug retards bacterial growth because folic acid is essential for proper cell function and growth.

Thymine is made from uracil and is essential for DNA synthesis. If the nucleotides of thymine are in low concentration or absent, DNA synthesis is slow or halted. Methotrexate inhibits the ability of cells to grow because it binds to the active site of an enzyme needed for thymine synthesis. Rapid cell growth requires DNA synthesis, thus rapidly dividing cancer cells are more adversely affected by methotrexate than are normal cells.

Another example of competitive inhibition is found in the action of the *sulfa drugs*. These drugs are effective against a number of bacteria. Bacteria use *p*-aminobenzoic acid to synthesize folic acid, which the bacteria use to make some coenzymes. The sulfa drugs, including sulfanilamide, resemble *p*-aminobenzoic acid sufficiently well to bind to the active site of dihydropteroate synthetase (Figure 14.14). While sulfanilamide is bound to this enzyme, no folic acid can be produced by the enzyme molecule. If enough of the enzyme is inhibited, insufficient folic acid is produced, and growth of the bacterial population is inhibited. Sulfa drugs do not affect the metabolism of mammals directly because mammals cannot make folic acid; it is obtained from their diets.

EXAMPLE 14.1

Ethylene glycol, $HOCH_2CH_2OH$, is the principal component of antifreeze. It is not toxic itself, but it is oxidized in the body to oxalic acid, HOOCCOOH, which is toxic. Ethanol, CH_3CH_2OH, is sometimes given to patients who have ingested ethylene glycol. Why?

Solution: Note the similarity in structure between ethylene glycol and ethanol. Because they are structurally similar, ethanol is a competitive inhibitor and binds to the first enzyme that oxidizes ethylene glycol. This competition reduces the rate at which ethylene glycol is bound, and, fortunately, the body slowly excretes the glycol, thus reducing the amount of oxalic acid formed. Interestingly, the enzyme that first binds ethylene glycol is alcohol dehydrogenase, the enzyme responsible for oxidizing ethanol. In this therapy, the normal physiological substrate, ethanol, serves as a competitive inhibitor of a nonphysiological substrate, ethylene glycol.

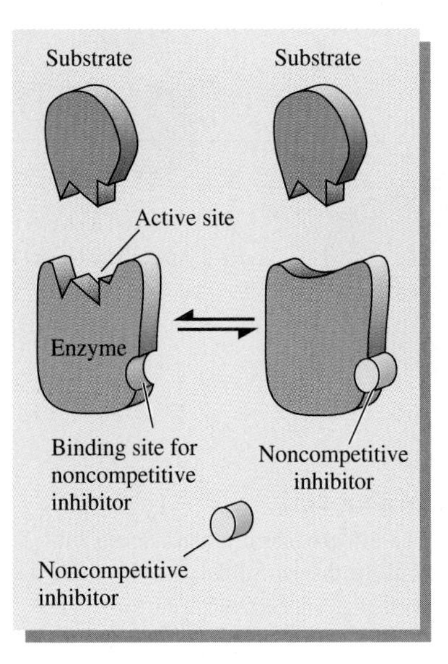

FIGURE 14.15
The interaction of an enzyme and a noncompetitive inhibitor. This type of inhibitor does not bind to the active site, but its binding causes the enzyme to become inactive. In this example, the inhibitor induces conformational changes in the enzyme that prevent the substrate from binding to the active site.

Some reversible inhibitors do not bind at the active site of an enzyme. These are **noncompetitive inhibitors** because they do not compete with the substrate. If they do not occupy the active site and prevent S from binding, how do they inhibit the enzyme? Noncompetitive inhibitors bind to the enzyme and cause a conformational change in it (Figure 14.15). In this new conformation, the enzyme is less active or inactive. It now binds S less efficiently or not at all, or the catalytic groups of the active site are no longer aligned properly for efficient catalysis.

NEW TERMS

reaction rate (velocity) The rate at which a reaction occurs, usually expressed in terms of substrate consumed per unit time or product formed per unit time

irreversible inhibitor Molecule that binds tightly to an enzyme and reduces or eliminates the activity of the enzyme

reversible inhibitor Molecule that binds temporarily to an enzyme and, while bound, reduces or eliminates the activity of the enzyme

competitive inhibitor Reversible inhibitor that binds to the active site of an enzyme and thus competes with the substrate

noncompetitive inhibitor Reversible inhibitor that binds to an enzyme at a location other than the active site and thus does not compete with the substrate.

TESTING YOURSELF

Rates of Enzyme-Catalyzed Reactions

1. Why is the rate of most enzyme-catalyzed reactions virtually zero at temperatures much above body temperature?
2. Compare the binding of competitive and noncompetitive inhibitors to enzymes.

Answers **1.** The enzyme is denatured at these temperatures and is therefore inactive. **2.** Competitive inhibitors bind to the active site of an enzyme, and noncompetitive inhibitors bind elsewhere on the enzyme.

14.5 Regulation of Enzyme Activity

♦ Enzyme activity can be altered by changing enzyme concentration or by altering the activity of the enzyme molecules.

The activity of enzymes is regulated in several ways. One way is to change the concentration of the enzyme. This can be accomplished by altering the rate at which the enzyme is synthesized or the rate at which it is broken down. Regulation of protein synthesis by control of gene expression was discussed in Chapter 13. Control of gene expression results in increased or decreased rates of enzyme synthesis, which increases or decreases the concentration of the enzyme. The rate of protein breakdown is also regulated. Body proteins are continuously degraded within cells, but the rate of degradation of some enzymes may vary with cellular conditions. Both synthesis and degradation affect the concentration of enzymes; regulation of these processes contributes to the regulation of metabolic activity.

CHEMISTRY CAPSULE

Blood Clotting

Blood clotting prevents blood loss from injuries, but clotting also contributes to heart attacks. How does this vital process occur? Blood contains several proteins called clotting factors that are zymogens, inactive enzymes. A cut or injury initiates sequential activation of these zymogens, resulting in the formation of an insoluble mass that closes the wound.

Injury stimulates release of clot-activating factors and exposes blood to nonphysiological surfaces that also activate the sequence. At each step another zymogen is activated to an active enzyme. Each step thus amplifies the process, and increasing amounts of zymogen are activated. The activation proceeds through several factors until fibrinogen, a soluble protein, is converted to fibrin, an insoluble fibrous protein. At the end of the sequence, active factor XIII cross-links fibrin molecules to form a massive molecular plug, which, along with platelets, blocks blood flow from the wound.

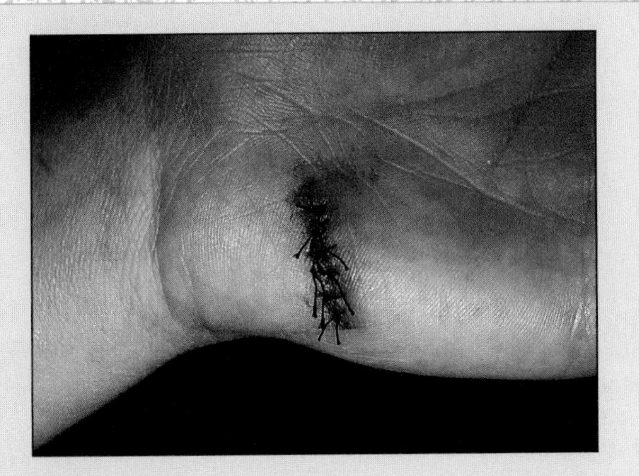

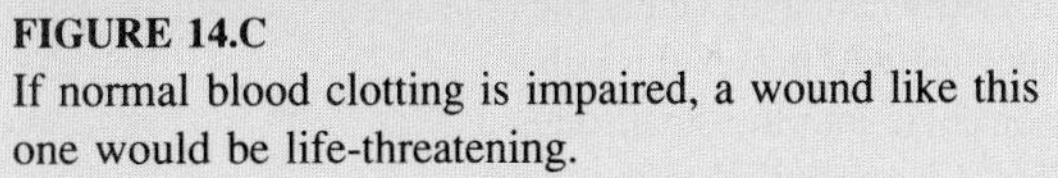

FIGURE 14.C
If normal blood clotting is impaired, a wound like this one would be life-threatening.

Enzyme activity can also be regulated in several ways that do not change the concentration of the protein. Three examples of this type of enzyme regulation are discussed here: (1) activation of zymogens, (2) reversible covalent modification of an enzyme, and (3) allosteric regulation.

Zymogens

Some enzymes are synthesized by the body in an inactive form called a **zymogen**. Zymogens can be recognized by the suffix *-ogen* or the prefixes *pre-* or *pro-*. Zymogens possess no enzymic activity when synthesized, but they gain activity when one or more peptide bonds are cleaved at some time following synthesis. Cleavage of these peptide bonds is followed by conformational changes to yield an active form. Examples include some of the digestive enzymes of the gastrointestinal tract and the enzymes responsible for blood clotting (see box).

A specific example is useful here. *Trypsin* is a digestive enzyme that cleaves specific peptide bonds of ingested proteins. Since dietary and body proteins are so similar, what prevents trypsin from breaking down the proteins of the pancreas where it is made? If trypsin were active when synthesized, it would immediately begin to work on the proteins of the pancreas. Instead, trypsin is synthesized as the inactive precursor *trypsinogen*, which is released into the small intestine when partially digested food enters from the stomach.

In the small intestine, trypsinogen is converted to trypsin by trypsin molecules already there or by the enzyme enterokinase. These enzymes cleave a peptide

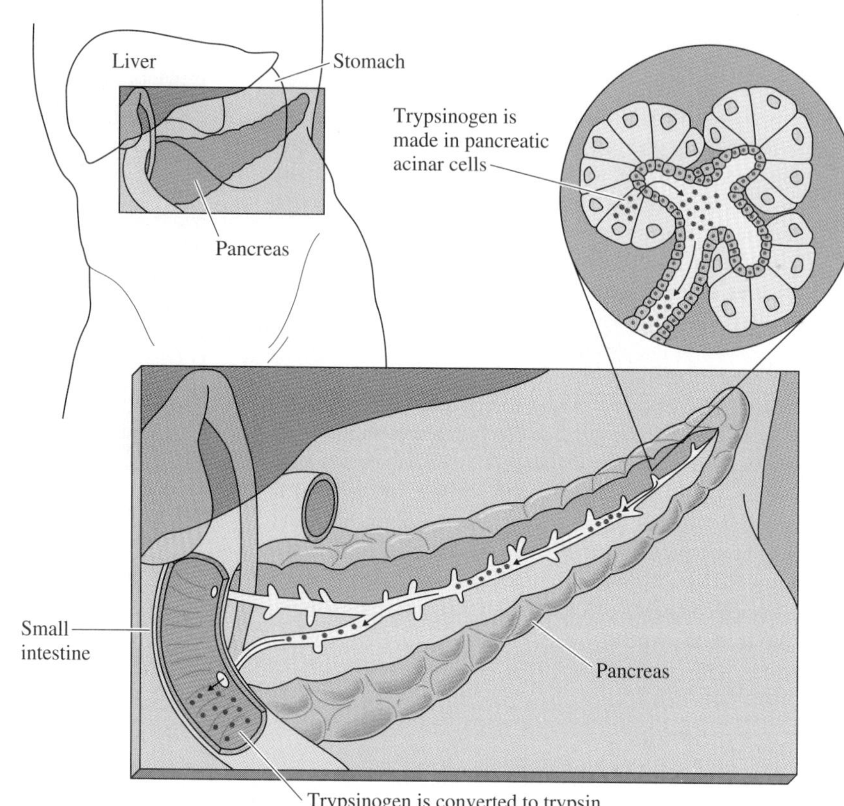

FIGURE 14.16
The conversion of trypsinogen molecules (blue dots) into trypsin (red dots). Trypsinogen is the inactive form of trypsin synthesized in the pancreas. It is released when food enters the small intestine and passes through a system of ducts into the small intestine. Here it is converted to the active form trypsin by the enzyme enterokinase or other trypsin molecules. If massive amounts of trypsinogen become activated to trypsin in the pancreas, trypsin will lyse proteins with resulting loss of cells and tissues. The serious and often fatal condition, acute pancreatitis, results.

bond near one end of the trypsinogen molecule. When freed of the small peptide, trypsin undergoes a conformational change to the active form (Figure 14.16). The activity of some other digestive enzymes is controlled in a similar manner.

Covalent Modification of Enzymes

♦ The activity of some enzymes can be turned on and off by covalent modification.

The activity of some enzymes can be reversibly increased or decreased by covalent modification of the enzyme. An example is the phosphorylation–dephosphorylation of the enzyme glycogen phosphorylase. This enzyme cleaves glucose molecules from glycogen and attaches a phosphate group to the glucose. This cleavage occurs on demand, that is, whenever more phosphorylated glucose is required by the muscles or liver (see Chapters 15 and 17). When phosphorylase has no phosphate groups covalently attached to certain serine residues in the enzyme, it has little enzymic activity. When the need for phosphorylated glucose increases, a series of events is initiated that results in the attachment of phosphate groups to these serines. When the serines are phosphorylated, glycogen phosphorylase is more active and phosphorylated glucose is rapidly formed from glycogen. The enzyme remains active until dephosphorylation returns it to the less active form (Figure 14.17). The activity of the enzyme is regulated by reversible covalent modification.

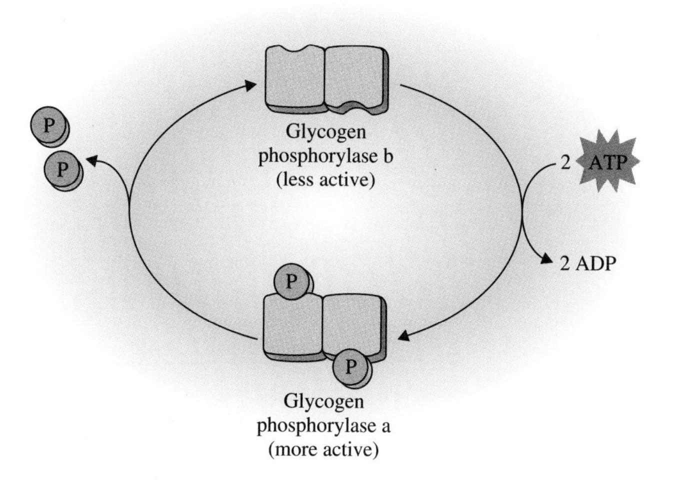

FIGURE 14.17
The reversible covalent modification of glycogen phosphorylase. When increased activity of glycogen phosphorylase is required in muscles, the enzyme is phosphorylated by transfer of phosphate from ATP to each polypeptide chain of this protein. When the activity is not needed, the phosphate is cleaved from the polypeptide chain, returning the enzyme to its less active form.

Allosteric Regulation of Enzymes

A third mechanism for the regulation of preexisting enzymes is **allosteric regulation**. In addition to the active site, some enzymes possess other sites that can bind certain small molecules found in the cell. These sites are called *allosteric sites*, or "other place" sites. These sites show the same high specificity of binding for their small molecules, called *effectors*, that the active site does for substrates. When an effector binds to the allosteric site, the enzyme undergoes a conformational change. In the new conformation the activity of the enzyme is different. If the effector is a **positive effector**, the enzyme has greater activity than if the effector had not bound. If the effector is a **negative effector**, then the enzyme is less active in the new conformation (Figure 14.18). Negative effectors are, in essence, normal noncompetitive inhibitors of the enzyme.

♦ Effectors binding to allosteric sites on some enzymes increase or decrease the activity of the enzyme.

Allosteric regulation of the first enzyme of a metabolic pathway allows for the efficient regulation of the entire pathway through a process called *feedback inhibition*. Consider the example shown in Figure 14.19. A pathway of three enzymes converts substrate S into product P. Product P is a negative effector of the first enzyme (E_1) of the pathway. As S is converted to P, the concentration of

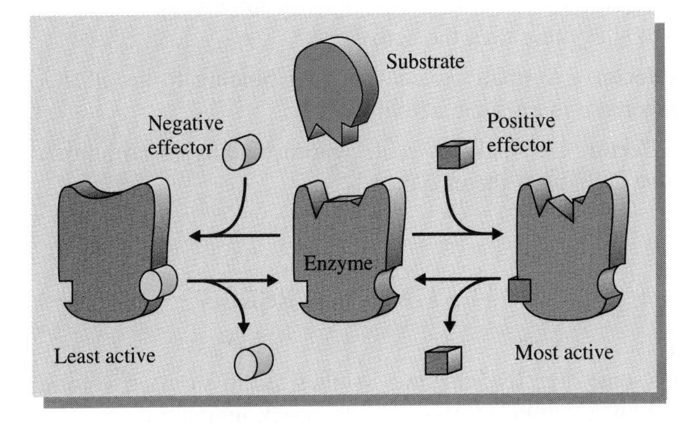

FIGURE 14.18
The regulation of allosteric enzymes by positive and negative effectors. Binding of a positive effector increases the activity of the enzyme, whereas binding of a negative effector decreases the activity. In this hypothetical enzyme, the effectors cause conformational changes in the enzyme that influence binding of the substrate.

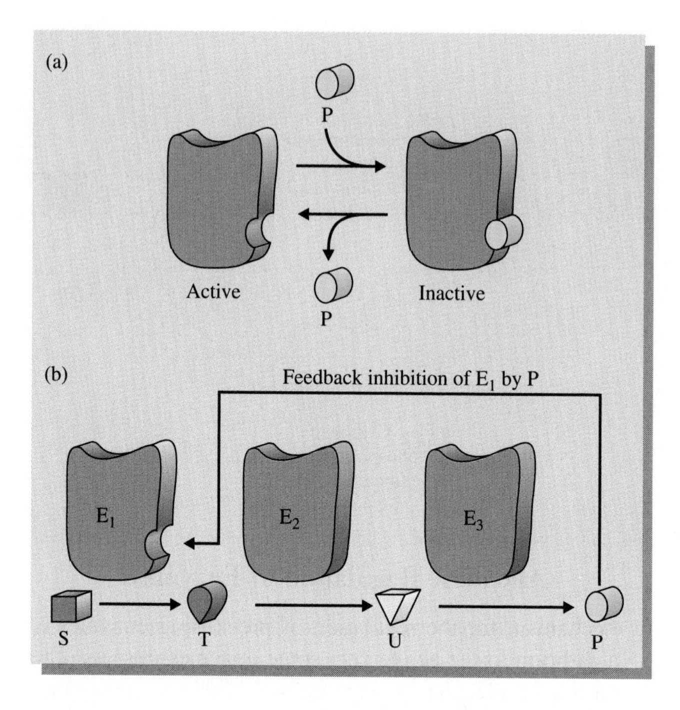

FIGURE 14.19
Feedback inhibition of a metabolic pathway. (a) P is a negative effector of the first enzyme of this pathway, E_1. (b) The activity of E_1 depends on the amount of P in the cell. If the concentration of P is relatively high, E_1 is relatively inactive (P feeds back and controls E_1). If a small amount of P is present, most of the E_1 molecules are active. This allows the synthesis of P to be controlled by the amount of P.

P increases and more of it binds to the allosteric sites of E_1. Fewer E_1 molecules are now active, so the rate at which P is produced is reduced. The process is readily reversible. If the concentration of P decreases within the cell, some P bound to E_1 will diffuse away. More of the E_1 enzyme molecules become active, and more P is formed. Allosteric control provides for rapid and sensitive regulation of metabolic activity.

NEW TERMS

zymogen An inactive form of an enzyme that is activated by cleavage of one or more peptide bonds

allosteric regulation Regulation of enzyme activity by the binding of small molecules to sites other than the active site

positive effector A small molecule whose binding to an allosteric enzyme causes an increase in enzymic activity

negative effector A small molecule whose binding to an allosteric enzyme decreases the activity of the enzyme

TESTING YOURSELF

Regulation

1. Why are some digestive enzymes synthesized in an inactive form?
2. Explain how a thermostat works by feedback inhibition.

CHEMISTRY CAPSULE

Enzymes In Medicine

Enzymes are essential to life, but they also play an increasingly important role in medicine. Enzymes are used in diagnosis (the determination of the cause and nature of a disease) and in treatment. The diagnosis and treatment of a heart attack provides an illustration of the role of some specific enzymes in medicine.

The muscle cells of the heart possess some enzymes that are not normally found in blood plasma. When a heart attack occurs, the reduced blood flow to the muscle cells results in the death and rupture of some cells. The contents of these broken cells enter the blood. Enzymes of heart muscle are now in the blood. If a blood sample is analyzed for these enzymes, the sample will be positive in a heart attack victim, but negative in someone suffering from some other ailment (see Figure 14.D).

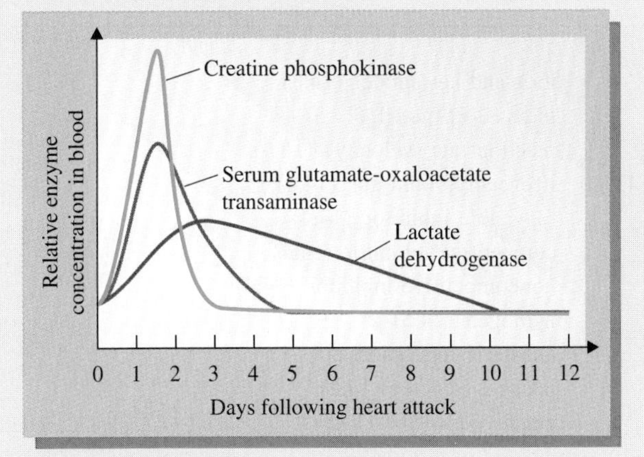

FIGURE 14.D
These three enzymes are normally absent from blood. Heart muscle cells are destroyed during and after a heart attack, and the enzymes of these cells are released into the blood stream. The presence of these enzymes in blood within these concentration ranges is one line of evidence a physician uses to diagnose a heart attack.

Another enzyme is used to treat heart attack victims. When a heart attack occurs, blood flow to a portion of the heart is reduced or stopped. Often this is due to a blood clot blocking an artery that is partially restricted by atherosclerotic deposits (see Figure 14.E). Intense discomfort may occur immediately, and with time more and more permanent damage occurs to the tissues deprived of blood. If the clot can be dissolved before permanent damage results, the probability of death and permanent disability is significantly lessened. Tissue-type plasminogen activator (TPA) is an enzyme that can be administered to heart attack victims. This enzyme converts a zymogen called plasminogen to the enzyme plasmin. Plasmin is the enzyme that normally breaks down blood clots in the body. Prompt treatment with TPA simply activates the normal clot-dissolving machinery of the body. If TPA is administered in time, blood flow to the heart muscle is restored before the muscle cells die.

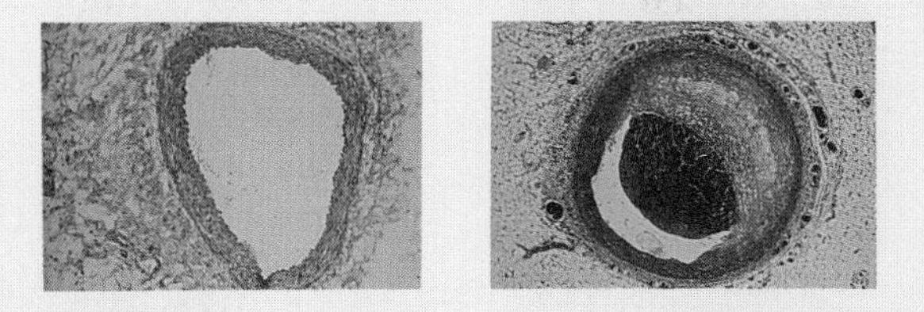

FIGURE 14.E
A comparison between a normal artery (left) and one nearly blocked with atherosclerotic plaque and calcified wall (right). Blood flow in a normal artery is unrestricted and free flowing. Normal blood flow is significantly reduced in the other artery and may be totally blocked by a blood clot. The atherosclerotic deposits include cholesterol and other lipids.

Answers **1.** If the enzymes were synthesized in an active form, they would hydrolyze the proteins, lipids, and carbohydrates of the cells of the pancreas and gastrointestinal tract. **2.** When a room gets too cold, an electric contact in the thermostat closes and activates the heater. The heater produces thermal energy in the room until the room temperature exceeds a set point in the thermostat. At this point the circuit opens, shutting down the heater. Heat, the product of the heater, effectively shuts down its own production.

Summary

All of the reactions of the body are collectively called *metabolism*. Catabolic reactions yield energy as they break down larger molecules to smaller ones. Anabolic reactions use energy and smaller molecules to build larger molecules. The reactions of the body are catalyzed by *enzymes*, which are either simple or conjugated proteins. Many enzymes can be recognized by the suffix *-ase* in their name. Enzymes bind substrate molecules and catalyze their conversion to products. Many enzymes show considerable specificity for their *substrates*. The catalytic activity of enzymes is influenced by substrate proximity, orientation, and strain. Enzymes may also act as acids or bases during a reaction. The rates of enzyme-catalyzed reactions are affected by the concentrations of substrate and enzyme and are influenced by temperature and pH. Enzymes can also be inhibited by *reversible* or *irreversible inhibitors*. The activity of an enzyme can be modified by changing the concentration of the enzyme or by altering its activity. *Zymogens*, covalent modification, and *allosteric regulation* are examples of enzyme regulation.

Terms

metabolism (14.1)
catabolism (14.1)
anabolism (14.1)
metabolic pathway (14.1)
enzyme (14.2)
-ase (14.2)
substrate (14.2)
cofactor (14.2)
coenzyme (14.2)
isozymes (14.2)
active site (14.3)
lock-and-key model (14.3)
induced-fit model (14.3)
reaction rate (velocity) (14.4)
irreversible inhibitor (14.4)
reversible inhibitor (14.4)
competitive inhibitor (14.4)
noncompetitive inhibitor (14.4)
zymogen (14.5)
allosteric regulation (14.5)
positive effector (14.5)
negative effector (14.5)

Exercises

Metabolism (Objective 1)

1. What are the two main branches of metabolism?
2. What reactions of the body are energy-yielding?
3. What part of metabolism consumes energy for synthesis of larger molecules?
4. What is a metabolic pathway?
5. Why are enzymes needed for metabolic reactions?
6. Classify the following reactions and pathways as either catabolic or anabolic. (Some of these pathways are covered in later chapters.)
   a. digestion
   b. DNA replication
   c. translation
   d. transcription
   e. glycolysis
   f. glycogen synthesis

Enzymes (Objective 2)

7. To which group of biomolecules do the enzymes belong?
8. In Section 14.2, a sentence begins "A catalyzed reaction requires less time to reach equilibrium. . . ." What does this statement imply about the equilibrium of an enzyme-catalyzed reaction?
9. Which of the following substances are enzymes?
   a. trypsin
   b. galactose
   c. tryptophan
   d. acetylcholinesterase
   e. catalase
   f. sucrase
   What clues do the names of these substances give you?
10. From the name of each of these enzymes, what is its substrate and what type of reaction does it catalyze?
    a. xylulose reductase
    b. glutamine synthetase
    c. xanthine oxidase

11. What is an enzyme's substrate?
12. What is a prosthetic group?
13. What name is given to an enzyme that lacks its prosthetic group?
14. What are small organic molecules that are needed for the activity of some enzymes called?
15. What nutrients are used to make some coenzymes?
16. You have isolated an enzyme from bacteria through a series of steps that removed all other proteins and small molecules. Your enzyme was active during the first steps of the isolation but appeared less and less active as its purity increased. In its pure form, it has no activity. Curiously, full activity is restored if a small amount of the ground-up bacteria is added. Explain these observations.
17. You have discovered an enzyme that is involved in the oxidation of crude oil. From the *Chemistry Capsule* on minerals and vitamins in metabolism, determine what coenzyme(s) might be involved in this reaction.
18. What is an isozyme?
19. Explain in general terms why there are different forms of lactate dehydrogenase (LDH).
20. What symbols are used to designate the different forms of LDH? What do these symbols mean?

Enzyme Specificity and Activity (Objective 3)

21. What is the active site of an enzyme?
22. For substrate–enzyme binding, compare the lock-and-key model and the induced-fit model.
23. What bonds or interactions are involved in the binding of a substrate to the active site of an enzyme?
24. It is said that a substrate is complementary to the active site of an enzyme. What factors make the active site and substrate complementary?
25. Describe or sketch the portion of the active site of a hypothetical enzyme that binds the dipeptide phenylalanyllysine with high specificity.
26. What effect does an enzyme have on the activation energy of a reaction?
27. How can the orientation of substrate molecules influence the activity of an enzyme?
28. Describe how the binding of two substrate molecules to an enzyme's active site enhances the activity of the enzyme. What term is used to describe this effect?
29. Could the concept of strained substrate molecules be integrated into the lock-and-key model of enzyme–substrate binding? Why or why not?
30. The side chains of some amino acids can act as general acids or bases during catalysis. List the amino acids that can donate H^+ and those that can accept H^+ during catalysis.

Reaction Rates (Objective 4)

31. Sketch the rate (velocity) of an enzyme-catalyzed reaction as a function of enzyme concentration.
32. Sketch the rate of an enzyme-catalyzed reaction as substrate concentration is changed from zero to a high level.
33. Explain why the shape of the graph of rate of reaction versus enzyme concentration is different from the graph of rate of reaction versus substrate concentration.
34. Why does a graph of enzyme activity versus temperature increase in a linear fashion until a precipitous drop is seen at higher temperatures?
35. Compare the activity of pepsin and trypsin in the pH range of 2 to 9. What is the physiological significance of these pH profiles?
36. Explain why a graph of the activity of an enzyme versus pH yields a bell-shaped curve.
37. Enzymes show varying rates with changes in pH and temperature. Are these factors normally important in the activity of the enzymes within the body?
38. Compare the binding of a reversible and irreversible enzyme inhibitor.
39. Compare the binding of a competitive and noncompetitive inhibitor.
40. What structural relationship exists between an enzyme's substrate and a competitive inhibitor?
41. How do sulfa drugs inhibit bacterial growth?
42. Succinate, $^-OOCCH_2CH_2COO^-$, is an intermediate in aerobic metabolism (see the TCA cycle in Chapter 15). The enzyme succinate dehydrogenase oxidizes succinate to fumarate, $^-OOCCH{=}CHCOO^-$. This enzyme is inhibited by malonate, $^-OOCCH_2COO^-$. What kind of inhibition is this?
43. Compare the structure of the substrate and inhibitor in Exercise 42. What similarities and differences may account for the type of inhibition that is observed?

Regulation (Objective 5)

44. Describe two ways cells can control the amount of an enzyme in the cell.
45. From its name, how can you recognize that a substance is a zymogen?
46. How are zymogens activated?
47. Why is trypsin not synthesized directly in the pancreas?
48. Give three examples of proteins that are synthesized as zymogens.
49. How is glycogen phosphorylase activated?
50. What regulatory advantage does covalent modification of enzymes have over regulation involving zymogens?
51. What does the term *allosteric* mean?
52. What is allosteric regulation?
53. Define positive and negative effectors.
54. What is meant by the term *feedback inhibition*? Does this form of regulation require a positive or negative effector?

Unclassified Exercises

55. Explain how a study of enzyme concentrations in blood can be used to diagnose a heart attack.
56. What use does the enzyme tissue-type plasminogen activator have in medicine?

BIOENERGETICS AND CATABOLISM

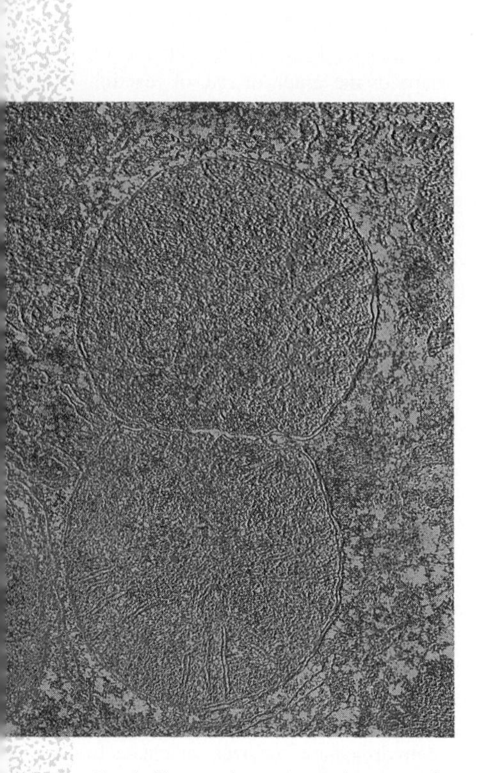

Catabolic reactions cleave larger molecules into smaller ones with the release of energy. Mitochondria, like those dividing here, are the major site of energy production in cells.

OUTLINE

15.1 Bioenergetics
15.2 Digestion, Absorption, and Transport
15.3 Carbohydrate Catabolism
15.4 Tricarboxylic Acid Cycle
15.5 Electron Transport and Oxidative Phosphorylation
15.6 Lipid Catabolism
15.7 Amino Acid Catabolism

OBJECTIVES

After completing this chapter, you should be able to

1. Describe the role of energy in maintaining the state of life.
2. Explain the central role of ATP in metabolism.
3. Describe how and where carbohydrates, lipids, and proteins are digested.
4. Describe the absorption and transport of digestion products.
5. Describe the central role of glucose in carbohydrate catabolism.
6. Explain the metabolic role of glycolysis and the aerobic and anaerobic fate of pyruvate.
7. Describe how the TCA cycle, electron transport, and oxidative phosphorylation provide energy under aerobic conditions.
8. Describe the catabolism of fatty acids.
9. Describe transamination, deamination, the urea cycle, and catabolism of amino acids.

The organic and biological chemistry section of this book began with a photograph of a woman climbing. The woman's image contrasts with the rock around her; there is no doubt that she and the rock are different. This contrast of body and rock can also represent the differences between living systems and the nonliving world. The woman differs structurally from the rock. The organic chapters explored the chemistry of carbon, a diverse chemistry that differs significantly from that of other elements. Later chapters examined biomolecules. These compounds have structures and properties that result in complex interactions. The climber's body form results from the structure and interactions of these molecules. Rock has a different set of structural features and interactions; rock has a very different form.

But there is more to living systems than structure, just as there is more to the image of the climber on the rock. The woman and rock not only differ structurally, they also differ in action. The rock is stationary, but the woman is moving; she is working her way up the rock. Similarly, a cell or organism must also do work to maintain life. Work requires energy. Energy is essential for life. A study of biochemistry requires more than a structural understanding of biomolecules, it also requires an understanding of bioenergetics.

15.1 Bioenergetics

Organisms must do work to maintain life. The cell is a highly organized system that includes many complex macromolecules. But the general tendency of the universe is to move with time to more random arrangements of smaller entities. This tendency is formally described as **entropy**. How does the cell maintain a complex state when entropy drives it towards smaller, more random arrangements? The cell does it through the use of energy. A constant supply of energy is needed to keep an organism from collapsing into inorganic ions and simple molecules like those in soil, water, and rocks. **Bioenergetics** is the study of energy in living systems.

Energy is defined as the capacity to do work and to affect change. Living organisms use energy to bring about three kinds of changes: (1) biosynthetic, or chemical, work, which is work used to build or sustain the molecules needed for life, (2) transport, which is work used to shuttle molecules and ions into, out of, or around a cell, and (3) work involving movement or motion. Cells obtain energy through chemical reactions. In Chapter 14, the reactions of the body were collectively called metabolism. Metabolism was in turn divided into two areas: catabolism and anabolism. Catabolic processes are generally *exothermic* (energy-yielding) and yield small molecules. Anabolic processes are generally *endothermic* (energy-requiring), and build more complex molecules. Anabolic reactions use energy that is provided by catabolism.

The reactions of the body produce or require energy, but how is useful energy obtained from exothermic reactions and how is it used to drive endothermic reactions? The answer to both of these questions is found in the concept of **coupled reactions**. An energy-yielding reaction can be coupled to an energy-requiring one. In essence, the energy given off by the exothermic reaction is made available to the endothermic one.

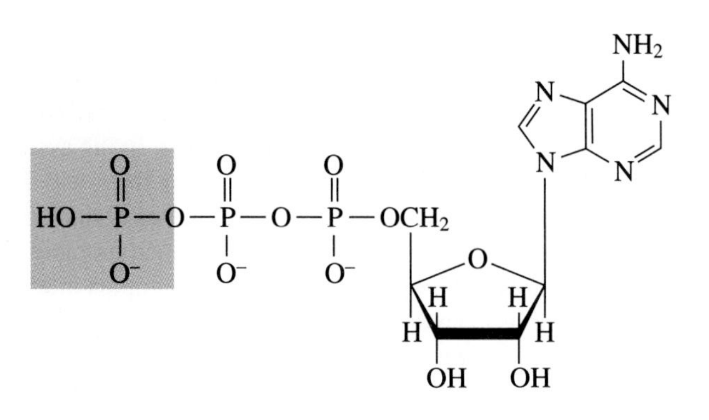

ATP is made by adding a phosphate (yellow-orange) to ADP.

Consider an example of a coupled reaction in metabolism. During a step of glycolysis (Section 15.3), a phosphoester bond in phosphoenolpyruvate (PEP) is hydrolyzed in an energy-yielding reaction:

$$\underset{\textbf{PEP}}{CH_2{=}\overset{\overset{\displaystyle OPO_3H^-}{|}}{C}COO^-} + H_2O \longrightarrow \underset{\textbf{Pyruvate}}{CH_3\overset{\overset{\displaystyle O}{\|}}{C}COO^-} + \underset{\textbf{Phosphate}}{H_2PO_4^-} + 14.8\ \text{kcal/mol}$$

In the same step of glycolysis, a molecule of **adenosine triphosphate (ATP)** is made from adenosine diphosphate (ADP) and inorganic phosphate in an energy-requiring reaction:

$$\text{ADP} + H_2PO_4^- + 7.3\ \text{kcal/mol} \longrightarrow \text{ATP} + H_2O$$

The enzyme pyruvate kinase catalyzes both of these reactions concurrently, that is, the reactions are coupled.

Coupled reactions can be expressed as a single equation. Chemical equations can be added in a manner similar to algebraic equations; terms common to both sides of the equation cancel. When two reactions are combined, both matter and energy for the reactions are summed. Combining the above reactions to form a single equation can be represented as:

$$CH_2{=}\overset{\overset{\displaystyle OPO_3H^-}{|}}{C}COO^- + H_2O \longrightarrow CH_3\overset{\overset{\displaystyle O}{\|}}{C}COO^- + H_2PO_4^- + 14.8\ \text{kcal/mol}$$

$$\text{ADP} + H_2PO_4^- + 7.3\ \text{kcal/mol} \longrightarrow \text{ATP} + H_2O$$

$$\textbf{Net:}\quad CH_2{=}\overset{\overset{\displaystyle OPO_3H^-}{|}}{C}COO^- + \text{ADP} \longrightarrow CH_3\overset{\overset{\displaystyle O}{\|}}{C}COO^- + \text{ATP} + 7.5\ \text{kcal/mol}$$

Water, phosphate, and part of the energy cancel out. Only 7.5 kcal/mol of energy are released because the rest was used to make ATP from ADP. Many of the reactions you will see in metabolism are coupled reactions.

The above example of coupled reactions was chosen for a reason. In that example, ATP was synthesized from ADP and inorganic phosphate (P_i). The energy for that synthesis was proved by an energy-yielding reaction of catabolism.

In one sense, energy became stored in the new bond in ATP. ATP has a very special role in the cell, serving as an energy reservoir for endothermic reactions. This is accomplished by coupling the hydrolysis of ATP with the energy-requiring reaction. When ATP is hydrolyzed, the same amount of energy is released that was required for its synthesis from ADP. The equation for the hydrolysis of ATP is:

$$\text{ATP} + \text{H}_2\text{O} \longrightarrow \text{ADP} + \text{P}_\text{i} + 7.3\ \text{kcal/mol}$$

◆ The bond between phosphates in ATP forms with loss of water. These bonds (there are two of them) are anhydrides of phosphoric acid and are sometimes referred to as high-energy bonds.

ATP hydrolysis provides the energy to drive most energy-requiring reactions of metabolism. Example 15.1 gives you the chance to examine the coupling of ATP to such an energy-requiring reaction.

EXAMPLE 15.1.

The phosphorylation of glucose to form glucose-6-phosphate is an early step of glycolysis. Direct phosphorylation of glucose by phosphate is an energy-requiring reaction:

$$\text{Glucose} + \text{Phosphate} + 3.3\ \text{kcal/mol} \longrightarrow \text{Glucose 6-phosphate} + \text{H}_2\text{O}$$

Couple this reaction to ATP hydrolysis to provide the energy that this reaction requires.

Solution:

$$\text{Glucose} + \text{Phosphate} + 3.3\ \text{kcal/mol} \longrightarrow \text{Glucose 6-phosphate} + \text{H}_2\text{O}$$

$$\text{ATP} + \text{H}_2\text{O} \longrightarrow \text{ADP} + \text{Phosphate} + 7.3\ \text{kcal/mol}$$

$$\text{Glucose} + \text{ATP} \longrightarrow \text{Glucose 6-phosphate} + \text{ADP} + 4.0\ \text{kcal/mol}$$

The energy-yielding hydrolysis of ATP provided the energy for the energy-requiring phosphorylation of glucose.

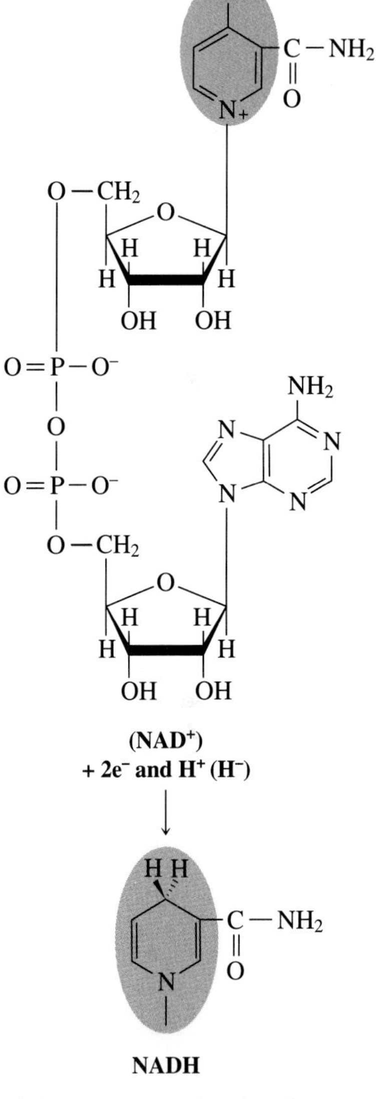

NAD⁺, shown here, is reduced to NADH. The nicotinamide ring in NAD⁺ (blue), accepts a hydride ion (H⁻) to yield the reduced form in NADH (reduced ring only shown in orange). The rest of the molecule is unchanged in the reaction.

There is another molecule that has a special role in metabolism, **nicotinamide adenine dinucleotide (NAD⁺)**. Just as ADP and ATP are involved in energy storage, NAD⁺ and its reduced form, NADH, are involved in removal and storage of electrons for oxidation–reduction reactions. NAD⁺ accepts two electrons and a hydrogen ion to become **NADH**:

$$\text{NAD}^+ + 2\,e^- + \text{H}^+ \longrightarrow \text{NADH}$$

Many reactions of catabolism are oxidative, and an oxidation must be accompanied by a reduction. In these oxidative reactions of metabolism, NAD⁺ is concurrently reduced. For example, in the tricarboxylic acid cycle (Section 15.4), malate is oxidized to oxaloacetate. That oxidation is coupled to the reduction of NAD⁺ to NADH:

$$\text{Malate} \longrightarrow \text{Oxaloacetate} + 2\,e^- + 2\,\text{H}^+$$

$$\text{NAD}^+ + 2\,e^- + \text{H}^+ \longrightarrow \text{NADH}$$

Net: $\text{Malate} + \text{NAD}^+ \longrightarrow \text{NADH} + \text{Oxaloacetate} + \text{H}^+$

The reduction of NAD^+ to NADH is an endothermic reaction; 52.6 kcal/mol is required for this reduction. When NADH is oxidized to NAD^+, 52.6 kcal/mol is released. NADH, like ATP, is a form of stored energy. You will see that this energy is used to synthesize ATP through oxidative phosphorylation (Section 15.5). ATP and NADH have key roles in catabolism.

♦ The reactions of catabolism yield energy and small molecules.

Catabolism is the part of metabolism that breaks down larger, more complex molecules into smaller and generally simpler ones. Catabolism of very different molecules such as glucose, fatty acids, and some amino acids yields the same products, although by very different reactions. For example, acetyl CoA is produced from all of the dietary molecules just discussed. These simpler product molecules are used by the body in one of two ways. They may be broken down further through exothermic reactions to yield energy that the cells then use as needed, or these smaller molecules may be used as building blocks for the synthesis of other molecules in the body.

Catabolic reactions are generally exothermic and oxidative. Catabolism yields smaller molecules with more oxidized carbon atoms; carbon dioxide is the final product of a number of catabolic pathways. Carbon in CO_2 has an oxidation number of $+4$, whereas the oxidation number of carbon in many biological molecules ranges from $+2$ to -3. In general, oxidative reactions are exothermic—as reduced molecules are oxidized, energy is released.

Catabolic reactions occur throughout the body. Although most of these reactions occur inside cells, digestion is extracellular. The rest of this chapter will discuss digestion, glycolysis, the tricarboxylic acid cycle, oxidative phosphorylation, electron transport, and several reactions and pathways involving lipids and amino acids.

NEW TERMS

bioenergetics The study of energy in living systems

entropy The tendency for the universe to go from more order to less order. Energy must be used to temporarily prevent this tendency in a living organism

coupled reaction A reaction that provides energy for an endothermic reaction by combining it with an exothermic one

adenosine triphosphate (ATP) A nucleotide that plays a role as an energy reservoir in living systems

nicotinamide adenine dinucleotide (NAD^+/NADH) A nucleotide that serves as an acceptor (NAD^+) and donor (NADH) of electrons in cellular reactions

TESTING YOURSELF

Bioenergetics

1. What is the source of the energy needed to drive endothermic reactions within the cell?
2. What name is given to the tendency for the universe to go from more order to less order with time?

CHEMISTRY CAPSULE

Measures of Energy: Enthalpy, Entropy, and Free Energy

Energy is the capacity to do work or affect change. Although there are several measures of energy used in chemistry, *enthalpy* (H) is the one you have been using. Changes in enthalpy, usually shown as ΔH, are related to energy gain or loss during a reaction. When the particles of a reaction lose energy, the enthalpy change (ΔH) is negative. The particles have lost energy to their surroundings. You have been calling these reactions exothermic. If the particles of the reaction have gained energy from their surroundings, ΔH is positive, the reaction is endothermic. Enthalpy is a useful measure of energy in chemistry, but it is not the only one.

Energy gain or loss is not the only energy factor that affects a reaction. Consider the spontaneity of a reaction, that is, the tendency for the reaction to occur. Many reactions that occur readily are exothermic; they give off energy. But some endothermic reactions are also spontaneous. Consider the melting of ice at room temperature as an example. This reaction is endothermic but spontaneous. The tendency for a reaction to occur involves another measure of energy called entropy. *Entropy* is related to randomness or disorder, and entropy increases during any spontaneous reaction.

Both enthalpy and entropy are involved in energy changes in reactions, but a chemist or biochemist does not need to measure both because there is a third measure of energy that is a combination of enthalpy and entropy. This measure is free energy (G). Free energy is used in biochemistry to predict available energy and the spontaneity of a reaction. Many of the reactions of metabolism are readily reversible. The direction that the reaction goes depends on the concentrations of the reactants and products of the reaction (LeChatelier's principle). If a reaction has a free-energy change (ΔG) that is negative, then the reaction is spontaneous. The reaction under these conditions is said to be *exergonic*. If free energy is gained by a reaction, then the reaction is *endergonic* and nonspontaneous. Free energy is a useful concept in biochemistry.

FIGURE 15.A
Entropy is a measure of the tendency for the universe to move toward increasing disorder. If you cover and shake these beakers of marbles, what will be the result? Is this consistent with entropy?

3. Characterize catabolic reactions with respect to (a) size of reactants and products and (b) oxidation state of reactants and products.

Answers **1.** The energy released in the exothermic reaction of ATP with water is often coupled to endothermic reactions in the cell. The energy originally comes from potential energy stored in chemical bonds in food molecules. **2.** entropy **3. a.** Catabolic reactions generally convert larger molecules to smaller ones. **b.** The reactants are generally oxidized during catabolism.

15.2 Digestion, Absorption, and Transport

In the gastrointestinal tract, large dietary molecules are converted to molecules that can be absorbed and utilized by the cells of the body. This process is called **digestion**, and it involves both mechanical and chemical factors. Chewing is a mechanical process that reduces food particle size, exposing more of the food to digestive enzymes. The chemical aspects of digestion include the hydrolysis

♦ Digestion is an extracellular process that prepares dietary molecules for absorption.

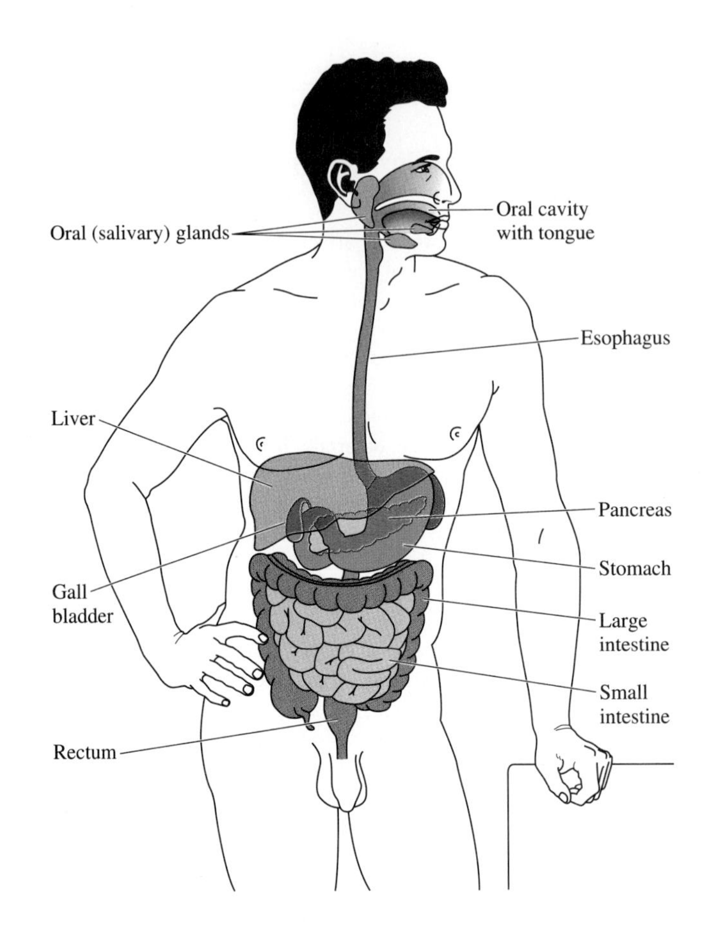

FIGURE 15.1
The human gastrointestinal tract. This organ system is responsible for the digestion and absorption of food.

of bonds in large molecules to form smaller ones, and the absorption of these smaller molecules into the body. The smaller products are absorbed by the mucosal cells of the small intestine. From the mucosal cells, the molecules enter the cardiovascular system and are distributed throughout the body. The cells of the body then absorb these molecules and use or store them. You can refer to Figure 15.1 as you read about digestion of biomolecules.

Digestion of Carbohydrates

The digestion of carbohydrates begins in the mouth. Food is chewed in the mouth and mixed with saliva, which contains the enzyme salivary **amylase**. This digestive enzyme catalyzes the addition of water to glycosidic bonds in the starches amylose and amylopectin. Salivary amylase hydrolyzes α 1$\rightarrow$4 glycosidic bonds of these starches, yielding smaller polysaccharides, oligosaccharides, and some glucose molecules (Figure 15.2). The effectiveness of salivary amylase is limited in two ways: (1) the amount of contact between the enzyme and the starches is limited because mixing of food and saliva is far from complete, and (2) the amount of time the amylase can work on the starches is limited because the amylase is inactivated by the acidic secretions of the stomach. The digestion

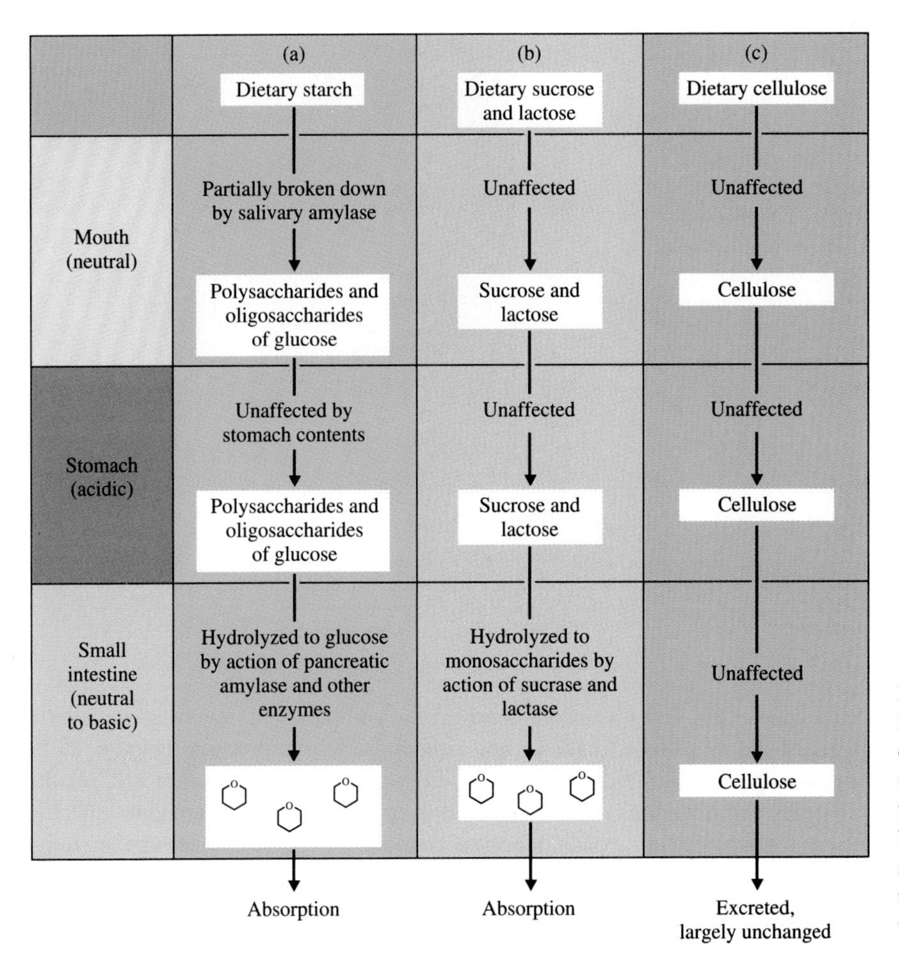

FIGURE 15.2
Digestion of carbohydrates. Dietary carbohydrates, including (a) starches, (b) sugars (shown as sucrose and lactose), and (c) indigestible complex carbohydrates (shown as cellulose), are subjected to a variety of conditions as they pass through the gastrointestinal tract.

of starch that begins in the mouth stops in the stomach where the amylase activity is lost. Dietary sugars and cellulose are not affected by amylase.

The chewed food that enters the stomach becomes mixed with gastric secretions. These secretions contain hydrochloric acid and are thus quite acidic; the pH of the stomach is less than 2. Many proteins, including salivary amylase, are denatured in this acidic environment. No specific enzymes for carbohydrate digestion are found in the stomach. The H^+ found in the stomach may catalyze the hydrolysis of some of the glycosidic bonds of poly- and oligosaccharides, but this makes no significant contribution to the breakdown of carbohydrates.

The partially digested contents of the stomach and small intestine are called **chyme**. Chyme entering the small intestine contains sugars and complex carbohydrates that were not altered in either the mouth or the stomach. It also contains the oligosaccharides and polysaccharides formed by the action of amylase on starch. Chyme is acidic, but the small intestine mixes chyme with secretions of the pancreas that contain bicarbonate. This base neturalizes the acid added by the stomach, changing the pH to between 7 and 8. The secretions of the pancreas also contain pancreatic amylase, an enzyme similar to the one in saliva. This enzyme completes the hydrolysis of the partially digested starches. Nearly all of

◆ The digestion of starches that began in the mouth is completed in the small intestine.

CHEMISTRY CAPSULE

Lactose Intolerance

Lactase is found in the young of all mammals, but many adults lack this enzyme and as a result are *lactose intolerant*. If these people eat lactose, the sugar present in milk and other dairy products, it is not digested in the small intestine but instead enters the large intestine. The presence of unabsorbed lactose in the large intestine alters normal water absorption, an alteration that can lead to diarrhea and dehydration. Furthermore, the large intestine contains numerous bacteria that can absorb and use the lactose. The lactose is an energy-rich addition to the supporting medium for the bacteria. This causes them to grow well and produce gases, which can lead to distress of the lower intestinal tract for the host.

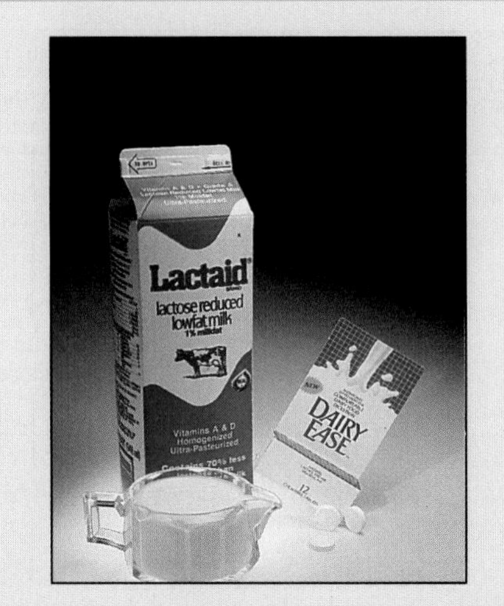

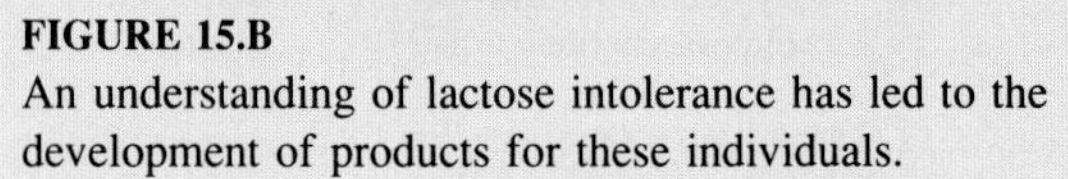
FIGURE 15.B
An understanding of lactose intolerance has led to the development of products for these individuals.

the exposed α 1→4 glycosidic bonds in the complex carbohydrates are hydrolyzed, leaving glucose, maltose, and isomaltose as the products (Figure 15.2). These sugars, like the dietary sugars, are water-soluble. Cellulose and related complex carbohydrates are not digested because humans lack enzymes specific for β 1→4 glycosidic bonds in polysaccharides. These substances and anything trapped within them are a major part of the fiber found in foods.

Dietary disaccharides, such as sucrose and lactose, and the disaccharides produced by starch digestion, are hydrolyzed to monosaccharides by several enzymes in the small intestine (Figure 15.2). The maltose and isomaltose produced by digestion of starches are hydrolyzed by reactions catalyzed by malt*ase* and isomalt*ase*, respectively. The hydrolysis of sucrose is catalyzed by sucr*ase*, and the hydrolysis of lactose is catalyzed by lact*ase*.

The final products of carbohydrate digestion are a variety of monosaccharides. Normally, large amounts of glucose are present in the chyme as well as some fructose, galactose, and other sugars. These sugars are absorbed by the mucosal cells of the small intestine. These cells possess a variety of specific transport systems in their cell membranes that move the small molecules produced by digestion into the cells. The absorbed sugars are then secreted by the mucosal cells into the abundant capillaries in the small intestine. Blood distributes the sugars throughout the body, making them available to all cells.

♦ Absorbed sugars are made available to all of the cells of the body.

Digestion of Protein

Saliva contains no enzymes that affect proteins. Protein digestion does not begin until food enters the stomach. The acidic environment of the stomach denatures proteins, leaving them in a partially or totally unfolded state. The peptide bonds of a denatured protein are more exposed to digestive enzymes than the peptide bonds of compact proteins. The peptide bonds of dietary proteins are cleaved by hydrolysis in a reaction catalyzed by the enzyme *pepsin.*

♦ The acidic environment of the stomach denatures proteins, which exposes more peptide bonds to pepsin.

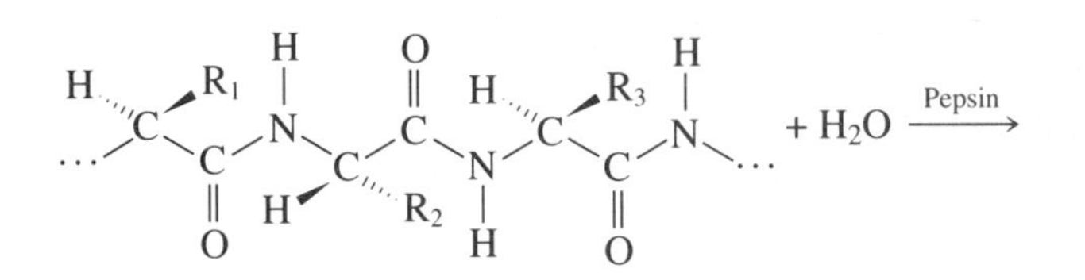

Polypeptide chain

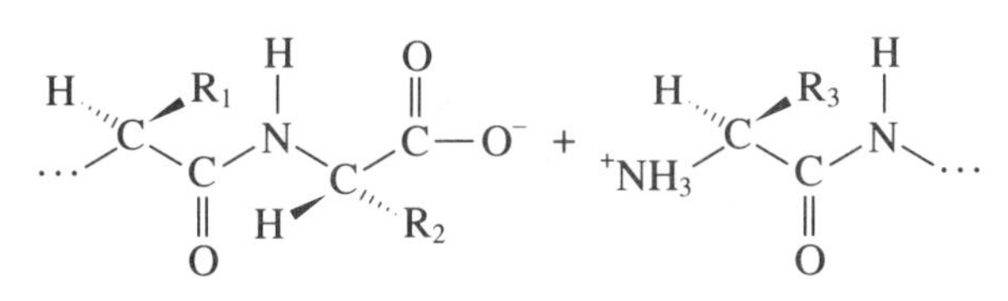

Smaller peptides

The hydrolysis of polypeptides is not completed in the stomach. Portions of the food are not adequately exposed to pepsin in the stomach. Furthermore, pepsin does not work equally well on all peptide bonds; pepsin shows some specificity for substrate. The chyme entering the small intestine contains a mixture of peptides, polypeptides, and free amino acids. Figure 15.3 summarizes the digestion of protein in the stomach and the rest of the gastrointestinal tract.

The secretions of the pancreas entering the small intestine contain several zymogens that are activated to **proteases** and **peptidases**, enzymes that catalyze the hydrolysis of proteins and peptides, respectively. Trypsinogen is converted to the enzyme trypsin by hydrolytic reactions catalyzed by trypsin itself or by the enzyme enterokinase. There is usually a little trypsin or enterokinase in the small intestine, so conversion of trypsinogen to trypsin always occurs. Trypsin catalyzes the activation of other zymogens, including chymotrypsinogen to chymotrypsin and procarboxypeptidase to carboxypeptidase. Collectively, these digestive enzymes and some others convert the peptides and polypeptides to free, water-soluble amino acids. The amino acids are absorbed by mucosal cells, which secrete them into the cardiovascular system for distribution throughout the body. Thus, like monosaccharides, amino acids are made available to all the cells of the body.

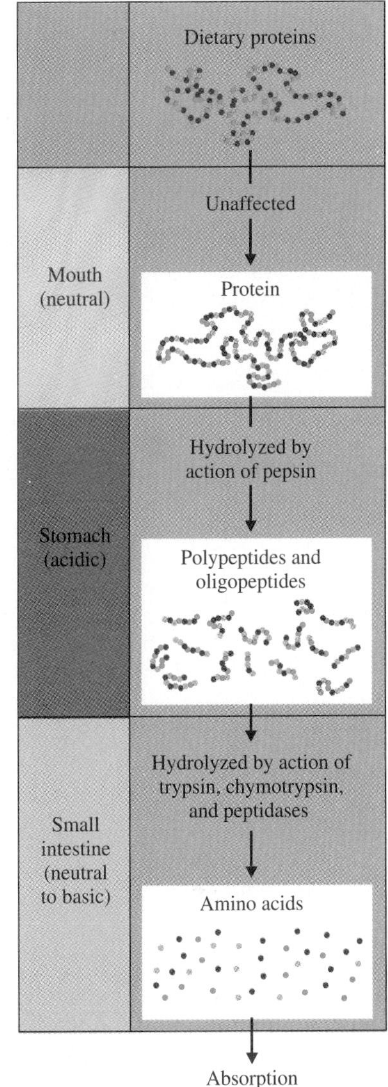

FIGURE 15.3
Digestion of dietary proteins occurs primarily in the stomach and small intestine. Because amino acids vary in size, shape, and polarity, there are several transport systems for the nonpolar, polar neutral, basic, and acidic amino acids.

Digestion of Lipids

Dietary lipids include triacylglycerols (fats and oils), cholesterol, and polar lipids. We will look at only triacylglycerols and cholesterol here. Digestion of

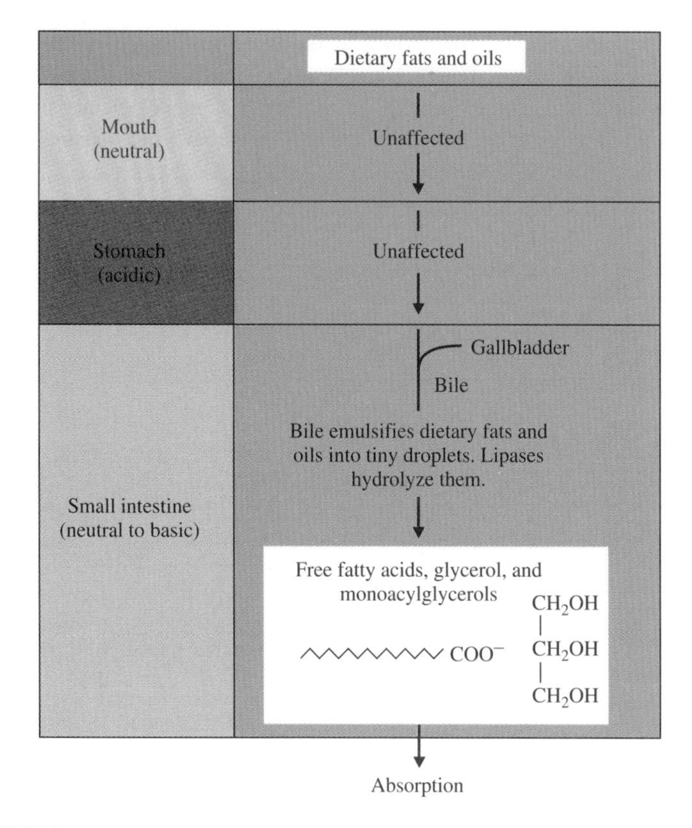

FIGURE 15.4
Digestion of fats and oils occurs in the small intestine through the action of both enzymes and bile.

triaglycerols does not begin until chyme enters the small intestine (Figure 15.4). Saliva and gastric secretions do not contain any significant amounts of enzymes that affect lipids. Cholesterol requires only absorption, whereas fats and oils must first be hydrolyzed. Digestion of lipids is complicated by their hydrophobic nature. Carbohydrates, proteins, and peptides are water-soluble or suspended in water so digestive enzymes can interact directly with these dietary compounds. But dietary lipids are water-insoluble and form droplets and globules in the gastrointestinal tract. Only lipid molecules on the surface of these droplets are exposed to enzymes; the vast majority have little chance to come into contact with enzymes because they are buried within the hydrophobic interior of the droplet.

This complication is eliminated by **bile salts**. These compounds are synthesized by the liver, stored in the gallbladder, and secreted into the small intestine as needed. See Figure 11.11 for the structure of two typical bile salts, taurocholate and glycocholate. Bile salts, like soaps, are amphipathic molecules, and they emulsify dietary lipid into many micelles. The micelles collectively have a much greater surface area than the ingested lipid droplets. People who lack a gallbladder should avoid a high-fat diet.

◆ The micelles formed from bile salts and triacylglycerols have large surface areas that are exposed to digestive enzymes.

The digestion of triacylglycerols involves hydrolysis of two or three of the ester bonds between the fatty acids and the glycerol. This hydrolysis is catalyzed by **lipases** synthesized and secreted by the pancreas.

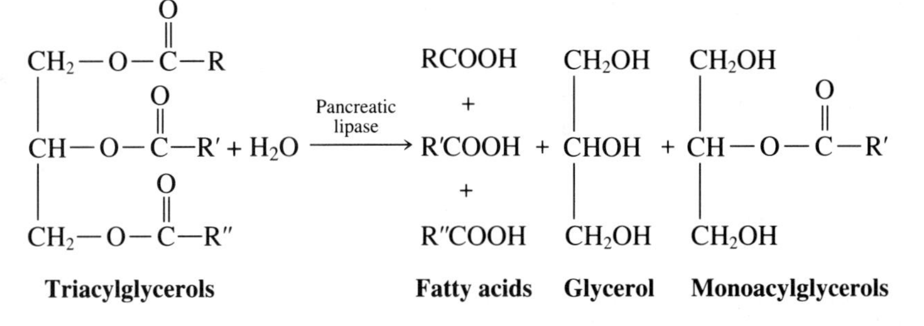

Dietary phosphoacylglycerols are digested by a phospholipase that is synthesized and secreted as a zymogen by the pancreas. This zymogen is activated by trypsin.

Cholesterol, monoacylglycerols, glycerol, and fatty acids are absorbed by the mucosal cells. In these cells, triacylglycerols are re-formed from fatty acids and from glycerol or monoacylglycerols. These triacylglycerols along with cholesterol are secreted into the lymphatic vessels of the small intestine. **Lymph** is a fluid similar in composition to blood but lacking red blood cells. Lipids are insoluble in lymph and blood because these fluids are principally water. In the mucosal cells and in the cardiovascular system, the lipids bind to specific proteins through hydrophobic interactions to form complexes called **lipoproteins**. Lipoproteins resemble micelles—they possess a hydrophobic interior of lipid coated with a hydrophilic surface of protein molecules (Table 15.1). Lipids are transported in blood and lymph as lipoproteins.

◆ Because lipids are water-insoluble, they must be transported as lipoprotein complexes.

The triacylglycerols in lipoproteins are broken down into fatty acids and glycerol, which are then taken up by the cells of the body. Adipose cells take up

TABLE 15.1
Typical Percentage Composition of Lipoproteins

Lipoprotein	Triacylglycerol (%)	Cholesterol (%)	Phospholipid (%)	Protein (%)	Average Density	Approximate Diameter (nm)
Chylomicrons	87	3–4	8	1.5–2.0	0.92–0.96	100–1000
Very-low-density lipoprotein (VLDL)	55	15	20	9–10	0.95–1.00	25–75
Low-density lipoprotein (LDL)	8–10	45	22	25	1.00–1.06	20–25
High-density lipoprotein (HDL)	1–5	18–20	21–30	45–55	1.06–1.21	7–12

CHEMISTRY CAPSULE

Cholesterol and Heart Disease

The role played by cholesterol in heart disease is not fully known, but studies of a human genetic disease called *familial hypercholesterolemia* have provided some valuable clues. People with two copies of the defective gene (homozygotes) die at an early age of heart disease. Liver transplants are a treatment for these individuals. People with one copy of the defective gene (heterozygotes) have a much higher risk of heart disease than the normal population. These individuals have elevated blood cholesterol levels because they have elevated levels of low-density lipoprotein (LDL). This elevated blood cholesterol is linked to an increased risk of heart attacks. The cause of increased LDL cholesterol appears to be due to a decreased ability of cells to take up LDL from the blood. The cells contain an abnormally low number of receptors for the LDL, thus less cholesterol is taken up by these cells and more cholesterol remains in the blood. These individuals are treated by depriving the body of cholesterol through restricted diet and by treatment with drugs to reduce both the uptake of dietary cholesterol and cholesterol synthesis. Under these conditions, the body makes more of the LDL receptor and uptake of blood cholesterol is stimulated. Whether the normal population benefits from reduced dietary cholesterol has been harder to prove.

much of the fatty acids and glycerol, and triacylglycerols are rebuilt and stored. When the body requires more energy, lipases within the adipose cell break down triacylglycerols and release free fatty acids into the bloodstream. These free fatty acids bind to the protein serum albumin by hydrophobic interactions. The fatty acids circulate as part of this protein complex, and they too are absorbed by cells as needed.

NEW TERMS

digestion Process that breaks down food molecules into smaller compounds that are absorbed by the body

amylases Enzymes that catalyze the hydrolysis of α $1 \rightarrow 4$ glycosidic bonds in starches

chyme Partially digested food in the stomach and small intestine

proteases Enzymes that catalyze the hydrolysis of peptide bonds in proteins and polypeptides

peptidases Enzymes that catalyze the hydrolysis of peptide bonds in peptides. The distinction between a protease and a peptidase is not clear-cut.

bile salts Emulsifying agents in bile that break down dietary lipid droplets into micelles

lipases Enzymes that catalyze the hydrolysis of ester bonds in fats, oils, and similar lipids

lymph Body fluid similar to blood but lacking red blood cells

lipoproteins Macromolecules composed of protein and lipid that transport lipids in blood

TESTING YOURSELF

Digestion

1. Where does the digestion of starches occur?
2. How do hydrophobic molecules, such as triacylglycerols, come in contact with the appropriate water-soluble digestive enzyme?

Answers **1.** Starch digestion begins in the mouth, but most of the digestion occurs in the small intestine. **2.** Dietary lipids are emulsified by bile salts. The micelles that are produced readily contact digestive enzymes in the gut.

15.3 Carbohydrate Catabolism

Carbohydrate digestion and absorption bring sugars into the blood from where they are absorbed into cells and used for energy or to build other molecules. For example, some galactose and glucose may be used to synthesize glycolipids or glycoproteins needed by the cell. Glucose in muscle or in the liver could also be used to synthesize glycogen, a polymeric storage form for glucose. Most of the dietary sugars, however, are used for energy production. Glucose is the most important monosaccharide and the only one maintained at approximately constant concentrations in the blood.

Glycolysis

The principal pathway for catabolism of glucose is **glycolysis**, which occurs in the cytosol of cells. It is a series of ten enzyme-catalyzed reactions that cleaves and oxidizes glucose to two molecules of pyruvic acid. At physiological pH, the pyruvic acid dissociates into pyruvate and H^+.

♦ Glycolysis is the principal catabolic pathway for sugars.

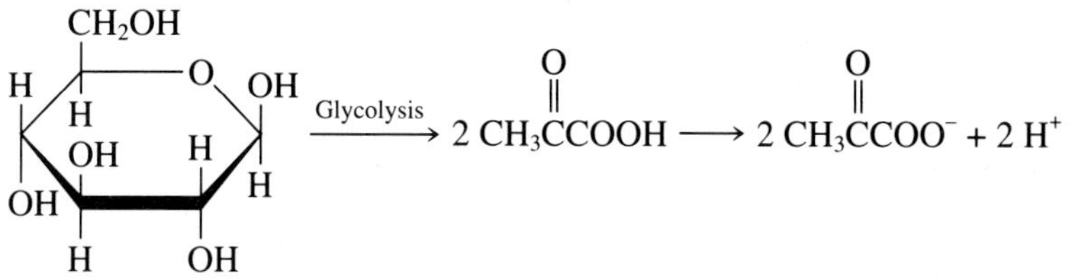

$$\text{Glucose} \xrightarrow{\text{Glycolysis}} 2\ CH_3\overset{O}{\overset{\|}{C}}COOH \longrightarrow 2\ CH_3\overset{O}{\overset{\|}{C}}COO^- + 2\ H^+$$

Glucose **Pyruvic acid** **Pyruvate**

Glucose catabolism is exothermic. A portion of the energy released is conserved by synthesizing adenosine triphosphate (ATP) from adenosine diphosphate (ADP) and phosphate. Furthermore, the oxidation that occurs during glycolysis is accompanied by the reduction of nicotinamide adenine dinucleotide from its oxidized form (NAD^+) to its reduced form (NADH). A representative equation for glycolysis is

$$\text{Glucose} + 2\ \text{ADP} + 2\ P_i + 2\ NAD^+ \xrightarrow{\text{Glycolysis}} 2\ \text{Pyruvate} + 2\ \text{ATP} + 2\ \text{NADH}$$

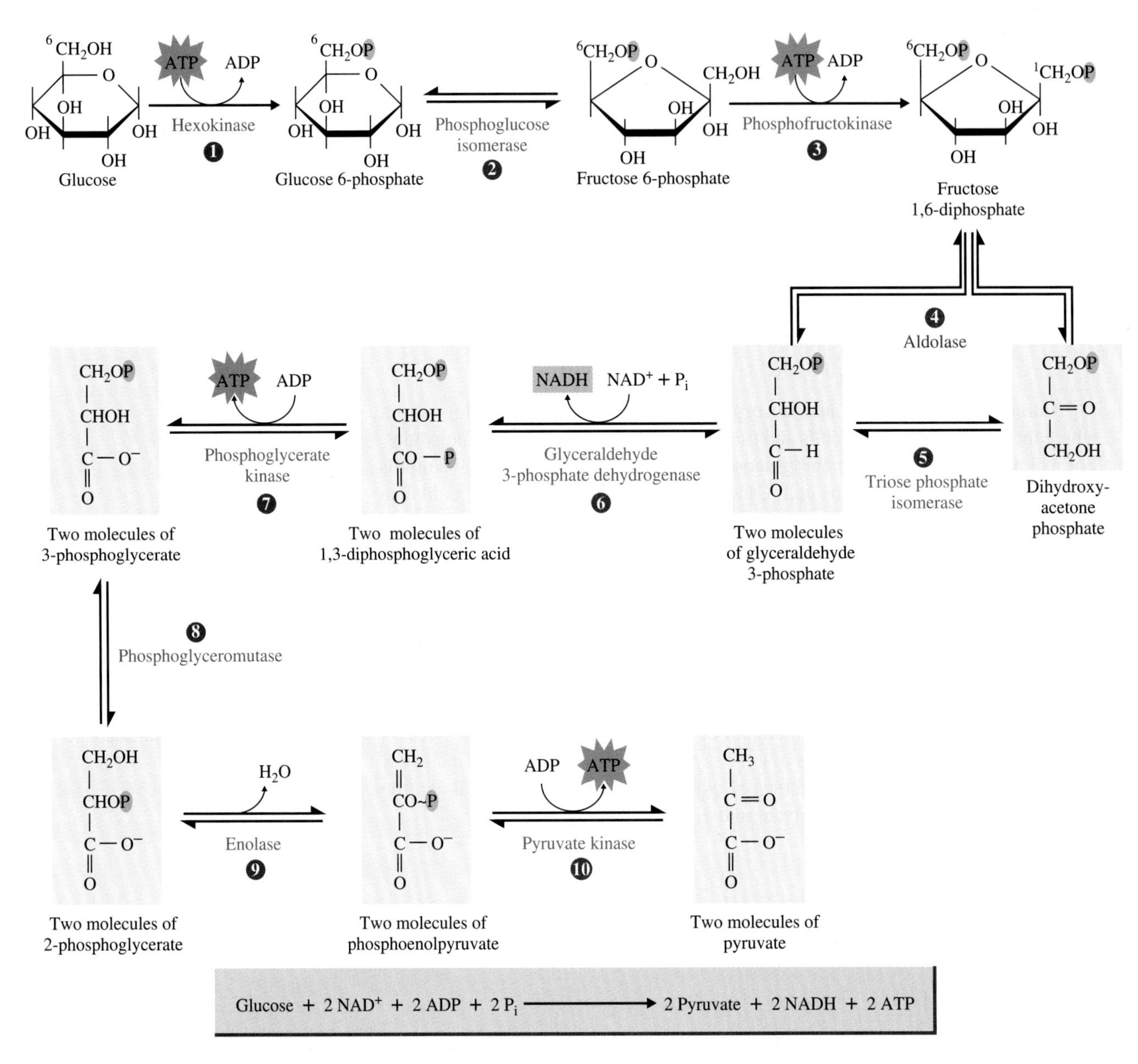

FIGURE 15.5
Glycolysis. This metabolic pathway converts glucose to pyruvate with production of energy. The steps of glycolysis are described in the text. Virtually all living organisms have the glycolytic pathway.

In the first step of glycolysis, a molecule of glucose is phosphorylated. The enzyme hexokinase catalyzes the transfer of a phosphate from ATP to the hydroxyl group on carbon 6 of glucose to produce glucose 6-phosphate (step 1, Figure 15.5). Note that even though glycolysis is a net producer of ATP, two ATPs are consumed in the early stages of glycolysis. Glucose 6-phosphate is a substrate of the next step of glycolysis, but the phosphorylation of glucose accomplishes something else. Glucose 6-phosphate is ionic, and ions do not readily cross cell membranes. Thus, phosphorylation of glucose traps glucose within the cell.

In the second step of glycolysis, glucose 6-phosphate is isomerized to fructose 6-phosphate in a reaction catalyzed by the enzyme phosphoglucose isomerase (step 2, Figure 15.5). Fructose 6-phosphate is then phosphorylated to fructose

1,6-diphosphate by transfer of a phosphate group from ATP (step 3, Figure 15.5). The current trend is to use *bis* instead of di. Thus, fructose 1,6-bisphosphate is a synonoym for fructose 1,6-diphosphate. This reaction is catalyzed by the enzyme phosphofructokinase. This is the second of two molecules of ATP that are consumed. The changes that occur in the sugars during the first steps of glycolysis are endothermic; ATP hydrolysis provides the energy to drive these reactions.

In the fourth reaction, the enzyme aldolase catalyzes the cleavage of fructose 1,6-diphosphate into two phosphorylated trioses, one molecule of glyceraldehyde 3-phosphate and one molecule of dihydroxyacetone phosphate (step 4, Figure 15.5). Next, dihydroxyacetone phosphate is isomerized into a molecule of glyceraldehyde 3-phosphate (step 5, Figure 15.5). This step is catalyzed by the enzyme triose phosphate isomerase. This enzyme and aldolase effectively convert fructose 1,6-diphosphate to two molecules of glyceraldehyde 3-phosphate. To balance glycolysis for 1 mol of glucose, each of the following steps should be multiplied by 2.

In the sixth step of glycolysis, glyceraldehyde 3-phosphate is both oxidized and phosphorylated. The enzyme glyceraldehyde 3-phosphate dehydrogenase catalyzes this complex reaction (step 6, Figure 15.5). The aldehyde group is oxidized to a carboxylic acid with concurrent reduction of NAD^+ to NADH. But the free carboxylic acid is not found because an inorganic phosphate becomes bonded to the carbon atom of the carboxyl group to form an anhydride bond. Just as the anhydride bonds of ATP are considered high-energy bonds, this mixed anhydride bond is also a high-energy bond. Since two glyceraldehyde 3-phosphates react, two NADH and two molecules of 1,3-diphosphoglycerate are formed for each glucose molecule. Phosphoglycerate kinase catalyzes the seventh step of glycolysis. 1,3-Diphosphoglycerate reacts with ADP to form ATP and 3-phosphoglycerate (step 7, Figure 15.5). The phosphate group that had been bonded to the carboxyl group is transferred to ADP. The high-energy mixed anhydride bond is broken in an exothermic reaction, but a portion of that energy is conserved in the endothermic formation of an anhydride bond in ATP. Two molecules of ATP and 3-phosphoglycerate are formed for each glucose. At this point, no net ATP synthesis has occurred because two ATPs were consumed earlier during the phosphorylation of glucose and fructose 6-phosphate.

The compound 3-phosphoglycerate is in turn converted to 2-phosphoglycerate by an isomerization catalyzed by the enzyme phosphoglyceromutase (step 8, Figure 15.5). The enzyme enolase catalyzes the dehydration of 2-phosphoglycerate into phosphoenolpyruvate (PEP) (step 9, Figure 15.5). The phosphoester bond in PEP is a high-energy bond. Two molecules of PEP are formed from a molecule of glucose. In the final reaction of glycolysis, PEP reacts with ADP in a reaction catalyzed by pyruvate kinase (step 10, Figure 15.5). Loss of phosphate from PEP to form pyruvate is highly exothermic, and again a portion of this energy is conserved through synthesis of a high-energy anhydride bond in ATP. Since two molecules of PEP are formed from each molecule of glucose, two molecules of pyruvate and two molecules of ATP are formed in this last reaction.

In summary, glycolysis oxidizes and cleaves one molecule of glucose to two molecules of pyruvate. Two molecules of NADH are formed. Two molecules of ATP are consumed with ADP as product, but later in the pathway four molecules of ATP are formed from four molecules of ADP. The pathway is a net producer of useful energy because two ATPs are now available to the cell.

♦ Glycolysis has a gross production of four ATPs, but the net ATP production is two.

$$\text{Glucose} + 2\ NAD^+ + 2\ ADP + 2\ P_i \longrightarrow 2\ \text{Pyruvate} + 2\ NADH + 2\ ATP$$

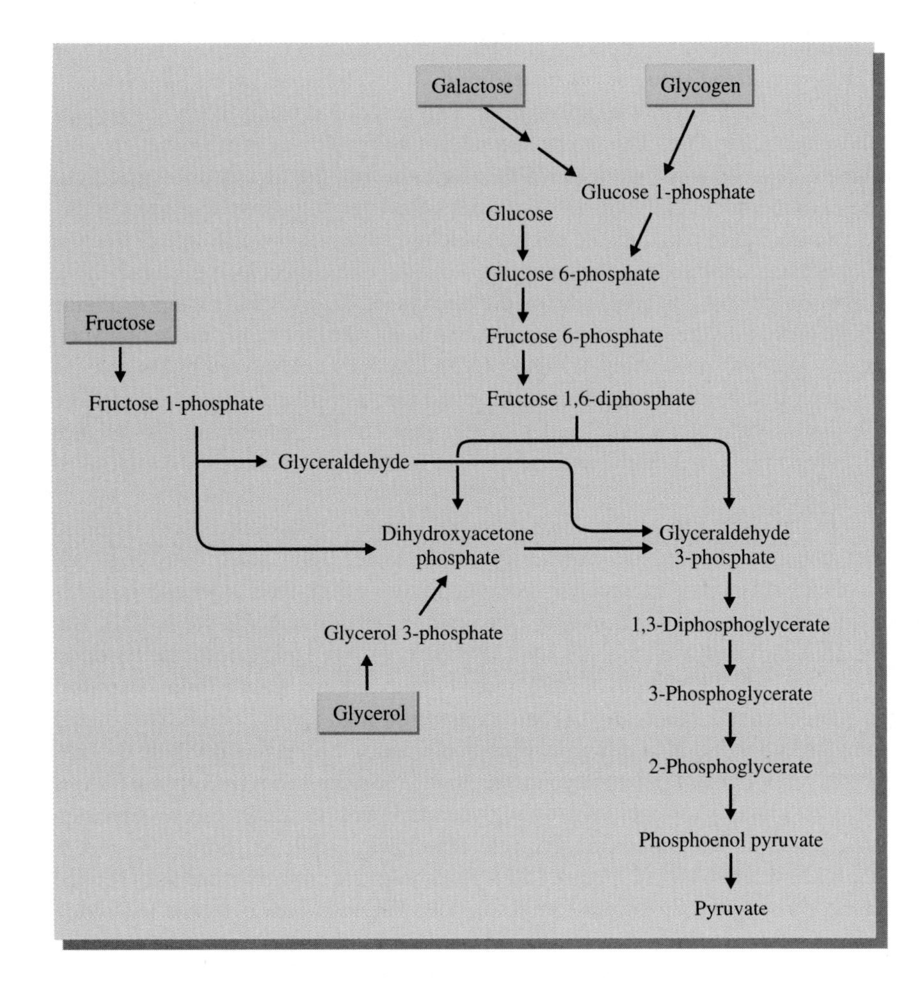

FIGURE 15.6
Other carbohydrates enter glycolysis in addition to glucose. This figure shows the entry points for galactose, glycogen, fructose, and glycerol.

Glucose is the principal compound that enters glycolysis, but it is not the only one. Other monosaccharides also enter glycolysis, though not necessarily at the same place as glucose. Figure 15.6 illustrates the entry point of some common carbohydrates.

Glycogen Catabolism

Glycogen, a polysaccharide found in the liver and muscle, serves as a storage form of glucose. When cellular energy is needed in muscle, glycogen reacts with inorganic phosphate in a reaction catalyzed by the enzyme glycogen phosphorylase, to yield glucose 1-phosphate.

$$\text{Glycogen}_n + \text{Phosphate} \xrightarrow{\text{Glycogen phosphorylase}} \text{Glycogen}_{n-1} + \text{Glucose 1-phosphate}$$

This highly regulated process is discussed in more detail in Chapter 17. The glucose 1-phosphate that is produced is converted to glucose 6-phosphate by a reaction catalyzed by the enzyme phosphoglucomutase. This molecule is an intermediate of glycolysis (Figure 15.6).

In the liver, the catabolism of glycogen often serves a different purpose. As in muscle, glucose 1-phosphate is produced from glycogen and inorganic phosphate and is isomerized to glucose 6-phosphate. This molecule is then cleaved to glucose and inorganic phosphate in a reaction catalyzed by glucophosphatase. Glucose is secreted from the liver into the blood as needed. Glycogen in muscle thus serves as a storage form of glucose to meet the immediate energy needs of muscle cells, whereas liver glycogen serves as a storage form of glucose for the entire body.

Pyruvate Catabolism

The pyruvate produced by glycolysis has two possible catabolic fates in mammals. If oxygen is available to the cell, *aerobic* catabolism will result in the oxidative decarboxylation of pyruvate to yield acetyl coenzyme A (acetyl-CoA) and CO_2 (Section 15.4). The acetyl-CoA, in turn, can be oxidized completely into carbon dioxide and water by the tricarboxylic acid cycle (see Section 15.4) and the electron-transport chain (see Section 15.5). These highly exothermic processes occur in the mitochondria. In these reactions, which are examples of **respiration**, the hydrogen atoms of the original substrate, glucose, are passed to an inorganic acceptor, oxygen. The NADH produced by glycolysis is also recycled to NAD^+ in the presence of oxygen (see Section 15.5).

◆ The catabolic fate of pyruvate depends on the presence or absence of oxygen.

If oxygen is not available to the cell, *anaerobic* catabolism occurs. Pyruvate is reduced to lactate in a reaction using NADH produced by glycolyis.

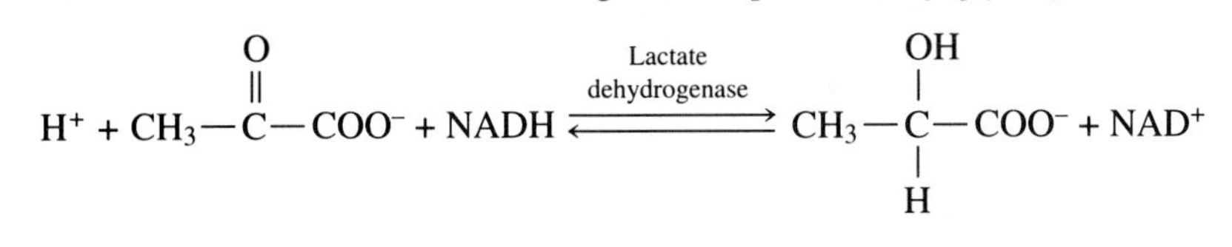

In this reaction, the hydrogen atoms derived from glucose are passed to pyruvate, an organic molecule. This process is a **fermentation**.

If NADH were not recycled to NAD^+ either in the presence or absence of oxygen, glycolysis would quickly come to a halt because cells contain only a small amount of NAD^+, which they continually recycle.

Yeast can also metabolize glucose in the absence of oxygen, but yeast recycle NADH and process pyruvate somewhat differently. In the absence of oxygen, yeast convert pyruvate first to acetaldehyde and carbon dioxide. The acetaldehyde in turn reacts with NADH to yield ethanol.

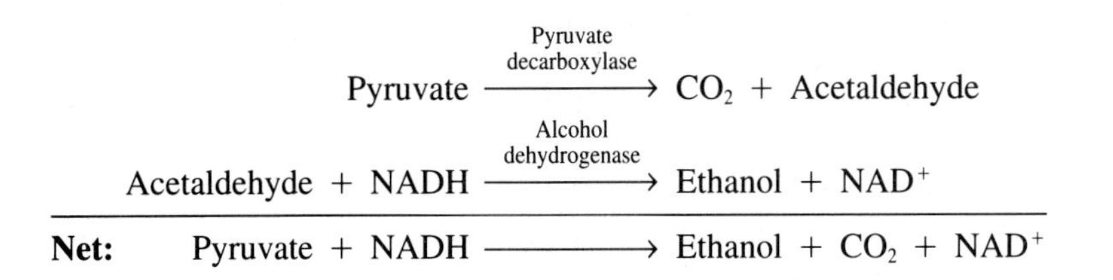

NAD^+ is recycled, and ethanol and carbon dioxide are produced. Since the acceptor of hydrogen atoms is an organic molecule, acetaldehyde, this is also a fermentation. This particular fermentation has been used by humans since antiquity—the baking, alcoholic beverage, and chemical industries use alcoholic fermentation as an industrial process.

NEW TERMS

glycolysis Cytosolic process that converts glucose to two molecules of pyruvate with the production of two molecules of ATP and two molecules of NADH

respiration Oxidation of a compound with transfer of electrons to an inorganic substance. Respiration using oxygen is the principal source of energy in the body.

fermentation Oxidation of a compound with transfer of electrons to an organic molecule. Formation of lactic acid from glucose in anaerobic muscle is a fermentation.

TESTING YOURSELF

Carbohydrate Catabolism

1. Compare the fate of pyruvate in aerobic and anaerobic catabolism.
2. Distinguish between a respiration and a fermentation.
3. Compare the net and gross production of ATP during glycolysis.

Answers **1.** During aerobic conditions pyruvate is converted to acetyl-CoA, which is catabolized further. During anaerobic conditions it is converted to lactate. **2.** In respiration, the electrons from the reacting molecule are accepted by an inorganic species, such as oxygen. In a fermentation, an organic molecule accepts the electrons. **3.** For each molecule of glucose, four ATP are produced during glycolysis. This is the gross production. Two ATP are consumed during glycolysis, so the net product is two ATP.

15.4 Tricarboxylic Acid Cycle

♦ Anaerobic catabolism of glucose is a very inefficient process for energy production.

Anaerobic catabolism of glucose, through glycolysis, yields two net ATPs. The useful energy saved in the synthesis of these two ATPs is about 14.6 kcal/mol of glucose. Complete oxidation of glucose to CO_2 and water yields 686 kcal/mol of glucose. The energy conserved by ATP synthesis in glycolysis is only a little more than 2 percent of the energy available when glucose is oxidized completely. Of course, glucose is not oxidized completely during glycolysis—much of the original chemical energy of glucose remains in the pyruvate molecules. The tricarboxylic acid cycle and electron transport coupled to oxidative phosphorylation are responsible for the complete oxidation of glucose with the concurrent release of large amounts of energy. This section discusses the role of the tricarboxylic acid cycle in catabolism.

If glucose is metabolized in the presence of oxygen, the pyruvate produced in glycolysis enters the mitochondrial matrix, where it is cleaved and oxidized to carbon dioxide and acetyl-CoA.

$$\text{Pyruvate} + \text{CoA} + \text{NAD}^+ \xrightarrow{\text{Pyruvate dehydrogenase}} \text{Acetyl-CoA} + CO_2 + \text{NADH}$$

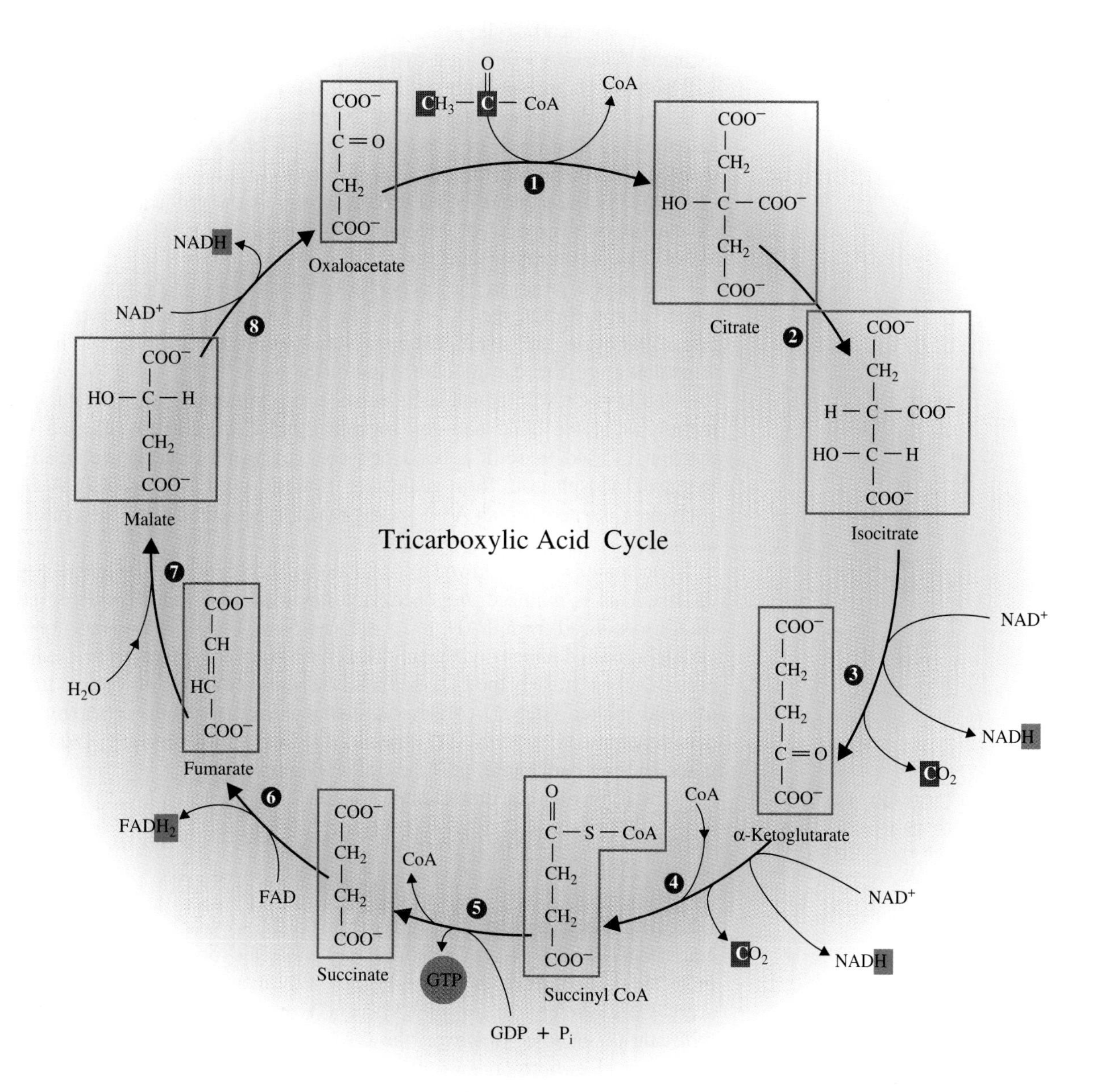

FIGURE 15.7
The tricarboxylic acid cycle. In this cycle, the carbon atoms (gray) of acetic acid (as acetyl-CoA) are oxidized to carbon dioxide. Energy is conserved as reduced NADH and $FADH_2$ (blue) and as GTP (purple). Steps 1 through 8 are discussed in the text.

The acetyl-CoA is then oxidized to carbon dioxide by the reactions of the **tricarboxylic acid** or **TCA cycle** (Figure 15.7). These oxidative reactions are highly exothermic. Some of the energy released by these reactions is conserved in the form of reduced and phosphorylated coenzymes. This cycle is also called both the *Krebs cycle*, named for Hans Krebs, the discover of many of the reactions in the cycle, and the *citric acid cycle*, named for the first product of the cycle.

In the first reaction of the TCA cycle, acetyl-CoA and oxaloacetate react to yield a molecule of citrate and a molecule of CoA (step 1, Figure 15.7). Citrate is one of two tricarboxylic acids found in this cycle that provide the cycle with its name. Citrate is isomerized by the enzyme aconitase to isocitrate (step 2).

Isocitrate is the substrate of isocitrate dehydrogenase, an enzyme that catalyzes the oxidation and cleavage of isocitrate to a molecule of CO_2 and α-ketoglutarate. During this oxidation, a molecule of NAD^+ is reduced to NADH (step 3). This step is the first of four oxidative steps in the cycle. The next step is an oxidative step that involves several substrates and products. In this reaction, α-ketoglutarate is oxidized, yielding CO_2 and succinyl-CoA. NAD^+ is reduced to NADH concurrently (step 4).

The acetyl group of the acetyl-CoA that entered the cycle contained two carbon atoms. So far, the TCA cycle has produced two molecules of CO_2. The rest of the cycle traps additional energy and returns the intermediate molecules to oxaloacetate, the starting compound of the cycle.

Succinyl-CoA is the substrate of succinate thiokinase, which catalyzes the hydrolysis of the bond between succinate and CoA. Energy released in this reaction is used to form a bond between guanosine diphosphate (GDP) and inorganic phosphate to form guanosine triphosphate (GTP) (step 5). A GTP is equivalent in energy to an ATP, so this reaction is equivalent to the synthesis of an ATP.

Succinate is next oxidized to fumarate by succinate dehydrogenase (step 6). As succinate is oxidized, the coenzyme flavin adenine dinucleotide is reduced from its oxidized form, FAD, to its reduced form, $FADH_2$. Fumarate contains a carbon–carbon double bond that undergoes the reactions typical of this functional group. In the next step, the enzyme fumarase adds a molecule of water to fumarate to yield malate (step 7). Malate is then oxidized to oxaloacetate by malate dehydrogenase (step 8). NAD^+ is reduced to NADH concurrently. Oxaloacetate is the starting compound, thus the cycle is complete.

♦ As acetyl-CoA is oxidized to carbon dioxide, coenzymes are reduced or converted to high-energy forms.

The following equation summarizes the TCA cycle:

$$\text{Acetyl-CoA} + 3\ NAD^+ + FAD + GDP + P_i \longrightarrow 2\ CO_2 + CoA + 3\ NADH + FADH_2 + GTP$$

Little available energy comes directly from the TCA cycle because only one ATP equivalent (the GTP) was produced per molecule of acetyl-CoA. Since one molecule of glucose yields two molecules of pyruvate and thus two molecules of acetyl CoA, two equivalents of ATP as GTP are produced in the TCA cycle per molecule of glucose. However, the NADH and $FADH_2$ produced in the TCA cycle and elsewhere contain significant amounts of energy that can be used for the synthesis of ATP. The processes of electron transport and oxidative phosphorylation are responsible for this synthesis of ATP.

NEW TERM

tricarboxylic acid, or **TCA, cycle** A cyclic pathway that oxidizes acetyl-CoA to CO_2, yielding reduced and phosphorylated coenzymes as the other products

TESTING YOURSELF

TCA Cycle

1. Which steps in the TCA cycle involve oxidation–reduction?

Answers **1.** Steps 3 (isocitrate dehydrogenase), 4 (α-ketoglutarate dehydrogenase), 6 (succinate dehydrogenase), and 8 (malate dehydrogenase). This can be recognized because dehydrogenases and the coenzymes NAD^+ and FAD participate in oxidation–reduction reactions.

15.5 Electron Transport and Oxidative Phosphorylation

Aerobic catabolism of glucose yields either 36 or 38 ATPs per glucose molecule, depending upon the fate of cytosolic NADH. Two ATPs were produced in glycolysis, and the two GTPs produced in the TCA cycle are equivalent to ATP, so four ATPs are accounted for. The remaining ATPs are synthesized by **oxidative phosphorylation**, a process in which ATP synthesis from ADP and phosphate is coupled to an oxidative process. Transport of electrons via an **electron-transport chain** within the mitochondria is intimately involved in this oxidation. This section describes electron transport and oxidative phosphorylation and accounts for the aerobic production of energy in the cell.

Oxidative reactions are generally exothermic, and the oxidation of NADH and $FADH_2$ to NAD^+ and FAD by oxygen are no exceptions.

$$NADH + H^+ + \tfrac{1}{2}O_2 \longrightarrow NAD^+ + H_2O + \text{Energy (52.6 kcal/mol)}$$

$$FADH_2 + \tfrac{1}{2}O_2 \longrightarrow FAD + H_2O + \text{Energy (43.4 kcal/mol)}$$

The formation of ATP from ADP and inorganic phosphate is endothermic; 7.3 kcal of energy is required per mole of ATP formed. Clearly enough energy is available during the oxidation of these reduced coenzymes to synthesize some ATP if the two processes are coupled.

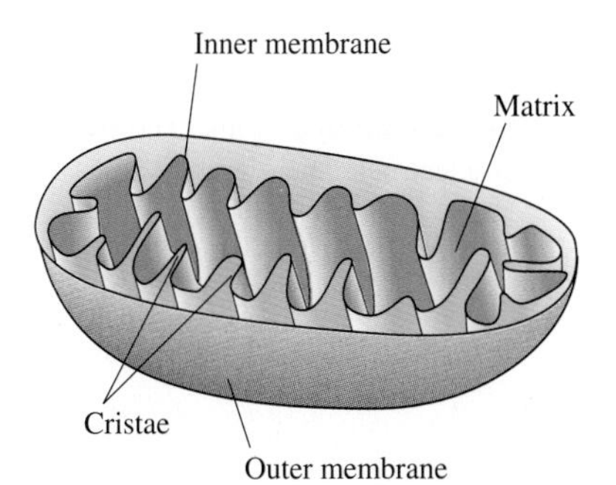

FIGURE 15.8
The matrix and inner membrane of a mitochondrion contain the important components of the TCA cycle (most are in the matrix), electron transport (inner membrane), and oxidative phosphorylation (inner membrane). The cristae are invaginations of the inner membrane that greatly increase its surface area.

Electron-Transport Chain

The mechanism by which NADH and $FADH_2$ are oxidized is understood in principle. When NADH is oxidized to NAD^+, two electrons pass from NADH to molecular oxygen through a series of electron carriers that are part of the inner membrane of the mitochondrion (Figure 15.8). Each of the carriers are alternately reduced and oxidized as they first accept then pass the electrons on to the next carrier of the electron-transport system. Ultimately, molecular oxygen accepts the electrons along with protons from solution to become water (Figure 15.9).

For each mole of NADH that is oxidized to NAD^+, over 52 kcal of energy is released. The **chemiosmotic theory** provides an explanation for how a portion of this energy is conserved. Energy is released as electrons flow from NADH through the carriers to oxygen. Some of this energy is used to pump protons (H^+) from the mitochondrial matrix to the outside of the mitochondrion. For each pair of electrons that flows from an NADH to molecular oxygen, three sets of protons

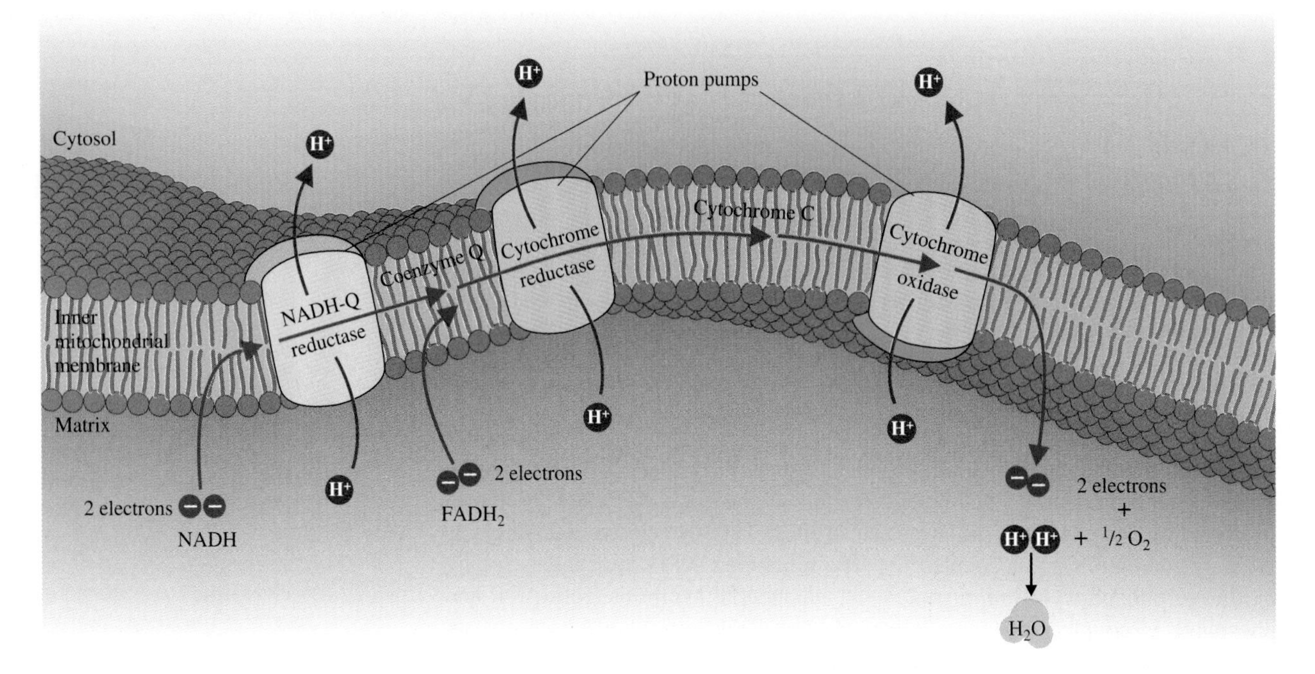

FIGURE 15.9
The electron-transport chain. As electrons pass from NADH to oxygen (blue arrows), they pass through three proton pumps. The pumps are protein complexes that use the energy of this electron flow to pump protons from the mitochondrial matrix into the cytosol (red arrows). The electrons from $FADH_2$ pass through only two proton pumps, thus fewer protons are transferred.

♦ The energy released when NADH is oxidized is used to establish and maintain an electrochemical gradient.

are pumped (Figure 15.9). This translocation of protons establishes a gradient that is both chemical and electrical in nature. Whenever a substance has a higher concentration in one region than another, a chemical gradient exists for that substance. If the substance is also electrically charged, as is a proton, then an electrical gradient also exists. Chemical and electrical gradients have potential energy. For example, it is the energy of a gradient that is responsible for the diffusion of particles from a region of higher concentration to a region of lower concentration. The electron-transport system of mitochondria is not just a transporter of electrons, it is also a proton pump that derives its energy from that flow of electrons. The pumping of protons in mitochondria is powered by a flow of electrons just as water pumped by an electric pump is powered by a flow of electrons (electricity). The chemical energy stored in the reduced state of NADH is converted to electric energy that is then used to establish a gradient possessing stored energy.

The oxidation of $FADH_2$ to FAD by oxygen also yields energy, but less than the oxidation of NADH: 43.4 kcal/mol versus 52.6 kcal/mol. This smaller quantity of energy pumps fewer protons. Figure 15.9 shows that the electrons from $FADH_2$ enter the electron-transport system at a different point from the electrons from NADH. When $FADH_2$ is oxidized, only two sets of protons are pumped. The energy from the oxidation of NADH drives three proton pumps, whereas the energy from the oxidation of $FADH_2$ drives two of the pumps.

Oxidative Phosphorylation

ATP is synthesized by the phosphorylation of ADP. When this synthesis is associated with the electron-transport system, it is called *oxidative phosphorylation*. It is observed that three ATPs are synthesized for each NADH that is oxidized, and two ATPs are formed for each oxidation of an $FADH_2$.

$$NADH + \frac{1}{2}O_2 + H^+ + 3\ ADP + 3\ P_i \longrightarrow NAD^+ + H_2O + 3\ ATP$$

$$FADH_2 + \frac{1}{2}O_2 + 2\ ADP + 2\ P_i \longrightarrow FAD + H_2O + 2\ ATP$$

This ratio of 3:2 is the same as the ratio of protons pumped when NADH and $FADH_2$ are oxidized. The amount of ATP synthesized is a direct result of the proton gradient, which in turn is a reflection of the amount of energy that was available to pump protons. The proton gradient is potential energy that can be used to do work. In oxidative phosphorylation, that energy is used to synthesize ATP from ADP and P_i.

♦ Energy stored in the proton gradient is used to make ATP from ADP and inorganic phosphate.

Within the inner mitochondrial membrane are multiple copies of a protein complex called ATP synthetase. In Figure 15.10 ATP synthetase is labeled F_0 and F_1. F_0 is imbedded in the membrane and appears to serve as a channel through which protons can pass from the outside of the membrane to the inside. As protons pass in a controlled fashion through this channel, a portion of the potential energy of the gradient is lost. F_1 uses some of that energy to synthesize ATP from ADP and inorganic phosphate (Figure 15.10).

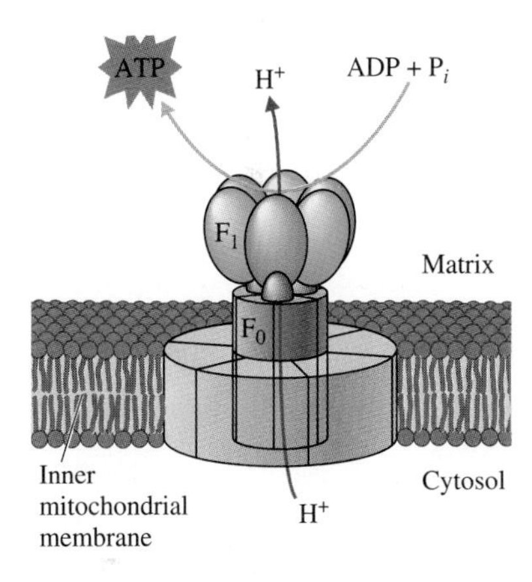

FIGURE 15.10
ATP synthetase consists of protein particles F_1 and F_0. As protons pass through F_0, the electrochemical gradient is reduced slightly. A portion of the energy made available from this proton flow is used to phosphorylate ADP to ATP. ATP synthetase is responsible for most of the cellular synthesis of ATP.

If the concentration of ATP is high within a cell, the concentration of ADP is necessarily low, because ATP is made from ADP and P_i.

$$ADP + P_i \rightleftharpoons ATP + H_2O$$

The aerobic synthesis of ATP, oxidative phosphorylation, is highly regulated; if little ADP is present in the cell, little oxidative phosphorylation occurs. If the concentration of ATP decreases, its rate of synthesis through oxidative phosphorylation increases. The gradient of protons is maintained in a similar way. If oxidative phosphorylation is occurring rapidly, then protons rapidly enter the mitochondrion. Electron transport must occur rapidly to maintain the proton gradient. Similarly, if NADH and $FADH_2$ are being rapidly consumed by the electron-transport system, then the TCA cycle will become very active to reduce the NAD^+ and FAD that is being formed. While the TCA cycle is very active, considerable acetyl-CoA is required, and the activity of glycolysis and pyruvate dehydrogenase or some other pathway must provide it in sufficient quantity. For convenience, the parts of metabolism are studied separately, but metabolism actually functions as a highly integrated, highly regulated process.

♦ Although metabolism is broken down into its components for easy study, it is actually a highly integrated process.

Analysis of the Energy from Glucose

Consider now the origin of the 36 or 38 ATPs that are formed from a glucose molecule. Glycolysis produces two ATPs and two NADHs. These NADHs are produced in the cytosol, but NADH cannot enter mitochondria directly. Cells of

TABLE 15.2

Two Alternatives for Aerobic Production of ATP from Glucose

Electrons passed to NAD^+ to form NADH	
From glycolysis =	2 ATP
2 NADHs from glycolysis. Electrons transported into mitochondria to form NADH. Thus, 2 NADH × 3 ATP/NADH =	6 ATP
2 NADHs from conversion of 2 pyruvate to acetyl-CoA. Thus, 2 NADH × 3 ATP/NADH =	6 ATP
2 acetyl-CoA molecules through TCA cycle yield 12 ATPs each. Thus, 2 acetyl-CoA × 12 ATP/acetyl-CoA =	24 ATP
	38 ATP
Electrons passed to FAD to form $FADH_2$	
From glycolysis =	2 ATP
2 NADHs from glycolysis. Electrons transported into mitochondria to form $FADH_2$. Thus, 2 $FADH_2$ × 2 ATP/$FADH_2$ =	4 ATP
2 NADHs from conversion of 2 pyruvate to acetyl-CoA. Thus, 2 NADH × 3 ATP/NADH =	6 ATP
2 acetyl-CoA molecules through TCA cycle yield 12 ATPs each. Thus, 2 acetyl-CoA × 12 ATP/acetyl-CoA =	24 ATP
	36 ATP

the body transport the electrons of cytosolic NADH into mitochondria by two different mechanisms: (1) the electrons are passed to NAD^+ within the mitochondria to form NADH or (2) the electrons are passed to an FAD to form $FADH_2$ in the mitochondria (Table 15.2). This is an important difference because $FADH_2$ yields one less ATP than NADH; two fewer ATPs are synthesized per glucose if the second transport mechanism is used.

♦ The amount of ATP formed from aerobic catabolism of glucose depends on how glycolytic NADH is recycled.

When pyruvate is converted to acetyl-CoA, an NADH is formed. Two NADHs and two acetyl-CoA molecules are formed per glucose molecule. When the two molecules of acetyl-CoA pass through the TCA cycle, three NADHs, one $FADH_2$, and one GTP are formed for each acetyl-CoA. After electron transport and oxidative phosphorylation, twelve ATPs are made per molecule of acetyl-CoA that passes through the TCA cycle, 24 per glucose molecule. In contrast, only two ATPs are formed per glucose by anaerobic catabolism of glucose. The amount of ATP formed from glucose catabolism is shown in Table 15.2

NEW TERMS

oxidative phosphorylation A process that synthesizes ATP from ADP and inorganic phosphate using energy stored in a proton gradient

electron-transport chain A series of electron carriers that transport electrons from the reduced coenzymes NADH and $FADH_2$ to oxygen. Energy released during this process is used to pump protons.

chemiosmotic theory A theory stating that the energy released by the flow of electrons along the electron-transport chain is used to establish a proton gradient. This high-energy gradient can be used to do work, including the synthesis of ATP.

TESTING YOURSELF

Electron Transport and Oxidative Phosphorylation

1. Twelve ATPs are obtained from each acetyl-CoA molecule that passes through the TCA cycle, yet no ATP is produced directly. Account for the 12 ATPs.
2. When 1 mol of NADH is oxidized to NAD^+, 52.6 kcal of energy is released. When 1 mol of ATP is synthesized from ADP and P_i, 7.3 kcal of energy must be added. If three ATPs are synthesized for each NADH oxidized, what is the efficiency of this coupling?
3. Proteins in the inner membrane of a mitochondrion pump protons from inside the mitochondrion to the outside to maintain a proton gradient. What is the source of the energy for this pumping?

Answers **1.** One GTP is produced and is equivalent to one ATP in energy. Three NADHs are produced, and when one NADH is oxidized by oxygen, enough energy is released to synthesize three ATPs. One $FADH_2$ is produced, and oxidation of this molecule by oxygen yields two ATPs. **2.** $[(3 \times 7.3 \text{ kcal/mol})/52.6 \text{ kcal/mol}] \times 100 = 42\%$. **3.** When NADH or $FADH_2$ is oxidized, electrons pass from it to oxygen. The energy released during this oxidation is used to pump protons.

15.6 Lipid Catabolism

Adipose tissue plays an important role in storing and releasing lipids, in much the same way as the liver stores glucose as glycogen and releases it as needed. Earlier it was stated that triacylglycerols are hydrolyzed in adipose cells to glycerol and fatty acids (see Section 15.2) The fatty acid molecules are released into the blood, where they bind to serum albumin, which is a transport protein for these hydrophobic molecules. The cells of the body absorb these free fatty acids as needed. The concentration of fatty acids in the blood is rather low, only a few milligrams per 100 mL, but they are constantly being absorbed by the cells of the body and replaced by new fatty acids from the adipose cells. Serum fatty acids are in rapid dynamic equilibrium with fatty acids in body and adipose cells. A significant amount of the energy needs of a human are met by catabolism of fatty acids.

◆ Free fatty acids bound to serum albumin are constantly available to the cells of the body.

The glycerol produced by hydrolysis of triacylglycerols is readily catabolized. Glycerol is a carbohydrate, not a lipid. Within cells, glycerol is readily converted to glycerol-3-phosphate. This is oxidized to dihydroxyacetone phosphate, which is an intermediate of glycolysis (see Figure 15.6).

Once inside the cell, the water-insoluble fatty acids are bound to a coenzyme A molecule that serves as a carrier. This reaction is an energy-requiring reaction in which concurrent ATP hydrolysis to adenosine monophosphate (AMP) and inorganic pyrophosphate (PP_i) provides the energy. Because two P_i molecules must be added to AMP to get back to ATP, hydrolysis of ATP to AMP is equivalent in energy to two high-energy bonds.

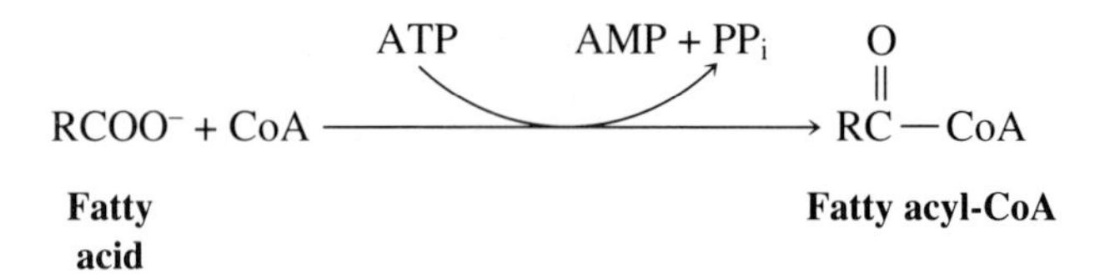

Fatty acyl-CoA molecules diffuse through the cytosol to mitochondria where they are split. The fatty acid part is transported into the mitochondrion bound to carnitine, which is $^{+}N(CH_3)_3CH_2CHOHCH_2COO^-$. Inside the mitochondrion the fatty acid is rebound to CoA.

Fatty acyl-CoA + Carnitine **(Cytosol)**	⟶	Fatty acyl carnitine + CoA **(Cytosol)**
Fatty acyl carnitine **(Cytosol)**	⟶	Fatty acyl carnitine **(Mitochondrion)**
Fatty acyl carnitine + CoA **(Mitochondrion)**	⟶	Fatty acyl-CoA + Carnitine **(Mitochondrion)**

Beta Oxidation of Fatty Acids

♦ Beta oxidation is the principal catabolic pathway for fatty acids.

Complete lipid catabolism is an aerobic process that occurs within mitochondria. The series of reactions that catabolize fatty acids constitute a cycle, or spiral, called **beta oxidation**. In each cycle, the fatty acid is oxidized and cleaved, yielding a smaller fatty acid, an acetyl-CoA, a NADH, and a $FADH_2$. In essence, beta oxidation catabolizes fatty acids just as glycolysis catabolizes sugars.

The compound palmitoyl-CoA (Figure 15.11) will be used to illustrate the steps of beta oxidation. In the first step, palmitoyl-CoA is oxidized to an unsaturated compound by removal of two hydrogen atoms. This oxidation is accompanied by the reduction of FAD to $FADH_2$ (Figure 15.11). This oxidation is similar to the oxidation of succinate to fumarate in the TCA cycle (see Section 15.4). The name *beta oxidation* signifies that the oxidation occurs between the β- and γ-carbons of the acyl-CoA.

The unsaturated product of the first step is hydrated by the addition of water to the double bond. This addition of water yields an alcohol, just as water was added to fumarate to make malate in the TCA cycle. The specificity of the enzyme ensures that the hydroxyl group is added to the β-carbon of the intermediate (step 2, Figure 15.11). In step 3, the β-hydroxyl group is oxidized to a β-keto group. This oxidation is accompanied by the reduction of NAD^+ to NADH. This intermediate has the same number of carbon atoms as the starting acyl-CoA, but the β-carbon atom has been oxidized from the oxidation number of an alkane to that of a carbon atom of a carbonyl group.

In the fourth step of beta oxidation, the intermediate is cleaved into two smaller molecules. The bond between the α- and β-carbon atoms is broken, and the β-carbon atom is transferred to a molecule of coenzyme A to form myristoyl-CoA, the CoA derivative of myristic acid. The rest of the intermediate leaves the reaction as acetyl-CoA. The product, myristoyl-CoA, is a compound identical to the starting material except that it has two fewer carbon atoms. These four steps of beta oxidation can be repeated on myristoyl-CoA to yield lauroyl-CoA, which

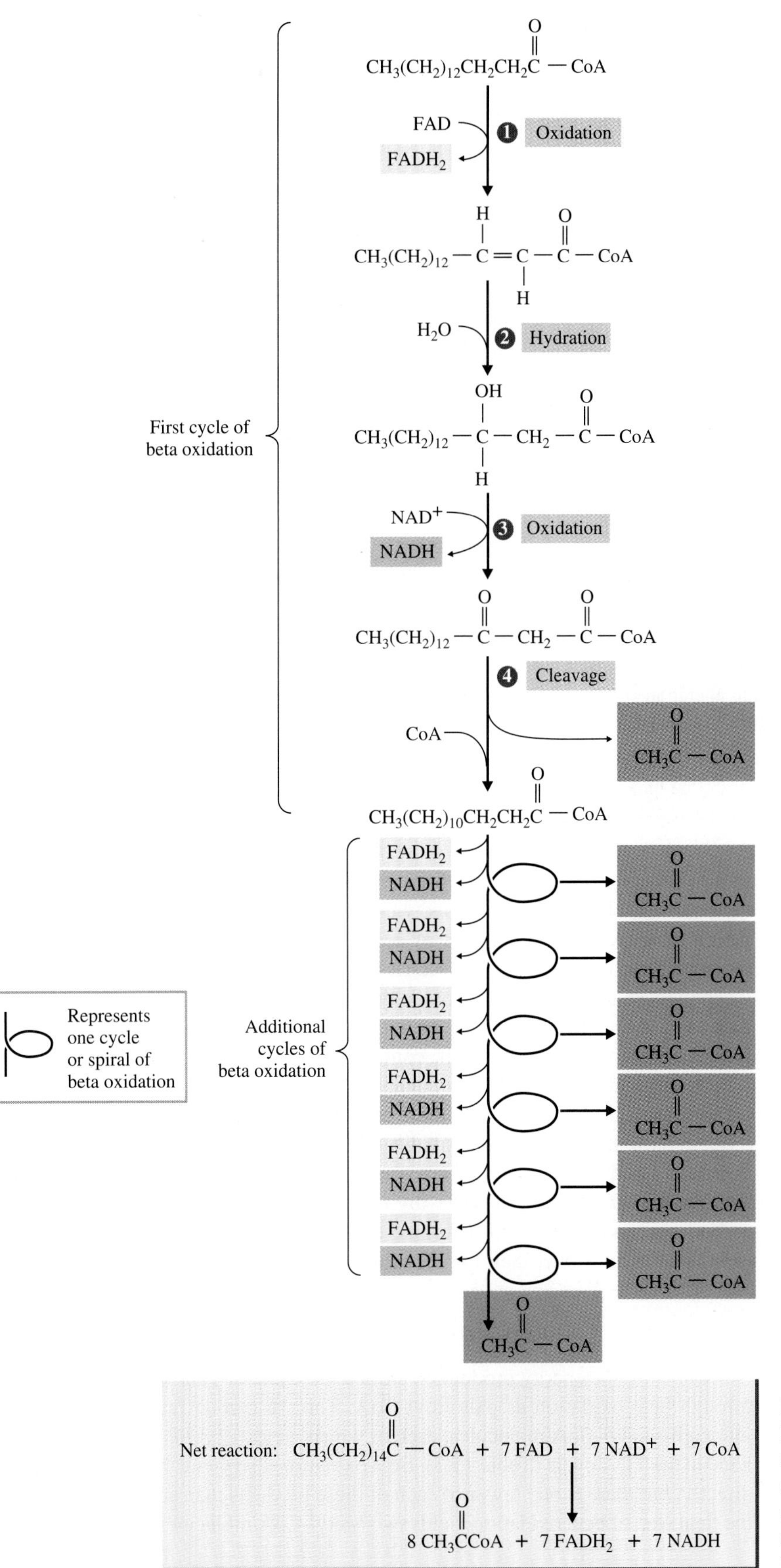

Net reaction: $CH_3(CH_2)_{14}C(=O)-CoA + 7\,FAD + 7\,NAD^+ + 7\,CoA \rightarrow 8\,CH_3C(=O)CoA + 7\,FADH_2 + 7\,NADH$

FIGURE 15.11
Beta oxidation of palmitoyl-CoA (palmitic acid). The specific steps are described in the text, but note that eight acetyl-CoA (red) are produced, one more than $FADH_2$ (yellow) and NADH (orange).

CHEMISTRY CAPSULE

Tay-Sachs Disease

One of the better known genetic disorders associated with lipid metabolism is *Tay-Sachs disease*. Gangliosides, which are glycolipids derived from sphingosine, are common in healthy brain tissue. Gangliosides are constantly being made and broken down in tissues. But in individuals with Tay-Sachs disease, the enzyme hexosaminidase A is missing. This enzyme is responsible for the cleavage of a sugar derivative from a specific ganglioside.

$$\text{Tay-Sachs ganglioside} \xrightarrow{\text{Hexosaminidase A}} \text{Hematoside} + N\text{-Acetylneuraminate}$$

Because this enzyme is not present, the concentration of Tay-Sachs ganglioside builds up in the brain. This accumulation results in paralysis, blindness, and early death, typically by the age of 4. Symptoms appear in the first several months of life. The patients typically have enlarged heads with doll-like faces and a characteristic alteration of the retina. No cure or treatment has been found for this disease at this time. However, the condition can be detected in embryos by testing for hexosaminidase A activity in amniotic fluid.

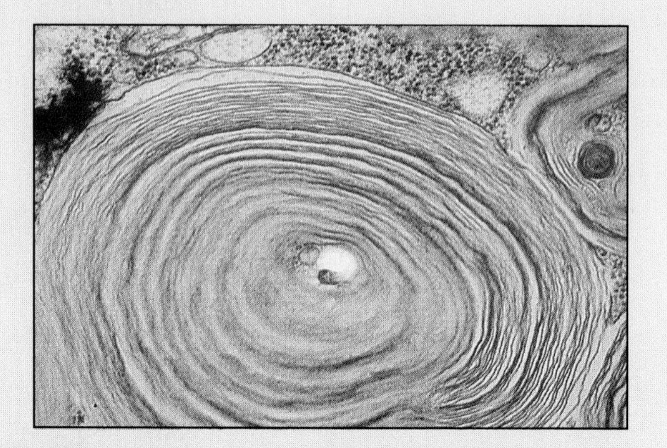

FIGURE 15.C
Degeneration of central nervous tissue from accumulated gangliosides causes Tay-Sachs disease.

in turn is beta-oxidized. Each subsequent acyl-CoA is oxidized and cleaved until the final step cleaves butyryl-CoA, yielding two molecules of acetyl-CoA. From palmitoyl-CoA, eight molecules of acetyl-CoA are produced along with seven molecules each of $FADH_2$ and NADH. Acetyl-CoA and NADH are also produced when glucose is catabolized aerobically. This is an important feature of catabolic pathways. Although the starting compounds may be very different, the products are generally one of a few small, common molecules found in the cell.

♦ Beta oxidation of fatty acids yields acetyl-CoA, NADH, and $FADH_2$.

Recall that the common fatty acids found in nature have an even number of carbon atoms. Beta oxidation of these compounds always yields acetyl-CoA. The unsaturated fatty acids are also oxidized by beta oxidation, but the product yield is different and a few additional enzymes are needed. Each double bond in a fatty acid reduces the yield of $FADH_2$ by one. An $FADH_2$ is produced whenever a carbon–carbon double bond is formed during beta oxidation; if the bond already exists in the molecule, then the reaction does not occur.

Energy Production from Fatty Acids

Consider the energy production from a molecule of palmitic acid that passes through beta oxidation as palmitoyl-CoA. This 16-carbon fatty acid is cleaved into eight acetyl-CoA molecules, each of which yields 12 ATP when it is metabolized in the TCA cycle (Table 15.3). Beta oxidation also yields NADH and $FADH_2$ directly, but there is one fewer of each of these products than acetyl-CoA because the final step of beta oxidation yields two acetyl-CoA molecules. These molecules

TABLE 15.3
Production of ATP from Palmitic Acid

8 acetyl-CoA × 12 ATP/acetyl-CoA =	96 ATP
7 NADH × 3 ATP/NADH =	21 ATP
7 $FADH_2$ × 2 ATP/$FADH_2$ =	14 ATP
Gross production of ATP from palmitic acid	131 ATP
Activation of fatty acid for entry into beta oxidation =	−2 ATP
Net production of ATP from palmitic acid	129 ATP

also yield ATP when they are oxidized (Table 15.3). But there is a cost associated with beta oxidation; an ATP was broken down into an AMP when the fatty acid was attached to coenzyme A. The hydrolysis of ATP to AMP reduces the net yield of high-energy bonds by two.

Fatty acids can also be oxidized at other atoms besides the β-carbon. Alpha oxidation yields fatty alpha keto acids, and omega oxidation yields dicarboxylic acids. These pathways are quantitatively less important than beta oxidation, but they do yield essential products.

Other Lipid Catabolism

Triacylglycerols are by far the most abundant class of lipids in the diet. Other saponifiable lipids are hydrolyzed to fatty acids that are catabolized by beta oxidation. There are specific catabolic pathways for many of the other components, but they are quantitatively less important and too numerous for a detailed study. The catabolism of cholesterol deserves comment because cholesterol seems to have such an important role in health. Small amounts of cholesterol are converted to steroid hormones, and some more is lost in the digestive tract. But the principal route for catabolism of cholesterol, about three fourths of it, is its conversion to bile salts. These compounds emulsify dietary lipid in the small intestine and are then reabsorbed from the small intestine. The reabsorption is not 100 percent efficient, however; some cholesterol must be continually converted to bile salts.

NEW TERMS

beta oxidation Metabolic pathway that oxidizes fatty acids to acetyl-CoA and reduced coenzymes

TESTING YOURSELF

Lipid Catabolism

1. Use Table 15.2 as a guide to determine the production of ATP by beta oxidation of stearic acid.

2. Triacylglycerols are composed of three fatty acids and a glycerol molecule. These components are catabolized in what pathways?

Answers **1.** Stearoyl-CoA yields 9 acetyl-CoA, 8 NADH, and 8 $FADH_2$. These yield 108, 24, and 16 ATP, respectively, for a total gross production of 148 ATP. Two ATP equivalents are required to activate the fatty acid, thus the net production is 146 ATP. **2.** The fatty acids are catabolized by beta oxidation. The glycerol is converted to dihydroxyacetone phosphate, which is an intermediate of glycolysis.

15.7 Amino Acid Catabolism

Amino acids derived from dietary protein are absorbed by the blood from the small intestines and join a pool of amino acids that circulate in the blood. A portion of these amino acids are used to synthesize proteins within cells of the body, but most diets provide more amino acids than are needed for protein synthesis. Excess amino acids are metabolized for energy or converted to other molecules.

Transamination of Amino Acids

♦ Catabolism of amino acids begins with the removal of amino groups via transamination.

The catabolism of amino acids ultimately requires 20 different pathways because the structures of the amino acids are so diverse. There are, however, several features of amino acid catabolism that are common to all of the amino acids. The first of these features is **transamination**. In this reaction, the α-amino group of an amino acid is transferred to an acceptor molecule, α-ketoglutarate (α-KG).

$$\text{Amino acid} + \alpha\text{-KG} \xrightarrow{\text{Transaminase}} \text{Keto acid} + \text{Glutamate}$$

In the cells of the liver, several transaminases are present. These enzymes collectively transfer the amino groups of all of the amino acids to α-KG. These transaminations effectively collect all of the nitrogen of the α-amino groups into a single molecule, glutamate, and generate keto acids that are the substrates for the specific catabolic pathways. The role of transamination in amino acid catabolism is shown in Figure 15.12.

Oxidative Deamination

The process of transamination collects nitrogen atoms as amino groups in glutamate. The nitrogen of these amino groups is not needed by the body and must be eliminated safely and efficiently. The next step leading to nitrogen excretion is **deamination**. In this reaction, the enzyme glutamate dehydrogenase catalyzes the deamination of glutamate to α-KG.

$$\text{Glutamate} + NAD^+ \xrightleftharpoons{\text{Glutamate dehydrogenase}} \alpha\text{-KG} + NADH + NH_3$$

This reaction recycles glutamate to α-KG and effectively converts what had been the amino groups of amino acids to ammonia (Figure 15.12). This does pose a potential problem because ammonia is toxic. Note that the deamination reaction

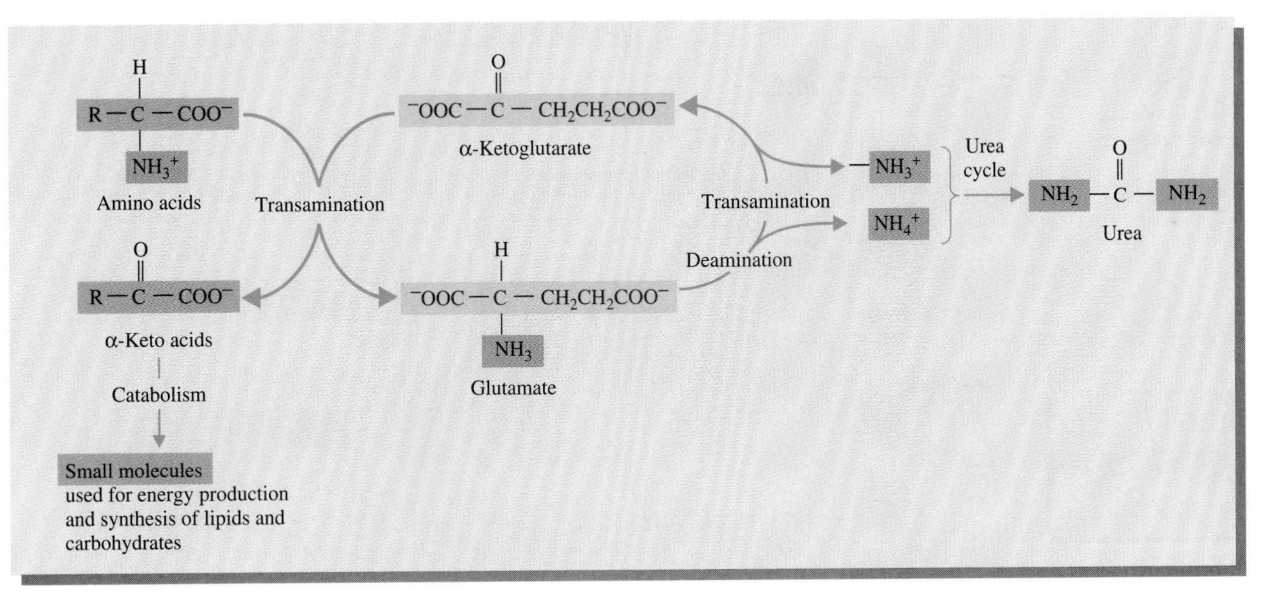

FIGURE 15.12
An overview of amino acid catabolism. Transamination yields alpha-keto acids that are metabolized to smaller molecules. The α-amino group from the amino acids combines with α-ketoglutarate to form glutamate. The amino groups of glutamate molecules are removed via transamination and deamination reactions. The nitrogenous products of these reactions are used to make urea.

is reversible. If the concentration of ammonia gets high enough, the reverse reaction will occur, and cellular α-KG will be converted to glutamate. Alpha-ketoglutarate is an intermediate of the TCA cycle, and if it is not available, the cell cannot carry out the aerobic metabolism that is responsible for most of the available energy in cells. Some fish and other aquatic species that have abundant environmental water available simply excrete ammonia directly into the environment before the concentration builds up to toxic levels. Other species, including mammals, convert the ammonia and other excess metabolic nitrogen to less toxic nitrogenous compounds that can be safely excreted.

Urea Cycle

The nitrogenous excretion product of mammals and some other groups is urea, which is synthesized in the **urea cycle** (Figure 15.13). The ammonia released from glutamate is combined with carbon dioxide to form *carbamoyl phosphate*. This endothermic reaction is coupled to the hydrolysis of ATP.

♦ The relatively nontoxic nitrogenous waste urea is synthesized in the urea cycle.

$$NH_3 + CO_2 + 2\ ATP \longrightarrow H_2N\overset{\overset{\displaystyle O}{\|}}{C}O\overset{\overset{\displaystyle O}{\|}}{\underset{\underset{\displaystyle OH}{|}}{P}}—O^- + 2\ ADP + P_i$$

Carbamoyl phosphate

Carbamoyl phosphate enters the urea cycle by reacting with ornithine to form citrulline. Citrulline reacts with aspartate to form a molecule of arginosuccinate. Note in Figure 15.13 that citrulline is bonded to aspartate through the amino group of aspartate. The nitrogen of this amino group and the nitrogen of ammonia are ultimately the nitrogen atoms that are found in urea. Arginosuccinate is cleaved to yield arginine and fumarate. The fumarate is recycled to aspartate,

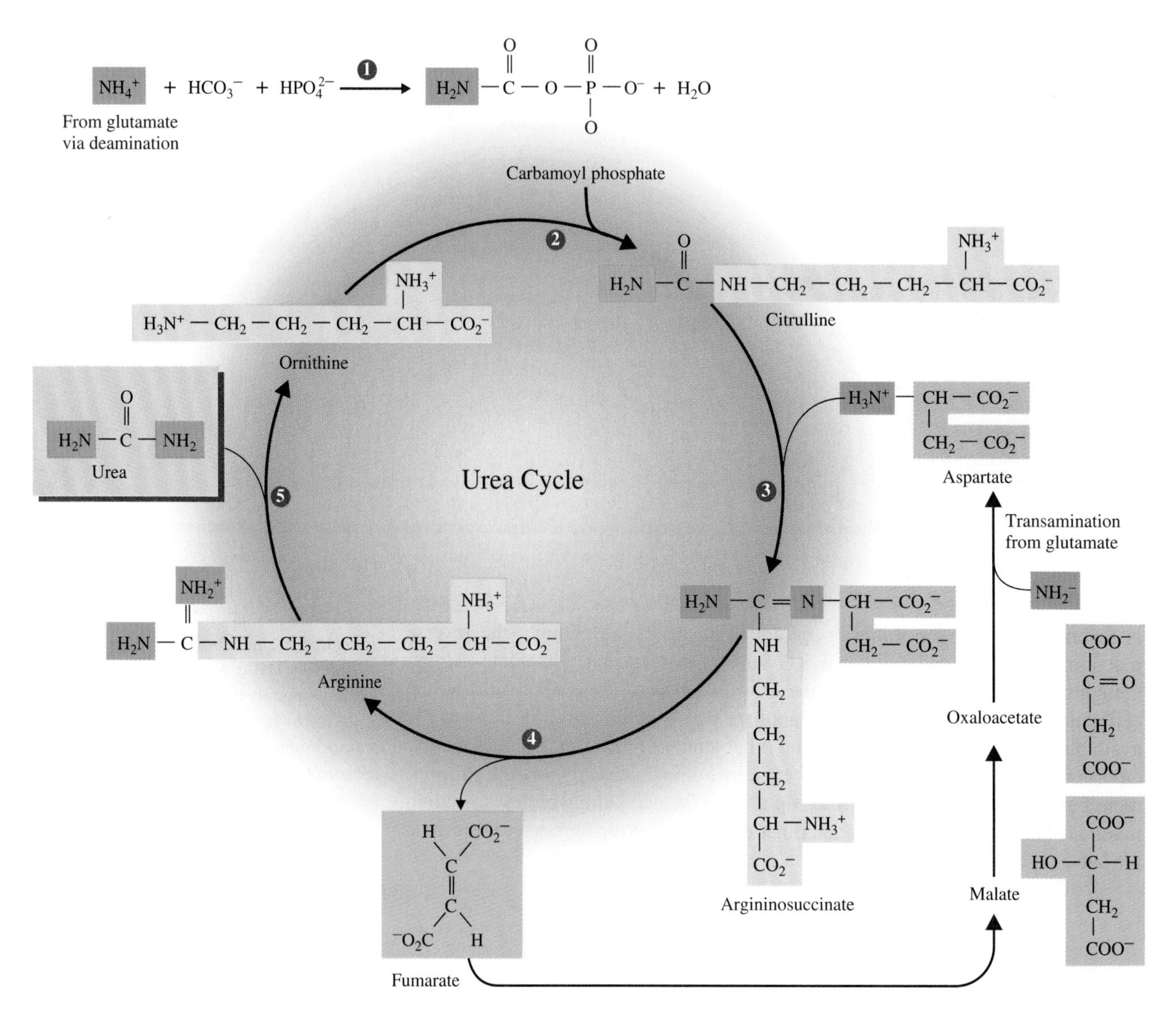

FIGURE 15.13
The urea cycle. The two nitrogen atoms of urea both come from glutamate, one via deamination to yield ammonium ion, the other via transamination to oxaloacetate, which yields aspartate.

with the amino group of aspartate coming from glutamate through a transamination reaction. Both of the nitrogen atoms of urea ultimately come from glutamate.

$$\text{Oxaloacetate} + \text{Glutamate} \xrightarrow{\text{Transaminase}} \text{Aspartate} + \alpha\text{-Ketoglutarate}$$

The arginine that is formed is cleaved to yield urea, the product of the cycle, and ornithine, which can combine with carbamoyl phosphate to repeat the cycle.

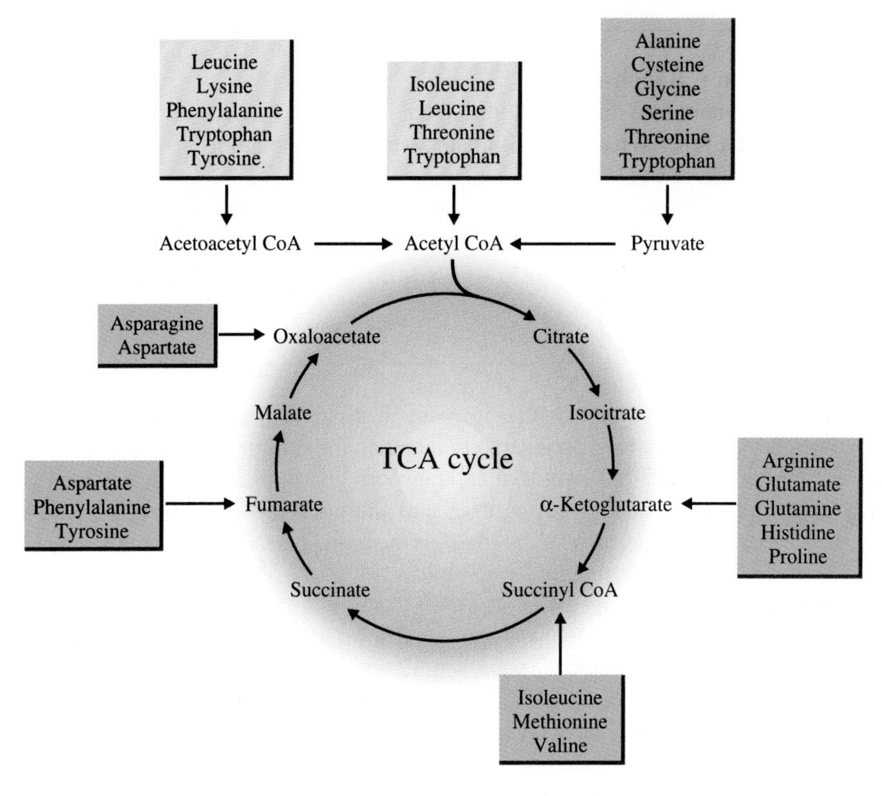

FIGURE 15.14
The products of amino acid transamination are keto acids. The carbon skeletons of the keto acids are broken down into several smaller metabolites, which include intermediates of the TCA cycle (green), pyruvate (orange), acetyl-CoA, and acetoacetyl-CoA (yellow). These molecules can be used for energy and to synthesize carbohydrates and lipids (see Chapter 16).

Catabolism of Keto Acids

The specific reactions that break down the keto acids formed from the 20 amino acids are too detailed to pursue. However, the products of these pathways are compounds you have seen before. Figure 15.14 shows these compounds and the amino acids they came from. Note that the amino acids are degraded to intermediates of the TCA cycle, pyruvate, acetyl-CoA, and acetoacetyl-CoA. The dozens of different carbon atoms of 20 amino acids appear in seven common metabolites.

♦ The various carbon skeletons found in the 20 amino acids are broken down into seven common biomolecules.

Pyruvate and the TCA cycle intermediates can be consumed for energy, but they can also be used to synthesize glucose and other carbohydrates through gluconeogenesis. This topic is discussed in the next chapter.

NEW TERMS

transamination Reaction that transfers amino groups from amino acids to α-ketoglutarate to yield keto acids and glutamate

deamination Reaction that removes the α-amino group from glutamate, yielding free ammonia and α-ketoglutarate

urea cycle Pathway that converts excess metabolic nitrogen to relatively nontoxic urea

TESTING YOURSELF

Amino Acid Catabolism

1. Urea contains two nitrogen atoms. What is the direct source of these atoms in the urea cycle? What is the ultimate source of these atoms?

Answers **1.** One comes from carbamoyl phosphate and the other comes from aspartate. They ultimately come from the α-amino groups of amino acids via glutamate through transamination reactions.

Summary

The *entropy* of the universe increases with time. To resist this tendency towards chaos, organisms use energy to maintain the complex state of life. Energy is used for synthesis of biomolecules, transport, and motion. Energy released in exothermic reactions is used to drive endothermic reactions through *coupled reactions*. *Adenosine triphosphate (ATP)* is the immediate storage form for energy in cells. *Nicotinamide adenine dinucleotide (NADH)* stores electrons resulting from catabolic oxidations.

The reactions of *catabolism* convert many large, complex molecules to a smaller set of simpler ones. These reactions are generally oxidative and energy-yielding. Catabolism occurs within cells, but digestion prepares dietary molecules for cellular absorption by hydrolyzing food molecules into smaller molecules. *Digestion* yields monosaccharides from carbohydrates, amino acids from proteins, and fatty acids and glycerol from triacylglycerols. These products of digestion are absorbed by mucosal cells and released into the circulatory system. Cells take in these nutrients as needed. *Glycolysis* is the principal catabolic pathway for sugars and yields pyruvate as the product. Glycolysis occurs in the presence or absence of oxygen.

Lactate is produced from pyruvate in the absence of oxygen. When oxygen is available, pyruvate is converted to acetyl-CoA, which is consumed in the *TCA cycle*. This cycle produces carbon dioxide, GTP, and the reduced coenzymes NADH and $FADH_2$. These reduced compounds are oxidized back to NAD^+ and FAD via the *electron-transport chain*. The energy produced by this oxidation is used to pump protons from the mitochondria to maintain an electrochemical gradient. The energy stored in this gradient is used by ATP synthetase to make ATP from ADP and P_i. Fatty acid catabolism yields acetyl-CoA, $FADH_2$, and NADH. These products are consumed in the TCA cycle and electron-transport chain. Amino acid catabolism begins with removal of amino groups via *transamination*. The nitrogen from these amino groups is used to make urea, and the carbon skeletons of the amino acids are catabolized to several intermediates of metabolism, including acetyl-CoA. Catabolism of sugars, lipids, and amino acids yields the same small set of metabolites.

Terms

bioenergetics (15.1)
entropy (15.1)
coupled reaction (15.1)
adenosine triphosphate (ATP) (15.1)
nicotinamide adenine dinucleotide (NAD^+/NADH) (15.1)

digestion (15.2)
amylases (15.2)
chyme (15.2)
proteases (15.2)
peptidases (15.2)
bile salts (15.2)
lipases (15.2)
lymph (15.2)
lipoproteins (15.2)
glycolysis (15.3)
respiration (15.3)
fermentation (15.3)
tricarboxylic acid, or TCA, cycle (15.4)
oxidative phosphorylation (15.5)
electron-transport chain (15.5)
chemiosmotic theory (15.5)
beta oxidation (15.6)
transamination (15.7)
deamination (15.7)
urea cycle (15.7)

Exercises

Bioenergetics (Objectives 1 and 2)

1. Energy is used to do what three kinds of work in an organism?
2. What is entropy?
3. How do cells maintain the highly organized state necessary for life?
4. What are the roles of catabolism and anabolism?
5. What is the central role of ATP in metabolism? Where does the energy come from to make ATP from ADP?
6. What role does NAD^+ play in metabolism?
7. Classify the following reactions and pathways as either catabolic or anabolic. (Some of these pathways are covered in later chapters.)
   a. digestion
   b. DNA replication
   c. translation
   d. transcription
   e. glycolysis
   f. glycogen synthesis
8. Briefly describe the concept of coupled reactions in metabolism.

Digestion (Objectives 3 and 4)

9. How does chewing aid digestion?
10. Only a small amount of starch digestion occurs before food enters the small intestine. Why?
11. What are the roles of the enzymes maltase, sucrase, and lactase?
12. Why do digestive proteases not break down the stomach and pancreas cells that synthesize them?
13. What is the function of the highly acidic environment of the stomach?
14. What is gained by the emulsification of dietary lipids? What compounds serve as emulsifying agents?
15. Describe why lipoproteins are needed for lipid transport in the blood.
16. List the major lipoproteins and the types of lipids they transport.
17. Complete the following table.

Food Class	Enzymes	Sites of Digestion	Products
Carbohydrates		Mouth and small intestine	
Proteins	Proteases and peptidases		
Lipids			Fatty acids and glycerol

18. Describe where and how monosaccharides are absorbed.

Glycolysis (Objectives 5 and 6)

19. What steps of glycolysis require energy? What is the source of energy for these reactions?
20. What steps of glycolysis yield energy? What molecule stores this energy? During glycolysis, how many of these molecules are formed per mole of glucose?
21. Rapid anaerobic catabolism of glucose may lower cellular and blood pH. Why?
22. Other monosaccharides besides glucose enter glycolysis. Where do these other monosaccharides enter into glycolysis?
23. Write equations for the conversion of glycogen to an intermediate of glycolysis.
24. What coenzymes are needed in glycolysis?

25. In the absence of oxygen, what is the net energy yield from glycolysis in ATP/glucose?
26. If no oxygen is present, the pyruvate produced by glycolysis is converted to what compound?

TCA Cycle, Electron Transport, and Oxidative Phosphorylation (Objective 7)

27. How is pyruvate converted to acetyl-CoA?
28. Compare acetyl-CoA, which is the reactant in the TCA cycle, to the carbon-containing products of the TCA cycle. From an inspection of these compounds, what do you think is the role of the TCA cycle?
29. What roles do NAD^+ and FAD serve in the TCA cycle and other metabolic pathways? How many of each coenzyme are involved in one cycle of the TCA cycle?
30. Besides catabolism of glucose by glycolysis, what other sources provide acetyl-CoA for the TCA cycle?
31. Compare the structures of GTP and ATP. Why are they comparable in energy?
32. The catabolism of acetyl-CoA by the TCA cycle yields enough energy to make how many ATPs?
33. What is meant by the expression "oxidative phosphorylation"?
34. What substance accepts the electrons lost by NADH during oxidation via the electron-transport chain?
35. How are NADH and $FADH_2$ oxidized in mitochondria?
36. Why does oxidation of NADH yield three ATPs but oxidation of $FADH_2$ yield only two?
37. The acceptance and donation of electrons is only one function of the electron-transport chain components. What is the other one?
38. Describe the chemiosmotic theory.
39. What enzyme uses the energy stored in an electrochemical gradient to drive the reaction that makes ATP? Describe the location and spatial arrangement of this enzyme.
40. What effects does a high concentration of ATP have on the activity of catabolic pathways?

Lipid and Amino Acid Catabolism (Objectives 8 and 9)

41. Beta oxidation refers to oxidation of what compounds?
42. What coenzymes are needed for beta oxidation of fatty acids?
43. What role does serum albumin play in fatty acid metabolism?
44. Fatty acids are not water-soluble. How are they transported in cells and mitochondria?
45. Compare the ATP yield from palmitic acid (16:0) and palmitoleic acid (16:1).
46. What are the two most common fates for dietary amino acids?
47. During amino acid catabolism, the amino groups of amino acids are transferred directly to what acceptor molecule?
48. From the perspective of waste disposal, what is accomplished by transamination during catabolism of amino acids?
49. Name the reactants, products, and enzyme for deamination. Since NAD^+ is converted to NADH, is this process oxidative or reductive?
50. How many moles of nitrogen, as ammonia and amino groups, are removed for each mole of urea that is synthesized?

Unclassified Exercises

51. Where does respiration occur in human cells?
52. The catabolism of glucose is often shown as

$$C_6H_{12}O_6 + 6\,O_2 \longrightarrow 6\,CO_2 + 6\,H_2O$$

Where in metabolism are the carbon dioxide molecules and water molecules produced?
53. Some cells obtain 38 mol of ATP per mol of glucose oxidized, and others obtain 36. What is the basis for this difference?
54. Where does oxidative catabolism of fatty acids, amino acids, and the products of glycolysis occur? What are the products of these reactions?
55. A long, hard hike up a particular mountain path requires 200 kcal of energy as ATP. Hydrolysis of 1 mol of ATP yields 7.3 kcal of energy. Calculate the number of grams of glucose needed to produce 200 kcal under aerobic conditions.
56. Solve Exercise 55 for anaerobic conditions. Do you think the body could generate this amount of energy using only anaerobic metabolism?
57. An ounce of a breakfast cereal provides 25 g of carbohydrate. If this were all glucose (which is a reasonable approximation, since the cereal is primarily wheat starch and sucrose), how many moles of ATP can be formed during aerobic metabolism? (*Hint*: Assume 36 mol ATP per mol glucose.)
58. Dietary carbohydrate has about 4 kcal/g. A mole of ATP is equivalent to 7.3 kcal when it is hydrolyzed. Calculate the percentage efficiency of ATP synthesis from this dietary carbohydrate.
59. You have learned that enzyme names often provide information about the enzyme substrate and the type of reaction that the enzyme catalyzes. Look at the first three reactions and enzymes of glycolysis (Figure 15.5), then translate the enzyme name into a brief sentence or phrase that summarizes the reaction and substrate.

The sun is the ultimate source of energy for living systems. Plants use solar energy to make biomolecules that are consumed by other organisms.

OUTLINE

16.1 Introduction to Anabolism

16.2 Photosynthesis

16.3 Biosynthesis of Carbohydrates

16.4 Biosynthesis of Lipids

16.5 Biosynthesis of Amino Acids

16.6 Biosynthesis of Nucleotides

OBJECTIVES

After completing this chapter, you should be able to

1. Define the term *anabolism* and describe its relationship to *catabolism.*

2. Describe the light and dark reactions of photosynthesis.

3. Describe gluconeogenesis and its metabolic role.

4. Summarize the synthesis of glycogen.

5. Describe the biosynthesis of fatty acids, cholesterol, and other steroids.

6. Determine the dietary or metabolic origins of the amino acids.

7. Describe the medical role of inhibitors of nucleotide synthesis.

16.1 Introduction to Anabolism

♦ Anabolic reactions are those involved in biosynthesis of molecules.

♦ ATP provides the energy for many anabolic reactions.

♦ NADPH provides the electrons needed for reductive anabolic reactions.

Digestion of foods yields a variety of small molecules that the body can then use for energy production (see Chapter 15) or synthesis of larger molecules. **Anabolism** is that part of metabolism that involves the synthesis of larger molecules needed by the body. To carry out synthetic activity, a cell must have the right enzymes (Chapter 14), energy, small molecules that serve as building blocks, and a source of electrons that can be used for reductions.

Many of the small molecules that are used for synthesis of larger molecules are acquired through diet and digestion or are produced by catabolism. You have already learned that dietary protein provides the amino acids needed for protein synthesis and that dietary glucose is needed to synthesize glycogen. Intermediates of metabolism may also serve as building blocks. For example, acetyl-CoA is used to synthesize fatty acids. Anabolism requires a pool of appropriate building blocks.

Energy for synthesis is provided primarily by adenosine triphosphate (ATP). Energy is provided by coupling energy-requiring synthetic reactions with energy-yielding reactions, such as ATP hydrolysis. Now is a good time to review Chapter 15 if you have forgotten the concept of coupled reactions.

The principal source of electrons and hydrogen ions (reducing power) is the coenzyme **nicotinamide adenine dinucleotide phosphate (NADPH)**. This molecule is identical to NADH except for the presence of an additional phosphate group (highlighted).

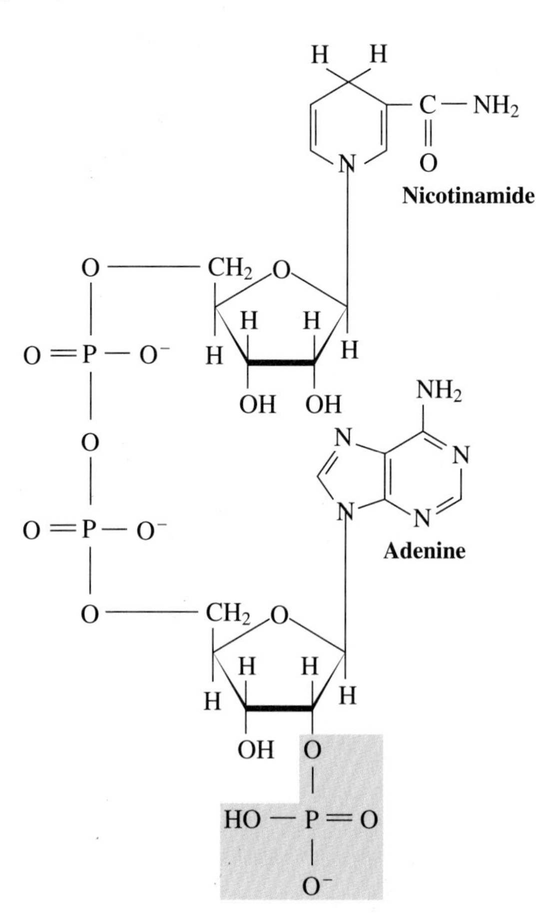

Like the very similar coenzyme NADH, NADPH alternately donates and accepts a pair of electrons and a hydrogen ion (a hydride ion).

$$NADP^+ + :H^- \rightleftarrows NADPH$$

NADPH must be in its reduced form to be used in biosynthesis. The formation of NADPH is discussed in Sections 16.2 and 16.4.

The enzymes involved in the oxidation–reduction reactions of catabolism use NAD^+/NADH, and those involved in anabolism use $NADP^+$/NADPH. This is another example of the order and regulation of metabolism. Each branch of metabolism has its own oxidation–reduction coenzyme that allows for efficient regulation of both anabolism and catabolism. Catabolism works best when the concentration of NAD^+ is high and the concentration of NADH is low. Anabolism works best when the concentration of NADPH is high. By maintaining two pools of oxidation–reduction coenzymes, both anabolism and catabolism can operate simultaneously and efficiently.

♦ Catabolism uses NAD^+/NADH, and anabolism uses $NADP^+$/NADPH.

Anabolism is a vast and complex field of study. This chapter concentrates on those examples that are of fundamental importance or are involved in human health.

NEW TERMS

anabolism The reactions of the body that synthesize larger, more complex molecules from smaller, simpler ones

nicotinamide adenine dinucleotide phosphate (NADPH) A coenzyme involved in reduction reactions in anabolism

16.2 Photosynthesis

Heterotrophic organisms are those that feed on other creatures because they cannot obtain energy and building blocks in any other way. Humans, other animals, fungi, and many bacteria are heterotrophic. **Autotrophic organisms**, or "self-feeders," can take simple inorganic molecules, such as water and carbon dioxide and ions such as ammonium ion, and use energy to convert them to whatever they need. Plants, algae, and some bacteria are autotrophic. Plants and algae use light as their energy source through a chemical process called **photosynthesis**. All life on earth ultimately depends upon photosynthesis; without photosynthetic organisms, all the heterotrophic organisms, including humans, would perish.

♦ Photosynthesis is the ultimate source of biomolecules on earth.

Photosynthesis in plants and higher algae occurs in chloroplasts (Figure 16.1) and can be summarized with a single equation:

$$6\ CO_2 + 6\ H_2O \xrightarrow{\text{Light}} \underset{\textbf{Glucose}}{C_6H_{12}O_6} + 6\ O_2$$

Carbon dioxide is reduced to yield glucose (or some other carbohydrate), and water is oxidized to oxygen. The photosynthetic reaction requires a great deal of energy, and light provides the needed energy. Although a single equation serves as a summary, the actual process is really many reactions. However, two major reactions are usually used to describe photosynthesis: the light reaction and the dark reaction.

CHEMISTRY CAPSULE

Synthesis of Vitamin D

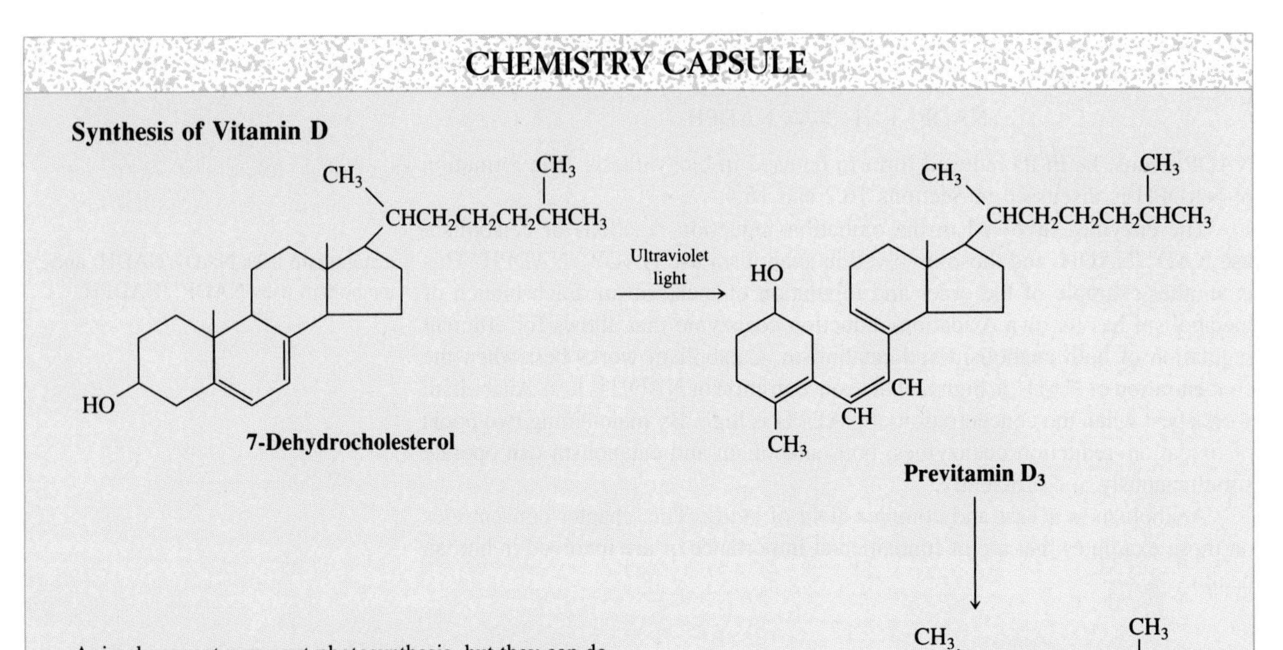

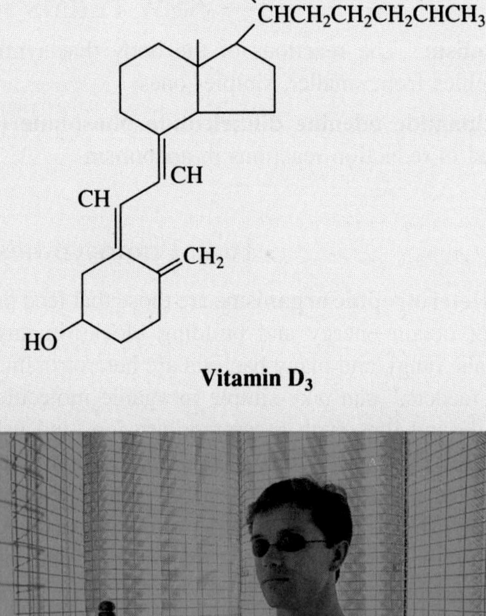

Animals cannot carry out photosynthesis, but they can do some photochemistry. That is, they can carry out some light-dependent reactions. The synthesis of vitamin D is an example. Vitamin D is made from 7-dehydrocholesterol, which in turn is made from cholesterol. (The synthesis of cholesterol is described in Section 16.4.) 7-Dehydrocholesterol is circulated throughout the body in the blood. When blood passes through capillaries in the skin, ultraviolet light is absorbed by 7-dehydrocholesterol. This absorbed energy cleaves a bond in the molecule, yielding previtamin D_3, which isomerizes to vitamin D_3.

The role of sunlight in the synthesis of vitamin D is why it is sometimes referred to as the "sunshine vitamin." The diet of people in polar and cold temperate zones must be fortified with vitamin D, since they may not be exposed to enough sunshine during the winter to make adequate amounts of vitamin D.

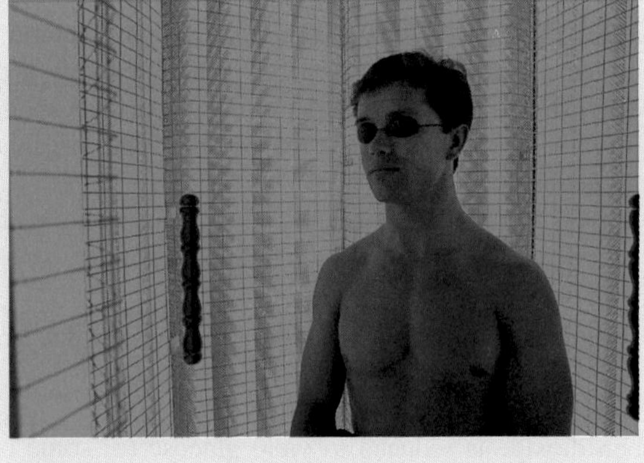

FIGURE 16.A
UV light, like that in this light therapy room, is needed to make vitamin D.

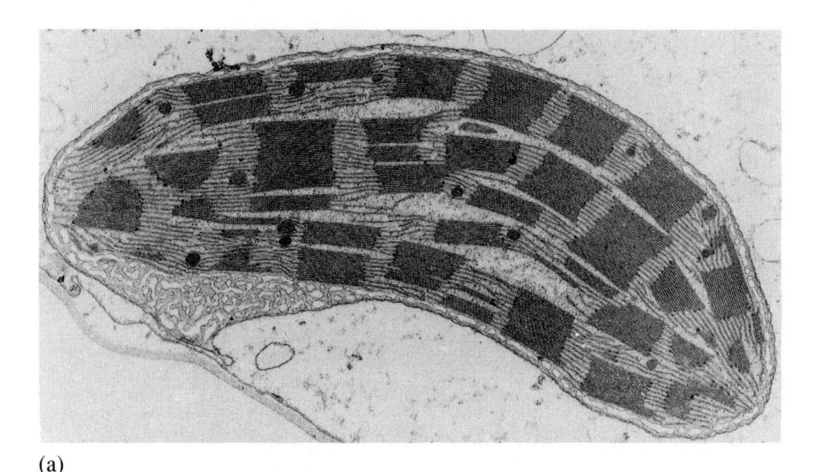

(a)

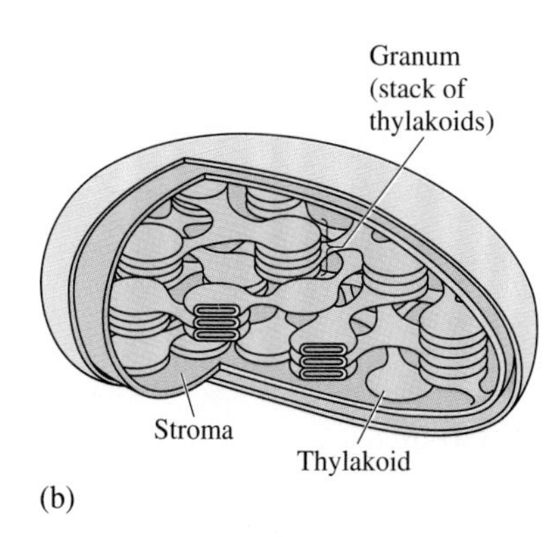

(b)

FIGURE 16.1
Chloroplasts are the site of photosynthesis in plants. (a) An electron micrograph, and (b) a diagram of a chloroplast. The stroma is the location of the dark reaction, and the grana, which consist of stacks of thylakoids, are the site of the light reaction.

The Light Reaction of Photosynthesis

The synthesis of glucose from carbon dioxide requires energy, provided as ATP, and the reduction of carbon dioxide. The reduction of carbon dioxide, like all reductions, requires electrons. In photosynthesis, water is the ultimate source of these electrons. But water is a very poor reducing agent because it does not readily donate electrons. The **light reaction** uses light energy to strip electrons from certain chlorophyll molecules, and then uses the electrons to convert $NADP^+$ to NADPH. The electron-deficient chlorophyll molecules regain electrons from water.

♦ The light reaction of photosynthesis provides energy (as ATP) and electrons (as NADPH).

The actual path of the electrons can be followed most easily by going backward through the light reaction (Figure 16.2). $NADP^+$ and H^+ gain electrons from a series of electron carriers that obtain electrons from a chlorophyll molecule in photosystem I. The transfer of electrons from photosystem I to the first of these carriers is endothermic. Energy from two photons of light is needed to strip two electrons from photosystem I. This leaves photosystem I electron-deficient. It in turn gains electrons from a series of electron carriers that get electrons from a chlorophyll molecule in photosystem II. Again, two photons of light are needed to strip two electrons from photosystem II. Photosystem II in turn strips electrons from water, leaving molecular oxygen. The net light reaction can be summarized by this equation:

$$H_2O + NADP^+ \xrightarrow{\text{Light energy}} NADPH + \frac{1}{2}\,O_2 + H^+$$

During this reduction of $NADP^+$, ATP is also made from ADP and inorganic phosphate. In the light reaction, electrons flow from water to $NADP^+$ through a

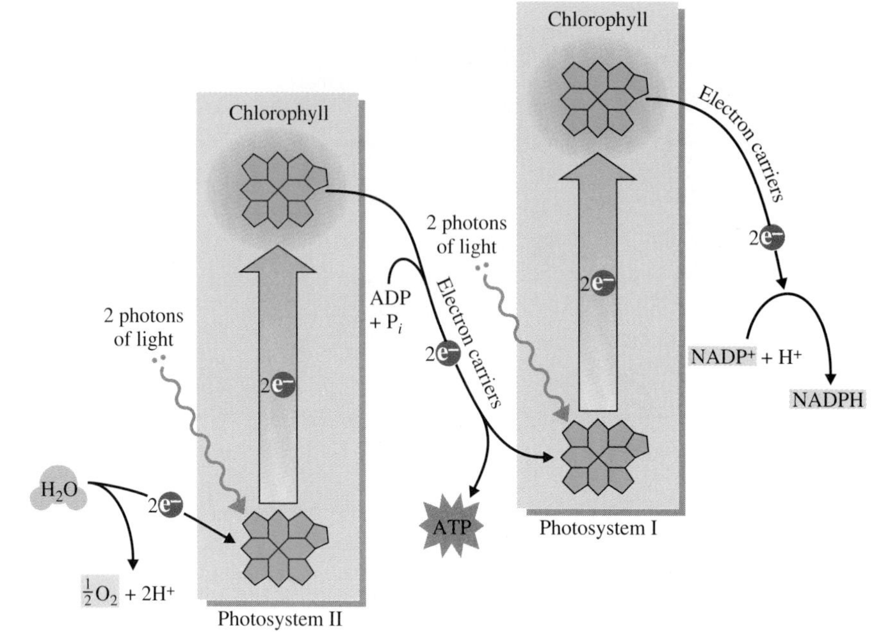

FIGURE 16.2
The light reaction of photosynthesis. Electrons from water pass through a series of electron carriers to $NADP^+$ (purple). Light provides the energy for this reaction. As electrons pass from photosystem II to photosystem I, protons are pumped to establish a proton gradient, which is stored energy. This energy can be used to synthesize ATP in a manner analogous to ATP synthesis in mitochondria.

series of electron carriers (Figure 16.2). Note that four photons of light are needed to move the pair of electrons from water to $NADP^+$. There is more than enough energy in four photons to reduce NADPH, however. A part of the energy is used to make some ATP. At least one ATP is made for each NADPH formed.

♦ A proton gradient across the chloroplast membrane is the source of energy needed for ATP synthesis.

During the light reaction, ATP is formed in chloroplasts in a manner similar to oxidative phosphorylation in mitochondria (see Chapter 15). As electrons pass from photosystem II to photosystem I, protons are pumped from the cytosol into the matrix of the chloroplasts. This establishes an electrochemical gradient similar to the one in mitochondria. When protons pass from the matrix to the cytosol, a portion of the energy is used to make ATP from ADP and inorganic phosphate. The chemiosmotic theory thus plays a role in oxidative phosphorylation in mitochondria and in light-dependent ATP synthesis in chloroplasts.

The Dark Reaction of Photosynthesis

The light reaction produces NADPH (the reducing source) and ATP (the energy), but it does not make any glucose directly. The synthesis of glucose occurs in the **dark reaction**. This reaction uses carbon dioxide and the products of the light reaction. The dark reaction, also called the *Calvin cycle*, involves several enzymes of glycolysis, some enzymes for the synthesis and isomerization of carbohydrates, and the single most abundant protein on earth: ribulose diphosphate carboxylase.

♦ Ribulose diphosphate carboxylase catalyzes the attachment of CO_2, an inorganic molecule, to ribulose diphosphate, an organic molecule.

The first step in the dark reaction is the attachment of carbon dioxide to the five-carbon organic molecule ribulose diphosphate (Figure 16.3). The product is an unstable six-carbon molecule, [C_6], that breaks into two molecules of 3-phosphoglycerate, an intermediate of glycolysis. Through a series of reactions

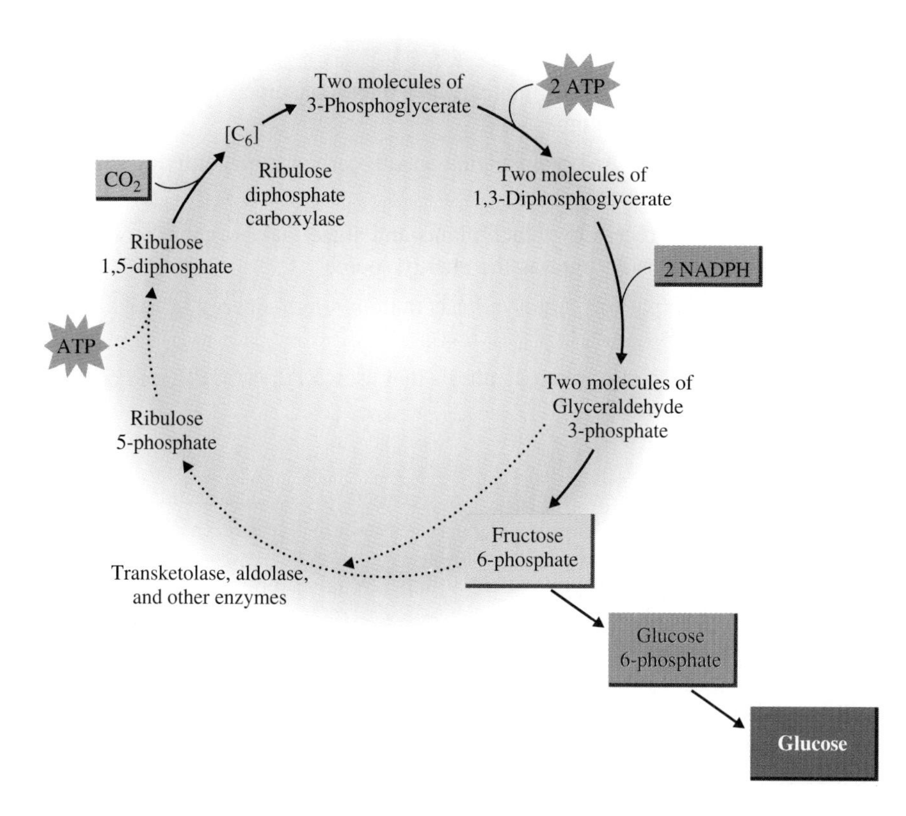

FIGURE 16.3
The dark reaction of photosynthesis. This reaction uses ATP (yellow) and NADPH (purple) from the light reaction to make carbohydrate (glucose) from CO_2. A portion of the 3-phosphoglycerate that is formed is used to make glucose, and the rest is recycled to ribulose 1,5-diphosphate.

that is partly a reversal of glycolysis, 3-phosphoglycerate is converted to glucose. Note that NADPH and ATP produced by the light reaction are needed for this synthesis.

In the equation for photosynthesis, glucose was shown to be made from carbon dioxide. Glucose is produced by the reactions shown in Figure 16.3, but only one of the six carbon atoms came from carbon dioxide; the remaining five already existed as ribulose diphosphate. Where do the other five CO_2 molecules enter into the picture? How is ribulose diphosphate formed? The whole process is balanced by multiplying the first step by six and adding several steps that synthesize sugars.

$$6\ CO_2 + 6\ \text{Ribulose diphosphate} \xrightarrow[\text{carboxylase}]{\text{Ribulose diphosphate}} 12\ \text{3-Phosphoglycerate}$$

$$\underset{\textbf{(inorganic)}}{6\ \text{C atoms}} + \underset{\textbf{(organic)}}{30\ \text{C atoms}} = \underset{\textbf{(organic)}}{36\ \text{C atoms}}$$

Twelve molecules of the product are formed in the preceding equation. Two of these product molecules, each containing three carbon atoms, are used to make glucose, the product of photosynthesis. Ten of them, collectively containing 30 carbon atoms, are rearranged and phosphorylated through the action of several enzymes to yield six ribulose diphosphate molecules, which can then participate in the dark reaction (Figure 16.3).

NEW TERMS

heterotrophic organism An organism that must consume organic matter to obtain energy and carbon atoms

autotrophic organism An organism that obtains carbon from nonliving sources such as carbon dioxide

photosynthesis A process by which plants and algae make organic molecules from carbon dioxide using light as the energy source

light reaction The part of photosynthesis that uses light energy to make ATP and NADPH

dark reaction The part of photosynthesis that uses ATP, NADPH, and carbon dioxide to make glucose

TESTING YOURSELF

Photosynthesis

1. What is the site of photosynthesis in a plant? What common organisms carry out photosynthesis?
2. Eighteen ATPs and 12 NADPHs are needed for the synthesis of one glucose molecule in photosynthesis. If an NADPH is equivalent to three ATP, compare the number of ATP required to synthesize glucose to the number produced during aerobic catabolism of glucose.

Answers **1.** The chloroplast. Plants, algae, and some photosynthetic bacteria. **2.** 54 ATP needed for synthesis, 36 or 38 produced during aerobic catabolism.

16.3 Biosynthesis of Carbohydrates

Digestion of carbohydrates yields large amounts of glucose and smaller amounts of other monosaccharides. These molecules are used for energy production (see Chapter 15), but they are also used to make oligosaccharides, polysaccharides, glycolipids, and glycoproteins within the body. Furthermore, the monosaccharides can be converted to other monosaccharides, and under certain physiological conditions, they can be made from other compounds within the body.

Gluconeogenesis

Although the diet provides large amounts of glucose to the body, there are circumstances that require the body to make glucose. This is accomplished by the process of **gluconeogenesis**, which literally means "new synthesis of glucose." The term does not refer to the conversion of other monosaccharides into glucose; instead, it refers to synthesis from smaller metabolic intermediates produced by catabolism. A summary of the molecules that can be used to make glucose is given in Figure 16.4. Note that the carbon atoms for glucose come from three sources: (1) lactate, (2) the glucogenic amino acids, and (3) glycerol.

◆ Glucose can be made from several small molecules that are produced from the catabolism of sugars, amino acids, and the glycerol of fats and oils.

Lactate is produced in oxygen-deficient muscle via glycolysis and reduction of pyruvate. Lactate diffuses into blood and is taken up by the liver. When

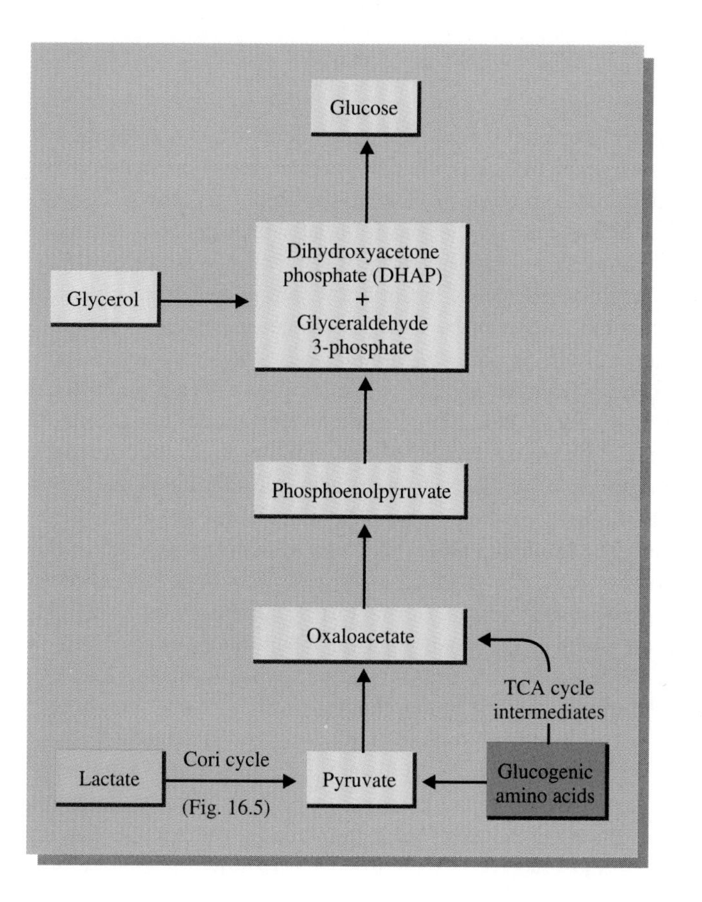

FIGURE 16.4
Glucose can be synthesized from several metabolites. Glycerol (yellow) is produced by catabolism of triacylglycerols; lactate (orange) is produced by anaerobic glycolysis. Catabolism of glucogenic amino acids (red), which is all of them except leucine, yields TCA cycle intermediates and pyruvate.

sufficient oxygen is available to the liver, the lactate is oxidized to pyruvate. This reaction is simply the reverse of the reaction that led to the production of lactate.

$$\text{Pyruvate} + \text{NADH} + \text{H}^+ \xrightleftharpoons[]{\text{Lactate dehydrogenase}} \text{Lactate} + \text{NAD}^+$$

Since this reaction is reversible, the concentrations of the species determine the direction of the reaction (Le Chatelier's principle). When lactate and NAD^+ concentrations are high, NADH and pyruvate are produced. There are several fates that are possible for the pyruvate: lipid synthesis, oxidation via the TCA cycle to produce energy, or conversion by gluconeogenesis to produce glucose. Much of this glucose is released into the blood. Exercising muscle converts glucose to lactate and is thus depleting its glucose stores. Under these conditions, muscle readily takes up glucose from blood. This anaerobic catabolism of glucose to lactate in muscle, resynthesis of glucose from lactate via gluconeogenesis in liver, and transport of glucose back to muscle constitutes a cycle called the **Cori cycle** (Figure 16.5).

Muscle cells produce two ATP during the synthesis of lactate. The Cori cycle removes lactate from the body by using it to make glucose. But in terms of energy, there is no such thing as a "free lunch" or breaking even. The synthesis of glucose from lactate requires four ATP and two GTP. More energy is required for gluconeogenesis than is derived from glycolysis. The reason for this is the irreversible nature of some steps in glycolysis: (1) the conversion of phosphoenolpyruvate and ADP to pyruvate and ATP, and (2) the phosphorylations of glucose

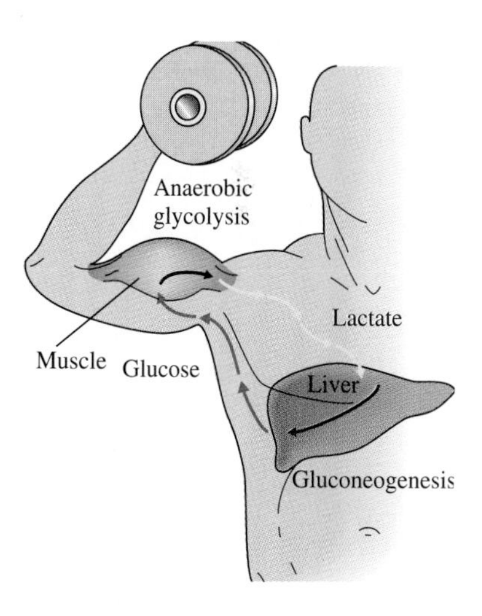

FIGURE 16.5
The Cori cycle. Lactate is produced from glucose by glycolysis under anaerobic conditions. The lactate leaves muscles, and is carried to the liver by blood where it is converted back to glucose.

and fructose 6-phosphate. Four ATP equivalents (as two ATP and two GTP) are required to bypass the phosphoenolpyruvate step. Furthermore, two ATP are consumed in the formation of two 1,3-diphosphoglyceric acid molecules. The phosphates of this acid are cleaved from later intermediates of gluconeogenesis; thus, the energy is not conserved as ATP. The net effect is that six ATP are required for gluconeogenesis but only two are gained from glycolysis.

Most amino acids can be used, in whole or part, to make glucose. These amino acids are called *glucogenic amino acids*. Two conditions lead to significant gluconeogenesis from amino acids: (1) fasting, and (2) a protein-rich diet. During fasts, the body breaks down body protein into amino acids. Small amounts of these amino acids are used to remake essential proteins, but the rest are used for energy and glucose synthesis. It may seem odd that under these conditions the body would bother to make glucose. Why not just use the amino acids for energy production? The brain and central nervous system use glucose as their principal energy source; thus, blood glucose must be maintained or dysfunction of the central nervous system occurs. During fasting, amino acids, through gluconeogenesis, provide much of the blood glucose. When much of the caloric intake of the diet is in the form of protein, it is also necessary to make glucose from amino acids to maintain adequate blood glucose levels.

Glycerol is the third source of glucose via gluconeogenesis. Glycerol is produced by catabolism of triacylglycerols. In animals, the fatty acids of fats and oils cannot be used to make glucose. During fasting, the amount of glycerol available to the body increases through increased catabolism of triacylglycerols. Amino acids and glycerol provide the carbon atoms for glucose during fasting.

Some of the specific reactions of gluconeogenesis are already familiar to you because they are reverse reactions of glycolysis. Figure 16.6 shows gluconeogenesis.

♦ The body can make all of the simple sugars that it needs.

The body can make all of the monosaccharides that it requires. Unlike lipids and amino acids, there are no essential sugars. Glucose is available from the diet or through gluconeogenesis. The other monosaccharides are obtained from the diet or made from glucose. The body possesses the necessary enzymes to make the required amounts of oligosaccharides and polysaccharides. Except for lactose produced during milk production, free oligosaccharides are uncommon; however, they are made as components of glycoproteins and some glycolipids.

Glycogenesis

Glucose and other monosaccharides are used as building blocks for the polysaccharides of the body. Each polysaccharide has its own pathway for its synthesis, but we will consider here only **glycogenesis**, the synthesis of glycogen. Glycogen is the principal storage form for glucose. Glycogen synthetase catalyzes the transfer of glucose molecules from uridine diphosphate glucose (UDPG) to the ends of the branches of glycogen molecules. Glucose cannot be added directly. The synthesis of the glycosidic bonds between glucose residues in glycogen requires energy. The energy needed for formation of this bond is found in the cleavage of the high-energy bond between the glucose and the UDP of UDPG. The reactions above add a glucose residue to the end of a chain in glycogen. A

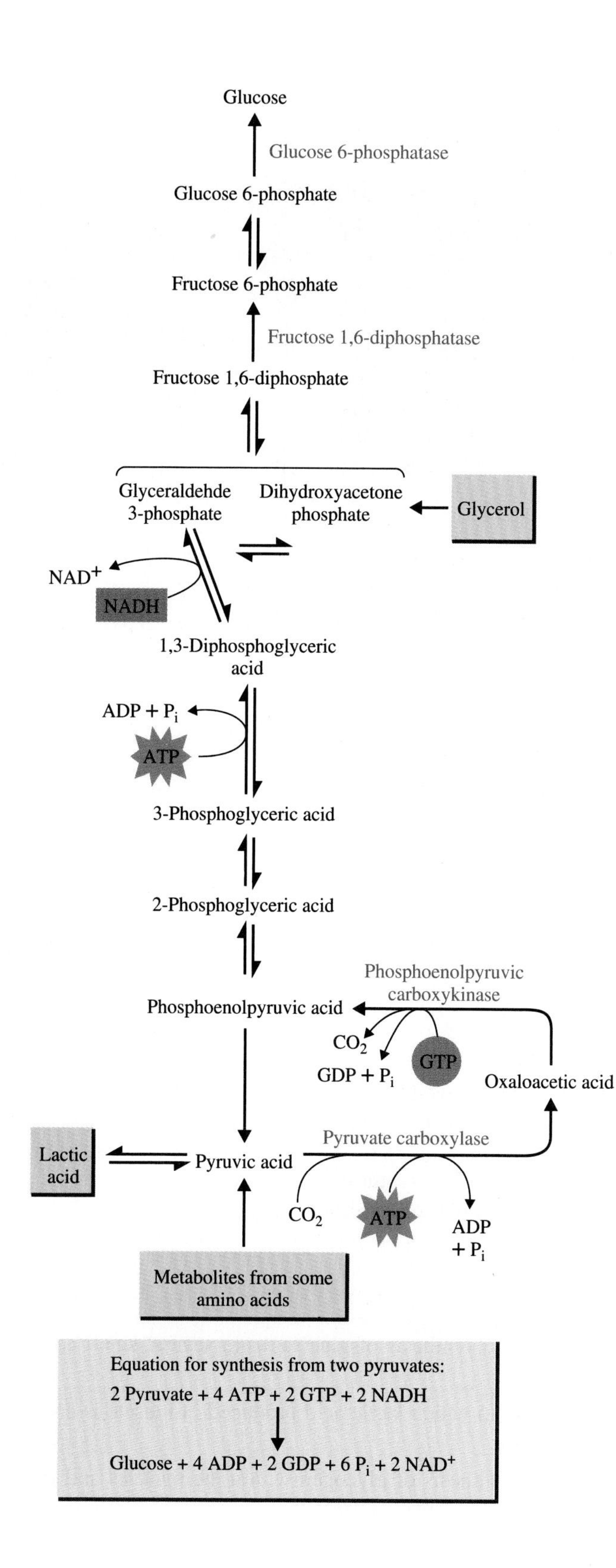

FIGURE 16.6
Gluconeogenesis includes many of the steps of glycolysis, but some of the steps of glycolysis are irreversible, and unique reactions exist for gluconeogenesis. Pyruvate carboxylase, phosphoenolpyruvate carboxykinase, fructose-1, 6-diphosphatase, and glucose 6-phosphatase are enzymes that catalyze reactions that are not shared by gluconeogenesis and glycolysis. Six high-energy bonds in ATP and GTP are cleaved per glucose synthesized by gluconeogenesis.

CHEMISTRY CAPSULE

Enzyme Modification Leads to Lactose Synthesis

Lactose is synthesized only by the mammary glands following birth. The cells of the body possess the enzyme galactosyl transferase, which transfers galactose residues onto appropriate acceptor molecules. Thus, a general reaction that occurs within all body cells is

Galactose + Acceptor molecule **(activated)** —(Galactosyl transferase)→ Galactosylated product

This enzyme alone cannot make lactose because glucose does not act as an acceptor molecule. But after a female gives birth, the concentration of the hormone prolactin increases in her body. In the mammary glands this hormone stimulates the production of another protein, α-lactalbumin, a protein that binds to galactosyl transferase. This complex has a different specificity than galactosyl transferase alone; it is a lactose synthetase. The complex catalyzes the attachment of galactose to glucose, yielding lactose.

Galactose + **(activated)** Glucose **(an acceptor molecule)** —(Lactose synthetase)→ Lactose **(galactosylated product)**

A common enzyme of the body is thus altered through allosteric regulation to yield a special-duty enzyme.

FIGURE 16.B
These are a few of the many formulas available as supplements or replacements for breast feeding.

specific branching enzyme is needed to add branching to glycogen. This enzyme cleaves an oligosaccharide from the end of a chain in glycogen and adds it internally to the chain through an α 1→ 6 bond. This increases the branching in the glycogen molecule. The synthesis and breakdown of glycogen are highly regulated processes that are discussed in Section 17.4.

NEW TERMS

gluconeogenesis The synthesis of glucose from several small common molecules of metabolism

Cori cycle A cycle in which lactic acid produced in muscle is converted back to glucose in the liver

glycogenesis The synthesis of glycogen within the muscles and liver

TESTING YOURSELF

Biosynthesis of Carbohydrates

1. Describe the Cori cycle and its metabolic role.
2. The central nervous system requires glucose as an energy source. During a prolonged fast, how does the body provide glucose to the brain?

Answers **1.** This cycle includes the production of lactate from glucose in anaerobic muscle, the transport of lactate to the liver in blood, the synthesis of glucose from lactate via gluconeogenesis in the liver, and transport of the glucose back to muscle. Lactate is recycled, rather than excreted. This is useful because lactate contains considerable energy that would be lost if it were excreted. **2.** The body breaks down cellular protein and uses many of the amino acids to make glucose via gluconeogenesis.

16.4 Biosynthesis of Lipids

As we saw in Chapter 11, the lipids are a diverse group of natural products that can be characterized as lipophilic (nonpolar) and hydrophobic (water-hating). In this chapter we will discuss the origins of some of these materials.

In general, separate pathways are used for anabolism and catabolism. This is also true for the lipids. Moreover, for fatty acids, the two processes are located in different parts of the cell.

Lipogenesis

Lipogenesis literally means "the synthesis of lipids," but it is more often used to mean the synthesis of fatty acids. Many fatty acids can be made by the body. Linoleic and linolenic acids are exceptions. In humans, the synthesis of fatty acids takes place in the cytosol of liver cells. Acetyl-CoA is the major precursor to the fatty acids and supplies two carbon fragments for assembling these acids. Any excess acetyl-CoA that does not pass through the TCA cycle for energy production may be used for fatty acid synthesis. This is the basic problem with weight control. If we do not use the acetyl-CoA for energy, it becomes stored as fat.

♦ Acetyl-CoA is the source of the carbon atoms for fatty acid synthesis.

Acetyl-CoA cannot pass directly through the mitochondrial membrane, but the acetate part of it is shuttled through the membrane by the pathway shown in Figure 16.7. This cycle not only transports the two carbon atoms of acetate from the mitochondria to the cytosol, but it also converts cytosolic NADH into NADPH as cytosolic oxaloacetic acid is converted into pyruvic acid.

How does the two-carbon acetate unit of acetyl-CoA become incorporated into a fatty acid? Fatty acids are synthesized by sequentially adding two carbon pieces to a two-carbon precursor molecule. The carbon atoms come from acetyl-CoA, but as you shall learn shortly, lipogenesis does *not* directly use acetyl-CoA. Acetyl-CoA is converted to malonyl-CoA and acetyl-ACP, which participate directly in fatty acid synthesis.

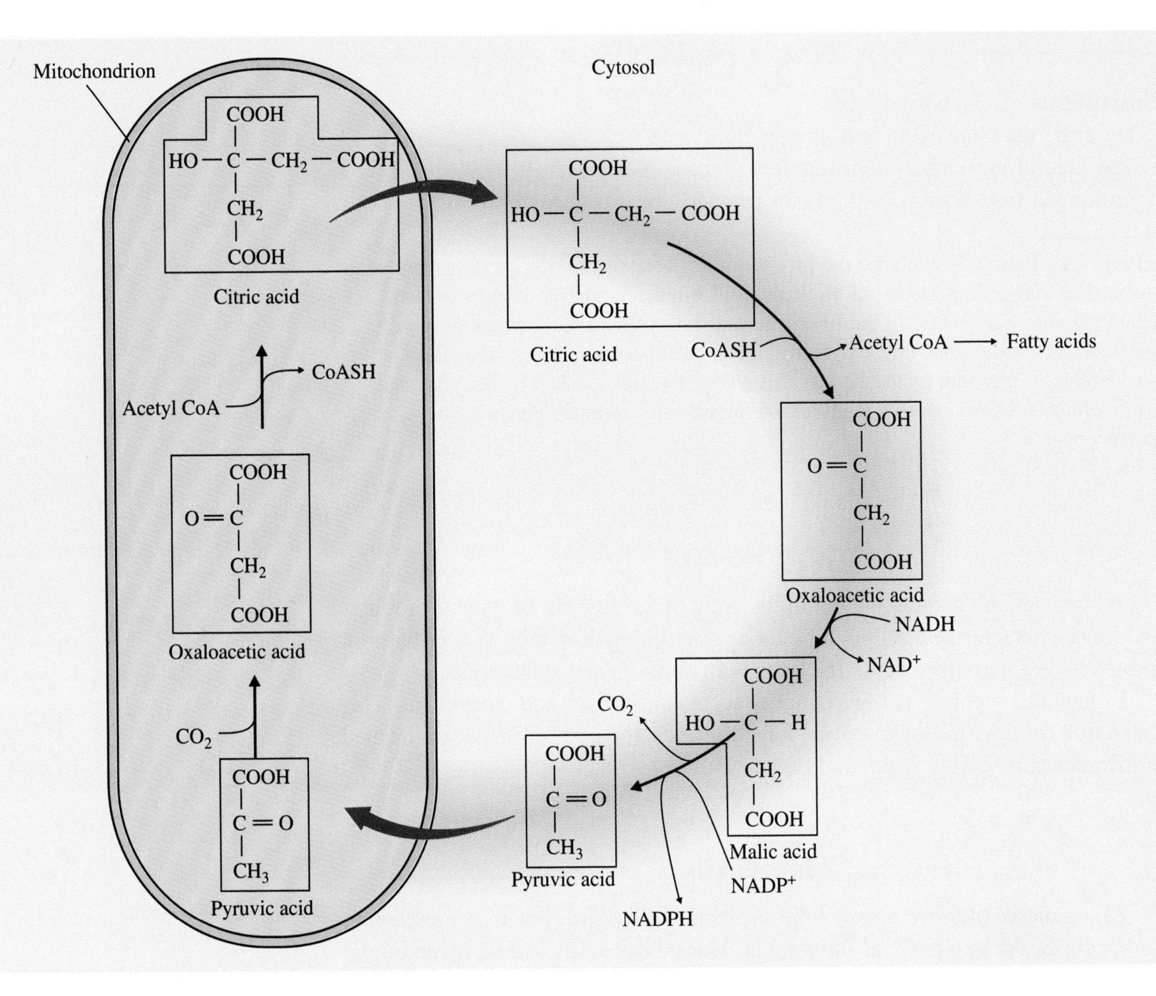

FIGURE 16.7
The transport of acetyl-CoA and reducing power from the mitochondria to the cytosol. Acetyl-CoA cannot pass through the mitochondrial membrane, but citric acid can. In the cytosol, the acetyl group in citric acid is transferred to CoA, leaving oxaloacetic acid as the other product. As oxaloacetic acid is metabolized to pyruvate, a reduction with NADH and an oxidation with $NADP^+$ occurs. This results in the conversion of cytosolic NADH into NADPH. The pyruvate that is made passes into the mitochondrion where the cycle repeats.

Let's examine the synthesis of palmitic acid, $CH_3(CH_2)_{14}COOH$, the saturated fatty acid with sixteen carbon atoms. This synthesis requires eight acetyl-CoA molecules to provide the sixteen carbon atoms. The acetyl group of one of these acetyl-CoA molecules is attached to the **acyl carrier protein (ACP)** to form acetyl-ACP.

$$CH_3-\underset{\underset{O}{\|}}{C}-CoA + HS-ACP \longrightarrow CH_3-\underset{\underset{O}{\|}}{C}-S-ACP + CoA$$

ACP is part of a large enzyme called *fatty acid synthase*. ACP serves as a carrier for the acetyl group and the growing intermediates during the synthesis.

The other acetyl-CoA molecules are converted to another intermediate, malonyl-CoA. This reaction is a carboxylation of acetyl-CoA with bicarbonate and requires energy provided by ATP hydrolysis.

♦ The carboxylation of acetyl-CoA is an endothermic reaction.

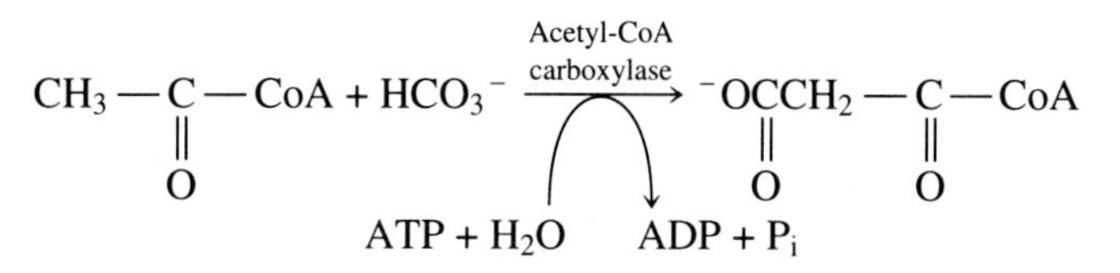

The malonyl group of malonyl-CoA is then transferred to ACP to yield malonyl-ACP.

The subsequent steps in the synthesis of palmitic acid are catalyzed by fatty acid synthase. A summary of this synthesis is shown in Figure 16.8. It is interesting to note that these reactions are biological examples of familiar organic reactions: reduction of a ketone to a secondary alcohol, dehydration of an alcohol, and hydrogenation of an alkene.

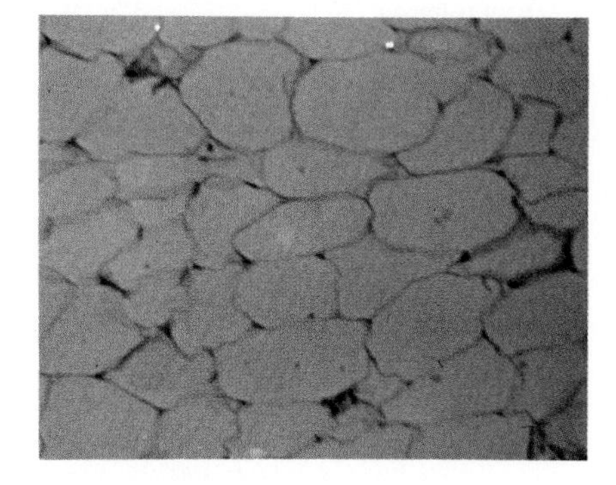

Adipocytes are a major site of fat synthesis and storage.

Humans synthesize other fatty acids in addition to palmitic acid. An additional enzyme system elongates some of the palmitic acid, two carbon atoms at a time, again using acetyl-CoA to provide the carbon atoms. The steps are very similar to those of the fatty acid synthase. Still another enzyme system introduces carbon–carbon double bonds to yield unsaturated fatty acids. This system cannot introduce double bonds beyond carbon atoms 9 and 10 of a chain. That is why linoleic acid and linolenic acid (Table 11.1) are provided only by the diet.

Note that NADPH was used in the synthesis of fatty acids just as it is used whenever reducing power is needed in synthesis. NADPH is produced during the cyclic transfer of acetyl-CoA from mitochondria to the cytosol, and by the **pentose phosphate pathway** (Figure 16.9). This pathway also yields pentoses that can be used by the body or recycled to hexoses.

Synthesis of Triacylglycerols and Phosphoacylglycerols

Fatty acids are components of the saponifiable lipids, notably the triacylglycerols and the phosphoacylglycerols. These compounds are esters that contain not only fatty acids but also glycerol. Triacylglycerols are formed in adipose cells by transferring the fatty acid (the acyl group) from CoA to yield the ester.

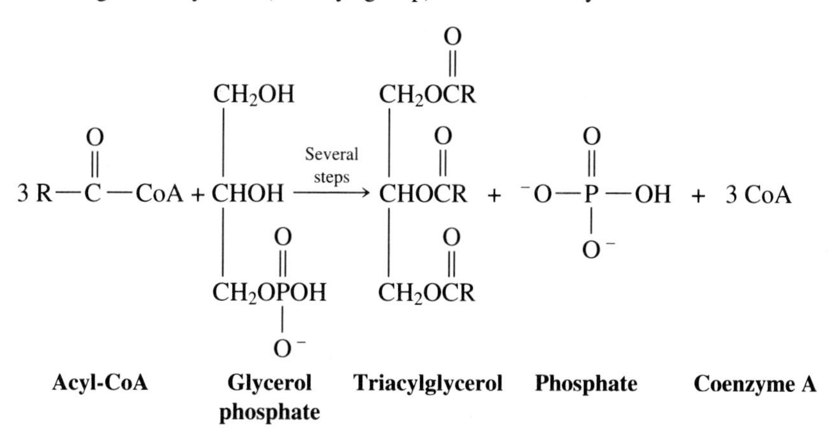

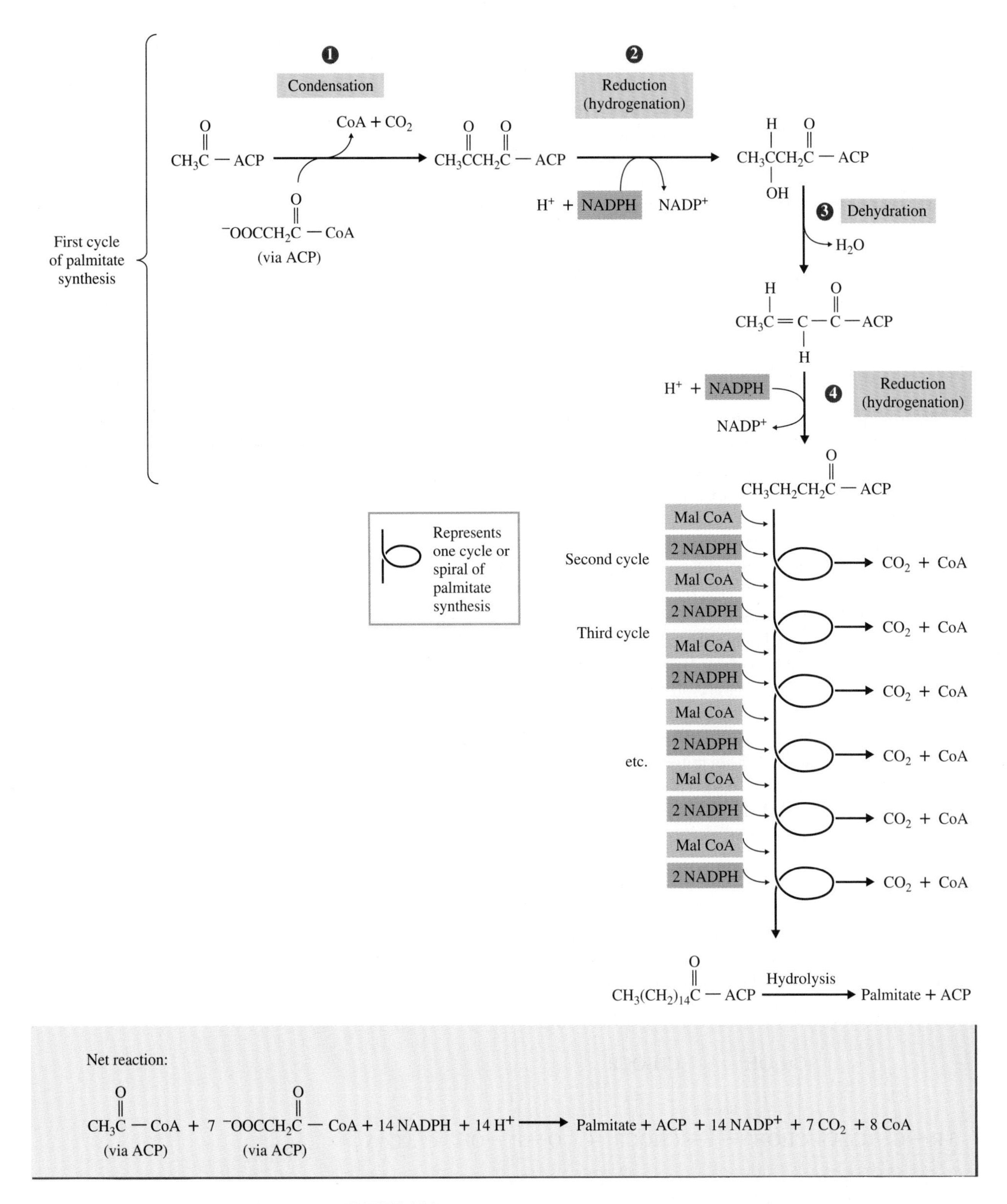

FIGURE 16.8
Fatty acid biosynthesis. Note that both acetate and malonate are attached to a carrier protein during the synthesis. This pathway is not strictly a cyclic pathway because the product is two carbon atoms longer than the substrate. The pathway more closely resembles a spiral.

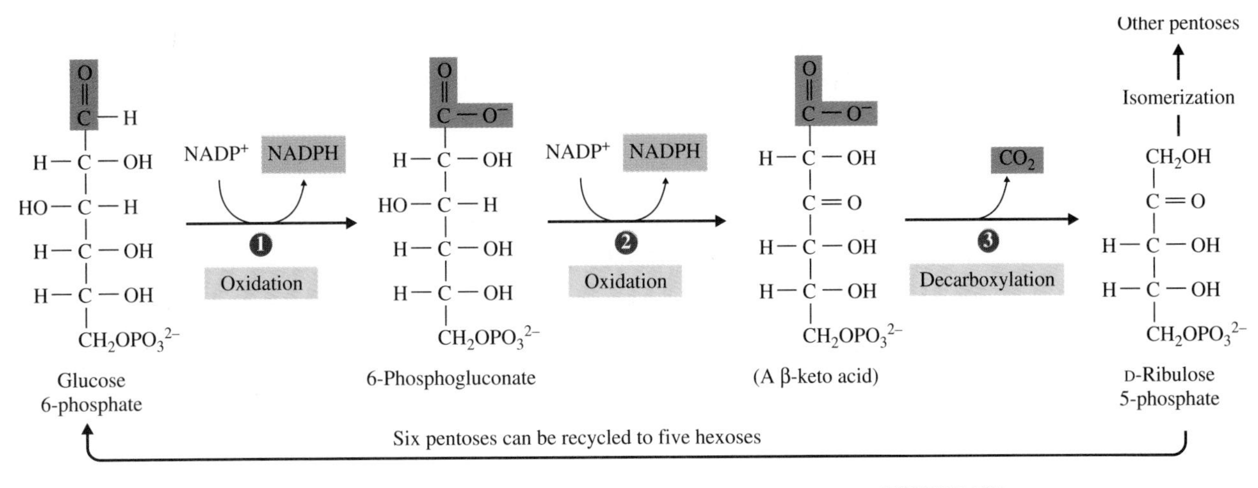

FIGURE 16.9
Pentose phosphate pathway. Glucose 6-phosphate is oxidized to a phosphorylated pentose yielding two NADPHs (purple) per mole. Pentoses can be made from these phosphorylated pentoses, or they can be recycled to glucose 6-phosphate.

Ester bond formation is endothermic, but the cleavage of the bond between the fatty acid and the CoA yields energy.

The synthesis of phosphoacylglycerols has some of the same features as the synthesis of triacylglycerols. Two acyl groups are transferred from acyl-CoA to glycerol phosphate to form a diacylglycerol phosphate (a phosphatidic acid) (see Chapter 11, Section 11.4). The phosphate is then cleaved from the molecule to yield a diacylglycerol. A molecule of phosphorylated choline or ethanolamine is then transferred from a carrier molecule (cytidine diphosphate, or CDP) onto the diacylglycerol to yield the phosphoacylglycerol (Figure 11.4).

Glycerol phosphate is needed for the synthesis of both triacylglycerols and phosphoacylglycerols. This compound is made by reducing one of the intermediates of glycolysis, dihydroxyacetone phosphate (DHAP).

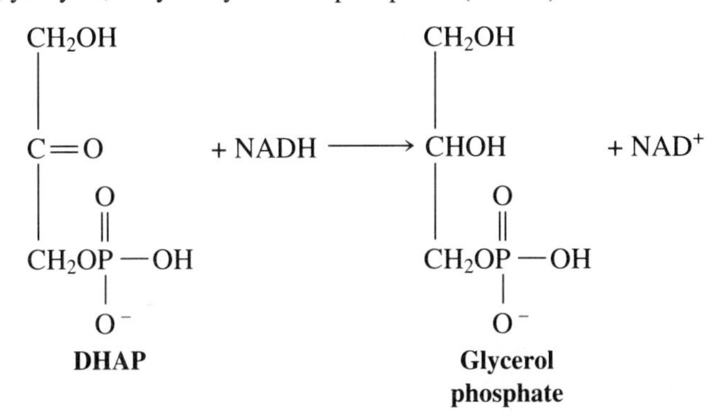

Biosynthesis of Cholesterol

Cholesterol biosynthesis has recently received much attention because blood cholesterol levels have been linked to heart disease. This synthesis occurs primarily in the liver, and the rate of synthesis is normally linked to the amount of dietary cholesterol. All of the carbon atoms of cholesterol are derived from the carbons of acetate in acetyl-CoA.

In two steps, three acetyl-CoA molecules are condensed into hydroxymethylglutaryl-CoA (HMG-CoA).

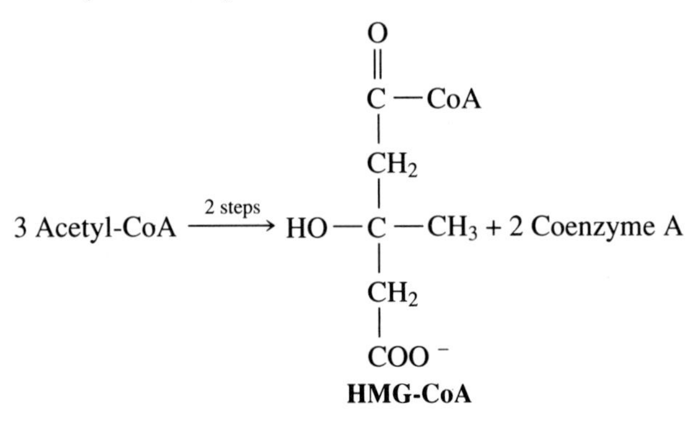

This compound is a key intermediate in metabolism because it can be used to make sterols or ketone bodies. If cholesterol is needed, some HMG-CoA is converted to mevalonic acid in a reaction catalyzed by the enzyme HMG-CoA reductase.

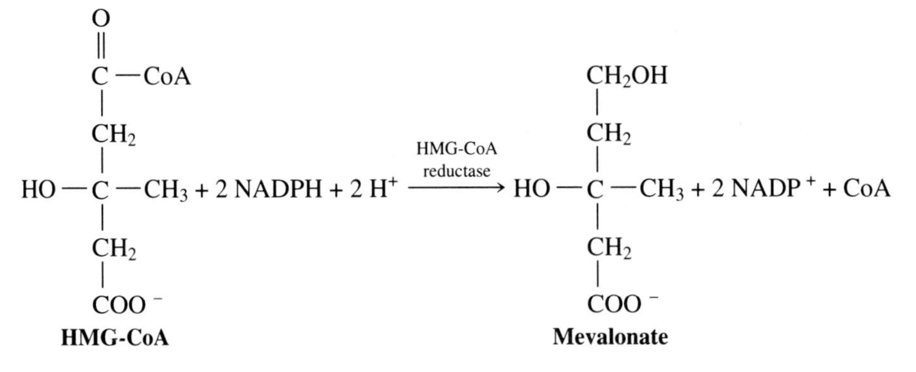

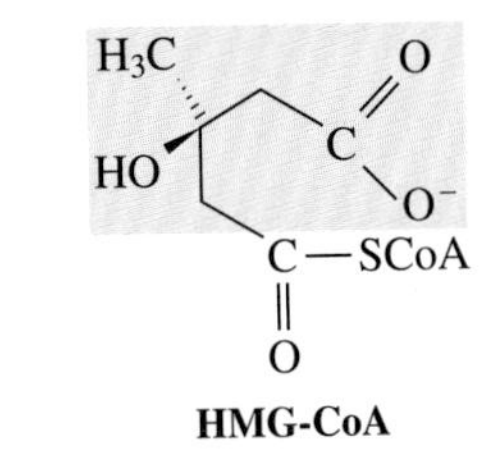

HMG-CoA

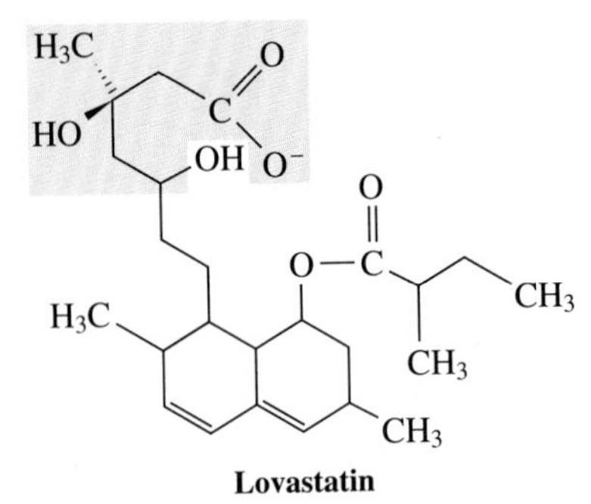

Lovastatin

This enzyme is involved in the regulation of cholesterol biosynthesis and is thus a target for drug treatment of individuals affected by high blood cholesterol. HMG-CoA reductase is competitively inhibited by lovastatin (mevinolin). This compound is used as a drug to help reduce blood cholesterol. The structures of HMG-CoA and lovastatin are shown in the margin, with their structural similarities highlighted.

Mevalonate is used by a variety of organisms to make such diverse compounds as terpenes and sterols. The terpenes are natural products that include a number of fragrances and oils. Mevalonate can be converted to isopentenyl pyrophosphate. Isopentenyl pyrophosphate is used to make a host of terpenes, but in mammals, six of these molecules condense to form squalene, an oil found in the secretions on the surface of skin.

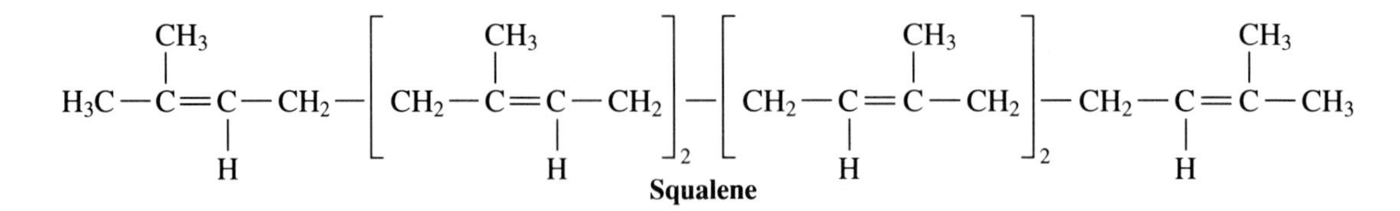

Squalene

Squalene is the precursor to some large terpenes and all of the steroids. Squalene is converted to a steroid ring system to form the sterol lanosterol, which is then converted to cholesterol (Figure 11.6). (Only the carbon skeleton is shown here.) The other steroids of the body are derived from cholesterol.

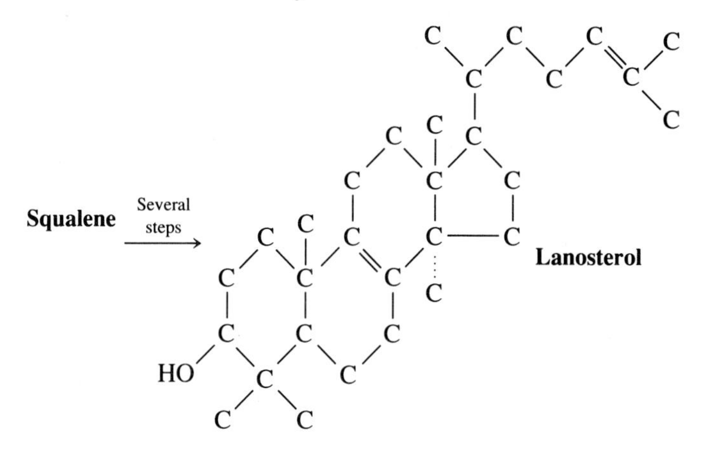

Synthesis of Ketone Bodies

HMG-CoA is also used to make the **ketone bodies**, acetoacetate, 3-hydroxybutyrate, and acetone, which provide a portion of the body's energy needs. In a single step HMG-CoA is converted to acetyl-CoA and acetoacetate.

♦ HMG-CoA is used for the synthesis of cholesterol and ketone bodies.

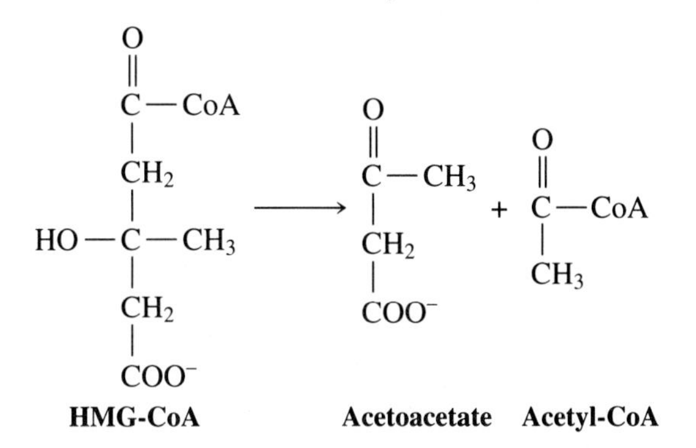

Acetoacetate is reduced to form 3-hydroxybutyrate.

$$\mathrm{CH_3\overset{\overset{\displaystyle O}{\|}}{C}CH_2COO^- + NADH + H^+ \longrightarrow CH_3\overset{\overset{\displaystyle OH}{|}}{C}HCH_2COO^- + NAD^+}$$

Small quantities of acetoacetate are decarboxylated to yield acetone.

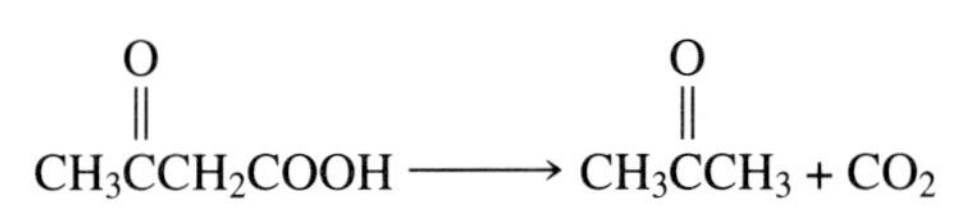

The ketone bodies are synthesized in the liver and circulate throughout the body at low concentrations in blood. Many organs use ketone bodies for fuel, and during fasting, their concentrations increase markedly.

NEW TERMS

lipogenesis The biosynthesis of fatty acids

acyl carrier protein (ACP) Protein involved in lipogenesis that carries the growing fatty acid and the molecules that condense to form the fatty acid

pentose phosphate pathway Series of reactions that effectively converts glucose to pentoses and NADPH

ketone bodies Fuel molecules in the blood that are made from acetyl-CoA via HMG-CoA

TESTING YOURSELF

Biosynthesis of Lipids

1. The carbon atoms of fatty acids come from what molecule?
2. What sources yield the NADPH needed to synthesize fatty acids and other reduced compounds?
3. Why does lovastatin decrease the biosynthesis of cholesterol?

Answers **1.** acetyl group of acetyl-CoA **2.** acetyl-CoA transport and the pentose phosphate pathway **3.** Lovastatin binds to the active site of HMG-CoA reductase, preventing HMG-CoA from binding. If HMG-CoA is not reduced to mevalonate, then cholesterol synthesis cannot proceed.

16.5 Biosynthesis of Amino Acids

♦ An essential amino acid cannot be made by the body.

The 20 common amino acids are required for protein synthesis. Diet normally provides these amino acids unless the amount or quality of dietary protein is too low. Some of the amino acids can be synthesized by the body, but others cannot. Recall that an amino acid that cannot be synthesized in adequate quantities is called an essential amino acid. The essential amino acids include the following:

Histidine	Lysine	Threonine
Isoleucine	Methionine	Tryptophan
Leucine	Phenylalanine	Valine

Plants and many bacteria synthesize all of the amino acids. It appears that humans, along with many other heterotrophic organisms, have lost the capability to make some of these amino acids. When a compound is readily available in the diet, an inability to synthesize the compound is not necessarily a disadvantage.

♦ Transamination of alpha keto acids provides some of the amino acids needed for protein synthesis.

Although the specific pathways through which humans make some of the nonessential amino acids are too long and complicated to present here, there are some common features that can be described. Some of the amino acids are made through transamination of the appropriate keto acid, a reaction that is essentially the reverse of a key step in catabolism of amino acids (see Chapter 15). Here are three keto acids and the amino acids produced by transamination.

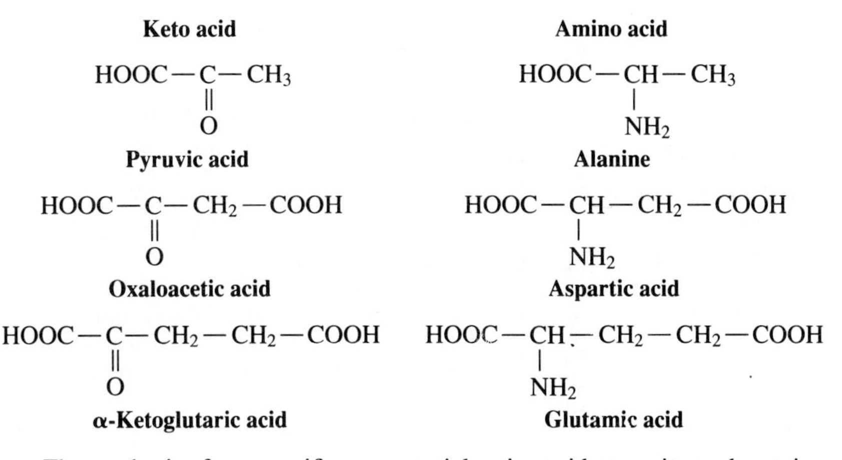

The synthesis of two specific nonessential amino acids, tyrosine and cysteine, requires an essential amino acid in the process. These amino acids can be made only from the essential amino acids phenylalanine and methionine. The phenyl ring of phenylalanine is hydroxylated to yield tyrosine.

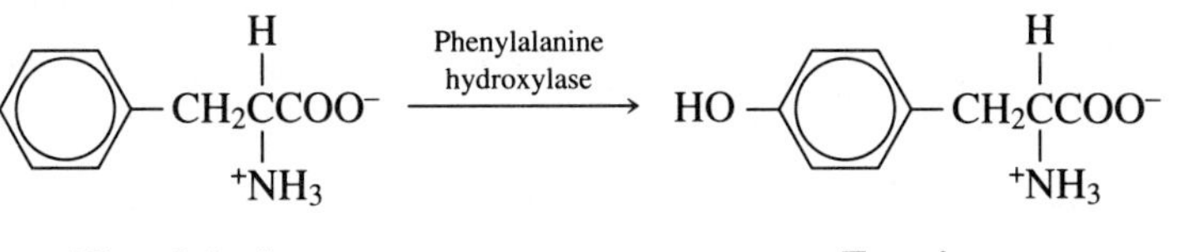

During the synthesis of cysteine, the sulfur atom is donated by methionine. The synthesis of an amino acid from an essential amino acid places greater dietary demand for those essential amino acids if the other amino acid is absent. For example, the amount of phenylalanine required in the diet is less if tyrosine is present than if it is absent. The same relationship exists between methionine and cysteine.

TESTING YOURSELF

Biosynthesis of Amino Acids

1. Why do mammals lack the ability to synthesize nearly half of the amino acids needed for protein synthesis? Is this a significant handicap for these organisms?
2. Show the reaction through which aspartate is synthesized in the body.

Answers **1.** They lack enzymes for the synthesis of these compounds. No. The diet of mammals normally includes all of the amino acids in sufficient amounts.

2.

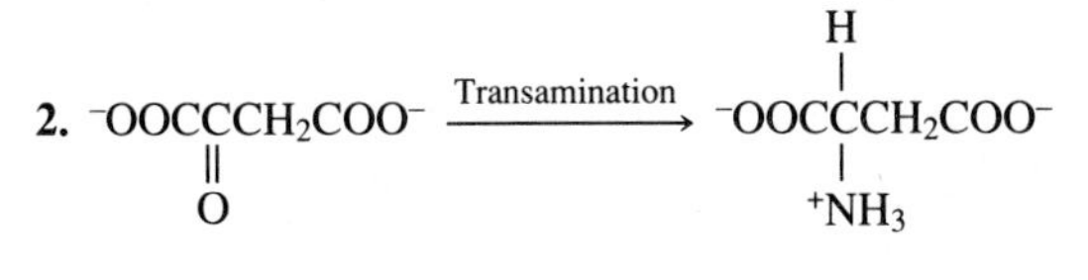

CHEMISTRY CAPSULE

Phenylketonuria (PKU)

Have you ever noticed the warning to phenylketonurics on the label of foods containing Nutrasweet? Nutrasweet is the brand name for aspartame, an artificial sweetener that contains phenylalanine. Phenylketonuria (PKU) is a genetic disease in which the patients are unable to properly metabolize phenylalanine to tyrosine because they lack the enzyme phenylalanine hydroxylase. If the diet contains more phenylalanine than is needed for protein synthesis, it will accumulate in cells and blood. If the concentration gets high enough, phenylalanine will convert to several phenylketo acids, including phenylpyruvic acid, which are then excreted in the urine. These compounds disrupt normal nervous system development in newborns. If the disease is untreated in a young baby, several severe developmental problems, including mental retardation, can result. Most newborns are tested for PKU. If they have this condition, they are given a diet with just enough phenylalanine for protein synthesis. Under these conditions, phenylpyruvate does not accumulate. Adult phenylketonurics should avoid excess amounts of dietary phenylalanine. The warning label on food packages and beverages containing Nutrasweet helps them to monitor their intake of this amino acid.

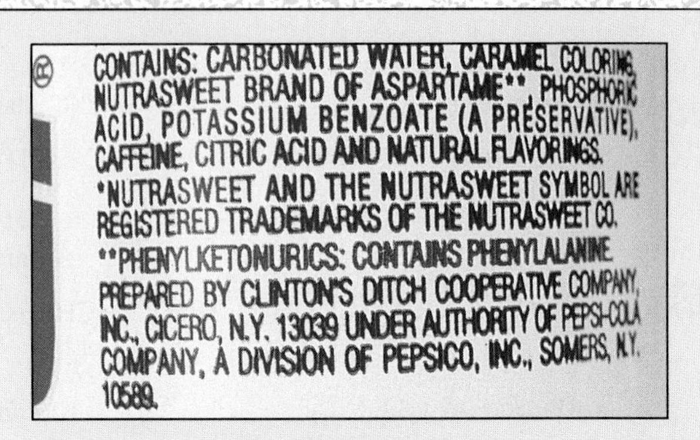

FIGURE 16.C
Foods and beverages containing aspartame provide a warning for phenylketonurics, who cannot metabolize the phenylalanine found in aspartame.

16.6 Biosynthesis of Nucleotides

♦ All needed nucleotides can be made by the body.

Nucleotides are involved in the transfer of hereditary information and numerous other reactions and are thus crucial to life. The nucleotides are all synthesized by the body; a dietary source is not needed. Again, each nucleotide has its own biosynthetic pathway that is too detailed for our purposes here. However, it is worthwhile to examine a part of one of these pathways because some anticancer agents are involved in this metabolism. Deoxythymidine monophosphate (dTMP) is formed from deoxyuridine monophosphate (dUMP) by the addition of a methyl group donated by the coenzyme methylene tetrahydrofolate (methylene-THF), which is derived from folic acid.

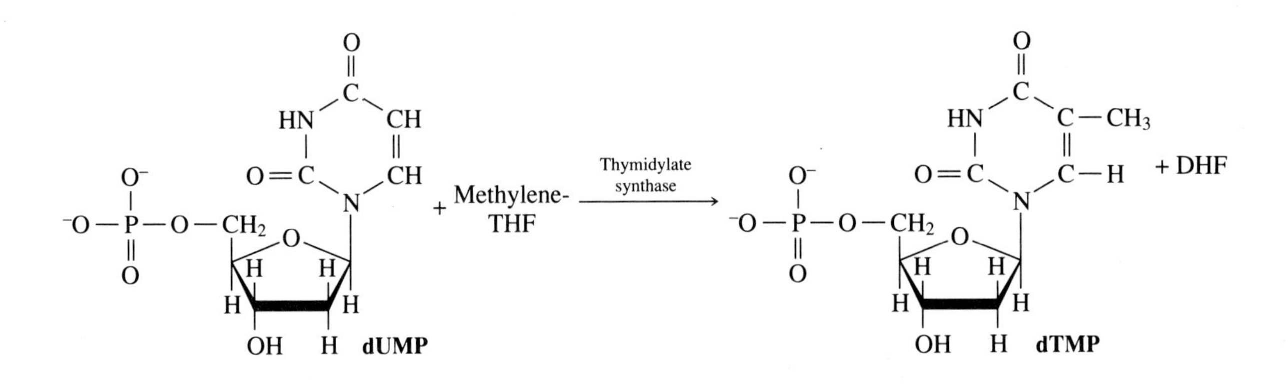

In the body, the anticancer drug fluorouracil is altered to form an irreversible inhibitor of thymidylate synthetase. When the appropriate amounts of fluorouracil have been administered, only a small amount of dTMP is formed. Other compounds, such as methotrexate, prevent the recycling of the coenzyme dihydrofolate (DHF) to methylene-THF (see Chapter 14, Section 14.4). If the concentration of methylene-THF is low, only a very little dTMP is formed.

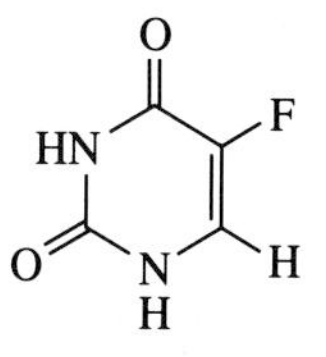

Fluorouracil

Why do these agents work as anticancer drugs? Cancer cells divide more rapidly than most normal cells. Cell division requires DNA replication, which in turn requires the deoxynucleotides. Inhibition of dTMP synthesis kills rapidly dividing cells. Many of the side effects of cancer chemotherapy can be traced to the loss of normal rapidly dividing cells. Hair cells die, and cells lining the gastrointestinal tract are affected as are the cells that divide to form blood cells. Hair loss, gastrointestinal problems, and suppressed immunity are, therefore, all side effects of chemotherapy.

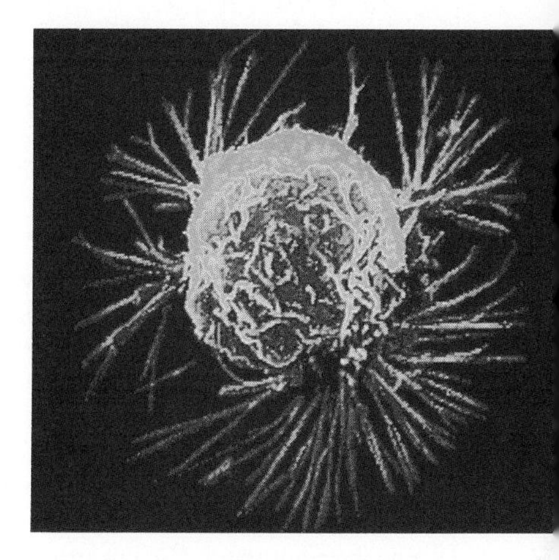

Anticancer drugs kill rapidly dividing cells such as this breast cancer cell.

TESTING YOURSELF

Biosynthesis of Nucleotides

1. How does fluorouracil kill rapidly dividing cells, such as cancer cells?

Answers **1.** This compound is metabolized to an inhibitor of thymidylate synthase, a key enzyme in the synthesis of deoxythymidylate. This nucleotide is needed for the synthesis of DNA, which must be synthesized *before* cell division can occur. If there is little or no DNA synthesis, there will be little or no cell division.

Summary

Anabolism is that part of metabolism that is involved in the synthesis of larger biomolecules. ATP provides much of the energy for these endothermic reactions and *NADPH* provides the electrons needed for reductions. *Photosynthesis* is the fundamental anabolic process because photosynthetic organisms provide the biomolecules needed by all other organisms. The *light reaction* of photosynthesis provides ATP and NADPH, whereas the *dark reaction* uses these compounds and carbon dioxide to make glucose. Glucose can also be made through *gluconeogenesis* from lactate, glycerol, and amino acids, but it cannot be made from fatty acids. Glucose is used to make the other monosaccharides, disaccharides, and glycogen that are needed by the body.

Lipogenesis is the process that synthesizes fatty acids in the body. ATP provides the energy, and NADPH produced in part by the *pentose phosphate pathway* provides electrons. Fatty acids combine with glycerol to make triacylglycerols, and with glycerol, phosphate, and several small molecules to make phosphoacylglycerols. Cholesterol is synthesized from acetyl-CoA. This synthesis can be inhibited by drugs in some people to help reduce high blood cholesterol levels. The body also synthesizes ketone bodies from acetyl-CoA. Only some of the 20 amino acids needed for protein synthesis are synthesized by the body. The rest are essential and are required in the diet. All of the nucleotides required by the body can be synthesized by the body. Rapidly dividing cells require more nucleotides for DNA synthesis, thus several inhibitors of nucleotide synthesis are used as anticancer drugs.

Terms

anabolism (16.1)
nicotinamide adenine dinucleotide phosphate (NADPH) (16.1)
heterotrophic organism (16.2)
autotrophic organism (16.2)
photosynthesis (16.2)
light reaction (16.2)
dark reaction (16.2)
gluconeogenesis (16.3)
Cori cycle (16.3)
glycogenesis (16.3)
lipogenesis (16.4)
acyl carrier protein (ACP) (16.4)
pentose phosphate pathway (16.4)
ketone bodies (16.4)

Exercises

Anabolism (Objective 1)

1. Compare anabolism to catabolism with respect to: (a) size of reactants and products, (b) whether the process is oxidative or reductive, and (c) whether the process is endothermic or exothermic.
2. Compare the roles and structures of NAD^+/NADH and $NADP^+$/NADPH. Why are there two coenzymes with such similar structures yet different biological roles?
3. In Chapter 14 you learned about the interaction of enzymes and their substrates. Use this type of interaction to explain why both NADPH and NADH are not substrates for the same enzymes.
4. What molecule provides electrons for reduction reactions in the synthesis of most biomolecules?
5. What molecule provides energy to drive many of the endothermic reactions of anabolism?

Photosynthesis (Objective 2)

6. How do heterotrophic organisms obtain organic compounds?
7. How do autotrophic organisms obtain organic compounds?
8. Write a balanced equation for the synthesis of 1 mol of glucose by the process of photosynthesis.
9. Make a table showing the products and reactants of the light reaction of photosynthesis.
10. Is the light reaction exothermic or endothermic? Where does the energy go to or come from?
11. Explain how the chemiosmotic theory is involved in the light reaction.
12. Compare the roles of the chemiosmotic theory in photosynthesis and oxidative phosphorylation.
13. How are ATP and NADPH used in the dark reaction of photosynthesis?
14. What enzyme is involved in "fixing" inorganic carbon dioxide into organic molecules?
15. Describe the flow of electrons from water to NADPH during the light reaction of photosynthesis.
16. Indicate how and where light is involved in the flow of electrons described in Exercise 15.
17. How many moles of photons are used to make the NADPH needed for the synthesis of a mole of glucose?
18. What is the most abundant protein on earth?
19. Briefly explain how a glucose molecule is formed as a product of the dark reaction and how ribulose diphosphate is recycled.
20. In photosynthesis, what compound is the ultimate source of electrons?
21. What molecule accepts carbon dioxide during the dark reaction of photosynthesis?

Biosynthesis of Carbohydrates (Objectives 3 and 4)

22. What is the role of gluconeogenesis in carbohydrate metabolism?
23. What molecules serve as precursors to gluconeogenesis? What is the product?
24. What is the normal fate for lactate produced in oxygen-deficient muscles? Where do these reactions occur? What is meant by glucogenic amino acids?
25. Name the intermediates formed as lactate is converted to glucose during gluconeogenesis. Where does gluconeogenesis take place?
26. Write a balanced word equation for glycogen synthesis. Formation of the glycosidic bonds in glycogen is endothermic. What is the source of energy for this reaction?
27. What relationship exists between galactosyl transferase and lactose synthetase?
28. What part of a triacylglycerol can be used for the synthesis of glucose?

Biosynthesis of Lipids (Objective 5)

29. What is lipogenesis?
30. Describe two sources of the NADPH used in anabolism.
31. Explain the transport of acetyl-CoA from the mitochondrion to the cytosol.

32. Compare the number of moles of acetyl-CoA, ATP, $FADH_2$, and NADH (NADPH) produced in the catabolism of palmitate to those consumed in the anabolism of palmitate. If acetyl-CoA and reduced coenzymes were used to make ATP, how much more ATP is needed to make palmitate than is produced from it?
33. What is the product of fatty acid synthase? How do humans make the other fatty acids?
34. How is glycerol phosphate synthesized in the body?
35. In Chapter 11 you learned that fatty acids nearly always have an even number of carbon atoms. Can you now explain why this is so?
36. What is the role of acyl carrier protein in fatty acid synthesis? What molecule serves as a carrier of acyl groups in other parts of the cell?
37. The pentose phosphate pathway produces what two products?
38. What role does cytidine diphosphate serve in lipid biosynthesis?
39. Describe the synthesis of lecithin.
40. What is squalene? What important class of biomolecules is synthesized from squalene?
41. What six-carbon molecule serves as a precursor to both cholesterol and the ketone bodies?
42. Name the ketone bodies and describe how two of them are synthesized from the other one.

Biosynthesis of Amino Acids (Objective 6)

43. What is the usual source of amino acids for animals?
44. Name the essential amino acids of humans.
45. How can transamination serve a role in both catabolism and anabolism of amino acids?
46. What is the corresponding keto acid for: (a) alanine, (b) aspartic acid, and (c) glutamic acid?
47. You have made a controlled diet for a population of mice. This diet is complete except for the lack of cysteine, a nonessential amino acid. The other 19 amino acids are present in the minimal amount needed by mice. The young mice grow slowly and show signs of malnutrition. What is wrong with your diet?

Biosynthesis of Nucleotides (Objective 7)

48. Chemotherapy for cancer patients has several common side effects. What is the cause of these side effects?
49. What is the role of methylene-THF in thymidine biosynthesis?
50. Are there any essential nucleotides? Why or why not?
51. Compare the structure of fluorouracil to that of uracil in dUMP. What is the relationship between this similarity and the mode of action of fluorouracil?
52. How does methotrexate inhibit cancer cells?

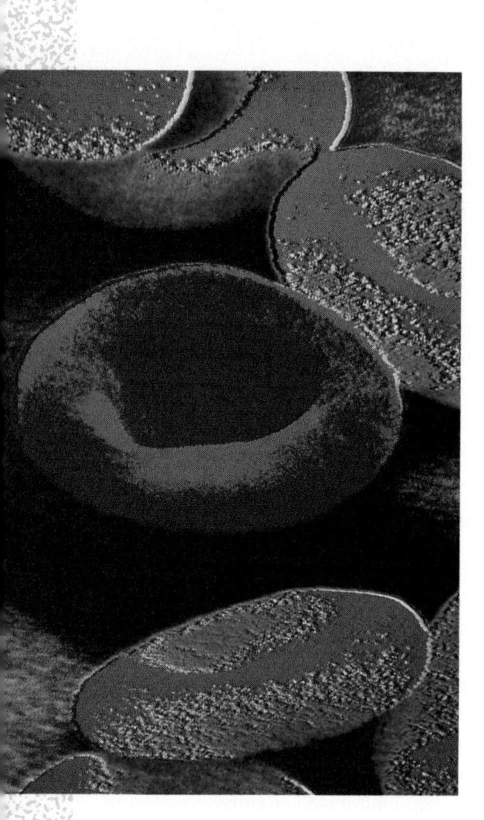

Red blood cells play a key role in the transport of gases.

BLOOD: THE CONSTANT INTERNAL ENVIRONMENT

OUTLINE

17.1 Gas Exchange
17.2 Removal of Wastes from the Blood
17.3 Regulation of pH
17.4 Circulation of Hormones and Their Role in Nutrient Maintenance
17.5 Blood and Immunity

OBJECTIVES

After completing this chapter, you should be able to

1. Explain the role of blood in gas exchange and transport.
2. Describe how filtration and reabsorption are involved in the elimination of wastes.
3. Describe the factors that influence and regulate blood pH.
4. Explain the mechanism of hormone action.
5. Describe the role of glucagon and insulin in the maintenance of nutrients in the blood.
6. Explain how antibodies contribute to the immune system.

Warm-blooded organisms maintain a more or less constant internal temperature. But temperature is only one of several factors that the body maintains within certain limits. Concentrations of nutrients, waste products, and oxygen are also closely regulated. **Homeostasis** is a state of dynamic equilibrium that maintains a constant internal environment in the body. This is accomplished through self-regulating processes that rely on feedback. An example of a homeostatic process is the regulation of body temperature. When the body is too warm, blood flow increases in the skin, which increases the rate of heat loss to the air. When the body is too cold, blood flow to the skin is reduced. Internal temperature is self-regulating, since body temperature influences the rate of heat loss.

Blood plays a major role in homeostasis because it is in contact with the interstitial fluids that surround all cells. Nutrient concentrations in the blood are maintained within a narrow range; the nutrients are thus available for ready absorption by cells. Cellular wastes diffuse from cells into the blood and are transported to excretory organs. Blood circulates hormones that help control and coordinate body activity. Blood also contains a variety of cells and proteins that neutralize or destroy foreign cells or chemicals. Many of these processes involve chemistry that you have already learned in this book. This chapter discusses the chemistry of blood.

♦ Blood plays an important role in the maintenance of a constant internal environment.

17.1 Gas Exchange

Gas Exchange in the Lungs

The cells of the body consume oxygen in the electron-transport chain and produce carbon dioxide, principally in the TCA cycle (see Chapter 15). Oxygen must be absorbed from the atmosphere by the lungs, transported by the blood to the tissues, and moved by diffusion from the blood into the tissues. Carbon dioxide produced by aerobic catabolism diffuses from the cells through the interstitial fluid into the blood. It is then transported by the blood to the lungs, from where it leaves the body.

Inhaled air that is relatively rich in oxygen passes through increasingly smaller passageways in the lungs to enter microscopic sacs called **alveoli** (Figure 17.1). Oxygen within the alveoli dissolves into the moist alveolar membranes. The concentration of oxygen within the alveolar membranes is higher than the concentration of oxygen in the blood that is flowing through the capillaries of the lungs. Because materials generally move from high concentration to low concentration, oxygen thus diffuses into the blood.

♦ Gas exchange occurs across the alveoli of the lungs.

Gas Transport in the Blood

The solubility of oxygen in plasma (the liquid part of blood) is rather low (0.3 mL/100 mL), and the body cannot transport enough dissolved oxygen in blood plasma to sustain life. This limitation is overcome by the presence of an additional oxygen-transport system. The red blood cells contain **hemoglobin**, an oxygen-binding protein. Hemoglobin greatly increases the capacity of the blood to transport oxygen. As the concentration of dissolved oxygen increases within the red blood cells, oxygen binds to hemoglobin molecules within the cells. Each hemoglobin molecule can bind four oxygen molecules, one to each of the four

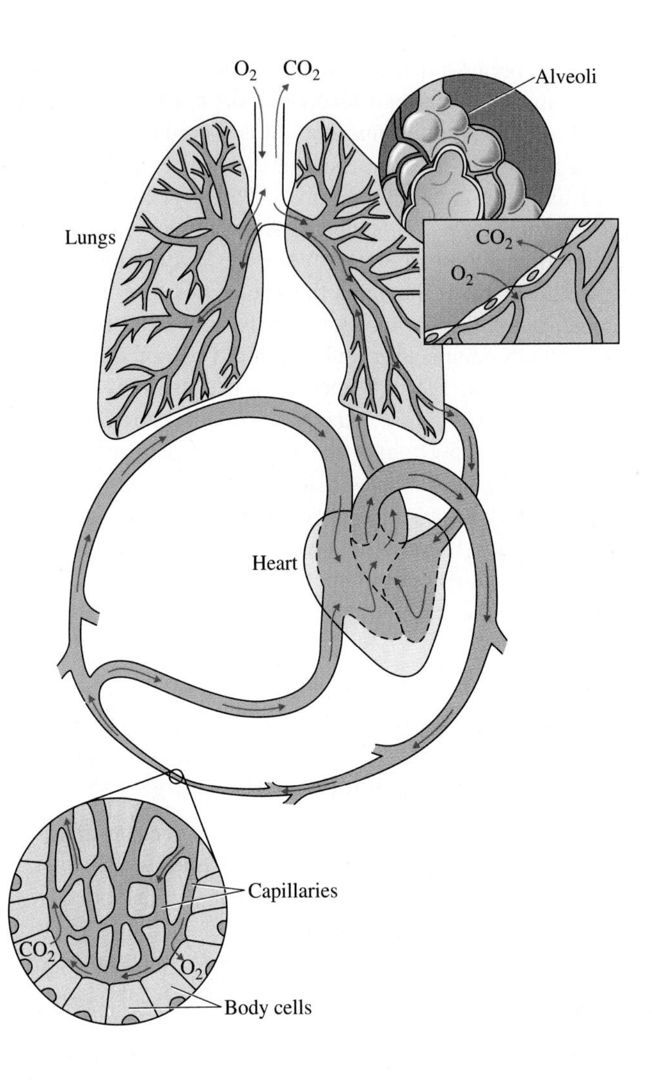

FIGURE 17.1
The movement of oxygen and carbon dioxide within the respiratory and circulatory systems. Oxygen (red arrows) moves from the lungs through the circulatory system to the tissues of the body. Carbon dioxide (blue arrows) flows in the opposite direction. These gases pass from one phase or system to another by diffusion.

subunits present in the protein. The binding of the first oxygen molecule to a subunit of hemoglobin causes a conformational change in that subunit, resulting in a change in shape of the other three subunits. In this new shape, the subunits bind oxygen molecules more readily. Thus, the binding of the first oxygen molecule to a subunit of hemoglobin enhances oxygen binding to the other three sites on the hemoglobin molecule. This type of binding is called *cooperative* binding. When blood leaves the capillaries of the lungs, most hemoglobin molecules have four oxygen molecules bound to them (Figure 17.2). The amount of oxygen in blood leaving the lungs is 20 mL/100 mL, mostly as oxygen bound to hemoglobin. This is far more than can be transported by a simple aqueous solution of oxygen.

♦ Hemoglobin greatly increases the oxygen-carrying capacity of blood.

Oxygen-rich blood is pumped to the tissues of the body. In these tissues, the concentration of oxygen is appreciably lower than the concentration in the blood. Oxygen diffuses from the blood into the surrounding fluids. This diffusion decreases the amount of dissolved oxygen in the blood and promotes the dissociation of oxygen molecules from hemoglobin. Like binding, the dissociation of oxygen

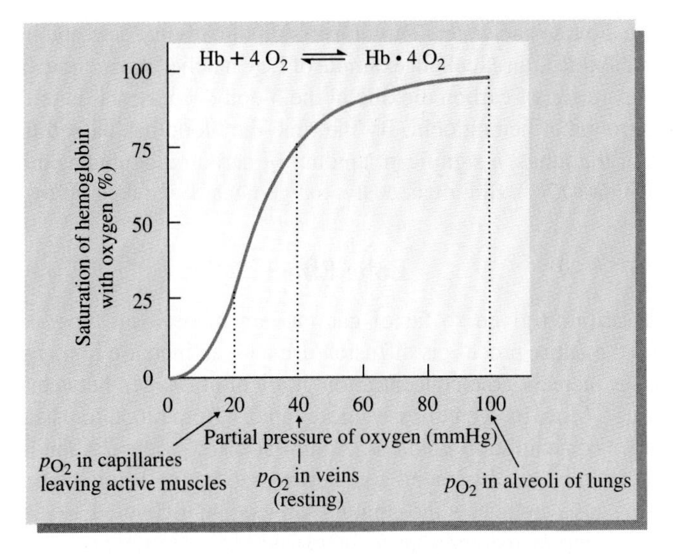

FIGURE 17.2
Hemoglobin (Hb) binds oxygen in the lungs and releases it to the tissues of the body. Blood leaving active tissues has less oxygen (p_{O_2} = 20 mm Hg), and hemoglobin is only 25 percent saturated with oxygen. In a resting individual, veins have more oxygen (p_{O_2} = 40 mm Hg), and hemoglobin is around 75 percent saturated. In lungs, there is more oxygen (p_{O_2} = 100 mm Hg), and hemoglobin is nearly 97 percent saturated with oxygen. Hemoglobin loads up with oxygen in the lungs and delivers it to tissues of the body.

from hemoglobin is cooperative; the first dissociation is quickly followed by others. This unloading of oxygen from hemoglobin ensures that the concentration of dissolved oxygen in the blood exceeds the concentration in the interstitial fluids. Adequate amounts of oxygen diffuse into the tissues as the blood passes through the capillaries. In a resting individual about 25 percent of the oxygen in blood is delivered to the tissues as the blood passes through. Considerably more is delivered to tissues that are metabolically active.

Gas Utilization and Production by Cells

The oxygen that diffuses into the cells of the body is used within the mitochondria for aerobic metabolism. Carbon dioxide is a product of aerobic metabolism and is only slightly soluble in body fluids. However, it reacts with water to form carbonic acid, which readily dissociates to yield bicarbonate ion.

$$\underset{\textbf{Carbon dioxide}}{CO_2} + H_2O \xrightleftharpoons{\text{Carbonic anhydrase}} \underset{\textbf{Carbonic acid}}{H_2CO_3} \rightleftharpoons \underset{\textbf{Bicarbonate ion}}{HCO_3^-} + H^+$$

The formation of carbonic acid is catalyzed by the enzyme *carbonic anhydrase* found within red blood cells. Some carbon dioxide also reacts with certain amino groups within hemoglobin molecules. Carbon dioxide in blood is thus transported in several forms: (1) as dissolved carbon dioxide, (2) as a complex with hemoglobin, and (3) as bicarbonate ion. Of the three, bicarbonate is the principal transport form. When venous blood arrives at the lungs, some of the dissolved carbon dioxide diffuses into the alveoli because the alveolar concentration of carbon dioxide is lower due to exchange with inhaled air. As the concentration of carbon dioxide decreases in the blood, carbonic acid dissociates to yield more carbon dioxide and water. This decreases the amount of dissolved carbonic acid, so bicarbonate combines with H^+ to form more carbonic acid, which yields even more carbon dioxide and water. This conversion of carbonic acid and bicarbonate into carbon dioxide is an excellent example of Le Chatelier's principle. In addition, as the concentration of carbon dioxide in the blood decreases, it is released from the amino groups of hemoglobin. By the time the blood has passed through the capillaries of the lungs, a significant amount of carbon dioxide has diffused into the lungs (Figure 17.1). Simultaneously, oxygen has been taken up by the blood.

Bohr Effect

The description given so far of gas movement between lungs and tissues implies that the entire process is diffusion driven—an increase in oxygen utilization in tissues increases the concentration gradient of oxygen between the blood and the tissues. More oxygen diffuses more rapidly from blood to tissues because of the steeper concentration gradient. A similar effect occurs at the lungs. This description is true, but hydrogen ion concentration (pH) and carbon dioxide concentration also influence the amount of oxygen delivered to tissues. The increased delivery of oxygen due to increased concentrations of hydrogen ion and carbon dioxide is called the **Bohr effect**.

Hemoglobin binds oxygen reversibly. At the relatively high concentrations of oxygen in the lungs, hemoglobin readily binds oxygen, but at the relatively low concentrations characteristic of tissues, hemoglobin releases oxygen. But hemoglobin also binds hydrogen ions (H^+) and carbon dioxide molecules. Carbon dioxide and hydrogen ions work like allosteric effectors to produce the Bohr effect. When they bind to hemoglobin, they induce a conformational change in the hemoglobin molecule to a form that binds oxygen molecules less tightly. Increased concentrations of these substances result in their increased binding to hemoglobin, which decreases the ability of hemoglobin to bind oxygen (Figure 17.3).

The physiological consequences of the Bohr effect are twofold: (1) increased delivery of oxygen upon demand and (2) increased binding of oxygen to hemoglo-

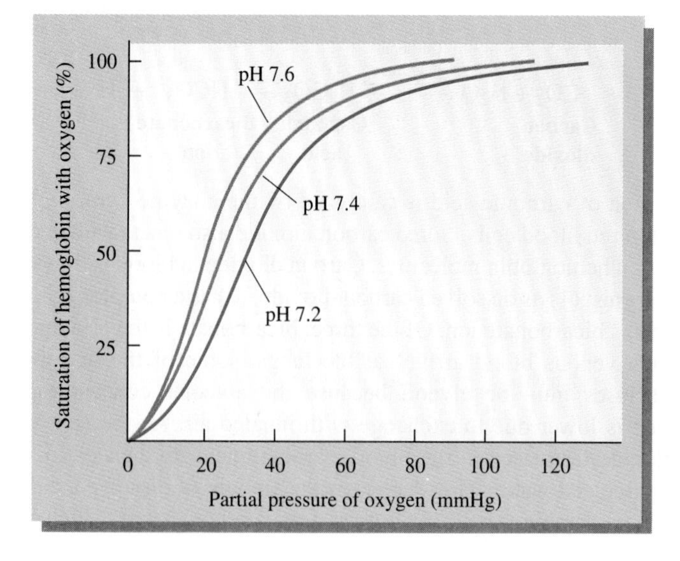

FIGURE 17.3
The effects of pH on oxygen binding to hemoglobin. Increased concentration of hydrogen ion (lower pH) reduces the binding of oxygen to hemoglobin. Since active tissue is more acidic (pH = 7.2) than less active tissue (pH = 7.4), more oxygen is delivered to active tissue than to less active tissue. Increased concentrations of carbon dioxide also reduce oxygen binding to hemoglobin.

bin in the lungs. When the metabolism of muscles increases during exercise, carbon dioxide is produced by aerobic metabolism and lactic acid is produced during anaerobic exercise. Both are acids and yield hydrogen ion in water. Thus, the pH of metabolically active tissue decreases during exercise. The hydrogen ion and carbon dioxide diffuse into blood where these molecules bind to hemoglobin. This binding promotes rapid unloading of oxygen. Thus, more oxygen is delivered to the locations where it is most needed, that is, to metabolically active tissues (Table 17.1). The liver removes lactic acid from the blood (see Chapter 16), and carbon dioxide is expelled in the lungs. Loss of these metabolites increases blood pH. Under these conditions, hemoglobin binds oxygen more tightly. The decrease in carbon dioxide concentration and the increased pH promote the oxygenation of hemoglobin in the lungs.

♦ Increased concentrations of H^+ and CO_2 in blood result in greater oxygen unloading from hemoglobin.

TABLE 17.1

Effects of pH on the Amount of Oxygen Delivered to Tissues

Tissue pH	mL of Oxygen Delivered/ 100 mL of Blood
7.2 (active tissue)	6.6
7.4 (normal)	5.0
7.6	3.0

NEW TERMS

homeostasis Maintenance of a constant internal environment

alveoli Tiny sacs in the lungs where gas exchange occurs

hemoglobin Protein in red blood cells that binds oxygen cooperatively. The presence of hemoglobin in blood greatly increases the capacity of blood for oxygen transport.

Bohr effect Increases in the concentration of carbon dioxide and hydrogen ions increase dissociation of oxygen from hemoglobin. Decreases in concentration have the opposite effect.

TESTING YOURSELF

Gas Exchange

1. Where is the site of gas exchange between the body and air?
2. Compare the oxygen-carrying capacity of whole blood and plasma.
3. What effect would alkalosis (increased blood pH) have on oxygen delivery to tissues?

Answers **1.** In the alveoli of the lungs. **2.** Plasma carries about 0.3 mL/100 mL, whole blood about 20.0 mL/100 mL. **3.** Less oxygen would be delivered (Bohr effect).

17.2 Removal of Wastes from the Blood

Carbon dioxide is a waste product that is circulated by the blood to the lungs, from where it leaves the body. Some water is also lost from the lungs. Sweat glands may also excrete significant amounts of water and salt under certain conditions. Most other wastes carried by blood are excreted by the kidneys. The physiological function of the kidneys is beyond the scope of this book, but several of the processes have chemical components that you have learned and are thus worth reviewing here.

The walls of the capillaries in parts of the kidneys are quite porous. As blood passes through the glomerulus, which is a cluster of capillaries, blood pressure forces water and solutes through the walls into Bowman's capsule, which funnels them into tubules. Cells and macromolecules, such as proteins, are too large to pass through the pores in the capillaries, thus the blood is filtered. Through **filtration**, large particles and cells are retained in the circulatory system and water and solutes pass out of it (Figure 17.4). Water, nutrients, and waste products are thus all filtered from the blood into the tubules of the kidney.

Nutrients do not remain in the kidney filtrate, however, The cells lining the tubules have active transport systems to reabsorb nutrients, such as glucose and amino acids. Water and ions may also be reabsorbed, but the actual amount reabsorbed is highly regulated. Normal kidney function can be lost through infection or environmental stress. Artificial kidney dialysis can be used to remove wastes from the blood.

♦ Diabetics sometimes excrete glucose because the concentration of blood glucose exceeds its renal threshold.

The ability to reabsorb nutrients in the kidneys is limited. If the concentration of a particular solute exceeds a concentration known as the **renal threshold**, then reabsorption will be incomplete and some of the solute will appear in the urine. This occurs with untreated diabetics. Following a meal, glucose is absorbed from the small intestine into the blood. The cells of diabetics do not take up glucose as rapidly as normal cells (Section 17.4). Glucose accumulates in the blood and may exceed the renal threshold for glucose, which is 180 to 200 mg/100 mL of blood.

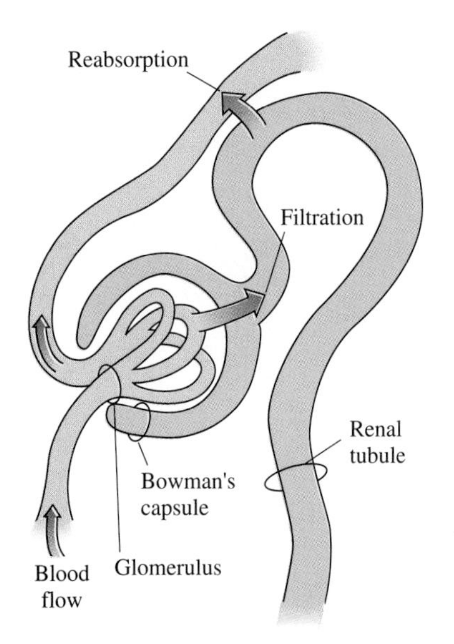

FIGURE 17.4
Filtration and reabsorption in a nephron of the kidney. Blood enters the glomerulus of the kidney and is filtered. The filtrate is channeled by the Bowman's capsule into the tubule. Cells lining the tubule reabsorb water and nutrients and return them to the blood.

NEW TERMS

filtration Process that separates large molecules from small ones by exclusion. Small molecules pass through pores, but large molecules and particles are retained.

renal threshold The maximum concentration of a substance that can be effectively reabsorbed by the kidneys. If the concentration exceeds this threshold, the substance will appear in the urine.

17.3 Regulation of pH

You learned in Chapter 12 that proteins have a native conformation that can be lost if the pH varies. Control of pH within a cell is critical, but because intracellular pH is influenced by extracellular fluids, control of the pH of these fluids is equally critical. If blood pH changes, the proteins of blood and cells may be denatured.

Buffering Systems in the Blood

The best way to control pH is with *buffers*. Buffers are solutions of a weak acid and its conjugate base. If hydrogen ion is lost from the system, the acid will give up hydrogen ions to replace those lost by the system.

$$HA \longrightarrow H^+ + A^-$$

This hydrogen ion is released from the acid to replace those lost from the system.

CHEMISTRY CAPSULE

Kidney Dialysis

Kidneys perform the vital function of clearing wastes and excess water and salts from the blood. If the kidneys fail, the constant internal chemical environment changes, and life-threatening problems arise. *Kidney*, or *renal*, *dialysis* is a treatment for kidney failure. In one form of dialysis, the patient's blood is passed through fine, semipermeable tubules that are bathed in a solution called the *dialysate*. This solution contains glucose, electrolytes, and other needed substances. Since both blood and the dialysate contain about the same concentration of these nutrients, there is no net diffusion between them. The dialysate does not contain waste products, and these materials diffuse from blood into the dialysate. Excess electrolytes can be removed by using a dialysate with the appropriate concentration of electrolytes. Excess water can be removed by making the dialysate hypertonic to the blood. A simpler form of renal dialysis takes advantage of the semipermeable nature of the membranes of the body. In this method, a dialysate is introduced into the abdominal cavity, where diffusion between the dialysate and blood can occur across the lining of the abdominal cavity. Both methods use simple physical properties—diffusion and osmosis—to accomplish what the kidneys can no longer do.

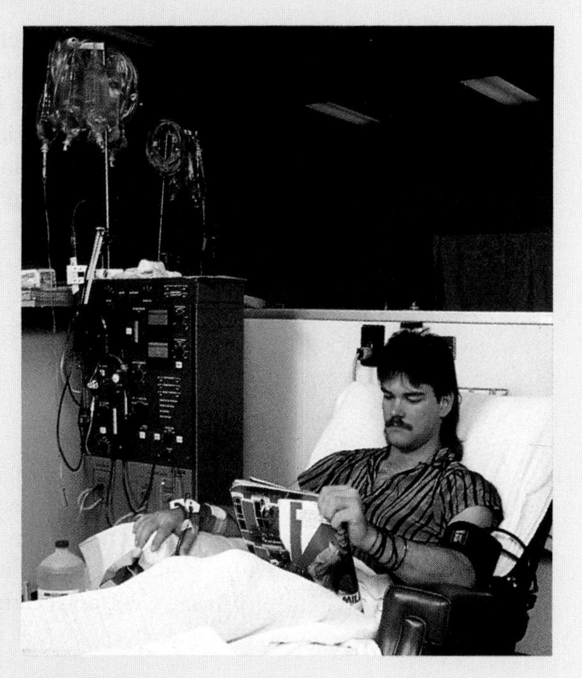

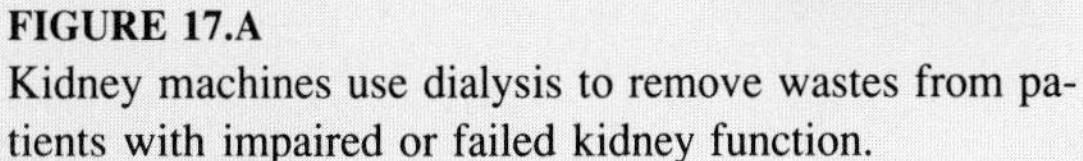

FIGURE 17.A
Kidney machines use dialysis to remove wastes from patients with impaired or failed kidney function.

If hydrogen ion is added to the system, then the conjugate base accepts the H^+ to become the acid.

$$HA \longleftarrow H^+ + A^-$$

This hydrogen ion added to the system is taken up by the conjugate base.

Collectively these two reactions provide protection against addition or loss of hydrogen ion.

♦ Buffers resist changes in pH because they can accept or donate hydrogen ions.

$$HA \rightleftharpoons H^+ + A^-$$

The acid and conjugate base of this system take up or give up hydrogen ions to prevent or minimize changes in pH.

Blood contains three potential buffering systems: (1) a bicarbonate system, (2) a phosphate system, and (3) a protein system. Recall that carbon dioxide produced during respiration combines with water to form carbonic acid, which dissociates to yield bicarbonate.

♦ Bicarbonate, phosphate, and proteins buffer blood.

$$\underset{\textbf{Carbon dioxide}}{CO_2} + H_2O \rightleftharpoons \underset{\textbf{Carbonic acid}}{H_2CO_3} \rightleftharpoons \underset{\textbf{Bicarbonate ion}}{HCO_3^-} + H^+$$

Because the concentration of bicarbonate is relatively high in blood, carbonic acid and bicarbonate are a conjugate acid–base pair that makes a significant contribution to the buffering of blood.

Phosphoric acid is a weak triprotic acid, that is, it can donate three hydrogen ions. Near physiological pH (pH 7), this acid exists mainly as either $H_2PO_4^-$ or HPO_4^{2-}. These two ions are an acid–base conjugate pair that contributes to blood buffering.

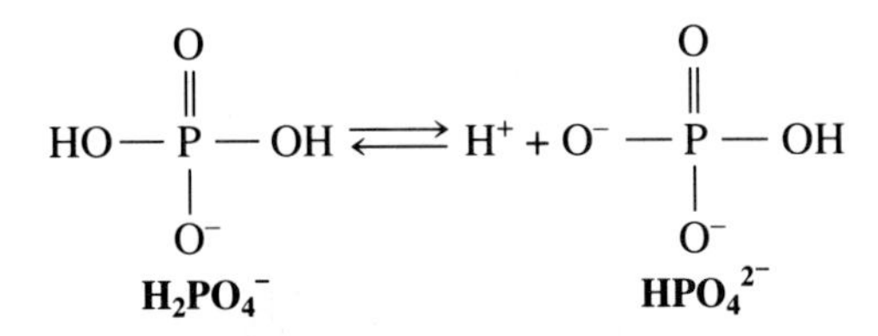

The third buffering system of the blood involves proteins. All blood proteins contribute to the buffering of blood, but hemoglobin, the most abundant blood protein, makes the largest contribution. All proteins possess some amino acids with basic or acidic side groups, and they also possess *N*-terminal amino groups and *C*-terminal carboxyl groups. These groups are potential hydrogen ion donors and acceptors that moderate blood pH.

Acidosis and Alkalosis

The body normally maintains blood pH within a narrow range from pH 7.35 to 7.45, but certain physiological states may push the pH beyond this range. If the pH drops below this range, the condition known as **acidosis** exists. If blood pH rises above this range, then **alkalosis** exists. Both of these conditions can be described as either respiratory or metabolic, depending upon the cause of the pH imbalance.

Respiratory acidosis occurs when carbon dioxide accumulates in the blood. This can occur through *hypoventilation*, which is abnormally low breathing rates and shallow breathing, resulting in reduced loss of carbon dioxide. Whenever the rate of carbon dioxide loss is below the rate of production, carbon dioxide accumulates. The increased carbon dioxide is accompanied by an increase in carbonic acid, which yields an increase in hydrogen ion and a corresponding drop in pH. *Hyperventilation* (breathing that is too rapid and deep) can cause *respiratory alkalosis* because carbon dioxide is lost at a rate greater than its production. Less carbon dioxide means less carbonic acid and less hydrogen ion. Loss of this acid means blood becomes more basic, that is, its pH increases.

Alterations in normal metabolism can cause *metabolic acidosis* and alkalosis. A principal cause of metabolic acidosis is *ketosis*, the presence of excess ketone bodies in the blood. The principal ketone bodies, acetoacetic acid and β-hydroxybutyric acid, ionize to yield hydrogen ion. If the concentration of ketone bodies is high enough, the buffering capacity of the blood is overwhelmed and the blood pH drops. *Metabolic alkalosis* is relatively uncommon, perhaps because the common body metabolites are not alkaline, but excess loss of stomach acid can yield metabolic alkalosis. Ingestion or intravenous feeding of excess bases can also raise blood pH.

NEW TERMS

acidosis Condition that exists whenever the blood pH drops below the normal range; can be metabolic or respiratory in origin

alkalosis Condition that exists when blood pH exceeds the normal range; like acidosis, it can result from respiratory or metabolic abnormalities

TESTING YOURSELF

Blood Buffers

1. What are the three principal buffering systems in blood?
2. a. What is respiratory acidosis?
   b. What is the actual source of the hydrogen ion?

Answers **1.** Bicarbonate, phosphate, and blood protein. **2. a.** Increased hydrogen ion concentration in blood due to an increase in blood carbon dioxide. **b.** Carbonic acid that forms from carbon dioxide and water.

17.4 Circulation of Hormones and Their Role in Nutrient Maintenance

Hormones are chemical messengers that regulate and coordinate the activity of the many different kinds of cells. Certain cells in endocrine glands produce hormones and secrete them into the blood where they are distributed throughout the body. The hormone molecules are then taken up by specific cells called *target cells*. The interaction between hormones and specific proteins of the cells results in an alteration of the cellular metabolism. Although there is considerable variation in the structures of hormones, they are classified into two main categories according to how they initiate a cellular response. Hormones of the first group actually enter the cell and exert their effect within the cell. This group includes the steroid hormones derived from cholesterol, the thyroid hormones, and some protein hormones, such as epidermal growth factor. Members of the second group bind to the surface of the cell and exert their influence by initiating a chain of events that results in metabolic changes. This group includes such proteins as glucagon and insulin, peptides, and some amino acid derivatives, such as epinephrine. Table 17.2 lists some important hormones in the body.

♦ The chemical messengers circulated in blood are hormones.

♦ Hormones may be steroids, proteins, peptides, or amino acid derivatives.

Hormones That Work within Cells

The hormones that enter a cell, such as the steroid hormones, diffuse into the cell and bind to specific receptor proteins. These complexes enter the nucleus where they bind at specific locations to the DNA–protein complexes located there (Figure 17.5). The binding of the hormone alters the expression of genes in that region of the DNA. Often the activity of the cell is altered for an indefinite time, and the cell requires several hours or more to respond to the hormone.

♦ Some hormones enter cells and bind to DNA, thus affecting gene expression.

TABLE 17.2

Some Important Hormones of the Body

Hormone	Endocrine Gland	Function
Hydrocortisone	Adrenal cortex	Controls a variety of metabolic processes
Estrogen	Ovary	Involved in secondary sex traits and female reproductive cycle
Progesterone	Corpus luteum	Involved in female reproductive cycle
Testosterone	Testes	Principal male sex hormone
Insulin	Pancreas	Regulates concentration of nutrients in blood
Glucagon	Pancreas	Stimulates release of glucose and lipid into blood
Epinephrine	Adrenal	Induces a variety of effects in response to environmental stimuli

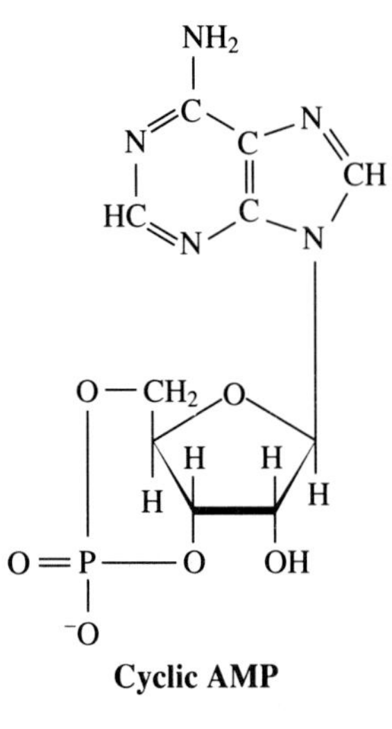

Cyclic AMP

Hormones That Work at Cell Surfaces

Members of the other group of hormones do not enter the cell. Instead, they bind to a specific receptor molecule in the cell membrane. The hormone–receptor complex serves as a signal to alter parts of the cell's metabolism. For example, the complex formed between the hormone epinephrin and its receptor triggers the activation of the enzyme adenylate cyclase to form **cyclic AMP** (3′, 5′-adenosine monophosphate) within the cell. Cyclic AMP (or cAMP) activates the enzyme *protein kinase*, which in turn alters the activity of some other key proteins within

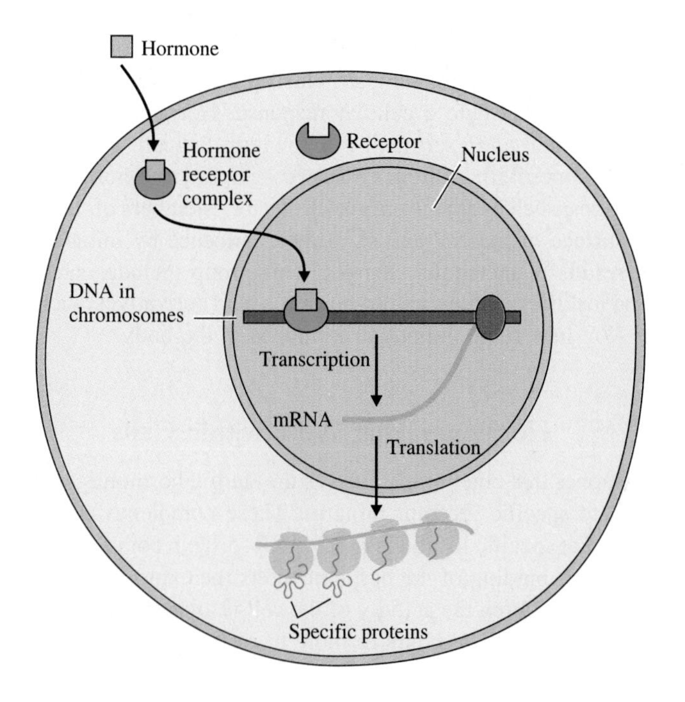

FIGURE 17.5
The entry and effects of a steroid hormone on a target cell. Hormone molecules in the blood stream (yellow) bind to a cytosolic receptor protein (green) at the cell membrane of the target cell. The hormone–receptor complex diffuses into the nucleus and binds at specific sites on DNA (blue) in chromosomes. This binding stimulates transcription of the DNA in that region, resulting in production of specific proteins (orange).

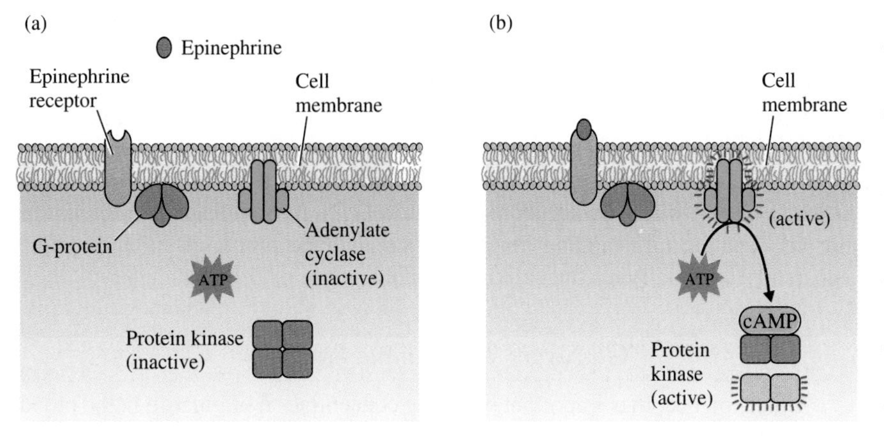

FIGURE 17.6
The interaction between a target cell and epinephrine (purple). (a) Target cells possess receptor molecules (orange) that bind the hormone on the outside surface of the cell. The membrane also has G protein (blue), and an enzyme, adenylate kinase, (green) that is inactive if hormone is not bound. (b) When epinephrine binds to its receptor, the receptor conformation changes, which activates G protein, which in turn activates adenylate cyclase. Adenylate cyclase catalyzes the formation of cAMP (yellow) from ATP. Cylic AMP binds to the enzyme protein kinase, which activates some of the subunits (pink) of this enzyme. (The role of protein kinase is shown in Figure 17.8).

the cell (Figure 17.6). Thus, the activity of the cell is changed. This process occurs rapidly, with a time frame measured in seconds, and is readily reversible. The hormone glucagon functions through this mechanism (see the next section).

Maintenance of Nutrients in the Blood

The concentration of nutrients in the blood is normally maintained within a narrow range. This ensures that cells have these nutrients available whenever they need them. Humans do not eat continuously. Nutrients in the blood do not simply reflect the concentration of nutrients in the gastrointestinal tract. This section provides an introduction to the mechanisms through which the body uses and regulates nutrients in blood.

Nutrient concentrations in blood are influenced by hormones. Although there are other hormones involved, the two primary hormones that influence blood nutrient concentrations are insulin and glucagon. **Insulin**, a "feast" hormone, signals an abundance of nutrients in the bloodstream. It is released when nutrients are available and need to be taken up, stored, or utilized. **Glucagon**, a "famine" hormone, serves as a signal that nutrients are scarce and should be released from storage and mobilized into blood to make them available to body cells. Let's first consider the feast hormone insulin.

♦ Insulin signals that an abundance of nutrients is present in blood.

♦ Glucagon signals that the concentration of fuel molecules is low and must be increased.

Insulin: The Feast Hormone

Shortly after a meal is eaten, glucose, lipids, and amino acids begin to enter the bloodstream through absorption from the small intestine. Depending upon the size and composition of the meal, absorption can continue for several hours. This absorption increases the concentration of glucose, lipids, and amino acids in the blood. There is an ample supply of nutrients available to cells.

Under these conditions, insulin is secreted by the beta cells of the pancreas, and secretion of glucagon is inhibited. Insulin stimulates anabolic processes. Cellular uptake of sugars and amino acids from blood is stimulated. Cellular synthesis of fatty acids, glycogen, and proteins increases. Insulin signals that

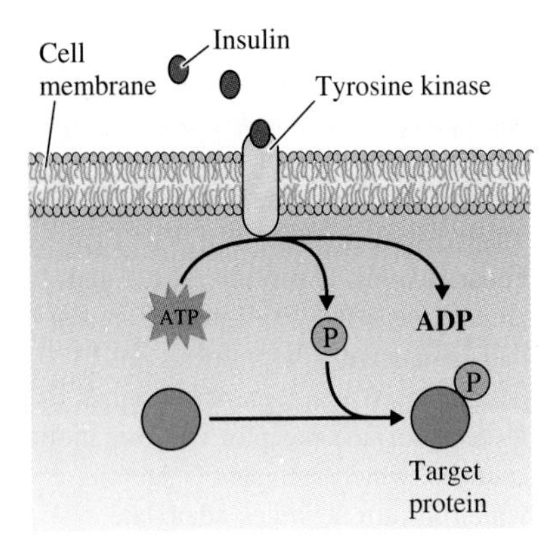

FIGURE 17.7
The activation of tyrosine kinase by insulin. Insulin (blue) binds to its receptor molecule, the enzyme tyrosine kinase (tan) in the cell membrane of the target cell. The binding of insulin activates tyrosine kinase. This enzyme catalyzes the transfer of a phosphate (P) (yellow) from ATP to target proteins (green) in the cell. Phosphorylation of these proteins changes their activity and thus alters the metabolism of the cell.

nutrients are abundant; the cellular equivalent of "eat, drink, and be merry" now occurs. The cells make needed proteins, and lipid and carbohydrate energy stores are replenished.

Insulin binds to insulin receptor proteins on the surface of target cells. This binding activates the receptor, which is an enzyme called *tyrosine kinase*. This enzyme, in turn, phosphorylates some of its own tyrosines and some tyrosines in several proteins within the target cells (Figure 17.7). Although specific details are not yet clear, insulin binding helps regulate anabolic processes in target cells, apparently by modifying the activity of one or more proteins.

Glucagon: The Famine Hormone

Glucagon becomes important when the concentration of glucose in the blood drops toward the lower end of the normal range. This can be thought of as a "famine" condition—nutrients for energy are becoming limited. At lower blood glucose concentrations, glucagon secretion from the pancreas is stimulated, and secretion of insulin is suppressed. The principal target cells for glucagon are liver cells and adipose cells. The ultimate effect of glucagon on these cells is the liberation of fuel molecules into the blood.

Consider the effects of glucagon on a liver cell. Glucagon binds to a specific receptor protein on the surface of the liver cell (Figure 17.8). This binding initiates a series of events that results in the activation of the membrane-bound enzyme adenylate cyclase. This enzyme converts ATP to cyclic AMP (cAMP). Cyclic AMP is released into the cell where it acts as a positive allosteric effector of the enzyme protein kinase. Cyclic AMP is often referred to as a second messenger. Glucagon serves as the first messenger, an extracellular messenger, and cAMP is the second intracellular messenger.

Activated protein kinase attaches a phosphate group to several proteins in the cell. The activity of these proteins is thus increased or decreased by covalent modification. In liver, the enzyme glycogen phosphorylase is activated by covalent modification (Figure 17.8). This enzyme catalyzes the cleavage of glucose from glycogen with attachment of phosphate to the glucose to yield glucose 1-phosphate. The equation for this reaction (which we saw in Chapter 15) is

$$\text{Glycogen } (n \text{ residues}) + \text{Phosphate} \xrightarrow{\text{Glycogen phosphorylase}} \text{Glycogen } (n - 1 \text{ residues}) + \text{Glucose 1-phosphate}$$

The phosphate is then cleaved from the glucose, which is then released into the blood. Glucagon thus causes an increase in blood glucose concentration by stimulating the release of glucose from liver glycogen.

Consider now the enzyme glycogen synthetase. This enzyme is responsible for the synthesis of glycogen from phosphorylated glucose. A moment's thought brings the realization that glycogen phosphorylase and glycogen synthetase have opposite effects. One is catabolic—it converts the macromolecule glycogen to a phosphorylated form of glucose. The other is anabolic—it takes a phosphorylated form of glucose and makes glycogen. The two should not have significant activity

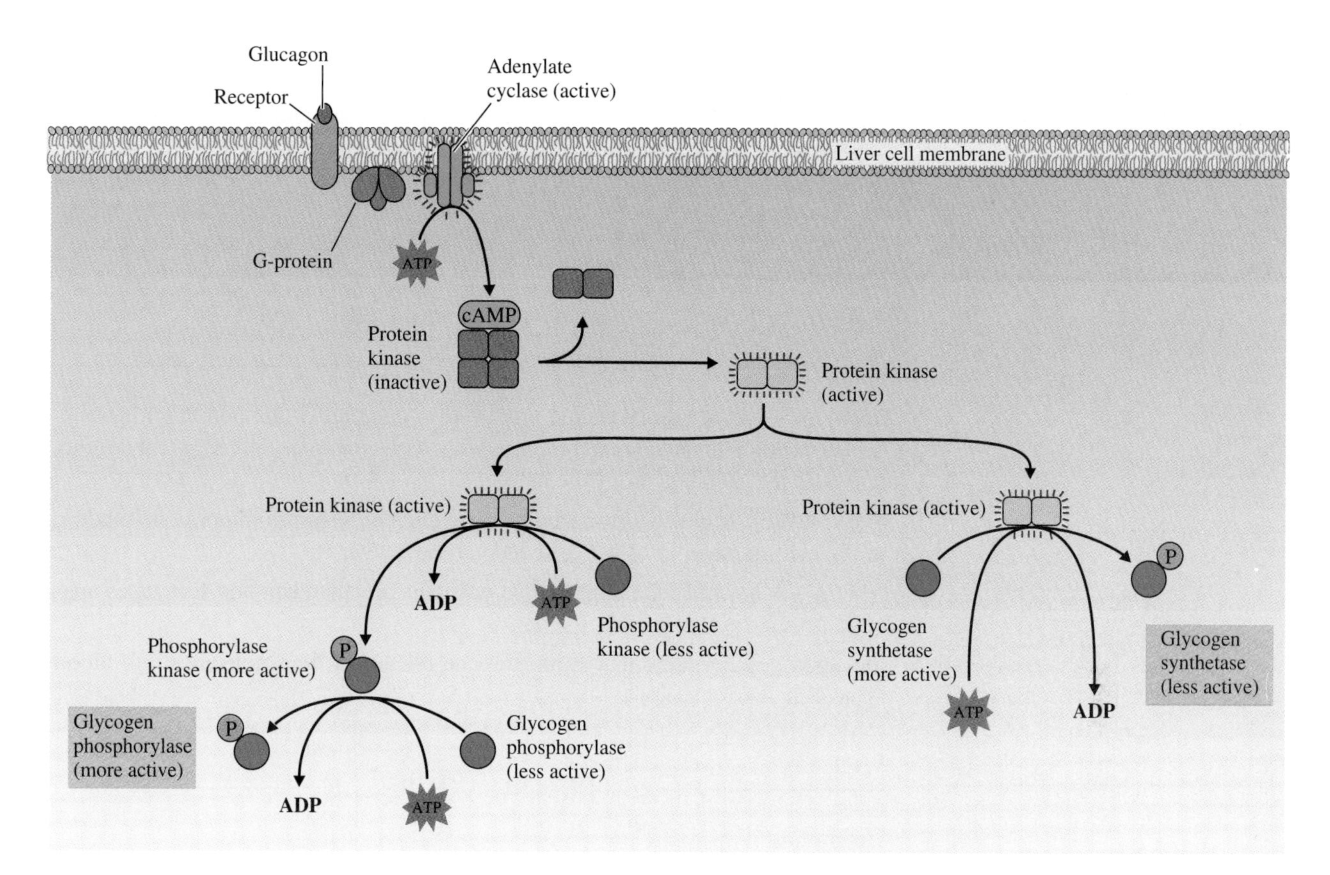

FIGURE 17.8
The effects of glucagon on glycogen metabolism in liver cells. Glucagon (purple), like epinephrine, activates protein kinase through a hormone receptor (orange), G protein (blue), adenylate cyclase (green), and cAMP (yellow) (see Figure 17.6). Protein kinase (pink) phosphorylates glycogen synthetase, which decreases glycogen synthesis, and it phosphorylates a second kinase (phosphorylase kinase), activating it. This enzyme catalyzes the phosphorylation of glycogen phosphorylase, activating it, which increases the rate of glycogen breakdown. In liver cells, protein kinase coordinates glycogen metabolism by simultaneously decreasing glycogen synthesis and increasing glycogen breakdown.

at the same time because the action of one, in essence, negates the action of the other. To work properly, one must be active while the other is relatively inactive.

The activities of glycogen phosphorylase and glycogen synthetase are regulated in concert. Protein kinases attach phosphate groups to both of these enzymes. But the effect on the catalytic activity of the enzymes is exactly opposite. Phosphorylation of the phosphorylase increases its activity; phosphorylation of the synthetase decreases its activity. One signal, the hormone, initiates a series of events that speeds up the release of energy molecules, a part of catabolism, and reduces the formation of glycogen, an anabolic process.

♦ Glucagon simultaneously stimulates glycogen breakdown and inhibits glycogen synthesis.

Lipid Maintenance

♦ Glucagon mobilizes glucose and lipids when they become low in blood.

Glucagon also affects fat cells. It binds to a receptor, and adenylate cyclase is activated to form cAMP, which activates protein kinase. Protein kinase, in turn, activates several proteins in fat cells including a lipase. The lipase hydrolyzes triacylglycerols to glycerol and fatty acids that are released into the blood. Glucagon, through its effects on the liver and adipose tissue, mobilizes glucose and lipid.

NEW TERMS

hormone Messenger molecule produced in one part of the body that is transported throughout the body in blood and that binds to target cells triggering events that alter the metabolism of the cell

cyclic AMP Molecule produced in some cells in response to hormone binding to the cell membrane; cAMP is a second messenger

insulin A hormone that signals that nutrients are abundant and stimulates anabolic activity in target cells

glucagon A hormone that stimulates the release of fuel molecules into blood as needed

TESTING YOURSELF

Hormones

1. Hormones must interact with their target cells to exert their effect. Describe the two types of initial interaction.
2. Why is cyclic AMP called a second messenger?
3. Describe how glucagon exerts opposite effects on glycogen breakdown and glycogen synthesis simultaneously.

Answers **1.** Some hormones bind to the target cell surface to exert their effects, others must enter the cell to exert their effects. **2.** Some hormones (first messenger) bind to the target cell surface and stimulate the synthesis of cAMP; cAMP then serves as a second, intracellular messenger. **3.** Glucagon stimulates the production of cAMP, which activates the enzyme protein kinase. This enzyme phosphorylates glycogen synthetase (which decreases its activity) and another kinase that phosphorylates glycogen phosphorylase (which increases its activity).

17.5 Blood and Immunity

The immune system protects against bacteria, viruses, parasites, and other foreign matter that enter the body. It is able to distinguish between self and nonself—molecules of the body (self) do not normally cause an immune response, but foreign (nonself) molecules do. The immune system has a chemical component and a cellular component. The chemical component includes numerous

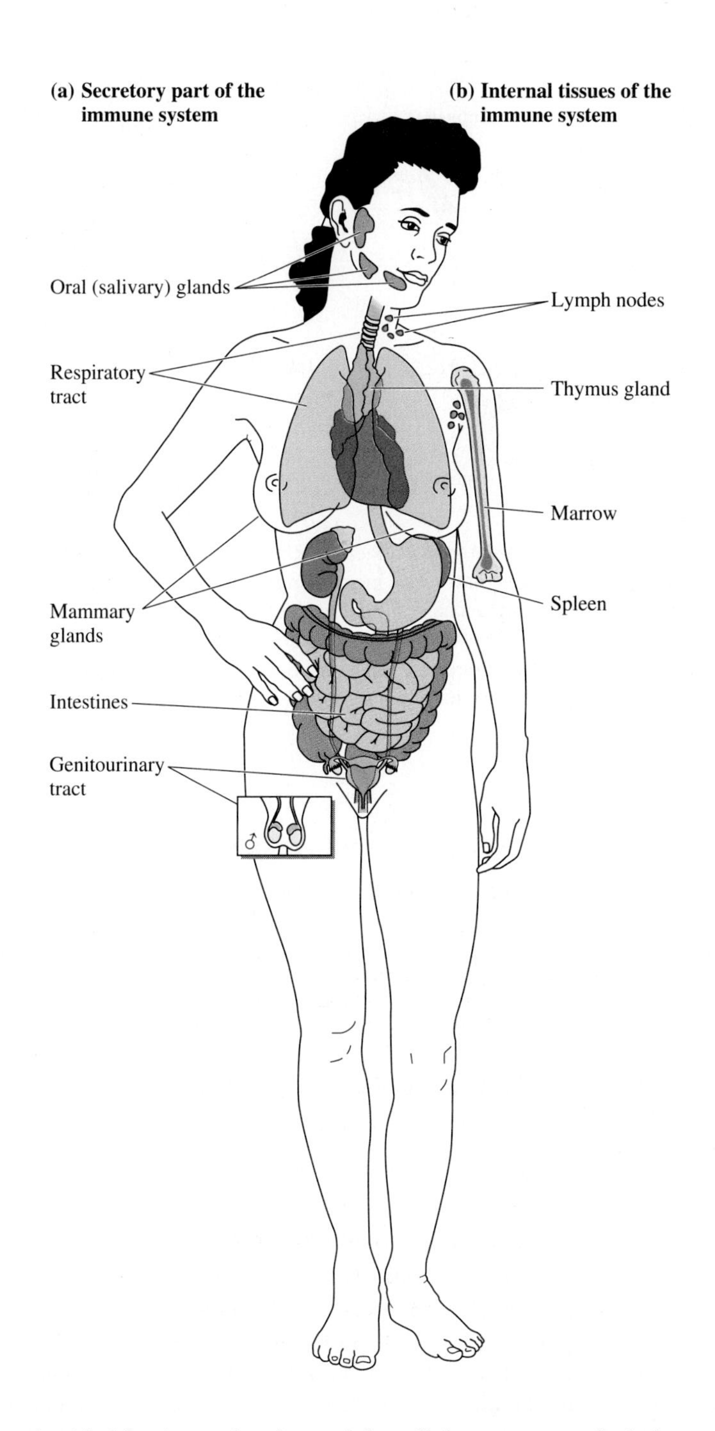

FIGURE 17.9
The immune system produces proteins and cells that protect the body from foreign matter. (a) Secretions of the immune system contain antibodies that reduce the chances of infection. (b) The internal tissues produce, modify, and distribute the cells and proteins of the immune system.

proteins that bind foreign molecules, and the cellular component includes a variety of cells, such as white blood cells, that are responsible for defending the body against invasion. There is considerable overlap in function and much interaction between them (Figure 17.9).

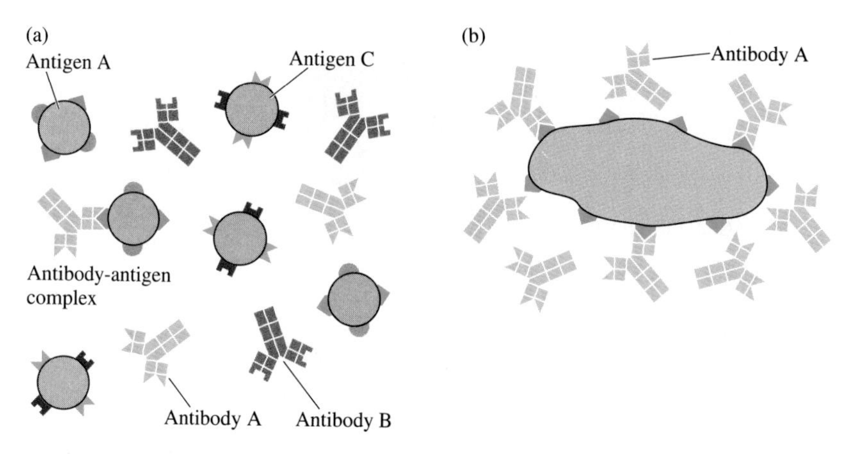

FIGURE 17.10
The binding of antibodies to antigens. (a) Antibodies bind to soluble antigens, usually proteins or polysaccharides, that are present in the body. Antibodies can only bind to antigen surfaces that are complementary to their binding site. Antibody A is binding to antigen A. (b) Antibodies can bind to the surfaces of bacteria, viruses, and parasites. Again, the surface must have a region that is complementary to the binding site of the antibody.

Antibodies and Antigens

♦ Antibodies are proteins that bind foreign molecules called antigens.

The principal components of chemical defense are the **antibodies**, which are also called *immunoglobulins*. These macromolecules are proteins that bind specifically to large foreign molecules called **antigens**. Antibodies bind antigens like enzymes bind substrates, but antibodies normally have no catalytic activity. Antigens may be free molecules, but they are often portions of the cell surface of an invading cell or part of the surface of a viral particle that has entered the body. Complexes formed between antibodies and antigens are readily recognized by cells of the immune system. The cells ingest the complex and destroy the foreign cell or molecule. Antibodies aid immune cells in clearing the body of foreign particles.

Each antibody can bind specifically to one kind of antigen. But because the body can potentially make many tens of millions of different antibodies that differ in their ability to bind specific antigens, virtually all foreign molecules will be an antigen to one or more antibodies (Figure 17.10). This ensures that antigen–antibody complexes form to aid in ingestion by various immune cells.

All possible antibodies are not present in the body at any one time. Instead they are formed by an immune response whenever a new antigen enters the body. The antigen binds to the surface of an immune cell called a *B lymphocyte*. This cell possesses surface proteins that can bind specific parts of the antigen. The binding of the antigen to the B cell causes two major changes in the cell: (1) it begins to divide very rapidly, yielding many copies of the cell, each of which is a clone of the original cell; and (2) it changes into a new cell type called a *plasma*

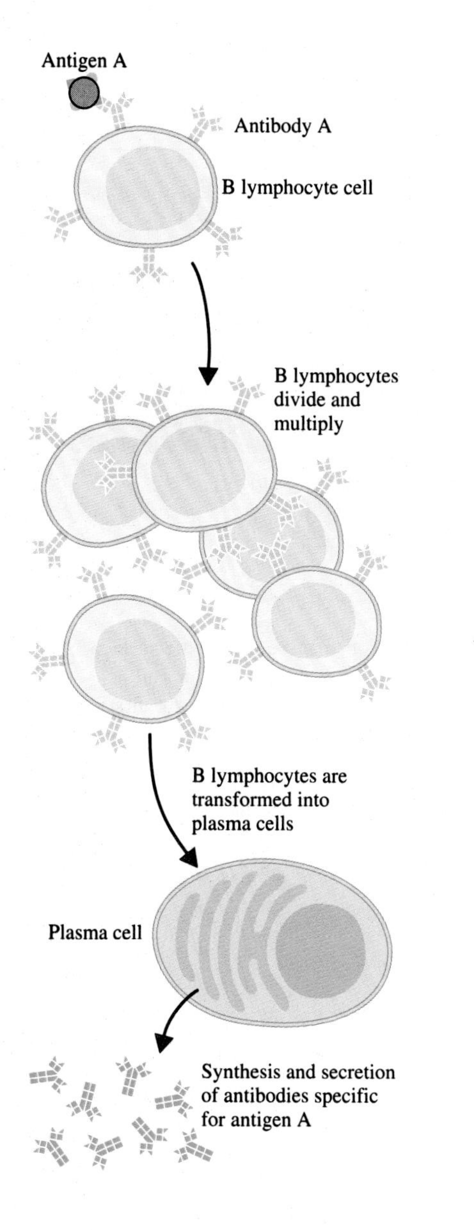

FIGURE 17.11
The immune response to an antigen. Antigens bind to complementary binding proteins on the surface of B lymphocytes. This binding stimulates the B lymphocyte (yellow) to divide and initiates changes in the dividing cells that transform them into plasma cells (orange). These plasma cells make and secrete antibodies specific for the antigen that bound to the B lymphocyte.

cell. Once developed, the plasma cells secrete antibodies that bind specifically to the antigen. The binding properties of the new antibodies are identical to the binding properties of the proteins on the B-cell surface. This ensures that only populations of cells that produce the proper antibodies are stimulated by the antigen (Figure 17.11).

The immune response just described takes 7 to 14 days to develop. The next time the organism is exposed to the same antigen, the response is much more rapid and massive. In a sense, the immune system has "memory" and thus responds more quickly to antigens it has previously encountered.

CHEMISTRY CAPSULE

ABO Blood Group

Blood grouping involves the immune system and the concept of self and nonself. The ABO blood system has four different blood groups: A, B, AB, and O (Table 17.A). Individuals with blood group A have the A antigen protein on the surface of their red blood cells. Those with group B have B antigen, and those with group AB have both. Individuals with Group O have neither. In an individual possessing the A protein or the B protein, the protein is not an antigen because it is a self-protein. If the A or B antigen is absent from red blood cells, the blood plasma contains antibodies (called *agglutinins*) to A or B (Table 17.A).

What would happen if group A blood were given to a group B individual? Since the individual with group B blood lacks the A protein, A is an antigen to that individual. The agglutinin (anti-A) in the recipient's blood plasma would bind to the A antigen on the donor red blood cells, causing clumping (agglutination), which would initiate a response that could have grave consequences.

Because the antigens of the ABO blood group can elicit agglutination in other individuals, it is necessary to match donor blood to recipient blood for transfusions. Group O blood is considered to be the *universal donor* because type O blood contains no AB antigens on the red blood cells and thus can cause no agglutination in the recipient. Type AB has both A and B antigens, thus anti-A and anti-B are absent. People with AB blood are thus *universal recipients*.

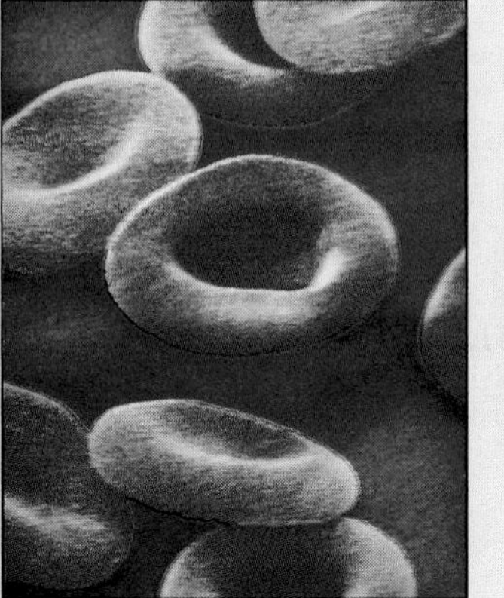

(a)

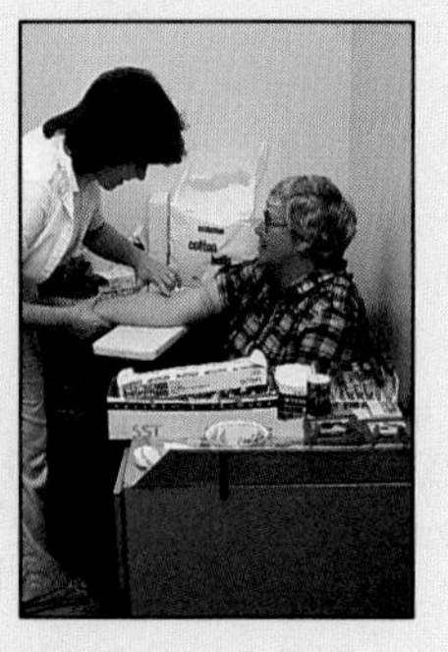

(b)

FIGURE 17.B
(a) Before blood transfusions can be given, the surface antigens of the erythrocytes of both the donor and acceptor must be known. (b) Drawing blood is the first of many steps in blood analysis.

Table 17.A
The ABO Blood Group System

	Blood Type			
Characteristic	**A**	**B**	**AB**	**O**
Antigen present on the red blood cells	A	B	Both A and B	Neither A nor B
Agglutinin normally present in the plasma	anti-B	anti-A	Neither anti-A nor anti-B	Both anti-A and anti-B
Incompatible with donor blood group	B, AB	A, AB	None	A, B, AB
Percentage in a mixed Caucasian population	41	10	4	45
Percentage in a mixed Black population	27	20	7	46

Interferons

Interferons are a group of proteins that interfere with the replication of viruses in cells; thus, they can also be considered as part of the chemical defense system. Interferons are produced by cells that are undergoing attack by viruses. The activity of viruses within body cells stimulates the cells to make and secrete interferons. Like some hormones, interferons bind to specific receptor molecules on the surface of other body cells. This binding stimulates the production of antiviral proteins within the cell that appear to protect the cell from viral attack (Figure 17.12). Interferons have recently attracted considerable attention in medical research because they also appear to affect the growth of some kinds of cancer cells.

♦ Interferons reduce viral replication in body cells.

Cellular Immunity and AIDS

The cellular component of the immune system involves elements of biology that have not been covered in this book. Immune cells include such cells as lymphocytes that function through molecular interactions with antigens and other cells and molecules of the body. They possess surface proteins that, through binding, recognize antigens, antibodies, and some surface proteins of other cells. Binding stimulates cellular activity that contributes to the defense of the body.

Acquired immune deficiency syndrome, or AIDS, results from the infection of a specific type of T lymphocyte by a virus called HIV (human immunodeficiency virus). The virus may lie dormant within the lymphocytes for years, but eventually, it multiplies within the lymphocyte and kills the cell by emerging from it. The virus infects other T cells and the cycle repeats. With time, the virus destroys most of these T lymphocytes. Since these cells play a vital role in controlling the immune response, the immune system functions poorly when these cells are lost.

When the immune system fails, the AIDS patient is vulnerable to a host of organisms that would normally be destroyed without injury or harm to the individual. These opportunistic infections include *Pneumocystis carinii*, a normally harmless organism that causes a type of pneumonia. The immune system also normally destroys cancer cells, so it is not surprising that cancers, such as Kaposi's sarcoma, an otherwise rare skin malignancy, appear in AIDS patients. The Center for Disease Control now predicts that AIDS infection will always result in death, unless a cure is developed.

How can AIDS be cured or an HIV vaccine be developed? The interactions between the HIV virus and the T lymphocytes are all molecular interactions. Specific binding between viral proteins and T-cell proteins must occur before infection can occur. The replication of the virus within the cell requires specific viral molecules interacting with host molecules. The search for a cure and a vaccine for AIDS requires a basic understanding of these molecular interactions. Many research projects studying AIDS are of a molecular nature. With ever-increasing frequency, biochemistry and molecular biology are being used to improve health care.

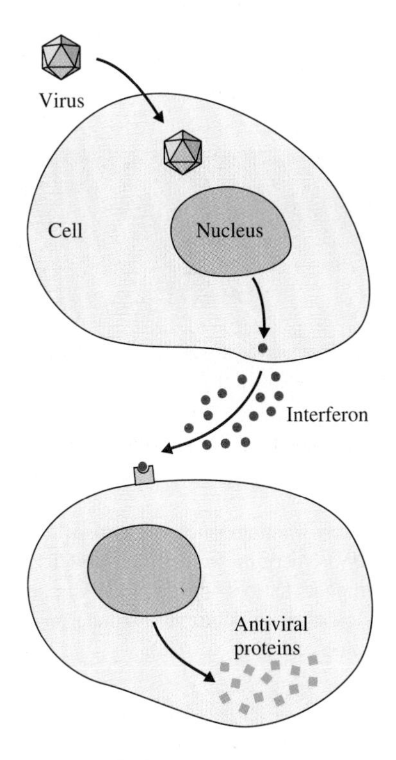

FIGURE 17.12
The antiviral action of interferons. Viral invasion of cells stimulates the synthesis and secretion of interferons. Even though the invaded cell may die, these proteins bind to noninfected cells and stimulate the production of antiviral proteins. If this cell is invaded by a virus, these antiviral proteins protect the cell by disrupting normal viral reproduction.

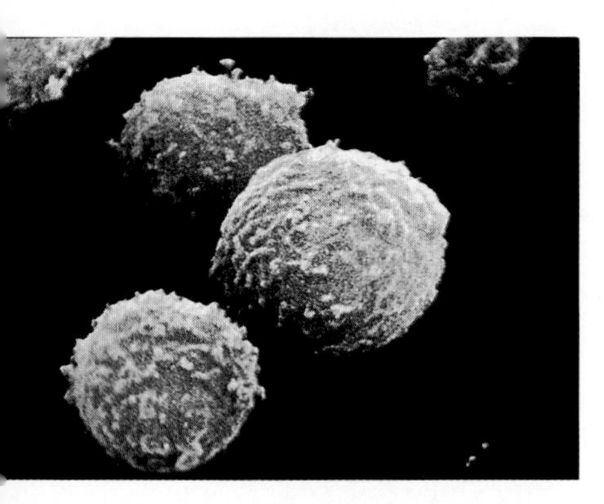

T-lymphocytes play important roles in immunity, including vital regulatory roles.

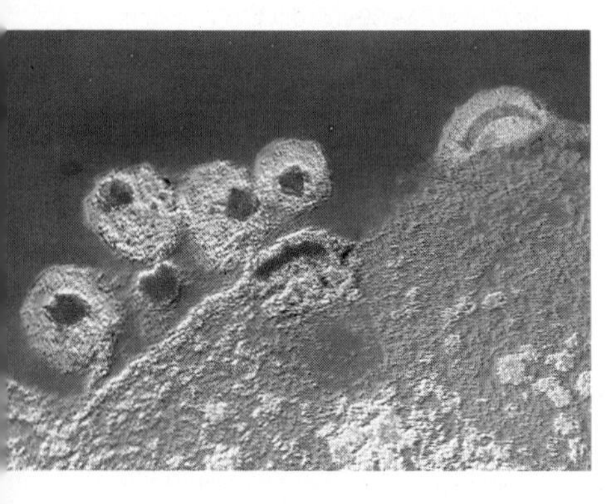

Human immunodeficiency viruses (HIV) emerging from an infected T4 lymphocyte. Loss of these T4 lymphocytes suppresses normal immune response and leads to AIDS.

NEW TERMS

antibodies Proteins produced by the body that bind foreign molecules and particles. These complexes are then readily taken up and destroyed by cells of the immune system

antigens Large foreign molecules within the body. These molecules cause an immune response that results in production of antibodies to the antigen

interferons Proteins produced by virally infected cells. They bind to other cells and stimulate an antiviral state in them

TESTING YOURSELF

Immunity

1. What are molecules called that are normally absent from the body and that cause an immune response?
2. What cells produce antibodies during an immune response?
3. What group of immune cells is infected by the AIDS virus?

Answers **1.** antigens **2.** plasma cells **3.** T lymphocytes

Summary

The body maintains a constant internal environment through *homeostasis*, which is a set of self-regulating mechanisms. The composition of blood is an important part of homeostasis. Oxygen and carbon dioxide are exchanged between the body and air in the *alveoli* of the lungs. These gases are transported by blood. Their movement into and out of blood is driven by diffusion. *Hemoglobin* greatly enhances the capacity of blood to carry oxygen; this molecule is adapted to readily take up oxygen in the lungs and deliver it to tissues. Wastes are removed from blood by filtration in the kidneys, while nutrients are reabsorbed.

Maintenance of a narrow pH range is critical in blood because denaturation of proteins occurs outside this range. Bicarbonate, phosphate, and proteins are the main buffers in blood. If the pH extends beyond this narrow range, *acidosis* or *alkalosis* will result. Hormones are chemical messengers in blood. *Insulin* serves as a signal that nutrients are abundant, so cells take in more nutrients. *Glucagon* signals that fuel molecules are scarce, and the liver and adipose cells respond by releasing more fuel molecules.

Blood also carries many components of the immune system. *Antibodies* are proteins that bind to foreign molecules to help with their removal. *Interferons* affect viral replication, thus they reduce the chance of viral infections. AIDS is a viral disease that destroys T lymphocytes involved in regulation of the immune response. When these cells are destroyed, the immune system fails.

Terms

homeostasis (17.1)	hormone (17.4)
alveoli (17.1)	cyclic AMP (17.4)
hemoglobin (17.1)	insulin (17.4)
Bohr effect (17.1)	glucagon (17.4)
filtration (17.2)	antibodies (17.5)
renal threshold (17.2)	antigens (17.5)
acidosis (17.3)	interferons (17.5)
alkalosis (17.3)	

Exercises

Gas Exchange (Objective 1)

1. What role does hemoglobin play in transporting oxygen? In transporting carbon dioxide?
2. The binding of oxygen to hemoglobin is said to be cooperative. What does this mean?
3. What is the role of carbonic anhydrase?
4. What are alveoli, and how are they involved in gas exchange?
5. How large might an organism be if an oxygen-transporting protein, such as hemoglobin, were not available to it? (*Hint*: Review Section 9.1 before answering this question.)
6. Examine the equation in Section 17.1 that deals with CO_2 in equilibrium with H_2CO_3 and HCO_3^-. Describe the effects of an increase in CO_2 on this equilibrium.
7. What term is used to describe the effects seen in Exercise 6?
8. Carbon dioxide is transported in the blood in what forms?
9. Describe the effects of increased acidity and increased carbon dioxide concentration on the oxygen-carrying capacity of hemoglobin. What is this effect called?
10. When blood reaches the lungs, carbon dioxide leaves the blood. What effect does this have on blood pH and the capacity of hemoglobin to bind oxygen?
11. List and describe briefly two ways in which vigorous exercise increases oxygen delivery to tissues.

Removal of Wastes (Objective 2)

12. What organs are responsible for removal of carbon dioxide? Urea? Water? Salt?
13. Describe the process of filtration and reabsorption in the kidney.
14. What is meant by renal threshold? What is the renal threshold for glucose?
15. The reabsorption of bicarbonate from the tubules in the kidneys is regulated. What effect would a decrease in the reabsorption of bicarbonate have on blood pH?

Regulation of pH (Objective 3)

16. List the three principal buffering systems in blood.
17. Describe how bicarbonate and carbonic acid contribute to the buffering of blood.
18. If the blood contained bicarbonate but no carbonic acid, would it be buffered?
19. Identify the acid and conjugate base for the bicarbonate buffer system in blood.
20. Which blood protein makes the greatest contribution to the buffering of blood?
21. What form of phosphate ion absorbs an H^+ at the conditions found in blood?
22. A patient is diagnosed as having ketosis. What effect may this have on blood pH? What name is given to this condition?
23. What effects can hyperventilation have on blood pH? What is this condition called?
24. What effects can hypoventilation have on blood pH? What is this condition called?
25. An individual has an infection of the gastrointestinal tract that causes vomiting and diarrhea. What effect could this have on blood pH?

Hormones (Objectives 4 and 5)

26. Hormones circulate throughout the body in the blood, and thus come in contact with all cells. Why are only some cells affected by any given hormone?

27. Describe how hormones that enter cells cause changes in the cell.
28. What hormones described in this text exert their effect without entering the target cell?
29. What hormones described in this text exert their effect from within the target cell?
30. Name three hormones that are steroids.
31. Name two hormones that are proteins.
32. A third group of hormones are derived from amino acids. Give an example.
33. How does protein kinase alter the activity of a cell? What activates protein kinase?
34. How is cyclic AMP formed in a cell?
35. Compare the structures of cyclic AMP and AMP. What structural difference results in binding of cAMP to protein kinase but no binding of AMP?
36. Why is insulin considered a feast hormone?
37. Why is glucagon considered a famine hormone?
38. What role does tyrosine kinase play in the effect of insulin on cells?
39. What effect does glucagon have on adipose cells? How does glucagon cause this effect?
40. Would you expect to find relatively high concentrations of insulin and glucagon in blood at the same time? Why or why not?
41. What term is used to describe the maintenance of the internal environment? What role does insulin have in this maintenance?

Immunity (Objective 6)

42. Describe how an interferon affects replication of a virus in body cells.
43. What happens to a B lymphocyte when an antigen binds to its surface?
44. How can an otherwise harmless organism kill an AIDS patient?
45. What are proteins called that bind foreign molecules in the blood?
46. What are proteins called that help inhibit viral replication?
47. The immune system is said to have memory. What does this mean?
48. The destruction of what cells results in the syndrome known as AIDS?
49. What agglutinins are found in the blood of an individual with group B blood?
50. Who could receive blood from a group B donor?
51. An individual with group B blood could receive blood from whom?

SOLUTIONS TO ODD-NUMBERED EXERCISES

CHAPTER R

1. The outermost electron of a neutral atom of sodium will be in a 3*s* orbital.

3. a. The orbital notation for a neutral atom of fluorine is $1s^2 2s^2 2p^5$.
   b. The orbital notation for a neutral atom of bromine is $1s^2 2s^2 2p^6 3s^2 3p^6 4s^2 3d^{10} 4p^5$

5. a. The orbital notation for the outermost electrons of a neutral atom of magnesium is $3s^2$.
   b. The orbital notation for the outermost electrons of a neutral atom of calcium is $4s^2$.
   c. The orbital notation for the outermost electrons of a neutral atom of strontium is $5s^2$.

7.

Li = +1	B = +3	O = −2	F = −1
Na = +1	Al = +3	S = −2	Cl = −1
K = +1	Ga = +3	Se = −2	Br = −1

   [Note that these elements are placed as they are found in the periodic table.]

9. LiCl

11. AlF_3

13. a. Nitrogen and oxygen are closest in electronegativity and will thus form the bond of highest covalent character.
    b. Potassium and oxygen are farthest apart in electronegativity and will thus form the bond of highest ionic character.

15.

```
   H H     H H
   ·· ··  ·· ·· ··
 H:C:C:O:C:C:H
   ·· ·· ·· ·· ··
   H H     H H
```

17. Bond type prediction: Carbon and

hydrogen	covalent
nitrogen	covalent
oxygen	polar covalent
fluorine	polar covalent
sulfur	covalent
chlorine	covalent
bromine	covalent

19. Cesium is the element farthest from carbon in electronegativity (0.8). The electronegativity difference is 1.9, an ionic bond.

21. a. The nitrogen–hydrogen bonds in ammonia are polar covalent (0.9 electronegativity difference).
    b.

```
   ··
 H:N:H
   ··
   H
```

    c. There are three pairs of bonding electrons.
    d. There is one pair of nonbonding electrons.
    e. The nitrogen atom will be tetrahedral
    f. The molecule has a dipole.

23. The hybridization of the nitrogen atom is sp^3.

25. a. Propane is a nonpolar molecule. It has no dipole.
    c. Ethyl alcohol is polar in the —OH region of the molecule.

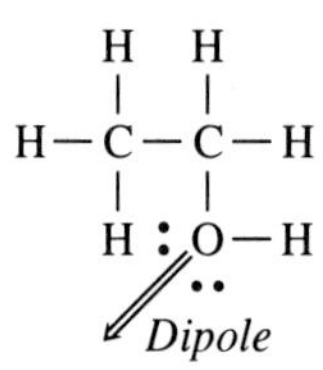

b. Methanol is polar in the — OH region of the molecule.

d. Methelethyl ketone is polar in the region of the oxygen.

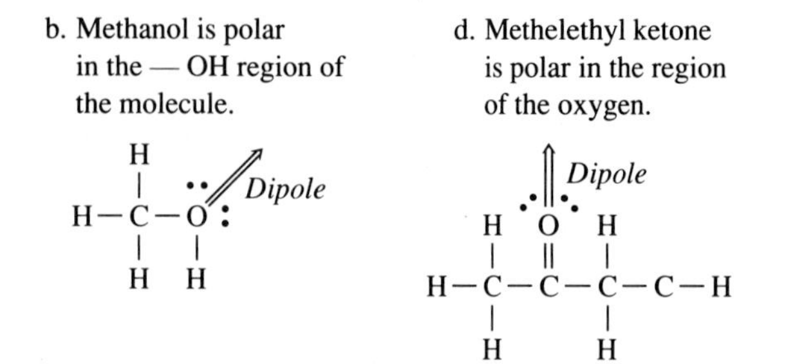

27. Methanol would be the most polar and thus the best solvent for the ionic compound sodium chloride.

29. Hydrogen bonding increases the effective mass of a molecule.

31. Molecules that hydrogen-bond show an increase in boiling point, as compared to molecules of similar mass that do not hydrogen-bond.

33. The outer electron configuration of a neutral atom of selenium will be s^2p^4.

35. A compound of sulfur and oxygen would probably have covalent (or polar covalent) bonding.

37. a. Covalent and polar covalent bonds exist in a ethylene glycol molecule.
    b. The oxygen atoms have unshared pairs of electrons.
    c. The carbon atoms have sp^3 hybridization.
    d. There are two dipoles in this molecule.
    e.

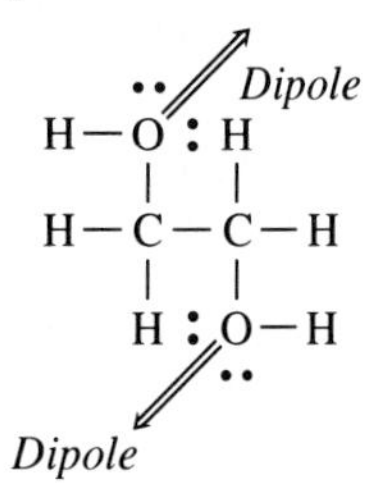

    f. The molecule will be water-soluble.

39. a. Methane: nonpolar
    b. Water: polar
    c. Ammonia: polar

41. Sodium chloride will not dissolve in methane.

43. Hydrogen bonding raises the boiling point of a compound.

45. Water expands as it freezes because the hydrogen bonds between water molecules tend to line them up in an ordered lattice; this configuration takes more space than the molecules require when in the liquid state.

CHAPTER 1

1. a. i
   b. ii
   c. ii
   d. iii

3. c

5. d

7. b

9. c.

CHAPTER 2

1. sp^3

3. "Straight chain" means a nonbranching carbon chain. Because of the tetrahedral arrangement required by sp^3 hybridization, a straight chain (180° bond angles) is not possible for more than two carbon atoms.

5. Two things need to happen: (1) Attractive forces (London for hydrocarbons and hydrogen bonding for water) have to be disrupted and (2) some new, *more stable* (lower-energy) attractive force between water and hydrocarbon has to result. This does not happen.

7. Branched hydrocarbons cannot line up as well as straight-chained hydrocarbons; thus, the already weak London attractive forces are even less between molecules. With less attraction between molecules, it is easier for them to become separated.

9. The atoms must be distributed so that the *number* of atoms, but not their connectivity, is the same.

11.

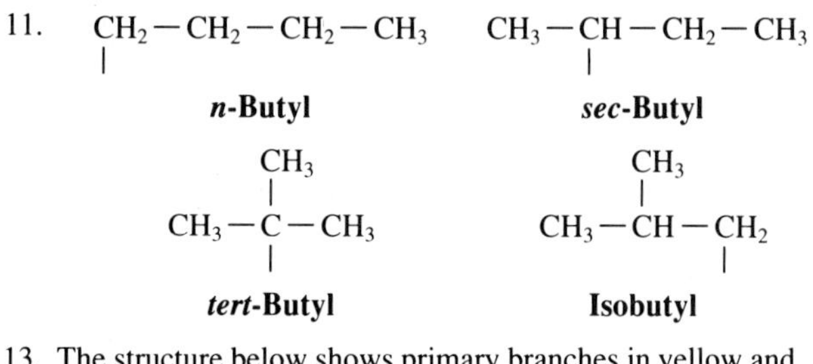

13. The structure below shows primary branches in yellow and secondary branches in red.

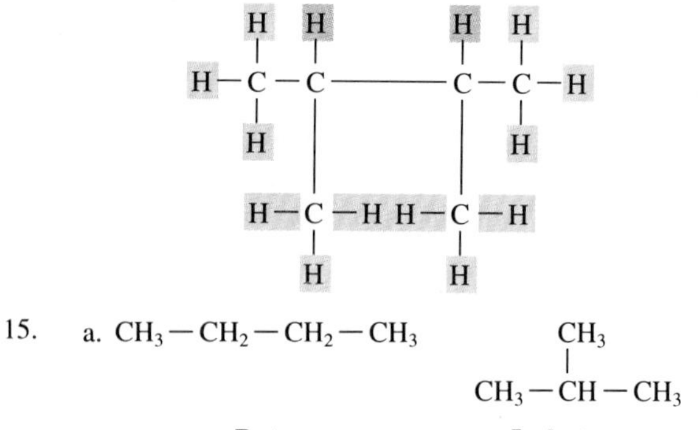

15. a. $CH_3-CH_2-CH_2-CH_3$ **n-Butane**; $CH_3-CH(CH_3)-CH_3$ **Isobutane**

b. $CH_3-CH_2-CH_2-CH_2-CH_3$

***n*-Pentane**

$CH_3-CH(CH_3)-CH_2-CH_3$ $\quad$ $CH_3-C(CH_3)_2-CH_3$

2-Methylbutane $\quad$ **2,2-Dimethylpropane**

c. $CH_3-CH_2-CH_2-CH_2-CH_2-CH_3$

***n*-Hexane**

$CH_3-CH(CH_3)-CH_2-CH_2-CH_3$

2-Methylpentane

$CH_3-CH_2-CH(CH_3)-CH_2-CH_3$

3-Methylpentane

$CH_3-CH(CH_3)-CH(CH_3)-CH_3$ $\quad$ $CH_3-C(CH_3)_2-CH_2-CH_3$

2,3-Dimethylbutane $\quad$ **2,2-Dimethylbutane**

17. H, H, CH_3 — C — C — CH_3, H, H

Note that the large groups are as far apart as possible.

19. Axial CH_3; CH(CH_3)(CH_3) Equatorial

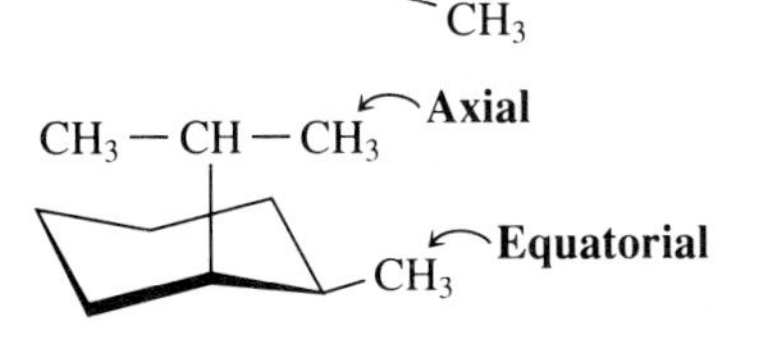

The larger group prefers to be equatorial.

21. a. 3,3,5-trimethylheptane
    b. 2-methylbutane
    c. 4-ethyl-2,2-dimethylhexane
    d. 2-methylhexane
    e. 2,2-dimethylbutane
    f. 2,2,4-trimethylpentane
    g. 3-methylpentane
    h. 2,2,3,3-tetramethylbutane
    i. 3-methylpentane
    j. 3,3,6-trimethylheptane, or 2,5,5-trimethylheptane—both have the same carbon chain length, and the same "score" when adding total numbers. Here we use the important rule that the name with the lowest first number takes priority.

23. a. 1,1-dimethylcyclopentane
    b. 2-ethyl-1,1-dimethylcyclooctane
    c. 1,1,4-trimethylcyclohexane
    d. isopropylcyclopropane
    e. isobutylcyclopentane
    f. *tert*-butylcyclohexane
    g. *sec*-butylcyclopropane
    h. 1,2-dimethylcyclobutane
    i. isopropylcyclononane
    j. *n*-propylcycloheptane
    k. 1,4-dimethylcyclooctane

25. a. 2-bromobutane
    b. 2,2-dichlorobutane
    c. 2,3-dibromopentane
    d. 4-fluoro-2,2,4-trimethylpentane
    e. 1,1,1-tribromo-2,2-dimethylpropane
    f. 2,2,5-trichloro-4-methylpentane
    g. 3,3-dibromopentane
    h. 1,1,1,3,3,3-hexachloropropane
    i. 2-chloro-2-methylpropane

27. The halogens increase in atomic number and atomic weight. There is an increase in London attractive forces.

29. a. *trans*-1-isopropyl-2-methylcyclohexane
    b. 1,3-dibromo-2-methylbutane
    c. *cis*-1-chloro-2-isobutylcyclopentane
    d. 3,4-dibromo-4-isopropylheptane
    e. 1,1-dibromo-2,2-dimethylcycloheptane
    f. *cis*-1,5-dibromo-1-isobutylcyclooctane
    g. 3,5-dibromo-4-methylheptane
    h. *cis*-3,4-dichloro-1,1-dimethylcyclopentane
    i. *cis*-1,6-dichlorocyclodecane

31. a. 3,5-dimethylheptane
    b. *cis*-1,2-dimethylcyclohexane
    c. 2,3-dimethylpentane
    d. 1,2-diethylcyclopentane
    e. 5,5-dibromo-2,2,3,3-tetramethylhexane

33. Complete combustion of any hydrocarbon gives carbon dioxide and water as products.

35. d

37. a.

39. c

41. e

43. a

CHAPTER 3

1. A hydrocarbon that contains double or (and) triple bonds.
3. sp^2
5. sp
7. The pi bond restricts rotation because in order to turn around the C—C sigma bond, the pi bond has to be broken.
9. In both uses the idea of "cis" means *on the same side*, whereas "trans" means *on the opposite side*.
11. The two carbons attached to each end of the triple bond must be linear with the triple bond. This results in four of the five bonds being linear. It is just too difficult for the fifth atom to make a ring.
13. a. 4-methyl-2-pentene b. 1-propene
    c. 1-methylcyclohexene d. 4,4-dibromocyclooctene
15.

a.

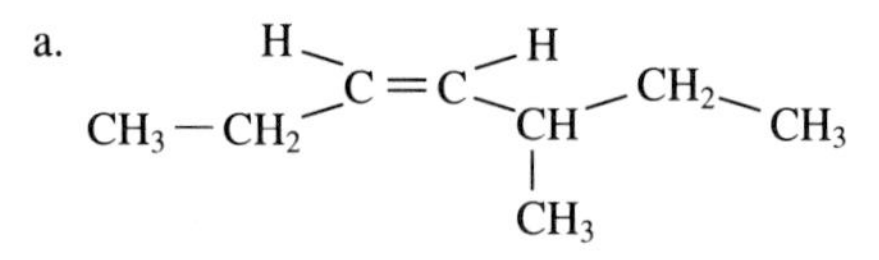

b.

$$H-C\equiv C-CH_2-CH_2-\overset{\overset{CH_3}{|}}{\underset{\underset{CH_3}{|}}{C}}-CH_2-CH_2-CH_3$$

c. CH$_3$ CH$_3$ CH$_2$CH$_3$ CH$_3$

d.

$$H-C\equiv C-\overset{\overset{CH_3}{|}}{\underset{\underset{CH_3}{|}}{C}}-CH_2-CH_2-CH_3$$

e.

$$CH_3-C\equiv C-CH_2-\overset{\overset{Br}{|}}{\underset{\underset{Br}{|}}{C}}-CH_3$$

f.

$$Cl-\overset{\overset{Cl}{|}}{\underset{\underset{Cl}{|}}{C}}-CH_2-CH=CH-CH_2-CH_2-CH_2-CH_3$$

g. CH$_3$ CH$_3$

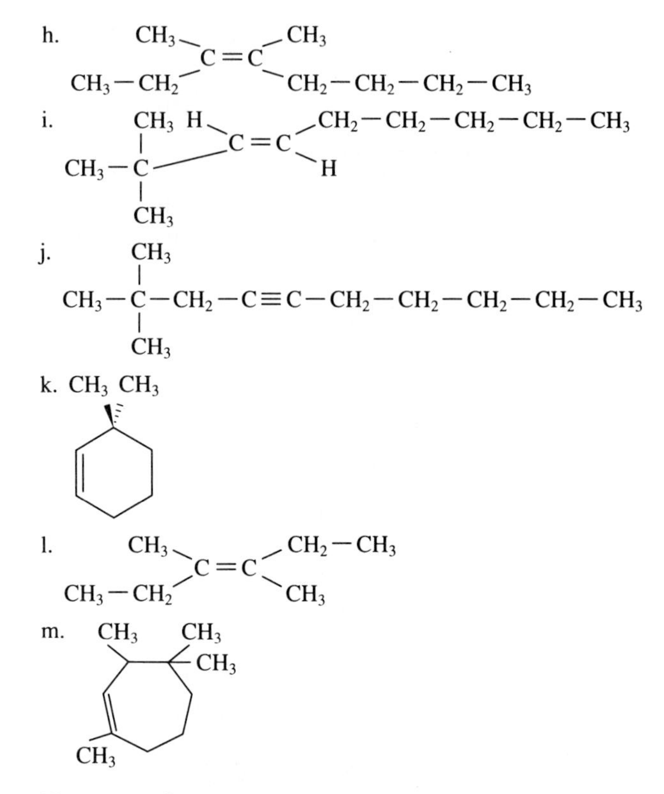

17. a. *trans*-2-pentene
    b. 4-methylcyclohexene
    c. 2,3-dimethyl-1-butene
    d. 3,4-dimethyl-*cis*-3-hexene
    e. 2-pentyne
    f. 1,3-dimethylcyclohexene
    g. 2-ethyl-4,4-dimethyl-1-pentene
    h. 4,4,5,5-tetramethyl-2-hexyne
    i. 4-methyl-1-hexene
    j. *trans*-5,6-dimethylcyclooctene
    k. 4-cyclohexyl-1-pentyne
19. Ethene has electrons in a pi bond that are easily accessed by acids.
21. Carbon dioxide and water.
23. An alkane
25. sp^2 to sp^3
27. 1
29.

a.

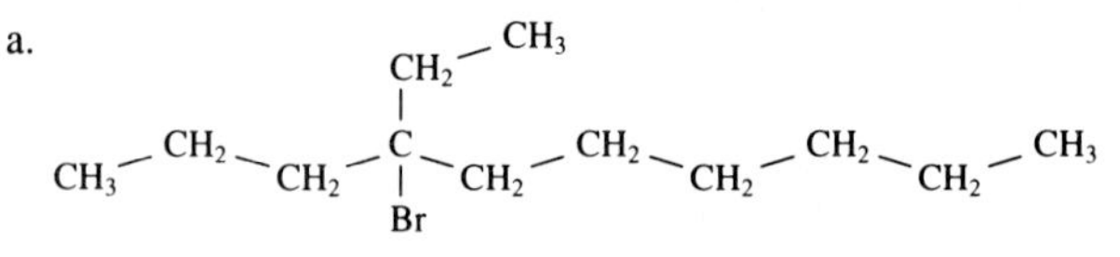

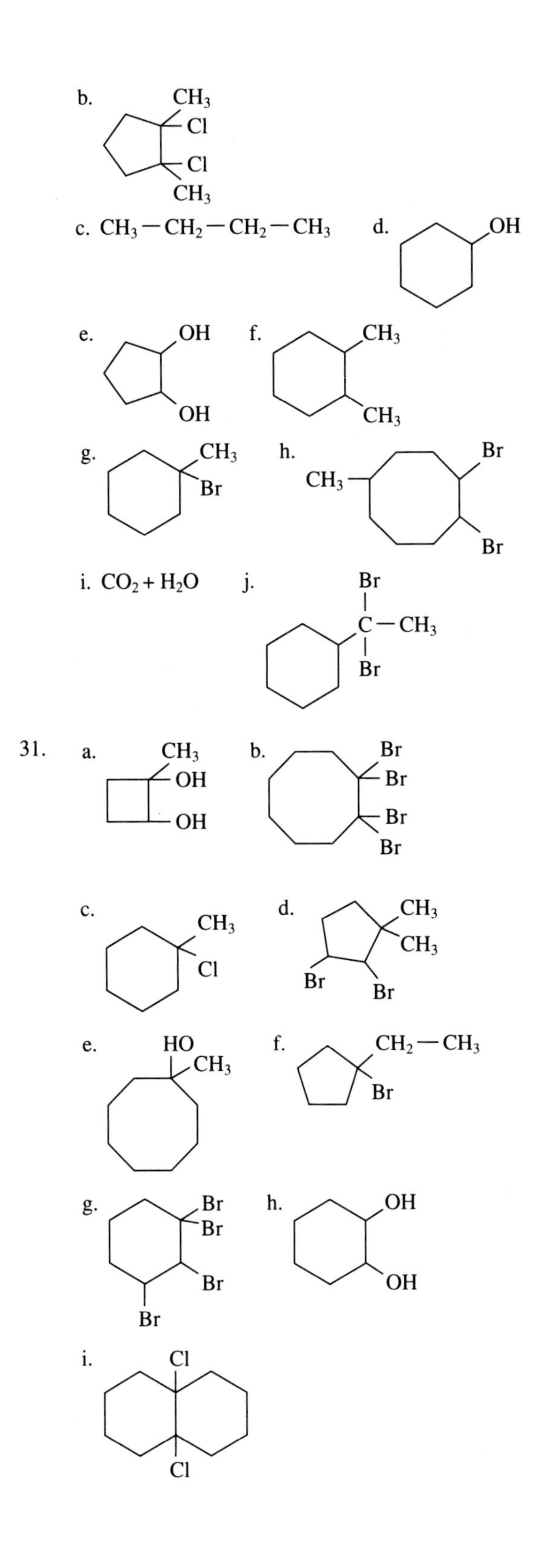

33. A complex molecule made up of repeating chains of a smaller unit, called a monomer

35.

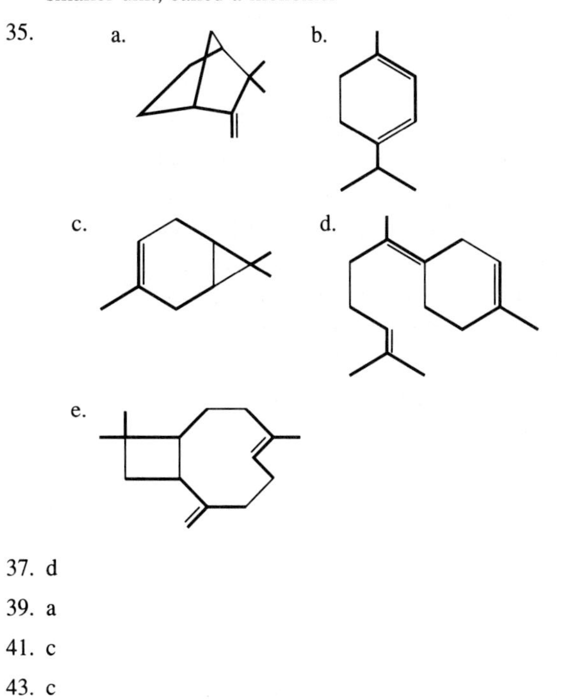

37. d

39. a

41. c

43. c

CHAPTER 4

1. Electrons that are not *localized* between two atoms in a bond. Refer to answer 3 below.

3. "a" represents delocalized electrons, whereas in "b" they are localized.

5.

7.

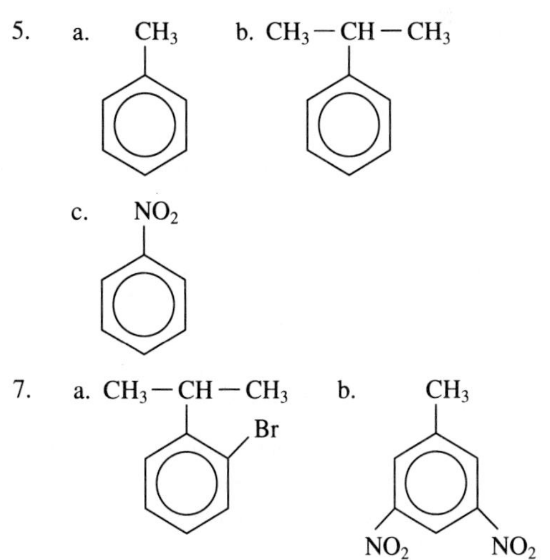

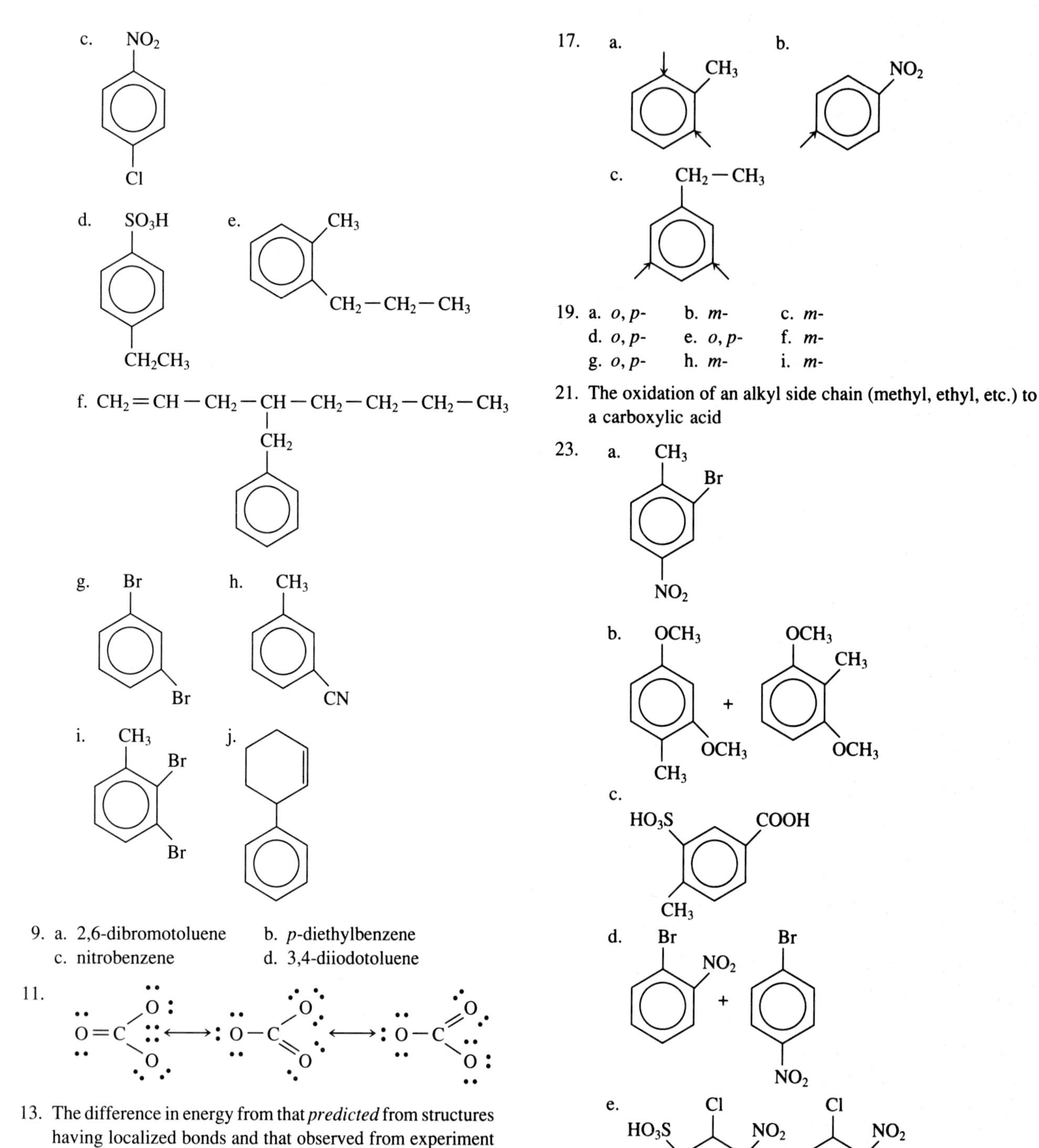

9. a. 2,6-dibromotoluene b. *p*-diethylbenzene
c. nitrobenzene d. 3,4-diiodotoluene

11. (resonance structures)

13. The difference in energy from that *predicted* from structures having localized bonds and that observed from experiment

15. If reagents are added to a double bond (assuming they were localized), the aromatic pi-system would be lost. The product would be considerably *less stable* than the reactant since the large aromatic resonance energy would be lost.

19. a. *o, p*- b. *m*- c. *m*-
d. *o, p*- e. *o, p*- f. *m*-
g. *o, p*- h. *m*- i. *m*-

21. The oxidation of an alkyl side chain (methyl, ethyl, etc.) to a carboxylic acid

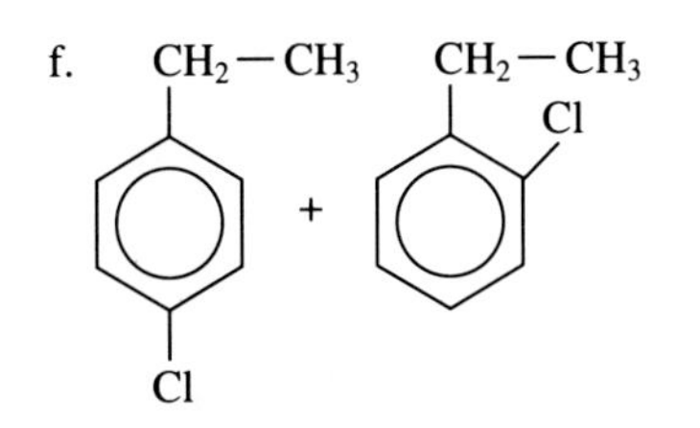

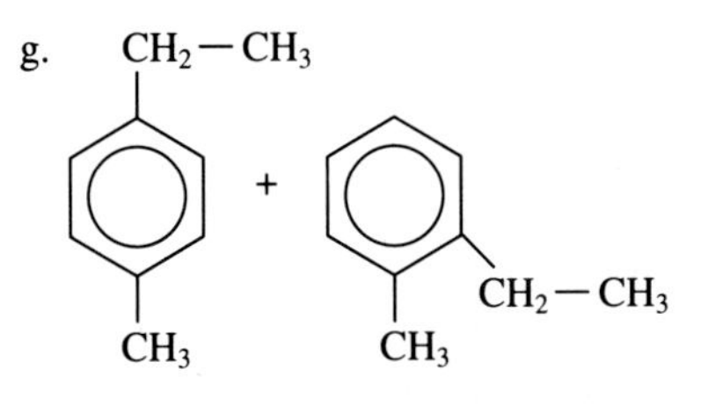

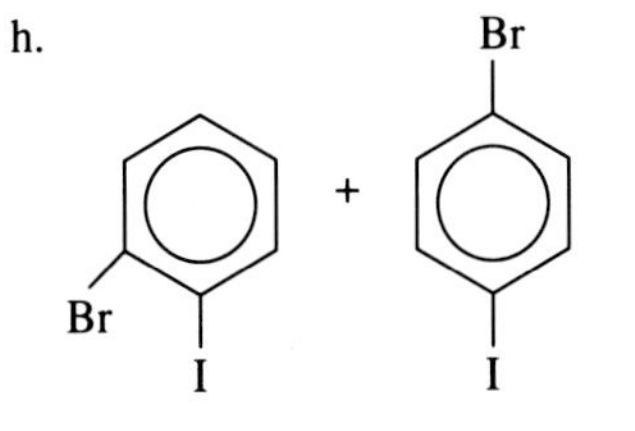

25.

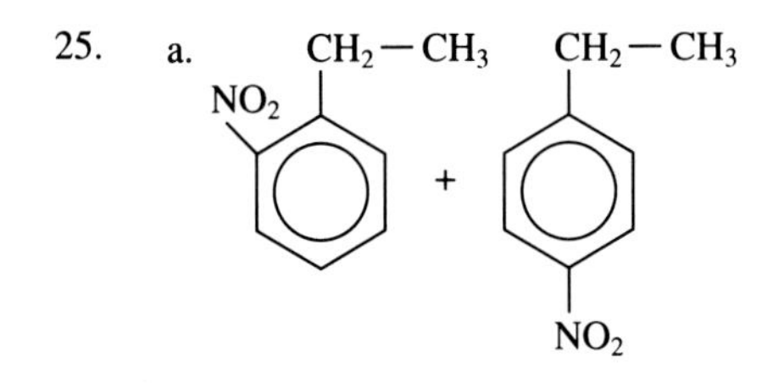

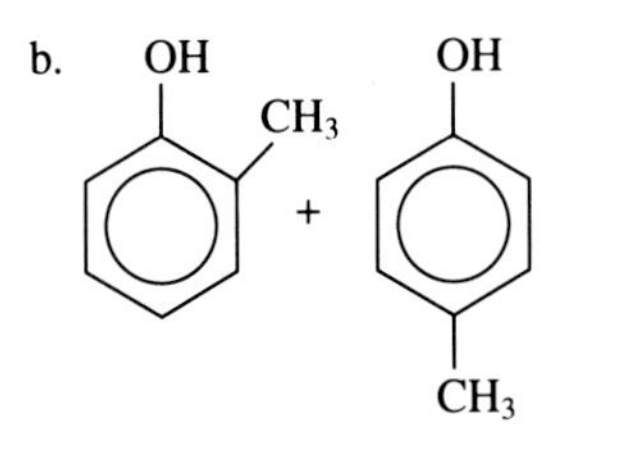

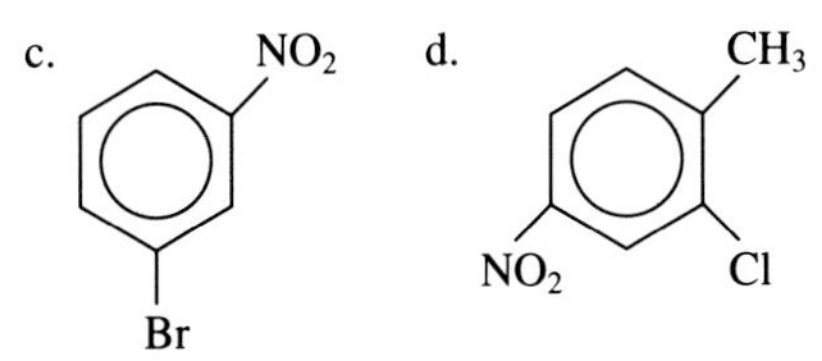

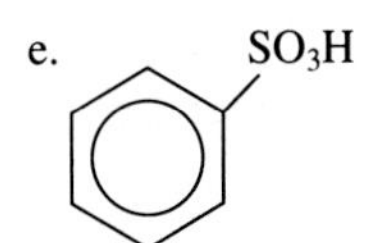

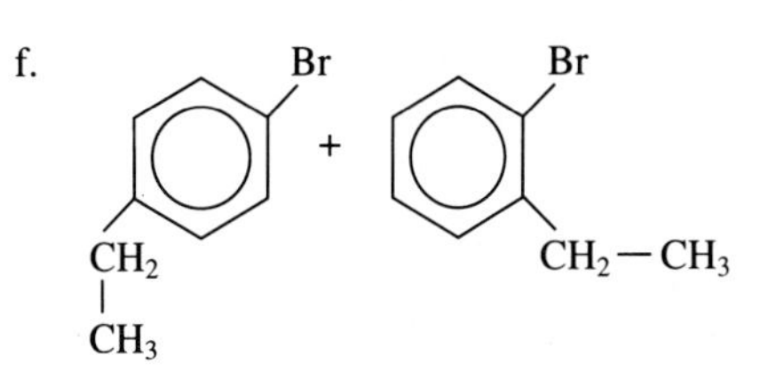

CHAPTER 5

1. Water participates in hydrogen bonding to a greater degree than methanol. This is a major factor in the boiling points of alcohols with low carbon numbers.

3. With only two carbon atoms, ethanol properties are dominated by the —OH group. Ethane, a simple hydrocarbon, will not have water solubility.

5.

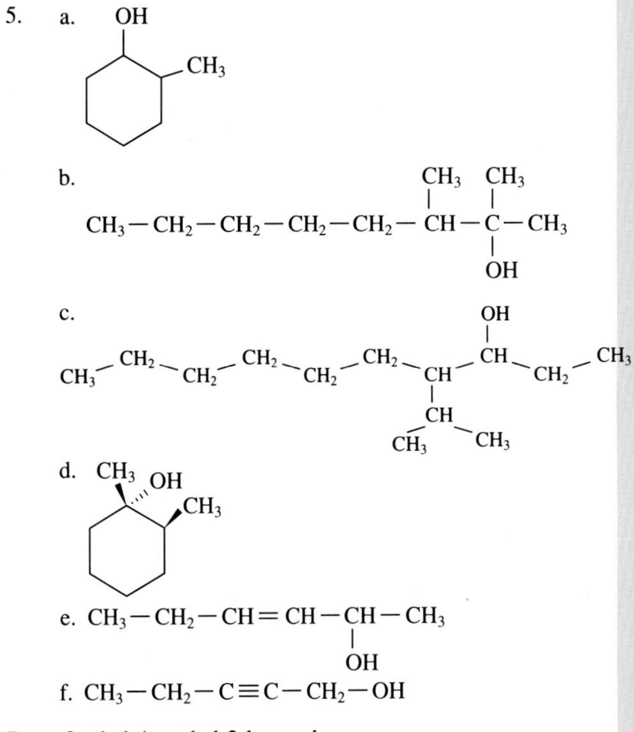

7. a 3-ethyl-4-methyl-2-hexanol
   b. 4-octanol
   c. 2,7-dimethyl-4-nonanol
   d. 4-methyl-4-octanol
   e. 4-ethyl-3,5-dimethyl-4-heptanol
   f. 4-octanol
   g. 2-isobutylcyclohexanol
   h. *cis*-1,2-cycloheptanediol
   i. 5,5-dimethyl-2-cyclopentenol

j. *cis*-2,8-dimethylcyclooctanol
k. 3-*tert*-butylcyclohexanol
l. 3,3-diisopropylcyclobutanol

9. a. 4,4-dibromo-2-pentanol
   b. 3-buten-2-ol
   c. 2,3-dichloro-1-pentanol
   d. 2,4-dimethyl-3-pentanol

11. 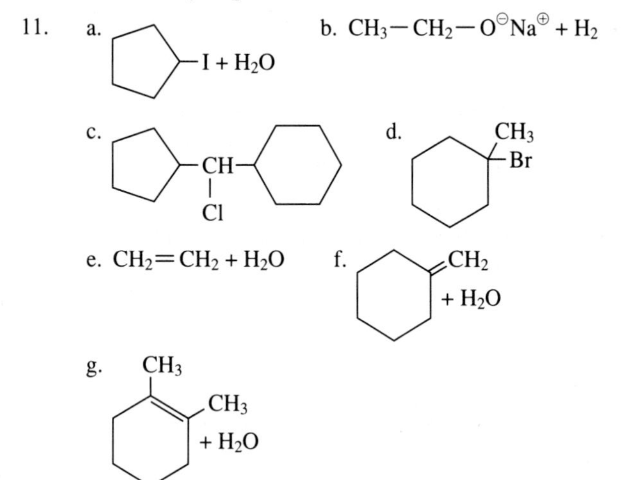

13. When the proton is removed from the oxygen of the —OH group, the resulting phenolate anion is resonance-stabilized.

15. There is no hydrogen bonding from the ether.

17. $CH_3-CH_2-CH_2-OH$ $\quad CH_3-\underset{\displaystyle OH}{\underset{|}{CH}}-CH_3$

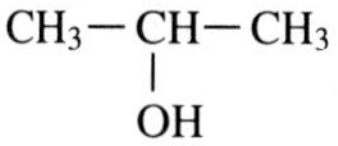

    $CH_3-O-CH_2-CH_3$

19. The sulfur is less electronegative than oxygen. This is expressed by a lower degree of hydrogen bonding.

21. a. $CH_3-S-Hg-S-CH_3$ b. CH_3-CH_2-SH

    c.  (ring with S—S) d. 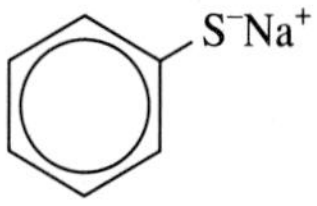 (benzene ring) S^-Na^+

    e. $CH_3-S-S-CH_3$ f. 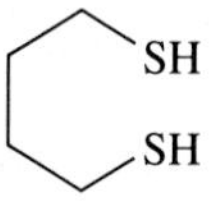 (SH, SH)

    g. $CH_3-CH_2-CH_2-S-Hg-S-CH_2-CH_2-CH_3$

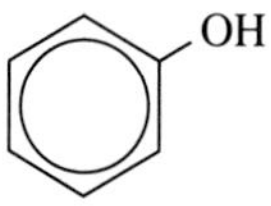

23. 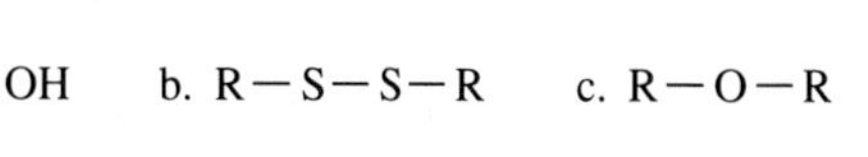

    c. $R-O-R$

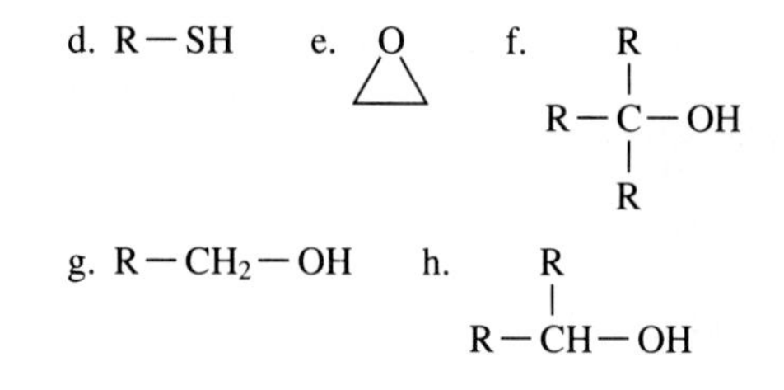

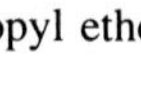

25. a. 3-methyl-2-cyclohexenol
    b. 2,4,5-trichlorophenol
    c. 2-isopropylcyclopentanol
    d. 2-chlorocyloheptanethiol
    e. 2,2-dibromo-3-pentyn-1-ol
    f. 4,4-dibromo-2-methyl-2-pentanol
    g. *trans*-1,2-cyclohexanediol
    h. *o*-nitrophenol
    i. *o*-ethylphenol
    j. *p*-bromophenol
    k. 2,5-dinitrophenol
    l. *o*-*tert*-butylphenol
    m. 2,3-dimethylcyclooctanol
    n. 3-isopropyl-2-*n*-propylcyclohexanol
    o. 1,3,3,4-tetramethylcyclohexanol
    p. 1-phenyl-1-pentanol
    q. 1,2-cyclohexanedithiol
    r. 2-methyl-1,2-butanediol
    s. 5,5-dimethyl-2-cylopentenol
    t. 3-butynethiol
    u. ethyl propyl ether
    v. *tert*-butyl isopropyl ether

27. a. cyclohexene
    b. 2-bromopropane
    c. 2-chloro-2-methylpropane
    d. 1-methylcyclohexene

29. c

31. a

33. a

35. a

37. b

CHAPTER 6

1. The dipole of the carbonyl compound is greater because of the difference in electronegativity between the carbon and the oxygen.

3. d, b, a, c

5. Alcohols have hydrogen-bonding interactions between molecules. These attractive forces are stronger than dipole–dipole interactions of the carbonyl compounds.

7. a. 2-bromopropanal
   b. 2,2-dimethylpropanal
   c. 3,3-dibromopentanal
   d. 2-propenal
   e. 2,2,2-tribromoethanal
   f. *o*-nitrobenzaldehyde
   g. 4-hydroxypentanal
   h. 3,3-dimethylbutanal
   i. acetaldehyde (or ethanal)

9. a. 3-methyl-2-butanone
   b. 2,2-dibromocyclopentanone
   c. 5,5-dibromo-4,4-dimethyl-3-hexanone
   d. *trans*-2,6-dimethylcyclohexanone
   e. 2-cycloheptenone
   f. 6-*tert*-butyl-2,2-dimethylcyclooctanone
   g. 3,5-dimethyl-2-hexanone
   h. 2,4,6-trimethyl-3-heptanone
   i. 3,5-dimethyl-2-hexanone

11. a. 3-methyl-2-butanone
   b. 2-methylbutanal
   c. 2-bromo-2-isopropylbutanal
   d. 4-methyl-3-hexanone
   e. 2,4-dimethylpentanal
   f. *trans*-4-hepten-3-one
   g. 2-ethyl-3-methylhexanal
   h. 3-ethyl-4-methyl-2-hexanone
   i. 2,3-dimethylpentanal

13. a. 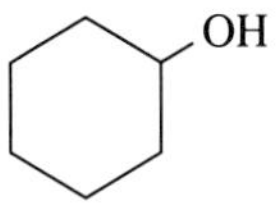 (cyclohexane)–OH b. 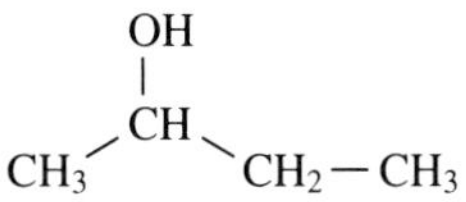 $CH_3-CH(OH)-CH_2-CH_3$

   c. $CH_3-CH(OH)-CH_3$ d. 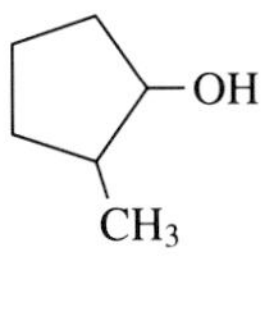 (cyclopentane with OH and CH_3)

15. a. 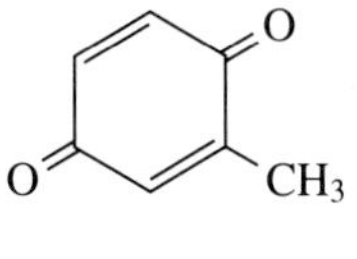 (quinone ring with =O, =O and CH_3) b. no reaction

   c. 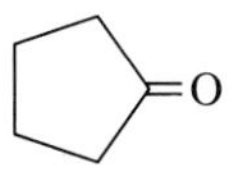 (cyclopentanone, =O) d. CH_3-CH_2-COOH

17.

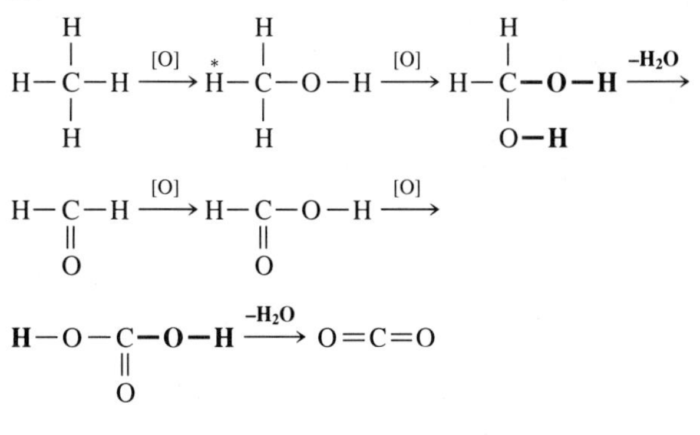

19. a. 2-methylhexanal
   b. 4-*sec*-butyl-2-methylheptanal
   c. 2,3,4-trimethylheptanal
   d. 2,3-dimethylhexanal
   e. 2-ethylbutanal
   f. 2-isopropylpentanal

21. There is a carbon–oxygen double bond. Because of electronegativity differences, the carbonyl group is polarized towards oxygen having "negative charge."

23. The dipole results in attraction between molecules of the type:

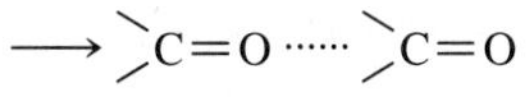

$$\longrightarrow \rangle C{=}O \cdots\cdots \rangle C{=}O$$

25. Both have a "double bond." Due to differences in electonegativity, the carbonyl is polarized towards oxygen.

27. a. 3,3-dibromoheptanal
   b. *trans*-3,4-dimethylcyclopentanone
   c. 5-phenyl-3-nonanone
   d. *m*-nitrobenzaldehyde

29. Because of the polarity of the carbonyl group, nucleophiles preferentially attack carbon over oxygen.

31. The generally unstable addition product of water to a carbonyl is shown here:

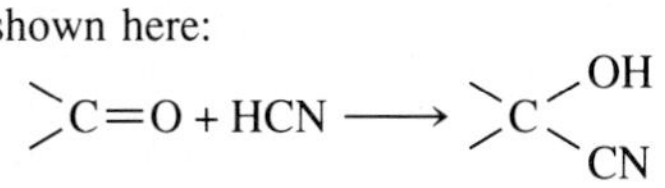

$$\rangle C{=}O + HCN \longrightarrow \rangle C(OH)(CN)$$

33.

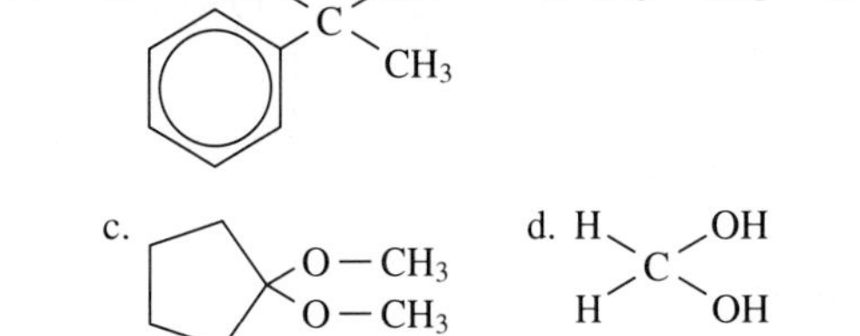

35. no reaction

37. a. carboxylic acid
    b. primary alcohol
    c. no reaction

39. b

41. d

43. b

45. b

47. a

CHAPTER 7

1. Acids can hydrogen-bond as well as dimerize, thereby greatly increasing the energy needed to separate molecules from one another. Alcohols have only hydrogen-bonding influences, but these are considerably more important than the dipole effects of the aldehyde carbonyl.

3. Carboxylic acids can form a reasonably strong dimer through hydrogen bonding

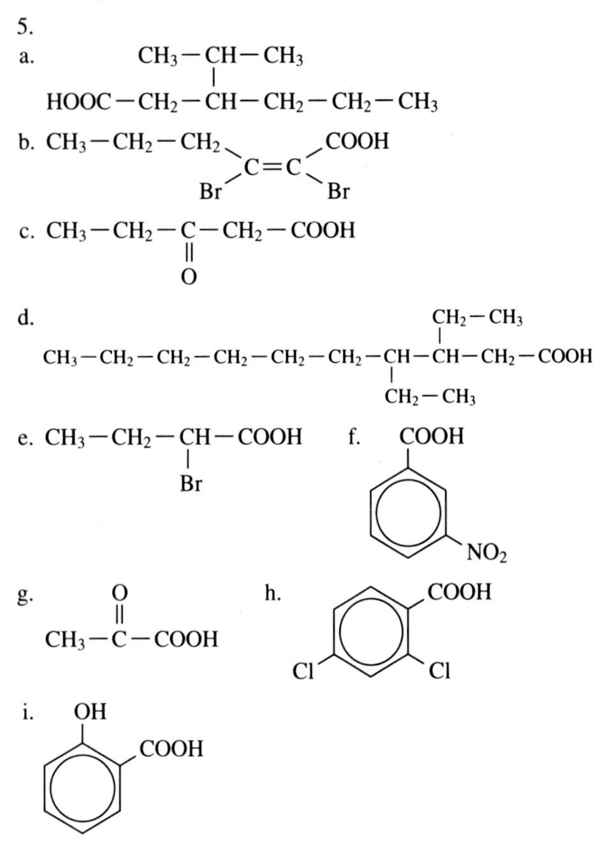

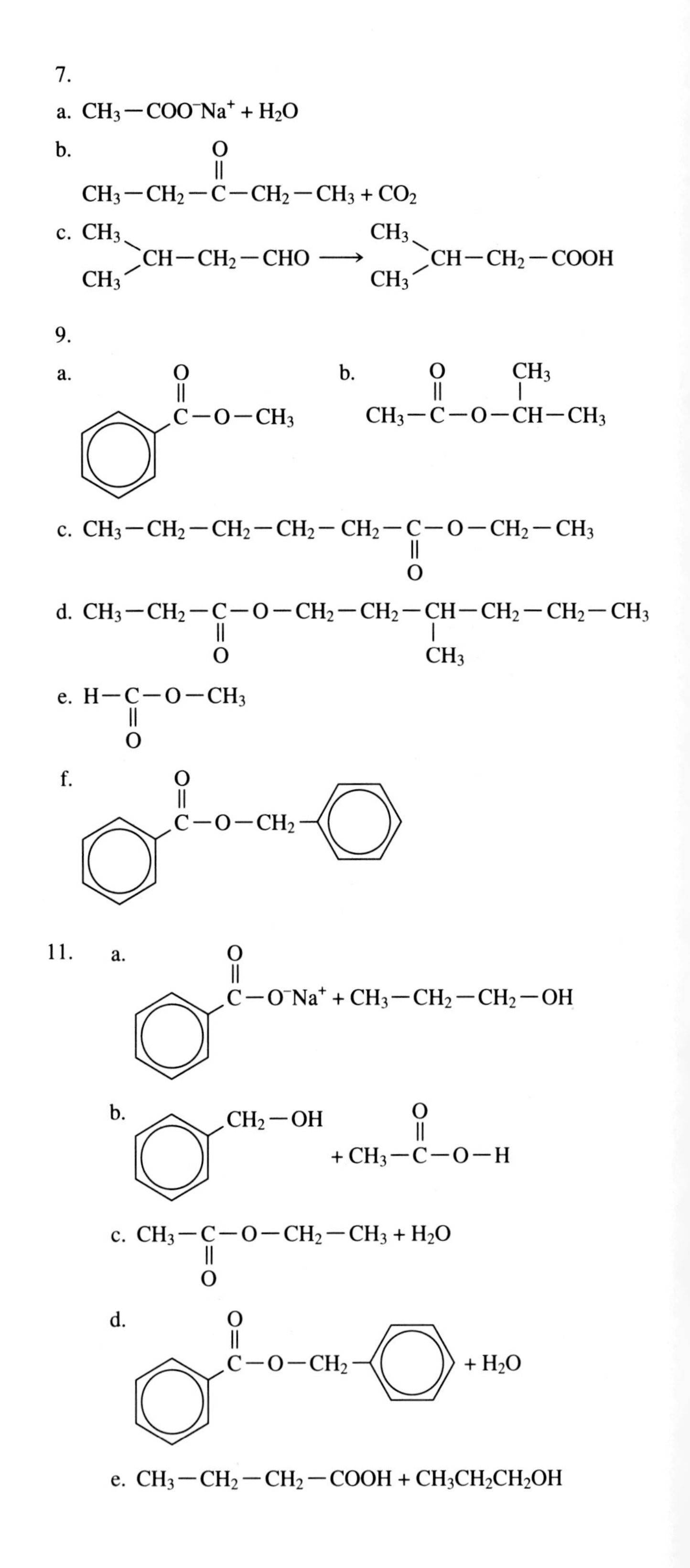

13. a. ethyl pentanoate
    b. 4,4-dichloro-7-methyloctanoic acid
    c. 6,6-dibromo-4-hydroxy-7-methyloctanoic acid
    d. *n*-propyl butanoate
    e. heptyl methanoate (heptyl formate)
    f. 3,4,5,5-tetramethyl-*trans*-3-hexenoic acid
    g. *n*-butyl benzoate

15.

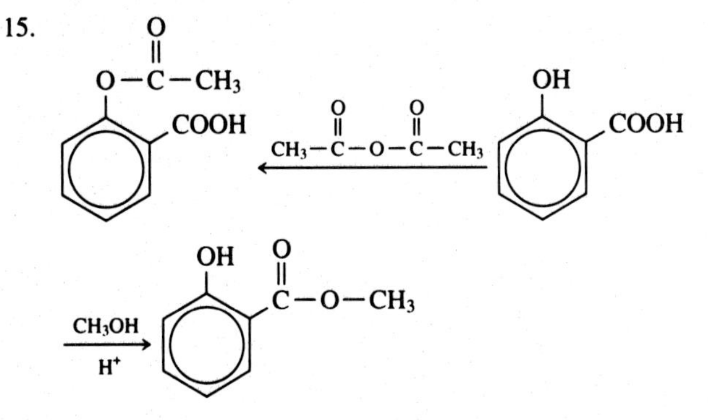

17. The detergent acts as a hydrocarbon and dissolves the hydrocarbon materials in oil paint. Again, this is an example of "like dissolves like."

19. sp^2

21. Hydrogen bonding is more important than hydrocarbon insolubility

23. a. 4,4-dimethyloctanoic acid
    b. *o*-nitrobenzoic acid
    c. *trans*-2-pentenoic acid
    d. 3-hydroxyheptanoic acid

25.

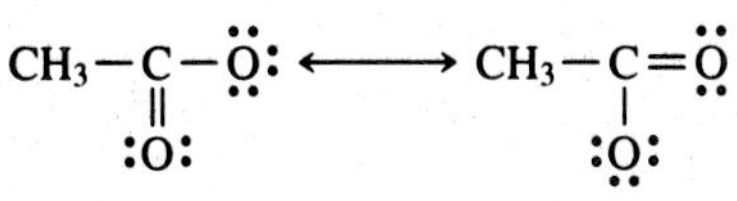

$$CH_3-\underset{\underset{:O:}{\|}}{C}-\ddot{\underset{\cdot\cdot}{O}}: \longleftrightarrow CH_3-\underset{\underset{:\ddot{\underset{\cdot\cdot}{O}}:}{|}}{C}=\ddot{\underset{\cdot\cdot}{O}}$$

27. 2-heptanone and carbon dioxide (decarboxylation of a beta keto acid)

29. carbon dioxide

$$CH_3—COOH + NaHCO_3 \longrightarrow CH_3—COO^-Na^+ + CO_2 + H_2O$$

31.

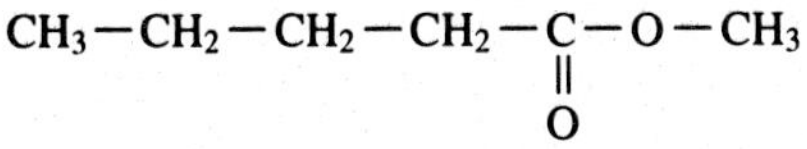

$$CH_3-CH_2-CH_2-CH_2-\underset{\underset{O}{\|}}{C}-O-CH_3$$

33. The hydrocarbon ends are able to interact with "greasy materials," and the ionic (or polar) ends are water-soluble.

35. d

37. c

39. a

41. c

CHAPTER 8

1. Amines have much weaker hydrogen-bonding interactions.

3. a. *N*-ethylaniline
   b. diethylamine
   c. isopropylamine
   d. *N*-ethyl-*N*-methylpropylamine (the longest chain takes priority)
   e. triethylamine
   f. *N*-methylethylamine
   g. *N*-methylpyrrolidine
   h. *N*-methyldiethylamine (*N*-ethyl-*N*-methylethylamine)

5. a. $CH_3-CH_2-\overset{\oplus}{N}H_2-CH_2-CH_3$ $Br^{\ominus}$

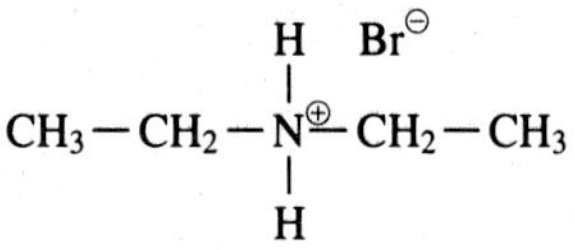

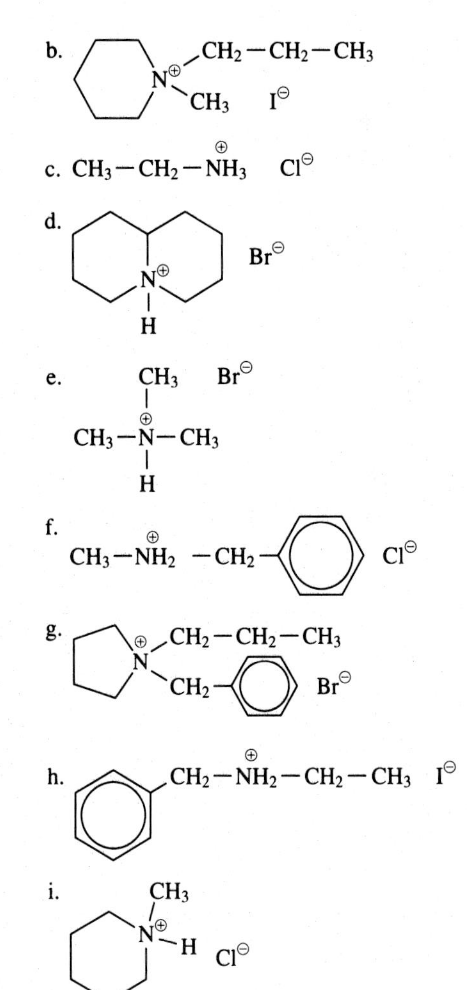

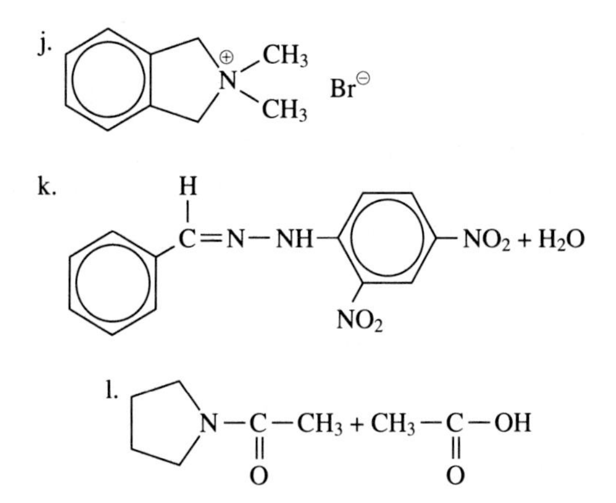

7. The lone pair of electrons are available for donation.

9. The hydrogen bonding is more important for alcohols.

11. The lone-pair electrons of nitrogen can, by resonance, be delocalized into the aromatic ring. Since they are less available on nitrogen, this reduces the Lewis basicity.

13. $R-CH(NH_2)-COOH$

15. Electrons on the nitrogen, by resonance, are delocalized into the carbonyl oxygen.

17. c

19. c

21. a

23. c

25. c

CHAPTER 9

1. The cell is the smallest unit of life that is capable of sustaining itself.

3. Faster transfer of nutrients and waste products to and from the cell.

5. Bacterium:

$$\text{Surface area} = 2\pi r^2 + 2\pi r h = 69\ \mu\text{m}^2$$

$$\text{Volume} = \pi r^2 h = 31\ \mu\text{m}^3$$

$$\frac{\text{Surface area}}{\text{Volume}} = 2.2\ \mu\text{m}^2/\mu\text{m}^3$$

Cell:

$$\text{Surface area} = 6 \times \text{length}^2 = 6.0 \times 10^4\ \mu\text{m}^2$$

$$\text{Volume} = \text{Length}^3 = 1.0 \times 10^6\ \mu\text{m}^3$$

$$\frac{\text{Surface area}}{\text{Volume}} = 0.06\ \mu\text{m}^2/\mu\text{m}^3$$

The bacterium has 2.2/0.06 = 37 times the surface area per unit volume.

7. Diffusion. (Some substances involve active transport.)

9. The cell membrane possesses transport systems that selectively move some substances across the membrane. Thus, this membrane helps control the contents of the cell.

11. They are the site of protein synthesis.

13. a. Block cell wall synthesis.
    b. Disrupt protein synthesis.
    c. Disrupt protein synthesis.

15. Carbohydrate (polysaccharide) and lipid.

17. Does not cause fever.

19. Probably not. The cell loses water and contracts. The cell wall only prevents the cell from swelling and bursting in hypotonic solution.

21. Animals, plants, fungi, true algae.

23.

Organelle	Animals	Plants	Bacteria
nucleolus	yes	yes	no
mitochondrion	yes	yes	no
endoplasmic reticulum	yes	yes	no
chloroplast	no	yes	no

25. DNA

27. Synthesis and transport of some biomolecules and cellular structures.

29. a. A mixture containing an equal amount of two enantiomers.
    b. Any two objects (like hands) that are nonsuperimposable mirror images of each other show chirality.
    c. A tetrahedral atom, usually carbon, with four different groups bonded to it.
    d. A compound containing chiral centers that is not chiral because of symmetry in the molecule.
    e. Stereoisomers with chiral centers that are not enantiomers.

31. Enantiomers

33. a, c

35. Measure several physical properties, such as boiling point. Most isomers have different values for these properties. If they are the same compound or enantiomers, they will have the same values. Then check optical activity with a polarimeter. If the samples are the same compound, the value will be identical, if they are enantiomers, they will rotate light in opposite directions by the same magnitude.

37. Chiral centers have four different substituents, so the chiral center in these compounds is the ring carbon atom with the dashed bond to the hydrogen. It is bound to the hydrogen, two different ring carbon atoms, and another carbon atom through the wedge bond.

39. d

41. a. 0
    b. 4
    c. 4 (Cis-trans isomerism can be ignored because the ring prevents the trans isomer of the carbon–carbon double bond.)
    d. 16

43. a. 4
    b. 4
    c. 0

45. Substance provided by food that the body must have for growth or maintenance.

47. The body conserves some of this energy as potential energy stored in chemical bonds of some molecules. The rest is lost as heat.

49. It is the fluid in which all other molecules and structures are either dissolved, dispersed, or suspended.

CHAPTER 10

1. A carbohydrate that cannot be broken down into a smaller unit (sugar) by simple hydrolysis.

3. This sugar contains a ketone functional group and has five carbon atoms.

5. a. D-glucose b. D-fructose

7. Glyceraldehyde; dihydroxyacetone

9.

Glucose:	carbon 1;	aldehyde
	2 through 5	alcohol (secondary)
	6	alcohol (primary)
Fructose:	1 and 6	alcohol (primary)
	2	ketone
	3 through 5	alcohol (secondary)

11. Switch the —H and —OH on carbon 5.

13. c

15. Both are aldopentoses and are identical at carbon atoms 1, 3, 4, and 5. But 2-deoxyribose has two hydrogen atoms as the substituents on carbon atom 2. Ribose has one hydrogen atom and a hydroxyl group on this carbon.

17. D- and L- glucose are enantiomers because they are nonsuperimposable mirror images. The other two are anomers.

19. In the ring structure that is typically illustrated, the hydroxyl group on carbon 1 points down, below the ring.

21. The cyclic forms of sugars are hemiacetals that form as shown in Figure 10.2.

23. Three. Eight.

25. There is no relationship between the D- and L- stereochemical families and the direction of rotation of plane-polarized light.

27. This is the change in rotation of plane-polarized light that is seen in a solution of a pure anomer as some of the anomer isomerizes to the other anomer.

29. A sugar that can reduce a metal ion to one of smaller oxidation number is called a reducing sugar. An aldehyde group must be present, or a keto group that can be isomerized to an aldehyde must be present.

31. Fructose is a ketose that can isomerize to an aldose. While it is an aldose, it is a reducing sugar.

33. No. They are not nonsuperimposable mirror images.

35. a. sucrose b. maltose c. lactose

37. The aldose is glucose; the ketose is fructose.

39. The glycosidic bond involves the anomeric hydroxyl group of both sugars. Because of this, it is not a reducing sugar.

41. a. α,β $1\rightarrow2$ b. α $1\rightarrow4$ c. β $1\rightarrow4$

43. Yes

45. The number of moles of monosaccharide formed per mole of carbohydrate when it is hydrolyzed in acid. Ten or fewer moles of monosaccharide are obtained from an oligosaccharide, more than ten from a polysaccharide.

47. In amylose, the glucose residues are linked by α $1\rightarrow4$ glycosidic bonds; in cellulose, these bonds are β $1\rightarrow4$.

49. *Similarities:* Both are unbranched polymers of monosaccharides connected by β $1\rightarrow4$ bonds.
    Differences: Cellulose contains glucose; chitin contains a modified glucose, *N*-acetylglucosamine.

51. Glycogen is a highly branched polymer of glucose. The ends of these numerous branches are sites where enzyme molecules can catalyze the release of glucose residues from the polymer.

53. It is found in connective tissue and helps protect and lubricate bone surfaces.

55. $$140{,}000 \text{ amu} \times \frac{1 \text{ glucose}}{162 \text{ amu}} = 864 \text{ glucose residues}$$

If your answer is 778 residues, you forgot to correct for the loss of water during glycosidic bond formation.

57. A. –f B. –d C. –a
D. –b E. –e F. –c

59. a. There are eight aldopentoses; ribose is shown here:

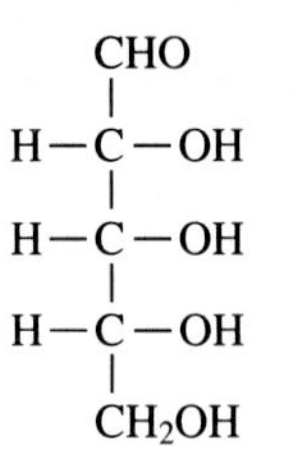

b. c.

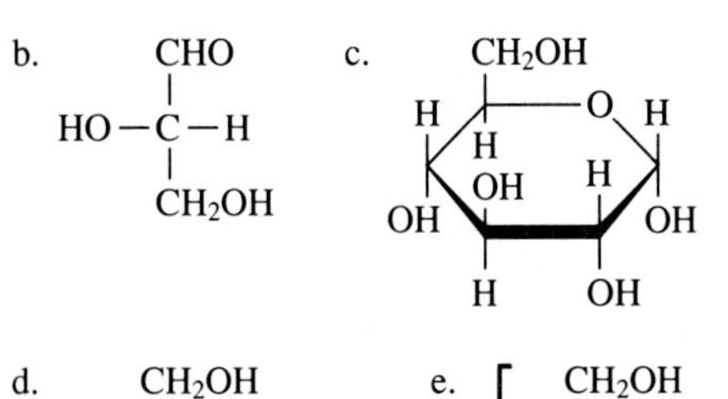

d. e.

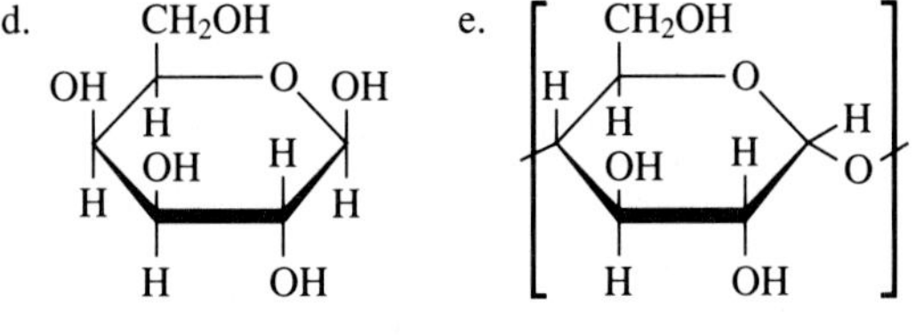

f.

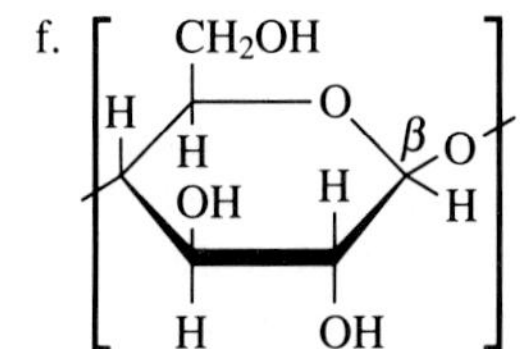

g.

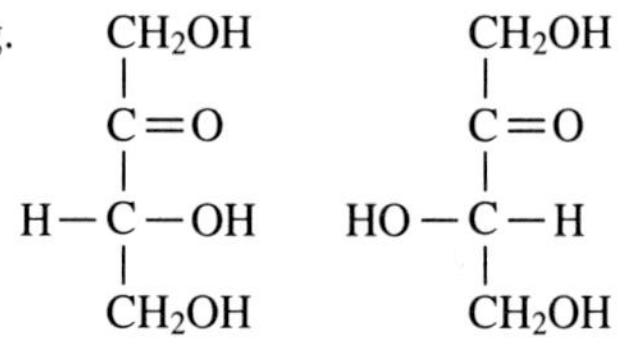

CH₂OH	CH₂OH
C=O	C=O
H—C—OH	HO—C—H
CH₂OH	CH₂OH

61. The appearance is concentration-dependent. A very dilute solution yields a slightly murky solution that is slightly green. A fairly concentrated solution yields a more obvious reddish precipitate, leaving little blue color in the mixture.

63. Oligosaccharide (tetrasaccharide)

65. Provide ample glucose to maximize synthesis of glycogen.

CHAPTER 11

1. They are generally soluble in nonpolar organic solvents and insoluble in water.

3. Triacylglycerols, phosphoacylglycerols, sphingolipids, glycolipids, some waxes

5. The nonpolar solvents, a, b, d

7. Their size; they have a long hydrocarbon-like tail.

9. It introduces a bend or kink in the nonpolar tail. It lowers the melting point.

11. a, d

13. a. $CH_3(CH_2)_5CH{=}CH(CH_2)_7COOH + H_2O \longrightarrow CH_3(CH_2)_6CHOH(CH_2)_7COOH + CH_3(CH_2)_5CHOH(CH_2)_8COOH$

b. $CH_3(CH_2)_5CH{=}CH(CH_2)_7COOH + H_2 \longrightarrow CH_3(CH_2)_{14}COOH$

c. $CH_3(CH_2)_5CH{=}CH(CH_2)_7COOH + I_2 \longrightarrow CH_3(CH_2)_5CHICHI(CH_2)_7COOH$

15. a. $CH_3(CH_2)_7CH{=}CH(CH_2)_7COOH + H_2 \longrightarrow CH_3(CH_2)_{16}COOH$

b. No reaction

c. $CH_3CH_2CHCHCH_2CHCHCH_2CHCH(CH_2)_7COOH + 3\ H_2 \longrightarrow CH_3(CH_2)_{16}COOH$

17.

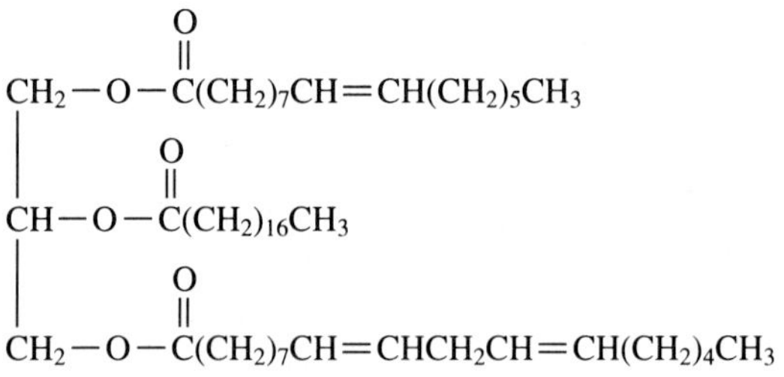

$CH_2{-}O{-}C(=O)(CH_2)_7CH{=}CH(CH_2)_5CH_3$
|
$CH{-}O{-}C(=O)(CH_2)_{16}CH_3$
|
$CH_2{-}O{-}C(=O)(CH_2)_7CH{=}CHCH_2CH{=}CH(CH_2)_4CH_3$

This is one of several isomers because each of the three acids could be connected to each of the three hydroxyl groups of glycerol.

19. This is the amount of iodine, in grams, that reacts with the carbon–carbon double bonds in 100 g of the fat or oil. The

iodine number is an indication of the amount of unsaturation present in the fat or oil.

21. The presence of double bonds in the fatty acids.

23. Antioxidants such as BHA and BHT can be added, or the oils used in the food can be partially hydrogenated.

25. 50,000 g × 0.20 = 10,000 g
10,000 g × (9 Cal/g) = 90,000 Cal
90,000 Cal × (1 day/2000 Cal) = 45 days

27. There are two possible structural isomers.

29. One

31. A molecule that has both polar and nonpolar regions. Membranes

33. Phospholipids are amphipathic molecules that can form the polar and nonpolar interactions needed for membrane structure; triacylglycerols are not amphipathic and thus cannot form these interactions.

35. A saponifiable lipid containing the base sphingosine

37.

$$
\begin{array}{l}
(CH_3)_3N^+CH_2CH_2OPO_2OCH_2 \\
\quad\quad\quad\quad\quad\quad | \\
\quad\quad\quad\quad\quad CH_2-O-\overset{O}{\overset{\|}{C}}(CH_2)_7CH=CH(CH_2)_7CH_3 \\
\quad\quad\quad\quad\quad | \\
\quad\quad\quad\quad\quad CH_2-O-\overset{O}{\overset{\|}{C}}(CH_2)_{16}CH_3
\end{array}
$$

Another structural isomer can be made by interchanging the fatty acids.

39. The structure of cholesterol is shown in Figure 11.6. The fused-ring system known as the steroid nucleus

41. Estrogen and progesterone. They play major roles in reproductive physiology, including the stimulation of ovulation by estrogen and the stimulation and maintenance of the uterine lining by progesterone.

43. They contain an aromatic ring.

45. Glycocholate and taurocholate. They emulsify dietary lipids in the digestive tract.

47. Aspirin inhibits the synthesis of prostaglandins that are involved in inflammation and pain.

49. 7-Dehydrocholesterol

51. a

53. Figure 11.13 illustrates a micelle. The polar head groups face out towards water, the nonpolar tails are clustered in the center.

55. For nonpolar molecules to dissolve in water, they must get between water molecules. This requires the breaking of relatively strong hydrogen bonds between the water molecules. The new bonds between nonpolar molecules and water are weak London forces. The replacement of stronger bonds by weaker ones is unfavorable and does not occur. Thus, the nonpolar molecules aggregate together.

57. The structure of a liposome is shown in Figure 11.14. They may gain use as a delivery system for drugs.

59. Amphipathic lipids and proteins are aggregated into a sheet-like structure held together by noncovalent bonds. The components are free to diffuse, thus the membrane is fluid.

61. c

63. A. – c B. – d C. – b
D. – a

65. a. Both are triacylglycerols. Oils have lower melting points due to more carbon–carbon double bonds or, less commonly, due to shorter chain fatty accids.
b. Both are saponifiable lipids derived from glycerol. Phospholipids (phosphoacylglycerols) are amphipathic, but fats are not.
c. Both are steroid female sex hormones, but they possess different physiological roles.
d. Both are nonsaponifiable lipids derived from unsaturated fatty acids. Both serve as signals or mediators, but the specific roles are different.

67. The following factors increase the risk of heart disease: the more obese an individual, the larger the dietary intake of fats and oils, the more saturated the dietary triacylglycerols, the greater the amount of blood cholesterol.

CHAPTER 12

1. Enzymes, structural, contractile, transport, protective, regulatory, storage

3. c

5. b

7. Two. The alpha carbon

9. Alanine, glycine, isoleucine, leucine, methionine, phenylalanine, proline, tryptophan, valine

11.

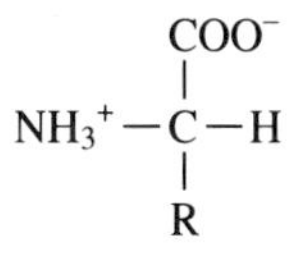

13. The basic amino acids have a basic group in the side chain; an acidic amino acid has a carboxyl group in the side chain. Lysine and aspartic acid are illustrated here.

COOH	COO^-	COOH	COO^-
^+H_3NCH	H_2NCH	^+H_3NCH	H_2NCH
CH_2	CH_2	CH_2	CH_2
CH_2	CH_2	COOH	COO^-
CH_2	CH_2	Aspartic Acid	
CH_2	CH_2		
NH_3^+	NH_2		
Lysine			

15. No. At the isoelectric point, the amino acid has both a positive and negative charge but is neutral. At more acidic pH, the amino acid has a net positive charge, at more basic pH, it has a net negative charge.

17.

COOH	COO^-	COO^-
^+H_3NCH	^+H_3NCH	H_2NCH
CH_2	CH_2	CH_2
COOH	COO^-	COO^-
pH 1	pH 7	pH 13

19. The C to N single bond actually has partial double-bond character due to resonance. Thus, rotation about that bond is restricted.

21.

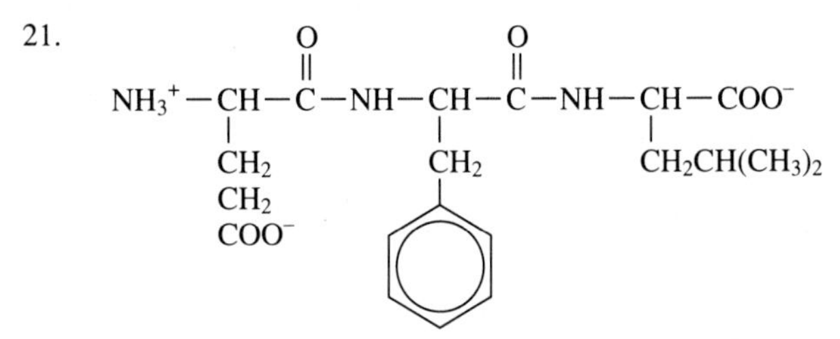

23. Alanylglycylvaline

25. Name	Structure	Biological Role
Oxytocin	See Figure 12.5	Contraction of smooth muscle in uterus and other target organs.
Vasopressin	"	Contraction of smooth muscle in arterioles and other target organs.
Enkephalin	"	Pain and pleasure perception.

27. *Primary:* the amino acid sequence
Secondary: the arrangement of nearby residues due to hydrogen bonding between atoms of the peptide bonds
Tertiary: the compact folding of the polypeptide chain back upon itself
Quarternary: the arrangements of subunits in an oligomeric protein

29. b

31. b

33. c

35. Figure 12.8 illustrates α-helix. Hydrogen bonding between atoms of peptide bonds

37. This oligomeric protein contains five subunits; two α subunits, two very similar subunits designated β and β′, and one σ subunit.

39. A simple protein contains only one or more polypeptide chains. A conjugated protein has, in addition, a prosthetic group.

41. The hydrophilic amino acids are on the exterior surface where their side chain can interact with the solvent water and polar and ionic solutes. The hydrophobic amino acids are in the interior where they form hydrophobic interactions.

43. b, d, e

45. $7^5 = 16{,}807$

47. *Positive:* lysine, arginine and histidine. *Negative:* aspartate and glutamate. This applies to solutions near pH 7.

49. At pH 6.13

51. If the proteins are more acidic, more of the ionizable side chains will be protonated. Thus, the protein will have a less negative or a positive charge. If they have a less negative charge, they will migrate more slowly. If they have a positive charge, they will migrate in the opposite direction.

53. Extremes of pH and heat denature proteins.

55. No. The amino acids can be bonded to each other in different orders, thus they can have different primary structures.

57. c

59. The rats will fare poorly on this diet. Because cysteine is lacking, methionine is needed for cysteine synthesis. Since a minimal amount of methionine is provided, there is not enough to meet a rat's needs for both methionine and cysteine made from methionine.

61. a. A carboxylic acid with an amino group on the α carbon. The basic building blocks of proteins. Any of the common 20 amino acids serve as an example.
 b. Amino acids that are in the same stereochemical family as L-glyceraldehyde
 c. A molecule possessing both a positive and negative charge. The 20 common amino acids exist as zwitterions.

d. Possessing both acidic and basic properties simultaneously. The amino acids are amphoteric.
e. The pH at which a charged molecule has no net charge. Each amino acid has its own unique pI.
f. A molecule composed of two amino acids linked by a peptide bond
g. The amide bond linking two amino acids
h. A macromolecule of more than 5000 mol wt that is made of amino acids linked by peptide bonds. Some have an additional group called a prosthetic group. Many examples can be given.
i. The amino acid sequence of a peptide or protein
j. The short-term arrangement of the polypeptide chain in a protein. Can be arranged into a spiral (α-helix) or into a sheet (β-pleated sheet).
k. The compact shape of most proteins resulting from the folding of the polypeptide chain.
l. The clustering together of nonpolar parts of a molecule that is in an aqueous solution
m. Covalent bond between the sulfur atoms of two cysteine residues in a protein. Contributes stability to the tertiary structure of a protein..
n. Loss of native conformation (shape) of a protein or other macromolecule

CHAPTER 13

1. Ribose is found in RNA; deoxyribose is found in DNA.
3. Deoxyribose has two hydrogens bonded to carbon atom number 2, ribose has one hydrogen and one hydroxyl group.
5.

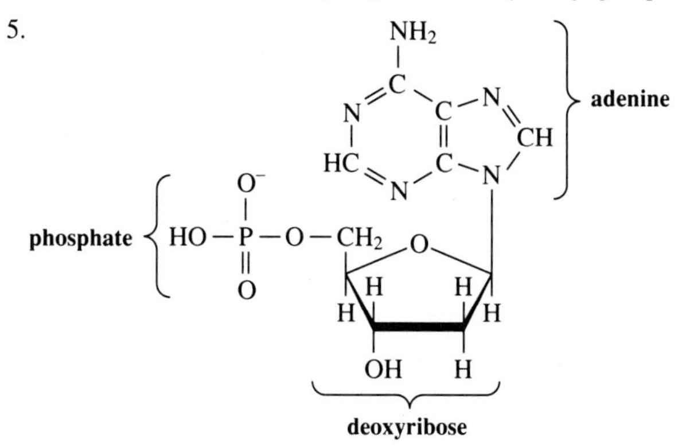

7. Uridine
9. Nucleotide
11. Adenine pairs with thymine through two hydrogen bonds; guanine pairs with cytosine with three hydrogen bonds.
13. Adenine, thymine, thymine, guanine, cytosine. Cytosine.
15. TCCGATGCATTCG
17. DNA polymerase
19. The new strand that is synthesized continuously is the leading strand. The new strand that is synthesized as a series of short (Okazaki) fragments is called the lagging strand.
21. The process of replication yields two new strands that are complementary to each of the original strands. The enzyme DNA polymerase synthesizes the new strand by sequentially attaching a new nucleotide whose base is complementary to the corresponding base on the original strand. Specific details are described in Section 13.3.
23. Each strand of the parent DNA molecule serves as a template for a new strand. Thus, each daughter molecule contains one intact parent strand and one new strand.
25. Ribose
27. DNA is usually the largest molecule in the cell. RNA is a polymer, but it is much smaller than DNA.
29. Promoters
31. A primary transcript is the RNA that is made by transcription of a gene on eucaryotic DNA. The primary transcript is modified to yield mRNA, which provides instructions for the synthesis of the protein during translation.
33. It is a portion of a eucaryotic gene between introns that remains after introns are cut out of the primary transcript as it is being modified to mRNA. *Exons* are *expressed* parts of the DNA.
35. At least 20 codons are needed because there are 20 amino acids. Codons of one or two bases only have 4 or 16 combinations of bases, respectively.
37. *Codon:* A sequence of three bases in mRNA that codes for a particular amino acid or that serves as a stop signal.
    Anticodon: A sequence of three bases in tRNA that pairs with a codon in mRNA. This pairing ensures that the proper amino acid is inserted into the growing polypeptide chain.
39. Codons are in mRNA, and thymine (T) is not found in mRNA.
41. Several sequences of codons work. Since there are two codons for phe, four for pro, and six for leu, there are $2 \times 4 \times 6 = 48$ possible answers. This is one of them: UUUCCUUUA
43. There is no unique set of codons for oxytocin; any of several sequences are correct. This is due to multiple codons for most amino acids. Your sequence should begin with either UGU or UGC, since these are the codons for cysteine.

45. Ribosomes

47. AUG with a set of bases preceding it serves as the start codon and insertion of methionine. AUG without those preceding bases signals insertion of methionine.

49. Posttranslational modifications may include hydrolysis of one or more peptide bonds, insertion of a prosthetic group, and aggregation of subunits.

51. After the enzyme catalyzes the formation of the bond between the nucleotide chain and the new nucleotide, the enzyme compares the new base with the complementary one on the old strand. If they are not really complementary, then the new nucleotide is cut out and the process repeats.

53. If a base changes in DNA, then a different base is present in the corresponding RNA, and a different codon now exists. Thus, a different amino acid may be inserted into the peptide chain, or the chain may terminate prematurely.

55. No. Since there is genetic variability in humans, there must be many possible sequences of DNA. They can, however, come up with a typical or consensus DNA.

57. The presence of the tryptophan in the diet repressed the synthesis of the enzymes needed for the tryptophan synthesis.

59. A bacterium grown on glucose does not make the enzymes needed to use lactose for energy. When the bacterium is given lactose, the lactose induces the synthesis of the enzymes needed to use the lactose. There is a lag between the time when lactose is added and the time when the enzymes appear.

61. They are used to cut the desired DNA from the donor DNA and to cut the carrier DNA (plasmid). They cut the DNA to yield sticky ends that allow the DNAs to come back together.

63. The DNA that corresponds to the gene is transcribed to yield RNA. This RNA is a functional mRNA in bacteria, and in eucaryotes, it is the primary transcript that is modified into an mRNA. The information in this mRNA is then used in translation to direct the sequence of amino acids in the polypeptide chain.

65. Proteins are responsible for most of the activities and characteristics of cells. Thus, if we understand the role of DNA in protein synthesis, we understand much of the role of DNA in controlling cell activity.

67. No. Since most amino acids have several corresponding codons, it is highly unlikely that a synthetic gene has exactly the same sequence as the real gene. Furthermore, introns are a factor if this was a eucaryotic gene.

69. The virus has single-stranded DNA as its nucleic acid.

CHAPTER 14

1. Catabolism and anabolism

3. Anabolism

5. They catalyze metabolic reactions, allowing them to proceed rapidly and efficiently at the temperature and conditions that exist within a cell.

7. Proteins

9. a, d, e, f

11. The molecule that bonds to the active site of the enzyme and is modified by the reaction that the enzyme catalyzes

13. Apoenzyme

15. Vitamins

17. NAD^+ and FAD

19. This enzyme catalyzes the formation of lactate from pyruvate and the reverse reaction, formation of pyruvate from lactate. The different forms of the enzyme have different responses to different concentrations of pyruvate and lactate, thereby allowing different tissues to carry out one of the reactions preferentially.

21. The location where the substrate(s) binds and where catalysis of the reaction occurs

23. Noncovalent interactions like hydrogen bonding, other polar interactions, and hydrophobic interactions

25. The size of the active site and the dipeptide should be similar. A portion of the active site may be nonpolar and complementary in size to the phenyl ring of phenylalanine. Another portion of the active site should fit the long side chain of lysine and have a negative charge or polar group to bond with the amino group at the end of the chain.

27. If the substrate molecules are oriented to place their reacting groups at the optimal distance and angle for reaction, then the activity of the enzyme is enhanced.

29. Probably not. Both the substrate and enzyme are normally in their most stable (least strained) forms. Thus, when substrate binds to the enzyme, it is not strained.

31. Your sketch should resemble the straight-line graph shown in Figure 14.8.

33. With constant amount of enzyme, the enzyme molecules become saturated with ever-increasing concentrations of substrate. This does not occur with additions of enzyme.

35. Pepsin is active around pH 2, trypsin is active around pH 7 to 8. Each is most active in the pH range where it is found: pepsin in the stomach (pH 2), and trypsin in the small intestine (pH 7–8).

37. No. Temperature and pH are normally maintained within a narrow range.

39. A competitive inhibitor binds to the active site of an enzyme. A noncompetitive inhibitor binds elsewhere on the enzyme.

41. They are competitive inhibitors of the enzyme dihydropteroate synthetase, which is involved in the synthesis of *p*-aminobenzoic acid. Without this product, the bacterium cannot make the essential cofactor folic acid.

43. The substrate and inhibitor both possess two carboxylate groups separated by a region of nonpolarity. The inhibitor cannot be oxidized to a carbon–carbon double bond because it has only one methylene group.

45. The name often has the suffix -ogen or the prefixes pre- or pro-.

47. Trypsin catalyzes the hydrolysis of proteins, and if it were active in the pancreas, it would catalyze the hydrolysis of pancreatic proteins.

49. It is covalently modified by the bonding of phosphates to specific serines. The activity of the enzyme increases when it is phosphorylated.

51. It means sites other than the active sites of enzymes.

53. A positive effector binds to an allosteric site of an enzyme. This binding causes a conformational change that increases the enzyme's activity. A negative effector binds to an allosteric site of an enzyme. This binding causes a conformational change that decreases the enzyme's activity.

55. Some enzymes that are normally absent in blood are released from heart tissue during and after a heart attack. Thus, the presence of these enzymes in blood indicates that a heart attack has occurred.

CHAPTER 15

1. Biosynthesis, transport, motion

3. Cells use energy to resist the tendency for the universe to move to ever-increasing disorder.

5. It is used for short-term energy storage. Some reactions of catabolism are exothermic, and a portion of the energy released in these reactions is used to synthesize ATP from ADP and inorganic phosphate.

7. *Catabolic:* a, e
   Anabolic: b, c, d, f

9. It reduces food particle size, which increases surface area; it mixes food with saliva, which contains the enzyme amylase.

11. These enzymes catalyze the hydrolysis of the disaccharides maltose, sucrose, and lactose, respectively.

13. It denatures proteins, which exposes the peptide bonds to digestive enzymes.

15. Lipids are insoluble in the aqueous environment of blood.

17.

Food Class	Enzymes	Site of Digestion	Products
Carbohydrates	Amylases, hydrolases for sugars	Primarily mouth and small intestine	Simple sugars
Proteins	Proteases and peptidases	Stomach and small intestine	Amino acids
Lipids	Lipases	Small intestine	Fatty acids and glycerol

19. The steps that involve hexokinase and phosphofructokinase. Hydrolysis of ATP

21. Lactic acid is the product of anaerobic metabolism and may accumulate under these conditions. This acid donates its hydrogen ion under physiological conditions, thus making blood or the cells more acidic.

23. $(\text{glucose})_n + \text{P}_i \rightarrow (\text{glucose})_{n-1} + \text{glucose 1-phosphate}$
    $\text{glucose 1-phosphate} \rightarrow \text{glucose 6-phosphate}$

25. Two

27. It is oxidatively decarboxylated in a reaction catalyzed by pyruvate dehydrogenease.

29. These coenzymes accept a pair of electrons and hydrogen ions during oxidation–reduction reactions. 3 NAD^+ and 1 FAD.

31. Both are nucleotide triphosphates; they possess identical phosphate–phosphate anhydride bonds. They are equivalent in energy because hydrolysis of an anhydride bond yields very similar products: a nucleotide diphosphate and an inorganic phosphate.

33. This is phosphorylation of ADP to ATP that accompanies the oxidation of the reduced coenzymes NADH and $FADH_2$.

35. Electron carriers of the electron-transport chain accept electrons from NADH and $FADH_2$, oxidizing them to NAD^+ and FAD.

37. They pump protons from the mitochondrial matrix to the cytosol.

39. ATP synthetase. A portion of it (F_0) is embedded in the inner mitochondrial membrane; the F_1 part projects into the mitochondrial matrix.

41. Fatty acids (as CoA derivatives)

43. It transports fatty acids in blood.

45. Palmitoleic acid yields two fewer ATP because it already has a carbon–carbon double bond. One fewer $FADH_2$ is produced from palmitoleic acid because of that double bond.

47. The α-amino groups of amino acids are transferred to α-ketoglutarate to form glutamate.

49. *Reactants:* glutamate, NAD^+
Products: α-ketoglutarate, NADH, NH_3
The enzyme is glutamate dehydrogenase.
The process is oxidative

51. In the mitochondria

53. The difference is due to the way the electrons of cytosolic NADH are brought into the mitochondrion of that cell. If mitochondrial FAD accepts them, then one fewer ATP is formed per cytosolic NADH than if mitochondrial NAD^+ accepts them.

55. (200 kcal) × (1 mol ATP/7.3 kcal)
× (1 mol glucose/36 mol ATP)
× (180 g glucose/1 mol glucose) = 137 g glucose

57. (25 g glucose) × 1 mol glucose/180 g glucose)
× (36 mol ATP/1 mol glucose) = 5.0 mol ATP.

59. *Hexokinase:* This enzyme phosphorylates a hexose (glucose).
Phosphoglucose isomerase: This enzyme converts one isomer of phosphoglucose to another.
Phosphofructokinase: This enzyme phosphorylates phosphofructose.

CHAPTER 16

1. a. Anabolism has small reactants and larger products; catabolism is the opposite.
   b. Anabolic reactions tend to be reductive; catabolic are generally oxidative.
   c. Anabolism is endothermic; catabolism is exothermic.

3. Since NADPH and NADH are structurally different (a phosphate on NADPH that is absent on NADH), they do not bind to the active site of the same enzyme.

5. ATP

7. They make them using carbon dioxide as a source of carbon atoms.

9.

Reactants	Products
H_2O	O_2
ADP	ATP
P_i	
$NADP^+$	NADPH

11. As electrons flow from photosystem II to photosystem I through an electron-transport chain, protons are pumped. This proton gradient is stored energy and can be used to do work.

13. ATP is used to phosphorylate intermediates of the dark reaction, and NADPH is used to reduce 1,3-diphosphoglycerate to glyceraldehyde 3-phosphate.

15. Electrons flow from water molecules to a chlorophyll in photosystem II, through electron carriers to a chlorophyll in photosystem I, through electron carriers to $NADP^+$.

17. Twelve NADPH are required, and four photons are needed per NADPH. Thus 48 moles of photons are needed.

19. Six carbon dioxide molecules combine with six ribulose diphosphate molecules to form twelve 3-phosphoglycerate molecules. Two of these are used to make glucose, the remaining ten are used to make ribulose 1,5-diphosphate.

21. Ribulose 1,5-diphosphate

23. Lactate, glycerol, glucogenic amino acids. Glucose is the product.

25. Pyruvic acid, oxaloacetic acid, phosphoenolpyruvic acid, 2-phosphoglyceric acid, 3-phosphoglyceric acid, 1,3-diphosphoglyceric acid, glyceraldehyde 3-phosphate (dihydroxyacetone phosphate), fructose-1,6-diphosphate, fructose-6-phosphate, glucose-6-phosphate. These reactions take place in the liver.

27. Lactose synthetase is formed by adding a modifying subunit to galactosyl transferase. Thus, they catalyze the same type of reaction but with different substrates.

29. The synthesis of fatty acids

31. Citrate leaves the mitochondrial matrix and enters the cytosol. It is cleaved to acetyl-CoA and oxaloacetate, which is recycled back into the mitochondrion.

33. Palmitate. Palmitate can be elongated and double bonds added to it.

35. Fatty acids are synthesized from acetyl-CoA, which contains two (an even number) carbon atoms.

37. NADPH and pentoses

39. A diacylglycerol reacts with cytidine diphosphate (CDP)-choline to yield phosphatidylcholine (lecithin) and CMP.

41. Hydroxymethylglutaryl-CoA

43. Dietary proteins

45. In catabolism transamination removes amino groups from amino acids, yielding keto acids that can be catabolized. In anabolism, amino groups are added to keto acids to synthesize amino acids.

47. The synthesis of cysteine requires methionine. If only a minimal amount of methionine is provided, there is not enough methionine in the diet to meet both the need for methionine and the need for synthesis of cysteine.

49. This coenzyme donates a methyl group to dUMP to form dTMP.

51. They differ only by the fluorine atom on the ring. Since fluorouracil resembles the uracil of dUMP, a metabolite of it binds (irreversibly) at the active site of thymidylate synthase.

CHAPTER 17

1. Hemoglobin binds oxygen, which greatly increases the oxygen-carrying capacity of blood. Hemoglobin also binds some carbon dioxide, facilitating its transport to the lungs.

3. This enzyme speeds up the reversible hydration of carbon dioxide to form carbonic acid.

5. Less than 1 mm in diameter, unless a mechanism is provided to deliver air to within a millimeter of all cells.

7. Le Chatelier's principle

9. Both factors decrease the oxygen-carrying capacity of hemoglobin. This is the Bohr effect.

11. Vigorous exercise decreases concentration of oxygen in the tissues and increases the acidity. Both of these effects result in increased dissociation of oxygen from hemoglobin in the tissue.

13. In filtration, blood pressure forces water and small solutes from the blood into the glomeruli of the kidneys. Water and nutrients are then reabsorbed from the tubule by active transport.

15. With less bicarbonate in blood, less H^+ may be taken up by bicarbonate and the blood may become more acidic.

17. Carbonic acid dissociates to bicarbonate and hydrogen ion whenever the concentration of hydrogen ion decreases. Bicarbonate reacts with hydrogen ion to form carbonic acid whenever the hydrogen ion concentration increases.

19. *Acid:* carbonic acid (H_2CO_3)
Conjugate base: bicarbonate (HCO_3^-)

21. HPO_4^{2-}

23. It makes it more alkaline (less acidic). Respiratory alkalosis

25. Loss of stomach acid (HCl) can increase blood pH due to metabolic alkalosis.

27. Steroid hormones bind to a receptor molecule within the cell. The complex diffuses into the nucleus and binds to specific protein–DNA complexes. This binding alters the rate at which the DNA is transcribed at that region.

29. Steroid hormones.

31. Insulin and glucagon

33. It phosphorylates certain proteins within the cell. When phosphorylated, these proteins have altered activity. Cyclic AMP activates protein kinase.

35. Cyclic AMP has its phosphate bound to both the 3′ and 5′ hydroxyl groups of ribose; AMP has it bound only to the 5′ hydroxyl group. The binding site on protein kinase only binds the cyclic form.

37. It is released into the blood stream when the concentration of nutrients (glucose) is low.

39. It stimulates catabolism of lipids in these cells. Binding of glucagon stimulates synthesis of cAMP, which activates protein kinase, which phosphorylates, thus activating, key enzymes needed for lipid catabolism.

41. Homeostasis. Insulin stimulates cells to uptake nutrients when they are present in above normal concentrations. This brings nutrient concentrations back towards the normal range.

43. The B lymphocyte is stimulated to both divide and change into plasma cells that secrete antibodies to the antigen.

45. Antibodies

47. The immune system responds more rapidly and more forcefully the second time (and each subsequent time) it is exposed to an antigen.

49. Anti-A

51. Individuals with B or O blood group

GLOSSARY

A

Absolute zero Temperature at which all molecular motion theoretically stops, measured as $-273.15°C$ or 0 K.

Accuracy How close the measured value is to the true value.

Acid Hydrogen ion donor; a molecule or ion that ionizes to give up one or more hydrogen ions.

Acid anhydride Derivative of an organic acid that is dehydrated, having the structure

```
R—C—O—C—R
  ‖   ‖
  O   O
```

Acid equilibrium constant (K_a) Equilibrium constant for a weak acid, which is a combination of the water concentration (considered to be constant) and the equilibrium constant.

Acidosis Condition that exists whenever the blood pH drops below the normal range; can be metabolic or respiratory in origin.

Activation energy Energy added to molecules so that they collide with sufficient energy to break chemical bonds.

Active site Site on an enzyme that bonds substrate and catalyzes the reaction to yield product.

Active transport Energy-requiring transport of substances across the membrane of a cell.

Acyl carrier protein (ACP) Protein involved in lipogenesis that carries the growing fatty acid and the molecules that condense to form the fatty acid.

Adenosine triphosphate (ATP) Nucleotide that plays a role as an energy reservoir in living systems.

Alcohol Functional group characterized by the general formula

R—O—H

Aldehyde Family of organic compounds formed when an alkyl group is placed on one of the carbon bonds of a carbonyl group and a hydrogen is placed on the other. The general structure is

```
Ar—C—H    R—C—H
   ‖         ‖
   O         O
```

The carbonyl group is always on a terminal carbon atom.

Aldose Monosaccharide containing an aldehyde group. The prefix *ald-* indicates the presence of an aldehyde group, and the *-ose* suffix signifies a carbohydrate.

Alkaloid Naturally occurring, basic, nitrogen-containing compound.

Alkalosis Condition that exists when blood pH exceeds the normal range; like acidosis, it can result from respiratory or metabolic abnormalities.

Alkane Family of hydrocarbons having only single carbon-to-carbon bonds characterized by the general formula C_nH_{2n+2}.

Alkene Family of hydrocarbon compounds having at least one double bond between carbon atoms in the molecular skeleton.

Alkyl group Hydrocarbon group made up of a hydrocarbon minus one of its hydrogen atoms. This group is named from the parent alkane by replacing the *-ane* ending with *-yl*. Alkyl groups exist only on paper as a tool for naming and for explaining organic reactions. Alkyl groups are often represented by the symbol R.

Alkyne Family of hydrocarbon compounds having at least one triple bond between carbon atoms in the molecular skeleton.

Allosteric regulation Regulation of enzyme activity by the binding of small molecules to sites other than the active site.

Alpha decay Radioactive decay involving emission of an alpha particle. The daughter product of alpha decay has two fewer protons and two fewer neutrons than the parent isotope.

α-helix Secondary structure of a protein in which the polypeptide backbone is arranged into a regular spiral shape.

Alpha (α) particles Relatively heavy particles having the mass of four hydrogen atoms that carry two positive charges and are fired out of some radioactive atoms.

Alveoli Tiny sacs in the lungs where gas exchange occurs.

Amide Nitrogen-containing compound derived from an acid and an amine with the general structure

```
R—C—N
  ‖
  O
```

Amine Organic compound in which a nitrogen is the central member of the functional group. Since nitrogen atoms have three bonds, only one of which connects to the parent hydrocarbon, it is possible to substitute additional alkyl groups on this nitrogen.

Amino acids Class of biological compounds whose members possess both an amino group and a carboxylic group. Amino acids are the building blocks of peptides and proteins. Most of the important natural amino acids are alpha-amino acids.

Amphipathic molecules Molecules with both polar and nonpolar regions.

Amphoteric molecule Molecule that functions as a base in the presence of an acid and as an acid in the presence of a base.

Amylases Enzymes that catalyze the hydrolysis of α 1$\rightarrow$4 glycosidic bonds in starches.

Anabolism Part of metabolism that uses energy to build larger, more complex molecules from simpler ones.

Anomeric carbon Carbon atom in the cyclic form of sugars that had been the carbonyl carbon of the open-chain form.

Anomers Pair of isomers that forms when a monosaccharide forms a ring. The oxygen attached to the anomeric carbon can have either a *beta* (β) ("up") or an *alpha* (α) ("down") orientation.

Antibodies Proteins produced by the body that bind foreign molecules and particles. These complexes are then readily taken up and destroyed by the cells of the immune system.

Anticodon Three-base sequence in tRNA that is complementary to a codon on mRNA.

Antigens Large foreign molecules within the body. These molecules cause an immune response that results in production of antibodies to the antigen.

Aromatic Class of ring compounds that have alternating double bonds and subsequent pi electron delocalization. An aromatic compound obeys the $4n + 2$ rule.

Arrhenius's model for an acid Idea proposed by Svante Arrhenius that acids are molecules that, when placed in water, ionize to produce hydrogen ions.

-ase Suffix used to indicate that the substance is an enzyme.

Atmosphere (atm) Normal atmospheric pressure, where 1.00 atm equals 760 torr.

Atom Smallest particle into which an element can be subdivided and still retain its chemical properties. It is a basic building block of matter. All atoms of an element are alike but differ from those of other elements in size, mass, and chemical reactivity.

Atomic mass Weighted average of the mass of all of the naturally occurring isotopes of a given element.

Atomic mass unit (amu) Basic reference for atomic mass measurements equal to 1/12 the mass of the carbon-12 atom. 1 amu $= 1.66606 \times 10^{-24}$ g.

Atomic number Number of protons (positive charges) in the nucleus of an atom. The atomic number is also the reference number used to specify the location of an element in the periodic table and to describe the number of electrons surrounding the nucleus of a neutral atom.

Autotrophic organism Organism that obtains carbon from nonliving sources, such as carbon dioxide.

Avogadro's number Number of atomic mass units that equal 1.00 g, which is 6.023×10^{23}.

B

Background radiation Radiation received from naturally radioactive elements in the atmosphere and the earth's surface and from cosmic radiation. About 65 percent of our annual radiation dose comes from background radiation.

Balanced chemical equation Chemical equation that indicates the balanced formulas and the relative quantity of each element or compound present before and after the reaction.

Base Hydrogen ion acceptor.

Benzene Simplest aromatic hydrocarbon.

Benzyl Alkyl group derived from toluene by loss of a hydrogen from the methyl group.

Beta decay Radioactive decay involving emission of a beta particle. The daughter product of a beta decay has one fewer neutron and one more proton than the parent isotope.

Beta oxidation Metabolic pathway that oxidizes fatty acids to acetyl-CoA and reduced coenzymes.

Beta (β) particles Negative, electron-sized particles fired out of a radioactive atom.

β-pleated sheet Secondary structure of a protein in which the polypeptide chain or chains are arranged into a sheetlike structure.

Bilayer Sheet-like structure made up of amphipathic molecules. The nonpolar tails form an interior core, and the polar head groups make the surfaces polar.

Bile salts Emulsifying agents in bile that break down dietary lipid droplets into micelles.

Binary compound Compound consisting of only two elements.

Bioenergetics Study of energy in living systems.

Bohr effect Increases in the concentration of carbon dioxide and hydrogen ions increase dissociation of oxygen from hemoglobin. Decreases in concentration have the opposite effect.

Bohr's planetary electron model Model proposed by Niels Bohr suggesting that electrons rotate around a nucleus in a set of fixed orbits like planets around the sun. The model predicts that light is given off when electrons jump from outer orbits to inner orbits.

Boiling point Temperature at which a substance in the liquid state is converted to the gaseous state. The gaseous form of the substance condenses into a liquid at the same temperature.

Bond strength Ability of one atom to hold onto another. Stronger bonds hold atoms together more tightly, resulting in higher melting points for their compounds.

Boyle's law Relationship stating that the product of pressure times volume for any gas sample is always a constant.

Brønsted–Lowry model for an acid Idea independently recognized by Brønsted and Lowry that an acid cannot donate a hydrogen ion unless there is another molecule—a hydrogen ion acceptor—present to accept the proton. The hydrogen ion acceptor is called a *base*.

Brownian motion Constant, random motion characteristic of colloidally suspended particles.

Buffer Mixture of roughly equal parts weak acid and the salt of the acid (its conjugate base) that maintains the pH of a solution reasonably constant with addition of moderate amounts of either acid or base.

C

calorie (cal) Unit of heat measurement equal to the amount of heat required to change the temperature of 1 g of water 1°C.

Calorie (Cal) Unit of energy used in nutrition. It is equal to 1 kcal or 1000 "small" calories.

Canal ray Stream of positive particles; positively charged atoms from which one or more electrons have been removed.

Carbohydrate Class of compounds consisting of polyhydroxy aldehydes and ketones and derivatives of these compounds.

Carbonyl Family of organic compounds composed of a carbon double-bonded to an oxygen. The two remaining carbon bonds may be connected to other atoms or alkyl groups.

Carbonyl group Functional group characterized by a carbon–oxygen double bond.

Carboxyl group Group of atoms composed of a carbonyl group with an —OH connected to one of the free carbon bonds. Carboxyls are the "trademark" of organic acids.

Carboxylate anion Resonance-stabilized anion resulting from removal of the acid proton of an organic acid.

Carboxylic acid Organic compound containing one or more carboxyl groups.

Carcinogenic Capable of inducing the formation of cancer cells.

Catabolism Energy-yielding part of metabolism that breaks down larger, more complex molecules into simpler ones.

Catalyst Compound that participates in a chemical reaction to provide an alternative reaction pathway of lower activation energy, thus increasing the rate of the reaction. The catalyst forms an intermediate compound with one reactant and is released following the second reaction.

Cathode ray Particles of negative electricity traveling through a vacuum tube, from cathode to anode. We know cathode rays today as the beam that sweeps across the face of a television tube to produce the picture.

Cell Basic structural unit of life; the smallest part of a living organism that can carry out life processes by itself.

Cell membrane Sheetlike structure that separates the interior of the cell from the external environment.

Cell wall Rigid structure that surrounds bacterial and plant cells. It protects the cell from rupturing in hypotonic solutions.

Cellulose Plant polysaccharide made up of β 1→4 linked glucose; the structural material in plants.

Celsius scale Temperature scale used worldwide in which water freezes at 0°C and boils at 100°C.

Centimeter (cm) Metric unit of length equal to 1/100 m (the decimal prefix *centi-* means times 1/100).

Chain reaction Series of very rapid reactions that occur among fissionable atoms if they are close enough together.

Charles' law Relationship stating that when the temperature of a gas sample is changed, its volume changes by an amount proportional to the change in absolute temperature.

Chemical change Change in the chemical composition of a substance that is usually not easily reversible and involves large changes in energy. Chemical change occurs as the result of a chemical reaction.

Chemical indicator Compound whose color is dependent on the hydronium ion concentration in its solution. Most indicators have just two colors and indicate whether the hydronium ion concentration is above or below a certain $[H_3O^+]$ value.

Chemical periodicity Cyclic or periodic repeating of chemical and physical properties with increasing atomic weight caused by the arrangement of electrons around atoms.

Chemiosmotic theory Theory stating that the energy released by the flow of electrons along the electron-transport chain is used to establish a proton gradient. This high-energy gradient can be used to do work, including the synthesis of ATP.

Chiral center Tetrahedral atom with four different substituents.

Chirality An object is chiral if it cannot be superimposed on its mirror image. This word is derived from the Greek word for "hand" because hands are chiral.

Chitin Structural polysaccharide found in arthropods similar to cellulose but has an amide in place of an —OH on carbon number 2 of the glucose residues.

Cholesterol Most abundant steroid in animals. An important membrane lipid with significant health implications.

Chromosomes Complexes of DNA and proteins found in the nucleus of eucaryotic cells. These structures carry genetic information.

Chyme Partially digested food in the stomach and small intestine.

cis- Prefix used to designate two similar groups on the same side of a molecule.

Codon Three-base sequence in mRNA that determines what amino acid is inserted into the polypeptide chain. Ultimately the sequence of codons in mRNA determines the primary sequence of proteins.

Coenzyme Name given to organic cofactors.

Cofactor Non-amino-acid portion of some enzymes.

Colligative properties Properties of solutions that are affected only by the concentration of the solute, not by its chemical identity.

Colloid Mixture in which the solute is evenly distributed but present in particles of sufficient size to reflect or scatter light.

Combined gas law Combination of Charles' law (compensating for temperature change) and Boyle's law (compensating for pressure change).

Competitive inhibitor Reversible inhibitor that binds to the active site of an enzyme and thus competes with the substrate.

Complementary bases Pairs of bases that hydrogen-bond to each other in nucleic acids: guanine pairs with cytosine and adenine pairs with thymine (or uracil in RNA).

Complex carbohydrates Digestible polysaccharides of the diet; mostly starches.

Compound Pure substance containing elements in definite and constant proportion. For example, sugar (sucrose) is always 42.1 percent carbon, 6.4 percent hydrogen, and 51.5 percent oxygen by weight. Its components can only be separated by chemical means.

Compound formula Formula of an ionic compound, which gives the identity and number of each atom in the smallest repeating unit of the compound.

Concentration Measure of the amount of solute per unit of solution, expressed as mass per unit volume, weight percentage, volume percentage, or moles per liter.

Condensed formula Condensed representation of a chemical structure that leaves out the vertical bonds and shows the whole structure set on one line, such as $CH_3CH_2CH_3$.

Conformation Three-dimensional shape of a molecule emphasizing the relative orientation of atoms in space.

Conjugate acid Base that has accepted a hydrogen ion from an acid and later gives it up to another molecule or ion of greater hydrogen-ion-attracting ability. In so doing, it acts as an acid.

Conjugate base Acid that donates a hydrogen ion to a base and becomes a negative ion that can later accept a hydrogen ion.

Conjugated polyene Polyene for which there are alternating double and single bonds.

Conjugated protein Protein that contains one or more prosthetic groups in addition to amino acids. (The groups may be organic or inorganic.)

Continuous spectrum Continuous band of all colors emitted by a hot solid.

Conversion factor Fraction in which numerator and denominator are in different units, but which are equal to the same quantity. The algebraic value of the conversion factor is always 1.

Coordinate covalent bond Covalent bond in which both electrons are contributed by one of the atoms.

Cori cycle Cycle in which lactic acid produced in muscle is converted back to glucose in the liver.

Coupled reaction Reaction that provides energy for an endothermic reaction by combining it with an exothermic one.

Covalent bond Chemical bond formed as two atoms share one or more pairs of electrons. The bond is due to the common attraction of each nucleus to the same pair(s) of electrons.

Crenation Shrinkage and dehydration of a cell that occurs when blood cells are placed in a solution of more than isotonic concentration and water moves out through the cell membrane.

Cubic centimeter (cm^3 or cc) Unit of volume equal to 1 mL; the preferred unit in medicine (for which the abbreviation "cc" is used).

Curie (Ci) Amount of radioactivity equal to 3.7×10^{10} disintegrations per second.

Cyclic AMP Molecule produced in some cells in response to hormone binding to the cell membrane; cAMP is a second messenger.

Cycloalkane Hydrocarbon compound with single carbon–carbon bonds, in which the skeletal carbons form a ring.

Cytosol Fluid interior of the cell, containing the ions and molecules that support life. Organelles are suspended in it.

D

Dark reaction Part of photosynthesis that uses ATP, NADPH, and carbon dioxide to make glucose.

Daughter nucleus (or **isotope**) Product of radioactive decay. The nucleus of a specific isotope is sometimes called a *nuclide*.

Deamination Reaction that removes the α-amino group from glutamate yielding free ammonia and α-ketoglutarate.

Decarboxylation Loss of CO_2. This reaction takes place very readily for beta keto acids.

Dehydration Formation of an alkene from an alcohol as a result of the loss of water.

Denaturation Process that results in the loss of a protein's native conformation and therefore its activity.

Density Mass per unit volume, commonly expressed in grams per cubic centimeter or grams per milliliter.

Deoxyribonucleic acid (DNA) Very large macromolecule found in the cell nucleus. This molecule is the carrier of genetic information.

Detergent Material that is a surface-active agent having a long hydrocarbon chain and a polar end. If the polar end is a carboxylate anion, the detergent is called a soap. Most other polar groups are often generically classified as detergents.

Dialysis Separation of colloidal materials from smaller ionic and molecular impurities by diffusion of impurity ions or molecules through a membrane that does not let the colloidal particles pass.

Diastereomers These are stereoisomers that are not enantiomers, that is, they are not mirror images of each other.

Diatomic molecules Stable molecules comprising two atoms of the same element. Oxygen (O_2), nitrogen (N_2), and chlorine (Cl_2) are examples of diatomic molecules.

Digestion Process that breaks down food molecules into smaller compounds that are absorbed by the body.

Dimensional analysis Method of converting a measurement from one unit to another by multiplying by an appropriate conversion factor. Conversion factors are always set up so that the desired unit is in the position (numerator or denominator) it should occupy in the answer. The unwanted unit then cancels as the conversion factor and the known quantities are multiplied. Since the conversion factor is equal to 1, the mathematical effect is that of multiplication by 1. The measurement does not change, only its unit.

Dipole An electric dipole is a region of unbalanced electric charge. A dipole is a characteristic of all polar bonds. Dipoles also affect the solubility characteristic of a molecule.

Dipole–dipole force Attractive force between two molecules as a result of the polarity of molecules.

Directing groups Groups already on a benzene ring that direct the position of attachment of electrophiles.

Distillation Process in which the solution is heated until the material of lower boiling point boils, leaves the solution, and is trapped and cendensed.

Disulfide Product of thiol oxidation, having the structure

$$R—S—S—R$$

Disulfide bridge Covalent bond between two sulfur atoms.

E

Electric charge Presence of excess positive or negative particles in or on an object. Objects with opposite electric charge attract each other, whereas objects of the same electric charge repel each other.

Electrical conductivity Ability of a material to transfer electric charge. A material that conducts electricity has either mobile electrons or mobile anions and cations.

Electrochemical series Listing of half-reactions in order of increasing tendency to gain electrons. Used to predict whether a given electron-transfer reaction will proceed. Any half-reaction can remove electrons from any half-reaction below it in the series.

Electrolysis Use of electric current to break molecules into their component elements.

Electrolyte Solution that contains ions and conducts electricity.

Electromagnetic radiation Form of wave motion comprising alternating electric and magnetic fields that travels at an extremely high speed (3×10^8 m/s).

Electron Light, negative particle that can be relatively easily removed from an atom. The mass of an electron is 1/1837 that of a hydrogen atom and its charge is -1.

Electron transfer Transfer of an electron from one atom to another, generally to achieve a stable electron configuration.

Electron volt (eV) Unit describing the energy of an electron.

Electron-dot method Graphic shorthand for writing the outer-group electron complement of an atom. Used principally for showing electron sharing in electron transfer but does not specify the orbitals involved.

Electronegativity Measure of an atom's ability to attract electrons. The type of chemical bond that forms between two atoms can be predicted by computing the difference in their electronegativities.

Electron-filling series System by which orbitals are filled with electrons, beginning with orbitals of lowest energy.

Electron-transport chain Series of electron carriers that transport electrons from the reduced coenzymes NADH and $FADH_2$ to oxygen. Energy released during this process is used to pump protons.

Electrophilic substitution reaction Reaction in which an electrophile (E^+) substitutes for an H on an aromatic ring.

Electrophoresis Technique that separates charged particles in an electric field; especially useful for detection of amino acids, peptides, or proteins.

Element Basic substances that cannot be broken down by chemical means to simpler substances. Each element is unique.

Empirical formula Simplest ratio of atoms in a compound, usually derived from experimental percentage composition data.

Enantiomers Pair of isomers that are mirror images of each other and are not superimposable; a pair of chiral molecules.

Endoplasmic reticulum Membranous organelle within the eucaryotic cell where some of the cellular synthesis and transport of materials occurs.

Endothermic Chemical reaction that requires a continual input of energy to keep going.

Energy Ability to do work. Energy is found in many forms, including heat.

Entropy Tendency for the universe to go from more order to less order. Energy must be used to temporarily prevent this tendency in living organism.

Enzyme Protein that has specific catalytic activity.

Epoxide Three-membered, oxygen-containing ring.

Equilibrium Condition in which the rate of a chemical reaction is the same in the forward and reverse directions. Concentrations of substances on the right and left are seldom equal at equilibrium.

Equilibrium constant Mathematical constant K_{eq} that enables the prediction of the predominant direction of a reaction.

Equilibrium equation Mathematical expression relating concentrations of reactants and products at equilibrium for a given reaction (*see* equilibrium constant).

Ester Functional group derived from an acid and an alcohol.

Ether Family of organic compounds formed when an oxygen atom serves as a bridge between two alkyl groups.

Eucaryote Organism with cells that possess a true nucleus and a number of other membranous organelles that are not present in the procaryotic cell.

Exothermic Chemical reaction that gives off energy.

F

Facilitated diffusion Transport of substances across the membrane of a cell in which proteins speed up the movement. The substances can only flow down their gradient.

Fahrenheit scale Temperature scale used in the United States in which water freezes at 32°F and boils at 212°F.

Fats Triacylglycerols that are solids at room temperature.

Fatty acids Carboxylic acids from biological sources that generally contain ten or more carbon atoms.

Fermentation Oxidation of a compound with transfer of electrons to an organic molecule. Formation of lactic acid from glucose in anaerobic muscle is a fermentation.

Fiber Indigestible carbohydrates of the human diet; mostly polysaccharides, such as cellulose, and complexes of polysaccharides.

Fibrous proteins Fiber-like proteins that are usually insoluble.

Filtration Process that separates large molecules from small ones by exclusion. Small molecules pass through pores, but large molecules and particles are retained.

First ionizing energy Energy required to remove one electron from a neutral atom.

Fluorescence Emission of light when an object is struck by another form of light, such as ultraviolet or "black" light.

Formula Abbreviated notation for a chemical compound that describes its composition in terms of the ratio of one ion to another. For example, the compound $MgCl_2$ has two chloride ions for every magnesium ion.

Formula weight Sum of the atomic weights of the elements constituting the simplest formula of an ionic compound.

Frameshift mutation Mutation resulting from insertion or deletion of a base from a DNA strand. The codon at that point plus all that follow are altered.

Frequency Number of waves that crosses a certain point each second, measured in waves per second, or hertz.

Functional group A particular combination and arrangement of atoms; when attached to a hydrocarbon, unique physical and chemical properties are given to the molecule.

Furanose Five-membered ring form of monosaccharides. The ring resembles the cyclic ether furan.

G

Gamma (γ) radiation Highly penetrating radiation with zero charge and zero mass that often accompany the emission of alpha and beta radiation by radioactive materials.

Gas State of matter in which the material assumes both the volume and shape of its container.

Geiger-Müller counter Gas ionization detector operated at high enough voltage that considerable multiplication occurs as electrons are collected within the tube. G-M counters are very sensitive to beta radiation and to alpha radiation if properly equipped. Also called a "Geiger detector."

Gene That portion of a DNA molecule that codes for a specific transcript. Since most transcripts become mRNA molecules, a gene can be considered the information in a DNA molecule that codes for one polypeptide chain.

Genetic engineering The branch of technology that manipulates genetic information to produce biological products, or organisms with permanently altered abilities.

Geometric structure Geometric form representing a molecule; carbon atoms are assumed to be at each vertex and hydrogens are not shown.

Globular proteins Spheroid, generally soluble proteins.

Glucagon Hormone that stimulates the release of fuel molecules into blood as needed.

Gluconeogenesis Synthesis of glucose from several small common molecules of metabolism.

Glycogen Animal equivalent of starch; the polysaccharide that is the storage form of glucose in animals.

Glycogenesis Synthesis of glycogen within the muscles and liver.

Glycol A dihydroxylic alcohol, that is, one with two —OH groups.

Glycoside Compound formed when a cyclic sugar is bonded to an alcohol through a glycosidic bond.

Glycosidic bond Bond between the anomeric carbon of a cyclic sugar and the —OH group of another sugar or an alcohol. This bond links sugars together in oligosaccharides and polysaccharides.

Glycolysis Cytosolic process that converts glucose to two molecules of pyruvate with the production of two molecules of ATP and two molecules of NADH.

Gram (g) Mass of 1 cm^3, or 1 mL, of water.

H

Half-life Time required for half of the radioactive nuclei present to undergo radioactive decay.

Half-reaction Oxidation or reduction component of a redox reaction, stated as a separate reaction.

Halogenation Addition of halogen to an alkene or alkyne. A halogen is added to both carbon atoms of the double or triple bond, and a pi bond is lost.

Halogenation reaction Reaction of an alkane with a halogen that is catalyzed by heat or light, in which a C—H bond is replaced by a C—X bond.

Heat Form of energy exhibited as molecular motion.

Heat of fusion Amount of heat that must be added to change 1 g of a material from its solid to liquid state or the amount that must be removed to change the material from liquid to solid. This value is 80 cal for water.

Heat of vaporization Amount of heat that must be added to change 1 g of a material from its liquid to its gaseous state. This value is 540 cal for water.

Hemoglobin Protein in red blood cells that binds oxygen cooperatively. The presence of hemoglobin in blood greatly increases the capacity of blood for oxygen transport.

Hemolysis Swelling and rupturing of a cell that occurs when blood cells are placed in a solution of less than isotonic concentration, and water moves in through the cell membrane.

Heparin Polysaccharide that functions as an anticoagulant.

Heterocycle Cyclic system having some atom other than carbon as a structural component of the ring. Some heterocycles are amines.

Heterotrophic organism Organism that must consume organic matter to obtain energy and carbon atoms.

Homeostasis Maintenance of a constant internal environment.

Homologous series Family of compounds differing only by the number of CH_2 groups in the formula. The alkane series is represented by the general formula C_nH_{2n+2}.

Hormone Messenger molecule produced in one part of the body that is transported throughout the body in blood and that binds to target cells, triggering events that alter the metabolism of the cell.

Hund's rule Electrons do not pair (occupy the same orbital) until all orbitals of that energy have at least one electron. Electrons are negative and stay as far apart as possible as long as possible.

Hyaluronic acid Complex polysaccharide found in connective tissue and in bone.

Hybridization Rearrangement of bonding and nonbonding electrons around an atom that occurs during bonding.

Hydrated ion Ion surrounded by water molecules.

Hydration Addition of water to an alkene or alkyne. An H is added to one of the carbon atoms of the double or triple bond, and an OH is added to the other carbon atom. A pi bond is lost.

Hydrocarbon Compound consisting only of the elements hydrogen and carbon. Hydrocarbons are almost always non-polar.

Hydrogenation Addition of hydrogen to an alkene or alkyne. An H is added to both carbon atoms of the double or triple bond, and a pi bond is lost.

Hydrogen bond Weak ionic bond between the negative region of a polar molecule and a positive hydrogen on an adjacent molecule.

Hydronium ion The ion H_3O^+ that exists in water solutions because the hydrogen ion (H^+) is not stable by itself.

Hydroquinone Common name for *p*-hydroxyphenol, an effective antioxidant.

Hypertonic As applied to blood, a solution that has an ionic concentration greater than that of blood.

Hypothermia Condition of lowered body temperature that causes a decrease in the chemical reactions that support body functions and life. The chemical reactions in the human body operate properly only in a narrow range of temperatures centered on 98.6°F (37°C).

Hypotonic As applied to blood, a solution that has an ionic concentration less than that of blood.

I

-ic Suffix indicating the higher of two possible oxidation states. For example, in ferr*ic* nitrate the iron is +3.

Ideal gas Theoretical gas composed of infinitely small molecules that behaves exactly as predicted by kinetic molecular theory.

Ideal gas law Equation relating gas volume to number of moles of gas present, temperature, and pressure:

$$PV = nRT$$

Imine Compound formed from the reaction of a carbonyl-containing compound and a primary amine.

Induced-fit model Model for substrate-enzyme interaction that states that the binding of substrate to enzyme causes a change in the shapes of one or both of the molecules. When bound to each other, the two are complementary.

Induction Regulation of gene expression involving the turning on of genes by the presence of a compound.

Inert gases Family of gases that are stable as individual atoms, all having a full complement of outer group *s* and *p* electrons. They are called the *noble gases.*

Insulin Hormone that signals that nutrients are abundant and stimulates anabolic activity in target cells.

Interferons Proteins produced by virally infected cells. They bind to other cells and stimulate an antiviral state in them.

Iodine number Index used to indicate the degree of unsaturation present in a fat or oil.

Ion Charged atom; an atom that has lost or gained electrons and no longer has the same number of electrons as the number of protons in its nucleus. Positive ions are called cations, and negative ions are called anions.

Ionic bond Attractive force between two oppositely charged ions.

Ionic crystals Group of positive and negative ions stacked in a regular manner.

Irreversible inhibitor Molecule that binds tightly to an enzyme and reduces or eliminates the activity of the enzyme.

Isoelectric point (pI) pH at which an amphoteric molecule, such as an amino acid or protein, has no net charge.

Isomer Variations of a particular compound having the same molecular formula but different arrangements of atoms and bonds.

Isotonic As applied to blood, a solution that has an ionic concentration equal to that of blood.

Isotope Term used to designate an atom of a specific mass number. Atoms of a given element all have the same number of protons in their nuclei (atomic number) but may have different numbers of neutrons.

Isozyme Structurally similar enzymes with similar catalytic activity and different metabolic roles.

J

Joule (J) SI unit for measurement of heat energy, equal to 0.239 cal.

K

Kelvin scale Temperature scale based on measurement of molecular motion. At zero Kelvin (absolute zero), all molecular motion theoretically ceases. Kelvin degrees are the same size as Celsius degrees, and the conversion is

$$K = °C + 273$$

Keto acids Organic acids that also contain a carbonyl functional group.

Ketone Family of organic compounds formed when an alkyl group is connected to each of the two remaining carbon bonds of the carbonyl group. The carbonyl group is not on a terminal carbon atom.

Ketone bodies Fuel molecules in the blood that are made from acetyl-CoA via HMG-CoA.

Ketose Monosaccharide containing a ketone group. Note the prefix *keto-* and the suffice *-ose*.

Kilocalorie (kcal) Unit of heat measurement equal to 1000 cal. In nutrition, the large Calorie is used instead of kilocalories.

Kilogram (kg) Basic SI unit of mass; 1000 mL of water has a mass of 1000 kg.

Kilometer (km) Metric unit of length equal to 1000 m (decimal prefix *kilo-* means times 1000).

Kinetic molecular theory Idea that molecules in gases are in constant motion.

L

Law of conservation of matter Matter (atoms) cannot be created nor destroyed by ordinary means.

Law of definite proportions Compound always contains the same elements combined in the proportions by weight.

LD_{50} Radiation dose that is lethal to 50 percent of the population being tested. The term LD_{50} is also used for toxic chemicals and drugs.

Le Chatelier's principle Method for easily predicting the effect of changes in reaction conditions on the equilibrium concentration of reactants and products.

Leukotrienes Lipids derived from arachidonic acid that are implicated in allergic responses.

Lewis acid Molecule or ion that can form a coordinate covalent bond with another species by accepting a pair of electrons.

Lewis base Molecule or ion that can form a coordinate convalent bond with another species by donating a pair of electrons.

Lewis theory Theory of electron arrangement in a molecule to accommodate eight electrons around most atoms (hydrogen has two electrons).

Light reaction Part of photosynthesis that uses light energy to make ATP and NADPH.

Line formula Simplified representation of a structural formula in which many of the C—H bonds are not shown. An example is

CH_3 /\ CH_3

Line spectrum Set of distinct spectral lines emitted by a hot gas sample.

Lipases Enzymes that catalyze the hydrolysis of ester bonds in fats, oils, and similar lipids.

Lipid Class of biomolecules characterized by insolubility in water and solubility in organic solvents.

Lipogenesis Biosynthesis of fatty acids.

Lipoproteins Macromolecules composed of protein and lipid that transport lipids in blood.

Liposome Structure consisting of an aqueous core separated from the external environment by a bilayer of amphipathic molecules.

Liquid State of matter in which the material has a definite volume but assumes the shape of its container.

Liter (L) SI volume unit equal to 1000 cm^3.

Lock-and-key model Model for substrate-enzyme interaction that states that the two molecules are complementary to each other before binding and fit together like a lock and key.

Lucas test Test for identifying whether an alcohol is primary, secondary, or tertiary by its rate of conversion to a chloroalkane.

Lymph Body fluid similar to blood but lacking red blood cells.

M

Markovnikov's rule Generalization used to account for the way an unsymmetrical reagent adds to an unsymmetrical alkene. The positively charged reagent (often H^+) adds to the carbon atom directly attached to the greater number of hydrogens.

Mass Measure of the *amount* of material, independent of gravitational field strength.

Matter Anything that has mass and occupies space.

Melting point Temperature at which a substance in the solid state is converted to the liquid state. The *freezing point* is the same temperature, with the change in the opposite direction.

Membranes Sheetlike structures in cells that separate the cell interior from the external environment and divide the cell interior into compartments.

Meso compound Compound with chiral centers that is nonchiral because of molecular symmetry.

Messenger RNA (mRNA) Ribonucleic acid that carries genetic information from the nucleus to the rest of the cell. The information is used to direct protein synthesis.

***Meta-* (*m-*)**—Prefix used to designate substituent position on a benzene ring. The *meta* position is second from the parent substituent.

Metabolic pathway Series of chemical reactions in the body that converts one substance to another distinctly different substance.

Metabolism Sum of all chemical reactions in the body that collectively sustain life.

Metallic bonding Type of chemical bond in which atoms share electrons by each contributing to a sea of mobile electrons that surround all of the atoms. This property allows metals to conduct electricity when in the solid state.

Metalloids Elements that share the properties of metals and nonmetals.

Metals Group of elements that are good electric conductors and are malleable.

Meter (m) Basic unit of length in the metric and SI measurement systems, originally defined as one ten-millionth of the distance from the earth's equator to the North Pole. It is now defined as the distance traveled by light in a vacuum during an interval of 1/299,792,458 s.

Metric system French measurement system from which evolved the SI. The metric system was the first measurement system to use the earth as a standard.

Micelle Aggregation of hydrocarbon materials having polar ends in the water phase.

Milliliter (mL) Common unit of volume used in science and medicine because of its convenient small size equal to 1/1000 L.

Millimeter (mm) Metric unit of length equal to 1/1000 m (the decimal prefix *milli-* means times 1/1000).

Mitochondrion Organelle of eucaryotes where oxygen-requiring energy production occurs.

Mixture Material that has no set proportion of its component substances and can be separated into its components relatively easily.

Moderator Material that slows neutrons. Slow neutrons are more readily captured by fissionable nuclei than are fast neutrons.

Molarity (M) Concentration of a solution in moles of solute per liter of solution. The preferred unit of concentration in chemistry.

Mole (mol) 6.023×10^{23} atoms or molecules. One mole of atoms or molecules has the same mass in grams as the atomic mass of an individual atom or molecule in atomic mass units.

Molecular collision theory Idea that for a chemical reaction to occur, molecules must (a) collide and (b) collide with sufficient energy to break chemical bonds. Changes in reaction rate occur if a change is made in any factor that determines collision rate or collision energy.

Molecular dipole Unbalanced distribution of electric charge across a molecule, giving one side a more negative charge and the other side a more positive charge.

Molecular formula Formula of a covalent compound, which gives the identity and number of each atom in the molecule.

Molecular weight Sum of the atomic weights of the elements constituting one molecule of a covalent compound.

Molecule Group of atoms bonded together in fixed proportions to constitute a compound. For example, two hydrogen atoms and one oxygen atom make a water molecule; two oxygen atoms make an oxygen molecule.

Mole method Use of the mole concept to compute masses of reactants and products involved in chemical reactions.

Monomer Smallest repeating unit from which polymers are made.

Monosaccharides Simple sugars, such as glucose; the smallest compounds that are carbohydrates. These are the basic units from which larger carbohydrates are made.

Mutagens Agents that cause mutations.

Mutarotation Change in specific rotation seen in sugar solutions that results from equilibrium between anomeric forms.

Mutation Damage to the genetic mechanism of a cell, causing it to reproduce in a different form. A permanent change in the base sequence of DNA that changes genetic information.

N

Native conformation Normal shape or conformation that a protein has in its biological setting performing its normal biological activity.

Natural radioactivity Spontaneous emission of nuclear particles and penetrating radiation by naturally occurring unstable atoms. The radioactive atoms are changed to new elements during the process.

Negative effector Small molecule whose binding to an allosteric enzyme decreases the activity of the enzyme.

Net ionic equation Chemical equation describing an ionic reaction that is written to identify those ions that *leave* the reaction via precipitation, formation of a gas, or formation of nonreactive covalent substance.

Neutron Nuclear particle with a mass of 1 amu and an electric charge of zero.

Nicotinamide adenine dinucleotide (NAD^+/NADH) Nucleotide that serves as an acceptor (NAD^+) and donor (NADH) of electrons in cellular reactions.

Nicotinamide adenine dinucleotide phosphate (NADPH) Coenzyme involved in reduction reactions in anabolism.

Nitrogenous base Basic, nitrogen-containing heterocyclic compound, the most common ones being adenine, guanine, cytosine, thymine, and uracil.

Noncompetitive inhibitor Reversible inhibitor that binds to an enzyme at a location other than the active site and thus does not compete with the substrate.

Nonmetals All elements that do not conduct electricity.

Nonpolar molecule Symmetrical molecule with no unbalanced electric charge.

Nonsaponifiable lipid Lipid that does not contain fatty acids and therefore cannot be saponified by base.

Normality For an acid, the concentration of a solution in moles per liter of reactable hydrogen ion. For example, a 1.0 M solution of HCl is 1.0 N, and a 1.0 M solution of the strong H_2SO_4 is 2.0 N.

Nuclear charge Number of protons in the nucleus, expressed as the atomic number of the element.

Nuclear fission Splitting of a heavy nucleus to produce two lighter "fission fragment" nuclei, a number of free neutrons, and a large amount of energy. Fission is triggered when an unstable nucleus absorbs a neutron.

Nuclear fusion Fusing of two light nuclei together to form a heavier nucleus. When very light nuclei are fused together, the product nucleus has less energy than the two "reactant" nuclei; considerable energy is released in the fusion reaction.

Nuclear waste Fission fragment nuclei are highly radioactive and extremely dangerous. They are referred to as "nuclear waste," or in the case of a weapons explosion, as "fallout."

Nucleoside Compound consisting of a nitrogenous base and a sugar.

Nucleotide Compound consisting of a nitrogenous base, a sugar, and one or more phosphate groups.

Nucleus Eucaryotic organelle that contains most of the DNA of the cell and directs cell division, growth, and maintenance of the cell.

Nutrients Substances required by the body that are provided by food.

Nutrition Branch of science that deals with the composition, preparation, consumption, and processing of food.

O

Octet rule Chemical stability is implied if an atom has a set of eight outer electrons (s^2p^6 electrons). Note that hydrogen does not obey this rule.)

Oils Triacylglycerols that are liquids at room temperature.

Oligosaccharides Carbohydrates that contain two to ten monosaccharides, such as table sugar and milk sugar (disaccharides).

Operon Set of genes in bacteria that work in concert. Genes within the operon are either turned on or off collectively.

Orbital Region in the space of an atom occupied by as many as two electrons.

Orbital notation Shorthand for indicating the orbital "addresses" of electrons. An atom with two $1s$ electrons, two $2s$ electrons, and a $2p$ electron could have its orbital population expressed as $1s^22s^22p^1$.

Organelle Structure within a cell that carries out one or more functions of the cell.

Organohalogen Organic compound in which a halogen (fluorine, chlorine, bromine, or iodine) has replaced one or more hydrocarbon hydrogens.

***Ortho-* (*o*-)** Prefix used to designate substituent position on a benzene ring. The *ortho* position is immediately adjacent to the parent substituent.

Osmosis Process in which two solutions separated by a semipermeable membrane attempt to equalize their concentrations by passing small molecules through the membrane. Biological systems use osmosis to pass water and some nutrients between cells.

Osmotic pressure Measure of the pressure driving the transfer of solvent molecules across a membrane from a solution of low solute concentration to a solution of high solute concentration. Osmotic pressure is the pressure that must be applied to the solution on the more concentrated side of the membrane to equalize the flow of solvent molecules in both directions.

-ous Suffix indicating the lower of two possible oxidation states. For example, in ferr*ous* sulfate the iron is +2.

Oxidation Loss of electrons in a chemical reaction. When iron rusts it changes its oxidation state from 0 to +3 by giving up three electrons in a two-step process, thus it is *oxidized.*

Oxidation number Charge predicted for an atom by the electron-transfer bonding theory, even if the molecule is covalent; or the number of electrons shared by the atom in a covalent or polar covalent molecule.

Oxidative phosphorylation Process that synthesizes ATP from ADP and inorganic phosphate using energy stored in a proton gradient.

P

***Para-* (*p*-)** Prefix used to designate substituent position on a benzene ring. The *para* position is across from the parent substituent.

Parent nucleus (or isotope) Unstable atom prior to radioactive decay.

Partial pressure Pressure, in torr or in millimeters of mercury (mm Hg), exerted by each component of a mixture of gases. The total of the partial pressures of all gases equals the total pressure of the gas.

Pentose phosphate pathway Series of reactions that effectively converts glucose to pentoses and NADPH.

Peptidases Enzymes that catalyze the hydrolysis of peptide bonds in peptides. The distinction between a protease and a peptidase is not clear-cut.

Peptide Compound consisting of amino acids linked by peptide bonds. Often the number of amino acids is indicated by prefixes, such as *di-*, *tri-*, or *oligo-*.

Peptide bond Amide bond between the amino group of an amino acid and the carboxylic acid group of another amino acid.

Percentage composition Percentage by weight of each element in a compound, computed from formula or molecular weights. It is equal to the number of atomic mass units of the element in question present in the formula, divided by the total formula or molecular weight.

Periodic table of the elements Table in which elements are arranged generally in order of increasing atomic number. The elements of each row constitute one period or cycle of chemical and physical properties, and the columns comprise elemental families of similar chemical behavior.

Perspective formula Representation of a chemical structure that conveys the three dimensions of a ball-and-stick model and has some of the simplicity of the structural formula.

pH Measure of hydronium ion concentration that equals $-\log[H_3O^+]$.

Phenol Any aromatic ring with an —OH group is classified as a phenol. Phenol is also the name for the simplest member of this class.

Phenyl Benzene ring when used as an alkyl group attached to a larger molecule.

Pheromone Naturally occurring compound produced by an organism for the purpose of chemical communication.

Phosphoacylglycerols Amphipathic molecules similar to triacylglycerols in structure, but with a substituted phosphoric acid in place of the fatty acid on the third hydroxyl group of glycerol.

Photon Packet of light waves, sometimes called a "quantum."

Photosynthesis Process by which plants and algae make organic molecules from carbon dioxide using light as the energy source.

Physical change Change that usually requires only moderate amounts of energy, is reversible, and results in no change in the chemical identity of the substance.

Pi bond (π-bond) Bond formed by overlap of unhybridized *p* orbitals of two adjacent atoms. No more than two pi bonds can exist between two adjacent atoms.

pK_a Logarithm of the acid ionization constant. The pK is the solution pH at which the acid transfers a hydrogen ion.

Planck's law Mathematical statement relating energy and frequency of electromagnetic radiation: the greater the frequency (and the bluer the color of light), the greater the energy: $E = h\nu$.

Plasmid Small pieces of DNA found in some bacteria; a common vehicle for introducing new genes into bacteria.

Plum pudding model Model of the atom representing it as a positive filling in which negative electrons ("plums") are floating. The design of the atomic model was similar to that of plum pudding popular in England, hence the name.

pOH Measure of hydroxide ion concentration that equals $-\log[OH^-]$.

Polar covalent bond Shared or covalent bond in which electrons are shared unevenly. For purposes of definition, bonds are considered to have polar properties when the electronegativity difference is 0.5 or greater.

Polar molecule Molecule with a molecular dipole.

Polyatomic ion Charged, covalently bound group of atoms that together act as an ion. Examples include the hydroxide group (OH^-) and the nitrate group (NO_3^-).

Polymer Complex compound resulting from the end-to-end union of a large number of smaller units (monomers).

Polypeptide Macromolecule containing many (ten or more) amino acids. This term is sometimes used to mean the chain in a protein.

Polysaccharides Carbohydrates that contain many monosaccharide units, such as starch and cellulose.

Polyunsaturated Molecules having more than one double or triple bond.

Positive beta decay (positron emission) Radioactive decay involving emission of a positron. The daughter product of positron decay has one fewer proton and one more neutron than the parent isotope.

Positive effector Small molecule whose binding to an allosteric enzyme causes an increase in enzymic activity.

Precision How close a set of measured values are to each other.

Primary carbon Carbon atom that is bonded to only one other carbon.

Primary structure Amino acid sequence in a protein or peptide.

Primary transcript First product of transcription. In higher cells it is modified to yield functional RNA molecules.

Principal quantum number Quantum number used to designate an electron's major electron group; its distance from the nucleus.

Procaryote Simplest and smallest cell type characterized by the absence of a true nucleus and other membranous organelles. Bacteria and blue-green algae are procaryotic organisms.

Prostaglandins Regulatory lipids derived from unsaturated fatty acids. Prostaglandins often have a cyclic portion based on cyclopentane.

Prosthetic group Group in a protein that is not an amino acid and that plays a major role in the activity of conjugated proteins.

Proteases Enzymes that catalyze the hydrolysis of peptide bonds in proteins and polypeptides.

Proteins Polymers of amino acids that have important roles in living systems.

Proton Nuclear particle with a mass of 1 amu and an electric charge of +1.

Pyranose Six-membered ring form of monosaccharides. The ring resembles the cyclic ether pyran.

Q

Quantum numbers Set of numbers used to designate an electron's major electron group, its orbital shape, and its orbital orientation.

Quaternary carbon Carbon atom that is bonded to four other carbons.

Quaternary structure Structure of an oligometric protein resulting from specific interactions among the subunits.

Quinone Common name for 2,5-cyclohexadiene-1,4-dione, the oxidation product of hydroquinone.

R

Racemic mixture Mixture that contains equal amounts of a pair of enantiomers and thus does not rotate plane-polarized light.

Radiation absorbed dose (Rad) Unit roughly equivalent to the roentgen, which describes the amount of energy deposited in tissue by x-rays or gamma radiation. A whole-body dose of 450 rad is fatal to 50 percent of the population.

Radioactive decay Process by which an unstable nucleus emits alpha or beta and gamma radiation.

Radiocarbon dating Determination of the age of an object by measuring the amount of radioactive carbon present relative to the amount in a similar living sample. Nuclear dating can also be based on other isotopes, such as thorium or uranium and lead.

Rancid Term used to indicate that a fat or oil is foul-smelling and bad-tasting due to the presence of volatile acids and aldehydes.

Reaction rate (velocity) Rate at which a reaction occurs, usually expressed in terms of substrate consumed per unit time or product formed per unit time.

Redox reaction Reaction that involves both oxidation and reduction.

Reducing sugars Sugars that have a free carbonyl group that can be oxidized. They are thus reducing agents.

Reduction Gain of electrons in a chemical reaction. When heated with carbon (coke), the iron Fe^{3+} in the iron ore hematite (Fe_2O_3) gains electrons and is *reduced* to metallic iron, Fe^0.

Relative atomic weight Average mass of a sample of atoms of an element, as compared with an accepted standard. The original standard was the lightest atom, hydrogen, assigned an arbitrary atomic mass of 1.000. The reference was changed recently to the most common type of carbon atom, which was assigned a mass of exactly 12 amu. (The term *atomic mass* is used interchangeably and is actually more accurate.)

Renal threshold Maximum concentration of a substance that can be effectively reabsorbed by the kidneys. If the concentration exceeds this threshold, the substance will appear in the urine.

Repression Regulation of gene expression involving the turning off of genes by the presence of a compound.

Residue Position within an oligomer or polymer; the residues of an oligosaccharide are occupied by monosaccharides.

Resonance energy Energy due to delocalization of electrons.

Respiration Oxidation of a compound with transfer of electrons to an inorganic substance. Respiration using oxygen is the principal source of energy in the body.

Reversible inhibitor Molecule that binds temporarily to an enzyme and, while bound, reduces or eliminates the activity of the enzyme.

Ribonucleic acid (RNA) Several kinds of acidic macromolecules found in the nucleus and other parts of the cell.

Ribosomal RNA (rRNA) Ribonucleic acid molecules found in ribosomes, the site of protein synthesis.

Ribosome Small organelle that is the site of protein synthesis.

Roentgen (R) Unit used to describe the amount of energy deposited in 1 g of air by x-rays or gamma radiation.

Roentgen equivalent man (Rem) Unit that equates radiation damage caused by alpha, beta, and gamma radiation. We are exposed to about 10 mrem/month from natural sources of radiation.

Rounding Process by which the appropriate number of significant figures are maintained after an arithmetic operation.

S

Salt Metal-containing product of an acid–base reaction.

Saponifiable lipid Lipid that is hydrolyzed by base to one or more fatty acids and one or more other molecules.

Saponification Basic hydrolysis of an ester to form the salt of the acid and the alcohol.

Saturated fatty acid Fatty acid that has no C to C double bonds.

Saturated hydrocarbon Hydrocarbon containing only single carbon-to-carbon bonds.

Scientific notation Process of writing a measurement or computed result as a number between 1 and 10 times a power of 10. It is particularly useful in rounding off and presenting results to the appropriate number of significant figures.

Scintillation Flash of light given off when an atom is struck by a nuclear particle or gamma ray.

Scintillation detector Detector used principally for detecting gamma rays. Photons striking the detector knock electrons from atoms; as these electrons fall back to lower energy positions in their atoms, light is given off. This light is observed by a photomultiplier tube, and the passage of the particle is recorded.

Secondary carbon Carbon atom that is bonded to two other carbons.

Secondary structure Helical or pleated structure of a protein that is due to hydrogen bonding between atoms of the polypeptide backbone.

Semiconservative replication Process that produces two DNA molecules from one. The process is semiconservative because each daughter DNA molecule receives one of the strands from the parent molecule.

Sigma (σ) bond Two electrons shared in overlapped orbitals on the axis *between* two covalently bonded atoms.

Significant figures Digits in a measured or computed number that are meaningful. Arithmetic operations sometimes artificially create additional digits. The result of multiplication or division operations may never have more significant figures than the least certain number used in the operation. The result of an addition or subtraction operation may never have more numbers after the decimal place than the least certain number used in the operation.

Simple protein Protein containing only amino acids.

Soap Detergent consisting of the salt of a long-chain fatty acid.

Solid State of matter in which the material has a definite shape and volume.

Solute Material that is dissolved in the solvent.

Solution Mixture of two or more substances (solvent and solute) that is clear and uniform throughout. Molecules or ions of solute are dispersed evenly throughout the mixture.

Solvent Pure substance in which another material (the solute) is dissolved.

Specific gravity Ratio of the mass of a sample of material to the mass of the same volume of water.

Specific heat Number of calories required to change the temperature of 1 g of a material 1°C.

Spectral line Single color of light emitted by a hot gas atom.

Spectrum Colors of light given off by an object.

Sphingolipids Group of polar membrane lipids characterized by the presence of sphingosine.

Stable electron configuration Complete complement of outer group *s* and *p* electrons. Often referred to as a noble-gas electron complement.

Standard temperature and pressure (STP) Conditions of 0°C (273 K) and 760 torr standardized worldwide.

Starch Readily digestible plant polysaccharide made of glucose residues bonded by α 1→4 and some α 1→6 glycosidic bonds.

Stem name First part of the element name that provides enough information to identify the element in a compound name. The stem name of sulfur, for example, is *sulf-*. Thus, the binary compound of hydrogen and sulfur has the name hydrogen *sulf*ide.

Stereoisomers Isomers that differ only in the arrangement of atoms in space.

Steroids Compounds possessing the steroid nucleus.

Stock method Oxidation states are indicated by Roman numerals immediately following the name of the element. For example, Fe^{3+} is iron (III).

Strong acid Molecule or ion with a greater tendency to give up hydrogen ions than H_3O^+.

Structural formula Representation of a structure that emphasizes the bond connection between atoms.

Subscript number Number placed below and after the symbol for an element in a chemical formula indicating the number of that element in the balanced formula for the compound. For example, the subscript "2" in H_2O indicates two hydrogens. If a subscript number is not given, it is presumed to be 1.

Substitution Mutation in which a single base has substituted for another in a DNA strand; a single codon is affected.

Substrate Reacting molecule that binds to an enzyme. The enzyme catalyzes its conversion to product.

Suspension Mixtures in which the solute particles are large enough that gravitational force eventually causes them to separate.

Système International d'Unités (SI) Internationally accepted measurement system based on physical rather than biological standards. The system is unique in its use of decimal prefixes to modify its basic units.

T

Temperature Measure of the average energy of motion of individual molecules or atoms.

Terpene Naturally occurring compound that contains isoprene units.

Tertiary carbon Carbon atom that is bonded to three other carbons.

Tertiary structure Compact, three-dimensional shape of globular proteins.

Tetrahedron Symmetrical geometric shape with four vertices and equal bond angles.

Thioester Fuctional group derived from an acid and a thiol.

$$R{-}\underset{\underset{O}{\|}}{C}{-}S{-}R'$$

Thiol Sulfur analog of an alcohol in which oxygen is replaced by sulfur.

Titration Process in which a measured volume of a known solution is reacted with a solution of unknown concentration to determine its concentration. The point at which the unknown is completely consumed is usually identified by a chemical indicator that changes color at the "end point."

Torr Unit of pressure equal to the pressure exerted by a standard column of mercury 1.00 mm in height.

Tracer Radioactive isotope used to follow the passage of the element of molecule in question through a biological or physical system.

Trans- Prefix used to designate two similar groups on opposite sides of a molecule.

Transamination Reaction that transfers amino groups from amino acids to α-ketoglutarate to yield keto acids and glutamate.

Transcription Process that synthesizes RNA molecules using a DNA molecule as a template.

Transfer RNA (tRNA) Ribonucleic acid molecules that transfer the correct amino acids into the protein that is being synthesized by ribosomes.

Translation Synonym for protein synthesis. Information stored in mRNA is used to direct the synthesis of protein.

Transmutation Process of conversion of one element to another by radioactive decay or nuclear bombardment.

Triacylglycerols Triesters of glycerol and three fatty acids.

Tricarboxylic acid (TCA) cycle Cyclic pathway that oxidizes acetyl-CoA to CO_2, yielding reduced and phosphorylated coenzymes as the other products.

U

Universal gas constant (*R*) Constant in the ideal gas law that relates pressure, temperature, and number of moles of gas present.

Unsaturated One or more double or triple bonds exist between carbon atoms in the molecule.

Unsaturated fatty acids Fatty acid that has one or more C—C double bonds.

Unsaturated hydrocarbon Hydrocarbon containing either double or triple carbon-to-carbon bonds.

Urea cycle Pathway that converts excess metabolic nitrogen to relatively nontoxic urea.

V

Valence-shell electron-pair repulsion (VSEPR) theory Idea that pairs of outer-group electrons move as far apart as possible. It is used to predict the shape of the bonds around an atom.

W

Water ionization constant (K_w) Product of hydronium ion and hydroxide ion concentration in water solutions, which is 1×10^{-14}.

Wavelength Length of one wave; mathematically equal to the wave velocity divided by the frequency.

Wave velocity Speed at which a wave moves through a material; it depends on the material and the type of wave, not the frequency of the wave.

Weak acid Molecule or ion with a lesser tendency to give up hydrogen ions than H_3O^+.

Weight Force a mass exerts downward on the earth, which varies with the gravitational field strength.

X

X-ray Highly penetrating ray produced as electrons slow up when they strike a material. Medical x-rays are produced today in a manner similar to that used by Roentgen.

Z

Zwitterion Dipolar ionic form of an amino acid that is formed by donation of an H^+ from the carboxyl group to the α-amino group. Because both charges are present, the net charge is neutral.

Zymogen Inactive form of an enzyme that is activated by cleavage of one or more peptide bonds.

PHOTO CREDITS

FRONTMATTER

Title Page: Richard Megna/Fundamental Photographs, New York, NY

unnumb. fig. p. xi: Richard Megna/Fundamental Photographs, New York, NY

unnumb. fig. p. xii: Mr. Johnny Johnson, Bruce Coleman, Limited

CHAPTER R

Chapter Opener: Bernard Giani/Agence Vandystadt/Photo Researchers

Figure R.1: W. L. Gore & Associates, Inc.

Figure R.2: Michael Keller, 1989 FPG International

Figure R.48: Dr. Frieder Sauer, Bruce Coleman, Limited

CHAPTER 1

Figure 1.A: Apatosaurus (formerly Brontosaurus) C. R. Knight Painting © American Museum of Natural History

Figure 1.B: American Petroleum Industry

Chapter Opener: NASA

CHAPTER 2

Chapter Opener: Atlanta Gas Light Company

unnumb. fig., p. 20: Bettman Archives

Figure 2.A: Source: Goddard Institute for Space Studies

Figure 2.B: Source: *Chemical and Engineering News*, April 1, 1991, p. 9.

Figure 2.C and Figure 2.E: *Chemical and Engineering News*, March 13, 1989, pp. 30–31.

Figure 2.17 (left): Animals, Animals © Mickey Gibson

Figure 2.17 (right): Animals, Animals © Kathie Atkinson

Figure 2.G: NASA

CHAPTER 3

Chapter Opener: Astrid and Hanns-Frieder Michler/Science Photo Library/Photo Researchers

unnumb. fig., p. 82: R. E. Litchfield/Science Photo Library/Photo Researchers

Figure 3.B: Giraudon/Art Resource, NY

Figure 3.14 (b): Gilbert Grant/Photo Reseachers

CHAPTER 4

Chapter Opener: © IBM, Peter Arnold, Inc.

Figure 4.C (left): Gemological Institute of America

CHAPTER 5

Figure 5.5: Boston Medical Library in The Francis A. Countway Library of Medicine

CHAPTER 6

Figure 6.4: E. R. Degginger

Figure 6.5: National Park Service

Figure 6.9: Thomas Eisner and Daniel Aneshansley

CHAPTER 7

Chapter Opener: Richard Megna/Fundamental Photographs, New York, NY

CHAPTER 8

Chapter Opener: Dr. Harold Rose/Science Photo Library/Photo Researchers

CHAPTER 9

Chapter Opener: Hank Morgan/Science Source/Photo Researchers

Figure 9.1 (b): Al Lamme, Phototake, New York City

Figure 9.1 (d): Astrid and Hanns-Frieder Michler/Science Photo Library/Photo Researchers

Figure 9.2 (b): Antoinette Ryter/Pasteur Institute

Figure 9.3: David M. Phillips/Visuals Unlimited

Figure 9.4 (b): Photo Researchers

Figure 9.5: Photo Researchers

Figure 9.6: Richard J. Feldman and Neil Patterson Publishers

Figure 9.12: Co Rentemeester/FPG International

CHAPTER 10

Chapter Opener: CDC/Science Source/Photo Researchers

Figure 10.3 (b): © 1992 Michael Davidson/Custom Medical Stock

Figure 10.7: © Irving Geis

Figure 10.9 (b) and (c): From Buckwalter, J. A., and Rosenberg, L. *Collagen and Related Research*, **3**, 489–504 (1983).

CHAPTER 11

Chapter Opener: David Parker/Science Photo Library/Photo Researchers

Figure 11.3: Animals, Animals, © David S. Fritts

CHAPTER 12

Chapter Opener: D. Cavagnaro

Figure 12.5: Division of Computer Research and Technology, National Institutes of Health/Science Photo Library/Photo Researchers

Figure 12.8 (b): © Irving Geis

Figure 12.B: Photo Researchers

Figure 12.C: Dagmar Fabricius/Stock Boston

CHAPTER 13

Chapter Opener: David M. Phillips/Visuals Unlimited

Figure 13.1 (top): Walter Chandoha

Figure 13.1 (bottom): Elyse Lewin/Image Bank

Figure 13.7 (b): © Irving Geis

Figure 13.7 (c): Robert T. Morrison and Robert N. Boyd, ORGANIC CHEMISTRY, fifth edition. Copyright © 1987 by Allyn and Bacon, Inc. Reprinted with permission.

Figure 13.8 (a): Dr. David Clayton

Figure 13.8 (b): Dr. Jack Griffith and *Scientific American*

Figure 13.21: Courtesy of Dr. Claude Martin and Dr. Françoise Dieterlen-Lievre

CHAPTER 14

Chapter Opener: Dr. F. Sussman/Peter Arnold Inc.

Figure 14.1 (a)(b): Zoltan Kanyo, Department of Biology, University of Pennsylvania

Figure 14.B: Andy Levin/Photo Researchers

unnumb. fig. p. 403: Carolina Biological Supply Company

Figure 14.C: Alex Bartel/Science Photo Library/Photo Researchers

Figure 14.E (left): Science Photo Library/Photo Researchers

Figure 14.E (right): Biophoto Associates/Photo Researchers

CHAPTER 15

Chapter Opener: Don Fawcett/Science Source/Photo Researchers

Figure 15.C: © 1992 IMS Creative/Custom Medical Stock

CHAPTER 16

Chapter Opener: Dwight R. Kuhn

Figure 16.A: © 1991 Custom Medical Stock

Figure 16.1 (a): L. K. Shumway/Photo Researchers

CHAPTER 17

Chapter Opener: Bill Longcore/Science Source/Photo Researchers

Figure 17.A: SIU School of Medicine/Bruce Coleman, Inc., New York

Figure 17.B (a): Bill Longcore/Science Source/Photo Researchers

Figure 17.B (b): Montagnier/Institut Pasteur/SPL/Science Source/Photo Researchers

INDEX

Numerals in *italics* indicate a figure, "t" following a page number indicates a table concerning the subject.

A

ABO. *See under* blood group
Acetal(s), 179
 formation of, 167–168, 179
 natural examples of, 170
Acetaldehyde, 10, 159, 162t
Acetaminophen, 191
Acetic acid, 2, *2*, *11*
Acetoacetate, 471
Acetone, 162t
Acetylcholinesterase, nerve gases and insecticides that inactivate, 405t
Acetyl–CoA, 202, 465, *466*, 475
Acetylcoenzyme A, 202
Acetylene, *64*, 71, 84–85
 electron-dot and hybrid orbital models for, *R-23*, R-23–R-24
Acetylsalicylic acid, 191
 See also Aspirin
Acid anhydride(s), 209, 210, 211
 reaction of, with alcohol, 209, 212
  with water, 212
 reaction of amines with, 225–226, 243
 fatty. *See* Fatty acid(s)
Acidosis, 486, 487, 498
cis-Aconitic acid, 198
Acquired immune deficiency syndrome, cellular immunity and, 497
Active transport, 318, *318*, 319
Acyl carrier protein, 466–467, 472
Adenine, 361
Adenosine diphosphate (ADP), oxidative phosphorylation and, 439
3′, 5′–Adenosine monophosphate, cyclic, 488, *489*, 492
 from glucose catabolism, 440, 440t
 oxidative phosphorylation and, 439
Adenosine triphosphate (ATP), 358, *418*, 418–419, 451
Adenosine triphosphate synthetase, 439, *439*
Adipocytes, *467*
Adrenocorticoid hormones, 310, *310*
Agglutinins, 496
AIDS. *See* Acquired immune deficiency syndrome
Alanine, 849, 326t, 473
Albinism, 380t
Alcohol(s), 9, *9*, 123, 133, 148
 artificial and naturally occurring, 128t
 boiling points of, 158, *158*
 dehydration of, 127–128, 133, 148
 formation of, 166–167
 hydrogen bonding and boiling point of, 123–125, *124*
 nomenclature of, 126–127
 oxidation of, 179
 and phenols, acidity of, compared, 136t
 primary, 132
  oxidation of, to acid, 195, 211
  secondary, and tertiary, testing for, 130–132
 reactions of, 127–132
 secondary, 132
 solubility of, *125*, 125–126
 special, 132–133
 structure and physical properties of, 123, *124*
 tertiary, 132
Alcoholic beverages, 131, 174
Aldehyde(s), 10, *10*
 boiling points of, 158, *158*
 common 162–163t
 nomenclature of, 159
 oxidation of, 180
 structure of, 157
 tests for, 172–174
Aldol, 159
Aldose(s), 270, 276
Alkaloids, 232–236, 237–238t, 242
Alkalosis, 486, 487, 488
Alkanes, 6, *7*, 17–35, 55, 63, *63*, 140
 boiling point and, 19, *19*
 bonding of, 44, *44*
 carbon–hydrogen and carbon–carbon bonds in, 18
 chemical reactivity of, 44–49
 chemical structures of, 21, *23*
 combustion of, 44–48
 conformations of, *41*, 41–42, *42*, 55
 first ten, names and formulas for, 22t
 general formula for, 21–22
 halogenation reaction of, 48–49
 homologous series and, 21–23
 hydrocarbons in, 20
 in medicine, 53–54
 nomenclature of, 20–25
 solubility in water, *19*, 20
 structure and physical properties of, 18–20
 tetrahedral shape of, 18
n-Alkanes, *25*
Alkene(s), 7, *7*, 63, *63*, 91
 addition of halogens to, 74–75, *75*
 addition of hydrochloric acid to, 76
 addition of hydrogen to, 74
 addition reactions of, 73, 107, *107*
 addition of water to, 76–77, 78
 boiling point and carbon number of, 65
 carbonyl group and, compared, 157, *157*
 chemical reactivity of, 72–82
 combustion reactions of, 73
 dehydration of alcohol to form, 129, 148
 hybridization and structure of, 64, 64–65, *65*
 lack of rotation of, about double bond, 65
 nomenclature of, *68*, 68–70
 oxidation reactions of, 80, *80*
 polymerization of, 81–82, 92
 polyunsaturated, 70, 86
 reaction with HX, 78
 reaction with nonsymmetric reactant, 77

Alkyl group(s), 7, *7*, 55
functional groups and, *9*
nomenclature of, 26–34
simple, 26–27, 27t
Alkyl halides, formation of, 130. *See also* Organohalogen(s)
Alkynes, 7, *1*, 63, *64*, 91
boiling points and carbon number of, 66–67, *67*
chemical reactivity of, 84–85
cis and trans isomers for, 67, *67*
hybridization and structure of, 66, *66*
nomenclature of, 71
Allosteric regulation of enzymes, *411*, 411–412, *412*, 414
Alpha amino acids. *See under* Amino acids
Alveoli, 479, 483, 498
Amide(s), 242
functional groups and, 12–13
hydrolysis of, 230–231, 243
as polymers, 231
reactions of, 230–231
structure of, 229–230, *230*, 231
Amine(s), 11–12, 12t, 222, 242
basicity of, 222–223, *223*, *224*, 243
boiling points of, 220, *220*
definition of, 11
heterocyclic, 221t, 240–242t
hybridization and structure of, 219–220
hydrogen bonding of, 220
nomenclature of, 220–222
and p*K*b values, 224t
primary, 219
reactions of, 222–228
reaction with acid anhydride, 225–226, 243
reaction with alkyl halide, 223–225, 243
reactions with carbonyl compounds, 226–227, 243
related to epinephrine, 236
secondary, 219
oxidation of, to nitrosamines, 228
special, 232–236
tertiary, 19
Amino acids, 232, 242, 326, 327t, 331, 348
α, 227, 323–331
acid–base properties of, 328–330
acidic, 328
amphoteric, 329, *329*, 331
basic, 328
biosynthesis of, 472–473
catabolism of, 446–450, *448*
glucogenic, 462
neutral polar, 328
nonpolar, 325
in nutrition, 330
polar, 328
proteins and, 335
structure of, 324–328, 325
transamination of, 446, *447*
Amino group, 11
Ammonia, 219, R-21, *R-21*, R-25, *R-25*
boiling point of, R-22t
Amphetamine, 236
Amphipathic molecules, 303, 304
Amygdalin, 165
Amylase(s), 422, 423–424, 428
Amylopectin, 283
Amylose, 283, *283*
Anabolic steroids, in athletics, 311
Anabolism, 391, 417, 453–477
Anesthetic(s), ether as, *142*, 142–143
Anethole, 114
Angel dust. *See* Phencyclidine
Anhydrides, 209
Animals, biological combustion of hydrocarbons by, 45–46, *46*
Anion, R-11
Anomers, 272, 276
Anthracene, 113
Antibiotics, 250
Antibodies, 494, *494*, 498
Anticancer drugs, 475
Anticodon, 374, 376
Antifreeze, 132
Antigen(s), 494, *494*, 498
immune response to, 494–495, *495*
Antioxidants, 176
Antiseptics, phenols as, 137
Apoenzyme, 393
Arginine, 328t
Argon, electron arrangement in, R-9
Aromatic compounds. *See* Aromatics
Aromatic hydrocarbons. *See* Aromatics
Aromatics, 98–121
bent, 113
as classification, 99, 101
fused–ring, 113
nomenclature of, 102–105
structure of, 99–101
See also Benzene(s)
Asparagine, 327t
Aspartame, 275, *275*
Aspartic acid, 327t
Aspirin, 191
Asymmetric carbon, 256
Atherosclerosis, 307
Atom(s),
behavior of electrons in, R-3, R-3–R-9
functional groups of. *See* Functional group(s)
Atomic nucleus(i), *See* Nucleus(i)
Atomic orbitals, 18, 66
Atropine, 235
Autotrophic organisms, 455, 460
Avery, Oswald, 354
Azulene, 113

B

B lymphocyte, 494
Bacteria, regulation of protein synthesis in, 381
Bacterial transformation, 354, *354*
Bakelite, 161
BAL. *See* British Anti-Lewisite
Barbiturates, 239, *239*
Base(s),
complementary, 359t, 361, 368
Beer, 131
Benadryl, 240t
Benedict's test, 173, 174, 278
Benzaldehyde, 104, 105, 162t
1, 2-Benzanthracene, 113
Benzene ring, alkyl side chain on, oxidation of, 196, 212
Benzene(s), 101
as aromatic compound, *100*, 100–101
disubstituted, 102–103, *103*
formula for, 99
monosubstituted, 102
polysubstituted, 103–104
resonance energy of, 100, *100*
structure of, 99
Benzenesulfonic acid, 102
Benzenol, 135

Benzpyrene, 113
1, 2-Benzpyrene, 142
Benzyl, 104, 105
Beta oxidation, 442, 445
BHA. *See* Butylated hydroxyanisole
BHT. *See* Butylated hydroxytoluene
Bilayer(s), 314, *314*, 318
Bile salts, 311, *311*, 426–427, 428
Biochemistry, R-3
  introduction to, 246–264
Bioenergetics, 417–420
Bioengineering, in future, 386–387
Biomolecules, 253–259
  size and functional groups of, 253–254, *254*
  stereoisomerism in, 254–259, *255*
Biosynthesis, of amino acids, 472–473
  of carbohydrates, 460–464
  of cholesterol, 469–471, 475
  of lipids, 465–472
  of nucleotides, 474–475
Birth control methods, R-2, *R-3*
Blood, 478–500
  artificial, 54, *54*
  buffering systems in, 484–486
  clotting of, 409, *409*
  in homeostasis, 479
  and immunity, 492–498
  maintenance of nutrients in, 489–492
  proteins in, 486
  removal of wastes from, 483–484
  transport of gases in, 479–481
Blood group, ABO, 496, 496t
Body, human. *See* Human body
Bohr's effect, 482–483
Boiling point(s),
  of alcohols and aldehydes, 158, *158*
  alkanes and, 19, *19*
  of alkenes, 65
  of ammonia, R-22t
  of carboxylic acids, 188, *189*
  of methane, R-22, R-22t
  of water R-22, R-22t
Bonding, chemical, R-10–R-16
  covalent, R-12–R-13
  electrons and molecular shapes, R-1–R-39
  ionic, R-10–R-11
Bonds. *See also* Hydrogen bonding
  covalent, *R-14*, R-14–R-15, R-16
    nature favors, R-18
  disulfide, 340, 341, *342*, 344
  glycosidic, 279, 280, 288
  ionic, R-11, R-14, R-14–R-15
  peptide, 331–333, *332*, 334, 348
  pi, 64–65, *65*, 90, R-23
  polar covalent, R-14, R-14–R-15, R-16
  sigma, 44, *44*, R-13
  type of, electronegativity and, R-13–R-14
Brevicomin, 170
British Anti-Lewisite, 146
Bromine, 75
  addition of, to alkene, 74–75, *75*
2-Bromopropanoic acid, 189
p-Bromotoluene, 104
Brønsted–Lowry acid, 4
Brønsted–Lowry base, 4–5
Brønsted–Lowry model for acid, 3
Bucky balls, 115, *115*
Buffer system,
  in the blood, 484–486
Buffers,
  pH and, 484
Butacote, 241t
1-3-Butadiene, 86, *86*
Butane, 23, 28, *28*
  conformations of, 40–42, *42*
  isomers of, and alkyl groups, 30t
  normal, 24
Butanedione, 162t
2-Butanone, 160
Butylated hydroxyanisole, 176
Butylated hydroxytoluene, 176
n–Butyl group, 28
sec–Butyl group, 28–29
tert–Butyl group, 129

C

Cadaverine, 221
Cadmium, texaphyrin, 111
Caffeine, 234
Calorie (Cal), 260, 261
Calvin's cycle, 458–459, *459*
Camphor, 163t
Carbamoyl phosphate, 447
Carbohydrate(s), 265–291, 268, 288
  biosynthesis of, 460–464
  catabolism of, 429–434
  classification of, 266–268
  complex, 283, 288
  definition of, 266, *267*
  digestion of, 422–424, *423*
  and nutrition, 284, *284*
Carbolic acid. *See* Phenol(s)
Carbon–carbon bond, 18
Carbon dioxide, greenhouse effect and, 46, *46*, *47*
  in blood, transport of, 481
  movement of, within body, *480*
Carbon–hydrogen bond, 18
Carbon tetrachloride, 12t
Carbon(s), of alkyl groups, 27–29
  anomeric, 272, 276
  chemistry of, R-2–R-3
  electronegativity of, 2–6, *3*
  in organic chemistry, 2
Carbonic acid, 481–482
Carbonic anhydrase, 481
Carbonyl, addition of electrophile and nucleophile to, 164, 179
  double-bond chemistry, 164–168
  reduction of, 167, 179
Carbonyl compounds, nomenclature of, 159–160
  preparation reactions of, 170–174
  reactions of, 164
  reactions of amines with, 226–227, 243
Carbonyl group, 179
  and alkene, compared, 157, *157*
  and compounds, 156–186
  functional groups and, 10
  physical properties of, 157–158
  structure of, 10, 157, *157*
Carboxyl group, 188, 192, 211
  functional groups and, 10, 11
Carboxylate, 296
Carboxylate anion, 198
  resonance stability of, 193, *193*
Carboxylic acid(s), 10, 11, 192, 211, 296
  acidity of, 193–194, 194t
  boiling points of, 188, *189*
  common names for, 190t
  and derivatives of, 187–217
  hydrogen bonding by, 188, *188*
  nomenclature of, 189–190
  preparation of, by oxidation, 195–196
  reactions and preparations of, 193–199

Carboxylic acid(s) (*continued*)
solubility of, 189, *189*
structure of, 188, *188*
Carcinogenesis, 378
Carcinogenic compounds, 141, 143
Carotene, 90
Carrots, as source of carotene, 90, *90*
Catabolism, 391, 417, 420, 450
Cation, R-11
Cell membrane, 249, 251, 262
Cell(s), 247–249
activity of, need for regulation of, 381
eucaryotic, 251–253, *252*
hormones working within, 487–488
of human body, 248, *248*
in multicellular organisms, 247–248
procaryotic, *249*, 249–250
size of, 247
surfaces of, hormones working at, 488–489, *489*
Cell wall, 250, 251, 262
Cellular immunity. *See under* Acquired immune deficiency syndrome
Cellulose, 284, 285, *286*
Cephalins, 304, *305*
Cerebroside, 304
Cetylpyridinium chloride, 225
CFCs. *See* Chlorofluorocarbons
Chain reaction, 52
Chair conformation, 43, *43*
Chemical bonding, R-10–R-16
Chemical industry, petroleum as source of, 4
Chemical stability, rule for, R-10
Chemiosmotic theory, 437–438, 441
Chemotherapy, side effects of, 475
Chiral center, 256, 259
Chirality, *255*, 255–256, 259, 262
Chitin, 284, 285, *286*
Chloral hydrate, 166
Chlordiazepoxide, 241t
Chlorinated hydrocarbons, 11, 12t
Chlorine, 11
electron distribution diagrams for, R-10, R-11
functional groups involving, 8t
3-Chloro-4-phenyl-3-heptene, 105
Chlorofluorocarbons (CFCs), ozone layer and, 52
Chloroform, 12t
Chloromethane, *11*, 12t
Chlorophyll, 111, *111*
Chloroplasts, 252, 455, *457*
2-Chloropropane, 76
Cholesterol, 307, *307*, 309, 313
biosynthesis of, 469–471, 475
catabolism of, 445
in foods, 309, *309*, 309t
and heart disease, 428
Chromosomes, DNA in, 364, *365*, 368
Chrysanthemic acid, 55
Chyme, 423, 428
Cinnamaldehyde, 162t
Circulatory system, movement of gases in, *480*
Citrate, 435–436
Citric acid, 198
Citric acid cycle. *See* Tricarboxylic acid (TCA) cycle
Civetone, 163t
Cocaine, 235
Codeine, 233
Codons, *372*, 372–373, 373t, *374*, *375*, 376
Coenzyme A, 395
Coenzyme(s), 393, 396
functions of, and vitamins, 394t
Cofactor(s), 393, 396
Collagen, 323
Combustion, of alkanes, 44–48
Competitive inhibitors, 406, *406*
Complementary bases, 359t, 361, 368
Complex carbohydrates, 283, 288
Condensed formula, 21, *23*
Conformations, of alkanes, 40–42, *41*, *42*, 55
of cycloalkanes, 42–43, *43*, 55
Conjugated polyenes, 86, *86*
Conjugation, 86
Contraceptives, oral, 85
Cori's cycle, 461, *461*, 464
Cortisol. *See* Hydrocortisone
Cortisone, 310, *310*
Coupled reactions, 173, 417–420, 450
Covalent bonding, R-12–R-13
Covalent bond(s), *R-14*, R-14–R-15, R-16
nature favors, R-18
Covalent molecules, shapes of, *R-18*, R-18–R-24
"Crack," 235
Crick, Francis, 359–360
Cristae, 253
Crude oil, 4
Cumene, 102
Cyanohydrin, 168, 179
as insect defense mechanism, 165, *165*
formation of, 164, 179
in laetrile, 165
Cyclamates, 275
Cyclic adenosine monophosphate, 488, *489*, 492
Cyclic AMP. *See* Cyclic adenosine monophosphate
Cycloalkanes, 35–39, 55
chemical reactivity of, 44–49
conformations of, 42–43, *43*, 55
general formula for, 36
geometric structure of, 36
in medicine, 54–55
nomenclature of, 36, 37–38, *38, 39*
plantar structure and ball-and-stick models of, 37t
Cyclobutane, 37t
Cycloheptane, *39*
Cyclohexane, 37t, 43, *43*
2-Cyclohexenethiol, 144
2-Cyclohexenol, 144
Cyclopentane, 37t
Cyclopentanone, 160
Cyclopropane, *36*, 37t, 141
as anesthetic, 54–55
Cyclopropanethiol, 144
Cyclopropanol, 144
Cysteine, 327t, 330
Cytosine, 361
Cytosol, 249, 251, 262

D

Dacron, 211
Darvon, 240t
Deamination, 446, 449
Decane, structural formula of, 32–33
Decarboxylation, of β-keto acids, 197, 198, 199, 212
Dehydration, of alcohols, 127–129, 133, 148
Deletion mutations, 378
Delocalized orbitals, 100

Demerol, 233
denaturation, 336, 344, 346
Deoxyadenosine monophosphate, 358
Deoxyribonucleic acid polymerase, in replication, and phosphate bonding, 366–367, *367*
Deoxyribonucleic acid (DNA), 249–250, 354, 355, 387
bacterial, *363*, 363–364
bonding of nucleotides to form, 360, *360*
changes in, mutagens and, 378
chromosomes in, 364, *365*
germinal, 379
hydrogen bonding in, 361, *362*, *363*
replication of, 364–368
and RNA, structural differences between, 368, 369t
somatic, 379
structure of, 359–364, *360*
higher levels of, *363*, 363–364
introduced changes in, *377*, 377–378
substitution in, 377–378
Deoxyribonuleotides, 357, *357*
Deoxyribose, *356*, 357
2-Deoxyribose, 273–274
Detergents, 203, 205, 207, 208, 208t, R-29, R-30
Diabetes mellitus, 174
Diacetylmorphine, 233
Diacylglycerols, 299
Diastereomers, 257, 259
Diazepam, 241t
2-2-Dibromopentane, 84
2,2-Dibromopropanoic acid, 190
Dicarboxylic acids, names for, 192
Dichloromethane, 12t
N-N-Diethylaniline, 221
Diethyl ether, *10*
Diffusion, in cell, 247
facilitated, 318, *318*, 319
Digestion, 421–428, 450
of carbohydrates, 422–424, *423*
of lipids, 425–428, *426*
of protein, 424–425, *425*
Dihydroxyacetone, 163t, 273
Dihydroxy aldehyde, 266
Dimercaprol, 146
Dimers, 188, *188*
1,1-Dimethyl-2-benzylcyclohexane, 104
Dimethylcyclobutane, isomers of, 38, *38*
Dimethyl ether, 140
4,4-Dimethyl-2-pentanone, 160
2,2-Dimethylpropane, 34
2,4-Dinitrobenzenesulfonic acid, 105
Dioxin, 143
Dipeptide, 333
Diphenhydramine, 240t
Dipole(s), R-16
electric, *R-24*, R-24–R-25, *R-25*, R-34
induced, 18, *18*
molecular, R-24–R-26
Dipole–dipole force, 157–158
Dipole–dipole interaction, R-28
Directing groups, 109, 109t, 112
Disaccharides, 281
dietary, 424
Disubstituted benzenes. *See under* Benzene(s)
Disulfide, 144–145, 148
reduction of, 145
Disulfide bond(s) 340, 341, *342*, 344
Disulfide bridge, 340, 341, *342*, 344
DNA. *See* Deoxyribonucleic acid (DNA)
Dopamine, 238t
Double helix, 359–361, *362–363*
Drug(s), synthetic, R-2, *R-3*

E

E^+, 107–108. *See also* Lewis acids
Electric dipole, *R-24*, R-24–R-25, *R-25*, R-34
Electrical charge, symmetrical distribution of, R-21, *R-21*
Electrolytes, R-30
Electron-dot method, R-8
Electron-dot model(s), determination of molecular for natural gas, R-20, *R-20*
Electron-dot structures, Lewis acid and base character from, 3–6
Electron dots, R-8–R-9
Electron energy diagrams, R-6, *R-7*
Electron filling series, R-6, *R-6*
Electron-pair repulsion, valence-shell (VSEPR), R-19
Electron-pair repulsion theory, R-33, R-34
Electron sharing, in chemical stability, R-12
Electron structure, of noble gases, R-10
Electron transfer bond, R-11
Electron-transport chain, *437*, 437–438, *438*, 440, 450
Electronegativity, and bond type, R-13–R-16, *R-14*
of carbon, 2–6, *3*
of elements, *R-13*, R-13–R-14
ionic character and, R-14, R-14t
Electrons, behavior of, and behavior of matter, R-3, *R-3*
in atoms, R-3–R-9
bonding, and molecular shapes, R-1–R-39
and orbitals, R-4–R-5, *R-5*
Electrophiles, 107–108. *See also* Lewis acids
reaction conditions and, 108t
Electrophilic substitution reactions, 107, 112, 116
Electrophoresis, 347, *347*, 348
Elements,
abundant, in earth's crust, 219
electronegativity of, *R-13*, R-13–R-14
found in living organisms, 7t
in periodic table, R-9
Emetine, 237t
Enantiomer(s), 256, 258, 259
Endoplasmic reticulum, 252, 253, 262
Endorphins, *333*, 334
Energy,
free, 260, *260*
resonance, 100, *100*, 116
use of, by living organisms, 417
Enkephalins, *333*, 334
Entropy, 417, 420, 450
Enzyme-catalyzed reactions, effects on pH on, 404
effects of temperature and, 402, 402–403, *403*
and reaction rate, substrates in, *401*, 401–402, *402*
Enzymes, 323, 390–415
active site of, 396–398, *397*
activity of, 398–399, *399*, *400*
allosteric regulation of, *411*, 411–412, *412*, 414

Enzymes (*continued*)
classes of, 393t
composition of, 392–393
covalent modification of, 410, *411*
definition of, 391–392
inhibition of, 404–408, *406, 407, 408*
in medicine, 413, *413*
in metabolism, 391
nomenclature of, 392
prosthetic group of, 393
rates of reactions and, *401*, 401–408
regulation of activity of, 408–412
specificity of, 396–398
Epinephrine, 236
amines related to, 236
interaction of, with target cell, *489*
Epoxide(s), 141, 143, 148
Ergotamine, 237t
Ergotism, 234
Erythrose, 269t
Ester(s), 207, 211
flavors and odors of, 201, 201t
formation of, 200, 212
functional groups and, 12–13
inorganic, 203
nomenclature of, 203–204
organic, 200–201
saponification of, 204, 208, 212
structure and reactivity of, 200–203
Estradiol, 113–114
Estrogen, 308
Ethane, 74
conformations of, 40, *41*
energies of, *41*
structural diagram and ball-and-stick models of, *63*
structural formula of, *6, 9*
Ethanethiol, 144
Ethanol, *9*, 77, 133, 159
Ethene, *63*, 64, R-23, *R-23*
Ether(s), *9*, 10, *10*, 123, 143, 148
as anesthetic, *142*, 142–143
common, 142t
nomenclature of, 140–141
special, 141–142
structure of, 140, *141*
Ethinyl estradiol, 85
Ethyl alcohol, 77, 133, 144, R-28, *R-29*
Ethylamine, 220
N-Ethylaniline, 221
5-Ethyl-2,8-dimethyldecane, 32–33, *33*
Ethylene, *63*, 64, 80, R-23, *R-23*
reaction with hydrogen, 74
Ethylene glycol, 80, 132
Ethyl group, *9*
Ethyl methyl ether, 140–141
Ethylmethylamine, 220
N-Ethylpyrrole, 221
2-Ethyltoluene, 103
Ethyne, 71
Eucaryotes, 251–253, *252*, 262
Eugenol, 105, 138
Exons, 371

F

Fabrey's disease, 380t
Facilitated diffusion, 318, *318*, 319
Fat(s), 299–300, 302
storage of, 302, *302*
Fat-soluble vitamins, 316–317
Fatty acid(s), 205, 208, 294–298, 319
in animals and plants, 295t
beta oxidation of, 442–444, *443*
biosynthesis of, 466–467, *468*
chemical properties of, 296–298
classification of, 294–296
energy production from, 444–445
essential, 301
halogenated, 297–298
physical properties of, 296
saturated, 294–295, 295t, *297*, 298
unsaturated, 295–296, *297*, 298
Fatty acid synthase, 467
Feedback inhibition, 411–412, *412*
Fehling's test, 173, 174, 278
Fermentation, 433–434
Fermentation of sugar, 131
Fiber, 285, 288
Fibrous proteins, 348
Filtration, 484, *484*
Flavin adenine dinucleotide (FAD), 395
Fluid mosaic model, of membranes, *315*, 315–318
Fluorine, 75
Fluorouracil, 475
Folic acid, synthesis of, sulfanilamide and, 407, *407*
Foods, 260, 261. *See also* Nutrition
Formaldehyde, 10, 161, 162t
Formalin, 161
Formic acid, 11
Formula(s), 21–22
Formulas, as replacement for breast feeding, *464*
Fossil fuels, 4
Frameshift, mutation(s), *377*, 378, 380
Franklin, Rosalind, 359
Free energy, 260, *260*
Freezing point, of water, R-33
Fructose, 269t, 274
Fuller, R. Buckminster, 115
Fumarate, 436
Functional group(s), 2
definition of, 3
and families of compounds created by, 8t
introduction to, 6–13
Furan, 221
Furanose(s), 272, 276
Fused-ring aromatic systems, 113

G

Galactose, 276
Galactosemia, 380t
Garlic, 147
Gas(es),
exchange of, in lungs, 479, *480*
noble, R-9
electron structure of, R-10
transport of, in blood, 479–481
utilization of, and production by cells, 481–482
Gastrointestinal tract, human, *422*
Gene(s), 372, 375
eucaryotic, insertion into bacterium, 385–386, *386*
expression of, regulation of, 381
Genetic code, *372*, 372–373, 373t
Genetic diseases, human, 380t
Genetic engineering, 385–387
future of, 386–387
process of, 385–386, *386*
Germinal DNA, 379
Globular proteins, 348
Glucagon, 489, 490–491, *491*, 492, 498
Glucogenic amino acids, 462
Gluconeogenesis, 460–462, *461, 463*, 464, 475

Glucose, 2, *2*, 162t, 266, 272–273, *273*, 274, 281
catabolism of, 429
diabetes tests and, 174
energy from, analysis of, 439–440
Glutamate, 446, 447
Glutamic acid, 327t, 473
Glutamine, 327t
Glutathionine, 334
Glyceraldehyde, 162t, 269t, 270, 273
Glycerine. *See* Glycerol
Glycerol, 132, 462
Glycerol phosphate, 469
Glycine, 326t
Glycocholate, 311, *311*
Glycogen, *283*, 283–284, 288
catabolism of, 432–433
metabolism of, glucagon and, 490, *491*
Glycogenesis, 462–464
Glycol(s), 132–133
Glycolipid, 304
Glycolysis, 429–432, *430, 432*, 434, 450, 461–462
Glycoside(s), 279, 280, *280*
Glycosidic bonds, 279, 280, 288
Goretex, R-2, *R-2*
Grain, 77, 133
Greenhouse effect, *46*, 46–47, *47*
Griffith, Fred, 354
Guaiacol, 138
Guanine, 361
Guanosine diphosphate (GDP), 436
Guanosine triphosphate (GTP), 436

H

Hair, "permanent waved," 145, *146*
Halogenation, 48–49, 55, 74–75
Halogenation reaction, 76
of alkanes, 48–49
Halogens, 11
Hardening, of oil to fat, 74
Health products, hydrocarbon-based, 53–55
Heart disease, cholesterol and, 428
Helium, R-10
electron arrangement for, R-6
α-Helix, *338*, 338–339, 344
Heme, 111, *111*, 343
Hemiacetal(s), 168
in monosaccharides, 271
Hemiketal(s), in monosaccharides, 271
Hemoglobin, 323, 343, 479, 480, *481*, 483, 498
Hemoglobin S, 344
Heparin, 286, *287*, 288
Herbicides, organohalogens as, 50t
Heredity, molecular basis of, 352–389
Heroin, *See* Diacetylmorphine
Heterocycles, 221, 222
Heterocyclic compounds, in medicine, 238, 240–242t
Heterotrophic organisms, 455, 460
Histidine, 328t
Holoenzyme, 393
Homeostasis, 479, 483, 498
Homologous series, 21–23, 55
Hormones, 487
adrenocorticoid, 310, *310*
of body, 488t
circulation of, nutrients and, 487–492
sex, 308, *308*
steroid. *See* Steroids
working at cell surfaces, 488–489, *489*
working within cells, 487–488
Human body, biological combustion of hydrocarbons by, 45
effects of temperature on, 403, *403*
hormones of, 488t
Hund's rule, R-6–R-7
Hyaluronic acid, 286, *287*, 288
Hybridization, R-22
of amines, 219–220
orbital, R-22–R-24
Hybridized orbitals, 18, R-34
Hydrate, 168
formation of, 166, 179
Hydrated ion, R-30, *R-30*
Hydration, 76, 77, 78, 297
Hydrazine(s), 227
reaction with carbonyl compound, 243
Hydrazones, 227
Hydrocarbons, 2, *6*, 6–7, *7*, 13
alkane. *See* Alkanes
aromatic. *See* Aromatics
biological combustion of, 45
chlorinated, 11, 12t
family, groups in, 20
saturated. *See* Alkanes
single-bonded, 17–61
strange, 69, *69*
unsaturated. *See* Alkene(s); Alkynes; Unsaturated hydrocarbons
Hydrochloric acid, as alkene reactant, 76
Hydrocortisone, 310, *310*
Hydrogen, electron arrangement for, R-6
electron-dot model for, R-21, *R-21*
Hydrogenation, 74, 297
Hydrogen atom, *R-12*, R-12–R-13
Hydrogen bond, R-32
Hydrogen bonding, R-32, R-33, *R-33*
of amines, 220
in deoxyribonucleic acid (DNA), 354, 355
Hydrogen cyanide, 165, *165*
Hydrogen gas, R-12
Hydrohalogenation, 78
Hydrophobic interactions, 341
Hydroquinone(s), 176, 178
oxidation of, 180
to quinones, 175–178
3-Hydroxybutanal, 159
4-Hydroxy-2-cyclohexenone, 160
Hydroxymethylglutaryl-CoA (HMG-CoA), 470, 471
Hypercholesterolemia, familial, 428
Hyperglycemia, 274
Hyperventilation, 486
Hypoglycemia, 274
Hypothermia, 403
Hypoventilation, 486

I

Ibuprofen, 191
Ice, frost patterns of, R-33, *R-33*
Imine(s), 226, 227, 228, 242
Immune response, to antigens, 494–495, *495*
Immunity, blood and, 492–498
cellular, and AIDS, 497
Immunoglobulins, 494
Indocin, 241t
Indomethacin, 241t
Induced dipoles, 18, *18*
Induced-fit model, 398, 399, *399*, *400*, 401

Induction, 381–383, 384
Inhibitor(s), 404–408, 414
 competitive, 406, *406*
 irreversible, 404–405, 414
 noncompetitive, 408, *408*
 reversible, 405–406, *406*, 414
Insecticides,
 inactivating acetylcholinesterase, 405t
 organophosphate, and acetylcholinesterase, 404–405, *406*
Insertion mutations, 378
Insulin, 489–490, *490*, 492, 498
 synthetic, production of, 385
Interferons, 497, *497*, 498
International Union of Pure and Applied Chemistry (IUPAC), 31
Introns, 371
Iodine number, 300, 302
Iodine test, 300
Ionic bonding, R-10–R-11
Ionic compounds, dissolving of R-29–R-31, *R-30*, *R-31*
Ionization, R-18
Ions, R-11
Irreversible inhibitors, 404–405, 414
Isoalkanes, *25*
Isobutane, 24, 29
Isobutyl group, 29
Isocitrate, 436
Isocitric acid, 198
Isoelectric point (pI), 329, 331
Isoleucine, 326t
Isomers, 23–25, 55
 for alkynes, 67, *67*
Isopropyl alcohol, 76–77, 78
Isopropylbenzene, 102
Isopropyl group, 28
Isotretinoin, 87–88
Isozymes, 393, 396

J

Jasmone, 163t

K

Kekulé, Friedrich, 99
Ketals. *See* Acetal(s)
Keto acids, 199
 α-keto acid(s), 199, 227
 β-keto acid(s), 198, 199, 211
  decarboxylation of, 197, 199, 212
 catabolism of, 449, *449*
α-Ketoglutarate, 446–447
α-Ketoglutaric acid, 198, 473
Ketone bodies, 472
 synthesis of, 471
Ketone(s), 10
 common, 162–163t
 functional groups and, *10*
 nomenclature of, 159
 structure of, 157
Ketose(s), 270, 276
Ketosis, 486
Kidney dialysis, 485, *485*
Kidneys, removal of wastes from, 483–484, *484*
Krebs cycle. *See* Tricarboxylic acid (TCA) cycle
Krebs, Hans, 435
Kwashiorkor, 347, *347*

L

Lactate, 460–461
Lactic acid, 172
Lactose, 281, *282*
 induction by, 382, *382*
 synthesis of, from enzyme modification, 464
Lactose intolerance, 424
Lactose repressor, 383
Laetrile, cyanohydrin in, 165
Lanosterol, 471
Le Chatelier's principle, 461, 481–482
Lecithin, 304, *305*
Leucine, 326t
Leukotrienes, *312*, 312–313, 319
Lewis acid(s), 12
 definition of, 4
Lewis acid–base chemistry, 6
Lewis base(s), 12
 definition of, 3–4
Lewis basicity, reactions of alcohols and, 127
Lewis theory, 3–6
Librium, 241t
Lidocaine, 235
Line formula, 21, *23*
Linear molecule, 66
Linolenic acid, 301–302
Lipases, 427, 428
Lipids, 292–321
 biosynthesis of, 465–472
 catabolism of, 441–442
 classification of, 293
 definition of, 293
 digestion of, 425–428, *426*
 and health, correlations of, 309
 maintenance of, 492
 nonsaponifiable, 293, 294, 306–313, 319
 polar, 303–304
 saponifiable, 293, 294
  of membranes, 303–306, 319
 sperm whales and, 302
Lipogenesis, 465, 472
Lipopolysaccharide, 250
Lipoproteins, 427, 427t, 428
Liposomes, 314, *314*, 318
Liquors, 131
Lister, Joseph, 137
Lithium, electron arrangement for, R-6
Lock-and-key model, 398, *399*, 401
London forces, 18, 65, 66–67, R-28
Lovastatin, 470
LSD. *See* Lysergic acid
Lucas reagent, 132
Lucas test, 132, 133
Lungs, exchange of gases in, 479, *480*
 removal of wastes from, 483
Lymph, 427, 428
Lysergic acid, 234
Lysine, *2*, 2–3, 328t
Lysozyme, 323, 396–398, *397*

M

Malnutrition, protein-calorie, 347, *347*
Maltose, 281, 282, *282*
Marasmus, 347
Markovnikov, Vladimir, 77
Markovnikov's addition reactions, 77–79
Markovnikov's rule, 77–79, 84
Matrix, 253
Medicine, enzymes in, 413, *413*
Medopren, 240t
Membranes, biological, *315*, 315–318
 saponifiable lipids of, 303–306, 319
Mendel, Gregor, 353

Menstrual cycle, 308, *308*
Meperidine, 233
Meprobamate, 240t
Mercaptans, 144
Meso compound, 258, 259
Mestranol, 85
Meta directors, 109
Metabolic acidosis, 486
Metabolic alkalosis, 486
Metabolic pathway, 391
Metabolism, 260, 391, 414, 417
 enzymes in, 391
 minerals and vitamins in, 394
Methadone, 233
Methamphetamine, 236
Methane, 18, 20–21, *21*, 123, R-21, *R-21*, R-25–R-26, *R-25*
 boiling point of, R-22, R-22t
 electron-dot model of, R-20, *R-20*
 electron filling series and electron-dot model for, R-20, *R-20*
 greenhouse effect and, 46, *47*
 solubility of, R-28
 structural formula of, 6, *6*, *7*, 9
Methanethiol, *9*
Methanol, *9*, 200
Methionine, 326t
Methotrexate, 406–407
Methoxyamphetamine, 236
Methylbenzene, 102
2-Methylbutane, 34
2-Methyl-1-butene, 69
2-Methyl-2-butenoic acid, 189
1-Methyl-1,3,5,7-cyclooctatetraene, 70
Methyldopa, 240t
Methyl ethyl ketone, 160
Methyl group, *7*, *9*
2-Methyl-3-heptyne, 71
trans-3-Methyl-3-octene, 70
N-Methylhexylamine, 221
2-Methylhexylamine, 221
4-Methyl-1,2-pentadiene, 70
2-Methylpentane, 31–32
Methyl salicylate, 105, 138
Mevalonate, 470
Micelle(s), 206, *206*, *207*, 208, 313, *313*
Miltown, 240t
Mineral oil, 53
Minerals, and vitamins, in metabolism, 394
Mitochondrion(a), 252, 253, *253*, 262
Molecular basis of hereditary, 352–389
Molecular dipole(s), R-24–R-26, *R-26*
Molecular polarity, R-19
 solubility and, R-28, *R-28*
Molecular shapes, determination of, from electron-dot models, R-19–R-22
 electron binding and, R-1–R-39
Molecules,
 amphipathic, 303, 304
 linear, 66
 nonpolar, R-25–R-26
 polar, R-26, R-34
  behavior of, *R-27*, R-27–R-31, *R-28, R-29, R-30, R-31*, R-34
 with polar and nonpolar parts, R-28–R-29, *R-29*
Monoacylglycerols, 299
Monomers, 81
Monosaccharides, 267, 269–276, 288, 424, 462
Monosubstituted benzenes. *See under* Benzene(s)
Montreal Protocol, 52
Morphine, 233
Mutagenesis, 376–379
Mutagens, 380
 changes in DNA and, 378
Mutarotation, 277–278, 279
Mutation(s), 376, 380
 frameshift, *377*, 378, 380
 as source of genetic variability, 379
 spontaneous, 376–378, *377*
Myoglobin, 323, *340*
 sperm whales and, *337*
Myricyl palmitate, 303
Myristic acid, 294

N

Naphthalene, 113
Native conformation, 336, 344, 346
Natural gas, electron filling series and electron-dot model for, R-20, *R-20*
Negative effector, 411, *411*, 412
Neon, electron arrangement in, R-9
Neo-Synephrine, 240t
Nerve gases, inactivating acetylcholinesterase, 405t
Nicotinamide adenine dinucleotide phosphate (NADPH), 454–455, 475
Nicotinamide adenine dinucleotide (NAD+), 395, 419, 450
Nicotinamide adenine dinucleotide, reduced form (NADH), 419–420
Nicotine, 236
NIH shift, 142
m-Nitrobromobenzene, 104
Nitrogen, 219
 electron filling series for, R-6, *R-7*
 functional groups involving, 8t
Nitrogenous base(s), 355–356, *356*, 358
Nitroglycerine, 203
Nitrosamines, oxidation of secondary amines to, 228
Nobel, Alfred, 203
Noble gases, R-9, R-10, *R-10*
2-5-Nonadiene, 70
Noncompetitive inhibitors, 408, *408*
Nonpolar molecules, R-25–R-26
Nonsaponifiable lipids, 293, 294, 306–313, 319
Nonsense strand, 370
Norepinephrine, 236
Norethindrone, 85, 310, *310*
19-Norethisterone, 85
Norethynodrel, 85
Novocain, 235
Nucleolus, 252
Nucleosides, 357, 358, 358t
Nucleotide(s), 355–358, 358t, 387
 biosynthesis of, 474–475
 bonding of, to form DNA, 360, *360*
Nucleus(i), 252, 253, 262
Nutrients, 260, 262
 classes of, 261
 maintenance of, hormones and, 487–492
 in blood, 489–492
 required by humans, 261t
Nutrition, 260–261, 262
 amino acids in, 330
 carbohydrates and, 284, *284*
 proteins in, 330, *330*
Nylon, 231

O

Octane, R-26
 structural formula of, *6*

cis-2-Octene, 70
Octet rule, R-11
Oil of Wintergreen, 191
Oil(s), 299–300, 302
 hydrogenation of, 74
Okazaki fragments, 367
Oligosaccharides, 267–268, 281–282, 288
Onions, 147
Operon, 382–383, 384
Optical isomers, 256
Orbital notation, R-7, *R-7*
Orbital(s), *R-4*, R-4–R-5, *R-5*
 atomic, 18, 66
 delocalized, 100
 hybridized, 18, R-34
Organelles, 247, 248
Organic acids, and acid derivatives, 211
 naturally occurring, 190t
Organic chemistry, R-2
 introduction to, 1–16
 writing equations in, 73–74
Organisms, common elements found in, 7t
 higher, regulation in, 384
Organohalogen(s), as pesticides and herbicides, 50t
 as solvents, 51t
 conversion of alcohol to, 130, 149
 functional groups and, 11, *11*
 medical uses of, 51t
 nomenclature of, 49
 reactions of amines with, 223–225, 243
Orinase, 240t
Ortho-para director(s), 108–109, 110
Oxaloacetic acid, 473
Oxidation,
 as alkene reaction, 80, *80*
 generalized scheme of, 170–171, 174
 preparation of carboxylic acids by, 195–196
 side-chain, 116
 simple model for, 195, *196*
Oxidation reactions, 110
Oxidation–reduction reaction(s), 170–172
 reversible nature of, 171–172, *172*
Oxidative deamination, 446–447
Oxidative phosphorylation, 437, 439, 440
3-Oxopentanoic acid, 190
Oxygen, cooperative binding of, 480, *481*
 delivery to tissues, pH, and, 483, 483t
 functional groups involving, 8t
 singlet, 111
Oxytocin, *333*, 334
Ozone layer, depletion of, 52, *52*

P

Palmitic acid. *See* Palmitoyl-CoA
Palmitoyl-CoA, beta oxidation of, 442–444, *443*
 production of adenosine triphosphate from, 445t
 synthesis of, 466–467
Pauling, Linus, 129, R-13
PCP. *See* Phencyclidine
Penicillins, 242t, 250, *250*
1,3-Pentadiene, 70
Pentane, 21
 isomers of, 24
n-Pentane, 34
2-Pentene, 70
Pentose phosphate pathway, 467, *469*, 472, 475
1-Pentyne, 71, 84
Pepsin, 425
Peptidases, 425, 428
Peptide bond(s), 331–333, 334, 348
 structure of, 332, *332*
Peptides, 333, 334, 348
 biologically important, *333*, 334
 structure and nomenclature of, *333*, 333–334
Perfluorodecalin, 54
Perfluorotripropylamine, 54
Periodic table of elements, R-9
Perspective formula, 21, *23*
Pesticides, organohalogens as, 50t
Petrolatum, 53
Petroleum, as source of chemical industry, 4
 hydrocarbon fractions from, 4t
pH,
 effects of, on enzyme-catalyzed reactions, 404, *404*
 regulation of, 484–487
Phenazopyridine, 241t
Phencyclidine, 234
Phenol(s), 110, 123, 137, 148
 and alcohol, acidity of, compared, 136t
 aromatic substitution reactions of, 137, 147
 common, 138t
 OH bond of, acidity of, 136–137
 reactions of, 136–137
 special, 137
 structure of, 135
Phenyl, 104, 105
Phenylalanine, 326t, 473, 474
Phenylbutazone, 241t
Phenylephrine, 240t
3-Phenylhexane, 104
Phenylketonuria (PKU), 380t, 474
2-Phenylphenol, 138
Phenyl propyl ether, 141
Pheromones, 89, 170, *170*
Phosphatidic acids, 304
Phosphatidylcholine. *See* Lecithin
Phosphatidylethanolamine. *See* Cephalins
Phosphoacylglycerols, 304, *305*
 synthesis of, 467–469
Phosphoester bond, 358
Phospholipids, 304
Phosphoric acid, 486
Photochemistry, 456
Photooxidation, 21
Photosynthesis, 455–460, *457*, 475
 dark reaction of, 458–459, *459*, 460, 475
 light reaction of, 457–458, *458*, 460, 475
Pi bond(s), 64–65, *65*, 90, R-23
pI (isoelectric point), 329, 331
Plasma cell(s), 494–495, *495*
Plasmid, 386, 387
Plasmin, 413
Plastics, disposal of, 82
β-Pleated sheet, 339, *339*, 344
Polar covalent bond(s), R-14, R-14–R-15, R-16
Polar lipids, 303–304
Polar molecule(s), R-26, R-34
 behavior of, *R-27*, R-27–R-31, *R-28*, *R-29*, *R-30*, *R-31*, R-34
Polarimeter, 256, *257*

Polarity(ies), molecular, solubility and, R-28, *R-28*
Polyenes, conjugated, 86, *86*
Polyethylene, 82
Polymerization, 81–82, 92
Polymers, amides as, 231
 common, 81t
Polypeptides, 333, 334
Polysaccharides, 268, 282–288
 storage of, 283–284
 structural, 284–285, *285*
Polysubstituted benzenes. *See* Benzene(s)
Polytetrafluoroethylene, 82
Polyunsaturated alkenes, 70, 86
Positive effector, 411, *411*, 412
Prednisolone, 310, *310*
Preservatives, antioxidants as, 176
Primary carbons, 27
Procaine, 235
Procaryote(s), *249*, 249–250, 251, 262
Progesterone, 163t, 308, 310, *310*
Proline, 325, 326t
Promoters, 370
Propanal. *See* Propionaldehyde
Propane, *21*, R-26
 isomers of, and alkyl groups, 30t
 structural formula of, *6*
 structure of, 28
Propanoic acid, 200
1-Propanol, 78
2-Propanol. *See* Isopropyl alcohol
Propene, 76, 78
Propionaldehyde, 160
α-Propoxyphene, 240t
n-Propyl alcohol, 77
n-Propyl group, 28
Prostaglandins, 312, *312*, 313, 319
Proteases, 425, 428
Protein kinase, 488
Protein(s), 333, 348
 and amino acids, 335
 blood, 486
 classification of, 348
 conformation of, 336, *336*, 344, 346
 conjugated, 343, 345
 denaturation of, 336, 344, *346*
 denatured, 403
 digestion of, 424–425, *425*
 fibrous, 348
 functions of, 323, 324t
 globular, 348
 macromolecular complexes of, 343
 in nutrition, 330, *330*
 primary structure of, 336, *337*, 344
 properties of, *345*, 345–347
 prosthetic group in, 343, 345
 quaternary structure of, 342–343, *343*, 344
 repressor, 383
 secondary structure of, *337*, 337–339, *338*, 344
 separation of, by electrophoresis, 347, *347*, 348
 simple, 343, 344
 structure of, 335
 synthesis of, 372
 tertiary structure of, *340*, 340–341, *341*, 344
Protein–calorie malnutrition, 347, *347*
Psilocybin, 238t
Purines, *356*, 356–357
Putrescine, 221
Pyranose, 272, 276
Pyrethrins, 55
Pyridine, 116, 222
Pyrimidines, *357*, 357
Pyrrole, 221, 222, *223*
Pyrrolidine, 222, *223*
Pyruvate, catabolism of, 433–434
Pyruvic acid, 172, 473

Q

Quaternary carbons, 27
Quinine, 232–233
Quinone(s), 176, 178
 bombardier beetle and, 178, *178*
 in everyday life, 177–178
 oxidation of hydroquinones to, 175–178
 reduction of, 180

R

Racemic mixture, 256, 259
Rancid, 300, 302
Reaction rate(s), 401, 408
Reducing sugars, 278, 279
Reduction,
 of carbonyl compounds to alcohols, 171, 174
Renal threshold, 484
Replication, 364–368, 387
 error in, mutations in, 376
 origins of, 365–366, *366*
 transcription and, 370–373
Replication forks, 366, *366*
Repression, 383, 384, *384*
Reserpine, 238t
Residue(s), 281, 282
Resonance energy, 100, *100*, 116
Resonance structures, 100
Respiration, 433, 434
Respiratory acidosis, 486
Respiratory alkalosis, 486
Respiratory system,
 movement of gases in, *480*
Restriction endonucleases, 386
Retin-A, 88
Reverse transcriptase, 386
Reversible inhibitors, 405–406, *406*, 414
Ribonucleic acid polymerase, 370, *370*
Ribonucleic acid (RNA), 354, 355, 368–369, 387
 and DNA, structural differences between, 368, 369t
 messenger (mRNA), 368, 369
 ribosomal (rRNA), 368, 369, 371
 transcription and, 368–369
 transfer (tRNA), 368, 369, 371, *371*
Ribonucleotides, 370
Ribose, 269t, 272, 273, 274
Ribosome(s), 249–250, 251
RNA. *See* Ribonucleic acid (RNA)

S

Saccharide, 267
Saccharin, 275
Salicylic acid, 191
Salol, 191
Saponifiable lipids, 293, 294
 of membranes, 303–306, 319
Saponification, of esters, 204, 208, 212
Saturated hydrocarbons. *See* Alkanes
Secondary carbons, 27

Semiconservative replication, 364, *365*, *366*, 368
Sense strand, 370
Serine, 327t
Serotonin, 238t
Sex hormones, 308, *308*
Sex pheromones, 170
Shells, R-4, *R-4*
Sickle cell anemia, 344, *344*, 379
Side-chain oxidation, 116
Sigma bonds. *See under* Bonds
Single-bonded hydrocarbons, 17–61
Singlet oxygen, 111
Soap(s), 205, 208, R-29, R-30
as detergent, 205–206, *206*, R-19, *R-19*
and hard water, 206
Sodium, electron distribution diagrams for, R-10, R-11, *R-11*
electron transfer from, to chlorine, R-11
Sodium bicarbonate, 136
Sodium hydroxide, 136
Sodium pentothal, 239
Solubility(ies), of alkanes, *19*, 20
of carboxylic acids, 189, *189*
and molecular polarity, R-28, *R-28*
of sugar, 277
Solvent(s),
organohalogens as, 51t
Somatic cells, 379
Somatic DNA, 379
Sorbic acid, 211
sp^2 hybridization, R-23
sp^3 hybridization, R-23
Spacesuits, NASA, R-2, *R-2*
Sphingolipids, 304
Sphingomyelin, 304, *306*
Sphingosine, 304, *306*
Squalene, 470, 471
Starch, 283, 288
Stereoisomers, 254, 259
in biomolecules, 254–255, *255*
of monosaccharides, 270–271
Steroid nucleus, 307, *307*
Steroids, 307–311, 313, 319
anabolic, in athletics, 311
effects of, on target cell, *488*
Structural formula, 21, *23*
Strychnine, 233
Substitution, 377–378, 380
Substrate(s), 392, 393, 414
in enzyme-catalyzed reactions, and reaction rate, *401*, 401–402, *402*
Succinate, 436
Sucrose, 281, *282*
Sugar(s), 269t
fermentation of, 131
mutarotation of, 277–278
optical properties of, 277
oxidation–reduction of, 278, *279*
properties and reactions of, 276–280
reducing, 278, 279
solubility of, 277
Sulfa drugs, 407
Sulfanilamide, and folic acid synthesis, 407, *407*
Sweeteners, artificial, 275

T

Target cells, hormones and, 487, *488*
Taurocholate, 311, *311*
Tay-Sachs disease, 379, 444, *444*
Temperature,
effects on body function, 403, *403*
effects on enzyme-catalyzed reactions, *402*, 402–403, *403*
greenhouse effect and, 46, *46*, 47
Terpenes, 87–88
sources of, 87, *89*
structure of, 87, *87*
and van Gogh, 88
Tertiary carbons, 27
Testosterone, 163t, 308
Tetrachloromethane, 12t
Tetrahedron(s), R-20, *R-21*
alkanes and, 18
Tetrahydrofolate, 395
Texaphyrin, 111
Thioester(s), 201–202, 208
formation of, 202, 212
Thiol(s), 123, 148
boiling points of, 144, *144*
nomenclature of, 144
oxidation of, 144–145, 149
reactions of, 144–146
reactions with mercury, 145–146, 149
special, 146–147, *147*
structure of, 144
Thiophene, 221
Threonine, 327t
Thujone, 88
Thymine, 361, 406–407
Thymol, 114, 138
Tissue-type plasminogen activator (TPA), 413
α-Tocopherol, 114
Tocopherols. *See* Vitamin E
Tolbutamine, 240t
Tollen's test, 173, 174
Toluene, 102, 110
Transamination, 446, *447*, 449, 450
Transcript, primary, 370, 371
Transcription, 369, 371, *370*, 371, 387
ribonucleic acid and, 368–369
Translation, 369, 372–375, *374–375*, 387
Translocation, 375
Transport, active, 318, *318*, 319
Triacylglycerol(s), *299*, 299–302, 319, 427–428, 445
biological roles of, 301–302
physical and chemical properties of, 300
synthesis of, 467–469
Tricarboxylic acid (TCA) cycle, 79, 434–436, *435*, 450
Trichloromethane, 12t
Triglyceride, 205, 299
Trimethylamine, 220
Trypsin, 409–410, *410*, 425
Trypsinogen, 409–410, *410*, 425
Tryptophan, 326t
repression of, 383, *384*
Tubocurarine, 237t
Tyrosine, 254, *254*, 327t, 330, 473
Tyrosine kinase, 490

U

Universal donor, 496
Universal recipients, 496
Unsaturated compounds, 87–89
Unsaturated hydrocarbons, 62–97. *See also* Alkene(s); Alkynes
structural diagrams and ball-and-stick models of, *63*
structure and physical properties of, 64–67
Uracil, 368
Urea cycle, *447*, 447–448, *448*, 449

Urodine, 241t
Uroshiol, 138

V

Valence-shell electron-pair repulsion (VSEPR), R-19
Valine, 326t
Valium, 241t
van der Waals forces, R-28
van Gogh, Vincent, 88
Vanillin, 114, 138, 163t
Vasopressin, *333*, 334
Velocity(ies), 401, 408
Vinegar, 174
Virulence, bacteria and, 354, *355*
Vision, processes in, *91*
Vitamin A, 88, 316, *316*, 317
Vitamin D, 316, 317
  synthesis of, 456
Vitamin E, 114, 316, *316*, 317
Vitamin K, 178, 316, 317
Vitamins, functions of coenzymes and, 394t
  minerals and, in metabolism, 394

W

Warfarin, 114
Water, addition to alkene, 76–77
  boiling point of, R-22, R-22t
  electron-dot model for, R-22, *R-22*
  freezing point of, R-33
  molecule, R-23, R-27, *R-27*
  molecules in ice, R-33
  solubility of alkanes in, 19, 20
  structure of, 123–124, *124*
Watson, James, 359–360
Waxes, 303
Whales (Sperm), and lipids, 302
  and myoglobin, *337*
Wilkins, Maurice, 359
Wines, 131

X

Xylocaine, 235

Z

Zwitterion, 329, 331
Zymogen(s), 409–410, 412, 414